Magnetism and Magnetic Materials – 1973

Part 1

AIP Conference Proceedings

Series Editor: Hugh C. Wolfe

Number 18, Part 1

Magnetism and Magnetic Materials – 1973
(19th Annual Conference - Boston)

Editors

C. D. Graham, Jr.
University of Pennsylvania

and

J. J. Rhyne
U. S. Naval Ordnance Laboratory

American Institute of Physics
New York   1974

Copyright © 1974 American Institute of Physics, Inc.
This book, or parts thereof, may not be
reproduced in any form without permission.

L. C. Catalog Card No. 59-2468
ISBN 0-88318-117-7
AEC CONF-731104

American Institute of Physics
335 East 45th Street
New York, N. Y. 10017

Printed in the United States of America

NINETEENTH ANNUAL CONFERENCE ON
MAGNETISM AND MAGNETIC MATERIALS
November 13 - 16, 1973
Boston, Massachusetts

Sponsored by
The American Institute of Physics
The Magnetics Society of the Institute of
Electrical and Electronics Engineers

In Co-operation with
The Metallurgical Society of the American Institute of
Mining, Metallurgical, and Petroleum Engineers
The Office of Naval Research
The American Society for Testing and Materials

The American Physical Society

The Conference is especially grateful to the
Office of Naval Research for its support of
the expenses of foreign and interdisciplinary
speakers under contract NONR (G)- 00004-74.

Contributions to the Conference from the following firms are gratefully acknowledged: Applied Magnetics Corp.; Arnold Engineering Co.; Audio Magnetics Corp.; Bell Laboratories; Ceramic Magnetics, Inc.; Colt Industries, Crucible Magnetics Div.; Computest Corp.; E. I. du Pont de Nemours and Co.; Eriez Manufacturing Co.; Ford Motor Co.; General Electric Co.; General Magnetic Co., Div. of Pemcor, Inc.; Honeywell; Infinetics, Inc.; International Business Machines Corp.; International Telephone and Telegraph Corp.; Magnetic Metals Co.; Magnetics, Div. of Spang Industries, Inc.; Pfizer, Inc.; Raytheon Company; RCA Laboratories; Rockwell International, Electronics Research Div.; Texas Instruments, Inc.; The Permanent Magnet Co., Inc.; Trans-Tech, Inc.; United Aircraft Corp.; Univac, Div. of Sperry-Rand; Westinghouse Electric Corp.

CONFERENCE ORGANIZATION

Conference
Chairman: F. B. Hagedorn

Steering
Committee:
- F. B. Hagedorn, Chairman
- M. K. Wilkinson, Secretary
- G. A. Fedde, Treasurer
- J. M. Lommel, Chairman-Elect
- D. I. Gordon
- A. Narath
- R. C. Byloff (IEEE)
- H. C. Wolfe (AIP)

Program
Committee: A. F. Mayadas and T. Penney, Co-Chairmen
- E. Banks
- P. I. Bonyhard
- W. J. L. Buyers
- S. H. Charap
- R. E. Dietz
- G. F. Dionne
- C. D. Graham
- F. B. Humphrey
- R. M. Josephs
- J. S. Kouvel
- D. L. Martin
- J. J. Rhyne
- M. Rubenstein
- R. S. Silberglitt
- R. E. Watson
- R. M. White
- M. Wortis

Local
Committee: K. Dwight and N. Menyuk, Co-Chairmen
- M. M. Antonoff
- R. W. Damon
- S. Foner
- J. B. Goodenough
- R. P. Guertin
- E. Schloemann
- B. Schwartz
- A. B. Smith
- J. Sokoloff
- H. J. Van Hook

Publications
Committee: C. D. Graham, Jr. and J. J. Rhyne

Exhibits: C & M Associates, P. O. Box 68, Maple Glen, Pa. 19002

CONFERENCE ADVISORY COMMITTEE

Chairman: W. D. Doyle

Term Expires
1973:
- H. Chang
- L. M. Corliss
- F. B. Hagedorn
- A. J. Heeger
- I. S. Jacobs
- A. Narath
- A. V. Pohm
- G. P. Rodrigue
- D. S. Shull
- J. K. Watson

Term Expires
1974:
- G. Bate
- R. F. Elfant
- G. Fedde
- A. Goldman
- B. Hershenov
- G. H. Lander
- K. Lee
- N. Menyuk
- G. T. Rado
- M. K. Wilkinson

Term Expires
1975:
- W. A. Baker
- W. D. Doyle
- D. I. Gordon
- E. M. Gyorgy
- V. Jaccarino
- S. Kern
- C. J. Kriessman
- J. M. Lommel
- R. L. White
- W. P. Wolf

Sponsoring Society Representatives:

- R. C. Byloff (IEEE)
- H. C. Wolfe (AIP)

Cooperating Society Representatives:

- G. Y. Chin (Met. Soc. of AIME)
- M. A. Garstens (ONR)
- D. H. Jones (ASTM Comm. A-6 on Mag. Mat.)

PREFACE

The Challenge for Our Future

The Conference on Magnetism and Magnetic Materials returned to Boston for its 19th annual meeting, having met there previously in 1967 and 1956. It seems appropriately traditional that Boston, the site of the second CM^3, should become the first city to host this meeting on three separate occasions. The outstanding success of the 19th CM^3 is also in keeping with tradition.

From a statistical viewpoint, CM^3 of 1973 had 703 registrants of which 51 were students. This number represents a modest increase over 1972 and is very slightly larger than the 1971 figure. Our registrants had the opportunity to choose from 34 invited and 270 contributed papers which were scheduled into 40 different sessions. The total number of papers was thus comparable with previous years, but the Program Committee used a scheme for organizing the sessions which was a departure from the traditions of recent M^3 Conferences. The organization of the sessions is retained in these Proceedings, where it may be seen that the invited papers were largely concentrated in sessions by themselves and that the invited and contributed sessions were run simultaneously.

There were also other breaks with tradition. The registration and reprint fees, for example, were both lower than in 1971 and 1972. But even so, the CM^3 goes into the final year of its second decade in a strong financial position, having reserves at the constitutional limit of \$10,000. Other aspects of the Conference also remain strong. Long range planning has established host cities and Local Chairmen through 1979, and active participation in the Steering Committee has been expanded substantially in the last few years. As the CM^3 approaches the traditional age of adulthood, all signs point toward stable maturity. It will be a challenge to avoid the lethargy that frequently accompanies middle age.

If this challenge is successfully met, it will be largely because the sage advice from past retiring chairmen has been taken. My recent predecessors have emphasized the necessity to change with the times, to remain constructively creative, and to infuse new blood into the conference organization. These sagacities possess continuing relevance as CM^3 concludes its second decade of bringing together scientists and engineers who are interested in basic and applied work in magnetism.

F. B. Hagedorn
Conference General Chairman
Bell Laboratories
Murray Hill, N. J. 07974

TABLE OF CONTENTS - PART 1.

Section 1.   SURFACES AND CATALYSIS

Magnetic Properties of Surfaces 1
    P. Fulde
Para- to Ortho-Hydrogen Conversion on Nonmetallic Magnetic Surfaces 8
    D.J. Scalapino, K.G. Petzinger
Experimental Relationships between Catalysis and Magnetism 19
    R.J.H. Voorhoeve
Magnetic Effects in Surface Reactions 33
    Harry Suhl

Section 2.   BUBBLE FILM GROWTH

Growth of Garnet Films by Liquid Phase Epitaxy 48
    B.F. Stein
Multiple Garnet Film Growth by LPE 63
    R.G. Warren, J.E. Mee, F.S. Stearns, E.C. Whitcomb
Rapid Epitaxial Growth of High Perfection Garnet Films 68
    W.A. Bonner
Characteristics of $Y_{3-x}Ca_x(Ge,Si)_xFe_{5-x}O_{12}$ Based Epitaxial Films for Magnetic Bubble Applications 69
    J.E. Geusic, D.H. Smith, L.G. Van Uitert, G.P. Vella-Coleiro
Growth Induced Anisotropy of $Y_{3-x}Lu_xFe_5O_{12}$: Dependence on Temperature and Composition 70
    E.M. Gyorgy, M.D. Sturge, L.G. Van Uitert
The Effect of Substrate Orientation and Growth Temperature on the Magnetic Anisotropy of $Eu_xY_{3-x}Fe_5O_{12}$ Films 75
    T.S. Plaskett, E. Klokholm, D.C. Cronemeyer
Misfit Strain in LPE $(Y,Eu)_3(Fe,Ga)_5O_{12}$ Garnet Films 80
    H. Makino, T. Hibiya, K. Matsumi
Perpendicular Anisotropy in $Gd_{1-x}Co_x$ Amorphous Films Prepared by RF Sputtering 85
    D.C. Cronemeyer
Neutron Irradiation of LPE Bubble Domain Garnets 90
    R.S. Sery, H.R. Irons

Section 3.   BUBBLE DEVICES

Progress in All-Permalloy Bubble Control Functions 95
    T.J. Nelson
Device Design and System Organization for a Decoder Accessed Magnetic Bubble Memory Chip 100
    P.I. Bonyhard, Y.S. Chen, J.L. Smith
Bonded Shift-Registers for Magnetic Bubble Memory Modules 105
    A.E. Feuersanger
Ni-Co Film with Large Magnetoresistance for Bubble Detection 110
    K. Asama, K. Takahashi, M. Hirano
Study on the Thick Film Chevron Detector 115
    T.T. Chen, P.K. George, L.R. Tocci, J.L. Archer
Experimental Confirmation of Field Access Device Modelling 116
    P.K. George, J.L. Archer
Field Access Bubble Propagation Circuits 121
    S.K. Singh
Laboratory Modeling of Bubble Propagation Systems 122
    C.H. Hsin, T.J. Matcovich, R.L. Coren
Propagation Characteristics of Parallel Bar Circuits 127
    G. Ng, W. Kinsner, E. Della Torre

Section 4.   RELIABILITY IN BUBBLE DOMAIN MEMORIES

Reliability Modeling: Application to Bubble Domain Memories 132
    W.G. Bouricius
Reliability of Magnetic Bubble Domain Memories 139
    D.H. Baird, C.F. Buhrer, J.J. Vytal, J.L. Archer, L.R. Toccio, O.D. Bohning
Long-Term Propagation Studies in Magnetic-Bubble Devices 140
    P.W. Shumate, P.C. Michaelis, R.J. Peirce
Error Rate Measurements in Bubble Circuits on Permalloy Coated YEu Garnet Films 152
    W.D. Doyle, W.E. Flannery, J.A. Coleman
Information Stability in Bubble Memory Elements 157
    K. Yoshimi, S. Fujiwara

Section 5.   WALL STRUCTURE IN BUBBLE MATERIALS

Observation of Strip and Bubble Domains in $PbFe_{12}O_{19}$ Using Lorentz Electron Microscopy — 162
    P. Dunk, G.A. Jones

LPE Garnet Films without Hard Bubbles — 167
    A.B. Smith, M. Kestigian, W.R. Bekebrede

Hard-Bubble-Free Garnet-Permalloy Composite Films — 172
    M. Takahashi, H. Nishida, T. Kobayashi, Y. Sugita

Static Hard Bubble Measurements on Several Epitaxial Garnet Film Compositions — 173
    D.H. Smith, A.A. Thiele

Interactions between Bloch Lines — 178
    A. Hubert

Domain-Wall Mass in Bubble Films — 183
    E. Schlömann

Controlled Compensation Walls and Compromise Compensation Walls in Garnet Films by the Silicon Annealing Technique — 188
    R.C. LeCraw, R. Wolfe

Magnetization Processes involving Planar Compensation Walls — 193
    O. Voegeli, E.B. Moore

The Exchange Constant of Gallium Substituted Iron Garnets — 194
    R.D. Henry, D.M. Heinz

Section 6.   WALL DYNAMICS IN BUBBLE MATERIALS

Domain Wall Dynamics in Low-Loss Garnet Films — 199
    P.J. Rijnierse, F.H. de Leeuw

Bubble Dynamics in Amorphous Magnetic Materials — 213
    M.H. Kryder, H.L. Hu

Dynamic Conversion Effects in Epitaxial Garnet Films — 217
    G.P. Vella-Coleiro

A Model for Dynamic Conversion in Bubble Domains — 222
    F.B. Hagedorn

The Influence of an In-Plane Field on Bubble Dynamics — 227
    R.M. Josephs, B.F. Stein

The Effect of a Constant In-Plane Magnetic Field on Magnetic Bubble Translation in $(YGdTm)_3(FeGa)_5O_{12}$ — 232
    D.C. Bullock

Temperature Variation of Magnetic Bubble Garnet Film Parameters — 237
    R.M. Sandfort, R.W. Shaw, J.W. Moody

Observation of Straight and Wavelike Domain Wall Motion in Bubble Films by High Speed Photography — 242
    T.M. Morris, A.P. Malozemoff

Observation of Domain-Wall Resonances by a Diffraction Technique — 247
    G.R. Woolhouse, P. Chaudhari

Section 7.   ALLOYS

NMR Study of the Electron Spin Density Near Iron Group Atoms in Cu — 252
    J.B. Boyce, T.J. Aton, C.P. Slichter

Determination of Moment Distributions and Hyperfine Fields at and Surrounding Transition Element Solute Atoms in Fe — 257
    M.B. Stearns

Single Crystal Study of Hyperfine Fields in Fe-3.2 at. pct. Mo — 262
    A. Asano, L.H. Schwartz

Charge Screening and the Magnetic Properties of the Ni-Sb and Ni-Ir Alloy Systems — 267
    J. Hudis, M.L. Perlman, R.E. Watson

Magnetic Impurity States in NiAl Intermetallic Compounds — 272
    J.R. Willhite, T. Yoshitomi, L.B. Welsh, J.O. Brittain

Conduction Electron Polarization in FCC Transition Metals — 277
    G.P. Huffman

Spin Fluctuations and Spin-Spin Interactions in Amorphous Metallic Alloys — 282
    F.R. Szofran, J.W. Weymouth, G.R. Gruzalski, D.J. Sellmyer, R. Ray, B.C. Giessen

Electron Spin Resonance of Metallic Si:P with Iron, a Variable Electron Concentration Alloy — 287
    T.A. Kennedy, R.T. Longo, J.H. Pifer

Anisotropic Kondo Resistance in Fe Doped $NbSe_2$ — 292
    R.C. Morris, B.W. Young, R.V. Coleman

NMR and Susceptibility Studies of Pt-Rh Alloys ... 297
   H.T. Weaver, R.K. Quinn
Paramagnon Effects Associated with Ni Impurities ... 301
   C.F. Eagen, A.I. Schindler
Effect of Strong Spin Correlations on the Magnetic Specific Heat and
Magnetization of Dilute PdFe Alloys ... 302
   N.C. Koon
Low Field Magnetic Susceptibility of $(Pd_{1-x}Ag_x)_{.99}Fe_{.01}$ ... 307
   J.I. Budnick, V. Cannella, T.J. Burch
The Low Temperature Specific Heat of Dilute Rare-Earth in Transition
Metal Alloys: Sc(Gd), Pd(Dy), and Pd(Gd) ... 312
   L.L. Isaacs
The Electrical and Magnetic Properties of Gold-Nickel Alloys ... 316
   J.R. Clinton, E.H. Tyler, H.L. Luo
Alteration of the Magnetic Properties of Au-Fe Alloys by Neutron
Irradiation ... 317
   R.J. Borg
Magnetic Order in Ni_3Mn Alloys ... 318
   C.E. Patton, G.L. Baker
Measurements and Trends of Hyperfine Fields in Heusler Alloys ... 319
   C.C.M. Campbell, W. Leiper
Analysis of Mossbauer ^{57}Fe Absorption in an F.C.C. Fe-Co-V Alloy ... 324
   G. Bambakidis, J.P. Cusick

Section 8.   PHASE TRANSITIONS

Magneto-Optical Studies of Metamagnetic Phase Transitions in $FeCl_2$ ... 329
   E. Y. Chen, J.F. Dillon, Jr., H.J. Guggenheim
Microscope Studies of the Mixed Phase Region of the Ising
Antiferromagnet Dysprosium Aluminum Garnet (DAG) ... 334
   J.F. Dillon, Jr., E. Y. Chen, W.P. Wolf
Magnetic Phase Boundaries and Spin Wave Renormalization in Two
Antiferromagnetic Compounds ... 335
   J.E. Rives, S.N. Bhatia, V. Benedict
Magnetic and Structural Transitions in NdS, DyS and ErS ... 340
   L.J. Tao, J.B. Torrance, F. Holtzberg
Magnetic Structure of FeI_2 and Phase Transitions in High Magnetic
Fields Parallel to the Spin Direction ... 341
   J. Gelard, A.R. Fert, P. Carrara
The Role of Harmonics in the First Order Antiferromagnetic to
Paramagnetic Transition in Chromium ... 342
   C.Y. Young, J.B. Sokoloff
The Proximity Effect for Very Weak Itinerant Ferro and Anti Ferro Magnets ... 347
   M. Kiwi, M.J. Zuckermann
Lattice Disorder and Magnetic Phase Transitions ... 351
   B.A. Huberman

Section 9.   MAGNETIC ORDER AND STRUCTURE

Magnetically induced Lattice Distortions in the Neptunium Monopnictides ... 352
   M.H. Mueller, G.H. Lander, H.W. Knott, J.F. Reddy
Magnetic Properties of UCu_5 and UNi_5 ... 357
   M.B. Brodsky, N.J. Bridger
Magnetic Properties of Gd-Rich Gd-Sm Alloys ... 362
   S. Arajs, D.L. Adour, E.E. Anderson, T.F. DeYoung, K.V. Rao
Magnetic Properties of $NpAl_3$ ... 366
   A.T. Aldred, B.D. Dunlap, D.J. Lam
A Study of Iron Impurities in the Linear Magnetic Systems TMMC and
$CsNiF_3$ ... 371
   P.A. Montano
Specific Heat of the Magnetic Chain TMMC Below 1.2 K ... 376
   H.W. White, K.H. Lee, J. Trainor, D.C. McCollum, S.L. Holt
Phase Transitions in $FeCl_2 \cdot 2H_2O$ in External Magnetic Fields: Mössbauer
Spectroscopy ... 380
   L.D. Kandel, M.A. Weber, R.B. Frankel, C.R. Abeledo

Mössbauer Study of Oriented YbCrO₃ Powder — 381
G.R. Davidson, B.D. Dunlap, M. Eibschütz, L.G. van Uitert

Magnetic Ordering and Hyperfine Parameters in the Linear Chain Compound RbFeBr₃ — 386
M. Eibschütz, G.R. Davidson, D.E. Cox

Magnetic Studies of a Canted Antiferromagnetic: $MnBr_3(CH_3)_3NH \cdot 2H_2O$ — 391
P.R. Newman, J.A. Cowen, R.D. Spence

Neutron Microscopy of Spin Density Wave Domains in Chromium — 396
J.B. Davidson, S.A. Werner, A.S. Arrott

Effects of Pressure on the Magnetic Ordering in Cr-Fe — 401
L.R. Edwards, I.J. Fritz

Generalized Neutron Polarization Analysis — 406
F. Mezei

Nuclear Polarization Dependence of Coherent Neutron Scattering From Holmium — 411
G.R. Little, R.A. Erickson

A Neutron Diffraction Study of Antiferromagnetic CoO with Nuclear Polarization from the HFS Interaction in the Region 0.35-4.2K — 416
D.A. Goer, R.A. Erickson

Neutron Diffraction Study of the Mn-Pd System — 421
G. Kadar, E. Kren, L. Pal

Magnetic Form Factor of Palladium at 4.2 K — 426
J.W. Cable, E.O. Wollan, G.P. Felcher, T.O. Brun, S.P. Hornfeldt

Antiferromagnetic Iron in Small γ-Phase Fe-Ni Crystals — 427
J.M. Crowell, J.C. Walker

Variation with Chemical Order of Spin Coupling in NiPt — 432
C.W. Chen, J.D. Greiner, R.W. Buttry

Temperature Dependence of the Co Hyperfine Field in $GdCo_2$ and its Relation to Magnetic Structure — 437
I. Wang, T.J. Burch, J.I. Budnick, J.J. Murphy, J.A. Cannon

Low Temperature Study of the $Co_{1-x}Fe_xSi$ Alloys — 438
P.H. Barrett, P.A. Montano, Z. Shanfield

Magnetic Properties of UC-UN Solid Solutions — 442
G.H. Lander, D.J. Lam, J.F. Reddy, M.H. Mueller

Section 10.   PHYSICS AND CHEMISTRY OF TRANSITION METAL COMPOUNDS

Demagnetization of Rare Earth Ions in Metals due to Valence Fluctuations — 447
M.B. Maple, D. Wohleben

Correlation between Lattice Constant and Magnetic Moment in 3d Transition Metal Alloys — 463
M. Shiga

Effects of Lattice Pressure on Sm Valence States in Monosulfide Solid Solutions — 478
F. Holtzberg

Magnetic Properties of Some Simple Metallic and Semimetallic Compounds of f-Electron Metals with Main Group Elements — 490
B. Stalinski

Studies on Rare Earth Cobaltites, $La_{1-x}Sr_xCoO_3$ and Related Systems — 504
C.N.R. Rao, V.G. Bhide

Section 11.   EXCHANGE: HYPERFINE EFFECTS

^{61}Ni Mössbauer Studies of Substituted Ni Spinels — 513
J.C. Love, F.E. Obenshain

Supertransferred Hyperfine Fields at Sb^{5+} in Insulating Ferrites: Effects of Local Order and Ion-Specific Properties — 518
B.J. Evans, L.J. Swartzendruber

A Mössbauer Study of Dy_2O_3 and the Ising Antiferromagnet $DyPO_4$: Relaxation, Anisotropy and Site Dependence — 523
D.W. Forester, W.A. Ferrando

Nuclear Magnetic Resonance in High Neel Temperature Garnets — 524
    H. Yokoyama
Magnetic Environment of Hydrogen in Fe from Muon Precession Measurements — 525
    N. Heiman, M.L.G. Foy, W.J. Kossler, C.E. Stronach
High Temperature EPR in Dense Solid and Molten Paramagnets — 529
    E. Dormann, V. Jaccarino
EPR Study of FeS_2:Ni, Co — 534
    R.N. Chandler, R.W. Bene
Evidence for Exchange Enhancement by Impurity-Band Electrons in $Sm_{1-x}Gd_xS$ — 535
    W.M. Walsh, Jr., L.W. Rupp, Jr., E. Bucher, L.D. Longinotti
Magnetic Properties of Gadolinium Chalcogenides with Varying Stoichiometry — 540
    W. Beckenbaugh, G. Güntherodt, R. Hauger, E. Kaldis, J.P. Kopp, P. Wachter
A Spherical Tensor Operator Description of the Exchange Splittings of Gd^{3+} in $GdCl_3$ — 545
    R.S. Meltzer, R.L. Cone
Interaction of U^{3+} Pairs in $LaCl_3$ — 550
    T.C.L.G. Sollner, R.N. Rogers

Section 12. AMORPHOUS AND DISORDERED MATERIALS

Exchange Interaction in Dilute Yttrium-Rare Earth Alloys — 555
    R.M. Nicklow, N. Wakabayashi
Magnetism in Amorphous Terbium-Iron — 563
    J.J. Rhyne, S.J. Pickart, H.A. Alperin
Amorphous Magnetic Materials — 578
    R.J. Gambino, P. Chaudhari, J.J. Cuomo
Theoretical Approaches to Spin Waves in Disordered Magnets — 593
    A.B. Harris

Section 13. AMORPHOUS, DISORDERED SYSTEMS

Magnetization Studies of an Amorphous Antiferromagnet — 594
    E.J. Friebele, N.C. Koon
Magnetic Behavior of Barium Titanium Silicate Glass — 595
    J.J. Santiago, H.W. Swenson, Jr.
Field Dependence of Magnetization in Random FCC Ferromagnets — 600
    P.M. Richards
Dilute Antiferromagnetic Systems in FCC and BCC Latices — 605
    H. Sato, R. Kikuchi
Random Magnetic Alloys — 610
    R.A. Tahir-Kheli, L.C.M. Miranda, S.M. Rezende
Studies in Exchange and Anisotropy Driven Amorphousness in Finite Size Spin Systems — 615
    J.D. Patterson, R.A. Tahir-Kheli
Moessbauer Studies in Non-Crystalline Magnets — 616
    F.J. Litterst, G.M. Kalvius, A.J.F. Boyle
On the Dilute Anisotropic Heisenberg Ferromagnet in the Coherent Potential Approximation — 621
    T. Horiguchi
Effects of Disorder on Ferromagnetism in Narrow Energy Bands — 626
    G.F. Abito, J.W. Schweitzer
Short-Range Order in Amorphous $GdFe_2$ — 631
    G.S. Cargill III
Mossbauer Effect Studies in Amorphous $TbFe_2$, $DyFe_2$, $HoFe_2$ and $ErFe_2$ — 636
    D. Sarkar, R. Segnan, A.E. Clark
Magnetic Properties of Amorphous GdCo Films — 641
    L.J. Tao, R.J. Gambino, S. Kirkpatrick, J.J. Cuomo, H. Lilienthal
Magnetic Properties of Amorphous Ni-P Alloys — 646
    D. Pan, D. Turnbull
Low Field Magnetic Susceptibility of Noble Metal-Transition Metal Spin Glass Alloys — 651
    V. Cannella, J.A. Mydosh

Section 14. THEORY

Variational Definition of Anderson's Ligand Field States N.P. Silva, T.A. Kaplan	656
Polar Singlet-Ground-State: Application to Wurster's Blue Perchlorate R.A. Bari	661
Persistence of Exchange Splitting in Metallic Ferromagnets above T_c J.B. Sokoloff	662
Finite Temperature Properties of the Hubbard Model: Phase Separation P.B. Visscher	663
Single Site Approximation in the Hubbard Model L.M. Roth	668
Magnetic Properties of the Hubbard Hamiltonian including the Resonance-Broadening Terms: Comparison to the CPA L.C. Bartel, H.S. Jarrett	673
Projection Operator Formalism: Application to the Hubbard Model A.J. Fedro, R.S. Wilson	678
Electron Correlation in a Narrow Band Metal; a New Method of Calculation S.P. Bowen	683
Determination of Error Minimized Green's Functions for the S-D Model P.E. Bloomfield, E.B. Brown	687
Microscopic Theory for the Fluctuation-Driven Phase Transition in Weak Itinerant Magnets K.K. Murata	692
Hydrodynamics of a Ferromagnetic Electron Gas with Spin-Orbit Coupling J. Tilley, A. Luther	697
Unified Itinerant-Localized Theory of Magnetism B.H. Brandow	702
Theory of Motional Narrowing of EPR Spectral Density of Magnetic Ions G. Reiter	707
Effects of Nearest Neighbor Four-Spin Correlation upon the Critical Properties of Spin-1/2 Heisenberg Ferromagnet T. Tanaka, L.F. Libelo	708
Some Consequences of the Non-Mixing Behaviour of the Four Spin Correlation Function in Heisenberg Paramagnets B. Foster, G. Reiter	713
Dynamic Form Factor and Nonequilibrium Behavior of the $S=1/2$ XY Model on Cubic Lattices M.H. Lee	714
Reciprocity Relations for Susceptibilities and Fields in Magnetoelectric Antiferromagnets G.T. Rado	719
Linear Momentum in the Rest Frame of a Ferromagnet and the Electrodynamic Consequences F.R. Morgenthaler	720

TABLE OF CONTENTS - PART 2

Section 15. MATERIALS SYNTHESIS AND CHARACTERIZATION

A Potential Pole Coil for High Field Measurement of Permanent Magnets — 725
    E. Steingroever

Magnetic Properties of Co-Fe-Nb Alloys for Remanent Reed Switches — 730
    M. Okada, M. Kassai, T. Sasaki, Z. Henmi

The Effect of Cobalt on the Formation of a Non-Magnetic Surface Developed in Grinding Carbon Steel — 735
    L.J. Swartzendruber, E. Siegel

Magnetic Properties of Internally Oxidized Copper — 740
    F.R. Fickett, D.B. Sullivan

Growing and Demagnetization Process of Single Crystal MnBi Thin Films — 741
    S. Honda, T. Kusuda

Magneto-Optics of Ceramic $YFeO_3$ and $FeBO_3$ — 746
    H.M. Kahan, D.P. Stubbs

Magnetic Properties of the New Magnetic Compounds Mn-X-Bi (X=Ni,Cu,Rh,Pd) — 747
    J.C. Suits, G.B. Street, K. Lee

Structural and Magnetic Properties of New $MnM_2^{III}Te_4$ Compounds (M^{III}=Al,Ga,In) — 748
    H.M. Kasper, E. Buchner

New Magnetic Oxides Involving Divalent Europium and Transitional Metal Ions — 749
    J.E. Greedan

Section 16. CRITICAL AND TRICRITICAL PHENOMENA

Critical Sound Absorption and Dispersion in EuO Near T_c — 754
    B. Golding, M. Barmatz

Critical Behavior of the Microwave Resonance Absorption Linewidth in $CrCl_3$ — 759
    T.G. Campbell, A.W. Lawson

Anisotropy and Temperature Dependence of the Zero Field Uniform Mode Relaxation Rate in MnF_2 Near T_N — 764
    A.M. Gottlieb, M. Feldman, M. Littman, P. Heller

Tricritical Behavior of the Blume-Capel Model — 769
    D.M. Saul, M. Wortis

Magnetic Equation of State of Dysprosium Aluminum Garnet (DAG) Near a Tricritical Point — 770
    A.T. Skjeltorp, R. Alben, W.P. Wolf

A Solvable Anisotropic Magnetoelastic Model — 771
    Y. Imry, D.J. Bergman, O. Entin-Wohlman

Scaling Hypothesis for a System Exhibiting a Critical Point of Order Four: Ising Planes with Variable Interplanar Interactions — 776
    T.S. Chang, A. Hankey, F. Harbus, H.E. Stanley

Section 17. SPIN WAVES AND CRITICAL PHENOMENA

Magnetic Excitations in Nickel and Iron — 781
    H.A. Mook, J.W. Lynn, R.M. Nicklow

$CrBr_3$: A Prototype Insulating Ferromagnet — 794
    R. Silberglitt

Heisenberg Ferromagnetism in Two Dimensions: an Experimental Study — 806
    A.R. Miedema, P. Bloembergen, J.H.P. Colpa, F.W. Gorter, L.J. de Jongh, L. Noordermeer.

Monte Carlo Studies of Magnetic Systems — 819
    D.P. Landau

Section 18. CRITICAL PHENOMENA

Crossover Phenomena, Critical and Tricritical Phase Transitions — 834
    E.K. Riedel

Asymptotic Scale Invariance in 4-ϵ Dimension — 849
    E. Brézin

Critical Behavior of Magnets with Dipolar Interactions — 863
    A. Aharony

Magnetic Phase Transitions in Alloys — 876
    A. Luther, G. Grinstein

Section 19. CRITICAL PHENOMENA AND LOWER DIMENSIONAL SYSTEMS

High-Temperature Expansions for Classical Systems — 878
    J.P. VanDyke, W.J. Camp

The Maximum of the Spin-Spin Correlation Function in the High-Temperature Critical
Regime 883
    M. Ferer
Exact Renormalization Groups for One-Dimensional Spin Systems 888
    D.R. Nelson, M.E. Fisher
Computer Simulation of a Discontinuous Phase Transition in the Two-Dimensional
One-Spin-Flip Ising Model 891
    T. Schneider, E. Stoll
Studies of the Dynamics of the Ising Model in a Transverse Field: One Dimension,
$T = \infty$ 892
    D.L. Huber, T. Tommet
Classical Heisenberg Chain in an External Magnetic Field 893
    M. Blume, P. Heller, N.A. Lurie
Propagating Modes in the Classical Heisenberg Chain 894
    J.S. Semura, D.L. Huber

Section 20.   TRANSPORT AND METAL-INSULATOR TRANSITIONS

Magnetic Susceptibility at a Peierls Transition 895
    P.A. Lee, T.M. Rice, P.W. Anderson
Hubbard Excitations and the Infrared Absorption Spectrum of (TTF)(TCNQ) 896
    J.B. Torrance, B.A. Scott, D.C. Green, P. Chaudhari
Linear Responses of a Magnetic Semiconductor and the Formation of a Bound
Magnetic Polaron 897
    P. Leroux-Hugon
Observation of the Localization of Electrons to Form Magnetic Polarons Near
T_c in EuS 902
    J.B. Torrance, F. Holtzberg
Photoemission Studies of Ni-Co Alloys having Large Anisotropic Magnetoresistance 903
    T.R. McGuire, W.D. Grobman, D.E. Eastman
Localized Ferromagnetic Polarons in Antiferromagnetic $Gd_{3-x}V_xS_4$ 908
    T. Penney, F. Holtzberg, L.J. Tao, S. von Molnar
Detailed Study on the Electronic Transition of Fe_3O_4 913
    S. Iida, M. Yamamoto, S. Umemura
Dragging of Domains by an Electric Current in Very Pure, Non-Compensated
Ferromagnetic Metals 918
    L. Berger
Contributions to the Electrical Resistivity from High-Degree s-f Scattering 923
    K. Ravishankar, M.J. Sablik, P.M. Levy, L.F. Uffer

Section 21.   OPTICAL PHENOMENA

Far-Infrared Spectroscopy of $TbPO_4$ 928
    J.L. Lewis, G.A. Prinz
Polarization of Magnon and Exciton Raman Scattering in $KCoF_3$ 929
    G.H. Johnson, D.B. Fitchen
Light-Scattering Study of Phonon-Magnon Coupling in the One-Dimensional
Antiferromagnet $FeCl_2 \cdot 2H_2O$ 930
    R.W. Kinne, J.F. Ryan, W.J. O'Sullivan, J.F. Scott
Light Scattering Study of Thermal Acoustic Magnons in Yttrium Iron Garnet 935
    J.R. Sandercock, W. Wettling
Infrared Light Scattering from a Coherent Spin Wave Packet of Controlled Width
and Energy Density 940
    J.K. Jao, F.R. Morgenthaler
Faraday and Kerr Effect of Bismuth Substituted Iron Garnets: Applications
and Physical Origin 944
    S. Wittekoek, T.J.A. Popma, J.M. Robertson
New Bi-Based Garnet Films for Magnetic-Bubble Devices with Magneto-Optic
Applications 949
    A. Akselrad, R.E. Novak, D.L. Patterson
Terbium Gallium Garnet for Faraday Effect Devices 954
    D.J. Dentz, R.C. Puttbach, R.F. Belt

Section 22.   DILUTE ALLOYS

Summary of Mössbauer Evidence for the Kondo Effect 959
    T.A. Kitchens, Jr., R.D. Taylor
Magnetization and Susceptibility of CuAu (Fe) 964
    J.B. Hadad, M.P. Sarachik
Magnetic Susceptibility of α-Phase Cu-Al and Cu-Al(Fe) 969
    F.B. Huck, W.R. Savage, J.W. Schweitzer

Co Hyperfine Fields in Spin-Fluctuation Alloys J.C. Bremer, J.A. Gardner	970
Thermopower Measurements of CuNi and CuNi(Fe) Alloys D.R. Zrudsky, A.B. Showalter	974
Dependence of Curie-Weiss Temperature on MN Concentration in Dilute ZNMN Alloys F.W. Smith	975
Magnetic Behavior of Mn Impurities in Silver J.C. Doran, O.G. Symko	980
Electron Spin Resonance of Gd in $Ce_xLa_{1-x}Ru_2$ K. Baberschke, U. Engel, G. Koopman, S. Hüfner	984
Electron Spin Resonance of Gd in $LuAl_2$ C. Rettori, D. Davidov, H.M. Kim, E.P. Chock	989

Section 23.   TOPICS IN APPLIED MAGNETISM

Integrated Magnetic Recording Heads: Review and Outlook J.P. Lazzari	990
Magnetic Wave Devices for Microwave Applications W.L. Bongianni	1005
Magnetic and Magnetoelastic Properties of Highly Magnetostrictive Rare Earth-Iron Laves Phase Compounds A.E. Clark	1015

Section 24.   BAND STRUCTURE

Photoemission Studies of Valence Bands and 4f Multiplet Structure in NdS, SmS, EuS, GdS, DyS and ErS D.E. Eastman, F. Holtzberg, J. Freeouf, M. Erbudak	1030
Optical Properties of GdS, GdSe, GdTe and LaS G. Güntherodt, P. Wachter	1034
X-Ray Photoemission Studies of Rare Earth Hard Magnets J.R. Cuthill, A.J. McAlister, N.E. Erickson, R.E. Watson	1039
One Electron Energy Levels of Iron Group Impurities in TiO_2 K. Mizushima, M. Tanaka, K. Asai, S. Iida	1044
Self Consistent Electronic Band Calculation for the Ferromagnetic (H.C.P.) Cobalt C.M. Singal, T.P. Das	1049
Electronic Structure and its Relation to the Magnetic Ordering of CrB_2 S.H. Liu, W.B. England, H.W. Myron	1054
Optical Absorption in a Model Transition Metal including Correlation J.C. Shaffer, A.J. Fedro	1055
A Band Model for Heusler Alloys A.A. Bahurmuz, M.J. Zuckermann	1059

Section 25.   SPIN WAVES AND RELAXATION

Spin Wave Renormalization in EuO C.J. Glinka, V.J. Minkiewicz, L. Passell, M.W. Shafer	1060
Spin Waves in γ FeMn: An Itinerant Antiferromagnet Y. Endoh, G. Shirane, Y. Ishikawa, K. Tajima	1065
A Mechanism for the Q-Dependent Anisotropies in the Rare Earth Metals R.J. Birgeneau, J.K. Kjems	1066
Neutron Measurement of Crystal Field Splittings in TmAs and TmBi H.L. Davis, H.A. Mook	1068
Thermal Conductivity and Spin-Phonon Interactions in $GdCl_3$ G.S. Dixon, J.J. Martin, N.D. Love	1073
Impurities and Higher Order Anisotropies in One-Dimensional Antiferromagnet $KCuF_3$ K. Okuda, H. Hata, M. Date	1078
Radiation Induced Broadening of the FMR and AFMR Uniform Modes S.M. Rezende, E. Soares, V. Jaccarino	1083
Nuclear Spin Lattice Relaxation in Ferromagnetic Ni_3Mn R.L. Streever	1088
T_1 Measurement of Mn^{55} NMR in MnF_2 A.R. King	1093

Section 26.   RECORDING, MEMORY DEVICES AND MATERIALS

Switching Dynamics of an Assembly of Interacting Anisotropic Ferromagnetic Particles R.F. Soohoo, K. Ramachandran	1098

Effects of Preferred Orientation on the Magnetic Properties of Co-Ni Thin Films	1103
K.Y. Ahn, K.N. Tu, A. Gangulee, P.A. Albert	
Improved Magnetic Metal-Alloy Particles	1108
R.J. Deffeyes, C.E. Johnson, Jr., J.H. Judy	
Anisotropy of Well Oriented γ Fe_2O_3 Magnetic Tape	1113
H.N. Bertram	
Optical Storage in Hot-Pressed Ferrimagnetic Spinels	1118
T. Coburn, R. Ahrenkiel, E. Carnall, D. Pearlman	
Cobalt-Platinum Films for Digital Magneto-Optic Recording	1122
D. Treves, J.T. Jacobs, E. Sawatzky	
Performance of a Beam Accessed Thermomagnetic Memory	1123
N. Minnaja, M. Nobile	
Dynamic Measurements of Thermomagnetic Reversal in Gd-Co Films	1128
H. Wieder	
In-Plane Bubbles in Continuous Metallic Films	1133
E.J. Torok, R.E. Lund, W.J. Simon	
Aging of Plated Wires Without a Hard Axis Field	1138
N. Goldberg	

Section 27.　　　　　　RARE-EARTH MAGNETS

A Review of the Binary Rare Earth-Cobalt Alloy Systems	1143
A.E. Ray	
Thermal Decomposition and Magnetic Behavior of Co_5Sm	1144
J.G. Smeggil, P. Rao, J.D. Livingston, E.F. Koch	
The Annealing Response of Sm-Co Magnets and its Dependence on Composition and Processing	1149
P.F. Weihrauch, D.K. Das	
Properties of Mischmetal-Cobalt Magnets	1154
D.V. Ratnam, M.G.H. Wells	
Influence of Samarium Content on the Structure and Magnetic Behavior of Co-Fe-Cu-Sm Permanent Magnet Alloys	1159
E.A. Nesbitt, G.Y. Chin, R.C. Sherwood, M.L. Green	
Thermal Stability and Temperature Coefficients of Sintered Samarium-Cobalt Magnets	1163
H.F. Mildrum, M.F. Hartings, K.J. Strnat	
Reversible and Irreversible Losses of Magnetization in $SmCo_5$-Magnets	1168
K. Bachmann	
A Co-Gd-Sm Permanent Magnet with a Zero Temperature Coefficient of Magnetization	1173
M.G. Benz, R.P. Laforce, D.L. Martin	

Section 28.　　　　　　RARE-EARTH MAGNETIC MATERIALS

The Magnetic Properties of RE Co_5-Single Crystals	1177
H.P. Klein, A. Menth	
Magnetic Properties of RE Co_5-Permanent Magnets	1182
A. Menth, H.P. Klein, J. Bernasconi, S. Strässler	
High Temperature X-Ray Studies of $SmCo_5$	1187
L.D. Jennings, D.R. Chipman	
Magnetostriction in Single Crystal $SmCo_5$	1192
D.A. Doane, C.D. Graham, Jr.	
The Magnetic Properties of Single Crystal $NdCo_5$	1197
W.G.D. Frederick, C.W. Searle, M. Hoch	
Time Effects in Co_5Ce Powders	1202
G. Roy, P. Gaunt	
Heat Capacity of $CeCo_5$, $PrCo_5$, $NdCo_5$, $SmCo_5$, and $GdCo_5$ from 5-300K	1207
D.A. Keller, S.G. Sankar, R. S. Craig, W.E. Wallace	
Magnetic Anisotropy of R_2Co_{17} Compounds (R-Er, Tm, Yb)	1212
K.S.V.L. Narasimhan, W.E. Wallace, R.D. Hutchens, J.E. Greedan	
Magnetic Measurements on Single Crystals in the $Y_2(Co_{1-x}Fe_x)_{17}$ and $Pr_2(Co_{1-x}Fe_x)_{17}$ Systems	1217
C.W. Shanley, R.S. Harmer	

Section 29.　　　　　　ANISOTROPY AND MAGNETOELASTIC EFFECTS

Magnetic Properties of MnBi Single Crystals	1222
W.E. Stutius, T. Chen, T.R. Sandin	
Anisotropy and Magnetostriction of Pr^{3+} and Nd^{3+} in YIG	1227
R. Krishnan, M. Rivoire	

Cubic Anisotropies in Some Mixed and Substituted Rare-Earth Garnets at 77°K 1232
    E.J. Heilner, W.H. Grodkiewicz
Contributions to the Magnetostriction of Ni^{2+} Ions in a Tetrahedral Site 1237
    A.J. Pointon, K.T. Lioliousis
High Temperature Mössbauer Study of RFe_2 Laves Phases 1242
    C.W. Kimball, A.E. Dwight, R.S. Preston, S.P. Taneja
Anisotropy Energy Measurements on Single Crystal $Tb_{.15}Ho_{.85}Fe_2$ 1247
    C.M. Williams, N.C. Koon
Magnetic Properties of $R_2Fe_{17-x}M_x$ (R=Y, Tm, Ho and M=Ni, Al) and $Y_{2-x}Th_xFe_{17}$ Compounds 1248
    K.S.V.L. Narasimhan, W.E. Wallace
Magnetostriction of Dy_2Co_{17}, Y_2Co_{17} and $Dy_xY_{2-x}Co_{17}$ Intermetallics 1253
    A.E. Miller, T. D'Silva, H. Igarashi, J. Shanley
Magnetic Anisotropy of the Heavy Rare Earths 1258
    M.S.S. Brooks, T. Egami
Magnetoelastic Effects in Erbium 1263
    W.C. Hubbell, K. Salama, C.L. Melcher, P.L. Donoho

Section 30.   MICROWAVE DEVICES AND MATERIALS

Tunable, Low Dispersion Magnon Tunnel Elastic Wave Transducers 1268
    J.T. Carlo, F.R. Morgenthaler
A Broadband Ferrite Limiter 1273
    S.S. Elliott
Magnetic Wave Propagation in YIG and Lithium Ferrite Slabs 1278
    N.D. Wilsey, C. Vittoria
Studies of FMR Linewidth in Thick YIG Films Grown by Liquid Phase Epitaxy 1279
    J.D. Adam, J.M. Owens, J.H. Collins
Approximate Method for FMR in Metals 1284
    A. Yelon, G. Spronken, T. Bui-Thieu, R.C. Barker, Y.J. Liu, T. Kobayashi
Coupled Uniform and Spin-Wave Modes in FMR 1285
    Y.J. Liu, R.C. Barker, A. Yelon
Electromagnetic Generation of Ultrasonic Waves in 3-d Transition Metals 1286
    M. Hanabusa, T. Kushida, J.C. Murphy
Low Field Resonance and Relaxation in Thin Ferromagnetic Films 1287
    T.P. Kehler, R.L. Coren
Spin Waves in Thin Films with a Diffusion Zone 1292
    C.S. Guenzer, C. Vittoria, H. Lessoff

Section 31.   SINGLET GROUND STATES

Evidence for Low Temperature Singlet Ground State in the Dense Kondo System $CeAl_3$ 1297
    J.V. Mahoney, V.U.S. Rao, W.E. Wallace, R.S. Craig, N.G. Nereson
Magnetic Ion-Lattice Interaction in Rare Earth Antimonides 1298
    E. Bucher, L.D. Longinotti, B. Lüthi, M.E. Mullen
Quadrupole Exciton Dispersion and the 151K Phase Transition in $PrAlO_3$ 1299
    J.K. Kjems, G. Shirane, R.J. Birgeneau, L.G. Van Uitert
Unitary Transformation Approach to the Singlet Ground State Problem 1300
    T. Egami, M.S.S. Brooks
Soft Modes in Singlet-Ground-State Systems 1305
    W.J.L. Buyers, T.M. Holden
The Ising Model With a Transverse Field 1306
    Y.L. Wang, J.W. Johnson
Weakly-Interacting Singlet-Triplet Magnetic Systems: Application to Copper Nitrate 1311
    J.C. Bonner, S.A. Friedberg
Singlet Ground State Effects in $NiSnCl_6 \cdot 6H_2O$ and Related Compounds 1316
    D. Meier, M. Karnezos, S.A. Friedberg
Binary Induced-Moment Crystal in a Mean-Field Model 1317
    E. Shiles, G.B. Taggart, R.A. Tahir-Kheli

Section 32.   INTERDISCIPLINARY AND APPLICATIONS

Effects on Plant Embryos of Strong Steady Magnetic Fields with and without a Gradient 1322
    P.W. Neurath, V.H. Neurath
Superparamagnetism in the Fungus, Phycomyces 1326
    K. Spartalian, N. Smarra, W.T. Oosterhuis
Magnetometric Measurements as Novel Means for Astronomical Observation 1330
    B.Z. Kaplan

Observation of Human Cardiac Bloodflow by Non-Invasive Measurement of Magnetic Susceptibility Changes J.P. Wikswo, Jr., J.E. Opfer, W.M. Fairbank	1335
Electromagnetic Propulsion for Magnetically Levitated Vehicle W.J. Harrold, R.S. Kasevich, C.H. Tang, N.P. Viens	1340
Suitability of the Rare-Earth Compounds $Dy_2Ti_2O_7$ and $Gd_3Al_5O_{12}$ for a Low Temperature (4K-20K) Magnetic Refrigeration Cycle D.J. Flood	1345
High Frequency Hysteresis Loops of High Voltage Ferrite Pot Core Transformers J.R. Asik	1349
A Magnetic Study of the Twist Viscosity in Nematic Liquid Crystals P.J. Flanders	1354
Fatigue Crack Growth Studies with Barkhausen Effect Type Measurements K. Schroder, J.C. McClure, Jr.	1355
Imitation Theory-Cooperative Phenomena in Social Systems E. Callen, D. Shapero	1360
Transferred Hyperfine Interactions in Heme Proteins R.G. Shulman, A.M. Mayer, S. Ogawa, T. Yamane, G.W. Kenner, K. Smith, A. Gonsalves	1361
Large Scale Applications of Superconductivity and Magnetism S. Foner, B.B. Schwartz	1362

Section 33.   SOFT MAGNETIC MATERIALS

(110) [001[Textured 0.80% Si-Fe for the Proton Synchrotron T. Wada, K. Takashima, M. Kawashima	1363
Domain Structure and Magnetic Losses in Minor B-H Loops of (100) [001] Crystals of 3% Si-Fe J.N. Sun, J.J. Kramer	1367
Magnetic Domain Studies in Iron-3 1/4 Weight Percent Silicon Transformer Sheet using the Scanning Electron Microscope D.E. Newbury, H. Yakowitz	1372
Ferromagnetic Domain Walls in Amorphous Iron Alloy D.K. Paul, J. Marti, L. Valadez	1377
Determination of the Domain Wall Energy from Initial Magnetization Curve in Ferrimagnetic Polycrystals M. Guyot, A. Globus	1382
Bloch Line Mobility of $\sim$ 48,000 cm/Oe-sec, Automatic Wall Placement, and some Dynamic Properties of Crossties L.J. Schwee, H.R. Irons, W.E. Anderson	1383

Section 34.   SURFACE AND SIZE EFFECTS

Surface Spin-Exchange-Scattering of Polarized Photo-Electrons from Ferromagnetic Polycrystalline Films M. Campagna, K. Sattler, H.C. Siegmann	1388
Attenuation Length of Hot Electrons in Ferromagnetic Ni D.T. Pierce, H.C. Siegmann	1393
Susceptibility Enhancement at the Surface of Paramagnetic Metals R.A. Weiner	1398
Observation of Surface Spinwave Modes in YIG Films: Effects of Annealing R.A. Turk, J.T. Yu, P.E. Wigen	1403
Exact Treatment of Non-Periodic Boundary Conditions for Heisenberg Chains K.H. Lee, Y.Y. Wang, H.W. White	1408
The Approach to Saturation in a Hard Ferrimagnet P.E. Clark, A.H. Morrish	1412

Section 1. Surfaces and Catalysis

MAGNETIC PROPERTIES OF SURFACES

P. Fulde
Institut Max von Laue-Paul Langevin,
(8046) Garching, Germany

ABSTRACT

The environment of a surface atom is of lower symmetry than that of a bulk atom. This can be very important in determining the magnetic properties of surfaces. We demonstrate this by investigating the surfaces of Van Vleck paramagnets and ferromagnets and of transition metals. Special attention is focused on the occurrence of surface magnetism (Van Vleck systems) and of spin dependent surface resonances (transition metals).

I INTRODUCTION

Surfaces of magnetic materials have obtained considerable attention over the last few years. Although most of the work done is theoretical in nature [1-4] there are also experimental results available [1,5,6] which indicate interesting surface effects. The theoretical work has been mainly concentrated on surfaces of Heisenberg- and Ising ferromagnets. There the surface effects result from a reduced number of nearest neighbors of a surface atom and from a possible change (usually decrease) of the nearest neighbor interaction at the surface. It is worth noticing that both models are insensitive to another important aspect of surfaces namely that surface atoms sit in an environment of lower symmetry than in the bulk. Therefore all those systems will show strong surface effects for which the symmetry of the local environment plays an important role in determining the magnetic properties of the bulk. In fact the changes at the surface due to the difference in symmetry can be so drastic that the magnetic properties of the surface may have nothing in common with the magnetic properties of the bulk. In order to demonstrate this we will consider two examples.

First we consider the surface of exchange induced Van Vleck ferromagnets (induced moment systems) [7-9] or paramagnets. In those systems the bulk magnetic properties depend on a competition between the nearest neighbor exchange interaction (favouring magnetism) and the crystalline electric field splitting (disfavouring magnetism). The crystalline field splitting of the (rare earth) ions is closely connected with the symmetry of their surrounding and is therefore completely different at the surface and inside the bulk. This way we may obtain magnetic surface layers on paramagnetic bulk samples.

The second example concerns surfaces of transition metals where the magnetism results from itinerant d-electrons. The magnetic properties of the bulk depend on the exchange interaction and on the density of states at the Fermi level. The latter varies with the number of holes in the d-band. In addition it is of importance among how many degenerate subbands a given number of d-holes has to be distributed. For example paramagnetic Ni has 0.3 d-holes per spin. They are sitting in a peak near the upper edge of the d-band which has predominantly t_{2g} character [10]. This is a consequence of the fcc symmetry of the Ni lattice. At the surface the symmetry is lower and the 3-fold degeneracy is lifted. As a consequence all the d-holes sit in one subband instead of three. This is certainly an important surface effect leading to strong surface potentials. Actually it turns out that there are additional changes taking place at the surface which are of equal importance. Together they lead to the possibility of spin dependent surface resonances which may drastically influence the magnetic surface properties.

II SURFACE OF VAN VLECK MAGNETS

We consider a lattice of interacting rare earth non Kramers ions in a cubic environment. The Hamiltonian of such a system is conventionally written as

$$H = \sum_i H_i^{CEF} - \frac{1}{2} \sum_{i,j} K_{ij} \underline{J}_i \underline{J}_j \qquad (1)$$

Here K_{ij} labels the interaction between ions i and j and $\underline{J}_i$ is the total angular momentum of the i-th ion. H_i^{CEF} is the crystalline electric field Hamiltonian acting on ion i. If the ion is inside the bulk it is given by

$$H_b^{CEF} = B_4^0 (O_4^0 + 5O_4^4) + B_6^0 (O_6^0 - 21O_6^4) \qquad (2)$$

Here the O's are certain polynomials in the angular momentum components J_+, J_-, J_z known as Stevens operator equivalents. The B_4^0 and B_6^0 are certain constants which can be calculated within the point charge model for the different (6, 8 and 12 fold) coordinations. Since the physical systems of interest are usually metallic compounds, there remains the question of the applicability of the point charge model [12].

Van Vleck magnets are characterized by a nonmagnetic ground state of the single ion in the crystalline field. For simplicity we assume that the ground state is Γ_1. An example of a crystal field level scheme is given in

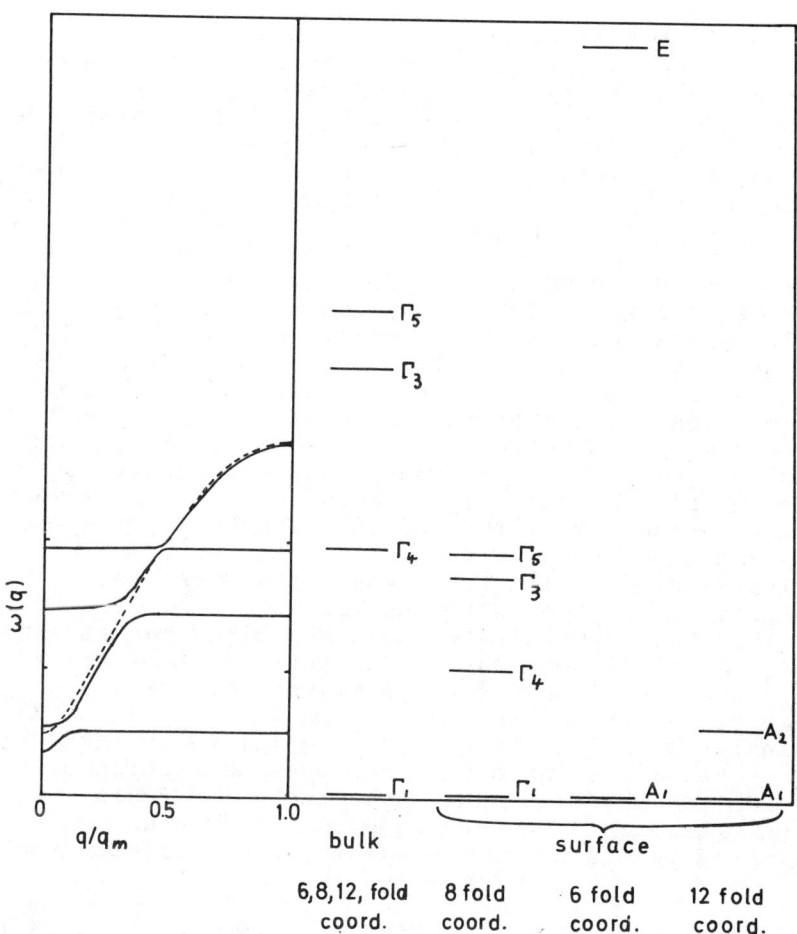

Fig. 1. Crystalline field level scheme for J = 4 ions in cubic environment for bulk and surface. Also shown is the bulk excitation spectrum for a temperature slightly larger than the bulk ordering temperature. (From Refs. 7 and 13.)

Fig. 1 where we have chosen $J = 4$ (Pr^{3+}). In the presence of the interaction K_{ij} the discrete energy levels are broadened into bands[13]. If the interaction is strong enough a magnetic phase transition may result.

For ions at a surface perpendicular to one of the principal axes H_s^{CEF} can be written as

$$H_s^{CEF} = B_2 O_2^0 + B_4' O_4^0 + B_4'' 5\, O_4^4 + B_6' O_6^0 - B_6'' 21 O_6^4 \qquad (3)$$

Again the B's can be calculated within the point charge model. The level scheme is mainly determined by the B_2 term which normally is an order of magnitude larger than the B_4 and B_6 terms except for the 8-fold coordination for which it vanishes. The results for the point charge model are shown in Fig. 1. For the 8-fold coordination the bulk level scheme is reduced by a factor of 1/2 at the surface. For the 6-fold coordination the separation of the lowest two energy levels is roughly 3 times that found in bulk while for the 12-fold coordination the splitting is smaller at the surface than in the bulk. In general however the parameters B_2, $B_4'^{(")}$, $B_6'^{(")}$ should be regarded as variables since the point charge model might be not a good approximation in some of the metallic compounds. Depending on the level scheme and the matrix elements between different levels at the surface one may have a higher ordering temperature at the surface than in the bulk. In favourable cases there may be magnetic ordering at the surface of a bulk probe which is paramagnetic even at $T = 0$. Of course there may be also situations in which the magnetization decreases at the surface as compared with the bulk.

It has been demonstrated that Van Vleck magnets are expected to show strong surface effects and that there is a good chance to observe surface magnetism. The above considerations, which were restricted to non Kramers ions in a singlet ground state can be generalized to magnetic ground states and to Kramers ions. The observation of spin polarized electrons in photoemission or field emission experiments on paramagnetic bulk probes would be a convincing way to prove the existence of surface magnetism.

III SURFACES OF TRANSITION METALS

Let us consider a transition metal in its paramagnetic state. The local density of states at a given lattice site i is described by the imaginary part of the electron Green's function $G(E, r_i, r_i)$. In general it will depend whether the atom is inside the bulk or part of the surface layer. We must therefore attach a surface potential to the surface layer such as to obtain the required changes. There are three distinct effects which contribute to the surface potential [14]. To demonstrate them we consider as a special example paramagnetic Ni. Inside the bulk paramagnetic Ni has 0.3 d-holes per spin and per Ni atom. The Fermi energy is situated in a high density of states peak near the upper edge of the d-band. The peak is primarily of t_{2g} character and therefore the 0.3 holes have to be distributed among 3 subbands. This picture is changed at the surface for the following reasons:

1. At a surface site the Ni wave functions have more space to extend than inside the bulk due to the smaller number of nearest neighbors. The renormalized atom approach can be used [14,15] to show that as a result the resonant d-level ϵ_ρ lies lower at the surface by ~ 0.15 Ryd as compared with the bulk. At the same time its width is reduced by approximately 1/3 due to the reduced number of nearest neighbors. As a consequence the number of d-holes drops from 0.3 per spin to approximately 0.2 per spin and atom.

2. In order to bring the Fermi level at the surface in agreement with the one inside the bulk charge will flow to the surface thus raising the position of the resonant d-level as well as the bottom Γ_1 of the conduction band. In doing the calculations it is found important to include screening of the bare Coulomb integrals as well as changes caused by the d-d and s-d Coulomb repulsion. It is found that this way the number of d-holes changes from 0.2 per spin to 0.1-0.15 per spin and atom.

3. While in the bulk the 0.3 d-holes per spin are distributed among 3 orbitals (t_{2g}) this is not true at the surface. Due to the lower symmetry of the surrounding of a surface atom the 3-fold degeneracy of the t_{2g} peak in Ni is lifted. For a surface of (100) orientation we obtain instead a singlet plus a doublet with the doublet being lower in symmetry. As a consequence at the surface all the d-holes sit in the singlet while the doublet is completely filled.

The effects considered in (1)-(3) can be described by attaching a surface potential to the surface layer. The latter can be thought of as a 2-dimensional "impurity" sheath surrounding the bulk probe. Of interest is the behaviour of the "impurity" sheath as the bulk becomes ferromagnetic. Especially one wants to know how big the magnetic moment is for the surface atoms and whether bulk and surface layer couple ferromagnetically or antiferromagnetically with each other. In a way the problem resembles the one of a (magnetic) impurity embedded in a ferromagnetic host [16,17] although the 2-dimensional character of the "impurity" sheath may not be neglected in a quantitative calculation.

Of importance is the concept of spin dependent surface resonances [14]. While in the paramagnetic state of bulk Ni the surface potential is spin independent this is changed if the bulk is ferromagnetic. The surface potential is then spin dependent and can be calculated self-consistently within the Hartree-Fock picture starting from the known potential for the paramagnetic bulk case. A spin dependent surface resonance occurs if the surface potential is strongly attractive for one spin direction and strongly repulsive for the other spin direction. In this case states may be pushed out of the d-band such

that they fall below the bottom of the band for one spin
direction and above the top of the band for the other
spin direction. This is illustrated in Fig. 2. Since

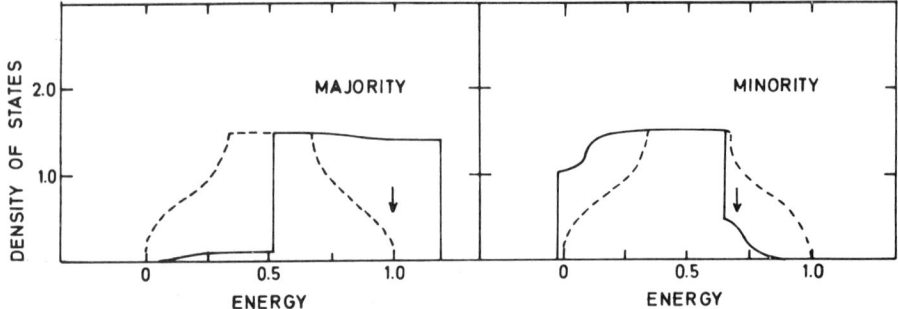

Fig. 2. Spin dependent surface resonance for a model-calculation attempting to describe ferromagnetic Ni. The Fermi level is indicated by an arrow and the dashed line indicates the bulk density of states (from Ref. 14).

those states overlap with the s-band they will be called resonant states. Their density of states reflects the 2-dimensional character of the "impurity" sheath. Spin dependent surface resonances incorporate features of both the (spin independent) Tamm's surface states as well as Friedel's virtual bound states. If spin dependent surface resonances are formed the coupling between bulk and surface moment can be antiferromagnetic. Model calculations which were performed for Ni suggest such an antiferromagnetic coupling.

Since there are various uncertainties in the parameters involved and approximations made in the model calculations they must be considered as a first attempt to describe the surface of Ni. The concept of spin dependent surface resonances however is of general interest and should have bearings on systems other than Ni.

IV CONCLUSIONS

By considering two specific examples we have demonstrated the importance of the lower symmetry of the environment of a surface atom as compared with the one of a bulk atom for the magnetic properties of surfaces. It was shown that it may lead to surface magnetism. Furthermore it may play an important role in the formation of spin dependent surface resonances and hence in a possible antiferromagnetic coupling between surface layer and ferromagnetic bulk.

REFERENCES

1. T. Wolfram, R. E. Dewames, W. F. Hall, P. W. Palmberg, Surface Sci. $\underline{28}$, 45 (1971).
2. D. L. Mills, Phys. Rev. $\underline{B3}$, 3887 (1971).
3. K. Binder and P. C. Hohenberg, Phys. Rev. $\underline{B6}$, 3461 (1972) and to be published.
4. M. J. Kaganov, Zh. Eksperim. i Teor. Fiz. $\underline{62}$, 1196 (1972) $\bigl/$ Sov. Phys. JETP $\underline{35}$, 631 (1972) $\bigr/$.
5. K. Sattler and H. C. Siegmann, Phys. Rev. Letters $\underline{29}$, 1565 (1972).
6. U. Gradmann, J. Appl. Phys. $\underline{40}$, 1182 (1969), and references cited therein.
7. I. Peschel and P. Fulde, Z. Physik $\underline{259}$, 145 (1973).
8. D. A. Pink, Phys. Rev. $\underline{B7}$, 239 (1973).
9. Y. Y. Hsieh and D. A. Pink, to be published.
10. L. Hodges and H. Ehrenreich, J. Appl. Phys. $\underline{39}$, 1280 (1968).
11. In the following recent review articles a complete list of references can be found citing the original work of Trammel, Bleaney, Cooper, Wang and others: P. Fulde and I. Peschel, Advan. Phys. $\underline{21}$, 1 (1972), R. J. Birgeneau, Proceedings of the 18th Annual Conference on Magnetism and Magnetic Materials 1972, American Institute of Physics Conference Proceedings edited by C. D. Graham and J. J. Rhyne, AIP, New York 1973.
12. K. C. Turberfield, L. Passell, R. J. Birgeneau and E. Bucher, Phys. Rev. Letters $\underline{25}$, 752 (1970).
13. I. Peschel, M. Klenin and P. Fulde, J. Phys. C$\underline{5}$, L 194 (1972).
14. P. Fulde, A. Luther and R. E. Watson, Phys. Rev. $\underline{B8}$, 440 (1973).
15. K. Levin, A. Liebsch and K. H. Bennemann, Phys. Rev. $\underline{B7}$, 3066 (1973).
16. J. Friedel in Proceedings of the International School of Physics "Enrico Fermi" Course 37 (1967), Academic Press, N.Y. 1967.
17. T. Moriya in Proceedings of the International School of Physics "Enrico Fermi" Course 37 (1967), Academic Press, N.Y. 1967.

PARA- TO ORTHO-HYDROGEN CONVERSION ON NONMETALLIC MAGNETIC SURFACES*

D. J. Scalapino
Department of Physics
University of California, Santa Barbara, Calif. 93106

K. G. Petzinger
Department of Physics
College of William and Mary, Williamsburg, Va. 23185

ABSTRACT

Recent theoretical and experimental results for the surface catalysis of para- to ortho-H_2 on magnetic materials are reviewed.

I. INTRODUCTION

According to the Pauli principle the wave function of an H_2 molecule must change sign when the two protons are interchanged. This implies that the states of the molecule with even rotational angular momentum J must have the proton spins in the singlet $I = 0$ state while states of odd J must have the proton spins in a triplet $I = 1$ state. The molecules with even J are called para H_2 while those of odd J are called ortho H_2. At temperatures large compared to the characteristic rotational temperature, the thermal equilibrium ratio of ortho to para H_2 approaches three due to the three possible nuclear states for the ortho molecules versus the one nuclear state for the para-molecules. At low temperatures, the equilibrium ratio of ortho to para concentrations approaches zero because the lowest ortho state has a finite rotational energy associated with $J = 1$. If the relative ortho-para concentration is not in thermal equilibrium, it relaxes toward equilibrium by interactions which must change both rotational and nuclear angular momenta. Here we will discuss certain theoretical and experimental aspects of this relaxation process when it occurs in a physiadsorbed H_2 layer on the surface of a magnetic material.

This "physical" or "low temperature" heterogeneous catalysis of ortho-para H_2 has been studied for many years. Our point is not to review this vast area of work, but to look at some features of it from the following point of view: the hydrogen molecule is a roving gradient magnetometer with reception bands in the infrared which can be used to explore the surface magnetization and adsorption potential near catalytic sites. The gradient nature of

---

*Research supported by AFOSR under Grant No. 71-2007.

the probe arises from the fact that a uniform magnetic
field cannot produce transitions which change the nuclear
spin. Furthermore, since the intraproton separation is
small compared to distances between the H_2 molecule and
the magnetic moments of the catalyst, only the gradient
part of the magnetization is important. Because the rotational angular momentum states of H_2 are coupled to the
nuclear spin states via the Pauli principle, the energy
transfers involve the infrared frequencies associated
with rotational energy differences rather than nuclear
spin state energies. Thus, the H_2 molecule is sensitive
to magnetic gradient fluctuations at certain narrow bands
in the infrared. These fluctuations may arise from the
relative motion of the H_2 molecule over a static magnetization and/or from the dynamics of the surface magnetization. Both the rotational motion and the translational
motion depend on the surface adsorption potential. Finally, as we will discuss at the end, this gradient
magnetometer is quantum mechanical so that one can modify
its behavior by changing its superposition of states with
an external magnetic field.

In Section II we begin by reviewing some of the features of ortho and para H_2 and discussing the effects of
the surface adsorption potential on the rotational levels.
The theory of the conversion is discussed in Section III
with an emphasis on the basic physical processes which
switch the probe molecule between para and ortho states.
In Section IV we conclude with a brief review of some experimental results. The reader will find that we have
still not really confronted the theory with experiment.
However, they are coming closer. Moreover, this work suggests the potential value of using H_2 ortho-para conversion
for the study of well-defined surfaces. Not only should
these be more susceptible to analysis, but one should be
able ultimately to use the temperature and magnetic field
dependence of the conversion as a probe of the surface.

II. THE ROTATIONAL LEVELS

For a free H_2 molecule, the rotational levels are
well approximated by the simple Hamiltonian

$$H_R = \frac{\hbar^2}{2I} \vec{J}^2 = B\vec{J}^2 \tag{1}$$

with $B = 86$ K. The eigen states of H_R are the spherical
harmonics Y_{JM} with $2J+1$ fold degenerate energy levels at
$BJ(J+1)$. Since the spherical harmonics change by $(-1)^J$
when $\theta \to \pi - \theta$, the even J levels must be associated with
the singlet nuclear spin state

$$X_s = \frac{1}{\sqrt{2}} (\alpha_1 \beta_2 - \beta_1 \alpha_2) \quad , \tag{2}$$

while the odd J levels have one of the triplet nuclear spin states

$$\begin{aligned} X_t^{(1)} &= \alpha_1 \alpha_2 \\ X_t^{(0)} &= \frac{1}{\sqrt{2}} (\alpha_1 \beta_2 + \beta_1 \alpha_2) \\ X_t^{(-1)} &= \beta_1 \beta_2 \end{aligned} \tag{3}$$

In this manner, the full wave function is odd under exchange of the protons. In thermal equilibrium, the ratio of the number of ortho molecules N_o^g to para molecules N_p^g in equilibrium in the gas phase is given by the ratio of their respective partition functions

$$\frac{N_o^g}{N_p^g} = \frac{3 \sum_{J \text{ odd}} (2J+1) e^{-BJ(J+1)/kT}}{\sum_{J \text{ even}} (2J+1) e^{-BJ(J+1)/kT}} \tag{4}$$

This is plotted as the "3D" curve in Fig. 1. Note that for room temperature this ratio of concentrations is essentially 3:1, at nitrogen (77 K) 1:1 and at low temperatures approaches 0:1.

The ratio of concentrations clearly depend upon the spectrum of the rotational eigen states of H_2. On a surface, the potential that an H_2 molecule experiences can hinder its rotation, changing the eigen states and hence the temperature variation of the equilibrium ortho to para surface concentrations. Sandler[1] was the first to observe this effect on TiO_2. He found that his observations at 90 K were in close agreement with a simple model in which the hindrance was approximated by assuming that the H_2 molecules on the surface behaved as two-dimensional rotors. In this case the wave functions are just $e^{iJ\theta}$ with $J = 0, \pm 1, \pm 2,...$ and energies BJ^2. As before the even J states (para H_2) go with X_s while the odd J states (ortho H_2) have X_t. In this case

$$\frac{N_o^s}{N_p^s} = \frac{6 \sum_{J=1,3,5,...} e^{-BJ^2/kT}}{1 + 2 \sum_{J=2,4,...} e^{-BJ^2/kT}} \tag{5}$$

This is plotted as the "2D" curve in Fig. 1. A separation factor S, defined as

$$S = \frac{(N_o^s/N_p^s)}{(N_o^g/N_p^g)} \qquad (6)$$

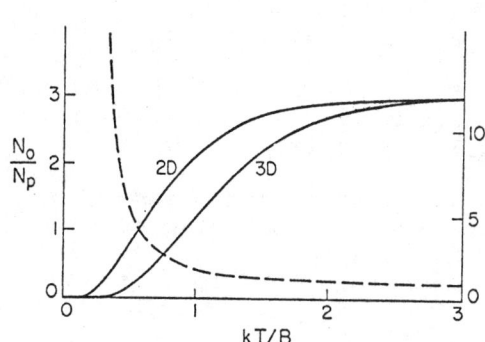

Fig. 1. Concentration ratios of ortho to para H_2 vs kT/B for free rotation (3D) and two-dimensional rotation (2D). The dashed curve is S, Eq. (6).

is plotted as the dashed curve in Fig. 1. Note that S at low temperatures S becomes large. This preferential adsorption of ortho forms the basis for separation of ortho from para H_2. There are a number of calculations[2,3] for various degrees of hindering so that the rotational states of the "typical" adsorbed H_2 molecules can be estimated from a fit of the observed separation factor $S(T)$. Similar measurements for D_2 and HD would provide further information on the rotational states of the probe on a given surface.

III. THE CONVERSION RATE

In this section we review a recent theoretical formulation of the conversion rate problem.[4,5] Let the protons of an H_2 molecule be located at $\pm \vec{d}/2$ relative to the center of mass $\vec{r}$ of the molecule. Then the interaction of the protons with a magnetic field is

$$\mathcal{H}_{int} = -\mu_p(\vec{I}_1 \cdot \vec{H}(\vec{r}+\vec{d}/2) + \vec{I}_2 \cdot \vec{H}(\vec{r}-\vec{d}/2)) . \qquad (7)$$

Expanding about the center of mass this becomes

$$\mathcal{H}_{int} = -\mu_p(\vec{I}_1+\vec{I}_2)\cdot\vec{H}(r) - \mu_p(\vec{I}_1-\vec{I}_2)\frac{\vec{d}}{2}:\vec{\nabla}\vec{H}(r) + \ldots \qquad (8)$$

The first term is a function of the total nuclear spin $\vec{I}$ and therefore cannot contribute to the transition matrix elements. The lowest order term just couples to the field gradient. The even terms in $\vec{d}$ vanish, and so the next higher order term is smaller by a factor $(d/2R)^2$ where R is some average distance between the H_2 molecule and a local spin. The magnetic field $\vec{H}(r)$ arising from paramagnetic electrons with moments $\mu_s \vec{S}_j$ located at positions $\vec{R}_j$ is

$$\vec{H}(\vec{r}) = \sum_j \frac{\mu_s}{|\vec{R}_j-\vec{r}|^5}\{\vec{S}_j|\vec{R}_j-\vec{r}|^2 - 3(\vec{R}_j-\vec{r})(\vec{S}_j\cdot(\vec{R}_j-\vec{r}))\} . \qquad (9)$$

Inserting this in Eq. (8) gives the basic interaction.

Now, proceeding in the usual way, the Born approximation for the transition rate can be expressed in terms of the correlation functions of the spins and the center of mass position of the hydrogen molecule. The reader is referred to Refs. 4 and 5 for these results. Here we outline a simple, approximate, way of obtaining the transition rate for some special cases which will emphasize the important physical ideas.

Suppose the surface has N_o adsorption sites and N_s of them contain a paramagnetic impurity with magnetic moment $\mu_s \vec{S}$. If R is the distance of the spin to the H_2 molecule, then the square of the matrix element for a J to J' transition is

$$k_{JJ'} \left(\frac{\mu_s \mu_p}{R^3}\right)^2 \left(\frac{d}{R}\right)^2 \frac{S(S+1)}{3} . \tag{10}$$

Here $k_{JJ'}$ is the form factor associated with the overlap of the rotational wave functions, $S(S+1)/3$ comes from the spin, and the remaining part is just the square of the field gradient.[6] Now if the hydrogen sits on a site for a typical time τ_H and the impurity spin relaxation time is τ_s, the frequency power spectrum of the interaction has the approximate form

$$\frac{\tau}{1 + (\tau\omega)^2} \tag{11}$$

with $\tau^{-1} = \tau_H^{-1} + \tau_s^{-1}$. The Golden Rule for the transition rate $\lambda_{JJ'}$, from a J para state to the J' ortho state is just: $2\pi/\hbar$ times the square of the interaction strength Eq. (10), multiplied by $\hbar^{-1}$ times the power spectrum evaluated at $\omega = (E_J - E_{J'})/\hbar = \omega_{JJ'}$

$$\lambda_{JJ'} = \frac{2\pi}{\hbar^2} k_{JJ'} \left(\frac{\mu_s \mu_p}{R^3}\right)^2 \left(\frac{d}{R}\right)^2 \frac{S(S+1)}{3} \frac{\tau}{1 + (\tau\omega_{JJ'})^2} \frac{N_s}{N_o} \tag{12}$$

The last factor N_s is just the probability that the H_2 molecule is on an active site. For isolated impurity spins, the spin relaxation time τ_s is sufficiently long that it does not contribute to the spectral weight.

In order to make contact with experiment, it is necessary to determine the number of para-ortho conversions per second. For the case in which the rate limiting process is the basic para-ortho conversion on the surface, rather than, for example, the surface adsorption-deadsorption step, the net number of H_2 molecules catalyzed per second can be written

$$k N \delta\theta \tag{13}$$

Here k is a rate constant which depends on the $\lambda_{JJ'}$

transition rates and the occupation of the surface rotational states in the ortho and para manifolds. N is the number of H_2 sites on the surface, and $\delta\theta$ is the non-equilibrium surface coverage factor

$$\delta\theta = |(N_p - N_p^s) - (N_o - N_o^s)|/N \qquad (14)$$

N_p^s and N_o^s are the surface equilibrium number of para and ortho H_2 respectively. For the usual case in which the surface rotational states of the separate para and ortho manifolds are in thermal equilibrium k is given by

$$k = \frac{\lambda_{01} + (\lambda_{21} + \lambda_{23})e^{-\beta\hbar\omega_2} + (\lambda_{43} + \lambda_{45})e^{-\beta\hbar\omega_4} + \cdots}{Z_p}$$

$$+ \frac{(\lambda_{10} + \lambda_{12})e^{-\beta\hbar\omega_1} + (\lambda_{34} + \lambda_{32})e^{-\beta\hbar\omega_3} + \cdots}{Z_o} \qquad (15)$$

Here Z_p and Z_o are the partition functions for the surface para and ortho rotational states. It is convenient to normalize k to k/λ with

$$\lambda = \frac{2\pi}{\hbar} \frac{N_s}{N} \frac{S(S+1)}{3} \left(\frac{\mu_p \mu_s}{R^3}\right)^2 \left(\frac{d}{R}\right)^2 \frac{1}{\hbar\omega_1} \qquad (16)$$

Then taking $S = 1$, $R \sim 2$ Å and ω_1 equal to the first rotational energy of a free H_2 molecule, the number of H_2 molecules converted per second Eq. (13) is approximately given by

$$kN\delta\theta \approx 10^2 \delta\theta (k/\lambda) N_s \qquad (17)$$

Thus, for a coverage $\delta\theta$ of order 1%, k/λ gives the number of H_2 molecules converted per surface spin per second.

A particular model of the surface sets both the rotational energies ω_1 and the parameter τ. From these the temperature dependence of k/λ can be obtained. This was done in Ref. 5 for the case of free rotation and three different translational modes: free two-dimensional translation, on-off site hopping, and diffusion. In Ref. 4 the case of two-dimensional rotation with the H_2 molecules diffusing over the surface was investigated. For the on-off site version discussed here, the site lifetime τ was parameterized as $\frac{a}{\omega_1} e^{b/t}$ with t the reduced temperature $T/172$. For physical adsorption one can estimate that a and b are of order 1. This corresponds to a vibrational frequency for the H_2 center of mass of order 10^{13}

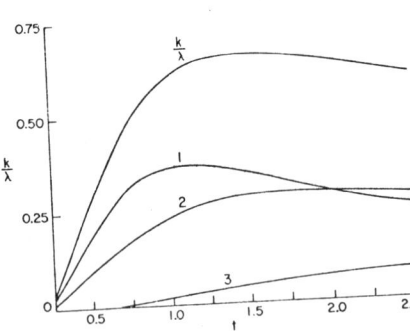

Fig. 2. The first three terms contributing to k/λ, Eq. (15), are plotted vs the reduced temperature $t = \frac{1}{172} T$ K. k/λ is obtained from the sum of these contributions.

and a hopping activation energy of order 350 cal/mole. The results[5] of calculating k/λ for the 1, $e^{-\beta\hbar\omega_1}$ and $e^{-\beta\hbar\omega_2}$ terms in Eq. (15) are shown in Fig. 2.

The basic temperature dependence of k/λ is determined by the effective spectral weight at the various rotational frequency differences as well as by the occupation of the different rotational levels. At low temperatures, where the dominant transition is between the $J=0$ and $J=1$ states, the spectral weight is small because of the slow hopping rate. Upon warming, the hopping rate increases pushing spectral weight out towards the ω_{10} frequency difference and leading to an increase in k/λ. However, at still higher temperatures, the higher rotational levels become populated leading to a relative decrease in occupation of the $J=0$, $J=1$ levels. The lack of spectral weight at the higher frequency differences $\omega_{J'J}$ cause a net decrease in k/λ as T continues to increase. This behavior is a general characteristic of a large number of models which have been studied.[5] However, the precise shape of $k(T)/\lambda$ should allow one to deduce information about the motional behavior of H_2 near the catalytic site.

In the case of a regular lattice of spins, several additional effects can occur. Ilisca[7] has pointed out that in this case the exchange coupling may be such that the energy for the rotational transition can come from the absorption or emission of a spin wave. Secondly, even when the energy transfer comes from the center of mass motion of the H_2 molecules, the static spatial correlations of the spins can effect the transition rate.[5] In the spin wave case, the basic modification of Eq. (12) is to replace the power spectrum $\tau(1 + (\tau\omega_{JJ'})^2)^{-1}$ by

$$\rho_s(\omega_{JJ'})n(\omega_{JJ'}) + \rho_s(\omega_{J'J})(1+n(\omega_{J'J})) \quad (18)$$

where $\rho_s(\omega)$ is the spin wave density of states and $n(\omega) = (e^{\beta\omega_s}-1)^{-1}$ is the usual Bose factor. The factor N_s/N_0 goes to a number order unity for this case. In the second case, static spin ordering plays a role. Here we consider the case of an antiferromagnetic system. In the ordered state, the local gradients are correlated with the spin directions. Assuming a simple site geometry and coupling, it can be shown[5] that the result for λ_{JJ}, given by Eq. (12) is modified by the replacement

$$S(S+1) \to S(S+1)\left(1 - \frac{\langle \vec{S}_o \cdot \vec{S}_\delta \rangle}{S(S+1)}\right) \qquad (19)$$

Here $\vec{S}_\delta$ is a near neighbor spin of $\vec{S}_o$ and $\langle \vec{S}_o \cdot \vec{S}_\delta \rangle$ is the static near neighbor correlation function. At temperatures well above the Néel temperature T_N, the correlation function $\langle \vec{S}_o \cdot \vec{S}_\delta \rangle$ goes to zero. However, as the antiferromagnetic correlations develop it becomes negative, and the conversion rate increases. For a simple Heisenberg exchange model, $\langle \vec{S}_o \cdot \vec{S}_\delta \rangle$ is proportional to the magnetic energy of the surface spins. Therefore, one might look for the temperature derivative of the conversion rate to vary as the surface specific heat nears T_N.

IV. SOME EXPERIMENTAL RESULTS

Experimentally the conversion rate can be determined by bringing a known non-equilibrium mixture of ortho and para H_2 in contact with a catalyst at temperature T and monitoring the time rate of approach of the gas concentrations towards equilibrium. Alternatively, the non-equilibrium mixture can be forced to flow past the catalyst and the rate constant determined by measuring the steady state change in concentration for a known flow rate. The ortho-para H_2 concentration is commonly determined by measuring the thermal conductivity of the gas.

Although the conversion rate predicted appears to be of the right order of magnitude, we have not made a detailed comparison.[8] We have been studying results for the rare earth oxides and dilute magnetic impurity systems such as ruby. Ideally one would like to have a set of experimental results in which the temperature dependence of the separation factor and the fractional coverage were known. In addition, an independent estimate of the effective number of spin sites should be available.

Recently Misono and Selwood[9] have observed a striking change in the rate constant for α Cr_2O_3 near its Néel temperature. Fig. 3 shows the variation of the rate constant with temperature. As the temperature is lowered through the Néel temperature, there is a dramatic increase in the conversion rate. It is presently not known whether this effect is due to spin wave interactions or to the static spatial ordering of the lattice. Other

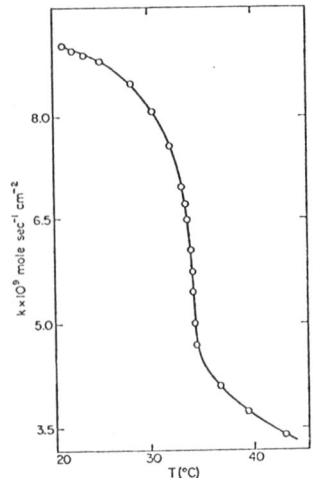

Fig. 3. Rate constant vs T for α Cr_2O_3 (courtesy P. W. Selwood)[3].

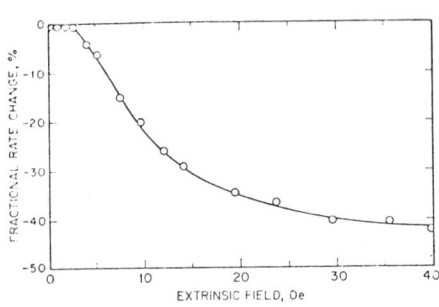

Fig. 4. Fractional change in the rate constant for Lu_2O_3 vs magnetic field (courtesy P. W. Selwood).

systems which exhibit ferromagnetic ordering are being studied.

Finally, Selwood[10,11] has also reported changes in the conversion rate in the presence of an external magnetic field. Fig. 4 shows the fractional change in the rate constant $(k(H) - k(o))/k(o)$ versus magnetic field for Lu_2O_3. Observable effects occur at magnetic fields of a few oersteds and saturation of the effect appears to set in at about 40 Oe. In addition to this "low field" effect Selwood has also observed increases in k over rare earths in fields of 10 kOe.[12] "High field" enhancements of k by factors of 2 occur for Yb_2O_3.

These low and high field effects are reminiscent of the influence of magnetic fields on luminescence involving triplet excitons.[13,14] The possibility of changes being produced by such small fields relative to kT/μ clearly depend upon the non-equilibrium nature of the problem. This suggests that the mechanism for the field effects depends on the existence of certain selection rules governing which of the spin triplet states can be converted to singlets by the gradient magnetization. We know, for example, in the case of free rotation in the absence of an external magnetic field, that of the nine $J=1$ ortho states only the five states with $J+I=2$ have matrix elements though $\mathcal{H}_{int}$, Eq. (8), to the $J=0$ para state.[5]

Ilisca and Legrand[4] found that for the two-dimensional limit of the hindered rotor, the transitions which conserve S_z and I_z are an order of magnitude stronger than those which involve changes in these quantum numbers. The external field changes the symmetry: in the low field region it competes with the nuclear dipole-dipole and $\lambda J \cdot I$ couplings which are of order 30 Oe, while in the high field region it may compete with the uniform part of the local paramagnetic impurity field. As the transitions are spread over more channels, the rate will increase even though the sum over the channels of the Golden Rule transition rates remains constant. However, what seems difficult to understand is that there is time for the spin precessions induced by the external field to alter the channel rates. With the recent observation by Eley, et al.,[15] that although a field of 2 kOe increases the rate of parahydrogen conversion on Nd_2O_3, it decreases the rate of orthodeuterium conversion, it is clear that the puzzle of the magnetic field effects is an intriguing one.

ACKNOWLEDGMENTS

The authors are greatly indebted to Professor P. W. Selwood for stimulating their interest in this problem, and for many helpful discussions and useful insights. One of the authors (DJS) wishes to thank Dr. P. Avakian and Dr. A. Suna of Du Pont for valuable insights into the effect of magnetic fields on triplet exciton processes. He would also like to thank Dr. P. Avakian for sending him a copy of an invited talk he gave on this subject at the XIth European Congress on Molecular Spectroscopy.[14]

REFERENCES

1. Y. L. Sandler, J. Phys. Chem. $\underline{58}$, 58 (1954).
2. D. White and E. N. Lassettre, J. Chem Phys. $\underline{32}$, 72 (1960).
3. T. B. MacRury and J. R. Sams, Mol. Phys. $\underline{20}$, 57 (1971).
4. E. Ilisca and A. P. Legrand, Phys. Rev. B $\underline{5}$, 4994 (1972).
5. K. G. Petzinger and D. J. Scalapino, Phys. Rev. B $\underline{8}$, 266 (1973).
6. For the case of free rotation of the H_2 molecules $k_{JJ'} = \frac{1}{6} (J + J' + 1)$.
7. E. Ilisca, Phys. Letters $\underline{33A}$, 247 (1970).
8. The calculated conversion rate agrees with some recent experiments on Y_2O_3 (P. W. Selwood, private communication, to be published.
9. M. Misono and P. W. Selwood, J. Am. Chem. Soc. $\underline{91}$, 1300 (1969).
10. P. W. Selwood, J. Catal. $\underline{22}$, 123 (1971).
11. K. Baron and P. W. Selwood, J. Catal. $\underline{28}$, 422 (1973).
12. P. W. Selwood, J. Catal. $\underline{19}$, 353 (1970).

13. R. C. Johnson, R. E. Merrifield, P. Avakian, and R. B. Flippen, Phys. Rev. Letters 19, 285 (1967).
14. P. Avakian, Proc. XIth European Congress on Molecular Spectroscopy in Tallinn, Estonia, USSR, 1973 (to be published in Pure and Applied Chemistry).
15. D. D. Eley, H. Forrest, D. R. Pearce, and R. Rudham, J. C. S. Chem. Commun. 1972, 1176.

EXPERIMENTAL RELATIONSHIPS BETWEEN CATALYSIS AND MAGNETISM

R. J. H. Voorhoeve
Bell Laboratories, Murray Hill, New Jersey 07974

ABSTRACT

Most catalysts are active because they contain transition metal ions, or are metallic transition elements. The reason for this is explained in a qualitative way. The relation between uncatalyzed chemical reactions and the elementary steps of a catalyzed reaction is indicated, with emphasis on the concept of the rate-determining step. Magnetic methods are of much use in the characterization of the particle size of ferromagnetic and antiferromagnetic catalysts, in the study of chemisorption, in the determination of the nature and number of active paramagnetic sites. The possible roles of magnetic ordering and of magnetic spin waves in the elementary steps of chemical reactions are discussed. Examples are given of changes in catalytic activity of ferromagnetic and antiferromagnetic catalysts correlated with the magnetic ordering at the Curie and Néel temperatures. The changes in the solid occurring at this critical temperature are not limited to the magnetic ordering but include also conductivity effects, changes in the defect structure, the lattice parameter and the nucleation rates of second phases. All of these may influence the rates of catalytic reactions.

INTRODUCTION

The correlation between the rate of chemical reactions and the magnetic field produced by external means or by paramagnetic-particles in the reaction medium has received active attention during a large part of this century. Recurrent increases in the level of interest at periods of about 20 years are a sure sign that there is an unsolved problem here. In the 1930's, J. A. Hedvall and his school, in the course of their work on the effect of phase transitions on the rate of chemical reactions, discovered that magnetic phase transitions may change the rate of the reaction in the neighborhood of the Curie temperature (Hedvall effect I). On occasion the temperature coefficient of the rate, i.e., the activation energy of the reaction, would be appreciably different below and above the Curie temperature (Hedvall effect II). In the 1950's these effects have been further investigated at the time when the electronic theory of catalysis was being developed, by Dowden,[2] Schwab[3] and coworkers. Especially Selwood[4] and his school have since that time elaborated the use of magnetic methods[5] in catalysis and the effects of intrinsic and extrinsic magnetic fields on the ortho-para H_2 conversion. The previous paper by Dr. Scalapino dealt with this reaction. Now, in the 1970's a further impetus is given by the theoretical interest in the magnetic properties of surfaces by Mills[6,7] and by Suhl.[8,9]

In the present paper, an overview will first be given of some concepts in catalysis. The effects of the presence and of the ordering of magnetic spins on the path and rate of reaction will be reviewed. Changes of the catalyst-reactants system which occur concurrent with magnetic ordering are examined. The experimental evidence for magnetocatalytic effects is critically discussed on that basis.

EFFECTS OF MAGNETIC FIELDS ON CHEMICAL REACTIONS

Direct evidence for the role of unpaired electrons is provided by the effects of magnetic fields on the rate and products of chemical reactions. The magnetic fields can be produced by external magnets (extrinsic fields) or by paramagnetic species in the reaction medium (intrinsic fields). The classic example of such effects is the catalysis of the ortho-para hydrogen conversion by paramagnetic species[10] and by combinations of solid paramagnetic catalysts and extrinsic magnetic fields of up to 18 KOe[11-13]. The rate of conversion may be field-enhanced by almost a factor two, due to an increase in the magnetic mechanism of o-p conversion. The latter depends on decoupling the proton spins in a strongly inhomogeneous field.[14,15] Field effects producing rate changes of about 30% or less at fields up to 80 kOe have been observed for several homogeneous catalytic reactions, e.g., for the isomerization of butene-2 under influence of iodine,[16] for radical pair reactions[17] and for reactions of biradicals.[18] The mechanism of these effects depends on the enhancement of the triplet-singlet transitions produced by equalizing the energies of these states in the magnetic field. Field effects have been well-established for reactions of large molecules such as enzymes.[19] An enhancement of the heterogeneous NH_3 synthesis over an iron catalyst by a field of 14-16 KOe has been reported.[20]

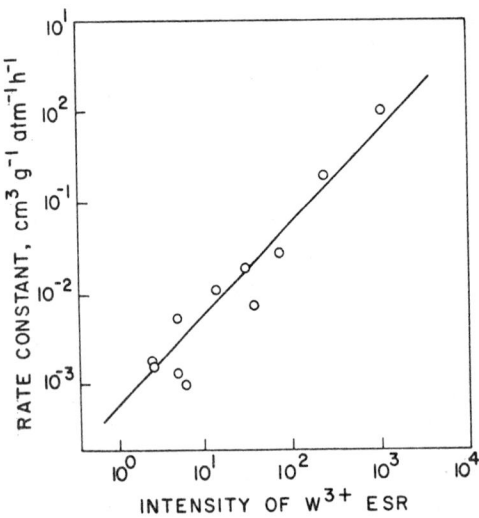

Fig. 1 Identification of W^{3+} as the active site in benzene hydrogenation over WS_2-based catalysts. After Voorhoeve.[25]

There is abundant experimental evidence for the importance of paramagnetic species such as transition metal ions in heterogeneous catalysis. Adsorption of reactant molecules on paramagnetic nickel particles has long been known to cause a large decrease of the magnetization, equivalent to the pairing of one or more electron spins in the bond formed between the molecules and the surface.[5,21] Measurements of the susceptibility and magnetization of catalysts are among the most important tools for characterizing magnetic catalysts dispersed on SiO_2, Al_2O_3 and other "carriers".[5,22] Relations between catalytic activity and the occurrence of paramagnetic ions in the surface have been established by ESR.[23,24] In a few cases, a quantitative relation between activity and the concentration of a specific ion in the surface of the catalyst has been found, e.g., for the hydrogenation of benzene on W^{3+} centers in the surface of WS_2 containing nickel[25] (Fig. 1). In the hydrogenation of ethylene, it was found that the activation energy E for the reaction decreased with decreasing distance between Cr^{3+} ions diluted with ZnO. The Cr-Cr distance was reflected by the antiferromagnetic coupling. A correlation between E and the Neel temperature T_N was found.[26] Of course, this may not be construed as a relation of cause and effect.

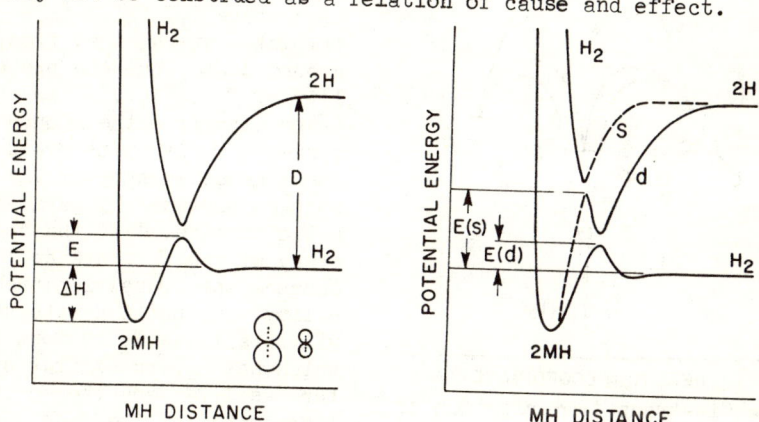

Fig. 2 Dissociative chemisorption of H_2 on a d-metal. Schematic potential energy diagram.

Fig. 3 Dissociative chemisorption of H_2 on s and d metals. Schematic potential energy diagram.

One of the most common questions asked about catalysis is how to explain the outstanding qualities of platinum as a catalyst. Somewhat more general, the question is: why are unpaired electrons of advantage and why should they be d-electrons, rather than s or p. As illustration, the class of reactions in which H_2 is a reaction partner is considered. In order to react, the H_2 molecule with a bond energy of D = 4.48 eV has to be dissociated (Fig. 2). The metallic catalyst assists in the dissociation by forming mainly covalent bonds with the hydrogen atoms. The potential energy diagram (Fig. 2) illustrates how the potential curves for 2H+M and H_2+M cross to provide for activated chemisorption of H_2. E is the activation energy for the adsorption. Inspection of the bond energies of MH bonds shows that $D(MH) = \frac{1}{2}(\Delta H+D)$ is 1-2 eV larger for metals with unpaired electrons such as Ni, Cu or Na than for metals with closed sub-shells such as Zn.[27] Stable chemisorption of H_2 is possible on the first group. However, H_2 chemisorption is a slow process on metals such as Na[28] or Cu, because the bonding by s electrons is more confined to short MH distances. In contrast, the spatial extent of d-orbitals leads to a smaller activation energy, even at the same well depth ΔH. This is illustrated in Fig. 3 by the s and d branches of the interaction potential, and it should be noted that $E(s) > E(d)$.

In addition to chemisorption of reactants, catalysts provide a stabilizing template for transitional stages in a chemical process, and here again the properties of d-orbitals are of advantage. By way of illustration, the cis-to-trans isomerization of butene-2 is considered. The carbon skeletons of the two isomers are given in Fig. 4. In the gas phase, the reaction may proceed through the transition state T_1, which is a triplet state at 1 eV from the ground state, but the probability of T_1 is low, since the transition from the ground state to T_1 is spin-forbidden. Inhomogeneous magnetic fields, such as provided by paramagnetic species, lift the degeneracy of the T_1 state and enhance the transition probability. Transition metals will stabilize T_1 to form T_2 which has a very similar structure as T_1,[30] and the activation energy of the reaction is thereby appreciably reduced. The availability in d-ions of occupied and unoccupied d-orbitals of the same energy is of importance for the catalysis of symmetry-forbidden reactions.[31]

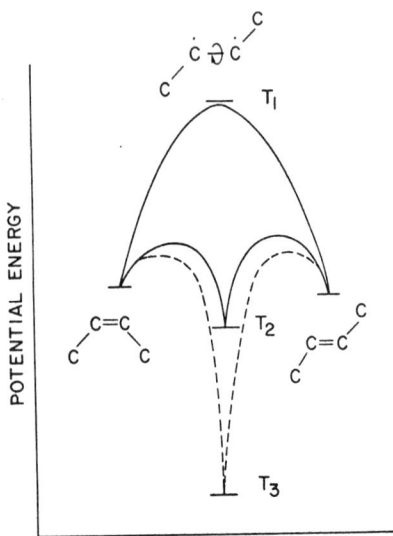

Fig. 4 Cis-trans isomerization of butene-2. Schematic potential energy diagram.

CHANGES OF THE CATALYST AND KINETICS OF THE TEST REACTION

In order to appreciate the following discussion of experimental evidence for magneto-catalytic effects, it is necessary to consider the steps which constitute the overall process which is measured. These are in general the preparation of the active site; the adsorption of reactants, chemical conversion in the adsorbed layer and desorption of the products. Schematically (Fig 5):

$$A_{gas} \rightarrow A_{ads} \rightarrow B_{ads} \rightarrow B_{gas} \qquad (1)$$

Any one of these steps may be slow compared with the others and it is the slowest step which is measured by the experiments. For example, in Fig. 4, T_2 is the transition state for a good catalyst on which all steps are of comparable rate. T_3, on the other hand is typical for a catalyst which bonds the intermediate too strongly, with the result that desorption becomes slow or impossible. In the last case, the catalyst is permanently changed by the process. This is the case for N_2O decomposition on Ni. In Fig. 5, the two catalysts are comparable, but the highest barrier for I is the chemical conversion, while for II the desorption of product B is slow. Changes from one type of kinetics to the other are possible on the same catalyst due to changes of the surface composition or of the temperature. These effects may interfere with the investigation of magneto-catalytic effects.

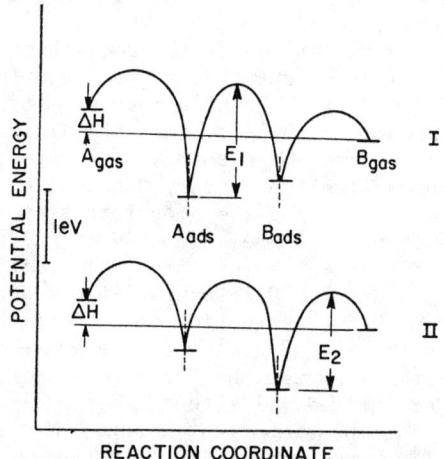

Fig. 5 Schematic potential energy diagrams for reaction (1) on two catalysts. I. The step $A_{ads} \to B_{ads}$ limits the rate. II. The step $B_{ads} \to B_{gas}$ limits the rate.

POSSIBLE EFFECT OF MAGNETISM ON THE CATALYTIC REACTION

Energy transfer. It is well known that a recombination reaction of two H atoms in the gas phase has a low probability due to the difficulty of removing the heat of formation H_2. In general, this energy can be transferred by emission of photons, as in chemiluminescence, or by the emission of an excited particle, or by transfer to a third body. The latter process is by far the most common on catalysts. Energy transfer to the solid can occur by exciting phonons or magnons. Since the energy to be transferred equals $E + \Delta H$ (Fig. 2), which is of the order of 1-2 eV, multi-phonon or multi-magnon excitation is required. From classical mechanics[31] it is known that phonon processes are very efficient for mass ratios between adsorbate and catalyst atoms close to one, but much less so for light atoms or molecules transferring energy to a heavy and hard metal. Therefore, it is expected that magnons will play a role primarily for light adsorbates such as H_2. This will be the subject of the following paper by Dr. Suhl.

Pre-exponential factor. The factor A in the usual rate equation, rate = f(pressures)A exp($-E/kT$), describes the probability of the process characterized by the activation barrier E. Alignment of the spins of an adsorbate species under influence of the intrinsic field of the paramagnetic atoms or ions in the catalyst surface may enhance A in the ferromagnetic and antiferromagnetic regions. On the other hand, an opposite effect on A is due to the fact that predominantly electrons of one spin polarization only can be accepted by a ferromagnetic lattice.[2] Inhomogeneous fields due to ferromagnetic particles may decouple the spins in biradicals such as T_1 in Fig. 4, thus substantially increasing the probability of singlet-triplet transitions and hence increasing A.[29] This, and the energy-transfer mechanism by coupling of the adsorbate-spin to the spin lattice[8,15] are probably the more important effects on A.

Activation energy. Contributions to the activation energy come from three sources. If a lattice spin has to be decoupled in order to bind with the adsorbate, an energy in the amount of $\sim n k T_c$ is required. Here, n is dependent on the coordination, T_c is the Curie temperature. In reactions where the adsorption is rate determining this will be added to the activation energy, whereas in reactions where desorption is rate determining the amount is subtracted from E. This effect is not more than 0.1-0.2 eV. An additional effect is due to the change of the work function beyond the Curie point. For nickel, this is 0.05 V between 350 and 450°C. Changes in the geometry of the surface, due to anomalous thermal expansion and the occurrence of magneto-striction will also have an effect on the value of E. In fact, the anomalous thermal expansion near the Curie point of nickel has been held responsible for the Hedvall effect I on this catalyst.[34]

Defect structure. The most important change which can occur together with the magnetic phase transition is a change of the defect structure of the surface, including both the kind and number of defects. This may lead to different reaction mechanisms, with concurrent changes in the activation energy, reaction orders etc. This would appear like a Hedvall effect II.

Nucleation of second phase. The solid state reactivity of many solids is clearly enhanced at magnetic phase transitions.[1,35] Important effects on the catalytic properties can be expected in catalysts prepared at low temperatures if the newly formed phase has better catalytic activity than the parent phase. As an example, Ni and NiO may coexist in catalysts as a result of such processes.

MAGNETO-CATALYTIC EFFECTS INVOLVING H_2^*

Many of the effects described by Hedvall[1] and his school were observed with hydrogenations of CO, unsaturated fats and the like. These reactions are quite complicated and the data will not be

---

*In the following examples, the transition temperatures T_N and T_c will be indicated in the figures by vertical lines.

discussed here. The same holds for newer work in which several catalyst preparations are compared.[36] The reproducibility of catalytic rate measurements is generally insufficient except where the same catalyst is being used in a wide temperature interval including the Curie or Néel point.

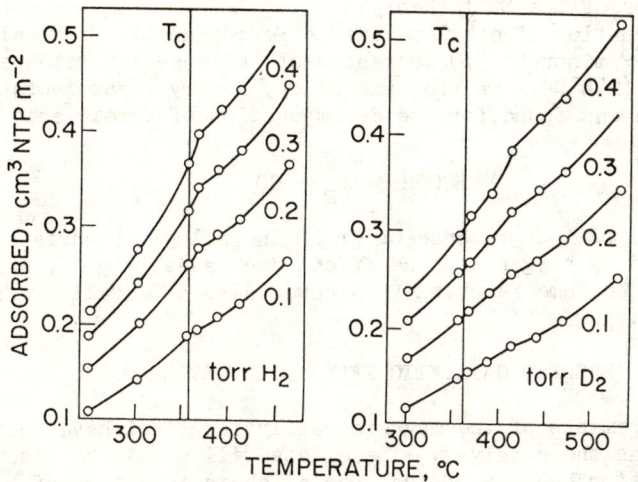

Fig. 6 Chemisorption of H_2 and D_2 on Ni. After Van Itterbeek et al.[34]

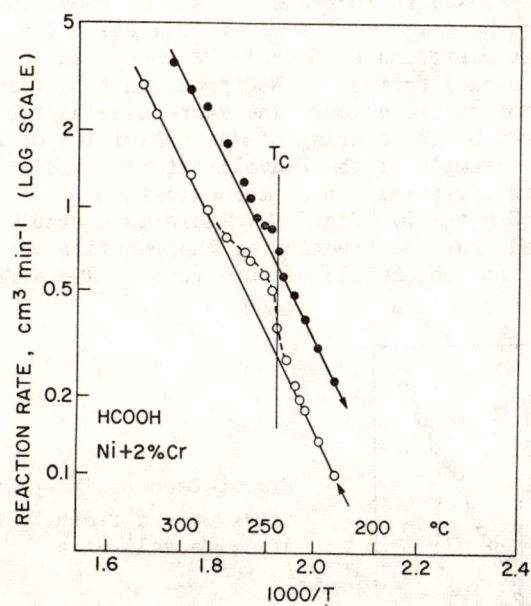

Fig. 7 Decomposition of formic acid on Ni 2% Cr alloy. Open data points: rising temperature. Solid data points: partially poisoned catalyst, decreasing temperature. After Schwab et al.[3]

Chemisorption of hydrogen and deuterium on nickel, measured in a static system, yielded adsorption isobares which show apparent anomalies for H_2 and D_2, especially at higher pressures. The anomalies appear close to the Curie point (360°C) for H_2, and at about 50°C higher temperature for D_2 (Fig. 6). The desorption of H_2 from Ni between 390 and 530°C has recently been measured and could be analyzed in terms of the Suhl theory.[37]

The hydrogenation of ethylene over a ferromagnetic Ni-Cr alloy catalyst showed[38] minor (< 5%) increases of the reaction rate near the Curie point (244°C). On the same alloy, a very clear increase in reaction rate was found for the decomposition of formic acid[3] (Fig. 7):

$$HCOOH \rightarrow H_2 + CO_2. \qquad (2)$$

However, because of the presence of Cr in the alloy the surface is quickly poisoned by oxygen and the effect decreases (Fig. 7, solid data points). The same reaction (2) showed also a Hedvall I effect on Ni-Cu alloys.[3]

MAGNETO-CATALYTIC EFFECTS INVOLVING O_2

Most measurements of the magneto-catalytic effect have been done with Ni or NiO as the catalyst. These data will mostly be discussed in the next Section because of the special characteristics of NiO.

The decomposition of nitrous oxide, N_2O, on Ni was studied in a flow system[38] and showed an increase in rate of about 10% near the Curie point. For the same reaction system, a steep increase in rate at the Curie point was found by Hedvall,[39] who noted that progressive oxidation of the nickel surface by N_2O produced NiO, which was a better catalyst for the reaction. The magneto-catalytic effect was eventually obscured by the overlap of the activities of Ni and NiO.

A very clear example of the Hedvall effect I was found by Parravano[40] in his investigation of the oxidation of CO over the perovskite $La_{0.65}Sr_{0.35}MnO_3$ (Fig. 9). This is a stable effect, and it was ascertained that the kinetics of the reaction was first order in $CO + O_2$ in the stoichiometric mixture (2:1). The effect of the

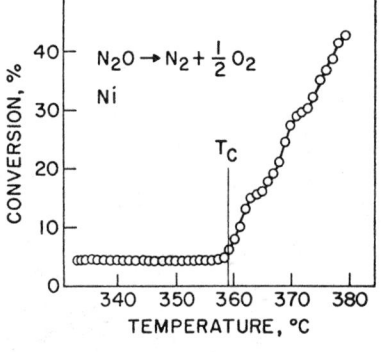

Fig. 8 Decomposition of nitrous oxide over a fresh Ni surface. After Hedvall et al.[39]

Curie interval is to decrease the reaction rate. Conflicting data have been obtained for the ferrite spinels.[38,41]

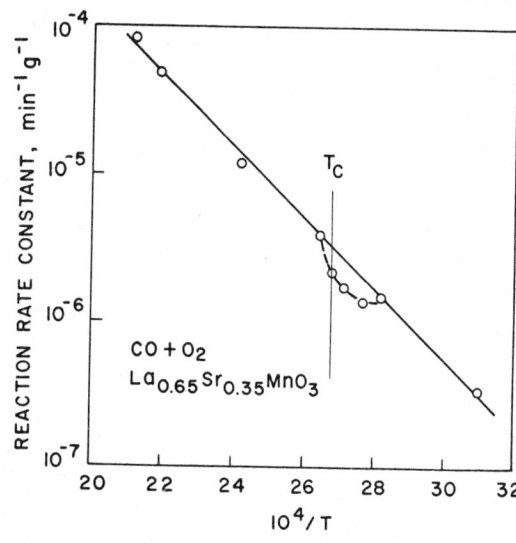

Fig. 9 Oxidation of CO with O_2 (2:1) over ferromagnetic $La_{0.65}Sr_{0.35}MnO_3$ powder. Curie point 100°C. After Parravano.[40]

"MAGNETO-CATALYTIC" EFFECTS ON NiO

The "magneto-catalytic" effects found on NiO catalysts provide an interesting example of the complexities of the subject. NiO has an antiferromagnetic spin lattice with a Néel temperature between 220 and 260°C, depending on the preparation procedure. The kinetics of the CO oxidation on NiO is strongly dependent on the CO:O_2 ratio in the gas phase,[42] with the oxidation state of the surface and the activation energy dependent on the ratio (Fig. 10). Changes of the apparent activation energy were observed near the Néel temperature of NiO. Studies with CO + O_2 (3:2) mixtures showed[43] anomalies in the reaction rate between 220° and 260°C, with the wider range at low firing temperatures of the NiO, when presumably the Ni^{3+} content is highest. In both studies it was observed that the electrical conductivity and the oxygen desorption rate changed at T_c.

Extensive studies of the oxidation of CO over NiO and of the electrical conductivity of NiO samples prepared in exactly the same manner as the catalysts have since been done.[44] These have established a complete parallel between the activation energies for the CO oxidation and the conductivity (Fig. 11). The catalysts were NiO prepared at 450°C and NiO thin films supported on silver. The electron distribution in the thin (< 1000 Å) NiO catalyst layer is changed by the proximity of the Ag support and also by UV irradiation. The catalytic activity was measured in UV light and in the dark. The experimental data are approximated very well with a model for the p-type semiconductor NiO in which below the Neel point (T_N) acceptor defects A_1 and a low concentration of donor defects D are present. Beyond T_N, more compensated acceptor defects A_2 and new defects D are formed, with the energy level for A_1 higher than for A_2. There

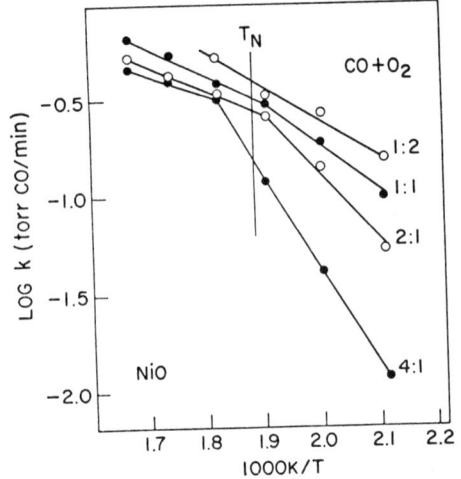

Fig. 10 Kinetics of the oxidation of CO over NiO, for various CO/O_2 ratios. After Bielanski et al.[42]

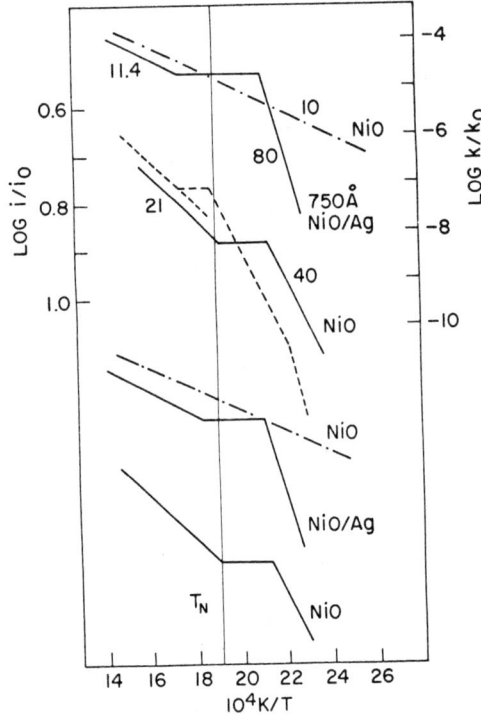

Fig. 11 Temperature-dependence of the conductivity i of NiO and of the rate constant k for CO oxidation over NiO and NiO/Ag catalysts. Top four curves: experimental data. — · — activity with UV lighting. — — — Conductivity, dark current. ——— Dark activities. Numbers next to curves give activation energies in kcal/mole. Lower three curves calculated on the basis of a defect model and band structure for NiO. The values of the reaction rate constant k and of the conductivity i are normalized with respect to the extrapolated values i_o and k_o at $1/T=0$. After Steinbach and Krieger.[44]

is beyond T_N a larger number of A_2 acceptors than of A_1 acceptors. The donor defects are fully ionized. The energy level diagram corresponding to Fig. 11, with all energies in kcal/mole, is given by

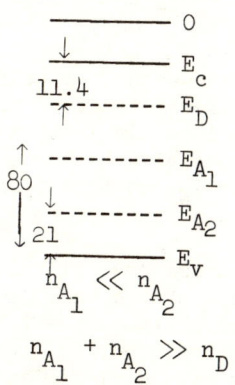

Fig. 12 Energy level diagram for NiO.44

$$n_{A_1} \ll n_{A_2}$$
$$n_{A_1} + n_{A_2} \gg n_D$$

The reaction rate is proportional to the number of holes in the valence band due to the formal reaction equation

$$CO + \oplus + NiO \rightarrow Ni^+ + CO_2 \qquad (3)$$

being the rate-determining step. Equation (3) is equivalent to breaking a covalent Ni-O bond. The dissociation of A_1 determines the concentration of holes at low temperature, whereas the dissociation of A_2 becomes the determining factor at higher temperature. In between, there are transition-ranges. For thin NiO layers on Ag, the dissociation of D becomes determining at

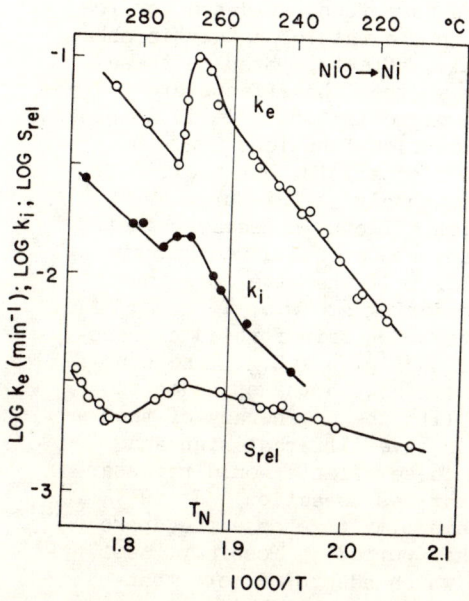

Fig. 13 Reduction of NiO with H_2 near the Neel temperature. k_e - overall reduction rate, k_i - rate of progress of the interface, S_{rel} - relative surface area of the interface. k_i and S_{rel} vertically displaced by an arbitrary amount.

high temperature. In UV light a Fermi level for defect electrons is established which is closer to E_v and this is reflected in the activation energy for the CO oxidation dropping to 10 kcal/mole without change in reaction mechanism. A detailed treatment of the reaction mechanism[44] is outside the scope of the present paper. The adsorbed oxygen species have been identified[45] as O_2 below room temperature, while at higher temperatures the sequence is O_2^-, O^- and O^{2-}. However, there is disagreement about the relative amounts at any temperature.[45,47] Both the defect structure and the conductivity vary strongly at the Neel point. It is felt that the magnetic transition is just one aspect of a rather drastic electronic reconstitution of the oxide at T_N, so that relating the catalytic properties to the ordering of the spins below T_N is extremely difficult.

An added complexity in the NiO system is the possibility of reduction to Ni. The kinetics of this process show a very pronounced Hedvall I effect near T_N, as measured by the overall production of H_2O from H_2 and NiO[48] (Fig. 13). Contributing to this are a modest anomaly in the rate of progress of the Ni-NiO interface, k_i, and an important anomaly in the nucleation rate of the new interface. This last anomaly is reflected in the curve for the relative interfacial area S_{rel}. Evidently the surface energy of a small ferromagnetic particle in a paramagnetic matrix is sensitive to the long range spin order in that matrix.

DISCUSSION AND CONCLUSIONS

Magnetic methods are of great value in identifying and counting paramagnetic ions active as sites in catalytic reactions; in measuring the bonding of adsorbates to paramagnetic catalysts; and in characterizing paramagnetic catalysts. Magnetic fields have been demonstrated to influence the rates of chemical reactions. A small number of experimental studies have found evidence for the effect of magnetic phase transitions on the rate of catalytic chemical reactions. These are all Hedvall I effects. Even in these cases it is still necessary to verify whether the effects are specifically due to the change in the magnetization. It is to be noted that other long-range spin correlation functions than the magnetization may have an effect on the catalysis, but their effects would not be observable as an anomaly at the Curie or Neel temperature. Two-spin correlation effects in nearly magnetic materials, as discussed by Suhl, must also be considered but are outside the scope of this paper. It is felt that the more important effects of the loss of spin ordering at T_N or T_c could be due to enhanced transition probabilities for spin-forbidden transitions, such as discussed in relation to Fig. 4. E.g., the coupling of the two spins of the triplet state with the spins on two adsorption sites in the lattice will lift the degeneracy of the triplet state if the two lattice spins have different sign and sufficiently long relaxation times. These effects would necessarily be rather specific for a particular test reaction. It is doubtful whether it is possible at this time to choose the appropriate catalytic test reaction for this purpose. Possibly, test reactions involving larger molecules which adsorb on more than one transition ion site and have conjugated double bonds are more likely to show effects of spin order than the small molecules used up till now. The solid state properties which change concurrently with the magnetic order have to be scrutinized for their effects on the catalytic reaction before a magneto-catalytic effect can be claimed.

ACKNOWLEDGMENTS

Thanks are due to Dr. H. Suhl and Dr. R. E. Dietz for very valuable discussions on this subject.

REFERENCES

1. J. A. Hedvall, Solid State Chemistry, Elsevier, Amsterdam,
2. D. A. Dowden, J. Chem. Soc., 1950, 242.
3. G.-M. Schwab and H. Goetzeler, Z. Physik. Chem. (Frankfurt), 2, 1 (1954).
4. P. W. Selwood, Magnetochemistry, Interscience, New York, 1956.
5. P. W. Selwood, Adsorption and Collective Paramagnetism, Academic Press, New York, 1962.
6. D. L. Mills, in Localized Excitations in Solids, Plenum Press, New York, 1968.
7. T. Wolfram and R. E. DeWames, in Progress in Surface Science, Vol. 2, Part 4, p. 233 (1972).
8. H. Suhl, J. H. Smith and P. Kumar, Phys. Rev. Lett., 25, 1442 (1970).
9. H. Suhl, unpublished lecture notes and private communications.
10. A. Farkas, Orthohydrogen, Parahydrogen and Heavy Hydrogen, Cambridge Univ. Press, 1935.
11. M. Misono and P. W. Selwood, J. Am. Chem. Soc. 90, 2977 (1968); 91, 1300 (1969).
12. P. W. Selwood, J. Catalysis 19, 353 (1970).
13. P. W. Selwood, J. Am. Chem. Soc. 92, 39 (1970).
14. E. Wigner, Z. Physik. Chem. (Leipzig), B23, 28 (1933).
15. E. Ilisca and A. P. Legrand, Phys. Rev. B5, 4994 (1972).
16. W. E. Falconer and E. Wasserman, J. Chem. Phys. 45, 1843 (1966).
17. R. Z. Sagdiev, K. M. Salikhov, T. V. Leshina, M. A. Kamkha, S. M. Sheim and Yu. N. Molin, JETP Lett. 16, 422 (1972).
18. G. L. Closs and C. E. Doubleday, J. Am. Chem. Soc. 95, 2735 (1973).
19. F. Figueras Roca, Ann. Chim. Fr., 2, 255 (1967).
20. G. G. Ivanov and G. P. Visokov, Zh. Prikladn. Khim. 25, 252 (1972).
21. G. A. Martin, Ph. De Mongolfier and B. Imelik, Surface Sci. 36, 675 (1973).
22. V. B. Evdokimov et al., Russ. J. Phys. Chem., 41, 1358 (1967); 44, 1435 and 1743 (1970); and references therein.
23. V. V. Voevodskii, in Proceed. Third Internat. Congr. Catalysis, Amsterdam 1964 (W. M. H. Sachtler, G. C. A. Schuit, P. Zwietering, editors), p. 88, North-Holland, Amsterdam, 1965.
24. R. J. Kokes, in Experimental Methods in Catalytic Research, (R. B. Anderson, editor), p. 436, Academic Press, New York, 1968.

25. R. J. H. Voorhoeve, J. Catalysis 23, 236 (1971).
26. H.-P. Walter, Z. anorg. allgem. Chem., 375, 214 (1970).
27. G. Herzberg, Spectra of Diatomic Molecules, Van Nostrand, Princeton, N.J., 1950.
28. J. Turkevich and T. Sato, in Proceed. Fifth Internat. Congr. Catalysis, Palm Beach, 1972 (J. Hightower, editor), p. 587, North-Holland, Amsterdam, 1973.
29. R. Ugo, in Proceed. Fifth Internat. Congr. Catalysis, Palm Beach, 1972 (J. Hightower, editor), p. 587, North-Holland, Amsterdam, 1973.
30. F. D. Mango and J. H. Schachtschneider, in Transition Metals in Homogeneous Catalysis (G. N. Schrauzer, editor), p. 223, Marcel Dekker, New York, 1971.
31. L. Trilling, Surface Sci. 21, 337 (1970).
32. R. A. Harman and H. Eyring, J. Chem. Phys. 10, 557 (1942).
33. A. B. Cardwell, Phys. Rev. 76, 125 (1949).
34. A. van Itterbeek, P. Mariens and O. van Paemel, Ann. Phys. 18, 136 (1943).
35. M. Daire and H. Forestier, in Proceed. Internat. Symp. Reactivity of Solids, Amsterdam 1960, p. 122.
36. D. Mehandjiev and G. Bliznakov, in Proceed. Third Internat. Congr. Catalysis, Amsterdam, 1964 (W. M. H. Sachtler, G. C. A. Schuit, P. Azietering, editors), p. 781, North-Holland, Amsterdam 1965.
37. L. A. Peterman, in Adsorption-Desorption Phenomena, F. Ricca, (editor), p. 227, Academic Press, London, 1972.
38. G.-M. Schwab and H. Goetzeler, Z. Physik. Chem. (Frankfurt), 4, 148 (1955).
39. J. A. Hedvall, R. Hedin and O. Persson, Z. physik. Chem. (Leipzig) B27, 196 (1934).
40. G. Parravano, J. Am. Chem. Soc. 75, 1497 (1953).
41. J. A. Hedvall and A. Berg, Z. physik Chem. (Leipzig), B41, 388 (1938).
42. A. Bielanski, J. Deren, J. Haber and J. Sloczynski, Z. physik. Chem. (Frankfurt) 24, 345 (1960).
43. A. Cimino, E. Molinari and G. Romeo, Z. physik. Chem. (Frankfurt) 16, 101 (1958).
44. F. Steinbach and K. A. Krieger, Z. physik Chem. (Frankfurt) 58, 290 (1968).
45. E. R. S. Winter, J. Catalysis 6, 35 (1966).
46. A. Bielanski and M. Najbar, J. Catalysis 25, 398 (1972).
47. J. Deren, S. Mrowec, J. Mater. Sci. 8, 545 (1973).
48. B. Delmon and A. Roman, J. Chem. Soc. FI 69, 941 (1973).

MAGNETIC EFFECTS IN SURFACE REACTIONS*

Harry Suhl
Univ. of Calif., San Diego, La Jolla, California 92037

ABSTRACT

Certain chemical reactions over solid substrates proceed at anomalous rates whenever the substrate carries anomalously large collective or quasi-collective fluctuations due, for example, to the proximity of a magnetic phase transition. Experimental evidence, and theoretical aspects of this effect will be discussed.

I. INTRODUCTION

Scattered through the literature on heterogeneous catalysis are references to anomalies in reaction rates near a phase transition of solid substrates. In most cases these observations, though suggestive, were not followed up in a systematic manner; only recently have studies begun; at least one of these concerns itself with a magnetic transition of the substrate.

How do these anomalies arise? A possible answer suggests itself upon a re-examination of the general theory of reaction rates as formulated by Kramers.[1] This theory treats a chemical reaction as a form of Brownian motion in the phase space of the N inert cores of the participating reagent atoms. The reacting system is represented by a point in the 3N dimensional space subtended by the coordinates of the cores. For each assignment of these coordinates, the system has a certain potential energy, which is thus a hypersurface $V(R_1 R_2 \ldots R_{3N})$ immersed in the 3N+1 dimensional cartesian space $\{\{\vec{R}\}, V\}$. That energy function is usually calculated in Born Oppenheimer approximation (in which the cores are held fixed, and the lowest energy of the outer electrons is calculated as a function of $\vec{R}$). Each minimum in the surface V corresponds to a chemical compound; a reaction corresponds to the motion of the representative point from one minimum to a neighboring one. For energetic reasons, the preferred path is close to the line of steepest ascent (and then descent) over the saddle point.

The propelling force in this motion (which is thus essentially one dimensional), as well as the "friction" needed to stabilize the reaction, are supplied by the fluctuating, random interactions of the reagents with a heat bath (the solid substrate in the case of heterogeneous catalysis). Furthermore, in most cases it is sufficient to consider this Brownian motion to be classical. The effective friction coefficient η characterizing this motion is related to the autocorrelation of the random force, whose principal effect is to impart a rapid series of small momentum changes to the particle. Kramers treats the problem as a transport process; because

---

*This research was supported by the Air Force Office of Scientific Research (AFSC) under grant no. AFOSR-72-2156 Mod. C.

the "collisions" involve small momentum changes, the appropriate equation governing this process is the Fokker-Planck equation[2] for the probability distribution f(R,P,t) for finding the particle near R with momentum P:

$$\frac{\partial f}{\partial t} + \frac{P}{M}\frac{\partial f}{\partial R} + F\frac{\partial f}{\partial P} = \eta \frac{\partial}{\partial P}(Pf + \frac{k_B T}{M}\frac{\partial f}{\partial P}). \quad (1)$$

where $F = -\partial V/\partial R$, k_B is Boltzmann's constant, T the temperature. No general solution being available, Kramers solves this equation for a steady state reaction ($\partial f/\partial t = 0$) in the limits of small and large η, arguing on the basis of a specific case that the socalled absolute rate theory of Eyring must hold in the intermediate range of values. Kramers' results are as follows: The reaction rate K (defined as the ratio of a specified current from one minimum in V to a neighboring one, divided by the number of particles in the first minimum needed to sustain this current) has the form

$$K = \upsilon \exp -\Delta V_B/k_B T \quad (2)$$

where ΔV_B is the height of the saddle point above the first minimum, and υ is given in terms of η by the formulae

$$\upsilon = \eta \frac{\Delta V_B}{k_B T} \quad \text{for} \quad \eta < \omega_A kT/\Delta V_B$$

$$\upsilon = \omega_A \quad \text{for} \quad \omega_A kT/\Delta V_B < \eta < \omega_B$$

$$\upsilon = \frac{\omega_A \omega_B}{\eta} \quad \text{for} \quad \eta > \omega_B \quad (3)$$

where $\omega_{A,B}^2 = \frac{2}{M}|\partial^2 V/\partial R^2|_{R=A,B}$ (See Fig. 1).

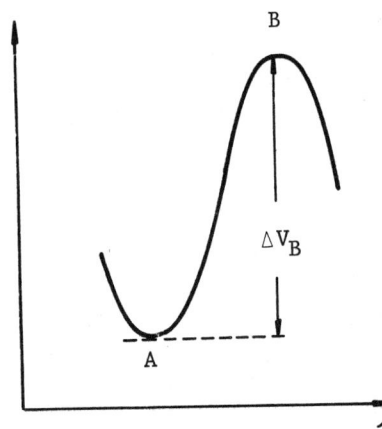

Figure 1

Note that in the intermediate range υ is independent of η. Most chemical reactions in the homogeneous phase are known or believed to be in that range.

We suggest that this may less frequently be true in the case of heterogeneously catalyzed reactions. In some of the experiments alluded to above, there is very often no or very little observed change in ΔV_B as the parameters of the problem are varied; yet the rate K undergoes large variations, suggesting that it is the prefactor that varies. Since ω_A is not likely to vary without an accompanying change in ΔV_B, one is driven to the conclusion that it is η that varies, and that one is either in the small or large η range. That η is very sensitive to changes in the substrate, particularly to changes in the fluctuation (or response) characteristic of the substrate follows from the fact that η is measure of the force flucutations exerted on the reagents by the substrate, and these fluctuations are directly related to the substrate response. For example, near the Curie point, of a magnetic substrate, spin flucutations (as well as the susceptibility) become large and, as described in an earlier publication,[3] this results in corresponding increases in η.

In support of the hypothesis that one may be outside the range of absolute rate theory, we note that one of the traditional reasons cited for the efficacy of catalysis in speeding reactions is that the presence of the catalyst lowers the barrier ΔV_B. This most likely is accompanied by a lowering in ω_B, and so the range

$$\omega_A k_B T/\Delta V_B < \eta < \omega_B$$

of applicability of absolute rate theory shrinks.

II. THE FRICTION COEFFICIENT IN TERMS OF ADPARTICLE RESPONSE FUNCTIONS

In this section we derive the appropriate friction coefficient η from the three basic responses of the reagents to an external perturbation in the form of a unifrom applied field: the density response, the current density response, and the force density response. The reason for doing this is that standard formulae for these exist which are model independent.

In Section IV these responses are then re-expressed in terms of substrate responses, in the limit of large adparticle masses.

We begin by writing down an equation of motion for the reduced density matrix of the adparticles alone. A simple form for such an equation can be found if it is assumed that the system as a whole, and in particular the subsystem of adparticles has associated with it two distinct timescales. There must be certain dynamical variables which vary much more slowly than others. (For example, the occupation numbers of adatom levels in the well responsible for the chemisorption is a case in point. Evidently these vary slowly compared with, for example, the charge fluctuations of a metal substrate.) A certain effective Hamiltonian H_s may be constructed which commutes with all these slow variables.[4] The density matrix ρ may be decomposed accordingly $\rho = \rho_s + \rho_f$, where $\rho_{s,f}$ vary slowly and fast respectively. The motion of ρ_s is dominated by H_s; the difference $H - H_s$ only causes Brownian motion, whose principal effect is to introduce irreversibility into the motion of ρ_s.

In addition we assume that $H_s = H_s^a + H_s^b + $ a very small coupling (we do <u>not</u> make such an assumption about the full Hamiltonian), where a and b refer to adparticles and bath respectively. Because of the large heat capacity of the bath, we finally write

$$\rho_s = \rho_s^a \rho_s^b$$

where
$$\rho_s^b \sim \exp -\beta H_s^b$$

Under these conditions arguments may be brought to show that the matrix elements of ρ_s^a satisfy the equation of motion

$$\frac{\partial}{\partial t}(\rho_s^a)_{nm} = \frac{1}{i\hbar}(H_s^a, \rho_s^a)_{nm} + [\sum_\ell w_{n\ell} e^{-\beta(\mathcal{E}_n - \mathcal{E}_\ell)}(\rho_s^a)_{\ell\ell} -$$
$$- \sum_\ell w_{n\ell}(\rho_s^a)_{nn}]\delta_{nm} - (1 - \delta_{nm})\sum_\ell (w_{n\ell} + w_{m\ell})(\rho_s^a)_{nm} \quad (4)$$

where the w are certain transition probabilities, and n labels the eigenstates of H_s^a with energies $\mathcal{E}_n$. For n = m, this equation becomes the familiar master equation. Henceforth we drop the subscript s. The equilibrium solution of (4) is

$$\rho_{eq}^a = \exp -\beta H^a / \operatorname*{Tr}_{ab} \exp -\beta H$$

On the other hand we know that

$$\rho_{eq}^a \sim \operatorname*{Tr}_b \exp -\beta H / \operatorname*{Tr}_{ab} \exp -\beta H$$

Thus

$$H^a = -\frac{1}{\beta} \log \operatorname*{Tr}_b \exp -\beta H \quad (5)$$

In general, H^a according to (5) will be a function of β. Only if its variation with β is very slow is H^a a useful Hamiltonian.* At sufficiently high temperatures (see footnote) this condition is presumably satisfied.

For small momenta we must have

$$H^a = \sum_{i,j=1}^{3N} \frac{1}{2M_{ij}} P_i P_j + V(Q_1 Q_2 \cdots Q_{3N})$$

where Q_i and P_i (i = 1 to 3N) are the cartesian coordinates and mo-

*We know from the work of Dagonnier[5] that at low enough temperatures the form (5) would give H^a an absurd temperature dependence. Dagonnier discusses the motion of an ion in a uniform medium. Because of translational invariance H^a can only have the form $P^2/2M_{eff}$, for small momenta P. The distribution $\exp -\beta P^2/2M_{eff}$ would give a mean square value $\langle P^2 \rangle$ proportional to $M_{eff}kT$. If M_{eff} were temperature independent, $\langle P^2 \rangle$ would go to zero. On the other hand the same mean square value calculated from the full Hamiltonian H remains finite at T = 0, due to zero point motion evoked by interaction with the medium. Hence M_{eff} would have to go as 1/T for low temperatures. Presumably, M_{eff} goes as $M_{eff}^1 + \langle P^2 \rangle_{zp}/k_BT$, and only for temperatures well above $\langle P^2 \rangle_{zp}/Mk_B$ does H^a make sense. We conjecture that in the present case, in which H^a has also a position dependent effective potential energy, the criteria for validity of H^a is $Mk_BT > |\langle P^2 \rangle_{zp}^{H^a} - \langle P^2 \rangle_{zp}^{H}|$ where the two zero point values on the right are calculated with H^a and the full H respectively.

menta, and the M_{ij} (generally functions of the Q) are effective masses.

In the large mass limit, in which the motion is classical, the effective potential is evidentally given by

$$\frac{e^{-\beta V(Q_i^o)}\hbar}{\int e^{-\beta V(Q_i)} dQ_i} = \frac{\text{Tr } \delta(Q_i - Q_i^o) \exp{-\beta H}}{\text{Tr } \exp{-\beta H}}$$

with the trace extending over <u>all</u> variables.

In this limit M_{ij} is diagonal, and the M_{ii} are equal, three at a time, to the bare adparticle masses. Since our ultimate objective here is not the determination of H^a but the determination of the dissipative processes, we shall for simplicity make the assumption that the M-tensor is diagonal and Q-independent. Thus

$$H^a = \sum_i \frac{P_i^2}{2M_i} + V(Q_i)$$

Instead of considering H, we consider

$$e^{iS} H e^{-iS}$$

where
$$S = \sum_{i_N} \log \left(\frac{M_i}{M}\right)(Q_i P_i + P_i Q_i)/2$$

$$M = \sum_{i=1} M_i$$

This transformation changes the kinetic energy to $1/2M \sum P_i^2$ (both in H and in H^a), and changes the Q_i to $\sqrt{M/M_i}\, Q_i$. All observables will be considered transformed in the same way, so their thermal averages will remain the same as those of the untransformed quantities.

In the new Hamiltonian,

$$H = \frac{1}{2M} \sum P_i^2 + V(Q_i)$$

we shall change coordinates from (Q_i) to arc length along the reaction path, and to a certain set of coordinates at right angles to the reaction path. This path over the saddle point can be characterized in several ways; a simple way is to note that it is the path of "slowest" steepest descent.* The change ΔV along a given arc length Δs drawn from a point Q_i on V is

$$\Delta V = \Delta s \cdot \sum \frac{\partial V}{\partial Q_i} n_i$$

where n_i are the direction cosines of Δs. Any path of steepest descent through Q_i is found by extremizing $\Delta V / \Delta s$ subject to $\Sigma n_i^2 = 1$. If we further seek the point Q_i at which the resulting ΔV is least we must extremize

$$\sum \frac{\partial V}{\partial Q_i} n_i$$

with respect to both n_i and Q_i, subject to $\Sigma n_i^2 = 1$ and V = const.

*The author is indebted to Dr. L. J. Sham for pointing this out.

The resulting equations are

$$\frac{\partial V}{\partial Q_i} - \lambda n_i = 0$$

$$\sum_i \frac{\partial^2 V}{\partial Q_i \partial Q_j} n_i - \mu \frac{\partial V}{\partial Q_j} = 0$$

or, combining these,

$$\sum_i^{3N} \frac{\partial^2 V}{\partial Q_i \partial Q_j} \frac{\partial V}{\partial Q_i} - \xi \frac{\partial V}{\partial Q_j} = 0 \qquad j = 1\ldots 3N$$

where ξ is a Lagrange parameter. These equations may be solved for the Q_i in terms of ξ, which gives the reaction path in parametric form.

We now change to the curvilinear coordinate system described by Corben and Stehle, and employed by Marcus[6] to discuss a three dimensional case. In fact, the following is a generalization of the work of Marcus. Define new coordinates q_i, of which the first q_o, is the arc length along the reaction path. In the hyperplane normal to the principal tangent at q_o, chose cartesian coordinates $q^1, q^2 \ldots q^{3N-1}$ along the principal normal, the binormal, and the successive higher normals (a total of 3N-1) to that curve. If the αth of these normal vectors $\vec{n}_\alpha$ has cartesian components n_α^i, we have

$$Q^i = Q_o^i(q^o) + \sum_{\alpha=1}^{3N-1} q^\alpha n_\alpha^i \ldots$$

Note that because the curve $Q_o^i(q^o)$ is orthogonal to V = const., V will depend only quadratically on the q^α for small q^α. The n_α^i are functions of q^o. They obey the Frenet-Serret equations[7]

$$\frac{d\vec{n}_\alpha}{ds} = -\varkappa_\alpha \vec{n}_{\alpha-1} + \varkappa_{\alpha+1} \vec{n}_{\alpha+1}$$

with $\varkappa_0 = \varkappa_{3N} = 0$, and $\vec{n}_o = \{dQ_o^i/ds\}$ the tangent vector. Here $\varkappa_1$ is the principal curvature and the remaining $\varkappa$ are the higher torsions.

The metric tensor of this coordinate system is diagonal, because of the ortho-normality of the n. We have, if α and β are not both = 0,

$$g_{\alpha\beta} = \sum_i \frac{\partial Q^i}{\partial q^\alpha} \frac{\partial Q^i}{\partial q^\beta}$$

$$= \delta_{\alpha\beta}$$

(recalling that $\partial Q^i/\partial q^o$ is the tangent vector n_o^i). On the other hand

$$g_{oo} = \sum_i \frac{\partial Q^i}{\partial q^o} \frac{\partial Q^i}{\partial q^o}$$

$$= \sum_i \left(n_o^i + \sum_{\alpha=1}^{3N-1} (-\varkappa_\alpha n_{\alpha-1}^i + \varkappa_{\alpha+1} n_{\alpha+1}^i) q^\alpha \right)^2$$

$$= (1-\varkappa_1 q^1)^2 + \sum_{\alpha=2}^{3N} (\varkappa_\alpha q^{\alpha-1} - \varkappa_{\alpha+1} q^{\alpha+1})^2 + \varkappa_2^2 (q^2)^2$$

$$= \theta^2 \text{, say}$$

$$= \det \| g_{ij} \| .$$

It follows that $g^{\alpha\beta} = \delta^{\alpha\beta}$ (α and β not both zero), $g^{00} = 1/\theta^2$. The Laplacian is[7]

$$\nabla^2 = \theta^{-1} \sum_{\alpha\beta=0}^{3N-1} \frac{\partial}{\partial q^\alpha} \theta g^{\alpha\beta} \frac{\partial}{\partial q^\beta}$$

Therefore $\frac{1}{2M} \sum_i P_i^2 = -\frac{\hbar^2}{2M} \frac{1}{\theta} \left[\frac{\partial}{\partial q^o} \frac{1}{\theta} \frac{\partial}{\partial q^o} + \sum_{\alpha=1}^{3N-1} \frac{\partial}{\partial q^\alpha} \theta \frac{\partial}{\partial q^\alpha} \right].$

Note that the curve $Q_o^i(q^o)$ is the projection of the reaction path onto the coordinate hyperplane. It is plausible (from consideration of simple 3-dimensional cases) to suppose that the curvatures $\varkappa$ of this curve vary only slowly with q^o. If that is the case, Schroedinger's equation $H^a\Phi = E\Phi$, with the substitution $\Phi = \theta^{-\frac{1}{2}} \Psi$, becomes

$$\left[-\frac{\hbar^2}{2M} \frac{\partial^2}{\partial q^2} + V'(q^o) \right] \Psi + \theta^2 \left[-\frac{\hbar^2}{2M} \sum_{\alpha=1}^{3N-1} \frac{\partial^2}{\partial q^{\alpha 2}} + v'(q^o, q^\alpha) - E \right] \Psi = 0 \quad (6)$$

where we have written

$$V(Q^i) = V(q^o) + v(q^o, q^1 \ldots q^{3N})$$

$$V'(q^o) = V(q^o) + V_c(q^o)$$

$$v'(q^o, q^\alpha) = v(q^o, q^\alpha) + V'(q^o)(1-\theta^{-2}) + v_c(q^o, q^\alpha)$$

Here

$$V_c(q^o) = \frac{\hbar^2}{2M} \left[\frac{3}{4} \varkappa_1^2 - \sum_{\alpha=2}^{3N-1} \varkappa_\alpha^2 \right]$$

and

$$v_c(q^o, q^\alpha) = \frac{\hbar^2}{2M} \left[\sum_{\alpha=1}^{3N-1} \frac{1}{4} (\partial \log \theta / \partial q_\alpha)^2 - \partial^2 \log \theta / \partial q_\alpha^2 \right] - V_c(q^o)$$

play the role of centrifugal effects due to curvatures (they vanish when $\varkappa = 0$). The point of the separation (6) is that all potentials in the second bracket vanish at $q^\alpha = 0$. For the small values of the q^α of interest (if V rises sufficiently steeply on either side of the reaction path) the second bracket is just a harmonic oscillator Hamiltonian, only terms up to order $q^{\alpha 2}$ being retained in the expansion of the potentials.*

Although the eigenstates of (6) have product form, (if the $d\varkappa^\alpha/ds$ are neglected) the density matrix $\rho(q^o q^\alpha)$ does not factor into $\rho(q^o)\rho(q^\alpha)$. Setting $\Psi = \Psi_1(q^o)\Psi_2(q^\alpha)$ we find that

$$\left[-\frac{\hbar^2}{2M}\frac{\partial^2}{\partial q^{o2}} + V'(q^o) \right]\Psi_1(q^o) = \alpha \Psi_1(q^o)$$

$$\left[-\frac{\hbar^2}{2M}\sum\frac{\partial^2}{\partial q^{\alpha 2}} + V''(q^o,q^\alpha) + \frac{\alpha}{\theta^2} \right]\Psi_2 = E\Psi_2$$

where α is a separation constant.

If β denotes the quantum numbers of the operation applied to Ψ_2, we see that

$$E = E(\alpha,\beta)$$

(rather than $E = E(\alpha) + E(\beta)$). From Liouville's equation for the matrix elements $\rho_{\alpha\beta,\alpha'\beta'}$ of ρ it then follows that an initial product form $\rho_{\alpha\alpha'}\rho_{\beta\beta'}$ will not remain a product in the course of time. We therefore write

$$H = H_1 + H_2 + \left(\frac{1}{\theta^2} - 1\right)H_1$$

$$= H_1 + H_2 + H_{12}$$

where

$$H_1 = -\frac{\hbar^2}{2M}\frac{\partial^2}{\partial q^{o2}} + V_o'(q^o)$$

$$H_2 = -\frac{\hbar^2}{2M}\sum_\alpha\frac{\partial^2}{\partial q^{\alpha 2}} + V'(q^o,q^\alpha)$$

and treat H_{12} as a perturbation. To see if it is small, we note that, in the case in which the levels of H_1 are taken to be harmonic oscillator levels of frequency ω_s, and those of H_2 have a typical frequency ω_T then since $1/\theta^{-2} - 1$ is of order $\varkappa^2 \langle q^{\alpha 2} \rangle$ (after displacements of the harmonic oscillators have been removed), the term

---

*The harmonic oscillators are actually displaced slightly (due to the linear term in $1/4(\partial \log\theta/\partial q_\alpha)^2$). This means that the actual reaction pthat is slightly shifted. Instead of $q^\alpha = 0(\alpha > 1)$ it is along $q^\alpha =$ displacement, which is a slow function of q^o. In the following we neglect this shift which easily can be taken into account if necessary (see Marcus (6)).

H_{12} is of order

$$\kappa^2 \frac{\hbar^2 \omega_s}{M\omega_T}$$

i.e. or order $\frac{\hbar^2 \kappa^2}{M}$. This is much less than the thermal energy (the average energy of $H_1 = H_2$), if the radii of curvature are much larger than a de Broglie wavelength. At usual temperatures of a few hundred degrees this is of order 10^{-8}cm, for hydrogen. From an estimate of the path of the reaction: chemisorbed $2H \to$ physisorbed H_2, this is marginally satisified. For larger mass atoms the condition is more easily satisfied. (It can be written $\gamma^2 \gg \hbar^2 kT/M$ where γ is a typical radius curvature or of torsion.)

We shall assume H_{12} to be small. Then ρ_a factors into $\rho(q^o)$ $\rho(q^1, q^2 \ldots q^{3N-1})$. Further, if the reaction proceeds slowly, we may assume that $\rho(q^\alpha)$ is a Boltzmann distribution.* Evidentally, we may then regard the q^α degrees of freedom ($\alpha > 0$) as part of the bath, and obtain an equation of the form (4) with ρ_a replaced by $\rho(q^o)$ and H_a by $H_1(q^o)$.

In many cases the Hamiltonian $H_1(q^o)$ will be such as to permit a reduction of this equation to a classical transport equation for the distribution function $f(q^o, p^o, t)$.

The quantum number n are then the cells around (q^o, p^o) themselves, and only the diagonal part of the master equation is needed. If $w_{nm} = w(q^o p^o, q^{o\prime} p^{o\prime})$ is taken to be zero unless $q^{o\prime} = q^o$ and w falls off so rapidly with $p^o - p^{o\prime}$, that only moments up to the second $\int (p^o - p^{o\prime})^2 w(q^o p^o, q^{o\prime} p^{o\prime}) dp^{o\prime}$, need be retained, the Fokker-Planck equation results:

$$\frac{\partial f}{\partial t} + \frac{p}{M} \frac{\partial f}{\partial q} - \frac{\partial V'(q)}{\partial q} \frac{\partial f}{\partial p} = \eta(q) \frac{\partial}{\partial p} (pf + Mk \frac{\partial f}{\partial p}) \quad (7)$$

where the superscript o has been dropped, as it will be in the following, where convenient.

The reason why a classical description for the motion along the reaction coordinate is frequently justified is the fact that the spectrum along this coordinate is essentially continuous. For example, the reaction in which the two chemisorbed hydrogens go to a form of H_2 is of this kind, if the adatoms are free to move in a "trough" along the surface.+

---

*The main effect of the slow q^o dependence of $\rho(q^\alpha)$ is then to convert the ΔV_b into $\Delta V_B - T \Delta S_B$, the activation force energy. (Here ΔS_B is the change in entropy due to changes in width of the "valley" between the minimum and the saddle point.)

+Strictly speaking, slow variation of $V'(q)$ is also required for the justification of (7). However, a further "coarse graining" in time of the Schroedinger equation, of the kind given by Goldberger and Watson[9], will give a slowly varying V', if the timescale of the reaction is long enough.

Finally, we indicate how $\eta(q)$ may be calculated directly by means of a "bootstrap" procedure as follows:
1. Obtain from the F.P. equation a relation between $\eta(q)$ and the density, current density, and energy density response of the system to a small force.
2. Calculate such responses from the Kubo formalism. This is normally done in cartesian coordinates.
3. Transform to curvilinear coordinates. The responses in the curvilinear system become linear combinations of those in the cartesian system.
4. Substitute the result in the relation between η and the responses.

A small force with cartesian components F_1^i ($i = 1 \ldots 3N$) applied to the assemblage of adparticles evokes responses in scalar, vector, and tensor quantities, such as density, current density, kinetic energy density, heat current, etc. For the first three of these we have, in cartesians,

$$R_1 = \sum_i X_i F^i \quad ; \quad J_1^i = \sum_j X_j^i F_1^j$$

$$K_1^{ij} = \sum_k X_k^{ij} F_1^k$$

Let lower case symbols denote these quantities in the Corben-Stehle coordinate system. We have, since F^i transforms like a vector

$$F^i = \sum_\alpha \frac{\partial Q^i}{\partial q^\alpha} f_1^\alpha$$

Hence, and since R_1 is a scalar and therefore equal to ρ_1,

$$\rho_1 = \sum_{i,\alpha} X_i \frac{\partial Q^i}{\partial q^\alpha} f_1^\alpha$$

$$= \sum_\alpha X_\alpha f^\alpha$$

with
$$X_\alpha = \sum_i X_i \frac{\partial Q^i}{\partial q^\alpha}$$

In the same way

$$j_1^\alpha = \sum_i \frac{\partial q^\alpha}{\partial Q^i} J_1^i$$

$$= \sum_{i,j} \frac{\partial q^\alpha}{\partial Q^i} X_j^i \frac{\partial Q^j}{\partial q^\beta} f_1^\beta$$

so that
$$X_\beta^\alpha = \sum_{i,j} \frac{\partial q^\alpha}{\partial Q^i} X_j^i \frac{\partial Q^j}{\partial q^\beta}$$

In the same way
$$X^{\alpha\beta}_{\ \ \gamma} = \sum_{ijk} \frac{\partial q^\alpha}{\partial Q^i} \frac{\partial q^\beta}{\partial Q^j} X_k^{ij} \frac{\partial Q^k}{\partial q^\gamma}$$

However, what is eventually needed is the gradient G^i of the kinetic energy density, which is the divergence of the energy density tensor:

$$G_1^i = \sum_j \frac{\partial}{\partial Q^j} (X_k^{ij}) F^k = \sum_k \phi_k^i F_1^k \text{ , say}$$

$$g_1^\alpha = \phi_\beta^\alpha f_1^\beta$$

and so $g_1^\alpha = \phi_\beta^\alpha f_1^\beta$, where

$$\phi_\beta^\alpha = \frac{\partial q^\alpha}{\partial Q^i} \left(\sum_{jk} \frac{\partial X_k^{ij}}{\partial Q^j} \right) \frac{\partial Q^k}{\partial q^\beta}$$

Now the matrix $|(\partial Q/\partial q)|$ has elements

$$\frac{\partial Q^i}{\partial q^o} = n_o^i + \sum_{\alpha=1}^{3N-1} \frac{dn_\alpha^i}{dq^o} q^\alpha$$

$$\frac{\partial Q^i}{\partial q^\alpha} = n_\alpha^i$$

and using the Frenet-Serret formulae we find that

$$\left(\frac{\partial Q}{\partial q}\right)^{-1} = \left(\frac{\partial q}{\partial Q}\right) = \left(\frac{\partial Q}{\partial q}\right)'/\theta^2$$

where the prime denotes the transpose. Hence

$$X_\alpha = \sum_i X_i \frac{\partial Q^i}{\partial q^\alpha} \quad ; \quad X_\beta^\alpha = \theta^{-2} \sum_{ij} \frac{\partial Q^i}{\partial q^\alpha} X_j^i \frac{\partial Q^j}{\partial q^\beta}$$

$$X^{\alpha\beta}_{\ \ \gamma} = \theta^{-4} \sum_{ijk} \frac{\partial Q^i}{\partial q^\alpha} \frac{\partial Q^j}{\partial q^\beta} X_k^{ij} \frac{\partial Q^k}{\partial q^\gamma}$$

and
$$\phi_\beta^\alpha = \theta^{-4} \sum_{ijk} \frac{\partial Q^i}{\partial q^\alpha} \frac{\partial Q^j}{\partial q^\gamma} \frac{\partial X_k^{ij}}{\partial q^\gamma} \frac{\partial Q^k}{\partial q^\beta}$$

Of particular interest X_o, X_o^o and X_o^{oo}, evaluated at $q^1 = q^2 = \ldots = 0$, since $\eta(q^o)$ may be expressed in terms of these quantities from the Fokker-Planck equation (7). At $q^1 = q^2 = \ldots = 0$, $\partial Q^i/\partial q^o = n_o^i$ the ith component of the tangent vector, and $\theta = 1$. Therefore

$$X_o = \sum_i n_o^i X_i \quad ; \quad X_o^o = \sum_{ij} n_o^i n_o^j X^{ij}$$

$$\varphi_o^o = \sum_{ij} n_o^i n_o^j \Phi_j^i \qquad\qquad X_o^{oo} = \sum_{ijk} n_o^i n_o^j n_o^k X_k^{ij}$$

Multiplying the linearized F.P. equation (7) by p and integrating over p we find the following relationship between the density $\rho_1 = \int f_1 dp$; $j_i^o \int \frac{p}{M} f_1 dp$ and $g_1^o = \frac{\partial}{\partial X} \int \frac{p^2}{2M} f_1 dp$

$$j_1^o M\eta = \left[f_1^o R_{eq}(q) + f_{eq} \rho_1 - 2 g_1^o \right]$$

where

$$f_{eg} = -\frac{\partial V'(q)}{\partial q} \quad \text{and} \quad R_{eq}(q) \sim \exp(-\beta V'(q))$$

Hence we have

$$\eta = \frac{1}{M} \left(\sum_{ij} n_o^i n_o^j X_j^i \right)^{-1} \left[R_{eq}(q) + f_{eq} \sum_i n_o^i X_i - 2 \sum_{ij} n_o^i n_o^j \Phi_j^i \right]$$

where the Q's in the response functions X have been re-expressed in terms of the q's, and the q^α have been equated to zero for $\alpha > 0$. Another, somewhat simpler expression for η may be obtained using the equation of motion linking the operators $\hat{R}$, $\hat{J}$ and $\hat{K}$ (see ref. (3)) and the force density operator $[\widehat{FR}]^i$.

These quantities are defined, in Q-representation, as

$$\hat{R}(Q) = \prod_{i=1}^{3N} \hat{\delta}(\hat{Q}^i - Q^i) = \prod_i \hat{\delta}^i$$

$$\hat{J}^i(Q) = \frac{1}{2M}(\hat{p}^i \hat{\delta}^i + \hat{\delta}^i \hat{p}^i) \prod_{k \neq i} \hat{\delta}^k = \hat{j}^i \prod_{k \neq i} \hat{\delta}^i$$

$$\hat{K}^{ij}(Q) = \frac{M}{2} \hat{j}^i \hat{j}^j \prod_{k \neq i,j} \hat{\delta}^k$$

$$[\widehat{FR}]^i = -(\partial \hat{H}/\partial \hat{Q}^i) R$$

and they satisfy

$$M \frac{\partial \hat{J}^i}{\partial t} + 2 \sum_j \frac{\partial}{\partial Q^i} \hat{K}^{ij} - [\widehat{FR}]^i = 0$$

when F^i is replaced by $F^k + f_1^k$, a similar equation holds for the small changes J_1 in H etc. When this equation is thermally averaged and f_1 is taken to be independent of time, there results, (since f_1^i is arbitrary)

$$2 \sum_j \frac{\partial}{\partial Q_j} X_k^{ij} = \Psi_k^i - \delta_k R_{eq} = 0$$

or

$$2 \Phi_k^i - \Psi_k^i - \delta_k^i R_{eq} = 0.$$

where ψ_k^i is defined by
$$[F^i\rho]_1 = \sum_k \psi_k^i f_1^k$$

Multiplying by $\frac{\partial q^\alpha}{\partial Q^i} \cdot \frac{\partial Q^k}{\partial q^\beta}$ and summing over i and k we get

$$2\varphi_\beta^\alpha - \psi_\beta^\alpha - \frac{1}{\theta^2} \frac{\partial Q^i}{\partial q^\alpha} \sum \frac{\partial Q^k}{\partial q^\beta} \delta_k^i R_{eq} = 0$$

which, at $\alpha = \beta = 0$ and for $q^\alpha = 0$ $\alpha > 0$, gives

$$2\varphi_o^o - R_{eq} = \psi_o^o$$

Substituting in the equation for η, we get

$$\eta(q) = \frac{1}{M} \left(\sum_{ij} n_o^i n_o^j X_j^i \right)^{-1} \left[f_{eq} \sum_i n_o^i X_i - \sum_{ij} n_o^i n_o^j \psi_j^i \right]$$

This result may be written

$$\eta(q) = \frac{1}{M} \left(\sum_{ij} n_o^i n_o^j X_j^i \right)^{-1} \sum_{ij} n_o^i n_o^j \left(X_j F_{eq}^i - \psi_j^i \right)$$

$$= \frac{1}{M} \left(\sum_{ij} n_o^i n_o^j X_j^i \right)^{-1} \sum_{ij} n_o^i n_o^j \left[\left(f^i - F_{eq}^i \right) \rho \right] / f_1^j$$

where F_{eq}^i are the components of $\vec{n}_o f_{eq}$ in cartesian coordinates.

In the limit of large mass, the arguments of ref (3) lead to the result

$$\eta(q) = \sum_{ij} n_o^i n_o^j \eta_{ij}(q)$$

where

and

$$kTM\eta_{ij}(q) = \text{Real part of } \int_0^\infty dt \langle \mathfrak{F}^i(0) \mathfrak{F}^j(t) \rangle$$

$$\mathfrak{F}^i = F^i - F_{eq}^i$$

and the time dependence is due to the electronic motion alone (i.e. the recoil of the adatoms is neglected). The thermal average is carried out with the representative point of the system fixed at q on the reaction path.*

III. THE FRICTION COEFFICIENT IN TERMS OF FLUCTUATIONS IN THE SUBSTRATE DEGREES OF FREEDOM

In the case of large M the preceding result takes a rather simple form. Suppose, for example, that the adatoms have virtually no free radical character, i.e. they completely share their electrons

*For a direct derivation in which M is very large to begin with, see forthcoming publication by W. Schaich (ref. 3).

with the substrate. Then spin forces are not involved. At this point it is necessary to write (ℓ,k) for the subscript i, where ℓ is the ℓth cartesian coordinate of the kth particle. Thus

$$F^i = \frac{\partial}{\partial Q^{k,\ell}} \left[\int V_k(Q^k - Q^b) R(Q^b) dQ^b \right]$$

+ direct interaction potential between adparticles.

Here $R(Q^b)$ is the charge density operator of the substrate, at position Q^b. Since the direct interaction between adatoms does not contribute to the fluctuating part of the force, we then find ($i = (k\ell)$; $i' = (k'\ell')$, the first index labels the particle, the second the cartesian axis)

$$kTM \, \eta_{ii'}(q) = \int \left[\frac{\partial V(Q^k - Q^b)}{\partial Q^{k\ell}} \frac{\partial V(Q^{k'} - Q^{b'})}{\partial Q^{k'\ell'}} \right] C(Q^b, Q^{b'}) \, dQ^b dQ^{b'}$$

where

$$C(Q^b, Q^{b'}) = R\ell \int_0^\infty \langle \delta R(Q^b, 0) \, \delta R(Q^{b'}, t) \rangle \, dt$$

is the time integral of the charge fluctuation correlation function of the substrate. It is convenient to write this result in terms of Fourier transforms. Then

$$\eta_{ii'} \sim \sum_k \int K^\ell K^{\ell'} V_k(K) V_{k'}(K') R_k^a(K \, Q^{\ell,k}) R_{k'}^a(K' Q^{\ell'k'})$$

$$C(K,K') \, dK, dK'$$

where $V_k(K)$ is the spatial Fourier transform of the interaction of the kth type of adatom core with the substrate, and $R_k^a(K,Q)$ is the Fourier transform of the density of the kth type of particle. $c(K,K')$ is the Fourier transform of C. All the friction coefficients thus have $c(K,K')$ as a "kernel." For crude estimates, one replaces c by its value for an infinite uniform substrate, in which case it becomes diagonal in K, and very simply related to an average over all wavenumbers of the imaginary part of the wavenumber dependent dielectric constant (and thus to the zero frequency resistivity (see ref. 3)).

In the same way, if the adatoms retain an electron spin S giving rise to some S·s coupling with the substrate spin density, C is replaced by a spin density-spin density correlation function, and in the infinite medium approximation, the η's are all related to the zero frequency limit of ω^{-1} Im $\chi(\omega,K)$ averaged over K with a weight depending on the Fourier transform of the coupling constant (see ref. 3). Here $\chi(\omega,K)$ is the spin susceptibility of the substrate at frequency ω and wavenumber K. In the case of ferromagnetic substrate, this quantity becomes larger near the Curie point T_c. Thus if a particular reaction away from $T = T_c$ was in the low η regime of Kramers, it will be enhanced near T_c; whereas a reaction that for T away from T_c was near the top of the Eyring range of η, will tend to be slowed down near $T = T_c$. The latter appears to be the case in the desorption of H_2 from thin nickel films, studied recently by Shanabarger.[8]

IV. CONCLUSIONS

It has been shown that the effective friction coefficient in the Brownian motion description of chemical reactions over surfaces depends on one or two simple response functions of the substrate, regardless of the complexity of the reaction or the number of reactants. Note that, although we have concentrated on the case of metal substrates, the same formalism applies more generally. It is only necessary to include in the charge fluctuations of the substrate those that are caused by the phonon excitation of the substrate ions. Anomalies are then to be expected where the ultrasonic attenuation, for example, has anomalies, e.g. near ferroelectric transitions. The notion that catalyzed reactions frequently fall outside the range of validity of Eyring absolute rate theory is not easily accepted by chemists, some of whom try to explain the observed anomalies in terms of changes of activation entropy, rather than changes of prefactor. This alternative is plausible only if structural changes in the surface occur near the transition but disappear both above and below the transition. If this is not the case it is hard to see how the entropic part of the activation free energy, which depends on the curvature of the effective energy surface away from the reaction path, should change. For that effective energy surface depends principally on the static configuration. It is true that there are corrections to it from virtual processes (e.g. spin wave emission and absorption below T_c and paramagnons above T_c) which are different above and below T_c. But if these were responsible for the anomalies, it is hard to see why the rates should return to their normal form once the transition is passed.

REFERENCES

1. H. A. Kramers, Physica $\underline{7}$, 284 (1940).
2. S. Chandrasekar, Rev. Mod. Phys. $\underline{15}$, 1 (1943).
3. E. G. d'Agliano, W. L. Schaich, P. Kumar and H. Suhl, paper presented at Nobel Symposium XXIV at Aspenasgarden, Lerum, Sweden, June, 1973. A more detailed treatment is given in two forthcoming publications by W. L. Schaich, and by E. G. d'Agliano, W. L. Schaich and H. Suhl.
4. H. Suhl and P. Kumar, Phys. Rev. $\underline{5}$, 4664 (1972).
5. R. Dagonnier, Adv. in Chem. Phys. $\underline{16}$, 1 (1969).
6. R. A. Marcus, J. Chem. Phys. $\underline{45}$, 4500 (1966).
7. C. E. Weatherburn, Riemannian Geometry and the Tensor Calculus, (Cambridge University Press, 1957), p. 95.
8. M. Shanabarger, private communication.
9. Goldberger and Watson, Collision Theory (John Wiley & Sons, 1964) pp. 494-498.

Section 2. Bubble Film Growth

GROWTH OF GARNET FILMS BY LIQUID PHASE EPITAXY

B. F. Stein
Sperry Univac, Blue Bell, Pa. 19422

ABSTRACT

A detailed discussion of the growth of garnet films by liquid phase epitaxy will be presented. The growth technique and the factors which either directly or indirectly affect the magnetic properties of the film will be discussed. Some of these factors are: oxide depletion which results from film growth; flux loss by volatization; and preferential oxide depletion produced by both non-unity and temperature dependent distribution coefficients. Quantitative estimates will be made of these effects and a control scheme to control film properties within narrow specifications will be proposed. This scheme has been used successfully to control film thickness, magnetization, and the bubble collapse field of $Y_{2.4}Eu_{0.6}Fe_{3.8}Ga_{1.2}O_{12}$ films. Growth of various compositions including those which support submicron diameter bubbles will also be discussed.

INTRODUCTION

Liquid phase epitaxy (LPE) is presently the most common technique for the growth of garnet bubble domain materials. One reason for this is the relative ease with which films can be grown with reasonable properties under diverse growth conditions. In a bubble domain memory large numbers of highly perfect films are needed whose magnetic properties are closely matched. Efforts to achieve these requirements in garnets of interest which contain more than one rare earth ion and either gallium or aluminum have revealed the complexity of LPE growth. We shall begin with a description of the general technique. Substrate selection and preparation, selection of an appropriate film composition, and determination of the physical and magnetic properties of the garnet film will be outlined rather than treated exhaustively.

We shall then discuss the effects of oxide depletion which results from film growth, flux loss by volatization, and preferential oxide depletion produced by non-unity and temperature dependent distribution coefficients on the magnetic properties of garnet films. A general scheme to control film properties within narrow specifications, which is based on compensating for the effects listed above, will be proposed. The results of applying this scheme to the growth of $Y_{2.4}Eu_{0.6}Fe_{3.8}Ga_{1.2}O_{12}$ films on $Gd_3Ga_5O_{12}$ (GGG) substrates will be presented.

Finally, we shall discuss the growth of several recently introduced compositions which promise high mobility and increased temperature stability as well as compositions which may be suitable for small bubbles.

GROWTH PROCESS

The growth process is that of isothermal growth from a supersaturated melt composed of garnet oxides dissolved in a lead oxide-boron oxide flux.[1-5] A non-magnetic garnet substrate is immersed in the melt at a growth temperature T_g which is below the saturation temperature T_s. The saturation temperature is the temperature at which LPE growth commences and is empirically determined for each melt. The substrate is held in the melt while being rotated at a constant rate for a specified time, during which a garnet film of uniform thickness is deposited on the substrate. Finally, the substrate is removed from the melt and cleaned of residual flux.

SUBSTRATE PREPARATION

The films which we will describe have been grown on GGG substrates. GGG is compatible with a large number of film compositions and it is now commercially available with low defect density. Since defects present in the substrate will also be present in the film, it is essential to begin LPE growth with highly perfect substrate material. The types of defects which may be found in Czochralski grown GGG have been reviewed in the literature.[6,7] We have placed the following criteria on the GGG boules: they must be free of a birefringent core, inclusions, and must have less than 5 defects/cm^2 over 85% of their central area. Wafers are either cut from the boule using a wire saw or are purchased already sliced. They are lapped in a slurry of 5 μ silicon carbide and water followed by a final Syton polish. The polished wafers are examined for defects using Nomarski interference contrast after being etched in 220°C H_3PO_4 for 2 minutes. This etch is sufficient to reveal both intrinsic boule defects and residual work damage which may have not been removed by polishing. Substrates are routinely polished which are free of both scratches and work damage.

Polishing, cleaning, and film growth are performed in Class 100 laminar flow hoods. We have found this to be a stringent requirement since any particulate contamination on the substrate will produce film defects which act as pinning sites for domains. The substrates are cleaned by boiling in an Alconox solution, rinsing a number of times in ultra high purity water, and then drying in a stream of filtered high purity nitrogen gas. The cleaned substrates are periodically examined to insure that they are free from contaminants.

MELT PREPARATION, APPARATUS, AND GROWTH PROCEDURE

All films have been grown in a melt containing a $PbO-B_2O_3$ flux. Melts are prepared by pre-melting the required garnet oxide and flux powders in a crucible at 1050°C. The constituents are first thoroughly mixed. The crucible is then filled and the pre-melt is performed. This process is repeated twice since the volume of the unmelted powders is greater than that of the crucible. The melt is heated to 1130°C for 16 hours to ensure complete dissolution of the oxides and homogenization of the constituents. No evidence of

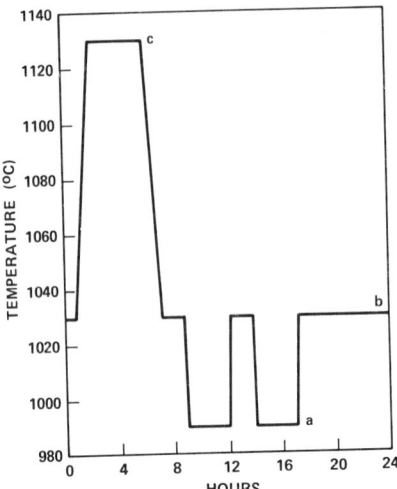

Fig. 1. Typical daily growth cycle for a melt which has $T_s = 1000°C$ and $T_g = 990°C$. Film growth takes place at a, the holding period is at b, and re-equilibration is at c.

undissolved material has been found after this treatment.

The furnace is composed of Kanthal windings and firebrick insulation and has a mullite muffle which insulates the crucible from the heating elements.

The furnace is suitably baffled so that the temperature variation over a 10 cm zone in the center is less than $\pm 0.5°C$. The temperature is controlled to $\pm .04°C$ during soaks and can be programmed at rates which range from 0.025 to 5°C/min. During programming at 1.5°C/min the deviation from the set point is $\pm 0.1°C$.

Films were grown isothermally using a continuous rotation rate of 60 rpm. A platinum substrate holder having four 0.1 cm diameter platinum legs which contact the substrate at points around the circumference is used. The substrates are lowered at a rate of 5 cm/min into the furnace and are held 0.6 cm above the melt surface for 3 min in order to achieve thermal equilibrium. Electron microscope studies using surface replicas[8] show no significant damage from exposure to flux vapor. The substrates are then immersed in the melt for typically 10 min during which film growth takes place. After the substrates are removed from the melt, they are held above the melt surface and rotated at 700 rpm for 30 sec to spin residual liquid off the substrate surface. The substrates are withdrawn from the furnace at the insertion rate. Solidified melt which adheres to contact points between the substrate and holder is dissolved using a 30% HNO_3 solution at 40°C for 10-15 minutes.

A daily temperature cycle is shown in Fig. 1. Up to six films have been grown consecutively without evidence of crystallization in the melt. Usually, however, the melt is cycled to 1130°C after four films have been grown to further insure that crystallization is prevented. The melt is held 30°C above the saturation temperature overnight with another re-equilibration cycle to 1130°C.

The LPE films are examined for visible defects using Nomarski interference contrast. The moving bubble raft described by Argyle, et al[9] is used to detect magnetic defects, although all magnetic defects which have been observed to date using this technique could be associated with physical defects visible by interference contrast. The density of magnetic defects which result from film growth ranges from 0-10 defects/cm^2. This low defect

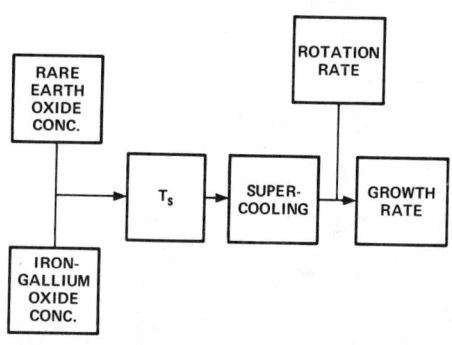

Fig. 2. Schematic representation of the relationship of fundamental growth parameters to the growth rate.

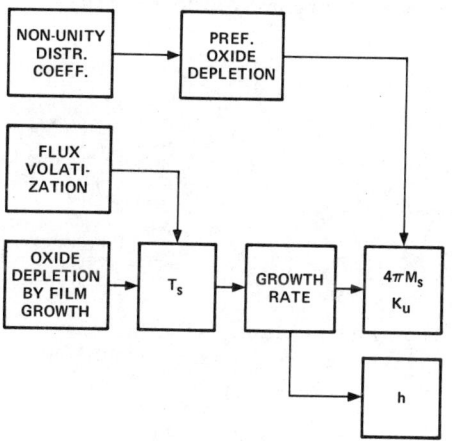

Fig. 3. Schematic representation of the influence of growth parameters on magnetic properties.

density can only be achieved if careful substrate preparation and cleaning procedures are strictly followed. Any relaxation of standards or breakdown of a step will increase the defect density, often by many orders of magnitude.

The magnetic properties are determined by the techniques described in detail by Josephs.[10] Routine measurements are made of the film thickness h, $4\pi M_s$, bubble collapse field, and bubble size. Generally, about 85% of the film area lies within one interference fringe ($0.13\,\mu$). Anisotropy and mobility measurements are also made periodically.

GROWTH PARAMETERS

There are complex relationships between the magnetic properties of an LPE film and its growth parameters. An overview of these relationships is given in Figs. 2 and 3. The order in which these parameters affect each other and ultimately determine the growth rate is shown in Fig. 2. The rare earth oxide and iron-gallium oxide concentrations are shown separately because their influence on T_s differs by an order of magnitude. The growth rate is a function of both the supercooling $\Delta T = T_s - T_g$ and the rotation rate. However, since all films were grown using a rotation rate of 60 rpm, only the supercooling will be considered as a variable in succeeding discussions.

Oxide depletion by film growth, flux volatization, and non-unity distribution coefficients will act to change the magnetic properties as shown in Fig. 3. Separate distribution coefficients can be defined for the rare earth atoms, which occupy the dodecahedral sites in the garnet structure, and for the iron and gallium atoms, which occupy the tetrahedral and octahedral sites. The principal effect on the magnetization arises from the large Ga distribution

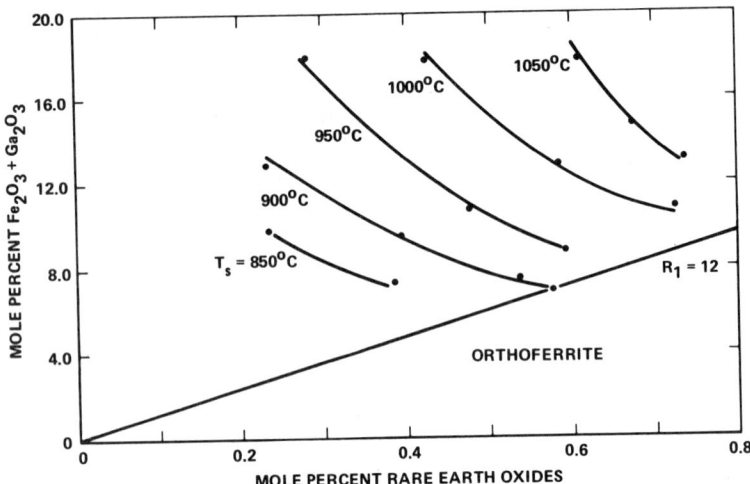

Fig. 4. Contours of constant T_S as a function of iron and rare earth oxide concentrations. (Note that mole percent oxide equals moles oxide/moles flux x 100).

coefficient. The distribution coefficient is defined as:

$$\alpha = \frac{\left(\frac{N_{Ga}}{N_{Ga} + N_{Fe}}\right)_{Film}}{\left(\frac{N_{Ga}}{N_{Ga} + N_{Fe}}\right)_{Melt}} \quad (1)$$

where the N are the molar amounts of iron and gallium. For a film composition $Ga_x Fe_{5-x}$ and a melt composition $Ga_\xi Fe_\eta$ the expression for α becomes

$$\alpha = \frac{\frac{x}{5}}{\frac{\xi}{\xi + \eta}} \quad (2)$$

This expression will be used later in a discussion of preferential oxide depletion.

SATURATION TEMPERATURE

Knowledge of the saturation temperature and its dependence on the garnet oxide concentrations is the starting point in an understanding of LPE growth and in the effects of repeated growth on

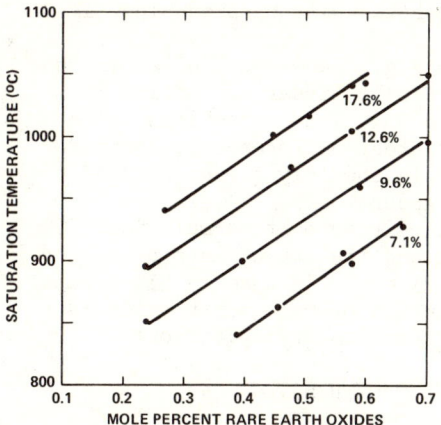

Fig. 5. T_S vs molar rare earth concentration for four $Fe_2O_3 - Ga_2O_3$ concentrations.

Table I. Corrections to be applied to T_S in Fig. 4 for the garnet compositions listed.

COMPOSITION	ΔT_s (°C)
$Y_{1.62} Eu_{0.78} Tm_{0.6} Fe_{4.07} Ga_{0.93} O_{12}$	0
$Y_{2.5} Sm_{0.5} Fe_{3.9} Ga_{1.1} O_{12}$	0
$Y_{2.5} Tm_{0.5}$ $Y_{2.3} Tm_{0.7}$ $Fe_{3.97} Ga_{1.03} O_{12}$ $Y_{2.1} Tm_{0.9}$	-5
$Y_1 Gd_1 Tm_1 Fe_{4.3} Ga_{0.7} O_{12}$	-5
$Y_{2.65} Gd_{0.35} Fe_4 Ga_1 O_{12}$	-5
$Eu_{0.8} Er_{2.2} Fe_{4.3} Ga_{0.7} O_{12}$	-30

film properties. There is a flexibility in the garnet system which allows garnets to be grown for a wide range of iron oxide to rare earth oxide ratios, R_1, which differ considerably from the stoichiometric 5:3 ratio. We have found that the garnet field extends from $R_1 = 12$ to at least $R_1 = 66$ without second phase precipitation. Below $R_1 = 12$ orthoferrite precipitation occurs in the YEu garnet system. The critical ratio for orthoferrite precipitation depends on the garnet composition and precipitation may occur for ratios as great as $R_1 = 20$ in garnets which contain Sm.[4]

Figure 4 shows in detail the dependence of the saturation temperature on the separate garnet oxide concentrations. In this figure T_s contours are plotted as a function of the rare earth oxide and iron-gallium oxide concentrations over the temperature range typically used in LPE growth.

The point to be noted from this figure is that T_s is much more sensitive to the rare earth oxide concentration than to the iron-gallium oxide concentration. This is reasonable since considerable iron and gallium oxides are used in excess of the stoichiometric value. The state of the excess iron and gallium oxide in the flux is not known, but it may form complex lead-boron-iron-gallium compounds.

T_S depends linearly on the rare earth oxide concentration as is shown in Fig. 5. The slope of the curves is 320°C/mole % rare earth oxide over the temperature range 850° to 1050°C. The dependence of T_S on the iron-gallium (Fe-Ga) oxide concentration is much weaker and non-linear and varies from 20°C/mole % Fe-Ga oxide at 8 mole percent Fe-Ga oxide to 10°C/mole % Fe-Ga oxide at 12-16 mole percent Fe-Ga oxide.

This weak dependence on the iron-gallium oxide concentration allows small additions of either iron or gallium oxides to be made to the melt to trim the magnetic properties without producing significant changes in T_s.

The constant T_s contours shown in Fig. 4 are not limited to the YEu system. They can be used with a simple linear correction in garnets containing Gd, Tm, Eu, Er, and Sm. The corrections to be applied are given in Table I. The greatest difference is found for Eu Er garnet, presumably because this is the only garnet studied which does not contain a substantial amount of Y.

The use of Fig. 4 considerably simplifies melt formulation. The desired T_s and oxide ratio are chosen and the appropriate oxide percentages are taken from the figure. R1 = 22 is used in YEu growth. It is low enough that a significant amount of rare earth oxide is present yet high enough to avoid orthoferrite precipitation.

GROWTH TEMPERATURE AND GROWTH RATE

Films can be grown over a large temperature range. The particular temperature chosen for a given composition depends upon the uniaxial anisotropy which is desired, the amount of lead incorporation in the film and the flux loss which can be tolerated. In general, both lead content and the uniaxial anisotropy decrease with increasing growth temperature, while the PbO loss by volatization increases.

The most obvious effect of lead incorporation is to increase the lattice parameter of the film. More subtle effects on anisotropy and domain motion may be present but have not been investigated to date. In order to minimize lead incorporation the films described here have been grown at the highest temperature possible consistent with sufficient uniaxial anisotropy to allow operation of bubble domain circuits. A saturation temperature of 1000°C has been used.

Within certain limits the growth rate is determined by the supercooling (at a constant rotation rate). Giess et al[2] have reported a linear relationship between growth rate and supercooling for ΔT as large as 30°C. We have observed similar behavior for ΔT up to 20°C. For greater ΔT spontaneous crystallization of the garnet phase has been frequently observed. Generally the first film grown even at large supercooling is free of crystallites. However, subsequent films grown without re-equilibration of the melt generally exhibit increased defect densities. In addition, the growth rate is considerably reduced because of competition between the mechanisms of film growth and spontaneous crystallization. To minimize the possibility of crystallization a supercooling of approximately 10°C is used. No garnet precipitation has been observed under these conditions in growth of over 300 films. The growth rate as a function of supercooling has been found to be:

$$G = 0.075 (T_s - T_g) \pm 10\% \; \mu/min \qquad (3)$$

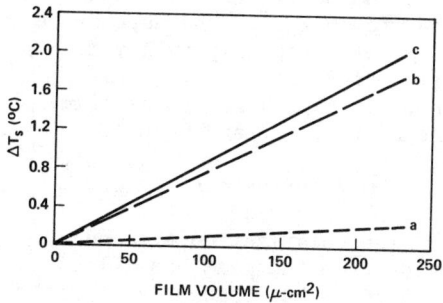

Fig. 6. Change in saturation temperature ΔT_s vs film volume for a 1010 g melt. Curves a, b, and c show the iron oxide, rare earth oxide, and total contributions, respectively.

Table II. Typical melt composition.

CONSTITUENT	WEIGHT (g)	MOLES
Y_2O_3	4.245	0.0188
Eu_2O_3	1.654	0.0047
Fe_2O_3	74.08	0.464
Ga_2O_3	12.17	0.065
PbO	900	4.03
B_2O_3	18	0.26
TOTAL WEIGHT = 1010 g		

EFFECT OF MELT COMPOSITION CHANGES ON MAGNETIC PROPERTIES

The mechanisms shown in Fig. 3 produce changes in the melt composition which affect the magnetic properties of the film. The magnetization is influenced by a non-unity distribution coefficient through preferential oxide depletion and influenced by oxide depletion and flux volatization through growth rate changes. These mechanisms have been enumerated by Hewitt et al[12] who presented a scheme to control the magnetic properties of the film. We shall treat the mechanisms in a quantitative manner and propose a somewhat different control scheme.

1. Oxide Depletion

A typical melt composition which is used in the growth of $Y_{2.4}Eu_{0.6}Fe_{3.8}Ga_{1.2}O_{12}$ at 990°C is given in Table II. The total weight of the melt is 1010 g. As the film grows, oxides are removed from the melt and T_s is reduced. The effect of oxide depletion on T_s for the melt given is shown in Fig. 6. Both the contributions of rare earth and iron oxide depletion are shown separately. The units of $\mu - cm^2$ are used for total volume since usually in film growth one or more films of a specified thickness and area are grown. For the 1010 g melt T_s decreases at a rate of 8.4×10^{-3} °C/$\mu - cm^2$. Of course this rate depends inversely on the melt size. The decrease in T_s described produces a decrease in the supercooling and in the growth rate if the growth temperature is held constant.

The growth rate can be expressed as a function of growth time by:

$$G = \frac{G_o}{1 + .075BAt} \quad (4)$$

where G_o is the initial growth rate, B depends on the molar oxide amounts used in the melt and is just the slope of the ΔT_s vs film

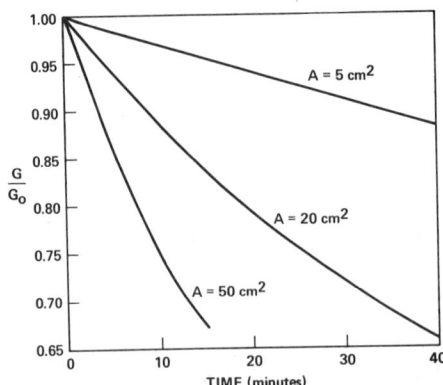

Fig. 7. Ratio of growth rate G to initial growth rate G_o vs growth time for several film areas in a 1010 g melt.

volume curve given in Fig. 6, A is the total film area and t is the growth time. The dependence of growth rate on growth time calculated from eq'n. (4) is shown in Fig. 7 for total film areas of 5, 20, and 50 cm^2.

2. PbO Volatization

The volatization of lead oxide from the melt has a less pronounced but cumulative effect on T_s. Lead oxide loss produces two competing effects. It concentrates the melt which raises T_s. It also decreases the lead oxide to boron oxide ratio, which lowers T_s.[12] Even though the effect becomes quite small as the melt size is increased, the net result is to increase T_s. For example, the increase in T_s per gram of PbO lost by volatization is 0.5°C, 0.08°C, and 0.01°C for melts of total weight 250 g, 1000 g, and 4000 g, respectively.

A typical daily temperature cycle for LPE growth has been shown in Fig. 1. Roughly, equal flux loss occurs during the holding period at 1030° and the equilibration period at 1130°C. The total flux loss from a 6.3 cm diameter crucible over 24 hours is 5 grams. For the 1010 g melt which was used this causes an increase in T_s of 0.4°C.

3. Growth Rate

Any change in the growth rate of the film will directly affect the magnetization in compositions which contain Ga. This occurs because Ga incorporation into the film is a strong function of growth rate. It is much more convenient to infer the gallium content of the film by measuring $4\pi M_s$ rather than using a direct technique such as electron microprobe analysis because the magnetization is extremely sensitive to Ga content. Up to at least 1.25 Ga formula units the relationship between Ga content and $4\pi M_s$ is linear with a slope of 900G/formula unit of Ga.[4] Figure 8 shows the saturation magnetization as a function of the growth rate for a garnet of nominal composition $Y_{2.4}Eu_{0.6}Fe_{3.8}Ga_{1.2}O_{12}$. Over the range of growth rates 0.3 to 1.2 μ/min there is a linear relationship between $4\pi M_s$ and G.

Under conservative growth conditions in which about 10°C supercooling is used, the growth time for a 6 μ film is 10 min. For a single film grown on a 5 cm^2 area (1" diameter) substrate the growth rate will decrease by 3% during the run for constant growth

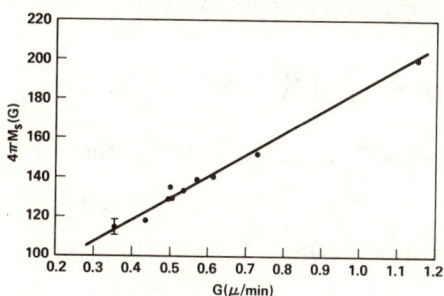

Fig. 8. Saturation magnetization vs growth rate for a $(YEu)_3(FeGa)_5O_{12}$ film.

Table III. Distribution coefficients of ions commonly used in LPE growth.

CATION	α
Y	1.00
Sm	0.86
Eu	1.00
Gd	1.10
Tm	0.90
Yb	0.95
Ga	1.6-2.0 [4]

temperature. The decrease in growth rate during the run given by equation (4) will cause a variation in $4\pi M_s$ through the film thickness. The average $4\pi M_s$ will be 1.5% lower than if the film were grown at constant supercooling. For a film which has a magnetization of 120G this effect would produce a variation of 3.6G through the film thickness. At constant growth temperature the average magnetization of two consecutively grown films would differ by 3%. In multiple growth this effect would be accentuated.

A possible remedy would be to increase the supercooling which, for constant film thickness, allows shorter growth times and a correspondingly smaller change in the growth rate. However, caution must be exercised since melt crystallization is enhanced under these conditions. We have chosen instead to decrease the growth temperature to maintain a constant supercooling and growth rate.

4. Non-unity Distribution Coefficient

The effect of a non-unity distribution coefficient is to preferentially deplete oxides relative to each other. Depletion of Ga relative to Fe will cause a gradual increase in $4\pi M_s$. Depletion of rare earths could cause changes in both $4\pi M_s$ and in the uniaxial anisotropy. The severity of the change will depend on both the distribution coefficient of the ion and on the sensitivity of $4\pi M_s$ to that ion. Table III gives the distribution coefficients of commonly used ions.[8] Ga is of particular interest since its distribution coefficient is both large and temperature dependent. For a film composition Ga_xFe_{5-x} and a melt composition $Ga_\xi Fe_\eta$, the change in $4\pi M_s$ is proportional to dx, the change in the molar amount of Ga. dx is given by the expression (recall equation 2)

$$dx = \frac{5\alpha - x}{\xi + \eta} d\xi - \frac{x}{\xi + \eta} d\eta + \frac{5\xi}{\xi + \eta} d\alpha \qquad (5)$$

where $d\xi$ and $d\eta$ are the molar changes of gallium and iron oxides, respectively, which result from film growth and ξ and η are the number of moles of gallium and iron oxide in the melt. Under usual growth conditions in which a small percentage of possible films are grown from a melt $\xi + \eta$ may be taken as a constant, N. With the temperature dependence of α shown more explicitly equation (5) becomes:

$$dx = \frac{5\alpha - x}{N} d\xi - \frac{x}{N} d\eta + \frac{5\xi}{N}\left(\frac{d\alpha}{dT}\right) dT \qquad (6)$$

For $\alpha = 2$, $x = 1$ and $\frac{d\alpha}{dT}$ given by Blank et al[4] as 1.342×10^{-3}, the net changes in $4\pi M_s$ are:

$$\delta (4\pi M_s) \text{ preferential oxide depl.} = 0.003 \text{ G}/\mu\text{-cm}^2$$

$$\delta (4\pi M_s) \text{ temperature dependence of } \alpha = 0.74 \text{ G}/^\circ\text{C}$$

The actual change in $4\pi M_s$ was found to be somewhat less than that expected. The difference in $4\pi M_s$ between the first and sixteenth films made without Ga_2O_3 addition but with a $0.6^\circ C$ temperature decrease during each run was 7.5G. The expected value was 10.5G. The difference may be due to the value of $\frac{d\alpha}{dT}$ which was used in the calculation. $\frac{d\alpha}{dT}$ was not determined for the melt used and may depend on the particular iron oxide and gallium oxide amounts present in the melt.

The strong dependence of T_s on the rare earth oxide concentration may present difficulties in compositions which contain ions with small distribution coefficients on dodecahedral sites. An example of this is the La ion whose distribution coefficient is 0.2[13-15] but is defined only over a narrow composition range. Because of the small value of α_{La} a large concentration of La_2O_3 would be needed in the melt in order to incorporate a significant amount in the film.

Since T_s is primarily a function of the total rare earth oxide concentration, the large concentration of La_2O_3 would raise T_s considerably over a composition containing Y_2O_3, for example. In order to maintain the saturation temperature at the usual value, the concentrations of other rare earth oxides would have to be reduced to keep their total concentration constant. This reduction in oxide available for film growth would accentuate oxide depletion effects and could cause changes in the anisotropy as well as in $4\pi M_s$.

Both distribution coefficient effects and flux volatization are common to compositions grown in the $PbO-B_2O_3$ flux system. Hiskes, et al[16] report that these effects are less severe in the $BaO-B_2O_3 - BaF_2$ flux system.

MELT CONTROL

Since the effects which we have described are caused either by oxide removal by film growth or by flux volatization the most straightforward technique of melt control would be to continu-

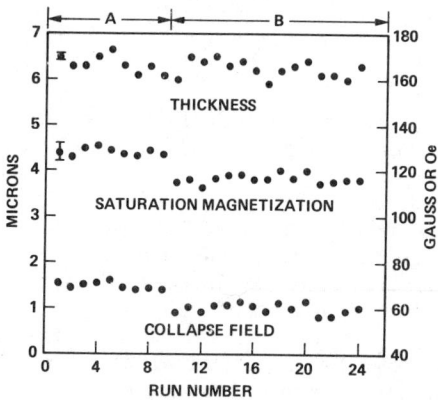

Fig. 9. Results of the melt control scheme. Two sets of films are shown which have $4\pi M_s$ = 128.5 and 116.5 G, respectively.

Table IV. Results of the melt control scheme for film sets A and B.

	A	B
h (μ)	6.26 ± 3%	6.26 ± 3%
$4\pi M_s$ (G)	128.5 ± 1.5%	116.5 ± 2.2%
ℓ (μ)	0.63 ± 3%	0.71 ± 2.5%
H_o (Oe)	70 ± 1.6%	59.8 ± 3.5%

ously add garnet oxides and flux to the melt. Unfortunately, equilibration of oxides added to the melt is too slow using present techniques so that continuous addition is not yet practical. The melt control scheme which we will propose includes both temperature changes during growth and periodic oxide additions to the melt. These additions are conveniently made during normal equilibration periods of the melt. The scheme described below has been successfully used to control the properties of 6 μ thick films grown on 2.5 cm diameter substrates. The bubble domain diameter is nominally 6 μ. The films were grown individually and a film was deposited on each side of the substrate. The control scheme used is:
1. Lower growth temperature during the run at a rate of $8.4 \times 10^{-3} °C/\mu\text{-cm}^2$ of film grown.
2. Add sufficient Ga_2O_3 to the melt during normal equilibration periods to compensate for $4\pi M_s$ increases.
3. Add sufficient garnet oxides to restore T_s to its original value after each ten films grown.
4. Add PbO to the melt daily to compensate for volatization losses.

The results of this control scheme are shown in Fig. 9 and in Table IV. Two sets of films which have different $4\pi M_s$ are shown. Those with $4\pi M_s$ of 128.5G (A) were grown over a two week period to accentuate the effects of flux volatization and provide a test of lead oxide addition.

Bubble domain memories will be composed of a large number of individual chips. The operating margins of the memory will depend on the overlap of the margins of individual chips, which to first order are determined by the bubble collapse field H_o. Further refinements in melt control which will reduce the run to run variations in H_o will undoubtedly be made as more subtle growth effects are understood. In situ monitoring of the growth rate or a practical technique of continuous oxide addition would be extremely valuable. An annealing procedure whereby the collapse field is trimmed to a desired value has been recently reported.[17]

Of course, the extent of the anneal which is necessary depends on the spread of as-grown H_0 values.

FUTURE LPE GROWTH

Two areas of future interest are the development of: 1) materials with improved magnetic and thermal properties and 2) materials which support small diameter bubble domains.

Several garnet compositions have been recently introduced which offer increased mobility and temperature stability.[18,19] These compositions contain Ca ions on dodecahedral sites and either Ge or Si ions on tetrahedral sites. Since the distribution coefficients of these ions are non-unity and small[13] oxide depletion effects will probably be present.

Micron or submicron diameter bubble domains are of interest because of the higher bit densities and data rates which can be achieved. The bubble diameter can be easily decreased by increasing $4\pi M_s$. The price which is paid for this is a decrease in $q = K_u/4\pi M_s^2$ and possible instability of the bubble domains. We have grown (111) YEu films with $4\pi M_s$ of approximately 420 G with $\ell = 0.10\mu$ and $q \simeq 1$ which support 1μ diameter domains.

An increase in anisotropy can be obtained by either introducing stress in a magnetostrictive composition or by increased pair ordering. Stress induced anisotropy was used in the growth of (100) YEu films on SmGG.[20] Giess et al[21] have reported submicron bubbles in garnets which contain no Ga and are matched to either (111) GGG or SmGG. Compositions can be used in which the number of pairs is greater than that found in many Ga substituted garnets. The absence of Ga should considerably simplify control of the magnetization.

CONCLUSIONS

LPE growth has been treated by enumerating the growth parameters and mechanisms which ultimately influence the magnetic properties of the garnet films. The saturation temperature has been shown to be dependent on the individual oxide concentrations which are present in the melt. Because of this dependence, the supercooling and thus the growth rate decrease as oxide is removed from the melt by film growth. The growth rate as a function of growth time is $G/G_0 = 1/(1+Ct)$, where the constant C is determined by the melt composition and the film area. The variation of G with time affects the Ga content and thereby the magnetization of the film. In addition, Ga has both a non-unity and temperature dependent distribution coefficient. This causes Ga to be preferentially depleted from the melt relative to Fe and increases $4\pi M_s$ while the overall incorporation of Ga into the garnet is temperature dependent. While these effects are most pronounced in compositions containing Ga they may also be present in newer compositions which contain other ions with distribution coefficients significantly different from one on both dodecahedral and tetrahedral sites.

A melt control scheme has been described which relies on both temperature changes and periodic oxide additions to the melt to compensate for the effects which have been described. This scheme has been applied to growth of YEu garnets. The parameters h, $4\pi M_s$, ℓ, and H_o have been controlled to about $\pm 3\%$ using this scheme.

ACKNOWLEDGMENTS

The author is grateful to J. Goodroe for his expert and enthusiastic growth of the LPE films. He especially wishes to thank R. Josephs for many valuable discussions about the LPE growth process and the experimental results, and for his comments on the manuscript. He thanks W. D. Doyle for his critical reading of the manuscript, H. B. Callen for valuable discussions, M. Kestigian for valuable discussions, J. O'Dowd for substrate preparation, and G. Moule for magnetic measurements.

REFERENCES

1. H. J. Levinstein, S. Licht, R. W. Landorf and S. L. Blank, Appl. Phys. Let. 19, 486 (1971)
2. E. A. Giess, J. D. Kuptsis and E. A. D. White, J. Cryst. Growth 16, 36 (1972)
3. S. L. Blank and J. W. Nielsen, J. Cryst. Growth 17, 302 (1972)
4. S. L. Blank, B. S. Hewitt, L. K. Schick and J. W. Nielsen, AIP Conf. Proc. No. 10, 256 (1973)
5. B. F. Stein and R. M. Josephs, AIP Conf. Proc. No. 10, 329 (1973). The pages in this article are out of order. The proper order is 329, 333, 330, 331, and 332.
6. G. A. Keig, AIP Conf. Proc. No. 10, 237 (1973)
7. J. W. Matthews, E. Klokholm and T. S. Plaskett, AIP Conf. Proc. No. 10, 271 (1973)
8. A. Baltz, private communication
9. B. E. Argyle and P. Chaudhari, AIP Conf. Proc. No. 10, 403 (1973)
10. R. M. Josephs, AIP Conf. Proc. No. 10, 286 (1973)
11. J. W. Nielsen, D. A. Lepore and D. C. Leo, Crystal Growth, Ed. H. S. Peiser (Pergamon, 1967) p. 457
12. B. S. Hewitt, R. D. Pierce, S. L. Blank and S. Knight, IEEE Trans. Magnetics (to be published)
13. M. Kestigian, private communication
14. α_{La} is uniquely defined only for compositions with less than 0.40 formula units of La. Compositions containing greater La amounts are not single phase.
15. G. P. Espinosa, J. Chem. Phys. 37, 2344 (1962)
16. R. Hiskes and R. A. Burmeister, AIP Conf. Proc. No. 10, 304 (1973)
17. D. H. Smith, F. B. Hagedorn and B. S. Hewitt, J. Appl. Phys. 44, 4177 (1973)
18. W. A. Bonner, to be presented at the 1973 Conference on Magnetism and Magnetic Materials, Boston, Mass.

19. J. E. Geusic, D. H. Smith, L. G. Van Uitert and G. P. Vella-Coleiro, to be presented at the 1973 Conference on Magnetism and Magnetic Materials, Boston, Mass.
20. T. S. Plaskett, E. Klokholm, H. C. Hu and D. F. O'Kane, AIP Conf. Proc. No. 10, 319 (1973)
21. E. A. Geiss, C. F. Guerci, J. D. Kuptsis and H. C. Hu, Materials Research Bulletin 8 (1973) (in press)

MULTIPLE GARNET FILM GROWTH BY LPE*

R. G. Warren, J. E. Mee, F. S. Stearns and E. C. Whitcomb
Electronics Research Division, Rockwell International
Anaheim, California 92803

ABSTRACT

Liquid phase epitaxy (LPE) film growth has been shown to provide device quality films utilizing several different flux systems. Using $PbO-B_2O_3$ system films have been grown of nominal $Y_{2.14}Eu_{0.56}Tm_{0.30}Ga_{1.1}Fe_{3.9}O_{12}$ composition simultaneously on eight 1-1/4 inch diameter gadolinium gallium garnet substrates. Each of these films has uniform film thickness and composition across the substrate, excluding small wafer edge effects. The thickness uniformity is within one interference fringe using the sodium D line which is 0.13μm. Films grown during one run, which involves up to eight films, have essentially the same magnetic characteristics. The physical arrangement of the equipment for this pilot line sized effort is described. Details of a horizontal substrate holder are discussed and the specifics for obtaining the film parameters are given.

INTRODUCTION

The bubble domain technology has advanced to a stage in its development where the availability and cost of films for devices and systems development are of paramount importance. Using the LPE system of garnet film growth[1] with horizontal substrate rotation[2] most workers deposit one film at a time or up to two as reported by Hewitt et al[3]. The main thrust of this work was that of increasing the number of films grown per each deposition run. The two major problems anticipated were thickness uniformity for each film and a ± 0.5 oe H_{coll} value for all films in a run. To this end many substrate holder designs were considered and tried. In the next section we will describe those experiments which lead to the present deposition scheme.

EXPERIMENT DETAILS AND RESULTS

The LPE growth station consists of a Kanthal wound vertical tube furnace with three zone control which has a maximum working diameter of 4 1/4 inches. The working zone of the furnace is controlled to $\pm 0.1°C$. For this work the melts were contained in 250 ml standard form pure platinum crucibles located at the geometric center of the furnace. The general furnace configuration was similar to that reported by others[3,5].

---

*This work was sponsored by the Air Force Materials Laboratory under Contract Number F33615-72-C-1299.

The films grown for this work had a nominal composition of $Y_{2.14}Eu_{0.56}Tm_{0.30}Ga_{1.1}Fe_{3.9}O_{12}$. The flux system used was the much reported 50:1 weight ratio of $PbO:B_2O_3$. The solute mole ratio was 0.09 and the Fe_2O_3/Ln_2O_3 mole ratio used was 20. This provides a saturation temperature (T_{sat}) of ≃920°C and deposition temperatures (T_{dep}) used were approximately 910°C with a resulting growth rate of the order of 0.3 μm/min. All of the substrates used for this work were 1 1/4 inch diameter GGG and were supplied by Crystal Technology Inc.

Since horizontal substrate rotation produces at least one film with excellent thickness uniformity, this was the area pursued. Our experience with this method of substrate rotation showed that the upper film would occasionally be fairly uniform in thickness but would more often have a bulls-eye appearance when viewed under monochromatic sodium lamp illumination. Each fringe represents a change of film thickness of 0.13μm. Films of this sort are typically thin at the center and thicker toward the edge. Holder configuration, substrate depth into the melt, melt size, and substrates being rotated off axis were all contributors to this undesirable film thickness nonuniformity. A holder was designed with four substrate positions, which using the back to back arrangement[3] could hold eight substrates. Figure 1 shows the four position holder on the left and a single position version on the right. Since the assembly is made from pure platinum, the wire used for its construction was rather large in diameter to obtain the mechanical rigidity required. The main holder sections, the finger arms, and substrate fingers are made of 0.080, 0.040, and 0.020 inch diameter wire respectively. The rotation rod attachment is 3/8 inch O.D. platinum tubing. The spacing between each substrate position was one centimeter. The substrates were held at three points by fingers 120 degrees apart, and the arms to which the fingers are attached permit the holder to accommodate several substrate sizes up to a maximum of 1 1/4 inch diameter.

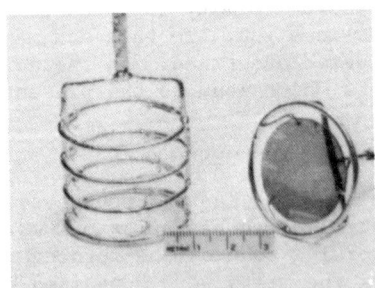

Figure 1. Photograph of Platinum Four Position and Single Position Substrate Holders.

Deposition experiments were made with the holders shown in Figure 1 both in the normal unidirectional rotation modes and under various rotation reversal conditions. Figure 2a is a photograph of the interference fringes of a film grown using the single position

holder. The film was grown using unidirectional rotation at a rate
of 60 rpm. In Figure 2a there are two sets of fringes. Those with
wider spacing covering essentially all of the wafer area are on the
backside or upper films. This is a variation of the bulls-eye
pattern. The other set of fringes, which are more intense, closer
together and with 120 degree periodicity are on the front side or
bottom film. The sweeping curves for this latter film show the
fluid flow pattern and indicate the direction of rotation. In
Figure 2b the fringe pattern also shows the films on each side of
the substrate but both films have very large single fringe areas.
This sample was made using the same single position holder and was
rotated at the same rate of 60 rpm as the films in Figure 2a, but
the direction of rotation was reversed each one revolution. The
photograph shows that the holder finger influence has essentially
been negated. The maximum edge effect is 1/8 inch around the pe-
riphery of the films on both sides of the wafer. Using a rotation
reversal rate of once each 10 revolutions the center of the bulls-
eye opens appreciably but not to the large single fringe area of
Figure 2b. The substrate rotation is provided by a shunt wound DC
motor with a worm gear reduction. For the reversal technique the DC
polarity is reversed at intervals by a preset clock. In this manner
the rate of reversal can be adjusted for differing rotation rates.
The actual reversal time is <0.02 sec and this is insignificant even
for a one second reversal period.

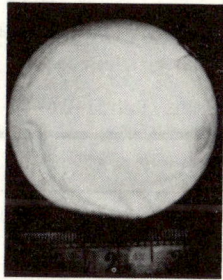 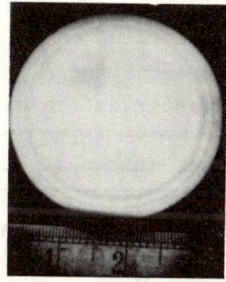

Figure 2a. Film Grown with   Figure 2b. Film Grown with
    Unidirection Rotation.   Rotation Reversal.

LPE garnet film growth using the four substrate position holder
with unidirectional horizontal rotation at 60 rpm produced two films
on each of the upper three and the top film of the bottom substrates
with an extreme bulls-eye appearance (small area with single fringe).
The bottom most film had the large single fringe area described by
Giess et al[2]. At rotation rates of $\geq$100 rpm this holder became such
an effective stirrer that the melt was spun out over the edge of the
crucible. Eight films grown on four G[3] substrates with the four
position holder each had a large single fringe area, as that shown
in Figure 2b, when the 60 rpm with rotation reversal each one revolution was utilized. With the back to back substrate arrangement
eight films were grown on eight G[3] wafers. Again Figure 2b is

representative of the uniformity of these LPE grown films. An intersubstrate position spacing of 0.5 centimeters using the same deposition scheme caused a deterioration of the large single fringe area.

Table I shows film characterization data for multiple dipped films. The data is presented such that the sample AU4 is the upper most film and DD4 is the bottom film in the stack. The columns of specific interest are growth rate and H_{coll} values for each film. For positions CU4 and DD4 the growth rates are the same and at these positions the H_{coll} values are within 0.2 oe. The temperature profile of the melt for this deposition run had a 1.6°C variation with a cooler top and also had a nonuniform temperature gradient in the region of films CU4 and DD4. The difference in the eight H_{coll} values can be directly related to melt temperature variation. The mean value of H_{coll} is 51.6 oe with a deviation of +4.8 and -3.2 oe. Several options are available for improving the $\geq \pm 0.5$ oe deviation of H_{coll} shown in the Table. These are:

1. Improved melt temperature profile
2. H_{coll} adjustment 4
3. Grouping of H_{coll} values for large numbers of films.

Uniform film thickness indicates uniform growth rate and uniform composition. The H_{coll} values for a single fringe area should be the same and this has been reported by others. This is also true when rotation reversal is used. H_{coll} values were determined for the sample shown in Figure 2b taken at the center and at four points 90 degrees apart and at a distance of 1 centimeter from the center for the films on both sides of the substrate. These values are given in Table II.

Table I Characterization of Multiple Dipped Films. Run 1-14-4

Film Position	Thickness	Growth Rate	Strip Width	H_{coll}
AU4	5.86μm	0.325μm/min	8.25μm	56.4 oe
AD4	5.76	0.320	8.30	54.9
BU4	5.53	0.307	8.30	51.7
BD4	5.42	0.301	8.55	52.6
CU4	5.39	0.299	8.70	50.0
CD4	5.34	0.297	8.84	49.3
DU4	5.29	0.294	8.55	48.4
DD4	5.38	0.299	8.70	49.8

Table II H_{coll} Uniformity for a Rotation Reversal Grown Film.

Position	Top Film	Bottom Film
0	50.7 oe	50.5
1	49.8	50.3
2	49.9	50.8
3	50.3	50.8
4	50.2	50.4

As can be seen, film growth using rotation reversal does result in very good uniformity of the magnetic characteristics across a film. The only film magnetic characteristic change observed for the reversal samples is an increase in H_{coll} caused by the slightly increased growth rate.

Using other substrates, which were 1 1/2 inch in diameter, the reversal technique produced thickness uniformity on both sides of the wafer with the edge effect being only 1/16 inch wide.

A brief investigation was made to study the specifics of growth for the multiple film dipping method. A simulated melt was used in a transparent container to observe the fluid flow characteristics with a rotation reversal situation. For simplification the single position holder, shown in Figure 1, was used and for unidirectional rotation the previously reported melt flow characteristics[5] were observed for the bottom portion of the melt. When rotation reversal was initiated, the rotation of the bulk of the fluid, as observed for unidirectional rotation, was stopped. Only a very slow "pumping" action was observed which appeared to have smooth motion, probably because the reversal rate was relatively high. A series of film growth runs were made with and without rotation reversal with the rotation rate varied from 10 rpm to 90 rpm. For the reversed rotation runs the sample rotation direction was reversed each one revolution. For a constant deposition time and $(T_{sat}-T_{dep})$ a plot of film thickness versus (rotation rate)$^{1/2}$ for each condition gave a straight line and, although offset, were essentially parallel. The rotation reversal samples were slightly thicker.

ACKNOWLEDGEMENTS

The authors wish to thank E. F. Grubb, T. N. Hamilton and R. Mendoza for film growth and characterization.

REFERENCES

1. H. J. Levinstein, S. Licht, R. W. Landorf and S. L. Blank, App. Phys. Let. <u>19</u>, 486, 1971.
2. E. A. Giess, J. D. Kuptsis and E. A. D. White, J. Cryst. Growth Vol 15, No. 5, 1972.
3. B. S. Hewitt, R. D. Pierce, S. L. Blank, and S. Knight, Intermag Conf. Washington, D. C., April 1973.
4. D. H. Smith, F. B. Hagedorn and B. S. Hewitt, J. App. Phys. <u>44</u>, 4177 (1973).
5. R. A. Burmeister and R. Hiskes, Semiannual Tech Rpt., Contract No. DAAH01-72-C-0996, January 1973.

RAPID EPITAXIAL GROWTH OF HIGH PERFECTION GARNET FILMS

W. A. Bonner
Bell Laboratories, Murray Hill, New Jersey 07974

ABSTRACT

Highly perfect films (>1 cm^2) of $\sim$200 gauss bubble domain materials are obtained by liquid phase epitaxy techniques using growth rates as fast as 2 μm/min. Rapid film growth combined with rotation of horizontally held substrates at $\sim$300 rpm and a low thermal gradient have been found to minimize problems related to precipitation of a second phase and compositional drift.[1] In this process relatively large rare earth concentrations are employed, requiring operation at a higher temperature. However, a greater working supersaturation can be achieved, permitting rapid film growth. Optimal conditions have been found for the systems $(Y,Lu,Ca)_3(Ge,Si,Fe)_5O_{12}$ and related materials containing other rare earth ions, $A\ell$ and/or Ga on {111} $Gd_3Ga_5O_{12}$ substrates. Films containing Ca^{2+} plus Ge^{4+} or Si^{4+} are of particular interest as the M^{4+} ions have a greater tetrahedral site preference than Ga^{3+} or $A\ell^{3+}$.[2] Outstanding bubble domain properties of these relatively high Curie temperature materials are reported by Geusic et al.[3]

1. B. S. Hewett, R. D. Pierce, S. L. Blank, and S. Knight; Proceeding IEEE MAG. 9, Sept. 1973 - to be published.
2. S. Geller, H. J. Williams, C. P. Espinosa and R. C. Sherwood; Bell System Tech. J. XLIII, 2, 1964.
3. J. E. Geusic, D. H. Smith, L. G. Van Uitert, G. P. Vella-Colerio, paper submitted to this conference.

CHARACTERISTICS OF $Y_{3-x}Ca_x(Ge,Si)_xFe_{5-x}O_{12}$ BASED EPITAXIAL FILMS FOR MAGNETIC BUBBLE APPLICATIONS

J. E. Geusic, D. H. Smith, L. G. Van Uitert and G. P. Vella-Coleiro
Bell Laboratories, Murray Hill, N. J. 07974

The growth and magnetic properties of epitaxial films of $(Y,RE)_{3-x}Ca_x(Ge,Si)_xFe_{5-x}O_{12}$ have recently been reported.[1] These compositions have Curie temperatures 70-100°C higher than Ca^{3+} and Al^{3+} substituted compositions of the same $4\pi M$. The (Ca-Si) and (Ca-Ge) substitutions do not adversely effect mobility; hence, in these garnet films high mobility ($\mu \sim$ 1400-2000 cm/sec-Oe) and a broad operating temperature range is simultaneously achieved. In a typical composition $Y_{1.64}Eu_{.1}Lu_{.3}Ca_{.96}Ge_{.96}Fe_{4.04}O_{12}$, which supports 6μ nominal bubbles, over the temperature range -10 to 100°C (see Fig. 1) the variation of strip width is less than 1μ and the bubble collapse field tracks within ±2 Oe the bias magnet which is used in the bubble mass memory module described by Michaelis and Bonyhard.[2] Preliminary studies of the propagation margins of this material at 0.6 MHz using a 26 micron period T-X test circuit (see Fig. 2) illustrates that extremely reliable operation over a broad operating temperature range is achieved at high rotating field frequencies. With these new garnet compositions, the operation of bubble devices at rotating field rates up to 1 MHz and over a temperature range -10 to 100°C can now be achieved.

REFERENCES

1. W. A. Bonner, J. E. Geusic, D. H. Smith, L. G. Van Uitert and G. P. Vella-Coleiro, Mat. Res. Bull. **8**, No. 10, 1223 (Oct. 1973).
2. P. C. Michaelis and P. I. Bonyhard, IEEE Trans. Magnetics MAG-9, 436 (1973).

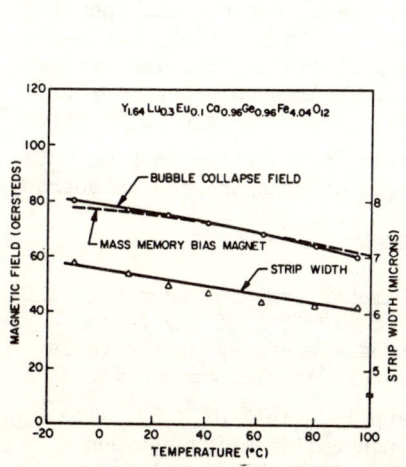

Figure 1

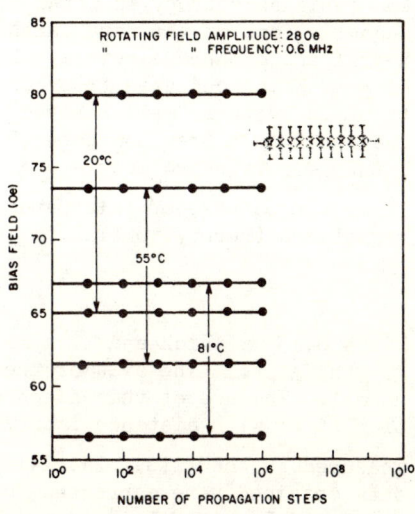

Figure 2

GROWTH INDUCED ANISOTROPY OF $Y_{3-x}Lu_xFe_5O_{12}$: DEPENDENCE ON TEMPERATURE AND COMPOSITION

by

E. M. Gyorgy, M. D. Sturge and L. G. Van Uitert
Bell Laboratories
Murray Hill, New Jersey 07974

ABSTRACT

We have measured the uniaxial growth-induced anisotropy in the growth plane of samples cut from the (110) facets of $Y_{3-x}Lu_xFe_5O_{12}$ with $0.4 \leq x \leq 2.7$, from 4K to 450K. Neither rare earth ion is magnetic. We find that at 77K and above the induced anisotropy can be roughly described by $K_u \sim K(1-x/3)x$ with $K \sim 8 \times 10^3$ ergs/cc at 300K. At 4K, K_u shows no regular dependence on x and the measured anisotropy at this temperature can reasonably be attributed to rare earth impurities. The temperature dependence of K indicates that the value of K is proportional to m_a^3, where m_a is the reduced magnetization of the octahedral sublattice.

INTRODUCTION

Non-cubic anisotropy is often observed in garnets that have only non-magnetic ions on the rare earth site and only Fe^{3+} on the tetrahedral and octahedral sites. For these garnets the usual theories[1] of growth induced anisotropy do not apply. Much of this anisotropy is neither "well behaved" nor reproducible. However, as was indicated in a previous paper,[2] the growth induced anisotropy of $Y_2LuFe_5O_{12}$ is reproducible and does not appear to be a "dirt effect". It is the purpose of this paper to extend the earlier work on this garnet and thereby to clarify the origin of this induced anisotropy.

EXPERIMENT

Discs 0.15 mm thick and 1.44 mm in diameter were cut from the (110) growth facets. The axis of the disc is parallel to the facet normal. The use of thin discs minimizes the effect of the variation of K_u with distance into the crystal.[3,4] The torque was measured in the (110) plane with an applied field of 7 kG. The data obtained were corrected for any non-collinearity between $\underline{M}$ and $\underline{H}$ and Fourier analyzed to obtain the first order anisotropy constant (K_1) and the uniaxial anisotropy constant (K_u).[4]

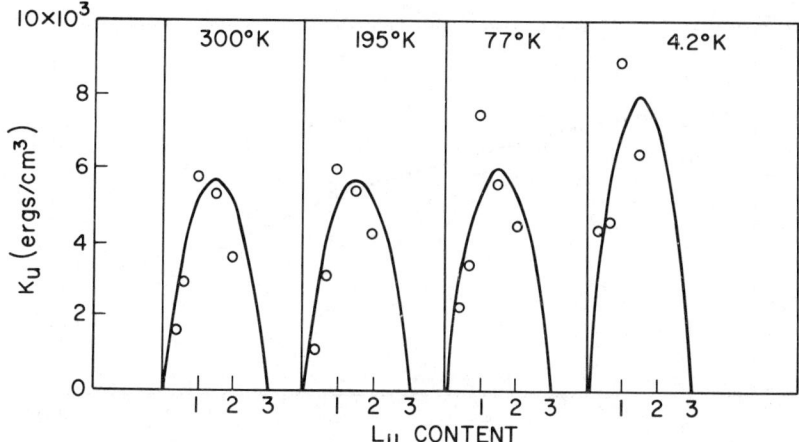

Fig. 1 The growth induced anisotropy as a function of Lu content for 4.2K, 77K, 195K and room temperature. The solid curve represents $K(1-x/3)x$ where x is the Lu content.

Since crystals from the same melt may have slightly different compositions, it is necessary to determine the Y:LU ratio of the actual samples used for the magnetic measurements. However, the samples used are too small for accurate chemical analysis. Therefore, the composition of each cylinder was determined from the measured lattice constant assuming that the lattice constant is a linear function of composition. Comparison of the rare earth ratio obtained from the lattice constant measurements to the ratio of rare earth ions in the melt shows that the distribution coefficient for Lu increases approximately linearly from 0.87 to 1.0 with increasing Lu content under the growth conditions employed.

RESULTS

The values of K_u at 4.2K, 77K, 195K and room temperature are plotted as a function of composition in Fig. 1. The solid line represents the curve $K(1-x/3)x$ where x is the Lu content in atoms per formula unit. In the (110) plane, K_u is given by $K_u = F(1\bar{1}0) - F(100)$. The positive value of K_u indicates that [001] is the easy axis in this plane.

The cubic anisotropy constant (K_1) at these temperatures is, with one exception, within 12% of the value of YIG.[5] The values of K_1 are scattered within this range and do not correlate with composition. The one exception is K_1 for $YLu_2Fe_5O_{12}$ (-2.9×10^4 ergs/cc) at 4.2K which may be compared to -2.48×10^4 ergs/cc given for YIG at this temperature. This deviation may be attributed to rare earth impurities. For example about 10^{-4} Sm ions per formula unit could

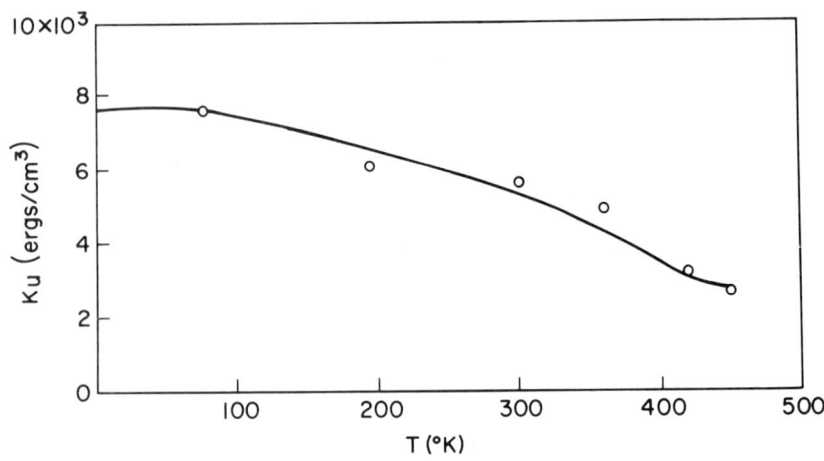

Fig. 2 The uniaxial anisotropy of $Y_2LuFe_5O_{12}$ as a function of temperature.

account for the difference in K_1 observed for YIG and $YLU_2Fe_5O_{12}$ at 4.2K.[6] It may be noted that K_u for this composition at 4.2K is also unexpectedly high (Fig. 1) and, as will be discussed later, this deviation from the expected value is also probably due to rare earth impurities. The presence of rare earth contamination at the levels postulated has not as yet been verified since the emission spectroscopic analytical technique used, is not sufficiently sensitive.

In Fig. 2, K_u for $Y_2LuFe_5O_{12}$ is shown as a function of temperature. The solid curve represents the temperature variation of m_a^3 where m_a is the reduced magnetization of the octahedral (a) sublattice.[7] The solid curve is normalized to fit the experimental data at 77°K and its theoretical significance will be discussed in the next section.

DISCUSSION

Although there is considerable scatter, the data shown at 77°K can be fitted reasonably well to $K_u = K(1-x/3)x$, where x is the Lu content (Fig. 1). This is the concentration dependence predicted by the accepted model of growth-induced anisotropy[8] in which there is preferential ordering of the rare earth ions among the different sites of the c sublattice. Experimental verification of this concentration dependence is given in Ref. 4 for $Y_{3-x}Sm_xFe_5O_{12}$ and $Gd_{3-x}Eu_xFe_5O_{12}$. The sharp increase in K_u below 77K, on the other

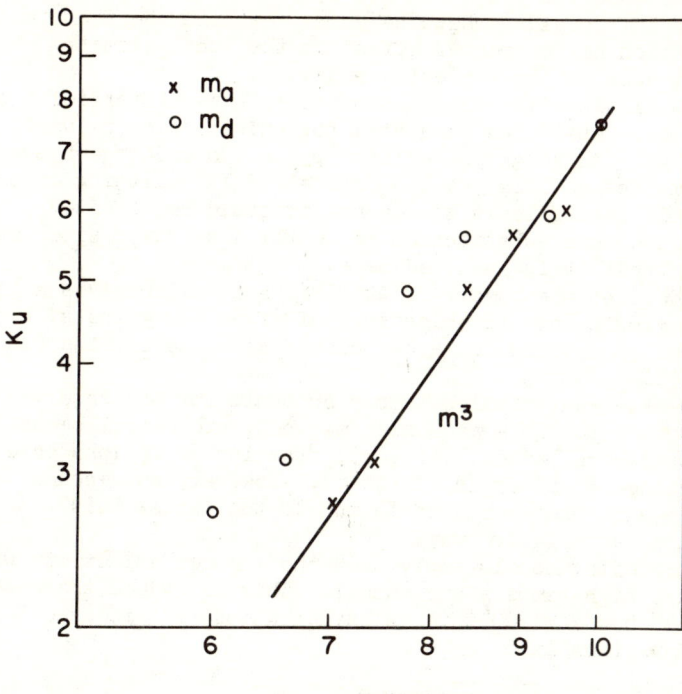

Fig. 3 A log-log plot of K_u as a function of m_d (circles) and m_a (crosses). The solid line shows that K_u is proportional to m_a^3.

hand, shows no regular dependence on x, and can reasonably be attributed to rare earth impurities.

The slow decrease in K_u with increasing temperature above 77K shown in Fig. 2 confirms that magnetic rare earth impurities are not important in this region, since their contribution should vary at least as fast as T^{-1}. It has previously been pointed out[2] that the model of Akselrad and Callen[9], which attributes K_u to crystal field effects at the tetrahedral Fe^{3+} sites, predicts the correct order of magnitude for K_u. In this model the ordering of nonmagnetic ions on the dodecahedral sites leads to a non-cubic induced anisotropy through the effect of these ions on the neighboring Fe^{3+} ions. This model also predicts that for $m_d \gtrsim 0.6$ the induced anisotropy should be proportional to m_d^3. m_a and m_d are the reduced magnetization of the octahedral (a) and the tetrahedral (d) sublattices respectively. A log-log plot of K_u versus m_d is shown in Fig. 3 (circles). It is seen that m_d^3 does not fit the data, and that a much better fit is obtained with m_a^3 (crosses). It appears that, although the c (rare earth) site is further from the nearest a site than it is from the nearest d site, the change in crystal field due to the substitution of Lu for Y on the c site is much larger at the a site. This could perhaps be connected with the fact that such a substitution has a greater effect on the local symmetry at the a site, than it does at the nearest d site.

The data will also fit the relation that K_u is proportional to $m_a m_d$. This is the dependence predicted for anisotropic (pseudo-dupolar) exchange. Although the anisotropy in the exchange between two Fe^{3+} ions may be as large as the crystal field splitting of the ground state[10] it is difficult to see how it could be affected appreciably by the mere substitution of Lu for Y at the c site, and we do not think that it is involved here.

We conclude that the theory of Akselrad and Callen gives a good account of the growth-induced anisotropy of mixed YLu garnets, but that the anisotropy appears to be associated with the a rather than the d sublattice.

While the Akselrad - Callen theory accounts for the observed anisotropy, another possible mechanism may be preferential occupation of Lu on the octahedral sites. (A small fraction of Yb ions have been observed in octahedral sites in YIG:Yb[11]). However, not enough is known about the site preference of Lu in YIG to discuss this possibility in a quantitative way.

All the preceding models imply that the induced anisotropy will disappear with a high temperature anneal. This is indeed the case. Annealing for 18 hrs at 1200°C in an oxygen atmosphere reduced K_u in all cases to less than 1×10^3 ergs/cc.

REFERENCES

1. A. Rosencwaig and W. J. Tabor, AIP Conf. Proc. 5, 57 (1972).
2. E. M. Gyorgy, M. D. Sturge, L. G. Van Uitert, E. J. Heilner and W. H. Grodkiewicz, J. Appl. Phys. 44, 438 (1973).
3. E. M. Gyorgy, J. F. Dillon, Jr. and J. P. Remeika, AIP Conf. Proc. 5, 680 (1972).
4. E. J. Heilner and W. H. Grodkiewicz, J. Appl. Phys. 44, 4218 (1973).
5. W. H. von Aulock, Handbook of Microwave Ferrite Materials (Academic Press, N.Y., 1965) p. 89.
6. We use here the value of K_1 at 4.2K (-4×10^7 ergs/cc) for $Sm_3Fe_5O_{12}$ given by R. Pearson (J.Appl.Phys. S33, 1236) (1962).
7. We use the sublattice magnetization data for pure YIG obtained by R. Gonano, E. Hunt and H. Meyer (Phys.Rev. 156, 521) (1967) and by A. M. van der Kraan and J. J. van Loef (Proc. Conf. on Applications of the Mössbauer Effect, Tihany. 1960 [Akademia Kiado, Budapest, 1971] p. 519).
8. H. Callen, Mat. Res. Bull. 6, 931 (1971).
9. A. Akselrad and H. Callen, Appl. Phys. Letters 19, 464 (1971).
10. R. C. Wayne and D. H. Anderson, Phys. Rev. 155, 496 (1967).
11. J. F. Dillon, Jr., J. P. Remeika and L. R. Walker, J. Appl. Phys. 38, 2235 (1967).

THE EFFECT OF SUBSTRATE ORIENTATION AND GROWTH TEMPERATURE ON THE MAGNETIC ANISOTROPY OF $Eu_xY_{3-x}Fe_5O_{12}$ FILMS

T. S. Plaskett, E. Klokholm and D. C. Cronemeyer

IBM Thomas J. Watson Research Center
Yorktown Heights, New York 10598

ABSTRACT

Garnet films of $Eu_xY_{3-x}Fe_5O_{12}$ of (100), (110) and (111) orientations were grown by the LPE technique from $PbO:B_2O_3$ fluxes at growth temperatures (T_G) between 744 to 884°C. The compositions of the films were between x = 1.0 and x = 2.1. The growth component (K_u^G) of the induced anisotropy for the (100) oriented films varied from negative to positive as T_G decreased. For the (111) and (110) films K_u^G was positive and varied slightly with T_G. The K_u^G for the (100) films for a constant T_G increased in magnitude with Eu content. The K_u^G cannot be explained entirely by the prevalent models. It is postulated that the K_u^G has a component dependent upon the Pb content of the film.

INTRODUCTION

In a previous paper[1] it was shown that the uniaxial anisotropy (K_u) in (100) oriented $Eu_2Y_1Fe_5O_{12}$ films deposited on $Sm_3Ga_5O_{12}$ substrates was dependent upon the growth temperature (T_G). At T_G above 800°C the K_u was negative while for T_G below 800°C the easy direction was normal to the plane and increased in magnitude as T_G decreased, reaching a value of 3×10^5 erg/cm^3 at 744°C. The compressive stress in the film increased when T_G decreased. This was attributed to an increase in the lattice parameter by Pb which was incorporated from the $PbO:B_2O_3$ flux.[2] The magnetostriction coefficient (λ) calculated from the stress vs K_u data was about 3 times greater than the value reported in the literature.[3] Since the value was unreasonably large, it was postulated that a growth component[4] of the induced uniaxial anisotropy existed which for (100) oriented films varied from negative to positive as T_G was lowered.

In this paper the anisotropy for the (111) and (110) oriented films as a function of growth temperature is reported. Also the effect of the Eu/Y ratio on the anisotropy for the (100) films is described.

EXPERIMENTAL

(a) <u>Film Growth</u> - The films were grown by the liquid phase epitaxy (LPE) "tipping" technique[5]. With this technique a saturated flux is tipped onto the substrate and growth occurs while the solution is cooled. The cooling rate used was about 4°C/min. The composition of the charge was 81.8 mole % PbO, 5.2% B_2O_3, 2.6% Eu_2O_3 + Y_2O_3 and 10.4% Fe_2O_3. The Eu/Y ratio was varied to provide films over a composition range of x = 1.0 to x = 2.1, where x is defined by $Eu_xY_{3-x}Fe_5O_{12}$. The mole ratio of Eu/Y in the film to that in the charge was about 0.66 for x = 2.1 and x = 1.6 and 0.76 for x = 1.1. The films were grown on (100), (111) and (110) oriented $Sm_3Ga_5O_{12}$ substrates. The T_G reported in this paper is the temperature at the start of growth and was varied between 744°C to 884°C. The growth time varied between 1 to 8 min., which provided a film thickness range of 0.4 to 2.0 μm. The film thickness was measured by optical interference using a Cary Scanning Spectrophotometer. The lattice mismatch between the film and substrate (Δa) was measured by an x-ray diffraction technique. The composition of the film was obtained by electron microprobe measurements.

(b) <u>Anisotropy</u> - The induced anisotropy in all samples was measured by a ferromagnetic resonance technique. Resonance fields were measured for the magnetic fields parallel and perpendicular to the film plane. For the (100) and (110) specimens the parallel resonance field was also measured as a function of crystal orientation in the film plane. From these data K_u was calculated using the convention that K_u is the difference between the anisotropy energy normal to the film plane and the anisotropy energy in the plane. For the (110) oriented films K_u is a function of the direction in the film plane. For this orientation K_u was calculated for the [001] and [1$\bar{1}$0] directions in the film plane. The magnetization (4πM) and a crystalline anisotropy (K_1) used for the calculation were interpolated from the bulk values given[3] for $Eu_3Fe_5O_{12}$ and $Y_3Fe_5O_{12}$.

The measured K_u was assumed to consist of a stress component (K_u^S) and a growth component (K_u^G), i.e.

$$K_u = K_u^G + K_u^S$$

To separate K_u^G from K_u^S, K_u^S was calculated from the Δa measurements. The magnetostriction coefficients (λ) used for the calculation for the various Eu concentrations were obtained by interpolation from bulk values given[3] for $Eu_3Fe_5O_{12}$ and $Y_3Fe_5O_{12}$.

RESULTS

The values of K_u^G for the various film orientations and Eu contents are shown in Fig. 1 as a function of T_G. For the (100) oriented films K_u^G changed from negative to positive as T_G decreased and the magnitude of the K_u^G increased as the Eu content increased. For the (111) oriented films K_u^G was positive and increased slightly as the T_G decreased. Similarly for the (110) oriented films K_u^G was positive and varied only little with T_G. Strip domains were observed in the (110) films but stable bubbles could not be formed.

The lowest T_G investigated was about 744°C. Below this the flux solidified during the growth cycle. The range of Eu content investigated was from x = 1.0 to x = 2.1. This is about the range of compositions that can be grown on $Sm_3Ga_5O_{12}$ substrates. The lattice parameter of the $Sm_3Ga_5O_{12}$ varied from 12.429 to 12.438 Å. The Δa between the film and substrate as measured was from + 0.0026 (tension) to - 0.0420 (compression). The value of Δa depended upon the growth temperature[1] and composition. As the growth temperature decreased the Pb content of the film increased as shown in Fig. 2 for the various orientations. The amount of Pb incorporated in the film was orientation dependent. For a constant growth temperature the Pb concentration was highest in the (110) films. The Eu/Y ratio was independent of T_G and the film orientation for the same initial charge composition.

DISCUSSION

The non-cubic magnetic anisotropy observed in garnets has been attributed to the growth process. This subject has recently been

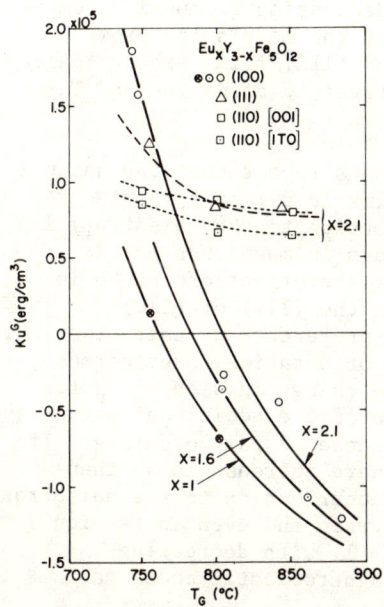

Fig. 1 - Growth component of the induced anisotropy (K_u^G) as a function of growth temperature (T_G) for various Eu contents and film orientation.

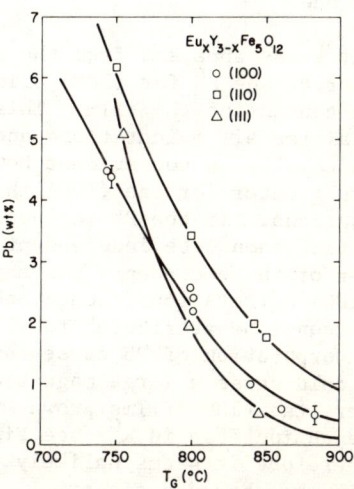

Fig. 2 - Pb content of the film as a function of growth temperature (T_G) for various film orientations.

reviewed.[4] According to the prevalent models, K_u^G is the result of a differential incorporation of rare-earth ions into essentially equivalent sites. The magnitude of K_u^G for mixed rare-earth garnets[6] depends upon the particular choice of rare-earth ions, the mole ratio of the rare-earth ions present and the growth orientation. The maximum K_u^G, as predicted from the models, occurs at about an equal mole ratio of the rare-earth ions and when their size difference is large.

Recently, the anisotropy energy of the Eu ions in $Eu_3Fe_5O_{12}$ was calculated and used to determine K_u^G in LPE films.[7] Reasonable agreement with experimental data for their (100) and (111) LPE films of $(EuY)_3(GaFe)_5O_{12}$ was obtained. It should be noted, however, that the reported value of K_u^G for the (100) film was negative. If the data shown in Fig. 1 for T_G at about 880°C is compared with calculation, reasonable agreement is also obtained. As shown in Fig. 1, K_u^G of the (100) films for this T_G is also negative. As T_G decreases and the amounts of Pb incorporated in the (100) films increases, K_u^G becomes more positive and reaches a value of about 2×10^5 erg/cm^3 at $T_G = 740°C$. As is shown by these data, K_u^G for (110) and (111) films also increases with decreasing T_G but not at the same rate as the (100) films. Furthermore, the compositional dependence of K_u^G for the (100) films does not show a maximum at $x = 1.5$ as predicted. The unusual change in sign of (100) K_u^G with decreasing T_G and its compositional dependence cannot be explained by the models for growth anisotropy. However, K_u^G for the (110) and (111) films are in reasonable agreement with the theories with respect to magnitude and T_G dependence.

It seems apparent from the data in this report that the unusual behavior of the K_u^G for (100) oriented films is directly related to the Pb content of the film. This contribution to the anisotropy is uniaxial for all orientations and increases in magnitude as T_G decreases. It is not evident however why the effect of the Pb is so much greater for the (100) than either the (111) or (110) orientations. If the Pb influences site-preference or pair ordering, the effect should be independent of film orientation. Furthermore, neither of these concepts can explain the change in sign of K_u^G for the (100) orientation. Hence, the source of the additional anisotropy cannot be attributed to site-preference or pair ordering. If the incorporation of Pb causes an inordinate increase in λ, then this would cause a large magnetoelastic contribution to the anisotropy. However, the (100) films grown with no strain and even in tension showed an increase in K_u^G (see Fig. 1, $x = 1$) with decreasing T_G and therefore it seems unlikely that a magnetoelastic model can explain the observed effect.

An attractive alternative is that the Pb causes a defect ordering similar to the oxygen vacancy induced ordering reported by Stacy and Rooymans.[8] The exact nature of this additional contribution to the anisotropy is not known at present and further work is in progress.

ACKNOWLEDGMENTS

We gratefully acknowledge the assistance of P. C. Yin, S. E. Blum and E. Mendel for the substrate growth and polishing, A. H. Parsons for film growth and characterization, J. D. Kuptsis and B. Chider for electron probe measurements, J. Karasinski for x-ray measurements and E. A. Giess for helpful discussions.

REFERENCES

1. T. S. Plaskett, E. Klokholm, H. L. Hu and D. F. O'Kane, AIP Conf. Proc. No. 10, p. 319 (1972).
2. E. A. Giess, B. E. Argyle, D. C. Cronemeyer, E. Klokholm, T. R. McGuire, D. F. O'Kane and T. S. Plaskett, AIP Conf. Proc. No. 5 p. 110 (1971).
3. L. G. Van Uitert, E. M. Gyorgy, W. A. Bonner, W. H. Grodkiewicz, E. J. Heilner and G. J. Zydzik, Mat. Res. Bull $\underline{6}$, 1185 (1971).
4. A. Rosencwaig and W. J. Tabor, AIP Conf. Proc. No. 10, p. 57 (1972).
5. L. K. Shick, J. W. Nielson, A. H. Bobeck, A. J. Kurtzig, P. C. Michaelis and J. P. Reekstin, Appl. Phys. Letters $\underline{18}$, 89 (1971).
6. H. Callen, Mat. Res. Bull. $\underline{7}$, 931 (1971).
7. M. D. Sturge, R. C. LeCraw, R. D. Pierce, S. J. Licht and L. K. Shick, Phys. Rev. B$\underline{1}$, 1070 (1973).
8. W. T. Stacy and C. J. M. Rooymans, Solid State Communications $\underline{9}$, 2005 (1971).

MISFIT STRAIN IN LPE $(Y,Eu)_3(Fe,Ga)_5O_{12}$ GARNET FILMS

H. Makino, T. Hibiya and K. Matsumi
Central Research Laboratories,
Nippon Electric Co., Ltd., Kawasaki, Japan

ABSTRACT

Misfit strain in epitaxial magnetic garnet films has been quantitatively investigated for $Y_{3-x}Eu_xFe_{3.9}Ga_{1.1}O_{12}$ deposited on (111) $Gd_3Ga_5O_{12}$ substrate by LPE techniques. By the X-ray double crystal method, measurements were carried out on the bending curvature, from which strain parallel to the surface, ε_x, was directly derived. For small misfits, ε_x is proportional to $\triangle a$, as expected from Besser's Region I model. However, in the region of $\triangle a \leq -0.02$ Å, it was confirmed that a distinct transient region exists between Region I and II, and some part of the elastic stress is relieved during the film growth. Degree of the elastic stress relief increases gradually with further increasing $|\triangle a|$. Similar conclusions were derived from rocking curve measurements using symmetric and asymmetric reflections.

INTRODUCTION

In magnetic garnet LPE films for use in bubble domain devices, it is important to know the substrate-film lattice misfit quantitatively, because misfit stress contributes to the stress-induced anisotropy and also affects the quality of epitaxial films[1-3]. Besser et al.[1] have presented a stress model for CVD films. They defined "Region I" as where the stress is caused only by room-temperature lattice misfit and "Region II" as where the stress depends on the difference of thermal expansion coefficients. Although Carruthers[4] proposed a modified stress model, which combines the concepts of Region I and II by introducing fractional stress relief (η), the boundary between Region I and II has not been clear. The purpose of this report is to clarify the presence of a transient region between Region I and II in LPE garnet films.

EXPERIMENTAL

In order to obtain garnet films with various misfits, a series of $Y_{3-x}Eu_xFe_{3.9}Ga_{1.1}O_{12}$ LPE films with $0.2 \leq x \leq 1.7$ were grown by tipping techniques on chemically polished (111) $Gd_3Ga_5O_{12}$ substrate from $PbO-B_2O_3$ fluxed melts whose compositions were similar to those reported by Shick et al.[5] Thickness of the film was measured by interferometric method. The backside of the specimen was polished by diamond grit.

"Stress-free" lattice constant of epitaxial film, a_f, was estimated from the expression: $a_f = 12.359 + 0.043 \cdot x$ (Å). This expression was found from the relation between Eu content x and lattice constant of standard polycrystals of $Y_{3-x}Eu_xFe_{3.9}Ga_{1.1}O_{12}$. The value of x in the epitaxial film was determined by electron probe microanalyzer, using sintered polycrystals as standards. Substituting 12.383 Å as the lattice constant of the substrate (a_s), $\triangle a$ (= $a_s - a_f$) was estimated.

Strain parallel to the film surface, ε_x, was obtained by measuring the radius of bending curvature, r, of the specimen. The radius of bending curvature of the substrate, onto which the film was deposited, was measured by X-ray double crystal diffractometer as a shift of the 888 diffraction peak position when a sample is moved parallel to the surface. Difference of lattice constants between substrate and "strained" film along the normal ($\triangle a^{\perp}$) and along the parallel ($\triangle a''$) to the (111) surface plane was measured using symmetric and asymmetric reflections. In the diffractometer, $CuK\alpha_1$ radiation is first diffracted from dislocation-free silicon single crystal. The 422 reflection with an asymmetric factor of 0.1 was chosen for the first crystal.

RESULTS AND DISCUSSION

In-Plane Strain and Fractional Stress Relief

Strain parallel to the film surface, ε_x, is given by[4]

$$\varepsilon_x = (1 - \eta)\triangle a/a_f + \eta \triangle \alpha \triangle T, \qquad (1)$$

where $\triangle \alpha \triangle T$ * is the thermal strain derived during the cooling process and η ($0 \leq \eta \leq 1$) is the fractional stress relief during film growth. The strain in the case of Besser's Region I and II is represented by substituting 0 and 1, respectively, for η in Eq. (1).

Since the film strain, ε_x, is partially relieved by bending, it is changed into ε_f, while substrate strain, ε_s, is introduced. Assuming that the substrate is sufficiently thicker than the film, these strains at the interface can be described as a function of radius of bending curvature[6]:

$$\varepsilon_f = -t^2/6rh, \qquad (2)$$

$$\varepsilon_s = t/2r, \qquad (3)$$

and thus, in-plane strain, which is expected to have occurred before bending, is given by

$$\varepsilon_x = \varepsilon_f - \varepsilon_s(a_s/a_f) \simeq \varepsilon_f - \varepsilon_s, \qquad (4)**$$

where: r = radius of bending curvature (r > 0, for compressive film

---

* $\triangle \alpha = \alpha_f - \alpha_s = +12.6 \times 10^{-6}/°C$ and $\triangle T = 900°C$.
** For Regions I and II and for their transient region, this relation is available by introducing some assumptions.

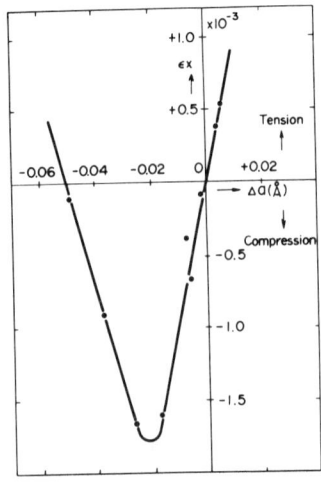

Fig. 1 In-plane strain obtained from bending curvature as a function of △a.

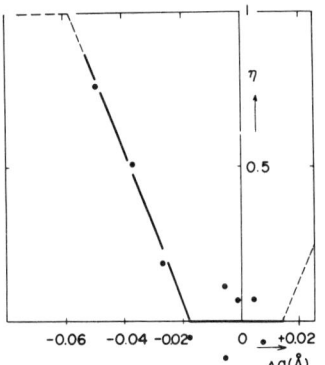

Fig. 2 Fractional stress relief as a function of △a.

strain), t=thickness of whole specimen, h= thickness of film.

By measuring t, h, and r, ε_x was obtained. t and h were 5-10μm and 200-400μm, respectively. r was 3-60m, depending on h, t and stress.

Figure 1 shows ε_x as a function of △a. Figure 2 shows obtained η values by substituting ε_x, △a and a_f into Eq.(1), as a function of △a. In the region of $-0.02\text{Å} \leq \triangle a \leq +0.015\text{Å}$, ε_x is proportional to △a. In addition, η is almost zero in the same region. This trend agrees with a model for the misfit strain without any elastic stress relief at the growth temperature. This region is almost in accord with Besser's Region I. Although η calculated by bending curvature measurement does not appear as exactly zero in Region I, this might have resulted from estimation error of ε_x and a_f during calculation process.

A maximum compressive strain is attained at $\triangle a \simeq -0.02$ Å and ε_x tends to change gradually from negative to positive with further increasing |△a|, and the condition of perfect Region II model ($\varepsilon_x \simeq 1.1 \times 10^{-3}$; tensile) is gradually approached. In such a region, η is not zero and tensile stress owing to thermal expansion difference is becoming more dominant over the compressive stress stemming from lattice misfit. The extrapolation of η values to $\eta = 1$ suggests that misfit strain in film would be explained by perfect Region II model over -0.06 Å. Film surface grown under such a condition was roughened and innumerable etch pits were revealed by immersion into hot phosphoric acid. This suggests the generation of the interfacial dislocations. The critical value of misfit ($\triangle a_c$) was found to be -0.02 Å, over which the elastic strain is partially relieved at the growth temperature.

According to Besser et al.[1], Region I and Region II are discontinuous at the boundary, however, present results reveal that they are not discontinuous, but that a distinct transient region exists between Region I and Region II.

Lattice Distortion in Epitaxial Film

According to the stress model, an epitaxial garnet film is an elastically distorted cubic structure and, therefore, its lattice planes $(hkl)_f$ are not parallel to the same $(hkl)_s$ planes in the undeformed substrate. In the case of asymmetric reflection, which uses such an inclined plane, the difference of peak positions of substrate and film in the rocking curve, $\Delta\theta$, results from not only the Bragg angle difference between them ($\Delta\theta_B = (\theta_B)_s - (\theta_B)_f$), but also the angular difference due to the nonparallelism of $(hkl)_s$ and $(hkl)_f$ planes ($\Delta\phi$). Then $\Delta\theta$ can be expressed as

$$\Delta\theta = \Delta\theta_B + \Delta\theta, \tag{5}$$

$$\Delta\theta_B = -\tan\theta_B [(\Delta a^\perp / a_s)\cos^2\phi + (\Delta a'' / a_s)\sin^2\phi], \tag{6}$$

$$\Delta\phi = -|\sin\phi\cos\phi|[(\Delta a^\perp / a_s) - (\Delta a'' / a_s)], \tag{7}$$

where θ_B is Bragg angle of hkl reflection for the substrate, and ϕ is the angle which the (111) surface plane makes with the (hkl) plane. It should be noted that Eqs. (5) to (7) are inaccurate, unless the rocking curve is measured under the condition that the diffraction vector lies perpendicular to an axis, around which the $(hkl)_f$ plane rotates from the $(hkl)_s$ plane, and that the absolute asymmetry factor value is less than 1. $\Delta a^\perp$ and $\Delta a''$ were then obtained in order to fit in the measured several $\Delta\theta$'s by the least square method. In Table 1, the results are listed for films with $x = 0.5$ and 1.7, repsectively.

On the other hand, the stress model predicts the lattice distortion of the film given by the following expressions:

$$\Delta a^\perp = \Delta a + [2\mu_f/(1-\mu_f)] \cdot [(1-\eta)\Delta a + \eta\Delta\alpha\Delta T a_f], \tag{8}$$

$$\Delta a'' = \Delta a - [(1-\eta)\Delta a + \eta\Delta\alpha\Delta T a_f]. \tag{9}$$

From the measurements of both $\Delta a^\perp$ and $\Delta a''$, we can determine η using Eqs. (8) and (9). The η values thus obtained are also listed in Table 1. They are in goods agreement with the results from the bending curvature measurements.

Figure 3 shows plotted measurements of $\Delta a^\perp$ versus Δa given only by symmetric 888 reflections. It is also found that $\Delta a^\perp$ is no longer proportional to Δa in the region of $\Delta a < -0.02$ Å. Figure 3 suggests that η would be nonzero when Δa exceeds a critical misfit, $\Delta a_c \simeq -0.02$ Å.

TABLE 1

	x=0.5	x=1.7
$\Delta a^\perp$ (Å)	+0.008	-0.049
$\Delta a''$ (Å)	+0.0004	-0.052
Δa (Å)	+0.004	-0.050
η	~0.	0.82

INDEX	$\Delta\theta$ (×10⁻⁴ rad)	
	x=0.5	x=1.7
8 8 8	-11.1	+66.4
4 6 6	- 5.6	+27.6
0 8 8	- 7.1	+38.9
12 2 2	- 7.5	+47.5
12 8 0	-11.8	
6 12 0	- 8.7	

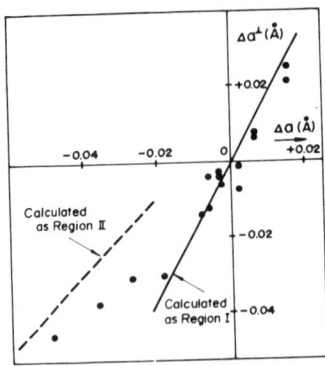

Fig.3 $\triangle a^{\perp}$ obtained from symmetric 888 reflections as a function of $\triangle a$.

CONCLUSION

It was experimentally confirmed that a distinct transient region exists between Region I and Region II in $a_f > a_s$ from several measurements, which were carried out independently using X-ray techniques. Values of η and $\triangle a_c$ obtained from these measurements were in good agreement with each other. In the transient region, η increases with increasing $|\triangle a|$. Dislocation etch pits were revealed by hot phosphoric acid etching for films grown under such a condition. These facts indicate that misfit dislocations are generated during the film growth in order to relieve some part of the elastic stress and film lattice no longer fits completely that of the substrate.

ACKNOWLEDGEMENT

The authors thank M. Maruyama, T. Ayusawa, K. Kokubu, T. Kawamura, J. Matsui and F. Yamamoto for useful discussions. They also thank T. Okada and other colleagues for their encouragement.

This work was, in part, supported by the Agency of Industrial Science and Technology of Japan.

REFERENCES

1. P. J. Besser, J. E. Mee, P. E. Elkins, and D. M. Heinz, Mat. Res. Bull., 6, 1111 (1971).
2. E. Klokholm, J. W. Matthews, A. F. Mayadas, and J. Angilello, AIP Conference Proceedings, No. 5 (17th Conf. on Magnetism and Magnetic Materials, Chicago, Nov. 1971) p. 105.
3. P. J. Besser, J. E. Mee, H. L. Glass, D. M. Heinz, S. B. Austerman, P. E. Elkins, T. N. Hamilton, and E. C. Whitcomb, ibid, p. 125.
4. J. R. Carruthers, J. Crystal Growth, 16, 45 (1972).
5. L. K. Shick, J.W. Nielsen, A. H. Bobeck, A. J. Kurtzig, P. C. Michaelis, and J.P. Reekstein, Appl. Phys. Lett., 18, 89 (1971).
6. See, for example, R. W. Hoffman, Physics of Thin Films, vol. 3, Eds. G. Hass and R. E. Thun (Academic Press, 1966) p. 211.

PERPENDICULAR ANISOTROPY IN $Gd_{1-x}Co_x$ AMORPHOUS FILMS PREPARED BY RF SPUTTERING

D. C. Cronemeyer

IBM Research Center, Yorktown Heights, N. Y. 10598

ABSTRACT

In a study of $Gd_{1-x}Co_x$ amorphous alloy films ($0.5 < x < 0.95$) prepared by rf sputtering onto glass substrates, the perpendicular anisotropy energy K_u as a function of composition and rf bias is derived from measurements of ferromagnetic resonance in the parallel and perpendicular configurations. This evaluation is facilitated by having magnetic moment data obtained with a force balance. With increasing rf bias, there is a monotonic increase of the Co concentration, whereas K_u at first rises and then appears to pass over a maximum. Starting from a $GdCo_2$ target, K_u has a maximum of 4×10^5 ergs/cm^3 for $x = 0.72$, dips to a low value near the compensation composition ($x = 0.795$), and increases again beyond this point. FMR values of K_u are roughly confirmed by analyzing the magnetization curves obtained with the magnetic field parallel to the plane of the samples. The approximate compositional range of interest for bubble domain applications ($K_u \geq 2\pi M_s^2$) seems to be $0.71 < x < 0.86$. The mechanisms of anisotropy in amorphous $Gd_{1-x}Co_x$ are discussed with reference to various possible models.

INTRODUCTION

The recent discovery that $Gd_{1-x}Co_x$ amorphous alloy films prepared in the range $0.7 < x < 0.9$ by rf sputtering, show uniaxial perpendicular anisotropy, and hence support cylindrical domains has excited considerable interest[1,2]. Crystalline compounds of nearly this composition (i.e. $SmCo_5$) had previously merited attention because of their permanent magnet properties[3,4]. In an amorphous system, there is an advantage over previous bubble materials because of the possibility of continuous adjustment of properties by variations of stoichiometry. The observation of bubble domains in $Gd_{1-x}Co_x$ by Lorentz electron microscopy, and by Kerr reflection optical microscopy[1,2] requires supporting evidence from the measurement of the magnetic properties. The usual criterion for magnetic bubble formation ($H_u \geq 4\pi M_s$), where H_u is the anisotropy field, and $4\pi M_s$ is the saturation magnetic moment, immediately dictated that information concerning these two particular magnetic properties was needed. This paper is devoted to the room temperature compositional and rf bias dependence of the perpendicular anisotropy K_u. The origin of this anisotropy is certainly an important scientific problem.

EXPERIMENTAL

The amorphous alloy material was deposited on 18 mm diameter glass discs; ~5 x 5 mm. samples were cut from the discs' centers. Sample compositions were obtained by the electron micro-probe (~1 at. %). Film thickness was obtained by means of a TALYSURF step height instrument. Most of the compositions studied were obtained from a $GdCo_2$ target rf-biased at up to 100 volts. The anisotropy field was obtained from an analysis of the ferromagnetic resonance data taken at 9.045 GHz using a rectangular H_{012} cavity. The sample itself was supported at the center of the cavity on a quartz fiber attached to a rotary mechanism. There may be interpretational difficulty with FMR in thick metallic films, since the magnetization due to the microwave field is non-uniform within the skin depth on the surfaces of the sample. In practice however, good resonances are usually obtained for the $Gd_{1-x}Co_x$ samples. In the evaluation of FMR measurements, both H_u, and the gyromagnetic ratio g, are obtainable from the usual equations for resonance in the perpendicular and parallel configurations, respectively:

$$\frac{\omega_\perp}{\gamma} = H_\perp + (H_u - 4\pi M_s) \qquad H_{dc} \perp film, \quad H_\perp \geq 4\pi M_s \qquad (1)$$

$$\frac{\omega_\parallel}{\gamma} = \left\{ H_\parallel \left[H_\parallel - (H_u - 4\pi M_s) \right] \right\}^{\frac{1}{2}} \qquad H_{dc} \parallel film, \quad H_\parallel \geq H_u \qquad (2)$$

where $H_\perp$ and $H_\parallel$ denote the resonance fields for a saturated state in the two configurations. The solutions for anisotropy field H_u, and for g are, respectively:

$$H_u = 4\pi M_s + r \left\{ H_\parallel \left[(1+r^2/4) H_\parallel + H_\perp \right] \right\}^{\frac{1}{2}} - (H_\perp + r^2 H_\parallel) \qquad (3)$$

$$g = \frac{h \omega_\parallel}{2\pi \beta} \left[\left\{ H_\parallel \left[(1+r^2/4) H_\parallel + H_\perp \right] \right\}^{\frac{1}{2}} - \frac{1}{2} r H_\parallel \right]^{-1} \qquad (4)$$

where $r = (\omega_\perp/\omega_\parallel) \simeq 1$, and $(h/\beta) = 714.6$ is Planck's constant divided by the Bohr magneton, for frequency $(\omega/2\pi)$ in GHz.

For this study, the magnetization was obtained by a force balance technique. The origin of this magnetization has been discussed previously[1-4]; it arises from anti-ferromagnetic coupling of the spins of Gd and Co atoms in the amorphous alloy[5]. It can be approximated: $4\pi M_s = |36,800 - 46,300 x|$, where x is the fractional Co concentration. The linear approximation is represented in Fig. 1 by the dashed line. This approximate form is derivable by setting the net magnetization M equal to the algebraic sum of constituent magnetizations, M_{Gd} and M_{Co} (strictly valid at T = 0°K); thus, $M = (1-x)M_{Gd} - xM_{Co}$. The compensation composition (M=0, x=0.795) fixes $(M_{Co}/M_{Gd}) = 0.258$, close to the value derived utilizing 7.1 μ_B and 1.72 μ_B as the moments of Gd and Co (that ratio being 0.242). A better theoretical approximation for the magnetization is shown as the solid curve (Fig. 1) which has been derived by R. J. Gambino[6]. This curve is fitted very well by a polynomial expression: $4\pi M_s = \left|\sum_{n=0}^{8} a_n x^n\right|$, where the constants $a_0 \ldots a_4$ are -24,220, 1.711×10^5, -3.220×10^5, 2.648×10^5, and -1.035×10^5, respectively. A further study of the magnetization appears in these conference proceedings[7]. The experimental magnetization data show considerable scatter as a function of x (Fig. 1), mainly because of inaccuracies in x, but also because of rare gas inclusion. In the subsequent analysis of the anisotropy, the raw values of x were corrected to fall on the theoretical curve. Without the adjustment, the anisotropy data appear to lose most of their coherence. The anisotropy field H_u is plotted in Fig. 2; the anisotropy energy $K_u = \frac{1}{2}M_s H_u$ is also shown (Fig. 3). The limiting values of H_u and K_u are indicated for 9.045 GHz, in accord with the specification given for eqs. 1 and 2 (Figs. 2 and 3). g-values for the Gd-Co alloy are plotted in Fig. 4, and are compared with a theoretical curve drawn using the Wangsness formula for an effective g (eq. 5):

$$g_{eff}(x) = g_{Gd}(S-R)/(S-\alpha R) \tag{5}$$

where $R = M_{Co}/M_{Gd}$, $S = (1-x)/x$, and $\alpha = g_{Gd}/g_{Co}$. For $0.795 < x < 0.812$, g is expected to be negative.[9] The region beyond x = 0.81 seems to be one of great measurement difficulty; large g-values are calculated, but measured values are lower, and frequently do not represent saturated resonances. The values of the anisotropy field H_u are expected to be very large for x > 0.81, and of course $H_\parallel$ must be very large also in order to achieve this condition. A few points obtained from the crude analysis of the static magnetization curves are also included in these graphs (square points) in order to demonstrate the trend of the anisotropy field, and to confirm the FMR values.

CONCLUSIONS

The interesting range in the Gd-Co system for bubble domains, with the present means of inducing anisotropy seems to be $0.71 < x < 0.86$. Outside the range mentioned above, the magnetostatic energy $2\pi M_s^2$ is apparently dominant. The magnitudes of the anisotropy energies found ($\sim 4 \times 10^5$ ergs/cm^3) are much smaller than those for crystalline Co[10] ($K_u = 1.4 \times 10^6$ ergs/cm^3, $H_u = 2000$ Oe), and for GdCo$_5$ ($K_u = 4 \times 10^7$ ergs/cm^3, $H_u = 2.7 \times 10^5$ Oe)[4]. Apparently, the anisotropy field derives from the rf bias applied to the film during the deposition process, since when the film composition is

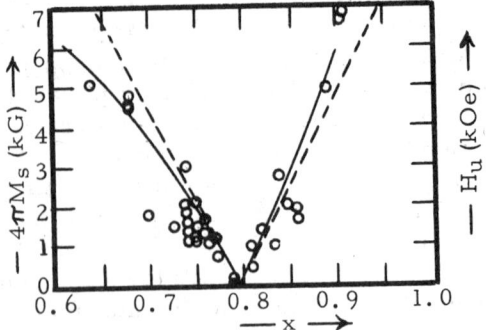

Fig. 1. The room temperature magnetic moment $4\pi M_s$ (kG) vs. atomic fraction x in $Gd_{1-x}Co_x$. The circles represent experimental data. The dashed line is a simple linear approximation, and the curve represents a better theoretical expression due to R. J. Gambino (see text for equation).

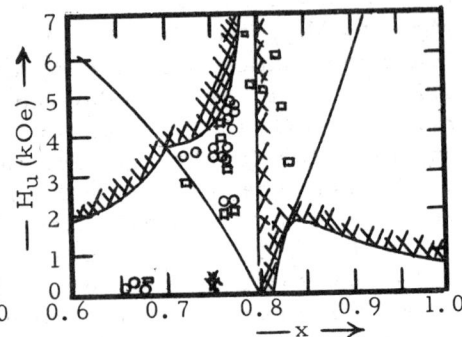

Fig. 2. The anisotropy field H_u(kOe) vs. x. The theoretical magnetization is also drawn to delineate where $H_u \geq 4\pi M_s$. The boundaries to H_u measurement at 9.045 GHz, and for a g-variation as given in Fig. 4, are shown cross-hatched. Circles denote FMR results, squares show those from magnetization.

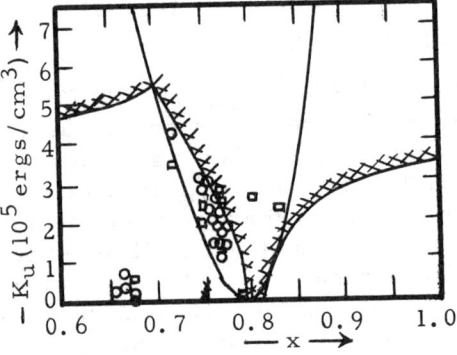

Fig. 3. The anisotropy energy $10^{-5} K_u$ vs. x. The boundaries to measurement of K_u at 9.045 GHz, and for a g-variation (Fig. 4) are cross-hatched. The magnetostatic energy is also shown to indicate where $K_u \geq 2\pi M_s^2$. Circles and squares relate to a $GdCo_2$ target, whereas x-x-x's refer to a $GdCo_3$ target.

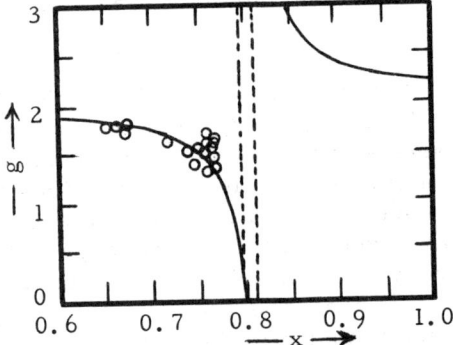

Fig. 4. The gyromagnetic ratio g as a function of x. The solid curve is calculated from the Wangsness formula with $\alpha = (g_{Gd}/g_{Co}) = (2.00/2.22) = 0.901$, $R = (M_{Co}/M_{Gd}) = 0.258$, for $M = 0$ at $x = 0.795$. The circles are experimental points. The vertical dashed lines show where $g < 0$.

essentially equal to the target composition (zero rf bias), the minimum magnetic anisotropy is found. One may consider the plots (Figs. 2 and 3) as parametric to applied rf bias relative to a $GdCo_2$ composition target. The points denoted by x-x-x's near x = 0.75

relate to a $GdCo_3$ target, however. In addition, the anisotropy in this amorphous alloy may be somewhat akin to the magnetocrystalline anisotropy related to alternate layers of Co and Co plus Gd in the 2:17, 1:5, and 2:7 monocrystal compounds[4]. This tendency for Co to coordinate with other Co's in sheets may introduce a layering type of anisotropy which would probably have little temperature dependence, as preliminary investigations show, in marked contrast to a steep temperature dependence of growth anisotropy in garnets[11]. It is expected that this basic layering tendency of the Co's, which may resemble pairing, would be enhanced by rf bias. The rf bias probably produces surface heating which causes just the right degree of pairing in the amorphous deposit. In addition, the rf bias provides a preferred orientation of rare gas ion damage tracks which may encourage pairing. The induced anisotropy effect is also modified by phenomena associated with the zero net magnetization at the compensation composition.

ACKNOWLEDGMENTS

It is a pleasure to thank J. C. Slonczewski, P. Chaudhari, R. J. Gambino, J. J. Cuomo, and R. Hasegawa for helpful discussion of this work; H. Lilienthal's careful measurements of the magnetic moment of these samples is greatly appreciated.

REFERENCES

1. P. Chaudhari, J. J. Cuomo, & R. J. Gambino, IBM J. Res. & Dev. 17, 66 (1973); Appl. Phys. Letters 22, 337 (1973); to be publ., Proc. Electronic Materials Conf., Las Vegas (Aug. 1973).
2. S. Herd & P. Chaudhari, Phys. Stat. Sol. (a) 18, 603 (1973)
3. G. Hoffer & K. Strnat, IEEE Trans. Magn. MAG-2, 487 (1966) & J. Appl. Phys. 38, 1377 (1967)
4. K. Strnat, G. Hoffer, J. C. Olson, W. Ostertag, & J. J. Becker, J. Appl. Phys. 38, 1001 (1967); K. Strnat, IEEE Trans. Magn. MAG-6, 182 (1970); K. Strnat & A. E. Ray, Z. Metallk. 61, 461 (1970)
5. J. Orehotsky & K. Schroder, J. Appl. Phys. 43, 2413 (1972); E. A. Nesbitt, J. H. Wernick, & E. Corenzwit, J. Appl. Phys. 30, 365 (1959); E. A. Nesbitt, H. J. Williams, J. H. Wernick, & R. C. Sherwood, J. Appl. Phys. 32, 342 (1961)
6. R. J. Gambino, unpublished.
7. L. J. Tao, et al., This Conference, paper 8E-3
8. R. K. Wangsness, Am. J. Phys. 24, 60 (1956)
9. S. Geschwind & L. R. Walker, J. Appl. Phys. 30, 163S (1959)
10. Z. Frait, Czech. J. Phys. 11, 360 (1961)
11. R. C. LeCraw & R. D. Pierce, AIP Conf. Proc. 5, 200 (1971)

NEUTRON IRRADIATION OF LPE BUBBLE DOMAIN GARNETS*

R. S. Sery and H. R. Irons
Naval Ordnance Laboratory, Silver Spring, Md. 20910

ABSTRACT

Five LPE iron garnets were irradiated in two steps to a total of 2×10^{15} n/cm^2 ($E_n > 10$ keV). Samples included GdEr-, GdY-, GdYTm-, GdYYb- and GdYLa-IG's. Pre- and post-irradiation measurements were made of ℓ, $4\pi M_s$, H_o, H_k and μ_w. A pre-irradiation simulation of ambient reactor conditions showed that the properties were not affected by these influences. Post-irradiation measurements revealed that negligible changes in properties had occurred. The estimated number of point defects produced in the samples is $\sim$ 1/10 of what would be required to cause threshold of damage property changes.

INTRODUCTION

Bubble domain devices for military and space applications may be subject to exposure to radiation environments. This work was done to determine what nuclear radiation effects might occur in bubble domain materials. Five samples of bubble domain materials were irradiated in the National Bureau of Standards Reactor. They were iron garnets (IG's) of GdEr, GdY, GdYLa, GdYTm and GdYYb (see Table I). The samples consisted of $\sim$ 5 µm thick LPE films on GGG ($Gd_3Ga_5O_{12}$) substrates 0.25 cm^2 by 0.05 cm thick.

EXPERIMENTAL

The samples were subjected to two irradiations. Only pre- and post-irradiation properties measurements were made. Because the samples were to be subjected to a combination of a hydrostatic pressure of 0.54 kgm/cm^2 (7.7 psi) and a 71°C ambient temperature during irradiation, they were first subjected to a 15-minute pressure-temperature test, i.e., immersion in distilled water at 71°C and 0.76 kgm/cm^2. None of the characteristic magnetic properties changed as a result of this simulation.

The irradiations were to 4×10^{14} n/cm^2 plus $\sim 10^7$ R and a total of 2×10^{15} n/cm^2 plus $\sim 10^8$ R, respectively, where the neutron energies were $E_n > 10$ keV. Pre- and post-irradiation measurements were made for five samples. Three controls were available due to the breaking of one sample, GdErIG, into two pieces (one became a control) and the GdYTmIG sample into three pieces (two were used as controls).

The properties measured included ℓ, $4\pi M_s$, H_o, μ_w and H_k. The characteristic length, ℓ, was determined, as outlined in a Monsanto report,[1] based on the theoretical analyses of Kooy and Enz and others.

*Work supported by DNA under subtask TB-055; samples from Monsanto.

Table I. Pre- and Post-Irradiation Values of Characteristic
Magnetic Properties of Bubble Domain Materials

	ℓ, μm (±10%)	$4\pi M_s$, G (±7%)	H_o, Oe (±3%)	μ_w, cm/secOe* (±25%)	H_k, Oe (±10%)
$Gd_{0.44}La_{0.04}Y_{2.52}Fe_{4.01}Ga_{0.99}O_{12}$					
Before (average)	0.66	118	48.5	590	400
After 4×10^{14} n/cm^2	0.70	120	48.5	920	390
After 2×10^{15} n/cm^2	0.66	125	52	950	400
$Gd_{0.8}Er_{2.2}Fe_{4.56}Ga_{0.44}O_{12}$					
Before (average)	0.80	184	62	220	1450
After 4×10^{14} n/cm^2	0.78	197	59	280	1370
After 2×10^{15} n/cm^2	0.76	180	54	200	1410
$Gd_{0.46}Y_{2.54}Fe_{3.95}Ga_{1.05}O_{12}$					
Before (average)	0.43	143	82	2000	110
After 4×10^{14} n/cm^2	0.425	142	81	2000	120
After 2×10^{15} n/cm^2	0.41	143	83.5	--	--
$Gd_{0.99}Y_{1.15}Yb_{0.86}Fe_{4.16}Ga_{0.84}O_{12}$					
Before (average)	0.43	163	98	1750	310
After 4×10^{14} n/cm^2	0.47	159	93	1750	290
After 2×10^{15} n/cm^2	0.44	155	93	1200	280
$Gd_{1.09}Y_{0.98}Tm_{0.93}Fe_{4.32}Ga_{0.68}O_{12}$					
Before (average)	0.63	169	86	560	1000
After 4×10^{14} n/cm^2	0.66	167	83	650	1020
After 2×10^{15} n/cm^2	0.67	169	84	650	970

*μ_w values vary with drive field used and changes > ±25% do not necessarily reflect changes caused by radiation.

From a micrograph an average value of P_o, the strip domain width, can be obtained directly. Tables based on the above analyses yield a value of ℓ/h corresponding to P_o/h, where h is the measured film thickness. Since we did not measure h directly, we used an average of the values determined by Monsanto (not measuring h directly at the point where P_o was determined probably increased slightly the magnitude of the experimental errors shown in Table I). Once ℓ/h is known, ℓ is easily determined.

In a similar manner $4\pi M_s$, the saturation magnetization value, is found from a similar table of values, i.e., $H_0/4\pi M_s$ vs P_0/h, which is also based on the methods mentioned in reference 1. H_0 is obtained by recording the field at which a bubble abruptly disappears when subjected to a slowly increasing biasing field.

The domain wall mobility, μ_w, was measured by the bubble collapse technique of Bobeck.[2] The basis of the method is to determine the pulse field required to change the bubble radius by a given amount in a given time. For convenience the radius is changed from the strip-to-bubble transition value r_{s-b} to the dynamic collapse value $r_c/2$ (r_c = static collapse value). An arbitrary pulse field of duration T which reduces the bubble radius to zero in the presence of a bias field r_{s-b} is determined. This procedure is repeated for several values of pulse duration T and a plot of 1/T vs H_p (the pulse field) is made. The slope of this curve $1/TH_p$ is then used to compute the mobility μ where $\mu = V/H_p = (r_{s-b} - r_c/2)/TH_p$.

Anisotropy Field (H_k). One measure of H_k is the minimum field required to align the magnetization along the hard anisotropy axis, which is approximately in the plane of the film. To measure H_k the film is oriented in an adjustable bias field so as to produce maximum second harmonic output from a hard axis pickup coil which is orthogonal to an easy axis RF drive coil (see Figs. 1(a) and 1(b). One advantage of detecting the flux change along the hard axis is that flux changes due to domain wall motion are mostly along the easy axis and hence are only weakly detected. The use of orthogonal drive and pickup coils and second harmonic detection follows the methods of Zappe[3] and Wichner[4] for thin film measurements.

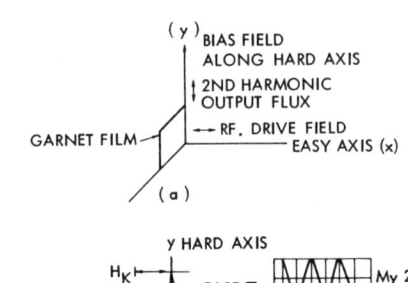
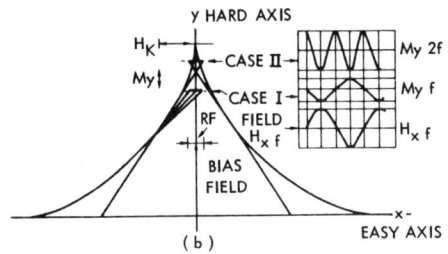

FIG. 1 GENERATION OF SECOND HARMONIC IN ANISTROPY FIELD MEASUREMENT

The cubic crystal anisotropy of the material complicates the measurement and causes $H_{k,matl}$ and the hard axis direction to vary slightly as the crystal is rotated about an axis normal to the film surface. The hard axis will be in the plane of the film every 60° as the film is rotated about its normal. For intermediate angles of rotation the hard axis is directed above or below the film plane by about 5°. Measurements were always made with the hard axis in the plane of the film.

The method can be understood in terms of the switching asteroid shown in Fig. 1(b), which is based on a rotational model for flux change in a material having an anisotropy energy of the form

$$E_A = K \sin^2 \theta$$

where K, the anisotropy constant, is equal to $(MH_k/2)$ and θ is the angle between M and the easy axis. Although domain walls are present in some phases of the measurement, they are neglected for the moment to simplify the analysis. For any value of field, M is found by drawing a tangent to the asteroid through the field point. For field points inside the asteroid there are two possible tangents and the field that was last applied outside the asteroid determines which tangent is to be used. The two cusps of the asteroid define the field boundary for irreversible flux switching. Fields within the asteroid cause only reversible flux changes. The directions of M at the peak values of the RF drive field are shown in Fig. 1(b) for two values of bias field. In Case I the time varying H vector does not cut both cusps of the asteroid and the resulting flux change along the hard axis is mostly at the same frequency as the drive field. In Case II the H vector cuts both cusps of the asteroid causing flux switching, and the flux change along the hard axis is a second harmonic of the drive frequency. The second harmonic signal will be a maximum when the bias field is along the hard axis and the magnitude of the bias is the minimum value that permits the total field vector to cut both cusps of the asteroid. For very small values of the RF drive field, the second harmonic will be a maximum when the bias field is equal to H_k. A typical excitation field of 1 Oe should produce an error of 2.4% if H_k is 500 Oe.

The above analysis neglects the effects of the multidomain state of the material. If the sample consists of domains in which the angle between M and the hard axis is much larger than the rotation caused by the RF field, and if the sample has a net M_x of zero, then the signals produced in oppositely directed domains cancel one another. To overcome this the hard axis field is reduced to about 0.8 H_k and a sweep field (not shown) is applied along the easy axis to produce single domain states on each side of the hard axis.

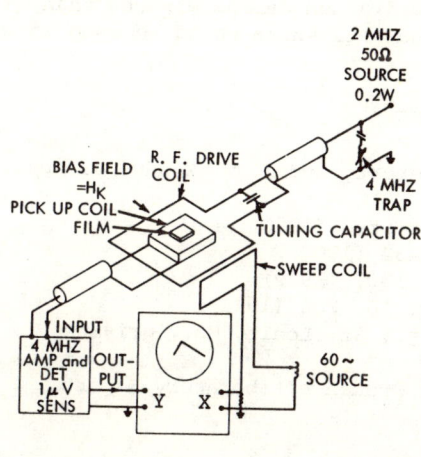

FIG. 2 ANISOTROPY FIELD MEASUREMENT APPARATUS

By adjusting the orientation of the sample about an axis normal to the film surface, signal peaks which are symmetrical about the zero value of easy axis field are obtained. The hard axis field is then increased until the two peaks merge. Although the signal is then very small, field values can be repeated within two percent for a particular orientation of the sample.

The hard axis field value as obtained above is shown in Table I as H_k. No correction for demagnetizing field was applied since the film was not in a single domain state.

A comparison of these H_k values with those obtained by the method of Josephs[5] on similar samples[6] (but not from the same batch) indicates that the values in Table I are low by about 5%. The experimental arrangement for measuring H_k is shown in Fig. 2.

RESULTS

Table I summarizes the results obtained for the five samples and includes values for the experimental errors. The way in which the magnitudes of these errors was derived is described elsewhere.[7,1] Some of the values for the mobility, μ_w, fall outside the ±25% range shown. This resulted, in part, from the use of different drive fields, H_D, for determining μ_w; this property varies with H_D for low H_D values. All of the before values recorded for the characteristic properties (excluding μ_w and H_k) and most of the after values are averages of two to four measurements. A table for pre- and post-test characteristic properties values for the controls, not included herein, shows that these properties exhibited no significant changes.

SUMMARY AND CONCLUSION

Two irradiations by neutrons to a total dose of 2×10^{15} n/cm^2, $E_n > 10$ KeV, plus $\sim 10^8$ R produced no observable changes in the characteristic magnetic properties tested for five representative bubble domain samples. Reactor environmental influences alone had no effect on these properties and, for three control samples, no detectable changes in properties occurred. A rough calculation shows that f, the fractional number of point defects introduced into the crystal lattices of the bubble domain films and substrates, was $\sim 5 \times 10^{-7}$. It is estimated that for changes in magnetic properties > 10% to occur f would have to be between 10^{-5} to 10^{-6}. Bubble domain materials per se should be less susceptible to radiation damage effects than, say, the associated semiconductor components which would be used in proposed bubble domain memory devices.

REFERENCES

1. R. W. Shaw, D. E. Hill, R. M. Sandfort, and J. W. Moody, J. Appl. Phys. 44, 2346 (1973).
2. A. H. Bobeck, Proc. Int. Conf. Ferrites (1970).
3. H. H. Zappe, J. Appl. Phys. 38, 1434 (1967).
4. R. Wichner, Rev. Sci. Instrum. 43, 1307 (1972).
5. R. M. Josephs, AIP Conf. Proc. No. 10, 286 (1972).
6. R. W. Shaw, Monsanto Research Corp., St. Louis, Mo., private communication.
7. R. S. Sery and H. R. Irons, Naval Ordnance Laboratory Report, NOLTR 73-33 (1973).

Section 3.   Bubble Devices   95

PROGRESS IN ALL-PERMALLOY BUBBLE CONTROL FUNCTIONS

T. J. Nelson
Bell Laboratories, Murray Hill, N. J. 07974

ABSTRACT

The purpose of this paper is to report progress in the design of permalloy overlay bubble circuits in which conducting paths in the permalloy carry the currents required to perform the control functions necessary in a major-minor loop organized magnetic bubble mass memory chip.

A versatile chevron-based building block has been found which can perform transfer or replication between chevron major and minor loops. A similar structure can be used to nucleate bubbles into a chevron propagation path. The all-permalloy control functions required in a mass memory chip have been demonstrated with 10 Oe bias field margins at 100 kHz in a 25 Oe rotating field. A memory chip using these functions would also require a non-magnetic conductor level which, though not directly involved in the control functions, is necessary to return their currents. This level would contain features gross by comparison with the permalloy features, and could be processed last, following deposition of a spacing layer on the bubble film and processing of the permalloy layer on top of that.

INTRODUCTION

Recently Bobeck and co-workers[1] described progress towards permalloy overlay designs which perform the necessary functions of generation, propagation, replication, annihilation, and detection of magnetic bubble domains in an adjacent magnetic garnet film without the addition of a nonmagnetic conductor level on the circuit. The advantages of one level circuit designs are fewer process steps in fabrication, coplanarity of all permalloy features, and absence of a registration process step between delineation of magnetic and nonmagnetic levels. The basic difficulty in one level bubble circuit design is that of locating conducting paths in the permalloy circuit which do not degrade the propagation margins.

The purpose of this paper is to report progress in the design of permalloy overlay structures which perform both propagation and current control functions. A versatile chevron-based building block has been found which can perform transfer or replication between chevron major and minor loops. A similar structure can be used to nucleate bubbles into a chevron propagation path. The design to be described is two level, strictly speaking, but the nonmagnetic conductor circuit features are gross by comparison with those in the permalloy level, and can be processed on top of it. The simplified processing made possible by all-permalloy control functions may permit these circuits to be fabricated with higher density or larger numbers of bits without the sacrifice in yield that would otherwise be expected.

AN ALL-PERMALLOY TRANSFER GATE

A two level "dollar sign" transfer gate which, when pulsed, diverts bubbles from one propagation path into another, has been characterized by Smith and Kish.[2] In their design, bubbles are transferred back and forth between a T-Bar major loop and the corners of T-Bar minor loops. The transfer gates are spaced two periods apart along the major loop, which is straight in the transfer region. The minor loops are therefore densely packed because adjacent propagation paths are one period apart. Because control is effected by the field of the current in the nonmagnetic level, that level must be carefully aligned with the permalloy level. Also, the conductor level must be processed before the permalloy level, which otherwise tends to shield the fields of the transfer current from the bubble film. This "conductor-first" processing, in turn, creates step-coverage and other problems in the permalloy level.

The design and operation of an all-permalloy transfer gate are shown in Fig. 1. Both the major and minor loops are composed of chevron elements. Because it is difficult to design a tight chevron corner, these transfer gates are spaced by three periods along the major loop. Adjacent minor loop propagation paths are, on the average, 1.5 periods apart, and the number of bubble positions per unit area is thus 2/3 of that achieved with two level circuits of the same 28 micron period. The transfer current flows in the straight permalloy path linking the major and minor loops. A gross nonmagnetic level, needed to return the currents, may be processed after the permalloy level.

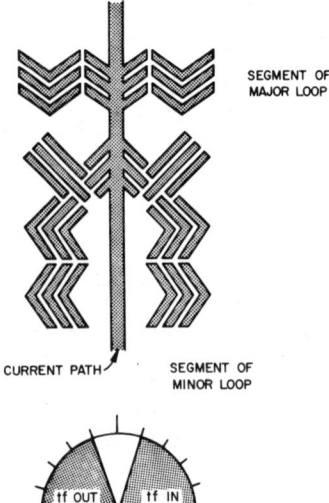
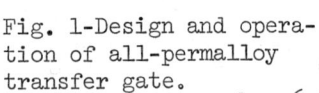

Fig. 1-Design and operation of all-permalloy transfer gate.

The test circuit constructed consisted of a major loop and ten minor loops with individually connected transfer conductors. The minor loops had identical corners on both ends: that is, the transfer conductors continued on through the loops. A wide gold stripe lying across the minor loops returned their transfer currents. The circuit was ion-milled in 4000Å permalloy deposited on a 0.8 micron SiO_2 spacing layer. The Sm-YIG[3] film thickness h, saturation induction $4\pi M$, and material length ℓ were

$$h = 6.5 \ \mu m, \ 4\pi M = 208G, \ \ell = 0.58 \ \mu m \ . \tag{1}$$

The circuit was tested at 100 kHz in a 25 Oe rotating field. The minor loop margins were limited on the low end by stripping of single bubbles in the corner, at 111 Oe, and on the high end by collapse of a bubble in a fully populated loop, at 121 Oe. The principle loss in margins is at the low end, and is caused by the presence of the conductor. It is hoped that decreasing the transfer conductor width from 6 microns to 5 microns or less may improve the propagation margins without significantly degrading the transfer function. Transfer-in was tested by gating on the rotating field and timing generation and transfer such that bubbles in successive bursts occupied adjacent positions in the minor loop. Bubbles were transferred into both the leading and trailing positions of the stream. Transfer-out was tested by generating three consecutive bubbles, transferring them into a minor loop, and then transferring any one of the three back out to the major loop. In this case, the bubbles remaining after a burst were collapsed by applying a pulse to the bias field control circuit before the next burst. In all testing, the results were inspected visually in a polarizing microscope between bursts at approximately 2 Hz repetition rates. With the phasing and currents shown in Fig. 1, the transfer margins equal or exceed the minor loop propagation margins. The pulse phasing shown exceeds the minimum required for transfer with reasonable bias field margins by 0.5 microseconds or 18 degrees on both ends.

The transfer pulse flows for almost half of a rotating field cycle in such a way as to block the passage of a bubble when it reaches the conductor. The bubble is protected from collapse, as the chevron becomes magnetically repulsive, by the proximity of the permalloy of the conductor and by the field produced by the current flowing in it. At the end of the transfer pulse, the opposing chevron has become magnetically attractive. The distance between the opposing chevrons, at their nearest approach, is only 6 microns, so the bubble can make the transition at low speed, relative to its motion in propagation. It is difficult to find room in the permalloy level for a current return path, but this can be achieved with gross features in a nonmagnetic conductor level. One way would be to place a wide gold stripe across the minor loops, as in the test circuit. The transfer gates would be electrically connected in parallel in contrast to the two level transfer design in which they are connected in series. While this does not change the power demanded of the electrical driving circuit, it does require higher current at a lower impedance level. This may be an advantage, for as described by Michaelis,[4] seven chips will be connected in series in a mass memory module. However, it will be necessary to insure that all transfer conductors within a given chip possess the same resistance so that they equally divide the current. It would also be possible to connect the transfer gates in series by using somewhat smaller features on the nonmagnetic conductor level. Because the transfer gates are spaced on 84 micron centers, the gold connections could still be considerably larger than the finest permalloy features.

ANNIHILATION AND REPLICATION

The remaining functions required in a mass memory design are replication, annihilation, detection and generation. All-permalloy detector designs have been discussed by Bobeck. In Fig. 2, which is the all-permalloy transfer gate, the chevron track receiving bubbles from the major loop could lead to the detector. This is necessary as

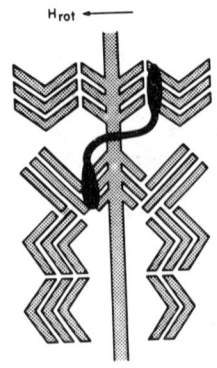

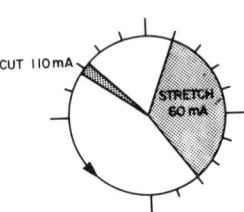

Fig. 2 - Operation of all-permalloy transfer gate in replicate mode.

a means of transporting bubbles to the detector, but can also be considered as the annihilate and replicate structure as well. Annihilation is performed simply by transferring unwanted bubbles into the detector. Since detection is the only function that might be required during annihilation, there is no penalty for this. Also shown in Fig. 2 is a pulsing scheme by which bubbles can be replicated, with one going to the detector and the other remaining in the major loop. The stretch pulse is turned on after the bubble has crossed the conductor. Its polarity is therefore opposite to that of the transfer pulse. About twice as much current was required to accomplish this stretching out, the result of which is shown in Fig. 2 when the rotating field is pointing to the left after the cycle containing the pulse. By applying a cut pulse (opposite in polarity to the stretch pulse) slightly before the field reaches the orientation shown in the figure, the strip is divided. Under the same test conditions quoted above for the transfer, replication was achieved over the entire propagation margins of the minor loops, i.e., from 111 to 121 Oe.

NUCLEATION AND GENERATION

Nucleation of magnetic bubble domains in the field of a short current pulse in a nonmagnetic hairpin shaped conductor has been reported by Nelson, Chen and Geusic.[5] Disk generators were seeded, and information bubbles were generated at 100 kHz data rates in their study. Nucleation is difficult in all-permalloy designs because the hairpin shape is not ammenable to propagation, and because the spacing required for propagation is large for nucleation. Moreover, permalloy has about five times the resistivity of gold or aluminum so significant heating may be expected at the current level required for nucleation. The low thermal conductivity of SiO_2 also contributes to the temperature rise in the permalloy, but this could be alleviated by using alumina as the spacing layer. Because of the thermal effect, the generator design shown in Fig. 3 is considered unsuitable for sustained generation at 100 kHz. The design can generate bubbles at low duty cycle, however; the current amplitudes and phasings required are shown in the figure for the same test conditions cited above. A bubble is first nucleated, with the field pointing up, by applying a short pulse making the stepped side of the conductor attractive. The bubble appears on the step, as shown in Fig. 3. The second phase of the process amounts to a transfer from the step into the chevron propagation path. The bias field operating range extended from 106 to 121 Oe, being several Oe better than the minor loops on the low end.

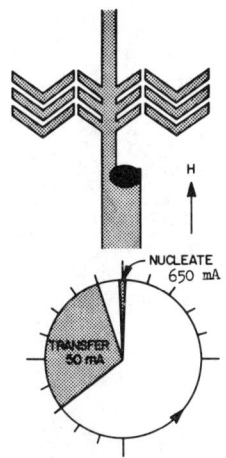

Fig. 3-Design and operation of all-permalloy nucleator.

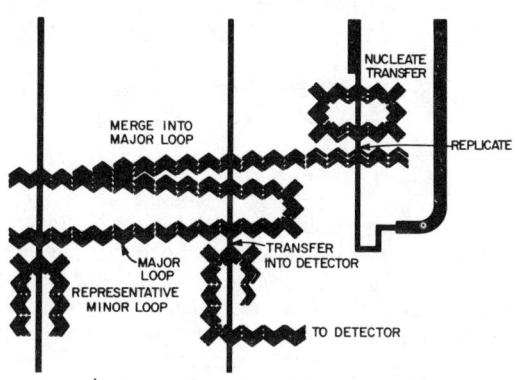

Fig. 4-Location of all-permalloy functions in mass memory chip.

Generation at the 100 kHz data rate could be achieved by making the recipient propagation path of the nucleator a short closed loop containing an integer multiple of three bubble positions. This generator structure would then be positioned with respect to the major loop so that once bubbles have been nucleated into it, they could be replicated onto the major loop as required. In fact nucleation and replication could be achieved with the same conductor, so that extra pads for the two functions would not be required.

CONCLUSIONS AND ACKNOWLEDGMENTS

The overall arrangement of the functions discussed above is shown in Fig. 4. Those other than transfer must be located at one end of the major loop, so that their pads nest in the way required in Ref. 3. The primary advantages of the design over previous efforts are that it permits permalloy-first processing and does not require a fine registration. The bias field margins obtained on experimental circuits are about 50% of the free bubble stability range. It is a pleasure to thank A. H. Bobeck, J. E. Geusic and F. B. Hagedorn for helpful discussions of this work.

REFERENCES

1. A. H. Bobeck, et al., IEEE Trans. Magnetics MAG-9, 474, 1973.
2. J. L. Smith, et al. IEEE Trans. Magnetics MAG-9, 285, 1973.
3. B. S. Hewitt, et al. IEEE Trans. Magnetics MAG-9, 366, 1973.
4. P. C. Michaelis and P. I. Bonyhard, IEEE Trans. Magnetics MAG-9, 436, 1973.
5. T. J. Nelson, et al., IEEE Trans. Magnetics MAG-9, 289, 1973.

DEVICE DESIGN AND SYSTEM ORGANIZATION FOR A DECODER ACCESSED MAGNETIC BUBBLE MEMORY CHIP

P. I. Bonyhard, Y. S. Chen and J. L. Smith
Bell Laboratories, Murray Hill, N. J. 07974

ABSTRACT

We have examined the feasibility of taking advantage of the superior access speed of the decoder organization of magnetic bubble memory chips, while retaining the wide operating margins and high degree of data integrity that have been achieved for major/minor loop organized chips. A semicircular bubble replicator has been designed and characterized that can be placed at the corner of T-Bar information storage loops. A novel decoding matrix, in which the half period wide control conductors simply retard the propagation of all bubbles other than the one selected, has been designed and operated. The bias field margins for both replication and decoding have been found to be substantially as wide as those for simple propagation. The difficulty in the initial design, that the output to be detected appears at different points depending on the selection address, is not a problem because of the incorporation of a multi-input chevron expander detector. A solution was found to the problem of erasing old information in connecting all the storage loops into one long shift register with a single write/erase point. This results in longer write access times which, however, need not be detrimental to system performance.

INTRODUCTION

Decoder accessed magnetic bubble memory chip designs have been reported.[1,2] We have examined the feasibility of taking advantage of the superior speed performance of the decoder organization, while retaining the wide operating margins and high degree of data integrity that have been achieved for major/minor loop organized chips.[3,4]

In addition to the basic bubble circuit functions, i.e., propagation, controlled bubble generation and detection, the initial design[1] stipulates at least two additional functions; bubble replication at 180° turns in propagating loops closely packed side by side, and means for decoding. Also, the detection scheme used must be capable of dealing with selected bubbles that emerge from the decoder at different physical locations depending on the loop of their origin. Finally, means must be found for selectively erasing stored data prior to writing new information. This last problem, unsolved in the initial reported design,[1] was subsequently solved by the use of "in-the-loop" decoding.[2] We anticipate that in-the-loop decoding, which stipulates a large number of gating operations performed even on parts of the stored data not selected for reading or rewriting at the time, may be detrimental to long term stored data integrity. Even though this anticipation is based only on our general understanding of the performance of bubble circuits rather than on any specific data, we preferred to solve the erase problem by modification of the initial design. We are, therefore, not considering the design of functions stipulated for in-the-loop decoding in this Paper. In the following sections we will discuss considerations relating to the design of the specific functions outlined above.

REPLICATION

A bubble replicator, suitable for the needs of the decoder organized chip, has been designed and characterized. The design is shown at the top of Fig. 1. The basic propagate elements are of the T-Bar type used in the major/minor loop organized chip design reported before,[3] but a novel half disk element is used in the 180° turns. Such turns exhibit good propagate margins and, in addition, permit bubble replication in a manner very similar to the operation of the disk type bubble generators.[5] Replication is accomplished by the application of a current pulse in a conductor (shown shaded) which splits the bubble on the half disk into two. After the split the leading bubble continues propagation around the corner, whereas the trailing bubble, i.e., the replica, transfers into the output track (to the right). The half disk replicator, however, exhibits margins superior to those of the disk generator by virtue of the "tail" added to the half disk which strengthens the magnetic poles that stretch the bubble to be split. Thus the bubble is stretched over a larger range of both the bias field and the drive field rotation phase. The corresponding pole strength of the opposite polarity is dissipated gradually along the wedge-shaped tail and, thereby, does not interfere with bubble propagation. Replicator operating data will be given at the end of the next section together with decoder operating data.

DECODING

In the initial design[1] decoding was accomplished by diverting unwanted bubbles into traps where they were subsequently eliminated. We have found it preferable to simply retard unwanted bubbles as they propagate along the parallel tracks following the replicators. Retardation is readily performed by the application of a current pulse in a conductor segment perpendicular to the propagate track.' Such a conductor segment and its placement relative to the permalloy propagate features is shown at the right hand end of the top of Fig. 1. Decoding is performed by the conductor segments connected into a binary matrix as indicated schematically in the middle section of Fig. 1. The effect of applying the appropriate pulse sequence to the select conductors is that only the selected bit emerges from the decoder on the proper cycle, whereas all other bits that may have started out on the same cycle are retarded by one or two cycles. For N parallel tracks $\log_2 N$ conductors and $2\log_2 N$ propagate cycles are required. Advantages of this design include simplicity, wide operating margins and good reliability.

The operation of an 8-track matrix has been demonstrated at 100 kHz propagation rate within a range of bias field of 112 Oe to

125 Oe, drive field values of 25 Oe and 30 Oe , range of replicate current of 60 mA to 100 mA with ±20° phase tolerance at 3 sec pulse width, range of decode current of 30 mA to 50 mA with ±40° phase tolerance at 5 sec pulse width. The garnet material, circuit processing, period, etc., were the same as reported in Ref. 3 and its companion papers listed there as references.

DETECTION

The detection problem mentioned in the Introduction can be readily solved by the use of a multi-input chevron expander detector[6] as indicated schematically in the right hand side of Fig. 1. Bubbles can laterally expand into long strips in a relatively small number of propagate cycles.[3] Thus, the multi-input detector serves conveniently as a logical OR gate spanning the large distances between tracks as well as providing a long bubble strip for a large magnetoresistive output. Of course, the detector output must be observed on the correct cycle and it may be followed by 2 cycles of "garbage" due to bubbles retarded in the selection matrix. In practice one can detect on every 4th cycle, which is not necessarily a disadvantage in a large memory, as outputs from different memory chips may be interleaved. For instance, the Motorola MC1544L/1444L sense amplifier chip, used in conjunction with our module design,[4] can select one out of 4 inputs. Thus, the same chip may be used to look at the outputs of 4 bridges in 4 consecutive cycles so that each amplifier senses bits at the basic bit propagation rate..

It is also possible to use chevron propagate elements in the decoder section, thereby performing some degree of expansion on the bubble while decoding. Thus, the total number of steps to the detector may be reduced. The decoding currents, however, must be increased as it takes more current to retard bubbles propagating on chevrons.

WRITING AND ERASING

Writing and erasing may be accomplished by connecting all the storage loops into one long "serpentine" shift register with a single write/erase point as indicated in the left hand side of Fig. 1. This, of course, results in long write access times. This, however, need not have any detrimental effect on system performance if a small fraction of the overall memory capacity is devoted to the buffering of memory input data. Consider, for example, a chip design with 128 replicate ports in the storage register. If the central processor were to spend half of its time in transferring data from and to the bubble memory then on the average 32 pages will be read and 32 pages will be dumped in the time period in which a given bit traverses the storage register. The average dumped page has to be buffered for half of this time period before it can be restored into its appropriate location in the bubble memory. Thus the average buffer capacity required would be only 16 pages, probably a tiny fraction of the total capacity of the bubble memory.

CONCLUSION

We have described the design of all additional circuit functions necessary for a decoder organized bubble memory chip. It seems clear that such a chip can be built using the same technology and having roughly the same margins and integrity as a major/minor loop organized memory chip. It would appear that at the presently typical 100 kHz propagate rate, when access times are in the milliseconds range, the performance advantage offered by the decoder organization would not justify the need for more conductors per chip and for more complex control pulse sequences. At higher bit densities and data rates, however, the decoder organization may become attractive.

REFERENCES

1. H. Chang et al., IEEE Trans. Magn. MAG-8, 214 (1972).
2. G. S. Almasi et al., AIP Conf. Proc. 5, 220 (1972).
3. P. I. Bonyhard et al., IEEE Trans. on Mag. MAG-9, 433 (1973).
4. P. C. Michaelis and P. I. Bonyhard, IEEE Trans. Magn. MAG-9, 436 (1973).
5. Y. S. Chen et al., IEEE Trans. Magn., to be published.
6. W. Strauss et al., AIP Conf. Proc. 10, 202 (1973).

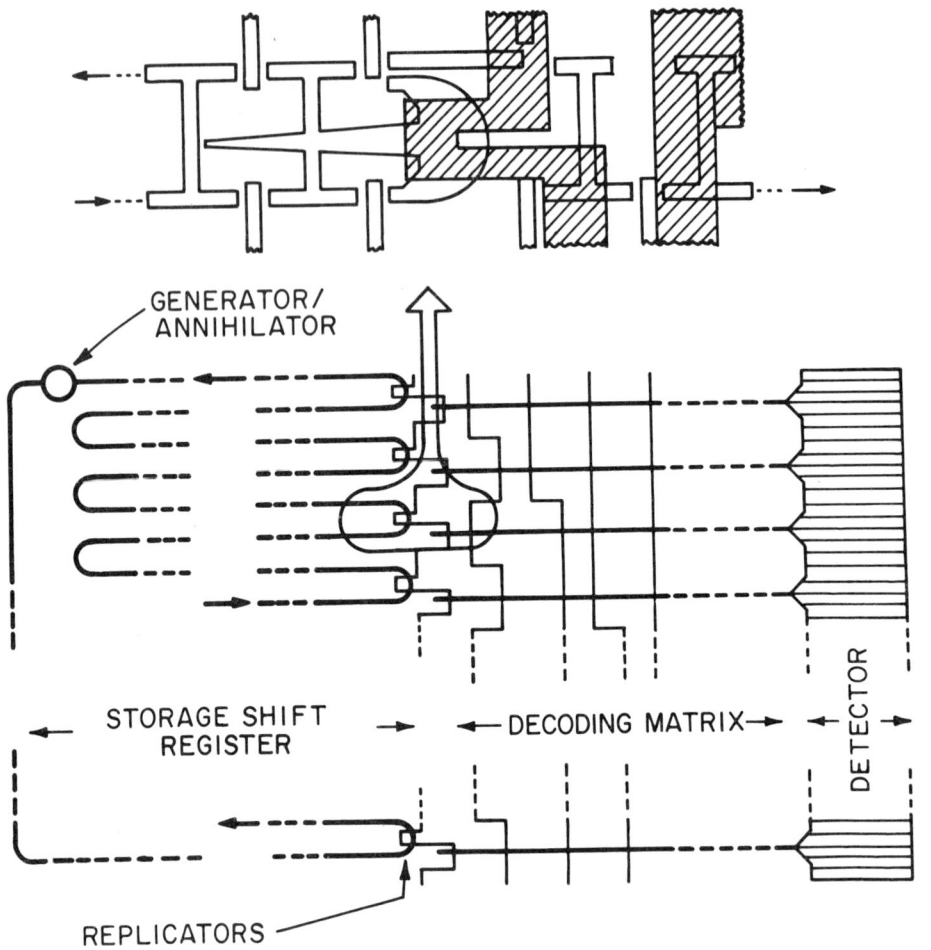

Fig. 1. Design of decoder-accessed bubble memory chip.

BONDED SHIFT-REGISTERS FOR MAGNETIC BUBBLE MEMORY MODULES

A. E. Feuersanger
GTE Laboratories Inc., Waltham, Mass. 02154

ABSTRACT

A garnet-to-permalloy circuit bonding technique has been developed that is used routinely in the assembly of shift-registers. In this approach thin-film circuits contain designed-in spacers for uniform crystal-to-circuit spacing to obtain uniform margins over the area of the circuit. The crystal is mated to the circuit by semiautomatic mechanical bonding using an optically transparent cement. Criteria for the selection of bonded shift-registers for incorporation on memory circuit boards have been established using a detailed test procedure on both circuits and crystals after ion implantation and dicing. Results of margin matching and yield figures for 28 register boards are given.

INTRODUCTION

Bubble memory circuits for systems under development are usually fabricated directly on epitaxial garnet films using an oxide layer as a spacer.[1] The bonded shift-registers described here are useful since they are assembled from prescreened circuits and crystals for experimental studies and for use in bubble memory systems. The objective was to assemble twenty-eight 1.1-kilobit shift-registers which operate in a single field module, each at a data rate of 66 kb/s and in the temperature range from 50 to 70°C. The prescreening allows improvements in shift-register yields specifically by separating the circuit yield and by nondestructive reuse of circuits or crystals. The assembly process has been refined to yield reproducibility and has resulted in shift-registers with well-matched margins for operation of multichip boards. The reliability of these shift-registers under various operating conditions is discussed in a companion paper.[2]

CIRCUIT AND CRYSTAL EVALUATION

The circuit designed for this bubble memory is a field-accessed shift-register with a 28-μm period Y-bar permalloy track of 3-μm line width prepared on 7059 glass. The interaction components are a seed-conductor loop generator, conductor loop annihilator switch and a matched pair of 5-bar longitudinal-stretch detectors. Preassembly circuit evaluation consists of high-resolution microscopic inspection and low-frequency (100 Hz) rotating field measurements of detector output, ΔV, vs. field amplitude, H_{rot}.

The crystals are LPE grown garnet wafers with a nominal composition $Y_{1.08}Gd_{0.72}Tm_{1.2}Ga_{0.8}Fe_{4.2}O_{12}$ and typical parameters $H_c = 60$ Oe, stripe width = 8.7 μm and epi thickness = 5.2 μm. Hard bubble formation is suppressed by implantation[3] of H_2^+ ions[4] at 60 keV and 1×10^{16} ions/cm^2. To avoid nonuniform implantation, a careful cleaning procedure is applied before implantation. The wafers are diced into 1.52-mm square chips by multiple wire saw cutting and each chip is scanned for major magnetic defects in a polarizing microscope equipped with a modulated bias field for defect decoration. The collapse field is measured for each chip to determine the implanted side of the crystal, the collapse field range, uniformity of the collapse field and minor defects by bubble-defect interaction. In this technique, a gradient is established in the crystal by the normal field component of a magnetic probe (100-μm Ni wire) driven by a conventional rotating field assembly. Crystals are easily flooded with bubbles by tuning the phase of the drive currents. An amplitude of 30 Oe and a frequency of 10 kHz are typical. Bubbles injected into a crystal by the probe are shown in Fig. 1. For collapse measurements the rotating field is turned off after filling the crystal. A sample of collapse measurements for several chips from the same wafer after implantation is shown in Fig. 2. Ranges of 1 to 2% are observed but generally the range is extended by nonuniform collapse over the crystal area. The unused edge region, first to collapse, is neglected in the measurements.

BONDING

System environment and practical bonding considerations determined the properties of the cement; i.e., optical transparency, easy nondestructive dissolution, rapid setting, high softening point, application of uniform layers a few thousand angstroms thick, and durability at temperatures in excess of 70°C. Among lens cements considered, Canada balsam has shown to fulfill all of these requirements. It dissolves to a clear bonding solution in only a few solvents, e.g., Xylene, Toluene and Trichloroethylene. The latter is preferred since it has the highest volatility necessary for rapid setting (approx. 10 seconds) in large-volume semiautomatic bonding. During assembly, it is required that the surfaces of both crystal and circuit are free of particles, processing remnants from cutting and growth defects. Crystal cleanliness, defects and flatness are checked by contacting the surface to be bonded with a flat and observing the wedge fringes produced in the air gap. To achieve a predetermined parallel spacing for uniform bubble-permalloy interaction (margins), we have designed into the circuit four 6-μm diameter spacers consisting of the highest levels in the circuit, as shown in Fig. 3. The spacers are located near the edge of the crystal and do not interfere with the operation of the circuit. Measurements of the spacer height show that we are near the desired coupling distance[5] of 4500 Å between propagation track and crystal surface. Margin measurements on shift-registers assembled with loads from 500 to 1600 gm did not show any variation from margins observed for shift-registers assembled at the normal load of 1050 gm. This indicates that the crystals rest on the spacers.

SHIFT-REGISTER EVALUATION

Bonded shift-registers are evaluated at low drive frequencies (5 to 20 Hz) by microscopic observation. Generator collapse field is used as a relative measure of uniformity in spacing of registers from the same crystal. High-frequency (66 kHz) bias field margins are determined, at a fixed drive field of 30 Oe, as the stable bias field limits on the propagation of 16 consecutive generated bubbles at 60°C. Write- and erase-all as well as detector output, typically 0.4 to 0.8 mV, are also evaluated. Yields of crystal chips

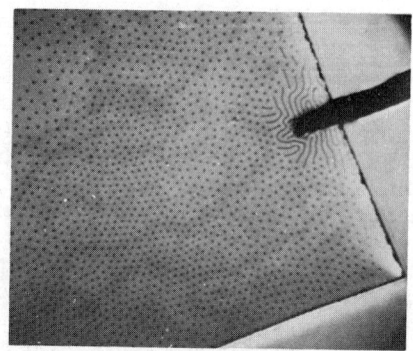

Fig. 1. Bubbles injected into epitaxial layer by a nickel probe. ($H_D = 0$; $H_B = 50$ Oe)

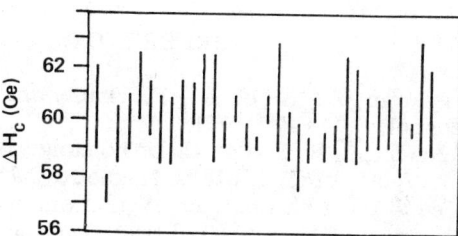

Fig. 2. Collapse range H_c of chips from the same garnet wafer after ion implantation.

passing magnetic scanning tests are in the range from 30 to 63%. Yield figures for registers assembled from pretested circuits and crystals and operating with all functions at high frequency are from 50 to 60%.

RESULTS AND DISCUSSION

Several multichip register boards have been assembled and operated in test systems. Results on a 28-register board are given here as an example. The distribution of bias field margins for 42 shift-registers from one crystal are shown in Fig. 4. The maximum margin observed is 7 Oe with a center at 73.5 Oe. The distribution peaks at a margin of 4.5 Oe and it is observed in a margin overlap diagram that the margin scatter is narrow. Since all margins are contained within the maximum device margin, it would appear that the scatter is primarily due to narrow margins rather than center shifting. The origin of this is believed to be in local variations of crystal perfection which have not been correlated with shift-register operational parameters so far. From the distribution shown in Fig. 4, 33 shift registers have a margin overlap of 3.5 Oe and 26 of 4.0 Oe. The results in margin matching show good reproducibility of the assembly procedure. Operation of shift-registers over periods of months has shown this technique to be a viable one for use in bubble memories.

ACKNOWLEDGMENTS

The author is indebted to J. Ramsey, who automated the bonding technique, to M. J. Urban for fabrication of the circuits, and to Dr. J. E. Mee, Rockwell International, Electronic Research Division, for providing the garnet crystals. Thanks are due to Dr. D. H. Baird for valuable discussions.

REFERENCES

1. J. P. Reekstin, A. G. Lehner, F. Vratny and G. W. Kammlott, J. Vac. Sci. Technol. $\underline{10}$, 847 (1973).
2. J. L. Archer, L. R. Tocci, O. D. Bohning, D. H. Baird, C. F. Buhrer and J. J. Vytal, "Reliability of Magnetic Bubble Domain Memories," presented at the 19th Conf. on Magnetism and Magnetic Materials (1973).
3. R. Wolfe and J. C. North, Bell Syst. Tech. J. $\underline{51}$, 1436 (1972).
4. J. C. Miklosz (Private communication).
5. Y. S. Chen and T. J. Nelson, IEEE Trans. Magnetics $\underline{8}$, 754 (1972).

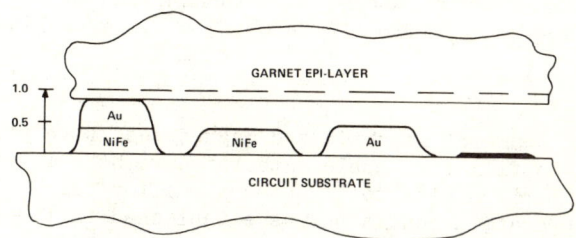

Fig. 3. Schematic representation of cross section through a bonded shift-register. Left to right: spacer, track element, gold conductor, detector. The horizontal scale is arbitrary.

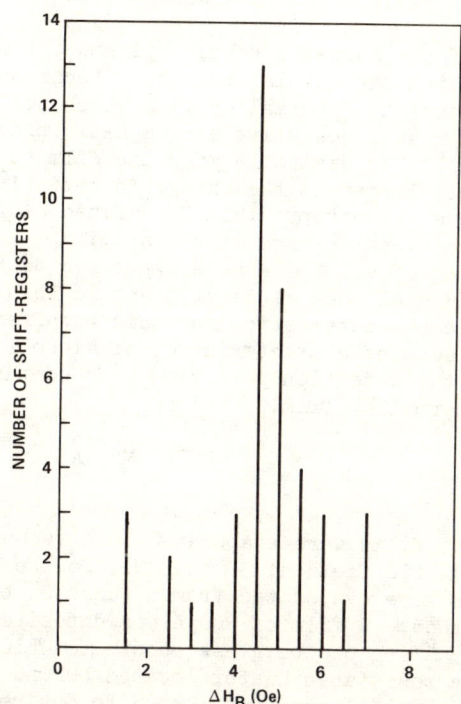

Fig. 4. Bias field margin distribution of 42 shift-registers. f = 66 kHz, T = 60 °C and H_D = 30 Oe.

Ni-Co FILM WITH LARGE MAGNETORESISTANCE FOR BUBBLE DETECTION

K. Asama, K. Takahashi and M. Hirano
FUJITSU LABORATORIES LTD. Kawasaki, Japan

ABSTRACT

Some of magnetic and magnetoresistive properties in Ni-Co film for bubble detector have been investigated in comparison with mostly used permalloy film. The optimum preparation condition was established in the film of 300 Å : the deposition temperature of 200 °C and a Ni content of 70 %. The fractional change in resistivity showed 3.8 % which was twice as large as permalloy film, and the low effective anisotropy field observed in this film was suitable for small bubble detectors. Using 6 μm bubble domain in $(EuEr)_3(GaFe)_5O_{12}$ and five-columns chevron stretcher, a signal voltage of 0.9 mV/mA was obtained, which was three times greater than that of our permalloy detector.

INTRODUCTION

Many of the magnetic bubble systems employ magnetoresistive detectors[1] with the bubble stretching technique[2,3]. At present most detectors consist of permalloy film which possesses a favorable combination of magnetoresistive and magnetic properties for this application[4]. The desirable magnetic film to improve the detection efficiency is larger in the change in resistivity, and smaller in the effective anisotropy field Hs defined as the field necessary to saturate fractional change in resistivity[5]. Here Hs depends on the demagnetizing effect and can be expressed as $Hs(w) = H_k + N \cdot 4\pi M_s$. Being examined on some kinds of magnetic thin films, 70Ni-Co revealed the suitable characteristics for this requirement. In this article we present some of characteristics of Ni-Co film and pertinent details on the detection performance in comparison with detectors made from permalloy film.

EXPERIMENTAL

Sample

All the films were evaporated in a vacuum of 2 to 3 x 10^{-6} Torr under a magnetic field of 30 oe. Ni, Co, Fe powders of the desired composition were evaporated from a tungsten boat by usual resistance heating method. A film of 300 Å was deposited at an evaporation rate of 2.5 Å/sec on the glass substrate which was heated by radiation from a resistance heater located behind substrate holder. A magnetoresistive element was formed to desired dimensions by ion milling method. The electrode was made from Au-Cr film (0.4 - 0.5 μm). For bubble detection a rectangular type detector was constructed with the five-columns chevron stretcher of 0.4 μm-thickness and 28 μm-period. The detector was positioned between chevron patterns. The dummy detector was used for the cancellation of the magnetoresistance signal due to the rotating field. An epitaxial garnet film was placed closely to the detector-propagation substrate. The bubble material used was $(EuEr)_3(GaFe)_5O_{12}$:$4\pi Ms$ was about 250 gauss and its bubble diameter was 6 μm.

Measurement

The fractional change in resistivity was defined as $\Delta\rho/\rho$ = (ρ easy - ρ hard)/ρ easy. Here resistivity was measured with the current along the easy axis. An inductive loop tracer was prepared for measuring coercive force Hc, and anisotropy field H_k. A value of Hs was measured as the field that the extraporation of $\Delta\rho/\rho$ in small drive field interesected to the saturation level of $\Delta\rho/\rho$. Ms of films was measured by torsion magnetometer[6]. Film composition was evaluated by atomic absorption spectroscopy. Film thickness was measured by the surface roughness meter (Talystep).

RESULTS AND DISCUSSIONS

The fractional change in resistivity $\Delta\rho/\rho$ and the change in resistivity $\Delta\rho$ for 70Ni-Co, 76Ni-Fe films deposited at a substrate temperature ranging from 120 °C to 280 °C were shown in Figure 1. A peak value was observed at the temperature of 200 °C for both films, and $\Delta\rho/\rho$ of 70Ni-Co was 3.8 % which was twice as large as 76Ni-Fe. The physical meaning of this substrate temperature dependence was not yet obvious. The influence of deposition temperature on Hc, H_k was examined. As can be seen in Figure 2, Hc became minimum at the temperature of 200 °C; 4 oe for 76Ni-Fe and 13 oe for 70Ni-Co.

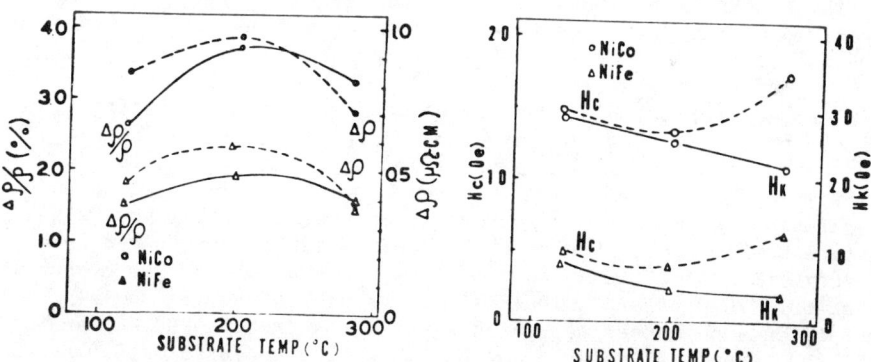

Fig. 1. $\Delta\rho/\rho$ and $\Delta\rho$ as a function of substrate temperature for film thickness of 300 Å.

Fig. 2. H_k and Hc as a function of substrate temperature for film thickness of 300 Å.

H_k showed the inverse proportion to the substrate temperature. From these results the following experiments were made on the specimen deposited at the optimum temperature of 200 °C.

Figure 3 showed the relation between film composition of Ni/Co, Ni/Fe and $\Delta\rho/\rho$. The maximum value was obtained at the composition of 70Ni-Co and 90Ni-Fe; 3.8 % and 2.9 %, respectively. The resistivity of Ni/Co film showed a maximum value of 25 $\mu\Omega$.cm at the composition of 70Ni-Co, and that of Ni/Fe decreased with Ni contents. The change in resistivity $\Delta\rho$ with respect to composition effect was resultantly described as in Figure 4; the maximum value of $\Delta\rho$ was obtained at the composition of 70Ni-Co and 76Ni-Fe. Namely $\Delta\rho$ of 70Ni-Co was twice as large as 76Ni-Fe. As the detection voltage is proportional to $\Delta\rho$ in a certain dimension of elements, 70Ni-Co detectors are superior to usual permalloy detectors.

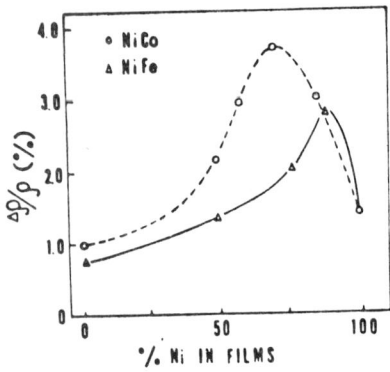

Fig. 3. $\Delta\rho/\rho$ vs. per-cent Ni in film thickness of 300 Å, deposited at the substrate temperature of 200 °C.

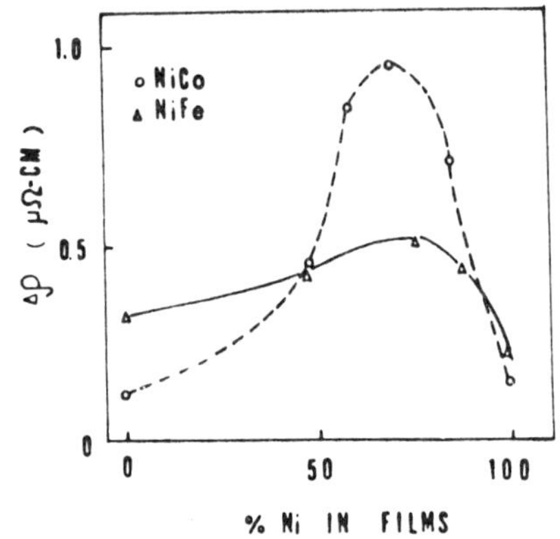

Fig. 4. $\Delta\rho$ vs. per-cent Ni in film thickness of 300 Å, deposited at the substrate temperature of 200 °C.

Another factor which must be considered will be the effective anisotropy field. The effective anisotropy field Hs in the width-wise direction of elements is expressed approximately as follows: $Hs = H_k + \ell/\ell+w \cdot t/w \cdot 4\pi Ms$, where ℓ, w, t : length, width and thickness of elements. To improve the detection efficiency Hs(w) must be smaller because a stray field from bubble domain can not overcome the effective anisotropy field in the case of garnet bubble detector. The authors calculated the value of Hs(w) as above-expression and further conducted experiments on both 70Ni-Co and 76Ni-Fe elements. The value of Ms, which was measured on 300 Å film by torsion magnetometer, was 600 emu/cc for 70Ni-Co and 1060 emu/cc for 76Ni-Fe. These values are rather different from the values obtained in the bulk material[7] ; 780 emu/cc for 70Ni-Co and 980 emu/cc for 76Ni-Fe. But it seems to be reasonable, if the accuracy in the measurement of thin films[6,8] is considered.

H_k of each film was 22 oe and 5 oe from Figure 2. $H_s(w)$ calculated in using these values was shown in Figure 5 in relation to the width of elements, where the length to width ratio was kept to a constant of 5. To compare it with the above calculated value, experiments were conducted on the specimen by changing the element width from 500, through 100, 50, 20, 5, to 3 μm, keeping the length to width ratio constant. The calculated and experimental results were rather in good agreement (Figure 5) and indicated that the value of Hs in 70Ni-Co film became smaller than that of 76Ni-Fe in the vicinity of 8 μm width, never-the-less large H_k.

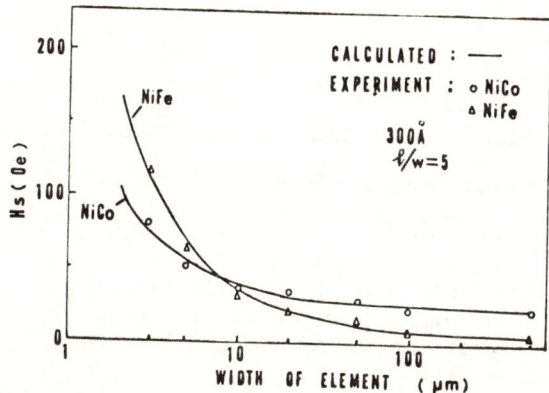
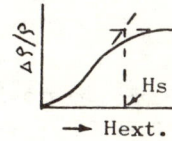

Fig. 5. Relation between Hs and width of element. The definition of Hs in measurements is indicated below.

At last the detection properties of these elements were compared at usual field access operation. The detector was a rectangular type of 6 μm x 50 μm x 300 Å, and its resistance was about 90Ω in both elements. The experimental result is summarized in Table 1. Here rotating field was 30 oe and bias field was 145 oe, using above-mentioned garnet. The operating frequency was 10 KHz and its typical signal form was shown in Fig. 6. It was found that the detection voltage of 70Ni-Co element was three times greater than that of permalloy detector. And the S/N ratio was more than 3. The resistance change $(\Delta R/R)_{static}$ of these detectors was measured as 2.8 %, 1.5 % by a field of 100 oe in the plane of detectors. The signal voltage change of 0.9 mV/mA and 0.3 mV/mA observed as a bubble passed

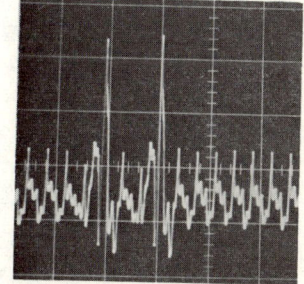

Fig. 6. A typical signal form of a 70Ni-Co detector. The vertical scale is 1mV/div and the operating frequency is 10 KHz. The detector current is 3 mA.

the detector corresponds to a ($\Delta R/R$) operating of 1.0 % and 0.33 %, respectively for 70Ni-Co and 76Ni-Fe detector. Above data were obtained on a chevron propagation pattern made from permalloy but the authors had confirmed that the results were the same if the pattern was made from NiCo.

Table 1. Detection performance of 70Ni-Co, 76Ni-Fe detector.

	70Ni-Co detector	76Ni-Fe detector
Detector resistance (Ω)	90	90
Detector ($\Delta R/R$)$_{static}$ (%)	2.8	1.5
Detector current (mA)	1	1
Detection voltage (mV)	0.9	0.3
Detector ($\Delta R/R$)$_{operating}$ (%)	1.0	0.33

Rotating field: 30 oe,   Bias field: 145 oe

CONCLUSIONS

A magnetoresistive detector made from 70Ni-Co film has been described which is superior to permalloy film in its change in resistivity, and is suitable especially for small bubble detection because of its low demagnetizing field.

ACKNOWLEDGEMENTS

We are grateful to H. Sasaki, Y. Nishimura, S. Hiyama for the encouragement, to K. Komenou, H. Nakajima for valuable discussions.

REFERENCES

1. G. S. Almasi, G. E. Keefe, Y. S. Lin, and D. A. Thompson, J. Appl. Phys. 42, 1268 (1971).
2. W. Strauss, P. W. Shumate, Jr., and F. J. Ciak, AIP Conf. Proc. 5, 235 (1972).
3. J. A. Archer, L. Tocci, P. K. George, and T. T. Chen, IEEE Trans. Mag. MAG-8, 695 (1972).
4. S. Krongelb, J. of Electronic Materials, Vol. 2 No.2 (1973).
5. K. Komenou, and K. Asama, National Conventional Record IECE, Japan, No.234 (1973).
6. C. A. Neugebauer, Phys. Review, 116, 1441 (1959).
7. R. M. Bozorth, Ferromagnetism, D. Van Nostrand Co.
8. J. F. Freedman, A. F. Mayadas, and E. Klokholm, IEEEE Trans. on MAG, Vol. MAG-5, NO.3, 170 (1969).

STUDY ON THE THICK FILM CHEVRON DETECTOR
T. T. Chen, P. K. George, L. R. Tocci and J. L. Archer,
Rockwell International, Anaheim, California, 92803

A theoretical model is presented for analyzing the magnetoresistance effect in permalloy structures as an aid to optimizing the operation of thick film magnetoresistive bubble detectors in field access circuits.

It has been demonstrated that in an unsaturated permalloy bar the average magnetization computed on the basis of a continuous two-dimensional magnetization function $M(x,y)$ is in good agreement with experiment. Using this two-dimensional model and further assuming the magnetoresistance to be linearly proportional to the magnetization such that: $\Delta\rho(x,y) = \frac{|\overline{M}(x,y)|}{M_s} \Delta\rho_s \cos\theta(x,y)$ EQ. 1

the magnetoresistance variation of an arbitrary two-dimensional structure may be computed as a function of the field orientation. Here $\Delta\rho(x,y)$ is the magnetoresistance change at point (x,y), $\Delta\rho_s$ the saturation magnetoresistance change and θ the angle between current $\overline{I}(x,y)$ and the magnetization $\overline{M}(x,y)$. Using this Eq. the magnetoresistance variation for a bar and an L-shaped structure placed in a rotating field and with and without the presence of a bubble field were calculated. The result for the L-shaped structure is shown in Fig. 1.

Experimental data was taken on different sets of serially connected etched bar and L-shaped elements. Fig. 2 shows results for 90° chevron elements (modified L-structure) connected as a two-level thick film detector. The measured ΔR vs. θ (angle between the rotating field and the current) with and without a bubble agrees qualitatively with the calculated results of Fig. 1 but ΔR shows a nonlinear dependence on the rotating field amplitude which is not predicted by the model. This nonlinear behavior can be taken into account in the model by establishing a proper domain distribution function related to the magnetization value which will, we feel, give good agreement between experiment and theory making the model useful in the optimization of magnetoresistive detector designs.

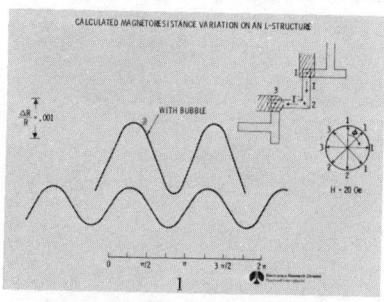

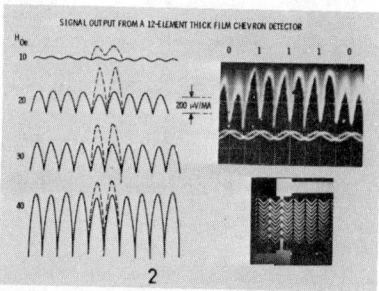

EXPERIMENTAL CONFIRMATION OF FIELD ACCESS DEVICE MODELLING

P. K. George and J. L. Archer, Electronics Research Div.
Rockwell International, Anaheim, California 92803

ABSTRACT

In order to experimentally verify the predictions of a two-dimensional field access model, two types of experiments have been performed on simple bar patterns. VSM measurements have been used to measure the average magnetization in bar arrays as a function of applied field and field angle relative to the major bar axis, and the results have been found to be in excellent agreement with theory. Effective z-component fields for several bar patterns have been measured with and without an in-plane applied field. Comparison with theory is complicated by the presence of an ion-implanted layer on the surface of the garnet which presumably affects the field of the bubble and, therefore, the outcome of the experiment. Calculations based upon the idealized bubble field give good qualitative and quantitative agreement with experiment, provided an effective spacing is used which takes into account the ion-implanted layer and consequent reduction in bubble field.

INTRODUCTION

In an effort to establish quantitative criteria for the design of the permalloy system in field access devices, a number of essentially similar device models have been proposed. Attempts to make comparisons between model predictions and experiment have not been successful in the past, principally because the approximations used in the modelling were either inadequate or simply incorrect. The recent solution of the two-dimensional modelling problem[1], however, allows a more critical examination of existing data to be made. In this paper we will show that (1) the magnetization process in permalloy bars can be predicted extremely accurately on the basis of a two-dimensional model, and (2) the magnetostatic well depths associated with a bubble trapped under a permalloy bar are in reasonable agreement with experiment. In an accompanying paper[2], the magnetoresistance of a bar and a chevron is considered in connection with detector simulation and is shown to be in good agreement with model predictions.

MODEL PREDICTIONS

Figure 1 summarizes what we consider to be the most important results that can be obtained for a typical bar ($2 \times 10 \times .5\mu^3$)

on the basis of two-dimensional modelling calculations. In Figure 1(a) the demagnetizing factor is shown as a function of position for a bar uniformly magnetized along its major axis. In general, the average demagnetizing factor for a bar calculated on the basis of Figure 1(a) is higher than that for the equivalent ellipsoid whether the structure is magnetized parallel or perpendicular to its major axis.

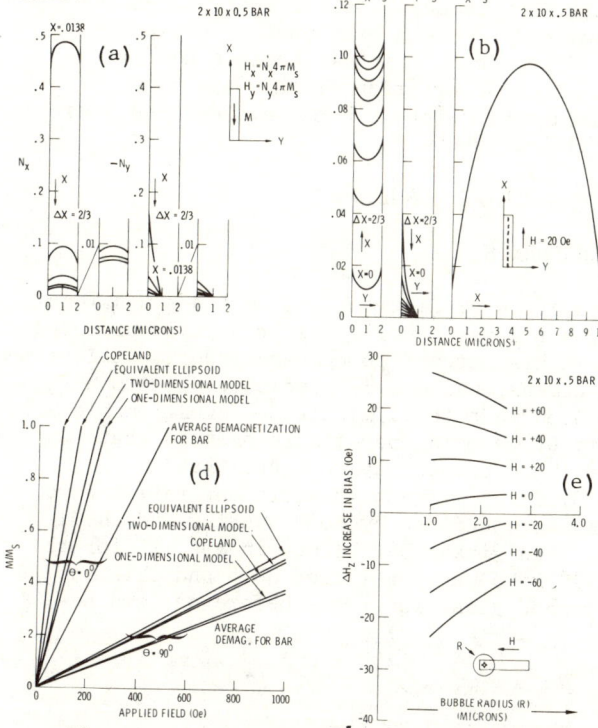

Fig. 1. Two-Dimensional Model Results for a $2 \times 10 \times .5\mu^3$ Bar

Figure 1(b) shows the resulting two-dimensional magnetization distribution obtained when a uniform 20 Oe in-plane field is applied along the major bar axis. The saddle-like shape of the magnetization is a result of the fact that the demagnetizing factor at any transverse cross section decreases as the bar edge is approached. In Figure 1(c) are presented the average x and y components of the magnetization as the in-plane field is rotated through 360°. In general, for a bar the ratio of maximum to minimum amplitude is less than that for the equivalent ellipsoid. The amplitudes shown in Figure 1(c) increase linearly with field and in Figure 1(d) are shown the linear portions of the predicted M-H curves ($H_c=0$) for $\theta=0$ and $\theta=90°$. For comparison purposes the results for various other models have been plotted on the same figure. As expected the results for the two-dimensional model fall between those for the equivalent ellipsoid and those for the average demagnetizing factor for the bar. The deviation between Copeland's[3] results and other work is apparently a consequence of the fact that he failed to include magnetic surface charge in his formulation. In Figure 1(e) is shown the reduction in bias field (ΔH_z) as a function of bubble radius incurred by a bubble as it moves to its equilibrium position under the end of the 2 x 10 x .5µ³ bar. The curve parameters are the in-plane field applied along the major bar axis either aiding (+) or opposing (-) the polarizing effect of the bubble. Typically the reduction in H_z increases with bar length and is independent of permalloy thickness (fixed length to width ratio) for fixed permalloy to garnet spacing. In Figure 1 and the accompanying discussion, we have summarized the important qualitative and quantitative predictions that can be made for a simple bar on the basis of the two-dimensional model. In the following section we present the results of two experiments which we believe confirm these predictions. Only a brief description of the experimental technique is given in each case principally because the techniques have been described in detail elsewhere.

EXPERIMENTAL RESULTS

(a) VSM-MAGNETIZATION MEASUREMENTS

To confirm the magnetization predictions of the two-dimensional model we have repeated the bar array experiments originally proposed and carried out by Doyle[4] in his attempt to compare experimental results with Copeland's one-dimensional model[3] predictions. A minor improvement in technique has been made to eliminate the questionable influence of the coercivity by choosing bar thicknesses so that their demagnetizing factors were considerably higher than any measured coercivity. Two different length to width bar ratios were considered and the patterns were defined by both sputter and chemical etching there being no significant difference in results for the dimensions used. Only 81-19 NiFe was considered here and this was sputter deposited on a polished glass substrate before pattern de-

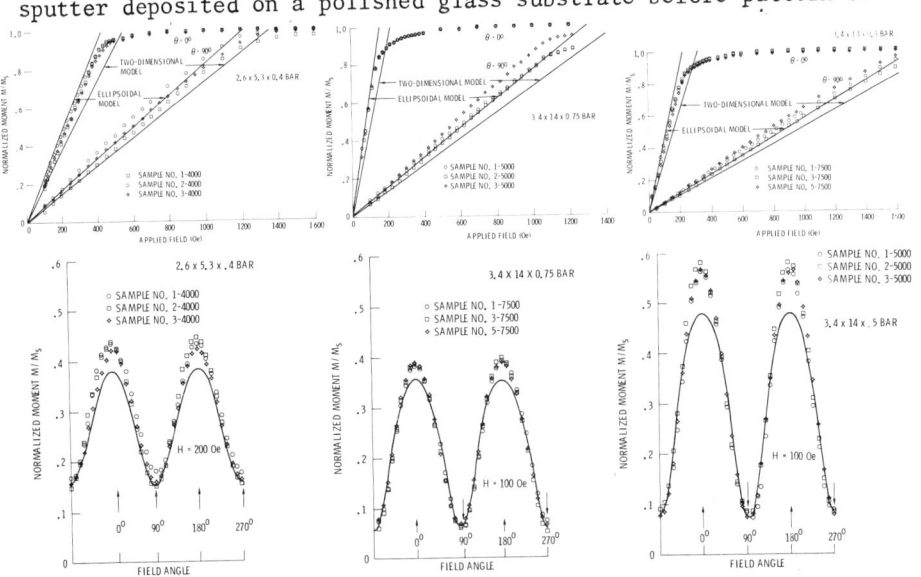

Fig. 2. VSM Magnetization Results

finition. All experiments were run on a PAR Model 155 vibrating sample magnetometer (VSM) which has the feature that the sample may be rotated in the field of the accompanying electromagnet. All data taken was reproducible and what we would consider to be extremely reliable.

Figure 2 presents the experimental VSM results along with the theoretical predictions based on the bulk value saturation magnetization of about 850 emu/cc which we found to be fairly close to the measured thin film value. The upper part of Figure 2 shows that the fit to the experimental M-H data for both easy and hard directions is extremely good for different thickness films and also when the length to width ratio is changed. The lower part reveals that the angular variation of the moment M is also predicted quite well by the two-dimensional model. On the basis of the uncertainties associated with the definition of the bars and the approximations made in the modelling we feel that the agreement reflected in Figure 2 is exceptional and that it provides partial justification for the use of the two-dimensional model in field access device modelling.

(b) MAGNETOSTATIC WELL DEPTH MEASUREMENTS

A bubble placed under the end of a permalloy bar suffers a reduction ΔH_z in effective bias relative to one placed in a permalloy free region[5]. The reduction in bias field may be measured at fixed bubble radius by experimentally measuring both the free and trapped bubble characteristics and subtracting to determined ΔH_z. Measurements of this type have been performed on two nearly identical YEuTmGaIG samples for three different bar geometries. The garnets were ion-implanted with Ne^+ at 80 KeV with the beam offset 5° from vertical resulting in specimens with less than 1 Oe bubble collapse field variation. The $4\pi M$ values were 157 and 160 respectively and collapse occurred at 71.3 Oe in one sample and 72.0 Oe in the other. Both samples were essentially 6μ thick and an SiO_2 layer was sputtered on top of them to provide an effective spacing of 1μ between garnet and permalloy. The coercivity of the permalloy was determined by measurements on a thin film monitor to be typically 1 Oe (except for one case for which it was 3.8 Oe) and the bar patterns were defined chemically (4000Å and 5000Å) and by sputter etching (7500Å).

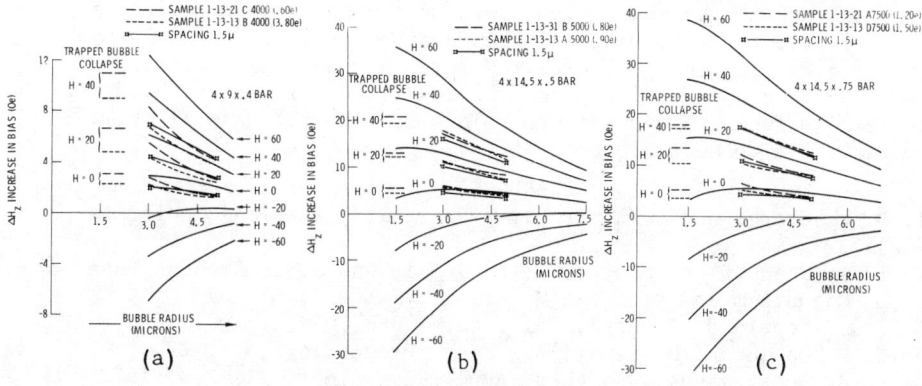

Figure 3 Magnetostatic Well Depth Results

The results shown in Figure 3 reveal what we believe to be agreement with the two-dimensional modelling predictions. The data exhibits some inconsistencies but on the whole suggest that the theoretical calculations over-estimate the bubble field which would be entirely consistent with the presence of the ion-implanted layer which has not been taken into account in the calculations. For the range of bubble radii considered both the experimental and theoretical results show an increase in well depth with decreasing radius, this effect being more pronounced as the in-plane field H (colinear with the bars) increases in magnitude. In making the comparison of Figure 3(a) it should be understood that experimentally (due to the shortness of the bar) the bubble equilibrium position shifts with varying bubble radius - no attempt was made to take this effect into account in our calculations. Both the experimental and theoretical results of Figures 3(b) and 3(c) tend to confirm the observation that the permalloy thickness is not an important factor in controlling the well depths. A rough estimate of the increase in garnet to permalloy spacing required to decrease the bubble field sufficiently to yield better agreement between experimental and theoretical results gives $\Delta z=.5\mu$ which is not unreasonable in view of the fact that the peak of the in-plane bubble field is reduced by about 33% only in the vacinity of the bubble radius and remains unchanged elsewhere. The result of increasing the spacing by $.54\mu$ is indicated on Figure 3 by the x's and light lines going through them.

CONCLUSION

The results presented here confirm that the two-dimensional model gives both qualitatively and quantitatively correct results for the magnetization and magnetostatic well depths for simple permalloy bars. Furthermore, measurements on bars of the variation of resistance with field angle reported elsewhere confirm the characteristic $\cos 2\theta$ behavior predicted theoretically. On the basis of these comparisons we would suggest that the importance of taking domain structure into consideration has yet to be proven and that the proposed two-dimensional continuum model provides a workable starting point to which switching fluctuations can later be added if necessary to account for possible margin narrowing and in the case of detector operation-switching noise. Our viewpoint is one in which the switching fluctuations play a minor role relative to the overall magnetization process which is determined by the demagnetizing effects of the pattern. This must be the case if reliable operation is to be achieved.

REFERENCES

1. P. K. George and J. L. Archer, Paper 13.7, Intermag Conference, 1973.
2. T. T. Chen, P. K. George, L. R. Tocci, and J. L. Archer, Conf. on Magnetism and Magnetic Materials,(Paper 2B-6) 1973.
3. J. A. Copeland, J. Appl. Phys. 43, 1905 (1972).
4. W. D. Doyle and M. Casey, AIP Conf. Proceedings, 10, 227 (1972).
5. P. K. George and T. T. Chen, Appl. Phys. Lett. 21, 263 (1972).

Field Access Bubble Propagation Circuits*

Shalendra K. Singh
Rice University, Houston, Texas

ABSTRACT

The theoretical model proposed by Archer et al[1] for analysis of field access magnetic bubble propagation circuits is extended to include rotating drive fields. This extension allows the study of bubble propagation along T-Bar patterns. Computer simulations are made to determine energy profiles for a bubble interacting with the drive circuit. The motion of the bubble along the propagation circuit is determined as a function of external drive field by assuming that the bubble follows the energy minima. Bubble velocity profiles can be calculated from the bubble position as a function of drive field orientation. This allows, for the first time, the comparison of Archer's theoretical model with the experimental observations. A study of the spacing between circuit elements of a T-Bar pattern reveals that the optimum value of this spacing should be a quarter of a bubble diameter. The factors which are primarily responsible for limiting propagation speed are determined and modifications to existing field access circuits are suggested to improve propagation at high frequencies. Extending the principles evolved from this computer simulation, a field access circuit of substantially different character from previously fabricated circuits was designed. This circuit shown below was built and tested, and showed an essentially uniform bubble velocity profile, one of the primary requirements of a high speed, wide margin propagation circuit. Results of experiments with this circuit are presented and relevant circuit parameters discussed.

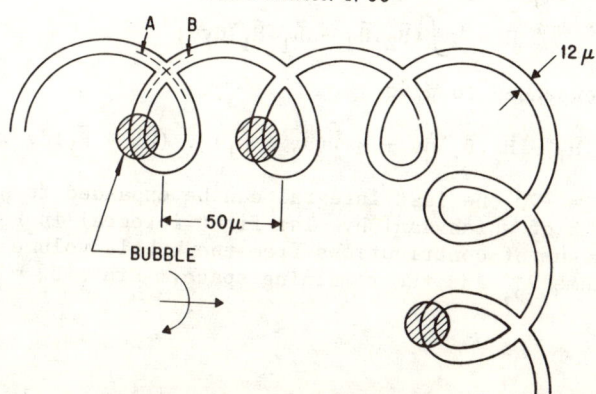

A NEW PROPAGATION CIRCUIT.

*Supported in part by NASA grant NGR 44-006-001

LABORATORY MODELING OF BUBBLE PROPAGATION SYSTEMS

C. H. Hsin*, T. J. Matcovich and R. L. Coren
Electrical Engineering Department, Drexel University,
Philadelphia, Penna. 19104

ABSTRACT

The use of scaled, field access bubble propagation systems has been proposed for examining the details of bubble conduction schemes[1]. It has been argued that such scaling cannot represent size dependent effects, cannot include the influence of bubble dynamics, and does not lend itself to including interactions between bubble and overlay. In this report we describe a large, appropriately scaled T-bar model which includes bubble interactions. It is shown that scaling effects are not significant, that bubble position is essentially determined by magnetostatic equilibrium and that the scale model can serve as a valuable tool in studying the design and extension of field access devices.

ANALYSIS

In this report we consider a configuration having a cylindrical magnetic domain in an unpolarized medium[2] which is infinite in the X-Y plane but finite in Z. We define the field H_1 to consist of the Z directed bias field, the planar (rotating) drive field, and the bubble field. The corresponding flux density is B_1. If a permalloy overlay is introduced the new field, H_2, differs from H_1 by H_p ($H_p = H_2 - H_1$), the field induced by the permalloy. The new flux density is B_2.

Since the energy of the first case is independent of bubble position, the bubble equilibrium can be explored by examining the energy difference

$$U = \tfrac{1}{2} \int (\bar{B}_2 \cdot \bar{H}_2 - \bar{B}_1 \cdot \bar{H}_1) dv , \tag{1}$$

which can be expanded to give

$$U = \tfrac{1}{2} \int (\bar{B}_2 \cdot \bar{H}_1 - \bar{H}_2 \cdot \bar{B}_1) dv + \tfrac{1}{2} \int (\bar{B}_2 + \bar{B}_1) \cdot (\bar{H}_2 - \bar{H}_1) dv . \tag{2}$$

Writing $H_2 - H_1 = -\nabla \phi$ the last integral can be expanded to a sum of integrals, both of which vanish. The first integral in Eq. 2 can be broken into a sum of contributions from the bubble volume V_B, the permalloy volume V_p, and the remaining space. Since $B_2 = \mu_0 H_2$ and

---

*Now at Burroughs Corp., EMSO Division, Piscataway, N. J. 08854

$B_1 = \mu_0 H_1$ outside of V_p and V_B the free space contribution vanishes. Substituting $B_1 = \mu_0 H_1$, $B_2 = \mu_0 (H_2 + M_p)$ in V_p and $B_1 = \mu_0 (H_1 + M_B)$, $B_2 = \mu_0 (H_2 + M_B)$ in V_B, and combining terms gives

$$U = \tfrac{1}{2}\int \bar{M}_p \cdot \bar{H}_1 \, dv_p - \tfrac{1}{2}\int \bar{M}_B \cdot \bar{H}_p \, dv_B = \tfrac{1}{2}U_p - \tfrac{1}{2}U_B \qquad (3)$$

U_p is the permalloy energy in the initial field, U_B is the bubble energy in the permalloy field. Minimizing U leads to $\delta U_p = \delta U_B$. The force exerted on the permalloy will be $-\nabla U_p$ and that on the bubble will be $-\nabla U_B$. Since, by Newton's third law these forces must be equal and opposite, then $\delta U_p = -\delta U_B$. From these results we have $\delta U_B = 0 = \delta U_p$, i.e. minimization of either the permalloy term[3,4] or the bubble term[1,5] can be used to determine equilibrium. We have chosen to use U_B for our analysis.

For the usual fields and materials we can take M_B to be uniform and vertically directed over the domain volume so that only the Z component of H_p enters into the expression for U_B. We then have $U_B = V_B M_B \langle H_{pz} \rangle$, where the brackets denote the mean value over the bubble volume. We see from this that minimizing U_B is equivalent to finding that configuration which minimizes the mean permalloy interaction field.

EXPERIMENT

The experimental arrangement consisted of the permalloy T-bar model, scaled approximately 400 times, which is described in reference 1. To that model we have now added a scaled bubble, consisting of a coil of 0.55" height, by 0.8" i.d. and 1.4" o.d. comprised of 420 turns of #26 AWG magnet wire. The bubble face was maintained 0.08" from the permalloy overlay. A current of 0.55A generated stray fields equal to those of a bubble with $4\pi M = 100G$. To establish the equivalency of this model to an actual bubble the vertical and radial stray fields were measured at different heights from the coil and were compared with the calculations of Druyvesteyn[6]. The comparison, shown on Fig. 1, is seen to be excellent.

The following procedure was adopted for the determination of $\langle H_{pz} \rangle$: with the bubble in position under the center line of the overlay and the planar drive field applied in a specific direction, the field H_z was measured along this center line at a distance of 0.29" above the overlay. (It is shown in reference 1 that this distance corresponds to the approximate average value of the permalloy vertical stray field over the bubble height.) The permalloy overlay was then removed and the bubble field H_{Bz} was measured over the identical path. Typical values are shown on Fig. 2. $H_{pz} = H_z - H_{Bz}$ was obtained by graphically subtracting the two curves. $\langle H_{pz} \rangle$ was taken to be proportional to the area under the resulting curve, measured over the bubble diameter. The bubble was then translated along the overlay axis and the process repeated at several points. These data yield $\langle H_{pz} \rangle$ as a function of bubble position so that the position yielding minimum field component can be found. Fig. 3 presents these results for the applied field at

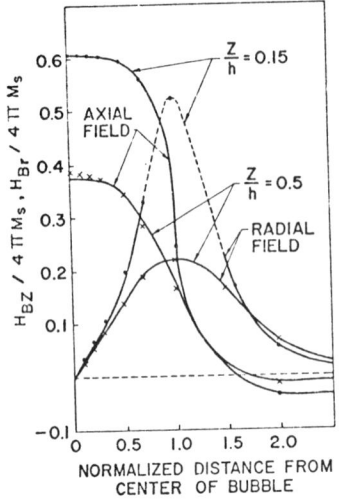

Fig. 1 Axial and radial fields from bubble model at two axial distances. Data are normalized to bubble axial length h=0.55" and mean bubble radius (.55"). Due to probe width radial data could not be taken near the coil windings for z/h=0.15. The x,o are the calculations of Druyvesteyn, and the smooth curves are measurements of the model.

290° to the overlay axis. At this drive field angle two equilibrium positions appear; by examining the approach to this angle it can be established that the bubble resides at point P. The barrier PQS does not vanish until the field angle exceeds 290°. It should be noted that the model used for this report is based upon the dimensions and properties of overlays and bubbles commonly used for orthorferrite materials. In a similar study corresponding to garnet materials the larger demagnetizing field (by a factor of 10) of the permalloy effectively eliminates the double minimum shown in Fig. 3.

By repeating the above process at successive field angles the translation of the bubble center can be determined as the field angle changes. This is shown on Fig. 4. Also shown on this figure are Rossol's stroboscopic data for a bubble driven at 21 kHz[7].

Fig. 2 Data used to calculate the permalloy interaction field. The applied field angle is 290°.

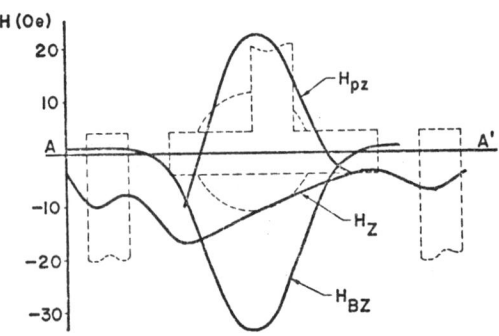

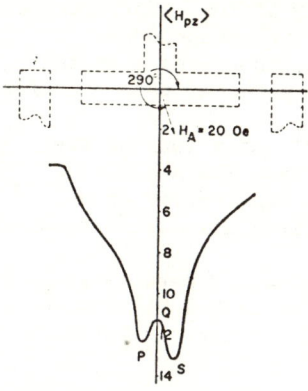
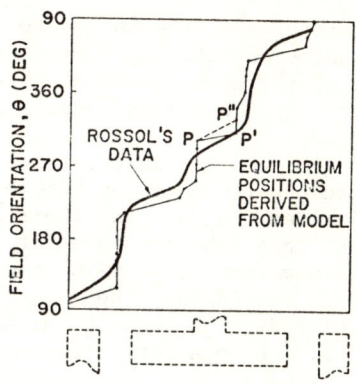

Fig.3 Mean permalloy interaction field as a function of bubble position, for a field angle of 290°.

Fig.4 Translational equilibrium position of magnetic bubble at each field angle. The rotating applied field is 20 oe.

DISCUSSION

Examination of Fig. 4 reveals an excellent agreement between the data from this study and that obtained from a stroboscopic study of an actual T-bar system. It may therefore be concluded that the major features of bubble propagation are realistically represented by this model.

The use of the scaled-up model is theoretically correct only if the magnetization is continuous, a condition which is known to be violated in actual overlays. The good agreement obtained here indicates that, at least with regard to the overlay external fields found here, the consequences of violating the continuity assumption are not significant.

As shown on Fig. 4, the motion of the bubble is a series of rapid jumps between energy minima which are determined almost exclusively by static considerations. Translation between these equilibria may be limited by bubble mobility in a way which has been calculated in reference 1. Illustrative of this correction is the adjustment made as the bubble leaves point P on Fig. 4; operating at 21kHz the bubble will follow path PP" rather than PP'. This modification makes the static curve more closely resemble the dynamic data. It is likely that better agreement with Rossol's data would be obtained if bubble size were allowed to vary in this study. Different field angles have different field profiles and, since the bubble diameter and shape depend upon the bias field, size variations do occur as a bubble propagates[9]. In addition, in low anisotropy materials M_B may change or rotate under the influence of these varying fields. Corrections for these effects would require tedious iterative procedures. In view of the good agreement obtained, it may be concluded that these effects are less significant than anticipated and they were therefore omitted.

Also omitted in this study are the known[1] variations of the lateral position of the bubble as it propagates. These shifts will alter $\langle H_{pz} \rangle$ and their determination would require area rather than line averages for $\langle H_{pz} \rangle$, an additional complexity which seems unwarranted at this time.

In conclusion, the closeness of our data to the data taken by Rossol indicates that the modeling procedure described here is a useful technique for simple modeling of bubble propagation systems.

REFERENCES

1. C. H. Hsin, T. J. Matcovich and R. L. Coren, AIP Conference Proceedings, 5, 244 (1971).
2. A. A. Thiele, Bell Sys. Tech. J. 46, 3287 (1969).
3. J. A. Copeland, J. Appl. Phys. 43, 1905 (1972).
4. P. K. George, and J. L. Archer, J. Appl. Phys. 44, 444 (1973).
5. K. Kempter, IEEE Trans. Magnetics 8, 746 (1972).
6. W. F. Druyvesteyn, D.L.A. Tjaden and J. W. F. Dorleijn, Philips. Res. Repts. 27, 7 (1972).
7. F. C. Rossol, IEEE Trans. Magnetics 7, 142 (1971).
8. C. H. Hsin, Ph.D. Dissertation, Drexel Univ., June 1973.
9. P. K. George and J. L. Archer, AIP Conference Proceedings, 10, 222 (1972).

PROPAGATION CHARACTERISTICS OF PARALLEL BAR CIRCUITS

G. Ng, W. Kinsner and E. Della Torre
Department of Electrical Engineering, McMaster University
Hamilton, Ontario Canada

ABSTRACT

Parallel bar circuits[1] of various aspect ratios have been tested and their operating margins determined. Particular attention has been paid to the design of turnarounds (corners) with respect to obtaining the same margins as for the propagating circuits. Three types of corners, i.e., the single, double, and triple bar corners of various width and aspect ratios have been tested. A study of various transfer gates, with different system applications, indicates satisfactory operation within wide margins. The permalloy-crystal spacing has been varied experimentally.

INTRODUCTION

The parallel bar circuit has been suggested[1] as a bubble propagating circuit, requiring only an oscillating transverse field. In order to incorporate this circuit into a major - minor loop organized memory, it is necessary to be able to turn corners and build transfer gates as well as to propagate in straight lines. This paper presents the operating margins for propagation circuits with various corners and gates obtained by quasi static measurements.

Measurements were carried out using 1 μm thick Permalloy circuits separated approximately 0.5 μm from platelets of $Sm_{.55}Tb_{.45}FeO_3$ that were $6\ell_M$ thick, where ℓ_M = 3.23 μm. A low frequency sinusoidal source was used for the transverse field in the propagation measurements. Fig. 1 illustrates the periodic Permalloy circuit used to measure the margins for linear propagation. The linear propagation was also measured on single loop circuits. The long and short bars were 6 mils and 2.5 mils long, respectively. Both were 1 mil wide, and the separation between them was 0.5 mil.

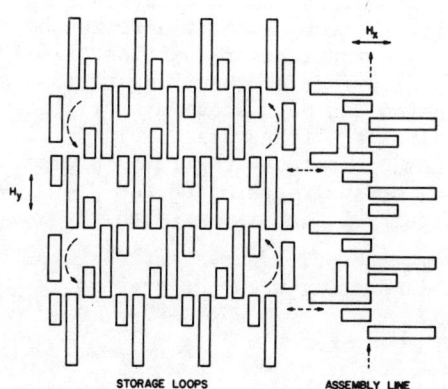

Fig. 1. Parallel bar bubble propagating circuit with corners and transfer gates. The field H_y propagates bubbles unidirectionally in the storage loops. The field H_x independently propagates bubbles in the assembly loop and is also used for transferring bubbles between the loops.

Since bubbles are stable under the ends of magnetized Permalloy bars, straight line motion of bubbles is possible only when the ends of the bars overlap the bubble path by approximately half a bubble diameter. In the circuits tested a slightly smaller overlap (0.2 mil) was used to minimize interaction between adjacent tracks as a compromise which still permitted nearly linear operation. The distance between the tracks in these circuits was three bubble diameters. It was found that this caused reduced margins due to interaction between bubbles of adjacent tracks propagating in opposite directions. It was calculated that a 20% increase in the track separation would overcome that problem.

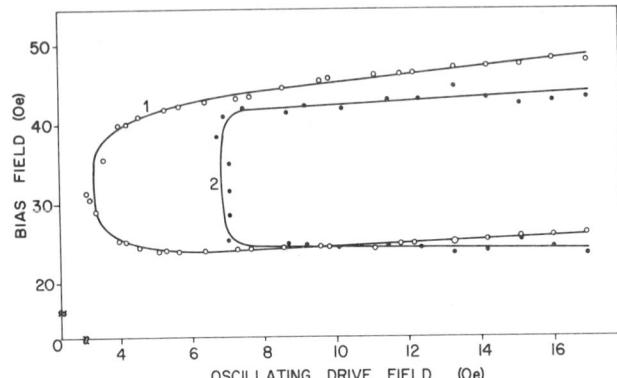

Fig. 2. Operating margins for linear propagation (1) symmetrical structure (2) a single symmetrical loop.

The operating margins for linear propagation are shown in Fig. 2. It is noted that the bias field margins are comparable to other transverse field propagating circuits such as T bar. The failure modes are similar with strip out and bubble collapse, determining the low and high limits on the bias field, respectively[2]. The important difference in the margins is that the parallel bar circuits can propagate with much lower drive fields. It is believed that this is due to the better defined poles and the better pole concentration as compared to T bar circuits or due to the shorter distance that the bubble has to propagate as compared to the other circuits. Circuits have been tested that operated with drive field as low as 3 Oe.

Analysis of the bubble motion under sinusoidal excitation indicated that the highest bubble velocities occur in the transition between the long bar and the short bar and thus this transition will limit the high frequency operation. If higher frequency operation is desired, a two step stairwave excitation can be used at a sacrifice of the ability to use tuned circuits for the transverse field drive coil[1]. This will result in greater demand on the drive transistors.

The influence of the Permalloy - crystal spacing on the propagating margins of the parallel bar circuits is consistent with the results reported previously[3,4], however, the pole concentration makes it less sensitive to the variations in spacing when the spacing is large (approximately 1 μm).

CORNER CIRCUITS

Three types of corner circuits were investigated. They utilized a triple, a double, and a single bar, respectively, to turn a bubble at a corner and are illustrated in Fig. 3.

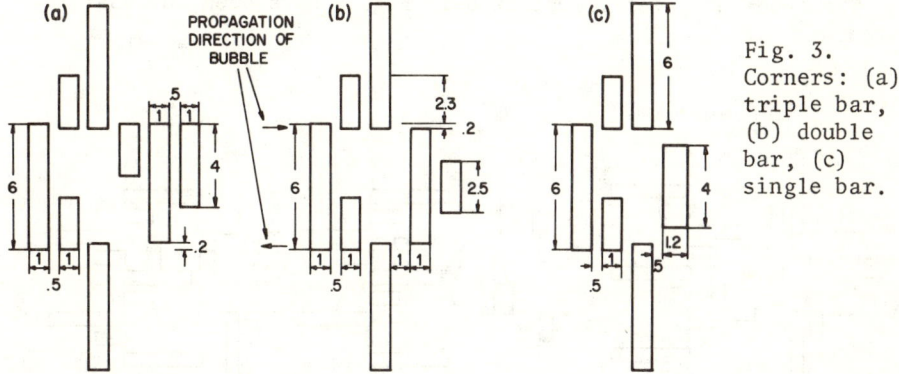

Fig. 3. Corners: (a) triple bar, (b) double bar, (c) single bar.

The single bar corner performs the turn essentially on a short bar. However, due to the distance the bubble has to travel, the short bar has to be made wider than the other short bars in order to obtain the aspect ratio of a short bar. Although a bubble can be turned from one track to another when the turning bar is symmetrically located between the tracks, better operating margins are obtained if the bar is placed asymmetrically. This results from the better interaction between the poles.

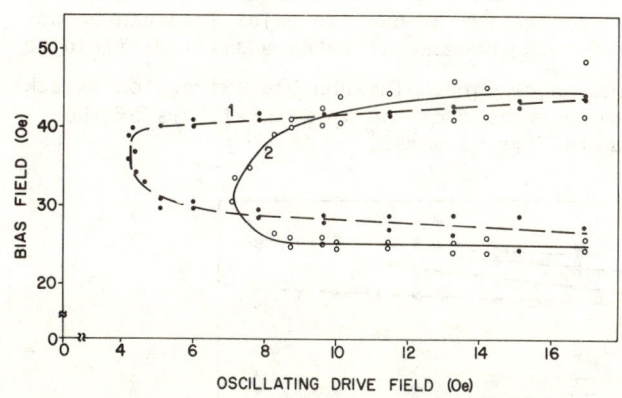

Fig. 4. Operating margins of the best two corners: (1) double bar, (2) single bar.

The operating margins of the corners are shown in Fig. 4. These circuits have not been optimized. Measurements of various corners indicate, however, that the separation between the turning bars and the end bars of the tracks is critical. It is noted that many other corners are possible; in fact, the transfer gates discussed below operate successfully as corners in the blocked state (i.e., when the H_y field is applied).

TRANSFER GATES

The types of gates illustrated in Fig. 5 have been successfully tested.

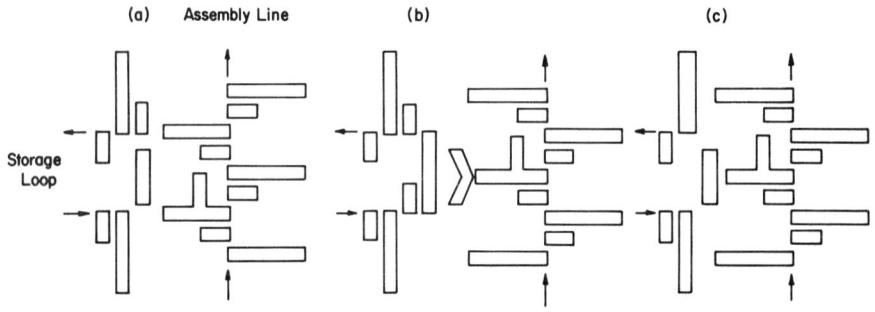

Fig. 5. Various transfer gates.

They all operate on the principle that a transverse field perpendicular to the propagate field causes a transfer. The memory can be organized with horizontal minor (storage) loops and a vertical major (assembly) loop or line. The field which transfers bubbles from the minor loops to the major loop can be used to circulate the bubbles around the major loop. Propagation around the major loop can be accomplished without transferring because it takes a small H_y field to move the bubbles from the major loop. This bubble extraction is achieved by the vertical protrusions from the horizontal bars of the transfer gates, as shown in Figs. 1 and 5.

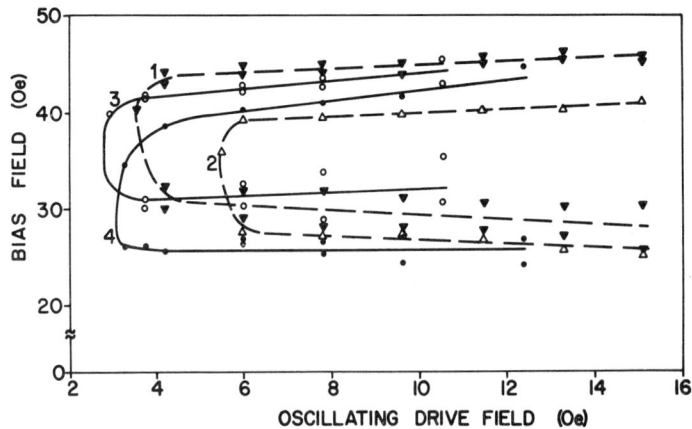

Fig. 6. Operating margins for transfer gate: (1) corner in Fig. 5a; (2) transfer in Fig. 5a; (3) corner in Fig. 5b; (4) transfer in Fig. 5c.

The transfer of a bubble from the major to minor loop is then completed by the H_x field. The direction of bubble transfer is determined by the polarity of the transfer field, H_x.

Fig. 6 shows the operating margins of transfer gates and associated corners. The operating margins shown in Fig. 7 refer to a transfer gate, a corner, and a symmetrical (periodic) propagating structure.

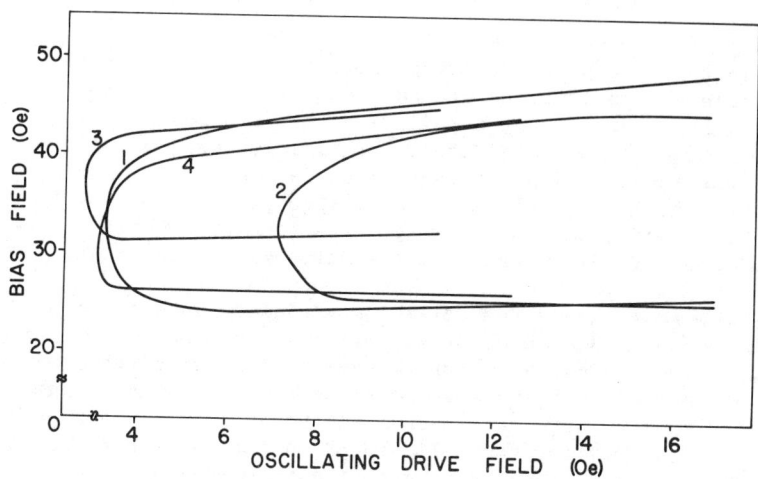

Fig. 7. Operating margins for (1) linear propagation; (2) single bar corner; (3) curve (3) in Fig. 6; (4) curve (4) in Fig. 6.

ACKNOWLEDGEMENTS

The authors wish to thank Bell Northern Research Laboratories, Ottawa, Canada for evaporating the Permalloy films and Bell Telephone Laboratories, Murray Hill, U.S.A. for supplying the crystals.

REFERENCES

1. E. Della Torre and W. Kinsner, INTERMAG Conf., Washington, April 1973, IEEE Trans. Magn., Vol. MAG-9, 298, (September 1973).
2. I. Danylchuk, J. Appl. Phys., 42, 1358, (March 1971).
3. Y. Kotera, R. Kinoshita, T. Namikata and Y. Nishimura, IEEE Tran. Magn., MAG-8, 673 (September 1972).
4. Y. S. Chen and T. J. Nelson, IEEE Tran. Magn., MAG-8, 754 (December 1972).

Section 4. Reliability in Bubble Domain Memories

RELIABILITY MODELING: APPLICATION TO BUBBLE DOMAIN MEMORIES*

W. G. Bouricius
IBM T. J. Watson Research Center
P. O. Box 218
Yorktown Heights, New York 10598

ABSTRACT

Beginning with future application requirements, the steps in the process of deriving useful reliability models for bubble domain memories are outlined. How one identifies failure modes, assigns failure rates, deduces the deleterious effects of failures, and finally derives and solves pertinent reliability equations is discussed. The use of reliability modeling in the design process is delineated, and finally the design of a specific bubble domain memory is discussed from the reliability viewpoint.

During the next decade the reliability requirements of data stores such as bubble domain memories, will increase by two orders of magnitude. The basic reason for these more stringent requirements is that the nature of the applications is rapidly changing. Computer systems with large data bases will become more integrated into overall operations and one will depend on getting the right data from the various stores involved in much the same way that one now depends on communications and electric power. Recognizing that everything fails occasionally, one must devise fall back strategies, emergency procedures, and so forth, whose cost is very much dependent on the probability, frequency, and duration of outages. Furthermore, for every application there is a threshold reliability and availability that must be exceeded before one attempts to implement it. A more subtle point is the following: even if the threshold requirement is met, the nature and complexity of the required backup emergency procedures constrains the system designers considerably.

Reliability modeling is like the elephant to the seven blind men. Depending on how you approach it, it can serve several different purposes. It can be employed to estimate servicing and maintenance requirements, inventory size, suitable applications, error checking requirements, etc. In this paper the main reason for employing reliability modeling is to improve the design, and as such it must be employed as part of the design process. It is extremely difficult to retrofit a design without excessive costs and delays, so unless reliability modeling is done concurrently with the design it loses most of its value. To improve the design means essentially to

* Supported in part by NASA contract NAS-8-26671

assess the trade-offs between reliability and other factors and make design changes accordingly. For example, one improvement comes after the identification of weak (i.e., unreliable) sections in the overall system. In some cases these weak parts require only slight modifications in their organizations to effect drastic improvements in reliability, and in other cases a major redesign is necessary. Often times a little bit of redundancy is the cheapest way to achieve the required reliability. Once the changes are made, the reliability modeling and calculations must be redone to get a new perspective and perhaps identify a new weak section that requires modification.

There are basically five steps in the reliability modeling procedure that must be taken in sequence and then reiterated in the case of a change in a design. These are: a) identification of failure modes and their distribution, b) assignment of values to the parameters in the component reliability equations, c) determination of the consequences of each type of failure as a function of the specific design, d) derivation of the system reliability equations, and e) calculations of system reliability. Each of these will next be discussed in the context of specific designs of a 10^8-bit bubble domain memory[1].

The memory consisted of several modules each containing 16 chips. Each chip contained 128 shift registers and each register consisted of 800 bits. Thus each module contained 1.6×10^6 bits. Preliminary reliability calculations were made for a straightforward simplex design, i.e. one without any standby spares, error correcting encoding, or any other form of redundancy. The results showed that one could not hope to achieve the reliability goal of .99 for a mission time of 2 months as desired by NASA. Therefore, error-correction encoding and associated circuitry would be mandatory. Two competing designs remained after some consideration was given to speed, power consumption, connections per chip, and reliability requirements. One of the two designs employed a bit-per-module organization which consisted of placing 72 modules in parallel and having 64 data bits plus 8 check bits read in or out simultaneously. This had the virtue that all faults up to and including failure of one of the driving coils supplying the rotating field would affect only one bit of the memory word. For this organization Hamming encoding was appropriate. The main drawback, for certain environments like outer space, was the power consumption.

The second competing design was a bit-per-chip organization which used 16 bits per chip to form the data word. In this design not all the bubbles had to move at the same time and hence power consumption was cut. On the other hand any malfunction on the module which affected more than one chip could no longer be corrected by a simple single error correction procedure. This drawback was compensated for by employing a 16-bit-adjacent group error correcting code[2] which corrects up to 16 simultaneously occurring bit errors in the group corresponding to a module. This second design required a greater amount of check bits and this increased the weight and volume. The final choice between the two designs depended on the results of the reliability modeling.

To identify the failure modes one had to examine all the elements composing the entire system and postulate failures at every location. On the basis of past experience, certain ones of all these potential faults could be eliminated. In the case of these particular memories the flip chip joints could be ignored in the reliability equations. In the following table of the bit-per-chip design the failure rate ratios were estimated from the geometry and complexity of the various magnetic elements.

Magnetics

Failure Rate Ratio	Failing Element	Numbers/chip
1	cell	128 x 800
10	bubble generator	128
10	bubble splitter	128
10	bubble annihilator	256
128	fan-in	8

All these affect a shift register and can cause therein a signal error.

Chip Electronics

Failure Rate Equivalent	Failing Element	Number/tetrad
1 discrete transistor	preamplifier	1
20 discrete transistor	sense amplifier	1
1 discrete transistor	write loop	1

The chips were arranged in groups of four that were oriented differently with respect to the rotating magnetic field. All four chips in a tetrad could be serviced by the same preamplifier, sense amplifier, and write loop. A failure in any of these three elements would affect all four chips.

Module Electronics

Failure Rate Equivalent	Failing Element	Number/module
1 discrete transistor & 1 transformer	clear & decode loops	8, 8
3 discrete transistors 1 transformer	field coils	2, 2

A failure in any of these elements affects the whole module. The failures were categorized according to their effects, since when the effects are the same the failure rates can be lumped in the reliability equations. Similar tables exist for the bit-per-module design.

Having identified the various failure modes the next step was to establish (or assume) a failure distribution. Many of the elements involved have an infant mortality increment on top of a constant failure rate. Since this is of short duration one could reasonably ignore it in the reliability equations. Similarly one could ignore the wear-out increase in the failure rate by assuming that it comes after the useful product life. Thus one was able to justify the use of constant failure rates, which resulted in exponential reliability equations for simplex organisms.

There still remained the problem of assigning numerical values to the failure rates. For the electronic elements this was relatively easy since field data was available. For the magnetics of bubbles, however, no data existed nor was it forthcoming in the foreseeable future. Therefore, it was decided to make a parameter study to help decide which of the two competing designs was preferable. Three values for individual cell failure rates were chosen. The middle value was set to that value which would yield the same failure rate for all the cells as for the rest of the memory system. Then this failure rate of 5×10^{-12} per hour was multiplied and divided by ten to get a reasonable range of values to use in the parameter study. At this point the implications of each type of element failure was reassessed and a discovery of an improper design was made. At a previous point in the design process an impedance mismatch occurred which was readily solved by designing 4 field coils to operate in series, and similarly with several clear and decoder loops. With such a design a driver failure could cause multiple group errors to occur which would not be correctable by the encoding employed. Upon discovering this fact the design was changed to employ pulse transformers to overcome the impedance mismatch. This incident illustrates the principal that the reliability modeling should coincide with the design as part of the design process.

It was now time to formulate the reliability equations.

Define the following:

$N = 6$ is the number of modules
$M = 16$ is the number of chips/module
$L = 128$ is the number of shift registers/chip
$O = 800$ is the number of cells/shift register

In the bit-per-chip design 16 bits came from one module to form one quarter of a data word, 4 modules were required for the whole data word and 2 modules for the check bits. The encoding used will correct any one of the $2^{16}-1$ possible error patterns coming from one module and also detect double errors coming from separate modules.

If either a module or a chip fails the error correction capability handles it, but any two distinct errors may only be detected. On the other hand a multiplicity of failures in shift registers can often occur without causing a double error because the likelihood is that they will occur in different words. Therefore, one can lump the failure rates of the module and the chip as follows:

$$\lambda_b = \lambda_{module} + M \times \lambda_{chip}$$

and call it a block failure rate.

Also let the shift register failure rate be

$$\lambda_s = \lambda_{cell} \times M \times O$$

Then define for mission time T

$$B = e^{-\lambda_b T} \quad \text{and} \quad S = e^{-\lambda_s T}$$

The memory reliability equation then is

$$R_{mem} = N(1-B) \; B^{N-1} S^{L(N-1)} +$$

$$B^N \left\{ S^{LN} + \binom{LN}{1}(1-S)^1 S^{LN-1} H_1 + \binom{LN}{2}(1-S)^2 S^{LN-2} H_2 + \ldots \right.$$

$$\left. + \binom{LN}{j}(1-S)^j S^{LN-j} H_j + \ldots \right\} \quad (1)$$

The first term of this equation consists of four factors; the number of ways one module could fail, the probability that one does fail, the probability that the remaining N-1 modules survive, and the probability that all of the shift registers in these (N-1) surviving modules also survive.

The second more complicated term represents the case where there are no failures due to λ_b but there may be shift register failures. Within the brackets the first term, S^{NL}, is the probability of no shift register failures. The second term consists of four factors. The first factor is a binomial coefficient representing the number of ways a single shift register can fail. The second factor is the probability of one failing. The third factor is the probability that the remaining NL-1 shift registers survive, and H_1 is the probability that the error correction circuitry can correct the errors resulting from the failure of the shift register. H_1 is of course unity since only one error can be caused by a single shift register failure and the encoding provides for single error correction. H_j for $j > 1$ is less than unity. The values of H_j up to $j = 14$ were calculated as described in appendix E of reference 1. The procedure employed was essentially the simulation of the following problem, expressed in classical terminology.

Given, an urn containing N x L balls, each ball having one of N colors and labelled from 1 to L. Removing j balls (with replacement), what is the probability that no two balls of different color have the same number? This corresponds to the physical case of: Given multiple shift register failures. What is the probability that they didn't occur in different modules at the same location, and hence that their associated errors will be correctible through single error correction encoding?

For the bit-per-module design an equation similar to Eq. (1) was also formulated. Then comparative computer runs were made to see the differences between the two competing designs.

In the following table the reliability of the error correction circuitry was included in calculating the mission time in months that the various magnetic domain bubble memory systems would have the indicated reliability. The three values in each triad correspond to λ_{cell} values of 5×10^{-13}, 5×10^{-12}, and 5×10^{-11} per hour

Mission time in Months for progressively increased Reliability.

System Reliability	Simplex Design	Bit/Module Design	Bit/Chip Design
0.5	7.79 1.53 0.17	11.9 1.51 0.15	42.4 19.9 6.62
0.6	5.88 1.13 0.12	9.07 1.11 0.11	34.4 16.4 5.47
0.7	4.11 0.79 0.09	6.56 0.78 0.08	26.9 13.1 4.39
0.8	2.57 0.49 0.05	4.26 0.48 0.05	19.2 9.75 3.30
0.85	1.87 0.36 0.04	3.17 0.35 0.03	15.2 7.98 2.73
0.9	1.21 0.23 0.02	2.1 0.23 0.02	10.9 6.04 2.11
0.95	0.59 0.11 0.01	1.05 0.11 0.01	6.01 3.73 1.38
0.99	0.11 0.02 --	0.21 0.02 --	1.36 1.10 0.52

Three conclusions are readily drawn from this last table. First, unless the λ_{cell} failure rate is less than a certain threshold value of approximately 10^{-12}/hr, no improvement occurs in going from a simplex design to the bit-per-module design that employs Hamming error correction encoding. The extra circuitry necessary to do single error correction plus the extra modules necessary for the check bits can balance out the advantages gained by having the single error correction capability. Second, the bit-per-chip design is more reliable in all cases than the bit-per-module design. Third, the bit-per-chip design is less vulnerable to reliability degradation as the value of λ_{cell} increases. At a reliability of .9 the mission time ratio in one case is 2.1 ÷ .02 or approximately 100 while in the other design this same ratio of degradation is 10.9 ÷ 2.11 or only 5. As mentioned before, this reliability improvement only came at a price in weight, size, volume, cost, etc. Nevertheless the trade-off is possible, and for certain applications is worth making.

In summary, the magnetic domain bubble memory system discussed here was designed for a specific application and maintenance environment with reliability modeling taking place concurrently with the rest of the design process, and not as an after thought for a possible retrofit. The results justified the extra effort involved.

1. G. S. Almasi, W. G. Bouricius, B. J. Canavello, W. C. Carter, F. A. Giess, R. J. Hendel, R. E. Houstmann, T. F. Jamba, D. C. Jessep, G. E. Keefe, F. F. Lee, P. N. Oliver, J. V. Powers, and L. L. Rosier, IBM Research Report, RC 3700, 1972.

2. W. G. Bouricius, W. C. Carter, E. P. Hsieh, D. C. Jessep Jr., and A. B. Wadia. IEEE Transactions on Computers Vol. C-22, pp. 269-275, 1973.

RELIABILITY OF MAGNETIC BUBBLE DOMAIN MEMORIES

D. H. Baird, C. F. Buhrer and J. J. Vytal
GTE Laboratories, Inc., Waltham, Mass. 02154

J. L. Archer, L. R. Toccio and O. D. Bohning
Rockwell International, Research Division, Anaheim,
Ca. 92803

ABSTRACT

As a test vehicle in evaluating bubble memory reliability, we have adopted a module consisting of 28 1100-bit shift registers operating in parallel at 66 kHz. Factors governing reliability include those defined by the shift register itself and those which depend also on its module environment. The register itself is characterized by the integrity of bubble propagation, and by the detector parameters of bubble signal level, noise level, and net 2ω signal resulting from imperfect cancellation between detector and "dummy" elements. With a register mounted on the memory plane, the magnitude of the induced $d\Phi/dt$ signal is also determined. With many registers mounted in a common drive and bias field, margin matching of the registers additionally affects overall reliability.

The discussion will draw on experience to date with the module described above. Characteristics of the individual shift registers will be presented, with emphasis on operating margins, and on signal and noise levels of the thin film Chinese character detector used. Comparison will be made with similar data for thick and thin film chevron detectors. Error rate data obtained for registers operated singly in the module will be discussed in terms of the measured register characteristics and compared with information on module performance with more than one register in simultaneous operation.

LONG-TERM PROPAGATION STUDIES IN
MAGNETIC-BUBBLE DEVICES

by

P. W. Shumate, P. C. Michaelis and R. J. Peirce
Bell Laboratories, Murray Hill, N. J. 07974

ABSTRACT

Extensive testing of magnetic-bubble circuits has been done where bubbles are propagated at frequencies up to 127 kHz and the long-term retention of data examined. From tests performed on small circuits in a microscope environment, mean-time-to-failure (MTTF) data have been taken. The effects on MTTF of garnet material parameters, ambient temperature, circuit design, ion-implantation dosage, circuit defects and operating frequency have been examined. A high-mobility material operated on a circuit modified as a result of these tests displayed near-infinite MTTF over a temperature range of at least 28°C-80°C. Tests were also performed on large circuits in a memory-module environment. The results of long-term ($\sim 10^{15}$ bubble steps) error-free operations show that very large MTTF's are realizable in actual devices. A method for estimating reliability from these data is described and possible failure mechanisms for propagated bubbles are discussed in view of the new data.

INTRODUCTION

Recent investigations have shown that the propagation of magnetic bubble domains on a permalloy circuit in a rotating magnetic field does not necessarily take place with 100% reliability.[1,2] During propagation, permanent modifications of the bubble data pattern occasionally take place through loss (collapse) of a bubble, self-replication of a bubble, or a one- or two-period displacement of a bubble relative to the rotating field. A behavior was observed which suggested that for any given number of steps N a probability $P(N) \leq 1$ exists that a successful propagation will occur. This probability is a complicated function of the combination of bubble material, the circuit design and the test parameters as well as of N. More extensive testing of this behavior has now been done and the results will be described here.

The data given in TEST CIRCUIT RESULTS were obtained from failure measurements made in a microscope environment on reentrant test circuits of ≤ 55 steps. The measurement technique and the tests made to examine its limitations are described in Ref. 1. Other data reported in MODULE RESULTS were obtained from 20 kb-capacity bubble circuits in a memory-module environment.[2,3] These module tests display a failure behavior similar to that observed in the small test circuits. However, because many functions other than propagation (e.g., generation, transfer and detection) are tested in the module, the numerical failure data are expected to be different from those of the microscope tests. The numerical data do indicate that the low-failure-rate predictions one can make for the large circuits based on the small-test-circuit results are realistic. How one can make these predictions is described in the DISCUSSION.

TEST-CIRCUIT RESULTS

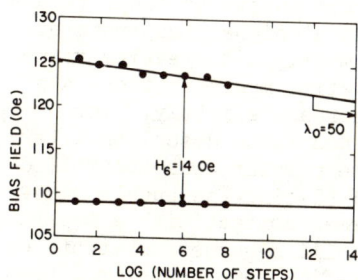

Fig. 1-Bias-field margins plotted vs. logarithm of number of steps propagated for material A operated on the circuit of Fig. 2a.

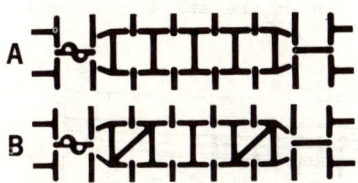

Fig. 2-(a) 11-step circuit described in text. (b) The circuit in (a) modified to smooth turns. A 23-step version of this circuit was also used.

The lines in Fig. 1 show failure data obtained using $Y_{2.6}Sm_{0.4}Fe_{3.8}Ga_{1.2}O_{12}$ garnet[4] (hereafter material A) in conjunction with a 29-μm-period test circuit. Parameters for material A are listed in Table I; the garnet was ion implanted with 2×10^{16} H+/cm^2 at 25 keV to suppress hard bubbles.[5] The circuit is a small-scale version of a design used for the 20-kb memory circuits and is shown in Fig. 2a. The testing parameters used here, and in all other tests unless otherwise specified, were: circuit-to-garnet spacing = 1.6 μm, drive field = 25 Oe at 100 kHz, and operating temperature = 25°C. In principle, the lines in Fig. 1 represent 50% failure loci; that is, if the circuit were operated for a number of steps N at a bias field H and the point (N,H) fell on the line, the probability of failure would be 50%. (The accuracy of this definition of the line is limited by experimental error.) If the point (N,H) falls inside (outside) the region bounded by the lines, the probability of failure is lower (higher). Therefore the lines approximate mean-time-to-failure (MTTF) for operation at any value of bias field. (More precisely, the 50%-failure loci represent the median times to failure MEDTTF, a parameter less commonly used than MTTF. MTTF differs from MEDTTF by ln2 if the failure mechanism is random--see DISCUSSION.)

Experience has shown that the appearance of the material-A data in Fig. 1 is typical for many materials and circuits. Specifically, the decrease in bias margin with increasing number of steps N is generally observed to be logarithmic, at least to 10^{11} steps. The best MTTF characteristic clearly is one where the bias margin shows no change with N and the margin itself is as wide as possible.

For comparing data such as these when experimental conditions are changed, it is convenient to introduce two parameters labeled H_6 and λ_0. The field H_6 in Oe is the width of the margin as recorded for 10^6 steps of propagation. The number λ_0 is the common logarithm of the extrapolated value of N where the bias margin vanishes. Therefore the best MTTF characteristic now can be described as having H_6 as large as possible and $\lambda_0 = \infty$. For the data in Fig. 1, these values are 14 Oe and 50, respectively.

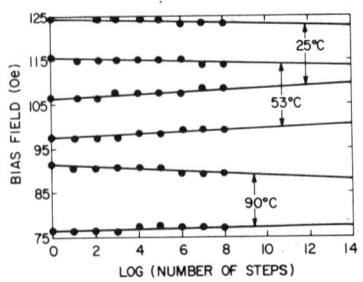

Fig. 3-MTTF data for material A operated on the circuit of Fig. 2a as a function of temperature.

Figure 3 shows the temperature dependence of MTTF data for a different sample of material A. It is seen that the margin shifts to lower values of bias field at higher temperatures (reflecting mostly a decreasing value of the saturation magnetization) and H_6 and λ_0 decrease slightly. For these tests and those described later, the sample temperature was controlled to ±0.5°C by placing the bubble material in thermal contact with a separate gadolinium-gallium-garnet substrate on which a thin-film SnO_2 heater had been chemically deposited.[6]

The bubble-material parameters influence the MTTF. For example, Fig. 4 shows MTTF data as a function of temperature for $Y_{1.6}Lu_{0.3}Eu_{0.1}CaFe_4GeO_{12}$ garnet[7] (hereafter material B). The properties of this high-mobility material are listed in Table I. It is seen from data taken out to 10^8 steps that λ_0 is infinite and H_6 is large (~15 Oe) over the temperature range 28°C-80°C.

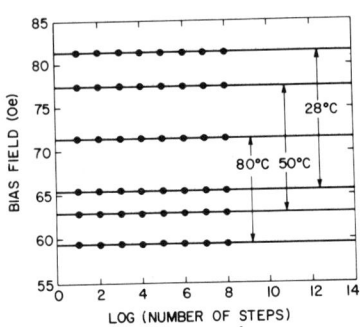

Fig. 4-MTTF data for material B operated on the circuit of Fig. 2a as a function of temperature.

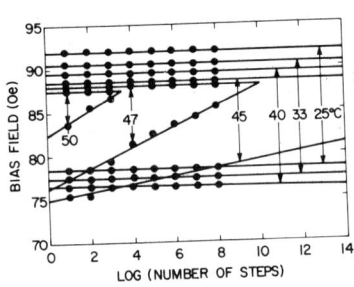

Fig. 5-MTTF data for material C operated on the circuit of Fig. 2a as a function of temperature

Figure 5 shows these data for a material similar to B but with a larger moment and smaller anisotropy field and coercivity. This material is $Y_{1.9}Sm_{0.1}CaFe_4GeO_{12}$ garnet[8] (hereafter material C) and its properties are also listed in Table I. Here the 25°C margin is wide and λ_0 is infinite but, at higher temperatures, the MTTF is short and is unsuitable for use in bubble devices.

143

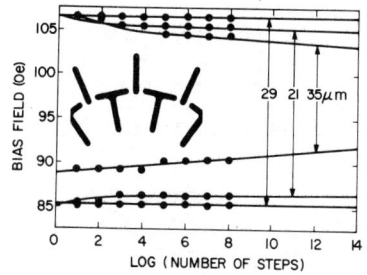

Fig. 6-MTTF data for material A operated on the circuit inset in this figure for three different circuit periods.

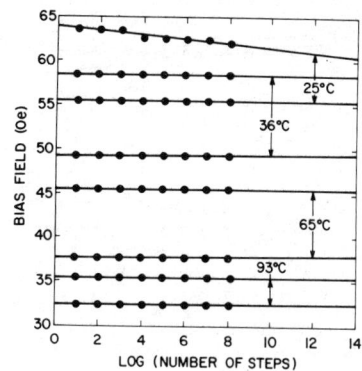

Fig. 7-MTTF data for material D (orthoferrite) operated on a circuit like Fig. 6 as a function of temperature.

The period τ of the permalloy can be optimized relative to the bubble diameter d to increase the MTTF. Figure 6 shows data for material A operated with three different circuit periods. T-Bar elements were arranged in a circular loop, a segment of which is shown in Fig. 6. It is evident that a period τ of 29 μm "tunes in" the best MTTF characteristic (H_6 = 21 Oe, $\lambda_0 = \infty$). The relevant parameters with which the 29 μm period can be compared are material length ℓ = 0.60 μm and epitaxial garnet thickness h = 6 μm.

A crystal of $Sm_{0.55}Tb_{0.45}FeO_3$ (hereafter material D), for which the properties are listed in Table I, was tested similarly but on a larger version of the circuit in Fig. 6 (τ = 178 μm). As seen in Fig. 7, at 25°C this orthoferrite displayed a failure characteristic similar to those seen for the garnets. As the temperature was first raised, the characteristic became better as H_6 opened to 9.2 Oe and λ_0 became infinite by 36°C. At higher temperatures λ_0 remained infinite although H_6 became smaller. In fact H_6 became smaller about twice as rapdily as a function of temperature as did, say, the lower side of the bias margin. This means that both the absolute and the relative margins became worse at temperatures higher than about 36°C.

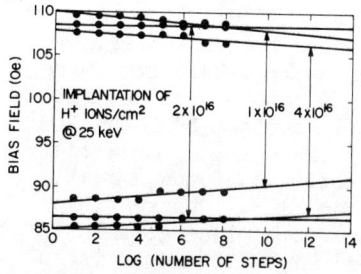

Fig. 8-MTTF data for material A operated on the circuit of Fig. 6 as a function H^+ ion implantation dosage.

MTTF is sensitive to ion implantation dosage. Figure 8 shows data for material A implanted with 0.5×, 1×, and 2× the "normal" dosage of 2×10^{16} H^+/cm^2 at 25 keV. The test circuit again is that shown in Fig. 6 with a period of 29 μm. One sees that the normal dosage gives the best MTTF characteristic (H_6 = 22 Oe and $\lambda_0 = \infty$). Figure 9 shows similar data for the same material but where Ne^+ implantation was used instead of H^+ implantation. Here the optimum dosage, now at 150 keV, is 2×10^{14} Ne^+/cm^2, giving

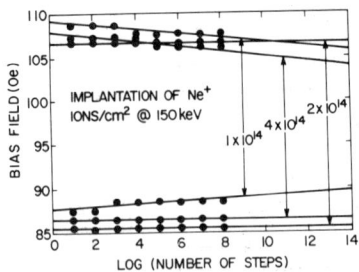

Fig. 9—MTTF data for material A operated on the circuit of Fig. 6 as a function of Ne+ ion implantation dosage.

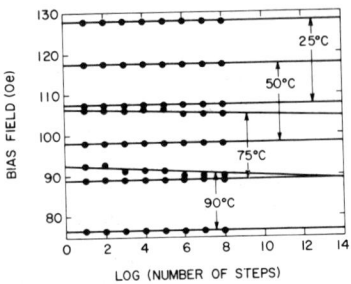

Fig. 10—MTTF data for material A operated on the circuit of Fig. 2b as a function of temperature.

H_6 = 21 Oe and λ_0 = ∞ for this sample.

The MTTF testing procedure is also useful for evaluating circuit design modifications. First it was noticed that λ_0 was increased by using more circuit elements (periods) for turning corners. For example, a smooth circle of T-Bars, as in Fig. 6, shows λ_0 = ∞ reliably at 25 Oe drive and 1.5 μm spacing. With the idea in mind of smoothing out sudden direction changes in 180° turns in the loop shown in Fig. 2a, the circuit of Fig. 2b was developed by J. L. Smith and P. I. Bonyhard. The idea was to alter the pole strength at such points in the turn as to smooth out the average field gradient in which the bubble moves. Indeed, slow-speed observations of bubbles propagating in the circuit of Fig. 2b show that the bubble moves very smoothly through the turns. For material A tested with 25 Oe drive at 100 kHz and a spacing of 1.5 μm, the circuit of Fig. 2b shows H_6 = 21 Oe and λ_0 = ∞. Figure 10 shows the temperature dependence for this circuit driven at 25 Oe and using material A.

Even these optimistic results can be misleading, however. In the circuits shown in Fig. 2 the conductor metallization for the transfer gate (l.h. side of photomicrographs) was omitted. Only the permalloy part of the circuit was tested. When the Al-Cu conductor was placed between the modified circuit and the garnet, the MTTF decreased (H_6 = 20 Oe and λ_0 = 80) but was restored again (H_6 = 20.5 Oe and λ_0 = ∞) when the drive field was increased to 30 Oe.

Certain types of fabrication problems in the permalloy circuit show up readily in MTTF tests. Figure 11 shows the characteristic for each of two loops similar to that shown in Fig. 2a. They were adjacent loops, manufactured simultaneously from the same metallization. The good loop shows the characteristic typical of those obtained using material A with this circuit (H_6 = 17 Oe, λ_0 = 68). The other loop failed at an element where there was found a small spurious piece of permalloy (see inset in Fig. 11). Observations made at 1 Hz verified the fact that failures took place at this defect site. For the defective loop, H_6 is only 4 Oe and λ_0 is 10. It should be pointed out that other types of defects may show up as a loss at either side of the margin or in a diminished H_6 with an infinite or near-infinite λ_0.

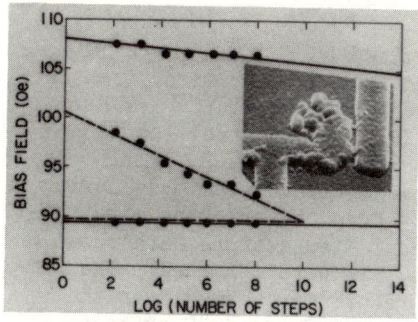

Fig. 11-MTTF data for material A operated on a defective circuit of the type shown in Fig. 2a. An SEM photomicrograph (1000x) of the defect is inset in the figure.

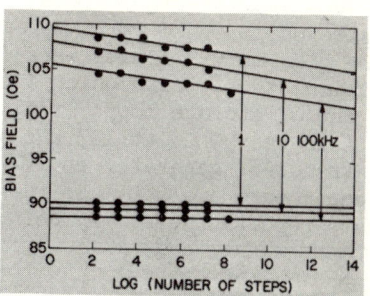

Fig. 12-MTTF data for material A operated on the circuit of Fig. 2a as a function of frequency.

MTTF can be dependent upon the frequency of propagation. The characteristic for material A operated on a circuit like that in Fig. 2a is improved as the frequency is lowered from 100 kHz, as seen in Fig. 12. When material C is similarly tested at 25°C, where already $\lambda_0 = \infty$ at 100 kHz, no change in the characteristic is seen. At 47°C, however, where $H_6 = 4.5$ Oe and $\lambda_0 = 10$, significant improvement is seen at lower frequencies. These data are shown

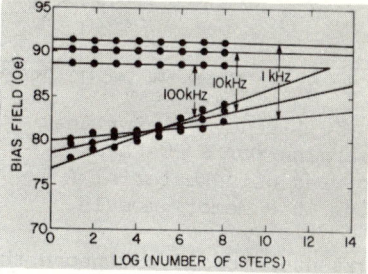

Fig. 13-MTTF data for material C operated at 47°C on the circuit of Fig. 2a as a function of frequency.

in Fig. 13 where, at 10 kHz, H_6 and λ_0 are increased to 8.5 Oe and 24, respectively, at 1 kHz H_6 and λ_0 are further increased to 9.5 Oe and 44, respectively.

MODULE RESULTS

The 20-kb-capacity circuits were tested with material A as follows. Two chips were mounted in adjacent positions on the same substrate with their detectors connected as adjacent arms of the same bridge circuit. The other two arms of the bridge were independent current sources. All observations of circuit failure, either hard (bubble information loss) or soft (read error), were made by electronic means using a specially designed bubble-memory exerciser. This exerciser operated the module in a read or a write mode, kept track of the bubble detector outputs and compared them with the original input patterns. Bubble circuit failure was defined as any error; i.e., read error, bubble collapse (or loss), or bubble strip-out. Read errors occurred when unexpanded or partially expanded bubbles passed through the chevron detector.[9] Therefore this condition was taken to be the upper margin limit although two or more Oe of additional margin existed before the stored information actually was destroyed.

The bias-margin limits were determined by short-term (2×10^6 propagation steps) tests which included bubble generation, transfer in and out of storage loops, replication and annihilation. The drive field was 29 Oe at 127 kHz and all of the control functions (i.e., transfer, generate, etc.) were optimized at the normal operating temperature of 36°C. No further adjustments were made during the tests reported here. By using this type of test, the margin vs. module-temperature data shown in Fig. 14 were obtained. The temperature was varied over the range -7°C to +80°C by means of thermoelectric devices in contact with the module. The temperature monitored was that of the substrate supporting the chips; the thermocouple was not in direct contact with either of the garnet chips. The dashed line in the figure indicates the permanent-magnet bias field variation with temperature. The bias-field margins were probed by an externally applied field which was used to aid or oppose the field from the permanent magnet.

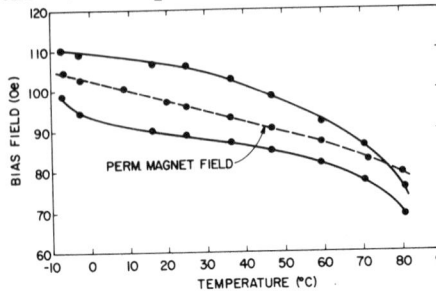

Fig. 14-Full read-cycle/write-cycle margins for a 20-kb chip manufactured using material A operating in a memory module.

All of the data for long-term propagation tests were obtained using a 127 kHz, 29 Oe drive field and the normal operating temperature of 36°C. Three long-term read-cycle tests were performed, each lasting 186+ hours which is the equivalent of 9×10^{10} field cycles. The first test was of a fully populated chip, $\sim 2\times10^4$ bubbles, with the bias field set 3.0 Oe above the lower bias margin. This test resulted in 10^{10} bubble output detections without an error for a total of 1.8×10^{15} bit-steps. This test was repeated with an information pattern consisting of 50% fully populated storage loops and 50% empty loops. The bias field here was 1.5 Oe above the lower bias margin. No errors were observed during the test. The third test consisted of a 50% populated chip also; however, this time the bias was set 1.5 Oe above the lower bias margin and a slowly varying (0.01 Hz) external bias field of ±1 Oe with triangular waveshape was applied. The purpose of this test was to determine if circuit performance could be degraded by a nonstatic bias field. No errors were observed during this test period either.

The lower half of the bias margin was chosen as the module test region because, by operating close to the lower margin limit, the largest operating temperature range is obtainable.

With the permanent-magnet bias field set as shown in Fig. 14, the temperature was lowered from +75°C to -7°C in approximately 40 minutes during a continuous read-cycle test without error.

DISCUSSION

The MTTF measurements have supplied a new criterion for evaluating materials and circuit designs. The measurements frequently differentiate among several available choices. In addition, the MTTF plots can be used to obtain estimates of error rates at any operating point.

For making error-rate estimates, it will be assumed that the ordinary statistical model[10] for a random failure mechanism can be applied. Indeed, detailed examination of the experimental data ($P(N,H)$ as a function of N and behavior of the standard deviations) indicates that the failure mechanism is random. Therefore it is assumed that the failure or hazard rate $h(H,\omega)$ at the bias field H and propagation frequency ω is constant and estimable from the data as

$$h(H,\omega) \equiv \frac{1}{MTTF} \approx \frac{\ln 2}{N_o(H)} \frac{\omega}{2\pi} \quad . \qquad (1)$$

That is, on the average the first failure occurs in propagating $N_o/\ln 2$ steps (in a time $2\pi N_o/\omega \ln 2$); $N_o(H)$ is the number of steps corresponding to 50% failure at H from the lines drawn through the experimental data points. The factor $\ln 2$ takes account of the fact that the $N_o(H)$ data, in principle, measure median time to failure rather than mean time to failure.

If a case such as that shown in Fig. 1 is considered, where the lower side of the margin is constant, then empirically for the top margin

$$N_o(H) = 10^{[(H_1-H)/m]} . \qquad (2)$$

Here H_1 is the y-intercept (N=1) of the collapse side of the margin and m is the absolute value of the slope of this boundary.

By further use of the model for random failures, the probability of failure-free propagation in moving any number of steps N at a bias H is

$$P(N,H) = \exp\left[-\frac{t}{MTTF}\right] = \exp\left[-\frac{N \ln 2}{N_o(H)}\right] . \qquad (3)$$

At the field H which corresponds to the intersection of the two sides of the margin (at $\log N = \lambda_0$), the failure rate is equally divided between possibly two mechanisms - one for the high-field failures and one for the low-field failures. Away from this point, however, only the dominant mechanism need be considered in reliability calculations. This is apparent because of the exponential dependence of $P(N,H)$. A factor of 10 in the number of steps attenuates the influence of one mechanism or the other by a factor of $e^{10} \sim 22,000$. When the low-field failures predominate, (2) is replaced by

$$N_o(H) = 10^{[(H-H_1)/m]} \qquad (4)$$

where H_1 and m are the intercept and slope, respectively, of the lower margin.

Now an example of the use of MTTF as stated in (1) will be considered. Equation (1) is implicitly related to all test parameters (e.g., temperature, circuit period, bubble material, drive field and frequency, as well as the circuit through $N_o(H)$). Suppose that all test parameters and the "character" of the circuit (design, period, processing) were to be kept the same but that the number of bits in the circuit were to be changed. That is, a failure prediction is required for a circuit with a capacity of B' bits (or total steps) based on the results obtained experimentally for a test circuit of different bit capacity B. Then it is reasonable to scale $h(H,\omega)$ in (1) and N in (3) by the ratio B'/B based upon the assumption that the failure probability for 1 bit moving N steps is the same as that for N bits moving 1 step. This assumption is reasonable since the product of the bits and the steps per bit is the total number of times the identical bubbles are exposed to the random failure mechanism. If the failure mechanism is <u>not random</u> then this assumption as well as those made for writing (1) and (3) are no longer valid. The capacity B of the test circuit did not appear in (1) because the test circuit was considered to be one unit, the entire unit failing as the result of failure of one or more of the B bits.[1]

Now (1) and (2) can be used to find a minimum acceptable value of λ_0. Suppose that data are taken for a 23-step circuit and that the extrapolation is to be made for a 1 Mb-capacity circuit which must operate at 100 kHz with an MTTF of 100 years (3.2×10^9 s). Thus

$$3.2\times 10^9 \text{ s} = (23/10^6)(N_0/10^5 \text{s}^{-1} \cdot \ln 2) \qquad (5)$$

from which $N_0 = 10^{19}$. This value of N_0 requires that the point (10^{19}, H) must fall on or within the collapse side of the margin for the 23-step circuit, where H is the bias field to be used for the operation of the large circuit. Next suppose that it is decided that this bias should be centered within the margins as measured for 10^6 steps; i.e., $H - H_{so} = 0.5 H_6$ if H_{so} is the low-field side of the margin for 10^6 steps. From congruent triangles constructed on the MTTF characteristic,

$$(\lambda_0 - \log 10^6)/H_6 = (\lambda_0 - \log 10^{19})/0.5 H_6 \qquad (6)$$

from which it is found that λ_0 must be greater than or equal to 32, regardless of the value of H_6. For the sake of reference, recall that the mass-memory circuit data in Fig. 1 had $\lambda_0 = 50$.

Something will now be said about possible failure mechanisms. Failures are not due to some simple mechanism like, for example, sample heating in the presence of a rotating field. If such were the case, thermal equilibrium would be reached in times far shorter than the tens of hours over which the logarithmic variation of the margin loss is observed.

When failure occurs at the high-field side of the margin, apparent collapse of a bubble or, much less frequently, a one- or two-period displacement of the bubble relative to the rotating field is observed. A mobility-related mechanism or a dynamic-conversion process may account for this. Failure at the low-field side of the margin is always by bubble strip out; i.e., a bubble becomes an elongated domain spread over more than one circuit period. A failure mechanism related to bubble-stability parameters may be the candidate here.

By comparing Figs. 3 and 4 it is obvious that the higher-mobility garnet B performed better than material A. This also applies to high-mobility material C (Fig. 5) below 40°C. The difference is apparent both in larger values of H_6 and in $\lambda_0 = \infty$. It is tempting to suggest that a failure mechanism exists which becomes more active as the phase lag between the bubble and the drive field increases, at which time the excess bias from the drive gradient is more likely to collapse the bubble. The frequency dependences of materials A (Fig. 12) and C (Fig. 13) bear out the excess-bias point of view to the extent that more bias range is available at the collapse side as the frequency is lowered. For material A, however, the slope of the collapse characteristic is unchanged at lower frequencies, thus indicating a failure mechanism still persists. (Note that the systematic shift in the low side of the Fig. 12 data could, indeed, be a heating effect and proportional to frequency.)

Dynamic conversion may account for the apparent collapse of bubbles in some cases. This is a process[11-13] in which a normal bubble's spin system is converted to a less-mobile configuration as it moves rapidly through a field gradient. The converted bubble may move erratically, leaving the propagation circuit and appearing in the tests reported here as if it had collapsed.

The othoferrite data (Fig. 7) imply that another collapse mechanism may be present. For this material, at 25°C, 200 kHz should be the limiting frequency of operation,[14] so substantial phase lags exist at 100 kHz. However, the bubble diameter increases with temperature and, unless the mobility increases faster to offset this change, the failures should be more evident at higher temperatures. This follows since the limiting frequency varies inversely with bubble diameter and thus is less than 200 kHz at temperatures above 25°C. Yet λ_0 becomes infinite at higher temperatures. As for hard bubbles (which move at an angle to the drive gradient[15]) or dynamically converted bubbles, neither has ever been observed in orthoferrites even in the highest speed tests.

Collapse failure mechanisms must also be consistent with the observed facts that λ_0 can be made infinite by increasing the drive field from 25 Oe to 30 Oe in one case or by tuning the circuit period to the material (Fig. 6). Also ion-implantation dosage (Figs. 8 and 9) affects the collapse margin (as well as the strip out margin). That an optimum dosage exists has been recognized,[16] and it may be that the long-duration experiments here are merely testing for the presence of occasional hard bubbles in a very sensitive manner.

At temperatures above 40°C, the poor performance of material C relative to B may possibly be related to bubble stability. Coercivity provides a stabilizing force for bubble domains while q (= anisotropy field/magnetization) is a measure of the stiffness of a bubble. At 50°C the coercivity and q of material C are 0.06 Oe and 3.2, respectively. The coercivity is an insignificant value while the q is low for most bubble materials. In contrast, the coercivity and q at 50°C for material C are 0.35 Oe and 6.3. This coercivity in not insignificant and q is a more typical value for bubble materials. Thus material B may be less susceptible to deformation in the propagate field gradients, hence no loss of margin at the strip-out side.

None of the above suggestions points to any obvious explanation for the frequency dependence of the strip-out side of material C in the range 40-50°C. It is necessary that further work be done to isolate and identify all of the failure mechanisms.

Finally, the memory-module results using material A are seen to be reasonably consistent with the tests made on small circuits. The module error rate does not increase over a very wide temperature range for the moderate propagation times examined. This agrees with the small-circuit results for material A. A change of bias field with temperature is required, of course, to track the margins. This is accomplished by proper choice of the bubble material and design of the module bias magnet.[7,17]

It is concluded that MTTF testing such as has been described here is a necessary and useful tool for designing and testing magnetic-bubble devices. Through its application, materials have been identified and circuits designed which display near-infinite MTTF's.

The authors wish to acknowledge the valuable contributions of F. B. Hagedorn, P. I. Bonyhard and J. L. Smith regarding the testing and circuit design. Thanks also go to D. E. Kish and W. J. Richards for constructing equipment for testing the memory module and to J. P. Reekstin for supplying the SEM photomicrograph in Fig. 11.

TABLE I: Properties of the materials used in these experiments as measured at 25°C.

	Magnetization (G)	Anisotropy Field (Oe)	Coercivity (Oe)	Material Length (μm)	Mobility (cm/s/Oe)	Strip Width (μm)	Thickness (μm)	q (Anis. Field / Magn.)
A	170	1540	0.20	0.60	250	5.5	6.0	9.1
B	140	1010	0.65	0.84	1400	7.4	7.4	7.2
C	165	700	0.45	0.62	1500	5.6	5.8	4.2
D	105	3200	0.05	4.0	1050	30.0	25.0	30.5

REFERENCES

1. P. W. Shumate and R. J. Peirce, Appl. Phys. Lett. <u>23</u>, 204 (1973).
2. P. C. Michaelis and P. I. Bonyhard, IEEE Trans. Magn. <u>MAG-9</u>, 436 (1973).
3. P. I. Bonyhard, J. E. Geusic, A. H. Bobeck, Y. S. Chen, P. C. Michaelis and J. L. Smith, IEEE Trans. Magn. <u>MAG-9</u>, 433 (1973).
4. S. L. Blank, B. S. Hewitt, L. K. Shick and J. W. Nielsen, AIP Conf. Proc. <u>10</u>, 256 (1973).
5. R. Wolfe and J. C. North, Bell Syst. Tech. J. <u>51</u>, 1436 (1972).
6. D. H. Smith, (unpublished).
7. W. A. Bonner, J. E. Geusic, D. H. Smith, L. G. Van Uitert and G. P. Vella-Coleiro, Mat. Res. Bull., October 1973; J. E. Geusic, W. A. Bonner, D. H. Smith, L. G. Van Uitert and G. P. Vella-Coleiro, Paper 1B-4 at this Conference.
8. J. W. Nielsen, S. L. Blank, D. H. Smith, G. P. Vella-Coleiro, F. B. Hagedorn, R. L. Barns and W. A. Biolsi, (unpublished).
9. W. Strauss, A. H. Bobeck and F. J. Ciak, AIP Conf. Proc. <u>10</u>, 202 (1973).
10. See, for example, A. M. Polovko, <u>Fundamentals of Reliability Theory</u> (Academic Press, New York, 1968).
11. G. P. Vella-Coleiro, F. B. Hagedorn, Y. S. Chen and S. L. Blank, Appl. Phys. Lett. <u>22</u>, 324 (1973).
12. F. B. Hagedorn, Paper 5B-4 at this Conference.
13. G. P. Vella-Coleiro, Paper 5B-3 at this Conference.
14. F. B. Hagedorn, AIP Conf. Proc. <u>5</u>, 72 (1972).
15. W. J. Tabor, A. H. Bobeck, G. P. Vella-Coleiro and A. Rosencwaig, Bell Syst. Tech. J. <u>51</u>, 1427 (1972).
16. R. Wolfe, J. C. North and Y. P. Lai, Appl. Phys. Lett. <u>22</u>, 683 (1973).
17. J. E. Geusic and L. G. Van Uitert, U.S. Patent No. 3,711,841, issued 16 January 1973.

ERROR RATE MEASUREMENTS IN BUBBLE CIRCUITS ON PERMALLOY COATED YEu GARNET FILMS

W. D. Doyle, W. E. Flannery and J. A. Coleman
Sperry Univac, Blue Bell, Pa. 19422

ABSTRACT

A quantitative method for describing the performance of a bubble circuit is presented. It is shown that two parameters are required to separately specify the margin and error rate. Data on the upper margin are presented as a function of in-plane rotating field, field shape, pulse sequence, film defects, the presence of hard bubbles and frequency for a typical integrated bubble circuit on a permalloy coated YEu garnet film.

The operating margins of magnetic field accessed bubble devices are described by the dependence of the limits of the dc bias field H_B, normal to the bubble wafer, on the amplitude of the in-plane rotating field, H_R.

No common method for defining margins has gained acceptance. Recently Shumate and Peirce[2] took a long step toward clarifying the situation by defining the margin for a particular measurement as the value of H_B for which errors of any kind in the data pattern were observed 50% of the time. They showed that H_B so defined decreased as the number of bubble steps increased and they reported qualitatively on the effect of different data patterns, rotating field magnitude and the number of stop-starts.

We have carried out related measurements on a 17 position minor loop (Fig. 1) in a major-minor loop configuration. The five layer integrated circuit consisted of a polished GGG substrate wafer, a 10.6μ thick $Y_{2.4}Eu_{0.6}Ga_{1.2}Fe_{3.8}O_{12}$ film with d = 5.9μ grown by liquid phase epitaxy,[3] ~100Å of evaporated NiFe for hard bubble suppression,[4,5] 9000 Å of sputtered SiO_2 and 3000 Å of sputtered NiFe from which the circuit with bar widths = 3.0μ was chemically etched. The hard bubble suppression layer was effective in reducing the free bubble static collapse range to < 1 Oe (83.4 - 84.2 Oe) for hundreds of bubbles generated over a wide range of field amplitudes and frequencies.

For the data reported here, the sample chips were bonded, circuit side down, to an identical GGG chip and two sets of ten turn coil pairs were wound directly on the chip. These coils allowed "continuous" operation for up to two seconds at 60 Oe at 250 KHz without serious distortion. The temperature rise in this time caused a change in the free bubble collapse field of 0.2 Oe. The field configuration for x rotating field cycles was always generated by (x + ½) cycles of sinusoidal current in one coil in quadrature with x cycles of sinusoidal current in the other. All the data were taken for the counter-clockwise direction of bubble motion with the bubble rest position at the end of the long bars.

The data reported here are for the upper limit of the bias margin only because the circuits were not protected by guard rails.[6] For every measurement, the loop was filled with 17 bubbles, H_B was adjusted to an appropriate value and H_R applied with the desired amplitude, frequency, number of cycles and burst sequence. When the sequence terminated, the number of missing bubbles was recorded as the number of errors and the process repeated a total of five times for each data point. In this way, a plot of H_B versus the average number of errors E for a particular set of conditions could be made. A family of such curves can be generated by repeating the measurements for increasing number of cycles and a partial set is shown in Fig. 2. Here N is the number of bubble steps defined as the product of the number of bubbles in the full loop (17) and the number of rotating field cycles. The data is for an extreme set of conditions, i.e., 300 KHz, five cycles/burst but the results are characteristic of all the measurements made to date. The error bars on E are the rms deviations for the five determinations. The most obvious difference between different sets of data is the relative displacement with N. All errors appeared to come from bubble collapse and no evidence of relative bubble shifting was obtained[2] during separate examinations of partially filled loops.

There is a considerable amount of information in these curves but for the present discussion we limit ourselves to the value of H_N, defined as the value of H_B for E = 1.0 on the straight line drawn through the data for a particular value of N. That is, H_N is the value of H_B required to obtain an error rate (average number of errors per bubble transfer) equal to N^{-1}. The dependence of H_N on N (Fig. 3) defines bubble circuit performance in a reproducible quantitative way. Presumably Shumate and Peirce's[2] definition of H_N corresponds to E = 0.5 and would be slightly smaller.

The dependence of H_N on N for all the data obtained to date can be described by

$$H_N = H_1 - \epsilon \log N \qquad (1)$$

where H_1 is the extrapolated value of H_N for E = 1 and ϵ is the decrease in H_N per decade of N. This is also consistent with the results of Shumate and Peirce.[2]

Fig. 3 shows the dependence of H_N on N continuous bubble steps at 150 KHz for two values of H_R. Above H_R = 28 Oe, H increased slowly as generally observed for quasi-static margins. Below H_R = 28 Oe, H_N dropped rapidly and meaningful data could not be obtained. Quasi-static measurements showed no margin below 20 Oe. Changing H_R from 76 Oe to 38 Oe caused a decrease in H_1 of 1.4 Oe and an increase in ϵ from 0.5 $\pm$ 0.05 Oe/decade to 0.65 $\pm$ 0.05 Oe/decade. The results at lower frequencies were qualitatively the same.

The value of H_1 was sensitive to the number of stop-starts especially, as reported earlier,[2] when the number of stop-starts approached the number of cycles. At 100 KHz and H_R = 57 Oe. H_1 increased by 4 Oe when the pulse sequence was changed from 5 cycles/burst (11,765 stop starts) to continuous (one stop-start.)

The value of ϵ decreased from 0.3 ± 0.1 Oe/decade to 0.2 ± 0.1 Oe/decade.

It was found that the data was extremely tolerant of the time at which the field shut off. No effect was observed even if the last cycle was terminated as much as $45°$ early. Relative phase between field components was also found to be uncritical. However, H_1 was very sensitive to overshoot on the last half cycle.

The six identical minor loops on the chip were examined at 100 KHz with $H_R = 66$ Oe. The value of ϵ was the same within experimental error in all six and the value of H_1 was the same in five of the six. In one loop, however, H_1 was reduced ~ 3 Oe. The dc margins, in which the criterion for failure is much more qualitative were the same except for a difference of ~ 1 Oe at one turn. An unidentified defect of $\sim 1\mu$ in diameter was contiguous with one bar of the suspected turn but the circuits were otherwise identical.

One chip in which the thin permalloy hard bubble suppression layer was omitted was examined. The scatter in the data prevented meaningful measurements except at very low frequencies. At 2.5 KHz, ϵ was > 1 Oe/decade and H_1 was reduced several oersteds below the dc margin.

The data for N continuous bubble steps at $H_R = 57$ Oe from 50 KHz to 250 KHz is shown in Fig. 4. It was found that up to 110 KHz, H_1 and ϵ were frequency independent with values of 86.7 ± 0.3 Oe and $0.15 \pm .05$ Oe/decade respectively. Above 110 KHz, ϵ increased sharply to $0.75 \pm .06$ Oe/decade and was frequency independent from 120 KHz to 250 KHz. Above 250 KHz the field distortion was serious and so the data has not been included in Fig. 4. However, ϵ did not change, at least up to 400 KHz. H_1 decreased slowly with frequency above 150 KHz.

It is possible to describe the operation of a bubble circuit in terms of H_1 and ϵ. It is apparent that at any particular phase of H_R there is a distribution in time of the net bias field which the bubble sees and/or the bubble collapse field. A change in the position of the center of this distribution affects H_1 while a change in the width affects ϵ. If the shape of the distribution is assumed to be gaussian for example, it is trivial to show that over the range of N we have measured, the calculated dependence of H_N on N is accurately described by Eq. (1). There is no difficulty in seeing how every parameter associated with a bubble circuit and its operation can affect H_1. It is harder to specify those fluctuating factors which give rise to ϵ. We can find no instrumental effect such as power supply noise, jitters in the rotating field or external noise large enough to explain the observed values of ϵ. This leaves as the source of the fluctuation either switching noise in the permalloy circuit or some phenomena in the bubble material itself.

Although the effect of H_R on bubble collapse and bubble velocity is quite complicated,[7,8,9] the dependence of H_1 on H_R (Fig. 3) simply seems to reflect the expected non-linear increase in the stray field from the nearly saturated permalloy as reflected in the dc margins. The slight decrease in ϵ with increasing H_R is consistent with reduced permalloy switching noise at higher drives.

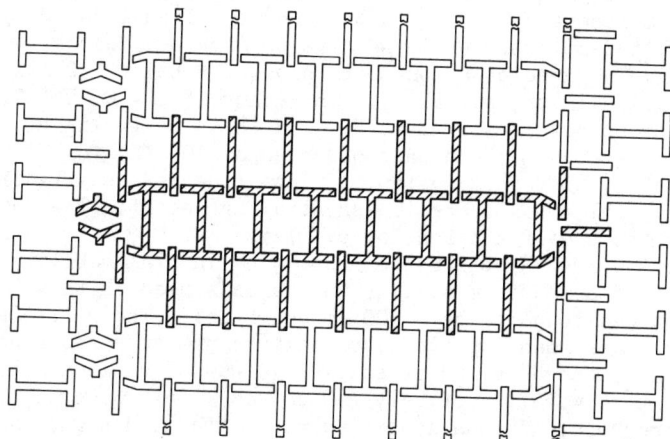

Figure 1. The 17 bit minor loop circuit (shown bold) under test. The bar width was designed to be 3.0μ.

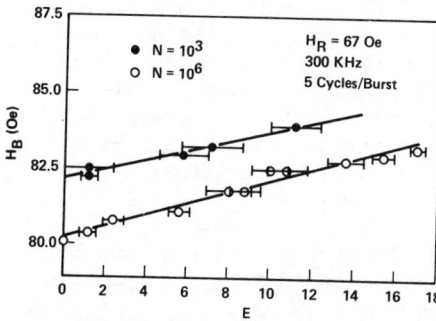

Figure 2. The dependence of the number of missing bubbles (errors) E in a single initially full 17 bit minor loop on the bias field H_B after N bubble steps.

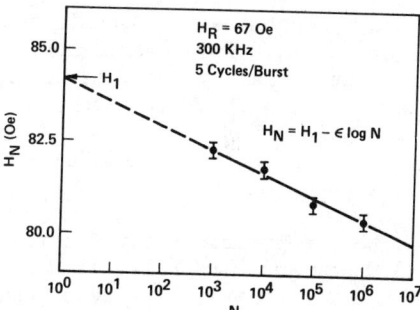

Figure 3. The value of H_N, the bias field at which the number of errors E = 1.0 as a function of N.

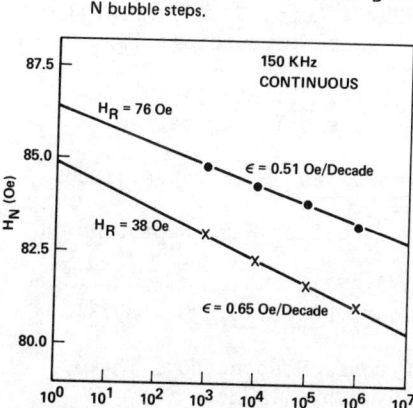

Figure 4. The value of H_N, the bias field at which the number of errors E = 1.0 as a function of N, for H_R = 38 Oe and 76 Oe.

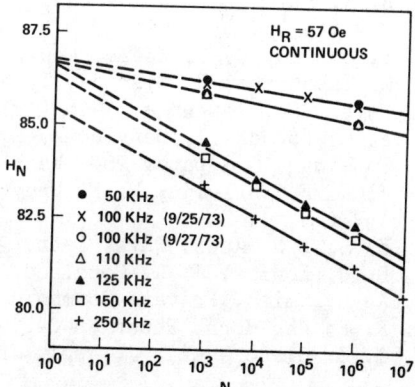

Figure 5. The value of H_N, the bias field at which the number of errors E = 1.0 as a function of N for several frequencies.

The data in Fig. 5 shows that below 110 KHz, H_1 and ϵ are independent of frequency. H_1 is determined by circuit design and operational parameters. While we have no direct evidence, we feel the most likely cause of a non-zero value of ϵ below 110 KHz is switching noise in the permalloy. The value of $\epsilon = 0.15 \pm 0.05$ Oe/decade which we have observed at $H_R = 57$ Oe is only slightly smaller than the value of 0.32 Oe/decade observed at 100 KHz for $H_R = 25$ Oe in a completely different system.[2] Since ϵ decreases slightly with H_R (Fig. 4), the results are in essential agreement. The one common element is the use of a permalloy overlay.

Above 150 KHz, H_1 begins to drop. The initial mobility of a permalloy coated YEu film similar to the ones used in this investigation was found to be ~ 700 cm/sec - Oe.[10] The roll-off at 150 KHz is lower than expected especially since comparable performance[1] has been obtained at smaller values of H_R in a material with an initial mobility of only 240 cm/sec - Oe. However, the roll-off frequency depends sensitively on circuit design,[1] and it is possible that the present circuit is not optimum.

The precipitous increase in ϵ above 110 KHz to a value which remains constant up to a frequency in excess of 250 KHz is unexpected. In this conference, Hagedorn[11] will describe a model postulated to describe the dynamic generation of wall transitions in a moving bubble. This should occur at a critical velocity estimated to be approximately the same as the saturation velocity V_0 calculated by Sloncezewski.[12] For the films used in this investigation, V_0 was calculated to be 2 - 4 M/sec. which would suggest a critical frequency of 100 KHz - 200 KHz. Our results are consistent with such a description.

We are grateful to the members of the Sperry bubble effort both at Sudbury and Blue Bell, all of whom contributed to the fabrication and testing of the chips described here.

REFERENCES

1. For a recent, comprehensive treatment, see Y. S. Chen, T. J. Nelson, J. Appl. Phys. **44**, 3306 (1973).
2. P. W. Shumate. R. J. Peirce, Appl. Phys. Lett. **23**, 204 (1973).
3. B. F. Stein, R. M. Josephs, AIP Conference Proceedings **10**, 329 (1973).
4. Y. S. Lin, G. E. Keefe, Appl. Phys. Lett. **22**, 603 (1973).
5. M. Takahashi, H. Nishida, T. Kobayashi, and Y. Sugita, J. Phys. Soc. Japan 34, 1416 (1973).
6. A. H. Bobeck, I. Danylchuk, F. C. Rossol and W. Strauss, Proc. Intermag 1973 paper 26-1 Washington, D. C.
7. J. W. F. Dorleijn, W. F. Druyvesteyn, Philips Res. Repts. **28** 133 (1973).
8. Y. S. Lin, Appl. Phys. Lett. **22**, 29 (1973).
9. D. C. Fowlis, J. A. Copeland, AIP Conf. Proc. **10**, 393 (1973).
10. A. B. Smith, Private Communication.
11. F. B. Hagedorn, Paper 5B-04 1973 M^3 Conf. Boston, Nov. 13-16.
12. J. C. Sloncezewski, J. Appl. Phys. **44**, 1759 (1973).

INFORMATION STABILITY IN BUBBLE MEMORY ELEMENTS

Koichi Yoshimi and Shozo Fujiwara
Central Research Laboratories,
Nippon Electric Co., Ltd., Kawasaki, Japan

ABSTRACT

A start-stop error in bubble memory elements and its stabilization methods have been described on $YFeO_3$ and Permalloy propagation circuits. The error rate largely increases with increasing the spacing between platelet and patterns. The phenomena are mainly caused by bubble interaction through Permalloy patterns and are also observed in YEu garnets. To improve the stability, several methods were examined; 1) Raising propagation pattern coercivity using NiFeCo films. 2) Arranging dot patterns near propagation patterns. 3) Applying DC in-plane bias field to the in-phase direction of start-stop operations. All three methods have been proven to be effective to increase the bias field margin.

INTRODUCTION

Recently, much attention has been paid to bubble domain memories. A 616-bit repertory dialer memory system was developed as a spring board to computer memories using $YFeO_3$, and the operating margin and reliability were evaluated. In the course of testing the memory, it was found that the written information was often partially destroyed after read operations. This information instability, which may be called start-stop error, poses a serious problem to fabricating non-volatile memories.

The phenomena of deteriorating operating margins have lately been reported by several authors. Copeland et al.[1] discussed random propagation failure and rephasing techniques. Bosch et al.[2] reported that the error rate increased significantly above 100 kHz. Shumate and Peirce[3] made "mean-time-to failure" measurements. These phenomena seem to be peculiar to garnet films. The start-stop problem, however, occurs both in garnet and orthoferrite materials. This paper describes the characteristics of the start-stop error and proposes methods to improve the operating margin decreased by the effect.

EXPERIMENTAL RESULTS

Bubble materials used in the experiments are mainly 60 μm thick $YFeO_3$ grown by the floating-zone technique. As a simple miniature memory element, 32-bit shift register patterns 1.5 μm thick were deposited on a glass substrate, as shown in Fig. 1. Spacing between the bubble platelet and overlay patterns was kept at about 8 μm with

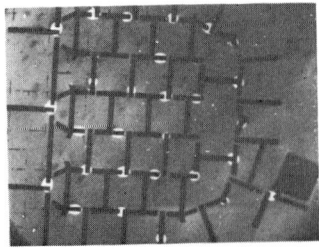

Fig. 1. Bubble configuration in YFeO$_3$ for Permalloy T-bar patterns in stopping state. Bit periodicity λ is 400 μm.

Canada balsam for most of the measurements. The drive field H$_R$ was synthesized with four sets of current pulses having half cycle pulse width. The start-stop error was found to be almost independent of the shape of drive field and to be essentially quasi-static. Therefore, the operating frequency was adjusted within one to 50 Hz. In the stopping state, no in-plane fields are applied, in contrast to the measurements made by Shumate and Peirce[3].

Figure 1 shows an extreme example of bubble configuration in Permalloy T-bar patterns after taking off the drive field at $\delta\phi = -90°$, where the phase difference $\delta\phi$ is defined as the angle between starting and stopping phase of the drive field. Coercive force H$_c$ and wall motion threshold H$_W$ of the patterns are 0.5 Oe and 0.1 Oe, respectively. Several bubbles moved away from predetermined positions at a relatively slow speed and reconfiguration of the bubble pattern took place. When a drive field is applied subsequently to this bubble configuration, several bubbles are prone to collapse.

According to the phenomenological point of view, a potential barrier must be raised to stabilize the bubble configuration in the absence of drive field. To improve the stability margin, the following methods were examined;
1) Raising the pattern coercivity, for example, using NiFeCo.
2) Arranging dot patterns near propagation patterns or changing the pattern shape.
3) Applying DC in-plane bias field in the direction of start-stop operations.

The error rate is defined as the relative number of error bubbles after 10 start-stop operational cycles, which consist of alternate 4-step clockwise drives and 4-step counterclockwise drives in order to avoid the effect of the generator pattern. In this paper, the error rate is measured in the 26-bit positions other than 7-bit positions near the generator (Fig.1). The error rate reaches an almost constant value after a few start-stop operations. It takes a minimum value when starting and stopping phase become 45° with respect to the direction of the bar pattern. Therefore, most of the measurements were done under this condition.

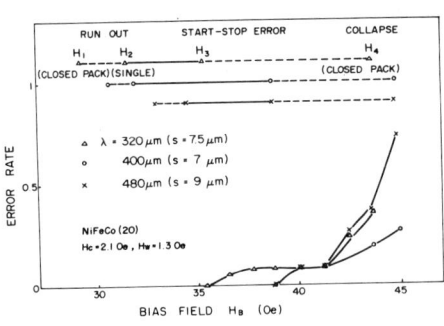

Fig.2. Error rates and start-stop margin in YFeO$_3$ for NiFeCo (64:16:20) patterns with bit periodicity λ as a parameter. H$_R$=42 Oe. $\delta\phi = -90°$

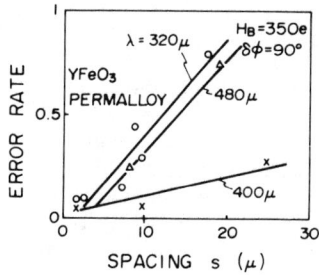

Fig.3. Spacing dependency of error rates in T-bar patterns as a parameter of λ.

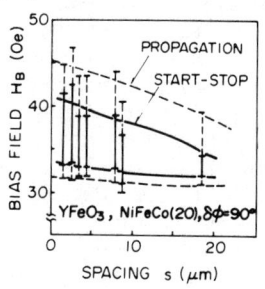

Fig.4. Operating margin in T-bar patterns as a function of spacing. $\lambda = 400 \mu$m, $H_R = 42$ Oe.

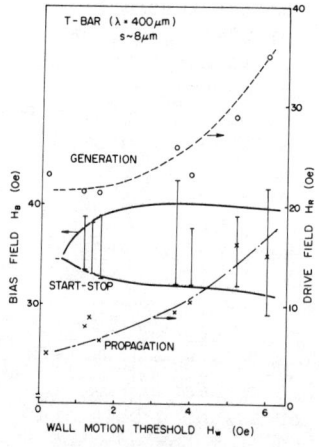

Fig.5. Start-stop margin and threshold drive field as a function of wall motion threshold H_W of NiFeCo patterns.

Figure 2 shows the typical error rates in NiFeCo patterns. The threshold bias field for normal propagation is denoted as H_1 for run-out in close packed bubbles, H_2 for run-out in a single bubble, and H_4 for collapse in close packed bubbles. Start-stop margins are H_2 to H_3. The error rate increases with increasing bias field and slightly decreases with increasing drive field. It becomes minimum at a certain bit periodicity λ ($\sim 400 \mu$m for YFeO$_3$).

For Permalloy patterns with small wall motion threshold, the start-stop margin is narrow and error rate is approximately proportional to the spacing as shown in Fig.3. For NiFeCo patterns, the start-stop margin also depends greatly on spacing, as shown in Fig. 4.

To determine the optimum H_C or H_W of the patterns, start-stop margin for $\delta\phi = -90°$ has been measured for various T-bar patterns consisting of differently Co-doped (0 ~ 30%) Permalloy films. Figure 5 shows the start-stop margins for $H_R=42$ Oe, and the threshold drive field for T-bar propagation and generation with a disk generator as a function of H_W. When H_W is larger than 1.5 Oe, margin increases significantly at the expense of a comparatively small drive field increment. Scatter of the collapse margin is thought to be caused by the fact that the magnetization process of the propagation pattern cannot be specified sharply only with H_W.

In-plane stray field was found to be a very sensitive factor with regard to start-stop error. Operating margins for $\delta\phi=0°$ exposed to the external DC in-plane field H_{DY} are shown in Fig.6, where positive H_{DY} sign means in the same direction as start-stop phase. In the case of Permalloy patterns (a), the start-stop margin vanishes at -0.5 Oe of H_{DY}, while the normal propagation margin is almost unchanged up to -5 Oe. For NiFeCo films (b), start-stop margins exist up to -2 Oe. The threshold value of H_{DY} is also improved up to as large as -2 Oe for the Permalloy patterns made up of small rectangular dots in the vicinity of the T-bar. Although the perpendicular stray field

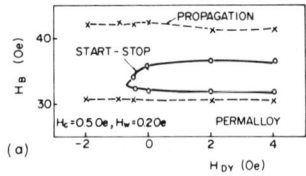

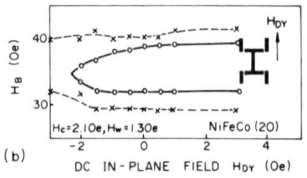

Fig.6. Operating margin in YFeO3 exposed to external DC in-plane field. $\lambda = 400 \mu m$, $H_R = 42$ Oe.

has been mainly considered to be an important factor deteriorating operating margin, one must bear in mind that in-plane stray field is also a serious factor and must be shielded. Figure 6 suggests that in-phase DC holding field can improve start-stop margin. In the early stages of developing the 616-bit repertory dialer memory utilizing YFeO3 and T-bar patterns ($\lambda = 320 \mu m$), the method of applying a 7 Oe in-plane DC field in stopping state was adopted to stabilize stored information. A 2.7 Oe overall worst case bias margin ($\pm 4\%$ of H_B) was obtained at 1 kHz and 2.2 Oe ($\pm 3\%$ of H_B) was obtained at 20 kHz for $H_R = 39$ Oe, including generation, detection and gating action of 25 conductor gates.

Start-stop error was also observed in $Y_{2.5}Eu_{0.5}Fe_{3.8}Ga_{1.2}O_{12}$ garnet films. Thickness h, magnetization $4\pi M_s$ and material length are $3.7 \mu m$, 147G and $1.05 \mu m$, respectively. In $40 \mu m$ period Permalloy T-bar patterns (thickness $t=0.4 \mu m$, $H_c=2.0$ Oe, $H_w=0.3$ Oe), start-stop margin was about 1 Oe for $\delta\phi = -90°$ using $2 \mu m$ spacing. On the other hand, the start-stop margins were 4 Oe for T-bar and 6 Oe for YY patterns with NiFeCo films ($t=0.36 \mu m$, $H_c=5.2$ Oe, $H_w=1.4$ Oe) using $2 \mu m$ spacing.

DISCUSSION

The optimum pattern condition is $\lambda/d=4$, where d is bubble diameter. According to the dipole approximation, direct interaction between two bubbles, converted into an equivalent field gradient, is estimated as 0.03 Oe for $4\pi M_s=100G$ and $h/d=0.5$. When a bubble is exposed to the field caused by semi-infinite bubble lattice, the value increases 5/3 times larger than that in two bubble system. In any case, the field is smaller than the coercive force of bubble materials (0.1 Oe) and Permalloy patterns. Furthermore, direct interaction fails to explain the fact that the error rate takes a minimum value at a certain bit periodicity.

In the absence of in-plane drive field, Permaloy patterns are magnetized only by bubbles. Indirect interaction between bubbles through Permalloy patterns should be appreciable. Calculations reported by Copeland[4] show that the magnetization distribution in a bar magnetized by a bubble is almost linear along the bar direction (see Fig. 6 in ref. 4). If linear approximation neglecting H_c, H_w is employed, the value of the field gradient across the bubble diameter is estimated as about 1 Oe at a point 1.5d from the edge of the bar (condition corresponding to $\lambda = 4d$) in terms of Copeland's results. In the case of small spacing between platelet and patterns, the field gradient is proportional to the spacing. The potential barrier when a bubble moves away from the edge to the center of a bar pattern is only about 0.2 Oe in the absence of a drive field, according to calculation reported by Archer et al[5]. The experimental data show that the potential barrier decreases with increasing bias field. The field gradient caused by indirect interaction can theoretically overcome the potential barrier. Thus, a model of indirect interaction qualitatively explains possibly the experimental data concerning spacing and bias field dependency.

The error rate for YY pattern was also examined and was found to be less than that for T-bar patterns. The reason is considered to be that, in YY patterns, all bit positions are separated and suffer from a lower interaction field through patterns.

To compare the data more quantitatively, further extended calculation must be conducted.

CONCLUSION

Information instability has been found to exist in the bubble memory elements reported here due to start-stop operations. The phenomena are thought to be caused by the indirect interaction between bubbles through Permalloy patterns. To improve the stability, several methods are proposed and found to be effective. Patterns employing NiFeCo films seem to be practical. In-plane stray field is an important factor besides the perpendicular stray field for bubble memory elements.

ACKNOWLEDGEMENT

The authors wish to express their appreciation to T. Okada and T. Furuoya for their encouragement and helpful comments, and to Y. Wada for valuable discussions.

REFERENCES

1. J.A.Copeland, J.G.Josenhans, and R.R.Spiwak, 1973 Intermag. Conf. 26-7.
2. L.J.Bosch, R.A.Downing, G.E.Keefe, L.L.Rosier, and K.D.Terlep, 1973 Intermag. Conf. 26-2.
3. P.W.Shumate and R.J.Peirce, Appl.Phys.Lett., 23, 204 (1973).
4. J.A.Copeland, J.Appl.Phys., 43, 1905 (1972).
5. J.L.Archer, L.Tocci, P.K.George, and T.T.Chen, IEEE Trans. Magnetics MAG-8, 695 (1972).

Section 5. Wall Structure in Bubble Materials

OBSERVATION OF STRIP AND BUBBLE DOMAINS IN $PbFe_{12}O_{19}$ USING LORENTZ ELECTRON MICROSCOPY

P. Dunk and G.A. Jones
University of Salford
Salford M5 4WT, England

ABSTRACT

The electron microscope has been used to investigate the domain structure of basally cleaved crystals of magnetoplumbite. After saturating a sample with an in-plane field, the remanent state contains extended arrays of bubbles and strip domains. Bubble domains have also been formed by the application of bias fields applied within the microscope. Some account of the wall structure is given including the presence of Bloch lines.

INTRODUCTION

Recently much work using Lorentz electron microscopy has been carried out on ferromagnetic metals. However due to preparation difficulties the technologically important ferrimagnetic oxides have not received attention to the same extent. There has been only one published contribution[1] to Lorentz microscopy concerned with magnetoplumbite, $PbFe_{12}O_{19}$. This oxide and more particularly the related garnets, are now very important potential bubble device materials. Examination by electron microscopy of any bubble material is valuable since it is the only method available which can resolve directly the wall structure of magnetic bubbles. The object of this paper is to report a modified preparation technique and some of the preliminary results obtained with the oxide magnetoplumbite. It is believed these studies will be relevant to garnets.

EXPERIMENTAL

Thin platelets of cleaved, basally oriented, magnetoplumbite were shaped ultrasonically into 3 mm. discs before being ground to a thickness of about 75 μm. The final stage of preparation used a chemical jet polishing technique. The polishing solution consisted of 75% orthophosphoric acid, 20% water and 5% concentrated sulphuric acid heated to 220 - 240°C. A few minutes polishing sufficed to perforate the disc and produce areas thin enough for electron microscope examination. Electron diffraction patterns showed the thinned material to be still of basal orientation.

Many of the studies were undertaken with a 1MeV microscope, the AEI - EM7. The accelerating voltage of one million volts allows greater areas of the specimen to be observed. As described previously[2] use of this microscope enables bias fields up to 20 KOe to be applied perpendicular to the sample. Specimens saturated in the basal plane were also studied. The Lorentz deflection of electrons in $PbFe_{12}O_{19}$ is small (4×10^{-5} rads) because of it's relatively low magnetization and hence conditions for domain wall observation are fairly critical. A small objective aperture (10 μm) was used, parallel illumination and a long exposure time (≃ 1 min). The domains are best seen with the microscope slightly out of focus, the so called Fresnel mode. Even under these conditions the domain walls are not visible on the viewing screen but only on photographic plates.

RESULTS

Fig. 1 is a micrograph, taken at 1 MeV, of a platelet of magnetoplumbite in the remanent state. The domain walls forming the strip lattice are clearly delineated and it will be seen that each consists of a bright and a dark component of contrast. In transferring from one domain to the next the magnetization may rotate across the wall in either of two possible senses and for a specimen with no previous magnetic history each is equally likely. Examination of the wall contrast in Fig 1 shows this to be the case. At certain points the magnetization rotation across the wall changes from one sense to the other, distinguishable on the micrograph by a reversal in contrast. In some cases this transition occurs over a very short distance: at other times it appears more gradual. It is envisaged that these discontinuities

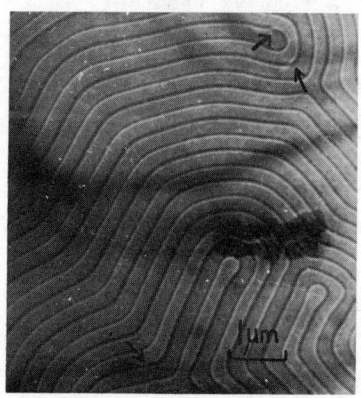
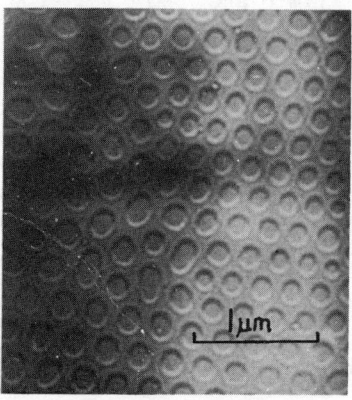

FIG 1
1 MeV micrograph of remanent strip array. Bloch lines shown by arrows.

FIG 2
100 KeV micrograph showing remanent bubble array after in-plane field of 17 KOe has been applied and removed.

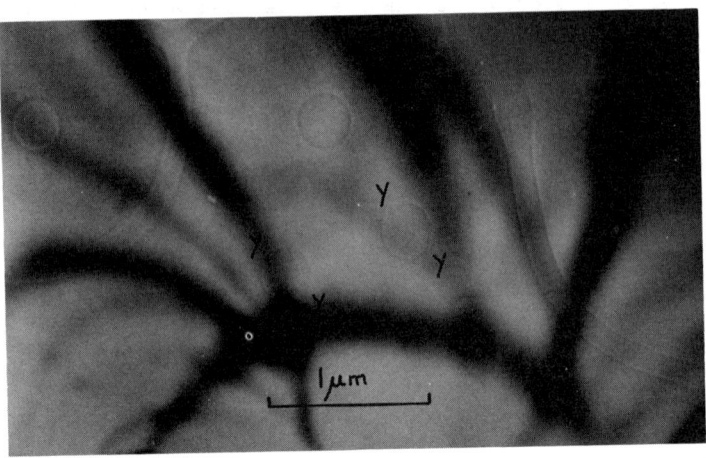

FIG 3
1 MeV micrograph near saturation with field applied along the c-axis Bloch lines shown at points marked Y.

are vertical Bloch (or Néel) lines.

It has been reported[3] that bubble domains may be created in hexaferrites by application of a saturating field ($\simeq 2K/M$) in the film plane. Fig 2 at 100 KeV is an extensive remanent bubble lattice created in $PbFe_{12}O_{19}$ by application and removal of a field of 17 KOe in the film plane. It will be seen that every bubble contains two Bloch lines situated diametrically opposite to each other. Moreover the disposition of the Bloch lines within a bubble is similar for the whole array of bubbles.

The final of this preliminary series of experiments was concerned with attempting to produce bubbles directly in the microscope by applying a perpendicular bias field in the normal way. Within a fairly restricted range of bias field this can be achieved as demonstrated in fig 3 at 1 MeV which shows part of an area containing bubble domains together with some narrow strips. The majority of bubble walls not shown on the micrograph contained no discontinuity but at least two, marked Y, possess two Bloch lines. The contrast is poor for the reasons discussed above.

DISCUSSION AND CONCLUSION

Kaczer and Gemperle[4] have given a comprehensive theoretical treatment for the strip domain structure of magnetoplumbite and quote an explicit formula for zero field strip width W, as a function of film thickness D, viz,

$$D = \frac{1.705 \, W^2 \, M_s^2}{\gamma}$$

where $M_s = 320$ gauss, $\gamma = 4.82$ erg. cm^{-2}. Adopting these values[5] and measuring W as 2800 Å (Fig 1), the film thickness observed at 1 MeV is about 3000 Å. This is in reasonable agreement with that predicted by the theory of Cape and Lehman[6] assuming a value of 375 Å ($= \gamma/4\pi M_s^2$) for the material parameter. As D tends to zero a thickness is reached where the crystal becomes a single domain saturated along the c-axis.[7]

The production of a honeycomb structure in both $PbFe_{12}O_{19}$ and $Ba\,Fe_{12}O_{19}$ (barium ferrite) has been reported[4,8] following the removal of a large in-plane field. In very thin films this process would be expected to lead to bubble formation as indeed happens, fig 2. There is no clear explanation for this behaviour although it is significant that all the bubbles as well as the neighbouring strips contain Bloch lines. If bubbles were first nucleated the planar field would tend to favour those containing two discontinuities;[9] as the field is decreased the bubbles will run out to the more energetically favoured strip domains. For some reason the bubbles in certain regions do not run out but persist at remanence. It is interesting to note that the application and removal of a saturating field along the c-axis in magnetoplumbite does not lead to bubble formation, the exact opposite to the behaviour of cobalt.[9]

The dynamic experiment producing bubbles, fig 3, resembles the classic work of Kooy and Enz[10] in barium ferrite. Unlike magneto-optical techniques however, the electron microscope reveals the detailed wall structures. Although fig 3 only shows a small area of specimen it was found that the majority of bubbles created in this way have no Bloch lines. Perhaps it is a significant fact in view of current bubble dynamics theory, that throughout this work no bubble containing more than two discontinuities was ever observed. After saturation the bias field was gradually reduced and an attempt made to determine the nucleation process. Only the gradual growth of strips was seen although this does not preclude the possibility that bubbles are formed at isolated sites[10] in thicker areas inaccessible to the electron microscope.

Having demonstrated the imaging capabilities of the microscope with respect to ferrimagnetic materials it is hoped to continue further the study of Bloch line behaviour.

ACKNOWLEDGEMENTS

One of the authors (Paul Dunk) would like to acknowledge the provision of a CAPS grant under the mutual auspices of S.R.C. and the Plessey Co. Ltd. Our thanks are also due to Mr. A. Marsh for his encouragement and assistance and to Dr. R.J. Fairholme, both of the Allen Clark Research Centre, The Plessey Co. Ltd. Towcester.

REFERENCES

1. P.J. Grundy, Brit. J. Appl. Phys., 16, 409, (1965)
2. G.A. Jones, P.J. Grundy, D.C. Hothersall and M.J. Goringe. Proc. Fifth. Eur. Cong. on Electron Microscopy, 356, (1972)
3. P.J. Grundy, Private Communication.
4. J. Kaczer and R. Gemperle, Czech J. Phys., B11, 510 (1961)
5. J. Kaczer and R. Gemperle, Czech, J. Phys. B10, 505 (1960)
6. J.A. Cape and G.W. Lehman, J. Appl. Phys. 42, 5732, (1971)
7. Z. Malek and V. Kambersky, Czech J. Phys., 8, 416, (1958)
8. H. Kojima and K. Got, J. Appl. Phys., 36, 538, (1965)
9. P.J. Grundy, D.C. Hothersall, G.A. Jones, B.K. Middleton and R.S. Tebble, I.E.E.E. Trans. Mag. Vol MAG 7 (1971) Phys. Stat. Sol. (a) 9, 79, (1972)
10. C. Kooy and U. Enz., Philips Res. Rep., 15, 7, (1960)

LPE GARNET FILMS WITHOUT HARD BUBBLES

A.B. Smith, M. Kestigian, and W.R. Bekebrede
Sperry Research Center, Sudbury, Mass. 01776

ABSTRACT

In previous studies of hard bubbles, these anomalous domains were found to exist in all unimplanted single-layer LPE garnet films at room temperature. We have prepared five garnet film compositions in which it is not possible to generate hard bubbles at room temperature. These compositions are $Y_{0.9}Gd_{1.2}Tm_{0.9}Fe_{4.6}Al_{0.1}Ga_{0.3}O_{12}$, $Y_{0.9}Gd_{1.5}Yb_{0.6}Fe_{4.5}Al_{0.5}O_{12}$, $Y_{1.0}Gd_{1.0}Er_{1.0}Fe_{4.5}Al_{0.1}Ga_{0.4}O_{12}$, $Y_{1.5}Eu_{0.75}Gd_{0.75}Fe_{4.4}Al_{0.5}Ga_{0.1}O_{12}$ and $Y_{2.0}Tm_{1.0}Fe_{4.1}Ga_{0.9}O_{12}$. All these films were prepared by standard LPE growth techniques and were **not** ion implanted or provided with any additional layers. The absence of hard bubbles in these films was inferred from the fact that all bubbles collapse within a 1.5% range of bias field even when the bubbles are generated by all the known techniques which normally produce hard bubbles. The dynamic behavior of these bubbles differs with the material composition. Some of this data indicates the presence of a small number of Bloch lines; however, the bubbles in one material all propagate at the same angle to a field gradient which seems to indicate the complete absence of Bloch lines. A common feature of all five compositions was that $q = H_k/4\pi M < 5$. This is apparently a necessary condition as is demonstrated by data we present on similar compositions with $q > 5$ which do exhibit hard bubbles. The effect of temperature on the behavior of these materials is also discussed.

INTRODUCTION

The properties of hard bubbles in LPE garnet films have been described by several authors.[1,2,3] The only single-layer LPE films reported to be free of hard bubbles at room temperature are those which have been ion implanted.[4] We would like to report on a number of unimplanted garnet films that do not exhibit hard bubble properties under quasi-static conditions at room temperature.

RESULTS

Our LPE garnet formulations that do not exhibit hard bubbles are listed in Table I. Also included are measurements of the properties of a typical film of each composition. The values given for thickness h, magnetization $4\pi M$, and characteristic length ℓ were all determined using standard techniques.[5] The quantity H_k is the value of in-plane field necessary to extinguish the stripe domains seen in a demagnetized film. (As discussed by Smith et al,[6] H_k is approximately equal to the anisotropy field 2 Ku/M.) The quantity q is the ratio of H_k to $4\pi M$. The quantity Δa is the mismatch between substrate and film as determined by x-ray measurements.[6] A positive number represents a film in tension. (Values of mismatch between + .0025 Å and - .0025 Å could not be determined with our apparatus and are listed in the table as $\cong$ 0.) The quantity μ is the bubble mobility whose measurement is discussed below.

The determination that the materials in Table I do not exhibit hard bubbles at room temperature was made on the basis of collapse field. The range of fields required to collapse all the bubbles in each material was less than 1.5% of the bias field. Of course, this determination is only meaningful if the proper techniques are used to generate the bubbles. We have used pulsed fields generated with a 5-turn, 4.5 mm-diameter coil which we find gives similar results to the somewhat different coil configurations described in the literature[1,7] for generating hard bubbles. We have employed a full range of pulse widths (0.025 - 25. μsec), pulse amplitudes (0 - 130 Oe), and bias fields with each sample. Only after numerous attempts had produced bubbles all with substantially the same collapse field did we conclude that hard bubbles could not be generated in a given material.

When a field gradient is applied to the materials in Table I by using the method of Vella-Coleiro and Tabor,[8] some of the bubbles generated by the above described pulsed-field technique move along the gradient while others do not. Some bubbles propagate at angles up to 70° but travel at approximately the same velocity as those which follow the gradient. They, therefore, appear to be "intermediate" bubbles with a few Bloch lines, but without the large numbers of lines that lead to the very low velocities reported[1] for really hard bubbles. In all but the YGdTm-garnet, bubbles can readily be found that move at angles ≤ 30° and these have been used for the mobility measurements reported in Table I. (In calculating all the mobility values in Table I, the velocity **component** along the gradient has been used.) When a field gradient is applied to the highest mobility material in Table I, YGdTm-garnet, **all** the pulsed-field-generated bubbles move at an angle 60 ±7°. Upon reversing the bias field, the bubbles move at the same angle on the other side of the gradient. These results seem to indicate that these pulsed-field-generated bubbles have no Bloch lines.[2] For comparison, we have also generated bubbles in this material by applying a large in-plane dc field. Presumably these bubbles all have just two Bloch lines[9] and should move parallel to the gradient.[2] We do observe such parallel motion if the field difference across the bubble is less than 3 Oe. With larger gradients these bubbles revert to propagation at the same angle as the pulse-field-generated bubbles.

One common characteristic of all the materials listed in Table I is that q ≤ 4.5. This seems to be a necessary condition for freedom from hard bubbles as may be demonstrated by preparing similar materials except with q > 4.5. Some examples are shown in Table II. (For convenience in comparison, we have numbered the entries to correspond with Table I, i.e., composition 1 in Table I is similar to composition 1 in Table II, etc.) In all films with q > 4.5, we have been able to readily generate hard bubbles. The parameter R in the final column of Table II expresses the "hardness" of the bubbles; it is the ratio of the largest collapse field observed to the collapse field for normal bubbles. (The normal bubble collapse field was determined by measuring bubbles which had been formed using an in-plane field.)

It is interesting to compare our results with the measurements taken by Henry et al[10] above room temperature. They found that no hard bubbles could be generated in $Y_{1.08}Gd_{0.72}Tm_{1.2}Fe_{4.2}Ga_{0.8}O_{12}$ above a critical temperature $T_H = 115^oC$. Since they did not attempt to correlate hard bubble properties with q, we have repeated their experiment with a sample of similar composition ($Y_{1.0}Gd_{1.0}Tm_{1.0}Fe_{4.4}Ga_{0.6}O_{12}$). We found the same behavior except that $T_H = 75^oC$. At this temperature the q of this material has dropped to 2.0 from its room temperature value of 7.2. This result indicates that hard bubbles could be generated in the materials of Table I if their q could be increased by means of a temperature change. We have observed such hard bubble generation in the YGdYb-garnet of Table I at temperatures below -5°C where $q \leq 6.1$. These measurements of the temperature dependence of hard bubble generation indicate a similar relationship between hard bubble generation and q as that exhibited by Tables I and II. However, the critical values of q noted in these experiments do differ somewhat from the $q \cong 5$ value that would be inferred from the tables.

In the course of these latter measurements, it was also noted that hard bubbles generated below T_H will remain hard if the sample is returned to room temperature without exceeding the collapse field. This is the same effect that Henry et al[10] observed at higher temperatures in the YGdTm-garnet sample mentioned above. Thus, it appears that it is only the generation of hard bubbles that is inhibited in $q \leq 4.5$ materials, not their existence.

It should be noted that hard-bubble suppression in LPE films can be achieved[11] by intentionally introducing compositional gradients which serve to suppress hard bubbles in a manner roughly analagous to the suppression achieved in multi-layer films. We do not believe that this mechanism is responsible for the results we have obtained since we know of no such gradients in our films. (Also, the above described temperature dependence of hard-bubble suppression would seem to indicate that some other mechanism is responsible.) Our films have been prepared using standard procedures and melt compositions which have been described in a previous publication.[6] The substrates were held vertical during growth in a few cases (#3 and 4, Table I; #3, Table II) and horizontal for all other samples.

The data presented in Tables I and II are representative of a much larger group of samples. We find that all samples with thicknesses greater than 3 μ obey the $q \cong 4.5$ requirement for hard bubble suppression indicated by these tables. The $q \cong 4.5$ criterion seems to apply regardless of the specific values of K_u and $4\pi M$. Also, there is no obvious dependence on whether the anisotropy is growth or strain induced. However, two samples which were less than 3 μ thick did show hard bubbles with q's less than 4. This dependence on thickness clearly requires further investigation as does the whole question of exactly how hard-bubble generation is suppressed in the materials of Table I. In explaining their observations, Henry et al[10] suggest that the appearance of a uniaxial in-plane anisotropy may be related to temperature-dependent hard-bubble suppression In the absence of any direct measurement of in-plane anisotropy in our materials, we made a qualitative check for the existence of a uniaxial in-plane anisotropy by observing bubble behavior. In all the materials in Table I except #1, when the bubbles become elliptical just before strip-out, the major axis of the ellipse always lies in one particular direction. (Unfortunately, this effect is not extreme enough to permit a precise determination of the crystallographic orientation of this direction.) The materials in Table II show no such evidences of uniaxial in-plane anisotropy.

CONCLUSION

One of the principal reasons for reporting these experiments is to correct the common impression that all unimplanted LPE garnet films exhibit hard bubbles at room temperature. We have also wished to point out the apparent correlation between hard-bubble generation and q. However, it should be emphasized that it is not yet clear whether q is the ultimate parameter for determining hard-bubble suppression since the controlling mechanisms are not yet definitely established.

REFERENCES

1. W.J. Tabor, et al, AIP Conf. Proc. 10, 442 (1972).
2. J.C. Slonczewski, A.P. Malozemoff, & O. Voegeli, AIP Conf. Proc. 10, 458 (1972).
3. H. Nishida, T. Kobayashi, & Y. Sugita, AIP Conf. Proc. 10, 493 (1972).
4. R. Wolfe, J.C. North, & Y.P. Lai, Appl. Phys. Lett. 22, 683 (1973).
5. R.M. Josephs, AIP Conf. Proc. 10, 286 (1972).
6. A.B. Smith, et al, AIP Conf. Proc. 10, 309 (1972).
7. H. Nishida, T. Kobayashi, & Y. Sugita, IEEE Trans. MAG-9, 517 (1973).
8. G.P. Vella-Coleiro & W.J. Tabor, Appl. Phys. Lett. 21, 7 (1972).
9. This experiment was suggested by R. Josephs and W. Doyle.
10. R.D. Henry, et al, IEEE Trans. MAG-9, 514 (1973).
11. F. Hagedorn, private communication.

ACKNOWLEDGMENT

The authors would like to thank W. Doyle and R. Josephs for useful discussions related to this investigation. We would also like to acknowledge the expert technical assistance of A. Doppler, F. Garabedian, W. Goller, and C. Ward.

TABLE I
PROPERTIES OF LPE GARNET FILMS THAT DO NOT EXHIBIT HARD BUBBLES

Composition	Δa (Å)	h (μm)	ℓ (μm)	4πM (G)	H_k (kOe)	q	μ (cm/sec/Oe)
1. $Y_{0.9}Gd_{1.2}Tm_{0.9}Fe_{4.6}Al_{0.1}Ga_{0.3}O_{12}$	-0.0065	5.6	0.29	249.	0.46	1.8	2700.
2. $Y_{0.9}Gd_{1.5}Yb_{0.6}Fe_{4.5}Al_{0.5}O_{12}$	+0.0038	10.4	0.36	186.	0.47	2.5	600.
3. $Y_{1.0}Gd_{1.0}Er_{1.0}Fe_{4.5}Al_{0.1}Ga_{0.4}O_{12}$	0	6.3	0.38	194.	0.64	3.3	440.
4. $Y_{1.5}Eu_{0.75}Gd_{0.75}Fe_{4.4}Al_{0.5}Ga_{0.1}O_{12}$	-0.020	9.2	0.18	327.	1.46	4.5	1000.
5. $Y_{2.0}Tm_{1.0}Fe_{4.1}Ga_{0.9}O_{12}$ *	-0.0054	11.2	0.37	215.	0.76	3.5	1130.

* This film was deposited on a Dy-substituted $Gd_3Ga_5O_{12}$ substrate having a lattice constant of 12.343. All other films were deposited on pure $Gd_3Ga_5O_{12}$. All substrates have a {111} orientation.

TABLE II
PROPERTIES OF LPE GARNET FILMS THAT DO EXHIBIT HARD BUBBLES

Composition	Δa	h	ℓ	4πM	H_k	q	R
1. $Y_{0.9}Gd_{1.2}Tm_{0.9}Fe_{4.5}Al_{0.1}Ga_{0.4}O_{12}$	-0.0043	3.7	0.82	138.	0.79	5.7	1.86
2. $Y_{0.9}Gd_{1.4}Yb_{0.7}Fe_{4.4}Al_{0.6}O_{12}$	+0.0092	5.0	0.80	142.	0.80	5.6	1.33
3. $Y_{1.0}Gd_{1.0}Er_{1.0}Fe_{4.4}Al_{0.1}Ga_{0.5}O_{12}$	-0.0043	8.7	1.20	106.	1.42	13.4	1.63
4. $Y_{1.5}Eu_{0.7}Gd_{0.8}Fe_{4.3}Al_{0.6}Ga_{0.1}O_{12}$	+0.0038	13.3	0.66	175.	1.65	9.4	1.20
5. $Y_{2.0}Tm_{1.0}Fe_{4.0}Ga_{1.0}O_{12}$ *	0	12.5	0.66	148.	1.40	9.4	1.12

* Same note applies as in Table I except that the lattice constant of this substrate is 12.350 Å

HARD-BUBBLE-FREE GARNET-PERMALLOY COMPOSITE FILMS*

M. Takahashi, H. Nishida, T. Kobayashi, and Y. Sugita
Central Research Laboratory, Hitachi, Ltd.
Kokubunji, Tokyo, Japan

In a previous publication, we reported hard bubble suppression by depositing thin permalloy films directly onto epitaxial garnet films[1]. Here, we summarize bubble behavior in these garnet-permalloy composite films.

Hard bubbles (including intermediate ones) are suppressed in the composite films for permalloy films thicker than about 60Å. This effect is probably attributed to the exchange coupling between garnet and permalloy. This interpretation is supported by the fact that permalloy, separated from garnet by a 100Å thick SiO_2 layer, failed to suppress hard bubbles. The same interpretation has been given by Lin and Keefe[2]. The collapse field, H_0, and runout field, H_2, increase with increasing permalloy thickness, t, up to $t \sim 300$Å. This is understood in terms of a decrease in magnetostatic energy. A bias field margin, $2(H_0 - H_2)/(H_0 + H_2)$, linearly decreases with increasing t, but the amount of the decrease is less than 10% of the original value for $t \leq 150$Å.

Wall mobility measurements by the bubble transport technique show a 50% increase in mobility in the composite film compared with that of normal bubbles in a single-layered film. It cannot be excluded, however, that the difference is due to the presence of a small number of Bloch lines in the bubbles in the single-layered film.

The permalloy deposition does not adversely affect bubble behavior in a memory chip. Bubbles can be propagated in T-bar circuits and detected by both permalloy and InSb detectors. Output signals are as high as those for single-layered films. Thus, the composite film is very promising for device applications since the deposition of permalloy is a very simple process. Furthermore, it should be noted that the hard bubble suppression by permalloy is likely the most suitable for amorphous materials such as GdCo.

We have also found that hard bubbles are absent in single-layered garnet films when their q (ratio of uniaxial anisotropy to shape anisotropy) is smaller than approximately 2. However, these films may not be suitable for device use because of spontaneous bubble nucleation and narrow bias margins. The reason for the absence of hard bubbles in such films is not clear at present.

REFERENCES

1. M. Takahashi et al., J. Phys. Soc. Japan **34**, 1416(1973) and to be published in Proc. of 1973 International Conf. on Solid State Devices (Tokyo).
2. Y. S. Lin and G. E. Keefe, Appl. Phys. Letters **22**, 603(1973).

* Supported in part by the Agency of Industrial Science and Technology of Japan.

STATIC HARD BUBBLE MEASUREMENTS ON SEVERAL EPITAXIAL GARNET FILM COMPOSITIONS

D. H. Smith and A. A. Thiele
Bell Laboratories, Murray Hill, N. J. 07974

ABSTRACT

Extensive hard bubble diameter measurements as a function of bias field and temperature have been made on several epitaxial garnet film compositions including $Y_{2.6}Sm_{.4}Ga_{1.2}Fe_{3.8}O_{12}$, $Y_1Eu_{1.85}Yb_{.15}Al_{1.2}Fe_{3.8}O_{12}$, $Y_{1.7}Eu_{.65}Tm_{.65}Ga_{1.2}Fe_{3.8}O_{12}$, $Y_{.9}Gd_{1.1}Yb_1Ga_{.9}Fe_{4.1}O_{12}$, $YGdTmGaFe_4O_{12}$ and $Y_{1.8}Ca_{1.1}Sm_{.1}Ge_{1.1}Fe_{3.9}O_{12}$. The data for each bubble were fitted to a detailed hard bubble model and the number of Bloch lines, N, was extracted. Comparing N with the bubble diameter at collapse provides broader support for the previously suggested idea of a maximum density of Bloch lines. The maximum density of Bloch lines (5 to 11 pairs/μm at 298°K) is found to be approximately linear in $Q = H_k/4\pi M_s$ with a slope $d\rho_m/dQ = 0.2$ to 1.4 pairs/μm for this group of compositions. Values of Q range from 2-5 at the highest temperature, T_H, where hard bubble generation is possible.

INTRODUCTION

A number of authors[1-4] have recently discussed the static properties of hard bubbles in mixed rare earth garnet epitaxial films. For bubbles having N pairs of Bloch lines, a modified domain wall energy expression,[1-3]

$$\sigma_w' = \sigma_w \left[1+(2N\delta_B/\pi d)^2 \right]^{1/2} \quad (1)$$

when substituted into the normal force equation,[5] has been used to fit bubble diameter, d, as a function of bias field, H, data and thus obtain a value for N. Here, σ_w and δ_B are the domain wall energy and wall width of a normal Bloch walled bubble, respectively. This modification can also be used to predict the collapse diameter for a hard bubble in which N is less than a critical value N_c. For this case, the collapse mechanism is the same as in normal bubbles[1,3] and the bubble collapses at a nonzero diameter. For $N \geq N_c$, the instability which leads to bubble collapse disappears;[1,3] the theory predicts that the bubble diameter will continuously approach zero with increasing field. However, all hard bubbles are found experimentally[1,2,4] to collapse from some nonzero diameter. A critical value for σ_w' (or equivalently a critical density $\rho_m = N/\pi d$ of Bloch lines) has been proposed[1-3] which makes the collapse diameter proportional to N once the critical density is reached. This is a result that is in general agreement with experiment.[1-3]

In the present work, our intent is to determine ρ_m for a number of different materials and to search for correlations with the other magnetic material parameters. In order to reduce the need to consider changes in properties from sample to sample of nominally the same composition, bubble data have been taken at a fixed location on a particular sample. Changes in the magnetic parameters have been achieved by altering the sample temperature.

EXPERIMENTAL

The static properties of circular hard bubbles have been measured for the garnet compositions listed in Table I at several temperatures between 250 and 340°K. The films were prepared by a liquid phase epitaxy technique.[6] The first 4 samples have monotonically decreasing effective uniaxial anisotropy field, and magnetic moment $4\pi M_s$ values while the last 2 exhibit increasing $4\pi M_s$ and decreasing H_k values with increasing temperature throughout the range 250-340°K.

Table I: Film Material Compositions and Curie Temperatures

Sample	Nominal Composition	T_c(°K)
1	$Y_{2.6}Sm_{.4}Ga_{1.2}Fe_{3.8}O_{12}$	398
2	$YEu_{1.85}Yb_{.15}Al_{1.2}Fe_{3.8}O_{12}$	402
3	$Y_{1.7}Eu_{.65}Tm_{.65}Ga_{1.2}Fe_{3.8}O_{12}$	405
4	$Y_{1.8}Ca_{1.1}Sm_{.1}Ge_{1.1}Fe_{3.9}O_{12}$	443
5	$Y_{.9}Gd_{1.1}YbGa_{.9}Fe_{4.1}O_{12}$	450
6	$YGdTmGaFe_4O_{12}$	437

The domains were viewed with a polarizing microscope having a temperature controlled stage. The bubbles were produced by cutting the demagnetized stripe domain pattern with a current pulsed, eliptically shaped wire loop with a 25 μm:125 μm aspect ratio. This loop, positioned with its major axis parallel to an array of stripe domains, approximated the hard bubble cutting conditions described by Nishida et al.[7] Current pulses of 2 amps, producing several hundred Oersteds of field in the plane of the film, and 1 μs duration were applied at a 10 Hz rate as the loop was moved along the stripes.

For each bubble within a group selected, d was recorded as a function of H on punched paper tape. After each increment in H, a charged capacitor was discharged into a small coil located beneath the sample support window. The coil subsequently oscillated at approximately 1 MHz with a decaying amplitude. The amplitude initially exceeded the coercive force and as it decreased the bubble approached its equilibrium diameter. Typical diameter measurement precision is about ±0.2 μm.

DATA REDUCTION

The film thickness, h, material length, ℓ_o, $4\pi M_s$, H_k and d as a function of H are input parameters to a computer program. The thickness is measured by optical interference and ℓ_o and $4\pi M_s$ are obtained from stripe width and normal bubble collapse field measurements.[8] A value of H_k is determined at room temperature using an optical magnetometer[9] while values at other temperatures are inferred from this measurement and the temperature dependences of ℓ_o and $4\pi M_s$. The computer calculates the wall force $\mathcal{F}_w$ from the force equation, Eq. (2), where $F(d/h)$ is the generalized radial force.[5]

$$\mathcal{F}_w = F(d/h) - \frac{d}{h}\frac{H}{4\pi M_s} \quad . \qquad (2)$$

In addition, the program fits a function similar to Eq. (3),

$$\mathcal{F}_w = \frac{\ell'}{h} = \frac{\ell_o}{h}\left[1+(2N\delta_B/\pi d)^2\right]^{-1/2} \qquad (3)$$

derived[1] from Eq. (1), to the data of a particular bubble to obtain an N value. The function[10] used contains a demagnetizing correction of the order of $1/Q$ compared to 1 inside the square brackets of Eq. (3). The function yields about 2.5% higher values of $\mathcal{F}_w$ than Eq. (3) for the bubble parameters of sample 1 at 298°K for N = 67 when d/h is between 0.2 and 1.2. Fitting Eq. (3) to the same data leads to $N \approx 63$. The difference is smaller than the precision of the experiment.

The fitting procedure is a vertical least-squares fit on a plot of $\mathcal{F}_w$ vs d/h. An example of this representation[2] for sample 1 at 298°K is shown in Fig. 1 where 3 bubbles with differing N are plotted and fitted to theoretical curves. Constant field loci can be determined on Fig. 1 with Eq. (2). The intercepts between these curves and the constant N curves fit to the data can be used to extrapolate the collapse diameter when the measured collapse field is available. These points are indicated by asterisks for the 3 bubbles in Fig. 1.

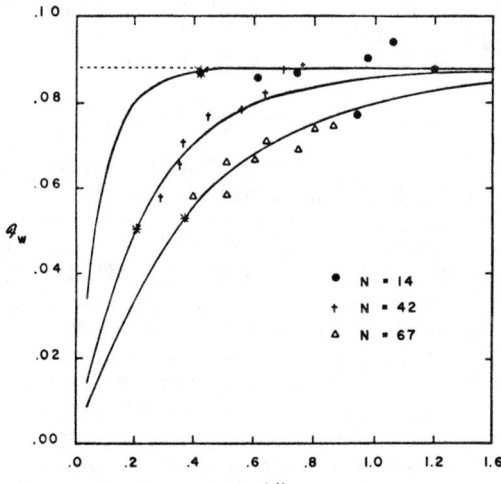

Fig. 1—$\mathcal{F}_w$ as a function of d/h for sample 1 at 298°K.

In Fig. 2, collapse diameters obtained by the extrapolation procedure are plotted[3] as a function of N for sample 1 at 298°K for various values of $N < N_c$, the change from a stable to an unstable solution to the modified force equation occurs at different values of d/h. This occurs when the second derivative of the total energy or (4) changes from positive to negative.

$$\ell' - \frac{d}{h}\frac{(\ell')}{(d/h)} - S_o(d/h) \quad . \qquad (4)$$

In (4), $S_o(d/h)$ is the radial stability function.[5] The locus of N and d/h corresponding to (4) = 0 is plotted as the solid curve on Fig. 2 for $N < N_c$. The largest $N = N_c$ for which (4) = 0 corresponds to the loss of unstable solutions for $N > N_c$. The experimentally observed collapse diameters decrease for $N < N_c$ in agreement with theory but for $N > N_i$ where $N_i \leq N_c$, the experimental collapse

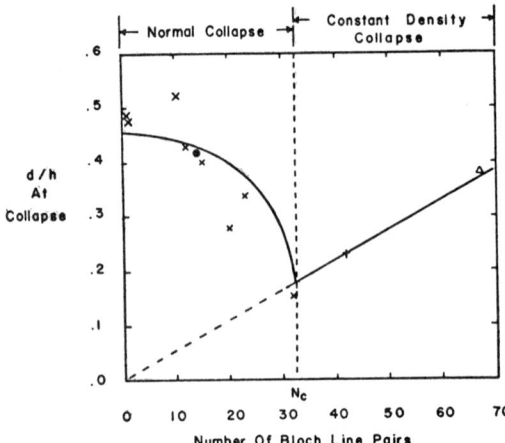

Fig. 2-Collapse diameters as a function of N for sample 1 at 298°K. The symbols ●, † and △ indicate the collapse points of the corresponding data of Fig. 1. Other bubbles are marked by X. Solid curves are calculated from the model described in the text.

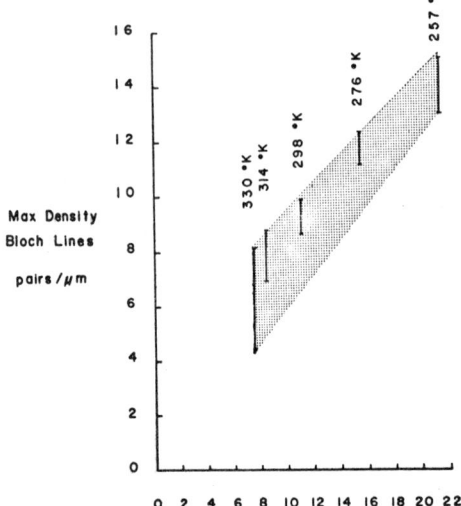

Fig. 3-Maximum Bloch line density as a function of Q for sample 1 at temperatures between 257 and 330°K.

diameters increase approximately linearly with N. The constant positive slope observed in Fig. 2 for $N > N_i \sim N_c$ is associated with a $\rho_m = 9.3$ pairs/μm. The fact that the maximum density line in Fig. 2 passes through the calculated normal collapse curve at N_c is fortuitous since two independent collapse mechanisms are involved. Measurements on other samples do not exhibit this fortuitous coincidence of N_i and N_c.

The maximum density of Bloch line pairs is found to increase almost linearly with increasing $Q = H_k/4\pi M_s$. An example is shown in Fig. 3 with data on sample 1 obtained over the temperature range 257 to 330°K. For these data $d\rho_m/dQ \sim 0.5$ pairs/μm.

CONCLUSION

Figures 1-3 illustrate the kind of data obtained on all samples. A summary of room temperature results is given in Table II for the 6 compositions of Table I. From H_k, ℓ_o and $4\pi M_s$, a value of exchange constant, A, has been computed and it is listed along with h, $4\pi M_s$ and Q values to characterize the samples used. Since ρ_m is approximately linear in Q, the room temperature value and $d\rho_m/dQ$ are listed for each sample. Finally the temperatures, T_H, above which it is not possible to generate hard bubbles[11] are listed. It should be noted that hard bubbles are stable above T_H, however.

At T_H, the Q values ranged from 2 to 5 for the samples measured. Sample 4 has the highest T_H (403°K) and it also has the lowest dQ/dT.

Table II: Properties of Samples Listed in Table I

Sample	h (μm)	A ergs/cm	$4\pi M_s$ Gauss	Q	ρ_m pairs/μm	$d\rho_m/dQ$ pairs/μm	T_H °K
1	6.5	1.1×10^{-7}	190	11.0	9.3	0.5	354
2	6.6	1.1×10^{-7}	185	11.2	11.0	0.9	358
3	6.0	2.3×10^{-7}	172	8.1	5.0	0.6	377
4	8.5	3.2×10^{-7}	107	6.0	4.4	1.4	403
5	4.4	3.4×10^{-7}	159	5.3	5.0	0.2	339
6	6.6	3.0×10^{-7}	200	7.0	7.0	0.95	338

ACKNOWLEDGMENTS

The epitaxial films used in this work were supplied by S. L. Blank, W. A. Bonner, B. S. Hewitt and S. J. Licht. Technical assistance was provided by E. R. Parshewski and R. J. Peirce. An unpublished version of Ref. 3 was graciously supplied by A. Rosencwaig.

REFERENCES

1. H. Nishida, T. Kobayashi and Y. Sugita, AIP Conference Proc. 10, 493 (1973).
2. W. J. Tabor, A. H. Bobeck, G. P. Vella-Coleiro and A. Rosencwaig, AIP Conference Proc. 10, 442 (1973); A. Rosencwaig, W. J. Tabor and T. J. Nelson, Phys. Rev. Lett. 29, 946 (1972).
3. A. Rosencwaig, Amer. Phys. Soc. Bull. 18, 398 (1973).
4. J. C. Slonczewski, A. P. Malozemoff and O. Voegeli, AIP Conference Proc. 10, 458 (1973).
5. A. A. Thiele, Bell Syst. Tech. J. 48, 3287 (1969).
6. H. J. Levinstein, R. W. Landorf, S. J. Licht and S. L. Blank, Appl. Phys. Lett. 19, 486 (1971).
7. H. Nishida, T. Kobayashi and Y. Sugita, IEEE Trans. Magnetics MAG-9, 517 (1973).
8. D. C. Fowlis and J. A. Copeland, AIP Conference Proceedings 5, 240 (1972).
9. P. W. Shumate, Jr., D. H. Smith and F. B. Hagedorn, J. Appl. Phys. 44, 449 (1973).
10. A. A. Thiele, to be published in J. Appl. Phys., Dec. 1973.
11. R. D. Henry, P. J. Besser, R. E. Warren and E. C. Whitcomb, IEEE Trans. Magnetics MAG-9, 514 (1973).

INTERACTIONS BETWEEN BLOCH LINES

A. Hubert[*]

IBM Research Center, Yorktown Heights, N. Y. 10598

ABSTRACT

The energy of a periodic arrangement of "winding" Bloch lines in a 180°-Bloch wall is calculated as a function of the distance between the Bloch lines and the material parameter $Q = K/2\pi M_s^2$. Due to a magnetostatic attraction and an exchange repulsion there results an equilibrium distance L_0 between the Bloch lines, which is of the order of $\sqrt{\pi A}/M_s$ for $Q \gtrsim 1$. The numerical results are compared with a uniform rotation model for the Bloch lines structure, which can be evaluated analytically. This ansatz proves to be a very good approximation for all Bloch line distances $L \lesssim L_0$. The uniform rotation model is therefore a reliable tool for the discussion of hard magnetic bubbles.

INTRODUCTION

Since the detection of Bloch lines (BLs) in the domain walls of bubbles[1,2,3] these Bloch lines have been shown to be responsible for many anomalous static and dynamic reactions of bubbles. In this paper we want to investigate the internal structure of BLs and especially their tendency to form stable clusters which has been inferred experimentally for example from stripe cutting experiments[2] and from an analysis of bubble collapse experiments.[4]

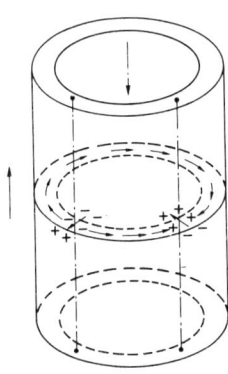

Fig. 1 A bubble with two BLs

Fig. 1 shows a schematic drawing of a bubble with two BLs. Since a single BL cannot exist in this geometry, the internal structure of BLs cannot be separated from their interactions. Qualitatively, one sees, that there will be a long range attraction due to their non-vanishing net magnetic charge, as indicated in Figs. 1 and 2. As the BLs approach each other, they will experience in addition an interaction due to exchange forces, which will be repulsive if the sense of rotation in both BLs is equal (as in Fig. 1), and attractive otherwise. BLs with opposite senses of rotation would therefore annihilate, and we will not consider them any further. In contrast, BLs of equal sign will have an equilibrium distance, which shall be calculated in the following.

We simplify the problem as far as reasonable: 1) We assume a planar domain wall instead of the circular wall of the bubble. 2) We assume a wall of infinite dimensions, thus neglecting the in-plane fields which deform the structure of the bubble wall near the platelet

[*]On leave from the Max-Planck-Institut für Metallforschung, Stuttgart, Germany.

surfaces. 3) Instead of a pair or a small number of BLs we assume a periodic array of BLs with period 2L. Under these conditions we calculate the energy per unit BL, subtracting the energy of the undisturbed Bloch wall. This BL energy should show a minimum at the equilibrium BL distance L_o.

THE UNIFORM ROTATION MODEL (URM)

Let the wall structure be described by the component of the magnetization direction perpendicular to the BLs $r(x,y)$ and the angle $\phi(x,y)$ defined in Fig. 2. Then the undisturbed Bloch wall in a uniaxial material with exchange constant A and anisotropy constant K is given by:

$$r = 1/\cosh(y/\sqrt{A/K}), \quad \phi = 0 \qquad (1)$$

If we replace the constant angle ϕ by a uniformly rotating angle, we come to a simple ansatz for a periodic BL structure:

$$r = 1/\cosh(y/\delta), \quad \phi = \pi x/L \qquad (2)$$

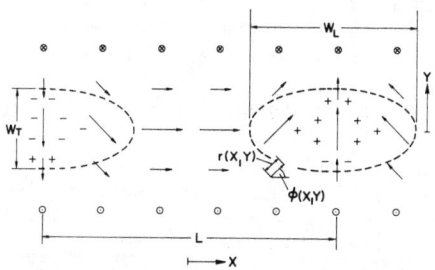

Fig. 2 Cross section through a periodic array of BLs.

which now contains two open parameters, the period 2L and the wall width parameter δ. This model has the advantage that its energy can be calculated analytically.

In the following we use $\sqrt{A/K}$ as the unit length and the exchange constant A as the unit for BL energies per unit length. Then one derives the following expressions for the surplus anisotropy and exchange energies of the BL relative to the undisturbed Bloch wall (1):

$$E_K = 2L(\delta - 1), \quad E_A = 2L(\frac{1}{\delta} - 1) + 2\pi^2 \frac{2\delta}{L} \qquad (3)$$

To discuss the magnetostatic or stray field energy we start from the density of the magnetic "charges" $\lambda = -\text{div } \vec{M}$, which we separate into a "σ-charge" $\lambda_\sigma = M_s r \sin\phi \cdot d\phi/dx$, which is symmetric with regard to the center plane of the wall, and a "π-charge" $\lambda_\pi = -M_s \sin\phi \cdot dr/dy$, which is antisymmetric. Integrating the self energy of these charges according to the laws of magnetostatics yields the simple result:

$$E_s = \frac{1}{Q} L\delta \quad \text{with} \quad Q = K/2\pi M_s^2 \qquad (4)$$

This is the same result as one would obtain neglecting the σ-charges and calculating the energy connected with the π-charges in the (localized) approximation which is justified for $L \gg \delta$. The non-local interactions between the σ-charges and between the π-charges at different BL positions cancel exactly.

This is somewhat surprising, since for widely separated BLs the non-local attraction between the σ-charges certainly dominates the interaction. However, as the BLs touch each other (and only this situation can be described by the URM), the density of the σ-charges must increase leading to a much weaker attraction than one would expect from the picture of magnetic line charges. But even with this change in character the remaining stray field energy (4) represents an attractive potential, which gives together with the exchange repulsion in (3) an equilibrium distance L_o. We calculate it by minimizing $E_K+E_A+E_s$ with regard to δ and L:

$$\delta_o = 1/(1+1/2Q) \quad , \quad L_o = \pi\sqrt{2} \cdot (Q/(1+1/2Q))^{1/2} \tag{5}$$

L_o as a function of Q is shown in Fig. 3. For a wide range of Q-values the equilibrium distance is thus predicted to be very near to $\pi\sqrt{2}$ times the BL width parameter $\sqrt{A/2\pi M_s^2}$.

For a given BL distance L the wall width parameter δ and the BL energy are calculated from the uniform rotation model to:

$$\delta(L) = (1+(1/2Q)+\pi^2/L^2)^{-1/2} \quad , \quad E_{BL}(L) = 4L[-1+1/\delta(L)]. \tag{6}$$

These results (Fig. 4) may be applied to the case when the bubble perimeter is smaller than the number of BLs times their equilibrium distance.

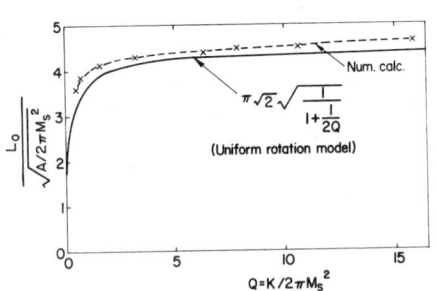
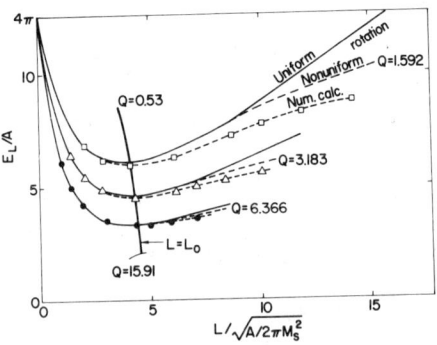

Fig. 3 The equilibrium BL-distance L_o as a function of Q.

Fig. 4 BL-energies as a function of their separation L.

ANALYTIC INVESTIGATION OF A NON-UNIFORM ROTATION MODEL

To investigate the significance of the URM we compare it with more general models. As a first step we replace (2) by:

$$\phi(x) = (\pi x/L) + \alpha\sin(2\pi x/L) \tag{7}$$

which for negative α describes more localized BLs separated by wider sections of more or less undisturbed Bloch walls. With this ansatz the anisotropy energy remains unchanged, the second term in the exchange energy is multiplied by $(1+2\alpha^2)$, and the stray field energy is calculated by means of a Fourier analysis to:

$$E_s = (L^2/\pi Q) \cdot G(\pi\delta/L) \quad , \tag{8}$$

with: $G(x) = x \cdot \sum_{i=0}^{\infty} [c_i^2 + 2(\tilde{c}_i^2 - c_i^2) F((2i+1)x)]$, $F(x) = x \cdot \sum_{k=0}^{\infty} (2k+1+x)^{-2}$ (8a)

$c_i = (\frac{\alpha}{2})^i \sum_{n=0}^{\infty} (\frac{\alpha}{2})^{2n} \frac{(-1)^n}{n!(n+i)!} [1 + \frac{(-1)^i \alpha/2}{n+i+1}]$, $\tilde{c}_i = \frac{c_i + \alpha(c_{i+1} + c_{i-1})}{2i+1}$, $c_{-1} = -c_0$

Minimizing numerically the total energy with respect to α and δ we get the dashed curves in Fig. 4. The main result is, that considerable deviations from the URM, as described by the parameter α, are found only for distances $L > L_o$. The position of the minimum is virtually not changed by the inclusion of the parameter α, thus justifying the URM for the range $L \lesssim L_o$.

It is interesting to compare the models (2) and (7) with the similar model of Rosencwaig et al.[5] Minor differences are that these authors assume also uniform rotation of the magnetization along the y-direction and that they introduce undisturbed Bloch wall segments in between the BL regions instead of the continuous ansatz (7). The essential limitation of their treatment is, however, that they use an approximate expression for the stray field energy, dropping the energy connected with the σ-charges and the non-local interactions of the π-charges, and thus loosing the attractive interactions between widely separated BLs altogether.

NUMERICAL CALCULATIONS FOR A MORE GENERAL MODEL

The true structure of a periodic array of BLs will not posses all the symmetries hitherto assumed. Certainly there will be some buckling of the wall contour in the regions of the BLs, due to the interactions of the σ- and the π-charges. Furthermore one has to allow for a non-uniformity of the wall width along the wall. Using such considerations, the following ansatz was chosen to give a better picture of the structure:

$r(x,y) = [1 + (1+\kappa)\sinh^2(a \cdot (y-b))]^{-1/2}$, $\phi(x,y) = \frac{\pi x}{L} + p_1 \cdot s \cdot v(p_2)$

$a = p_3 + p_4 \cdot c \cdot u \cdot v(p_5)$, $\kappa = p_6 + p_7 \cdot c \cdot u$, $b = p_8 \cdot c \cdot u \cdot v(p_9)$ (9)

$c = \cos \frac{2\pi x}{L}$, $s = \sin \frac{2\pi x}{L}$, $u = (1 + p_{10} c)^{-1/2}$, $v(p) = 1 - p + \frac{p}{1 + p_{11} y^2}$

This model includes the simpler models of the previous sections if all parameters but p_3 and p_1 vanish. The buckling is described by the function $b(x,y)$, the nonuniform wall width by the function $a(x,y)$, while the function $\kappa(x)$ describes a deviation from the wall profile of the undisturbed Bloch wall.

The energies which result from this ansatz were integrated numerically, using a rectangular lattice and replacing the continuous charge density by a step function which is constant in each lattice cell. The total energy was then minimized with regard to the structure parameters p_i and the results are shown in Figs. 3-6.

The deviations from the URM are most clearly visible from a plot of the BL width $W_L = \pi/(d\phi/dx)_{max}$ as a function of L (Fig. 5). For the URM W_L is equal to L by definition. Strong deviations from this law, approaching a more or less constant value — as one expects for isolated BLs — appear for distances $L > L_o$.

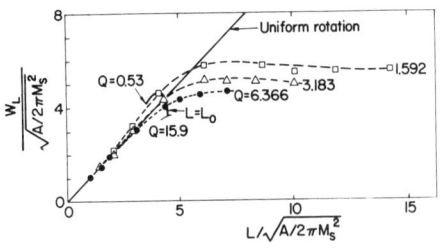

Fig. 5 BL-width W_L as a function of the BL-separation L.

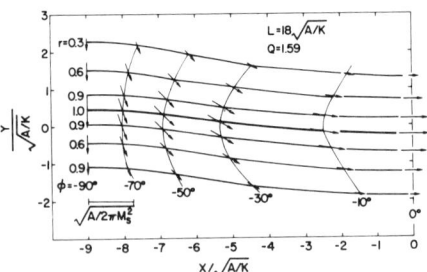

Fig. 6 Diagram of a numerically calculated BL-structure for $L \gg L_o$

CONCLUSIONS

The comparison of the more general calculations with the URM gives some insight into the mechanisms of the stray field energy. Due to the cancellation of non-local energy terms in the range $L \lesssim L_o$ the remaining energies are essentially of a local character. This leads to the conclusion, that even for non-periodic arrangments of BLs or for curved domain walls the interactions in a condensed BL array are mainly given by local nearest neighbor interactions. The results obtained in this paper should therefore also apply to BLs in bubbles. Slonczewski[4] used the equilibrium BL distance of Fig. 3 in his analysis of bubble collapse experiments and he found very good agreement with experiments. It should also be possible to observe BL interactions directly in the electron microscope.[1] In this case it will be essential to avoid in-plane magnetic fields which have been excluded in this analysis. Especially fields parallel to the wall would make a periodic array of BLs unstable.

ACKNOWLEDGEMENTS

The author appreciates many interesting discussions with J. C. Slonczewski and A. P. Malozemoff.

REFERENCES

1. P. J. Grundy, D. C. Hothersall, G. A. Jones, B. K. Middleton, R. S. Tebble, Phys. Stat. Sol.(a) 9, 79 (1972).
2. W. J. Tabor, A. H. Bobeck, G. P. Vella-Coleiro, A. Rosencwaig, Bell Syst. Techn. J. 51, 1427 (1972).
3. A. P. Malozemoff, Appl. Phys. Lett. 21, 149 (1972).
4. A. P. Malozemoff, J. C. Slonczewski, Int. Conf. Magnetism, Moscow 1973, Proceedings; J. C. Slonczewski, to be published.
5. A. Rosencwaig, W. J. Tabor, T. J. Nelson, Phys. Rev. Lett. 29, 946 (1972).

DOMAIN-WALL MASS IN BUBBLE FILMS

Ernst Schlömann
Raytheon Research Division, Waltham, MA 02154

ABSTRACT

The low velocity wall mass is calculated assuming that the film thickness is much larger than the width of Bloch lines. The method consists in first determining the wall-structure function $\phi(\zeta)$ [ϕ = angle between magnetization vector at middle of wall and wall plane, $\zeta = z/c$, z = distance from midplane of film, $2c = h$ = film thickness] for the static case and then solving a linear differential equation for the difference between the dynamic and the static wall structure functions. The dynamic wall structure may be characterized as a non-uniform rotation of the static wall magnetization which is concentrated primarily in the vicinity of "critical points" near the film surfaces. The local mass density is also concentrated near these points. Averaged over the film thickness the calculated mass is considerably larger than the value expected for an untwisted wall in an infinite medium.

INTRODUCTION

The internal structure of domain walls in bubble films differs from that expected in an infinite medium in that the orientation of magnetization at the midplane of the wall is not constant, but varies with the distance from the midplane of the film.[1-3] It has been shown theoretically that this "twisting" of the wall structure lowers the static wall energy,[3] but the effect is relatively small under the conditions encountered in practice. Slonczewski and collaborators have developed a theory applicable to moving walls[1,2] which shows that the effect of the wall twisting on the dynamic behavior is very strong, leading to velocity saturation at a level much lower than would be expected in bulk material. The Slonczewski theory[1,2] does not apply to slowly moving walls, and can not easily be adapted to this case. It is the purpose of the present paper to summarize the result of theoretical calculations of the wall structure at low velocities, and to derive numerical values for the low-velocity effective mass of domain walls. It will be shown that the wall mass is substantially larger for films than it is for bulk crystals. Some preliminary results on this subject have previously been published.[4]

The theory applies primarily to stripe domains in the demagnetized state (i.e., no bias field). By way of approximation the results can also be used for other domain structures such as isolated bubble domains, and bubble lattices.

The approach used in the present paper is suitable when the film thickness $h = 2c$ is very large compared to the width of Bloch lines

δ_ℓ. The latter is larger than the wall width by a factor of $\sigma^{-1/2}$ with $\sigma = 4\pi M_o / H_a$ (M_o = saturation magnetization, H_a = anisotropy field).

The wall structure can generally be characterized by the dependence of the angle ϕ between the wall and the magnetization at the midplane of the wall upon the distance z from the midplane of the film; or equivalently upon $\zeta = z/c$ (see for instance Fig. 1 of Ref. 4). For static stripe domains and $h/\delta_\ell \gg 1$ the dependence of ϕ upon ζ has recently been shown[5] to be approximately given by

$$\phi_o = \sin^{-1}[f(\zeta|\rho)] \qquad 0 \le \zeta \le \zeta_o$$
$$= \frac{\pi}{2} \qquad \zeta_o \le \zeta \le 1 \qquad (1)$$

with antisymmetric continuation for $\zeta < 0$. Here

$$f(\zeta|\rho) = \tanh^{-1}\left[\frac{\sinh(\pi\rho\zeta)}{\sinh(\pi\rho)}\right] \qquad (2)$$

and ρ is the aspect ratio of the stripe domains, defined as $\rho = c/d$, where d is the stripe width. ρ is a unique and known function of the ratio of material length ℓ to film thickness h.[6,7] The parameter ζ_o gives the location of the "critical point" of the wall structure, defined as the point at which $\phi_o(\zeta)$ first reaches $\pi/2$. It is mathematically defined by the condition $f(\zeta_o|\rho) = 1$.

Figure 1 shows the static wall structure function $\phi_o(\zeta)$ for several values of the aspect ratio ρ. The case $\rho = 0$ corresponds to the case of an isolated domain wall, and $\rho = 0.25$ corresponds approximately to the film thickness preferred for device applications.[8] The static wall energy of stripe domains and the velocity-momentum relationship at high velocities are further discussed in Ref. 5.

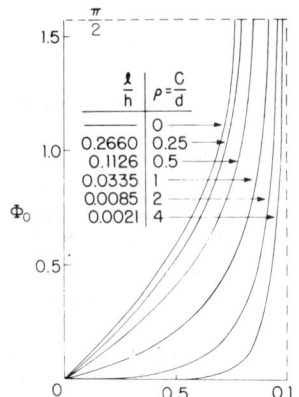

Fig. 1. Static wall structure $\phi_o(\zeta)$ for various domain aspect ratios ρ.

THEORY OF THE LOW-VELOCITY WALL MASS

For an infinite medium with uniaxial anisotropy the wall mass is known to be given by the "Döring formula"[9]

$$m_o = (1/2\pi\gamma^2)(K/A)^{1/2} \qquad (3)$$

where γ is the gyromagnetic ratio and K and A are the uniaxial anisotropy constant and the exchange stiffness constant. In the following the reduced notation defined in Ref. 3 is used and the calculated mass m is expressed in terms of the ratio m/m_o.

In order to determine the dynamic wall structure and hence the wall mass we first express the (reduced) wall energy $\widetilde{E}$ as a functional of the wall structure $\phi(\zeta)$. The function $\phi(\zeta)$ is then determined by minimizing the negative Lagrange function[10]

$$-\tilde{L} = \tilde{E} - 2\tilde{v}<\phi> \quad (4)$$

where $\tilde{v}$ is the (reduced) wall velocity and $<\phi>$ is the average of $\phi(\zeta)$. This variational principle is equivalent to the principle of minimizing the energy $\tilde{E}$ subject to the constraint that $<\phi>$ is constant.[2] After $\phi(\zeta)$ has been determined the mass ratio can be calculated as $m/m_0 = \sigma<\phi>/\tilde{v}$.

For $\sigma \ll 1$ the variational principle $\delta L = 0$ yields a nonlinear differential equation for $\phi(\zeta)$

$$-\tilde{c}^{-2}\phi'' + \sigma \cos\phi \, [\sin\phi - f(\zeta|\rho)] = \tilde{v} \quad (5)$$

where the prime denotes differentiation with respect to ζ and $f(\zeta|\rho)$ is given by Eq. (2). Assume now that this equation has been solved for the static case ($\tilde{v} = 0$) and that the solution is $\phi_s(\zeta)$. The dynamic wall structure can then be expressed as $\phi = \phi_s + \delta$. For low velocities $\delta \ll 1$, and $\delta(\zeta)$ therefore obeys the linear differential equation

$$-\tilde{c}^{-2}\delta'' + \sigma F(\zeta)\,\delta = \tilde{v} \quad (6)$$

where

$$F(\zeta) = \cos^2[\phi_s(\zeta)] - \sin^2[\phi_s(\zeta)] + \sin[\phi_s(\zeta)]\,f(\zeta|\rho) \quad . \quad (7)$$

If $\delta \ell \ll h$, the solution of Eq. (6) can be expanded in powers of $(\tilde{c}^2 \sigma)^{-1}$ since $h/\ell = (2/\pi)\,\tilde{c}\,\sigma^{1/2}$. Taking only the first term the mass ratio is then

$$m/m_0 = <1/F(\zeta)> \quad . \quad (8)$$

Thus $m_0/F(\zeta)$ can be interpreted as a local mass density.

Of particular interest is the behavior of the function $1/F(\zeta)$ near the critical points. Consider first the limiting case where $\delta\ell/h \to 0$. Under these conditions the static wall structure $\phi_s(\zeta)$ is given by the function $\phi_0(\zeta)$ defined by Eq. (1) and depicted in Fig. 1. It can readily be shown from Eq. (7) that in this case $F(\zeta)$ vanishes linearly at the critical points. The vicinity of the critical points, therefore, gives a divergent contribution to the mass ratio [see Eq. (8)]. This means that a more accurate solution of the static wall structure is required.

REFINED CALCULATION OF THE STATIC WALL STRUCTURE

A better approximation to the static wall structure can be obtained by modifying the previously defined function $\phi_0(\zeta)$ [see Eq. (1)] in the vicinity of the critical points. Calculations have been carried out with the trial function

$$\phi_s(\zeta) = (\pi/2) - b(\zeta_2 - \zeta)^\lambda \quad , \quad \zeta_1 \le \zeta \le \zeta_2 \quad (9)$$

assuming $\phi_s = \phi_0$ outside the interval $\zeta_1 < \zeta < \zeta_2$. Here b, λ, ζ_1 and ζ_2 are variational parameters to be determined from the condition of minimal energy. They are subject to the side condition that

$\phi_s(\zeta_1) = \phi_0(\zeta_1)$. Thus there are effectively three independent variational parameters.

The variational calculation yields $\lambda = 2.34$, but for simplicity $\lambda = 2$ has been used in the numerical calculations described below. Figure 2 shows the behavior of the wall structure function in the vicinity of the critical point for the single-wall case ($\rho = 0$) and different ratios of h to $\delta\ell$. In the limit $h/\delta\ell \to \infty$ the wall structure is accurately described by $\phi_0(\zeta)$ [see Eq. (1)], but for typical experimental conditions the actual structure differs considerably from $\phi_0(\zeta)$ near the critical point.

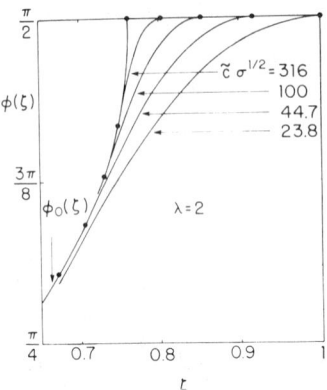

Fig. 2. Static wall structure in the vicinity of the critical point for various values of the ratio of film thickness to Bloch-line width.

RESULTS AND DISCUSSION

Figure 3 shows a representative plot of the local mass density $1/F(\zeta)$ versus distance from the midplane of of the film, as calculated from Eqs. (8) and (9). The ratio $h/\delta\ell$ is only moderately large for the case shown. Thus the method of approximation used in the calculation is not very accurate. This is apparent in Fig. 3 in that the two branches of $1/F(\zeta)$ are noticeably separated at $\zeta = \zeta_1$. It may be seen that the wall mass is largely concentrated in the vicinity of the critical point and that the mass ratio is significantly larger than unity.

Figure 4 shows the calculated mass ratio as a function of h/ℓ (where ℓ is the material length) for various values of σ. The end points of the curves correspond to the conditions at which the method used in the calculations becomes inapplicable because the point ζ_2 in the variational trial function for the static case reaches the film surface.

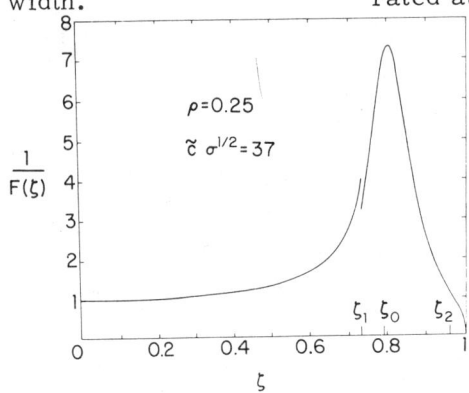

Fig. 3. Local mass density (on a reduced scale) for a film thickness in the preferred range ($h/\ell \simeq 4.8$).

Experimental observation of domain wall resonance in platelets and films containing stripe domains has been reported by Vella-Coleiro et al.,[11] Argyle and Malozemoff,[12] and Moody et al.[13] Vella-Coleiro et al.[11] used garnet platelets with an h/ℓ ratio near $10^4 - 10^5$. The wall mass inferred from their experiments agrees well with Eq. (3). This is consistent with the present theory since the theoretically expected m/m_0 ratio is very close to unity under these conditions.

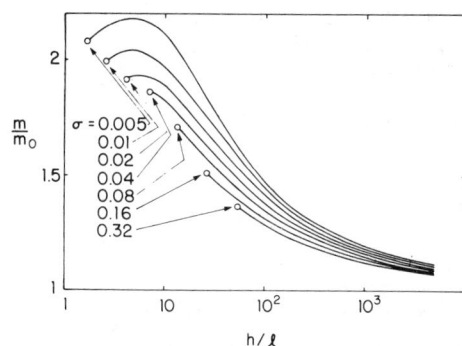

Fig. 4. Dependence of the mass ratio (for normal walls) on the film thickness for given material parameters.

In the experiments of Argyle and Malozemoff[12] and Moody et al.[13] the h/ℓ ratio was in the range of 2 to 20. The wall masses inferred from these experiments are generally larger (by a factor of about 2 - 10) than the theoretical value based on the Döring formula. In this qualitative sense the experimental data agrees with the theory described in the present paper. A detailed comparison is not possible at this time, partly because the theory is only marginally applicable to most of the data and partly because of an excessive scatter in the experimental data, which is not presently understood.

In the Argyle-Malozemoff experiments[12] it was observed that the resonance spectrum consisted of two distinct peaks. This may be attributable to the fact that the wall mass is not uniformly distributed, so that higher order modes of vibration can be excited.

ACKNOWLEDGEMENTS

It is a pleasure to thank J. Newell, L. Spiniello, and B. Matsinger for their help and J.C. Slonczewski, B.W. Argyle, A.P. Malozemoff and R.W. Shaw for stimulating discussions.

REFERENCES

1. B.E. Argyle, J.C. Slonczewski, and A.F. Magadas, AIP Conf. Proc. No. 5, 175 (1972).
2. J.C. Slonczewski, J. Appl. Phys. 44, 1759 (1973).
3. E. Schlömann, Appl. Phys. Letters 21, 227 (1972); J. Appl. Phys. 44, 1837 (1973); 44, 1850 (1973).
4. E. Schlömann, AIP Conf. Proc. No. 10, 478 (1973).
5. E. Schlömann, J. Appl. Phys., Dec. 1973 (to be published).
6. C. Kooy and U. Enz, Phillips Res. Rep. 15, 7 (1960).
7. R.W. Shaw, D.E. Hill, R.M. Sanford, and J.W. Moody, J. Appl. Phys. 44, 2346 (1973).
8. A.A. Thiele, Bell Syst. Techn. J. 50, 725 (1971).
9. S. Chikazumi, "Physics of Magnetism," J. Wiley and Sons, Inc., New York, 1964, Sect. 16.5.
10. E. Schlömann, J. Appl. Phys. 43, 3834 (1972).
11. G.P. Vella-Coleiro, D.H. Smith and L.G. Van Uitert, J. Appl. Phys. 43, 2428 (1972).
12. B.W. Argyle and A.P. Malozemoff, AIP Conf. Proc. No. 10, 344 (1973).
13. J.W. Moody, R.W. Shaw, R.M. Sandfort, and R.L. Stermer, Paper 17-4 at 1973 Intermag Conf. and private communication.

CONTROLLED COMPENSATION WALLS AND COMPROMISE COMPENSATION WALLS IN GARNET FILMS BY THE SILICON ANNEALING TECHNIQUE

R. C. LeCraw and R. Wolfe
Bell Laboratories, Murray Hill, New Jersey 07974

ABSTRACT

The Si annealing technique has been used on Ga-substituted LPE bubble garnet films to reduce $4\pi M$ through zero and produce controlled, well-defined compensation walls. A Si film was deposited on a garnet film, patterned photolithographically, and annealed at 525°C for 40 hours in O_2. This produced light and dark Faraday contrast regions separated by compensation walls which were independent of the bias field H_0. In the course of annealing, a new type of domain was observed. After 20 hrs. of annealing, the walls between the light and dark areas became fuzzy at $H_0 \sim 80$ Oe. At ~ 120 Oe they abruptly began expanding over large areas forming a grey (or 3rd color) domain corresponding to $\sim 0°$ Faraday rotation. Our model is as follows: At 20 hrs., under the Si, a compensation plane ($4\pi M = 0$) is produced near the middle of the film and approximately parallel to the surface. In small H_0 a "compromise compensation wall" perpendicular to the surface separates the light and dark areas. As H_0 is increased >80 Oe, the wall changes its shape to conform to the zero $4\pi M$ plane producing a true compensation wall. The two regions above and below this plane tend to cancel in rotation and create the new kind of grey domain. A new type of mobile grey bubble is also described.

INTRODUCTION

The development of the silicon annealing technique for fine scale localized control of the magnetization, $4\pi M$, in LPE bubble garnet films[1] has made it possible to reduce $4\pi M$ through zero and produce controlled, well-defined compensation walls. In the simplest case, such a wall is a plane through the film thickness separating two regions in one of which the tetrahedral iron moment, M_{tet}, predominates, whereas in the other the combined rare earth and octahedral moment, M_{oct}, predominates. The first observation of such a wall was reported by Mee[2] in Ga doped YIG with a compositional gradient. Similar walls were observed and named compensation walls by Krumme and Hansen[3,4]. However, because of the thickening of their walls in an applied field we believe that what they observed was not a true compensation wall but a "compromise compensation wall", which we shall explain below.

The Si annealing technique used to produce the controlled compensation walls discussed in this paper consists of depositing a thin film of Si on a bubble garnet film and annealing in oxygen, or an inert gas, at temperatures from 500-700°C. The Si annealing creates oxygen vacancies at the Si-garnet interface which diffuse through the garnet film. These oxygen vacancies greatly increase the redistribution rate of Ga ions between octahedral and tetrahedral sites. Thus in a material in which M_{tet} dominates the magnetization, $4\pi M$ is decreased under the Si. If enough Ga is present, $4\pi M$ can be reduced through zero and reversed.

In the experiment described here, a 1500 Å film of Si was deposited on a film of $Y_{2.5}Sm_{0.5}Fe_{3.7}Ga_{1.3}O_{12}$, patterned photolithographically and annealed at 525°C for 40 hrs. in O_2. (Much shorter times would be required at higher temperatures.) After annealing the remaining Si was removed. The Ga content in this sample was sufficient so that $4\pi M$ under the Si was reversed. Thus with the sample magnetically saturated, light and dark Faraday contrast regions were produced separated by compensation walls whose position and thickness were essentially independent of the applied field H_0. (Figure 1). The light areas are those from which the Si had been removed before annealing so that M_{tet} still dominates the magnetization. In the dark areas, previously covered by Si, $4\pi M$ was reduced through zero and M_{oct} dominates the magnetization. The contrast is observed because the Faraday rotation at room temperature is dominated by M_{tet}.[5] (At 4.2°K, the rotation is dominated by M_{oct}[6]). The compensation walls between the light and dark areas are well localized due to the steep gradients of $4\pi M$. When $H_0 = 0$ simple stripe domains can exist in the larger light areas. At $H_0 = 50$ Oe the pattern in Fig. 1 is obtained and remains essentially unchanged for all H_0.

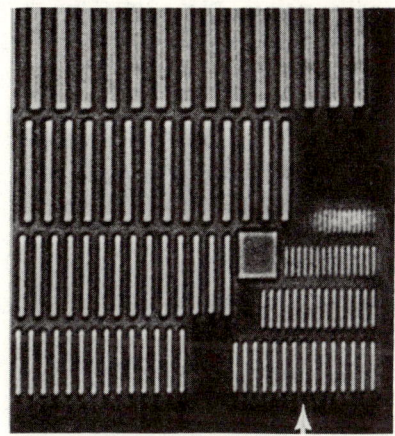

Fig. 1. Controlled compensation walls in a 4.2 μm thick film of $Y_{2.5}Sm_{0.5}Fe_{3.7}Ga_{1.3}O_{12}$. In 200 Oe bias field, the Faraday contrast shows that the dark areas around the bars (2μm thick at arrow) have passed through compositional compensation point after 1500 Å of patterned Si was annealed at 525°C for 40 hrs. in O_2 and then removed.

In the course of these experiments, a new type of domain was observed. After 20 hrs. of annealing at 525°C in O_2 of a sample from the same wafer as that used in Fig. 1 and covered entirely with 1500Å of Si, the domains appeared as in the left side of Fig. 2 for $30 < H_0 < 80$ Oe. At ~80 Oe the walls between the light and dark areas became fuzzy. As H_0 was increased further to 120 Oe, the walls abruptly began expanding over large areas, forming a grey (or 3rd color) domain. The Faraday rotation in the grey domains was found to be very nearly zero. As H_0 was

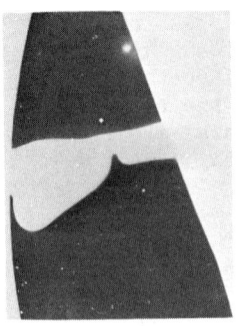
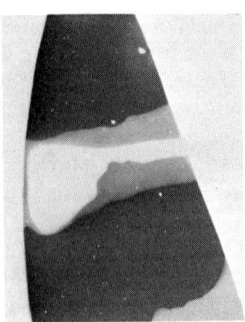

50 Oe   200 Oe

Fig. 2. A piece from the same wafer as that of Fig. 1, but with the Si covering the entire sample and annealed at 525°C for 20 hrs. in O_2. The walls separating the light and dark areas on the left are "compromise compensation walls". On the right the new grey domains are clearly seen. This is the state shown schematically in Fig. 3c. The sample is ~3.5 mm high.

increased still further to the maximum available of 420 Oe, the grey domains spread over almost the entire sample.

Our model for the grey domains, based on earlier unpublished observations by J. F. Dillon, Jr. on GdIG, is as follows: At 20 hrs., under the Si, $4\pi M$ was reduced below zero in the upper half of the film while still remaining normal in the lower part. Thus a compensation plane ($4\pi M = 0$) was produced near the middle of the film and not quite parallel to the film surface as shown in Fig. 3a. The wall shown here is a "compromise compensation wall". This minimum energy wall, perpendicular to the film surface, is associated with the compensation phenomenon but is not along the $4\pi M = 0$ plane. Its position is such that it minimizes the magnetostatic energy. The position of this wall moves rapidly with changing temperatures, because the position of the $4\pi M = 0$ plane is temperature dependent. As H_O is increased the compromise compensation wall tilts (Fig. 3b). This corresponds to the fuzziness which occurs at $H_O = 80$ Oe. As H_O is increased further, the wall changes its shape (Fig. 3c) to minimize the total magnetostatic and wall energy. At still larger H_O it conforms to the $4\pi M = 0$ plane, producing a true compensation wall (Fig. 3d). The two regions above and below this plane tend to cancel in rotation and create the new kind of grey domain. As the garnet film was thinned by chemical etching, the position of the compromise compensation wall moved in a manner consistent with this "two-layer" model of the magnetization.

The apparent increase in the "compensation wall" thickness with applied field observed by Krumme and Hansen[3] can be interpreted in terms of the same mechanism which explains these grey domains. The true compensation plane in their (111) platelet could be approximately parallel to a (112) growth plane which makes an angle of 19° 28' with the (111) surface. At low fields the wall which they observed could be a compromise compensation wall perpendicular to the surface. At high fields the domain wall conforms to the sloping compensation plane, giving rise to the observed spacial variation to the Faraday rotation and the apparent widening of the wall of 250µm in their 100µm thick platelet.

The grey domains were subsequently seen under conditions which

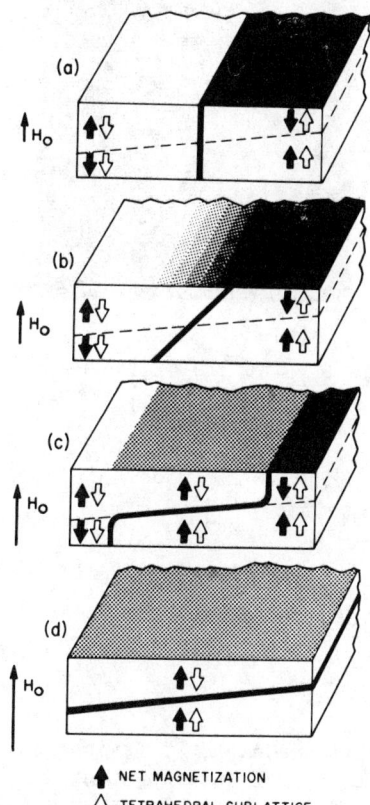

Fig. 3. Model for the "compromise compensation wall" and "grey domain". The dashed line delineates the compensation plane ($4\pi M = 0$). In low fields, 3(a), a "compromise compensation wall" separates the light domain in which the reverse magnetization (upper layer) predominates, from the dark domain in which the normal magnetization (lower layer) predominates. As H_o increases the wall tilts and becomes fuzzy, 3(b), and then breaks into a grey domain, 3(c) as it follows the compensation plane. In larger fields, 3(d), magnetostatic energy is greater than the wall energy and a true compensation wall is formed.

did not involve Si annealing. In a film of approximate composition $Y_{2.5}Sm_{0.5}Fe_{3.65}Ga_{1.35}O_{12}$, annealed at T = 1100°C in O_2 (with no Si) to adjust $4\pi M$ to approximately zero, they were seen over the range 46° < T < 56°C. Also on a sample of $Y_{1.67}Sm_{0.13}Ca_{1.2}Fe_{3.8}Ge_{1.2}O_{12}$ with a small negative $4\pi M$ at room temperature, they were seen at T ∼ 112°C. (Fig. 4). In these latter two experiments, a small compositional gradient normal to the film surface was produced during the LPE growth of the films, and not by Si annealing.

A new kind of mobile bubble is shown in Fig. 4. In the applied bias field of 80 Oe, a higher local field was produced by the Permalloy wire held near the surface of the garnet and a localized grey domain was formed. This bubble is in the upper half of the film, bounded by a vertical 180° wall around its circumference and a true compensation wall at its bottom. It could be moved in either the dark or light areas by moving the probe.

CONCLUSIONS

Using the Si annealing technique, sharply defined compensation walls can be made in any desired pattern in bubble garnet films containing Ga or Al. Such a pattern as shown in Fig. 1 may be useful in magnetooptic applications using a transmitted beam or a guided wave within the film. Controlled compensation walls have also been produced independently by Krumme and Hansen and were used in a thermomagnetically written magnetooptic memory application.[7] The new grey domains and mobile grey bubbles, associated with the compensation walls, may find application in future bubble devices.

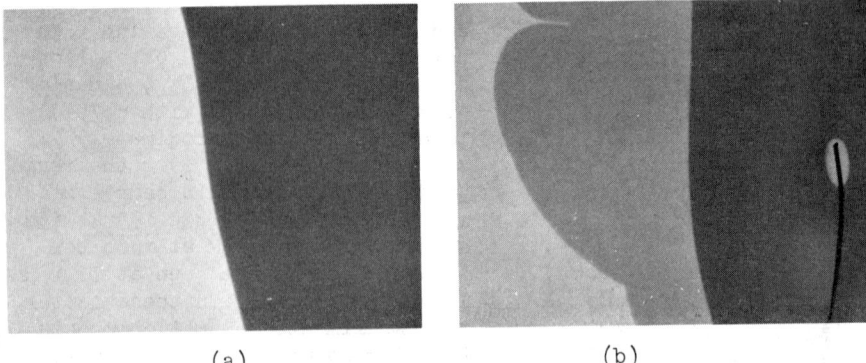

Fig. 4. A film of $Y_{1.67}Sm_{0.13}Ca_{1.2}Fe_{3.8}Ge_{1.2}O_{12}$ at 112°C with H_0 = 25 Oe in (a) and 80 Oe in (b). In (b) a large grey domain is seen, corresponding to Fig. 3c. A new kind of mobile grey bubble is also shown, produced by the increased local field from the Permalloy wire near the film surface. The section shown here is 2mm wide.

ACKNOWLEDGMENTS

We should like to thank J. F. Dillon, Jr., F. B. Hagedorn, and R. D. Pierce for many useful discussions in developing the model for the grey domains, and S. L. Blank for providing the garnet films.

REFERENCES

1. R. C. LeCraw, P. A. Byrnes, Jr., W. A. Johnson, H. J. Levinstein, J. W. Nielsen, R. R. Spiwak and R. Wolfe, IEEE Intermag Conf., Wash., D. C. 1973, to be published; R. Wolfe, R. C. LeCraw, J. C. North, International Conf. on Magnetism, Moscow, Aug., 1973.
2. C. D. Mee, Contemp. Phys. 8, 385 (1967).
3. J.-P. Krumme and P. Hansen, Appl. Phys. Lett. 22, 312 (1973).
4. J.-P. Krumme and P. Hansen, J. Appl. Phys. 44, 3805 (1973).
5. R. C. LeCraw and H. Matthews, unpublished.
6. H. Matthews, S. Singh, and R. C. LeCraw, Appl. Phys. Lett. 7, 165 (1965).
7. J.-P. Krumme and P. Hansen, to be published.

MAGNETIZATION PROCESSES INVOLVING PLANAR COMPENSATION WALLS

O. Voegeli & E. B. Moore

IBM Corporation, GPD

San Jose, California 95114

Annealing of chemically deposited $Gd_3Fe_5O_{12}$ films[1] on single crystal $Gd_3Ga_5O_{12}$ substrates at temperatures above 800°C causes appreciable cation cross-diffusion between film and substrate[2], as is evident from Auger composition analyses across the film-substrate interface. The resulting Ga concentration profile through the film thickness determines the position of a compensation plane, which separates the film into two layers, one being above and the other below compensation temperature. Depending on the direction of magnetization in each layer, there exist, as a function of a field applied along the film normal, four different states of magnetization. Recordings of Faraday rotation versus applied field show antiparallel magnetization at low fields and parallel magnetization at high fields. The latter state requires the formation of a planar compensation wall[3] about the compensation plane, whose energy, in combination with the field interaction energy of each layer, determines the equilibrium state at any given field value. As the position of the compensation plane changes with temperature, so do the magnetic moments of the two layers and different sequences of states occur during magnetization reversal at different temperatures. Respective sequences and associated transition field values are accurately predicted by assuming that a transition between two successive states occurs when the energy difference between these states equals some coercive energy barrier. The height of this barrier is proportional to the thickness of material whose magnetization reverses during the transition.

1. E. A. Giess and R. M. Potemski, IBM Tech. Disclosure Bull. 9, 960 (1967), 10, 852 (1967)

2. R. M. Josephs, IEEE trans. Magnetics 3, 553 (1970); B. F. Stein, J. Appl. Phys. 3, 1262 (1970)

3. P. Hansen & J. P. Krumme, API Conf. Proc. 10, 423 (1972).

THE EXCHANGE CONSTANT OF GALLIUM SUBSTITUTED IRON GARNETS

R. D. Henry and D. M. Heinz
Electronics Research Division, Rockwell International,
Anaheim, Ca. 92803

ABSTRACT

Ferromagnetic resonance techniques have been used to measure the exchange constant of gallium substituted iron garnets. The experimental results are compared to values calculated from Neel molecular field theory. Wall energies for bubble domain compositions are shown to agree with these measured values but not the values of the exchange constant calculated from $A = A_{YIG}\left(T_N/560°K\right)$.

INTRODUCTION

This investigation was initiated by an attempt to reconcile bubble domain wall energies obtained in different ways. An expression for the wall energy (σ_w) of a uniaxial material is

$$\sigma_w(A, K_u) = 4\sqrt{AK_u} \qquad (1)$$

where A is the exchange constant and K_u is the uniaxial anisotropy. The value of K_u is normally measured using techniques similar to that described by Shumate [1] or using ferromagnetic resonance. The value of A up to now has been calculated using a simplified molecular field theory. For magnetic garnet films, this is an approximation since the cubic anisotropy K_1 is ignored. A computer program, which utilizes both K_u and K_1 by integrating the anisotropy energy through a 180 degree rotation of the magnetization vector in a Bloch wall, has been used to provide a more accurate calculation of $\sigma_w(A, K_u, K_1)$ [2]. In contrast with the above method, the magnetization (M) and characteristic length (ℓ) are often obtained in characterizing bubble domain films using the Fowlis and Copeland [3] method. The wall energy is then obtained from

$$\sigma_w(M, \ell) = 4\pi M^2 \ell. \qquad (2)$$

Agreement between the two methods has been poor and this disparity has been traced to the value of A used with K_u (or K_u effective) in Eq. (1).

FILM GROWTH

For these studies, the garnet films were grown by the isothermal liquid phase epitaxy dipping technique [4]. The melts were prepared from 99.9 percent or purer Ga_2O_3, Gd_2O_3, La_2O_3, Tm_2O_3 and Y_2O_3, Reagent Grade Fe_2O_3 and PbO, and Purified B_2O_3. The molar ratios of Fe_2O_3 to rare earth and yttrium oxides were 20, and PbO to B_2O_3 were 16. Solute concentrations of 9.0 mole percent produced saturation temperatures of about 890C. Using a substrate rotation rate of 60 RPM, growth of garnet films on gadolinium gallium garnet (GGG) wafers was carried out at 5 to 10 C below the saturation temperature to give growth rates of 0.1 to 0.2 μm/ minute.

Following film growth, the thicknesses were measured using a standard infrared interference technique. Film-substrate lattice mismatch was measured by X-ray double crystal rocking curves to be between 0.005 and 0.008Å. Neel temperature measurements indicate the gallium distribution of the as-grown films to be representative of an equilibrium temperature near 1000C, in agreement with garnet film annealing studies [5].

Initially, FMR measurements were made on a thin film of the bubble material composition $(YGdTm)_3Ga_{0.8}Fe_{4.2}O_{12}$. However, because only two exchange modes were observed, interpretation of the spectra and surface boundary conditions was difficult. The linewidths of $Y_3(GaFe)_5O_{12}$ are an order of magnitude less than those of $(YGdTm)_3(GaFe)_5O_{12}$ so that the observation of several higher order exchange modes is possible. Since the lattice mismatch between $Y_3(GaFe)_5O_{12}$ and GGG is great enough to cause films to craze, substitution of a large ion for some yttrium ions was used to increase the lattice constant of the film. In as much as a narrow linewidth was desirable for these measurements, lanthanum was chosen to control the film-substrate lattice mismatch rather than another rare earth ion. Thus, this study was carried out on films of $(YLa)_3(GaFe)_5O_{12}$.

THEORY

According to the Neel molecular field theory of ferrimagnetic materials for sublattices aligned anti-parallel, the exchange constants are related to the molecular field coefficients (N_{ij}) by

$$A_{ij} \propto N_{ij} \tag{3}$$

and the Neel temperature is related to N_{ij} by [6]

$$T_N = -1/2[C_aN_{aa} + C_dN_{dd}] + 1/2[(C_aN_{aa} - C_dN_{dd})^2 + 4C_aC_dN_{ad}^2]^{1/2}, \tag{4}$$

where C_i is the effective Curie constant of the ith sublattice. Hence if one assumes that $N_{ad} \gg N_{aa}$, N_{dd} then it follows that the exchange constant of a gallium substituted material will be given by

$$A = A_{YIG}\left(T_N/560°K\right) \tag{5}$$

Equation (5) has been used [7] in calculating A for use in Eq. (1).

One can experimentally obtain values of the exchange constant from measurements of spin-wave spectra and measurements of the sample magnetization. For an infinite film the dispersion relation for spin-waves of wave vector k_z is

$$H_0 = \omega/\gamma + 4\pi M - H(K_u) - H(K_1) - \frac{2A}{M} k_z^2 \tag{6}$$

where ω is the frequency, γ the gyromagnetic ratio, H_0 the resonance field, and $H(K_u)$ and $H(K_1)$ the effective field of the uniaxial and cubic anisotropies respectively.

For complete spin pinning on both film surfaces k_z is given by

$$k_z = \frac{n\pi}{h}, \quad n = 1, 3, 5, 7, \ldots \tag{7}$$

with h the film thickness. For complete pinning on one surface and free precession on the other surface k_z is given by

$$k_z = \frac{n\pi}{h}, \quad n = \frac{1}{2}, \frac{3}{2}, \frac{5}{2}, \frac{7}{2}, \ldots \tag{8}$$

Of course another possible boundary condition corresponds to free spin precession on both surfaces and then only the uniform mode is observed ($k_z = 0$).

For the compositions grown and used for these experiments the Neel temperatures were 507K, 455K and 369K. The boundary conditions that existed after film growth would make an interesting study in itself, but time did not permit such an investigation. The film with $T_N = 560K$ appears to have boundary conditions corresponding to Eq. (7) as grown. Interestingly enough, the other two films had only the uniform mode as grown and we had to create the boundary conditions corresponding to Eq. (8) by post growth treatment. For the $T_N = 455K$ film, a 0.3 μm polishing slurry was used with an ultrasonic cleaner to damage the outer surface of the film to produce surface pits. The demagnetizing fields from the pits then produced spin pinning. When this technique was used with the $T_N = 369K$ film, still only one mode was observed. For this film it was necessary to ion implant the outer surface to create spin pinning and hence higher order modes.

EXPERIMENTAL RESULTS

An example of a spin-wave spectra for the $T_N = 455K$ at room temperature is shown in Fig. 1. T_N was measured using ferromagnetic resonance techniques.

The value of the exchange constant for this film obtained from Fig. 1 and vibrating sample magnetometer measurements is $A_m = 2.03 \times 10^{-7}$ erg/cm. This is quite different from the value obtained from Eq. (5),

$A_c = 3.33 \times 10^{-7}$ erg/cm.

Values of the exchange constant at room temperature for the three compositions are shown in Fig. 2. The exchange constant calculated from Neel molecular field theory and assuming that $N_{ad} \gg N_{aa}$, N_{dd} is shown by the dashed curve. The value of the exchange constant for YIG is from Henry et al[8]. Obviously Eq. (5) should not be used in calculating the exchange constant of substituted iron garnets.

DISCUSSION OF RESULTS

We have used the measured values of A (A_m) and values of A (A_c) calculated from Eq. (5) to calculate $\sigma_w(A_c, K_u)$, $\sigma_w(A_m, K_u)$, $\sigma_w(A_c, K_u, K_1)$ and $\sigma_w(A_m, K_u, K_1)$ for several different bubble compositions. These are compared in Table I to the values obtained for $\sigma_w(M, \ell)$. In all cases the best agreement is obtained between the values calculated for $\sigma_w(A_m, K_u, K_1)$ and $\sigma_w(M, \ell)$. Due to uncertainties inherent in the measurements of film thickness, strip width and collapse field, the values of $\sigma_w(M, \ell)$ probably have an accuracy of no better than ±10%. However, some general features are apparent in Table I. The values of $\sigma_w(A_c, K_u, K_1)$ and $\sigma_w(A_c, K_u)$ are always larger than the values of $\sigma_w(M, \ell)$. Interestingly enough the values of $\sigma_w(A_m, K_u, K_1)$ and $\sigma_w(A_m, K_u)$ are always less than or equal to $\sigma_w(M, \ell)$; but the differences in this case are all within experimental error.

The deviation from a simple linear relationship between A and T_N is in agreement with the results of Dionne[9]. In the work of Dionne it was found that the molecular field coefficients N_{ij} are dependent upon the gallium substitution. Even though Nad > Ndd, Naa for YIG, this is not necessarily the case even for moderate gallium substitutions. Dionne's results indicate that canting of the sublattices begins with small substitutions. The results of Fig. 2 appear to be in agreement with this.

In summary, the use of A_m in calculating wall energies leads to results consistent with values of wall energies obtained from measurements of bubble properties.

ACKNOWLEDGMENTS

The authors gratefully acknowledge helpful discussions with P. J. Besser in the course of this work and film growth by E. F. Grubb and F. S. Stearns. This research was supported by the Air Force Materials Laboratory under Contract F33615-73-C-5017.

REFERENCES

1. P. W. Shumate, Jr., D. H. Smith and F. B. Hagedorn, J. Appl. Phys. 44, 449 (1973).
2. G. R. Pulliam and F. A. Pizzarello, AIP Conf. Proc. 10, 413 (1973).
3. D. C. Fowlis and J. A. Copeland, AIP Conf. Proc. 5, 240 (1972).
4. H. J. Levinstein, S. Licht, R. W. Landorf and S. L. Blank, Appl. Phys. Lett. 19, 486 (1971).
5. D. H. Smith, F. B. Hagedorn and B. S. Hewitt, J. Appl. Phys. 44, 4177 (1973).

6. A. H. Morrish, The Physical Principles of Magnetism (John Wiley & Sons, Inc., New York, 1965).
7. L. G. Van Uitert et al, Mat. Res. Bull. **6**, 1185 (1971).
8. R. D. Henry, S. D. Brown, P. E. Wigen and P. J. Besser, Phys. Rev. Lett. 28, 1272 (1972).
9. G. F. Dionne, J. Appl. Phys. **41**, 4874 (1970).

Table I. Calculated Wall Energies

T_N^d (K)	K_u (10^{-3} erg/cm^3)	Ref	K_1^c (10^{-3} erg/cm^3)	A_c (10^7 erg/cm)	A_m (10^7 erg/cm)	$\sigma_w(A_c, K_u)$ (erg/cm^2)	$\sigma_w(A_m, K_u)$ (erg/cm^2)	$\sigma_w(A_c, K_u, K_1)$ (erg/cm^2)	$\sigma_w(A_m, K_u, K_1)$ (erg/cm^2)	$\sigma_w(M, \ell)$ (erg/cm^2)
396	6.90	a	0.60	2.93	1.32	0.18	0.12	0.18	0.12	0.13
399	2.04	a	1.08	2.95	1.35	0.10	0.07	0.11	0.07	0.07
400	14.3	b	0.60	2.96	1.36	0.26	0.18	0.27	0.18	0.21
402	8.04	a	4.50	2.97	1.38	0.20	0.13	0.21	0.14	0.15
416	4.88	b	1.12	3.08	1.54	0.16	0.11	0.16	0.11	0.12
418	9.40	b	0.18	3.09	1.55	0.22	0.15	0.22	0.16	0.17
438	2.88	a	1.55	3.24	1.80	0.12	0.09	0.13	0.10	0.11
450	18.85	b	0.18	3.33	1.95	0.32	0.24	0.32	0.25	0.27
451	6.40	a	2.20	3.33	1.97	0.19	0.14	0.19	0.15	0.15
459	43.4	a	0.38	3.39	2.08	0.49	0.38	0.49	0.39	0.40
480	8.98	a	3.35	3.55	2.39	0.23	0.19	0.24	0.19	0.19

a. J. W. Moody, R. M. Sandfort and R. W. Shaw, Advanced Research Project Agency Order No. 1999, Final Report "Magnetic Bubble Materials," Contract DAAH01-72-C-0490, August 11, 1972 and Semi-Annual Technical Report "Magnetic Bubble Materials," Contract No. DAAH01-72-C-1098, February 11, 1973. Data are average values of three to five samples.

b. This Study.

c. K_1 estimated for europium, gallium and aluminum content based in part on F. Euler, B. R. Capone and E. R. Czerlinsky, IEEE Trans. Mag. 3, 509 (1967), F. B. Hagedorn, W. J. Tabor and I. G. Van Uitert, J. Appl. Phys. 44, 432 (1973), E. M. Gyorgy, et al, J. Appl. Phys. 44, 438 (1973) and L. G. Van Uitert, et al, Mat. Res. Bull. 6, 1185 (1971).

d. The film compositions sequentially were $Y_{2.4}Eu_{0.6}Ga_{1.26}Fe_{3.74}O_{12}$, $Y_{2.52}Gd_{0.44}La_{0.04}Ga_{0.99}Fe_{4.01}O_{12}$, $Y_{2.54}Sm_{0.46}Ga_{1.17}Fe_{3.83}O_{12}$, $Eu_2Y_1Al_{0.9}Fe_{4.1}O_{12}$ $Y_{2.5}Gd_{0.5}Ga_{1.0}Fe_{4.0}O_{12}$, $Y_{2.525}Eu_{0.475}Ga_{1.0}Fe_{4.0}O_{12}$, $Y_{1.15}Gd_{0.99}Yb_{0.86}Ga_{0.84}Fe_{4.16}O_{12}$, $Tm_{1.20}Y_{1.08}Gd_{0.72}Ga_{0.79}Fe_{4.21}O_{12}$, $Gd_{1.09}Y_{0.98}Tm_{0.93}Ga_{0.68}Fe_{4.32}O_{12}$, $Er_{2.38}Eu_{0.62}Ga_{0.64}Fe_{4.36}O_{12}$ and $Er_{2.2}Gd_{0.8}Ga_{0.44}Fe_{4.56}O_{12}$.

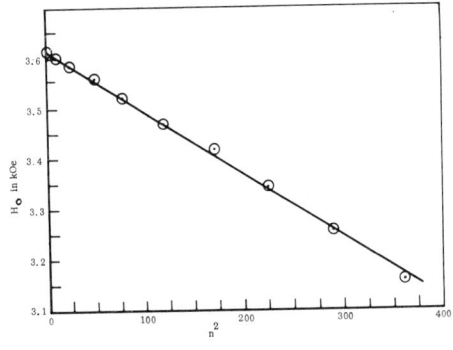

Fig. 1. Spin Wave Resonance Field Versus the Square of the Branch Number for a $Y_{2.89}La_{0.11}Ga_{0.78}Fe_{4.22}O_{12}$ Film ($T_N = 455K$)

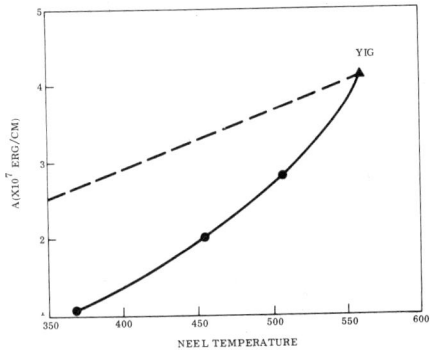

Fig. 2. Measured Exchange Constant Versus Neel Temperature

DOMAIN WALL DYNAMICS IN LOW-LOSS GARNET FILMS

P.J. Rijnierse and F.H. de Leeuw
Philips Research Laboratories, Eindhoven,
The Netherlands

ABSTRACT

Experimental techniques for the study of domain wall motion in bubble materials are discussed. New results are presented concerning the motion of a single plane wall in an epitaxial (YLa)(FeGa) garnet film with low damping ($\alpha \approx .01$). The accessible drive field region for linear wall motion is extended by application of an in-plane field H_{yo} normal to the wall. Linear wall motion is observed for velocities up to a peak velocity increasing linearly with H_{yo}. Beyond this region a strong non-linearity is observed: after a transient with a high wall speed, the velocity falls to a much lower, nearly drive-field independent value which again increases linearly with H_{yo}. It is also observed that the dynamic coercive field in the non-linear mode is 1 Oe higher than in the linear mode. A model of Bloch-line dominated wall motion may explain the observed phenomena.

INTRODUCTION

We first review briefly the available experimental techniques for the study of domain wall motion in bubble materials. Then new observations on domain wall motion will be presented, including an apparent increase in dynamic coercive force in the non-linear regime. The observed phenomena are interpreted in terms of Slonczewski's formalism[1]. Finally, the consequences of the observed domain wall behaviour for the interpretation of wall motion experiments are briefly discussed.

EXPERIMENTAL TECHNIQUES

Up to about 1969, most studies on domain wall motion were made on "picture frame" specimens[2] cut from bulk single crystals. However, the proposed use of magnetic "bubble" domains triggered off a great effort in the development of materials in the form of thin platelets or thin epitaxial films. With the possibility of visual observation of the domains, afforded by materials in this form, many of the doubts inherent in previous measurements due to a lack of knowledge concerning the exact domain configuration can be avoided. In fact, most present-day investigations rely on magneto-optical detection of domains by means of the Faraday effect, either visually or by photometric measurement. We will not attempt here an exhaustive survey of all methods proposed to date. Let us instead briefly discuss which methods are most suitable for extracting a given kind of information. Measurements both on bubbles and on plane walls are currently in use.

In measurements on single bubbles one has employed time-of-flight measurement in pulsed fields of short duration, relying on visual observation of the wall travel. In radial motion, the bubble collapse method[3] has the disadvantage of a strong variation of the drive field on the wall. This necessitates a careful analysis[4,5] and there is some uncertainty as to the exact conditions for dynamic bubble collapse[6]. An added complication, common to all bubble measurements, is the occurrence of "hard" bubbles[7,8]. We will not discuss these here, since their motional behaviour has been surveyed recently[9,10]. In the case of translational motion, the fields acting on the bubble wall do not vary strongly. Vella-Coleiro and Tabor[11] have proposed a compensation for the change in bias field felt by the bubble when moving in a gradient field. Another proposal is the "Permalloy disk" technique[12] which will certainly yield the true velocity of a bubble, but for which the actual drive field may be difficult to evaluate. Bubble translation measurements seem more relevant to the question of data rates in bubble devices than radial bubble motion; in the end, testing with a realistic propagation structure may provide the most practical data.

The description of plane walls is simpler, but in the usual configuration with the uniaxial anisotropy axis normal to the sample plane they are unstable against sinusoidal distortions[13] unless a sufficiently strong gradient field (cf. the configuration in Fig. 1) is applied. One consequence is the fact that upon application of a constant drive field, the actual field felt by the wall is a function of position, and the spatial dependence of the gradient field must be known to allow interpretation of the experimental data. Here it is an advantage to measure the instantaneous wall position, since then both field and velocity can in principle be deduced. A more serious aspect is the fact that the minimum field gradient required increases very strongly[13] with M_s, and for many bubble materials it is impossible to supply an external gradient field of sufficient strength. In such cases, one has to make use of the gradients felt by walls in a stripe domain pattern. Here the restoring forces on the walls have to be deduced from static measurements of the stripe width as a function of the field, or calculated[14] from known values of film thickness, magnetiza-

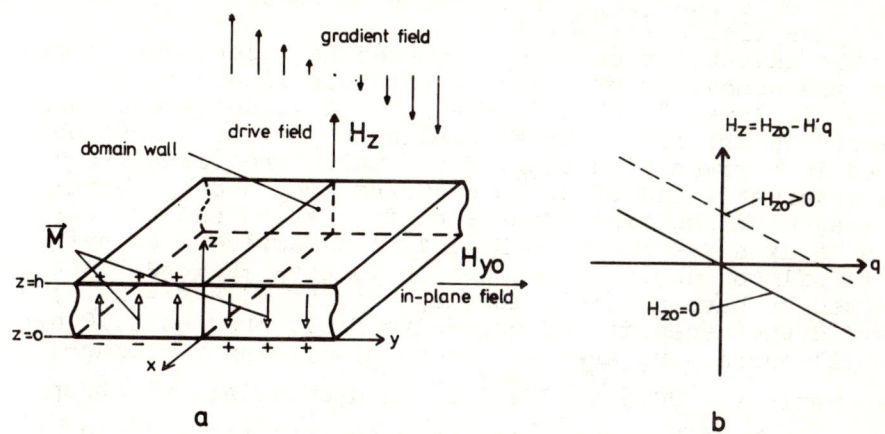

Fig. 1. a. Definition of coordinate axes and field directions
b. Drive field as a function of wall position.

tion and zero field stripe width. Since for practical applied fields the very high field gradients encountered imply very small excursions, measurements are made on a domain pattern rather than on a single wall. Sometimes such a domain pattern is spatially unstable, and small in-plane fields may be needed to stabilize it[15].

For our experiments, we have chosen the configuration of a single wall in a known external field gradient, employing photometric measurement of the instantaneous wall position. To this end, we have selected a specimen with a low magnetization and high anisotropy, which allow us to produce a plane wall in a gradient field in the kOe/cm range.

EXPERIMENTAL RESULTS

Domain wall motion often has a simple character: upon application of a constant drive field H_z, the velocity v reaches a stationary value v_s after a short time, v_s and H_z being related by

$$v_s = \mu(H_z - H_o) , \qquad (1)$$

where μ is the wall "mobility" and H_o is a dynamic coercive field. Generally H_o differs from the static coercive field, and Eq. (1) often does not hold for fields H_z very close to H_o. However, the linear relation (1) is sometimes valid in a limited field range only: for higher fields non-linear behaviour is observed, such as a change in slope of the v_s versus H_z curve, a velocity saturation[3,16] or a peak in v_s followed by a reduced velocity for higher fields[17,18].

The investigation to be reported here sets out to examine the behaviour of walls in this non-linear region. In what follows, "low" and "high" field values are always referred to the field where non-linearities are first observed in a given situation. Plane walls are established in a gradient field of up to 6 kOe/cm provided by permanent magnets. The experimental configuration and the coordinate axes are shown in Fig. 1. The sample is an epitaxial film of nominal composition $Y_{2.9}La_{0.1}Fe_{3.8}Ga_{1.2}O_{12}$, selected to have a fairly low magnetization ($4\pi M = 68$ G) to avoid the necessity of excessive field gradients. Other material parameters are: thickness $h = 3.9$ μm, gyromagnetic ratio $\gamma = 20.6 \times 10^6 s^{-1} Oe^{-1}$ and uniaxial anisotropy energy $K = 5500$ erg cm^{-3}, implying an anisotropy field of 2000 Oe. The exchange constant is estimated to be $A = 3 \times 10^{-7}$ erg cm^{-1}, and hence the wall width parameter $\Delta = (A/K)^{\frac{1}{2}} \approx .74 \times 10^{-5}$ cm.

An external dc magnetic field H_{yo} in the plane of the sample and normal to the domain wall ("in-plane field") was applied for the following reason. Models of moving domain walls, discussed in the next section, predict the occurrence of a peak velocity v_p beyond which uniform translation of the wall as a plane structure is impossible. This velocity is determined by the maximum field available for the spins in the wall to precess around, and one expects that v_p is increased when an extra precession field is provided. At the same time these models predict that Eq. (1) continues to hold with the same mobility μ. Thus an increase in v_p leads to an extension of the field region in which linear wall motion is possible to fields which are no longer close to the coercive field. In the ortho-ferrites such an extra field of internal origin is supplied[19] by a strong orthorhombic component of the anisotropy; in our case we have applied an external field[17] by displacing the specimen in the z-direction (cf. Fig. 1) to a position where $H_y \neq 0$, since a gradient $\partial H_z/\partial y$ implies a gradient $\partial H_y/\partial z$ of equal magnitude.

Wall position changes are observed visually or photographically in a polarizing microscope. They are also detected as intensity changes, caused by the Faraday effect, of the light passing through the sample between nearly-crossed polarizers by a fast photomultiplier followed by a broadband amplifier. The optical signal is calibrated in terms of wall displacement by comparison with the photographic observations. Field pulses are applied by a small 12-turn pulse coil close to the speci-

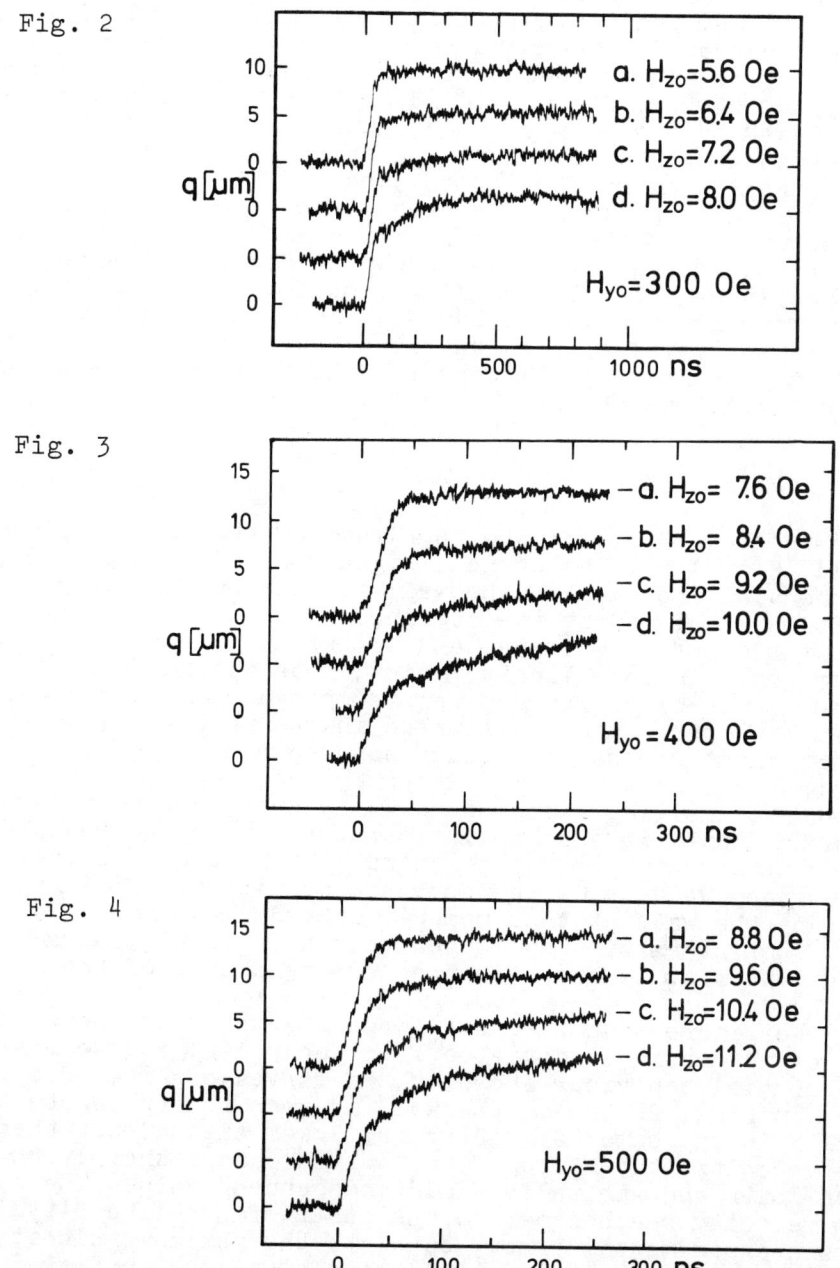

Fig. 2-4. Optical signal (wall position) as a function of time for different combinations of drive field and in-plane field.

men, with a coil constant calibrated as 40 Oe/A and a rise time of 8 ns, the maximum field amplitude being 16 Oe. The pulse duration is chosen to be roughly ten times the wall transit time and the repetition frequency varies accordingly up to 100 kHz. The amplified photomultiplier output is detected[20] by a slowly scanned sampling oscilloscope whose output is averaged with an integration time of 1.2 sec. The overall time resolution is 5 ns.

Typical experimental curves of the optical signals, corresponding to the wall position q along the y-axis, as a function of the time t are shown in Figs. 2-4, showing the motion of the wall for different applied fields H_{zo} and different in-plane fields H_{yo}. For low applied fields H_{zo} (Figs. 2.a, 3.a, 4.a) the behaviour is similar to that predicted by Eq. (1) allowing for the fact that the actual field H_z felt by the wall is equal to

$$H_z = H_{zo} - H'q, \qquad (2)$$

where $H' = -\partial H_y/\partial z$ is the magnitude of the linear gradient field (Fig. 1.b). The optical signal shows an essentially exponential approach of the wall to its new equilibrium position, modified by the combined effects of wall inertia[21], field coil inductance and measuring electronics. Many experiments in this linear region, made with different applied and in-plane fields, were compared with curves calculated[22] from a model discussed in the next section, and taking into account the field and electronics delays. All of these curves were well fitted by one single set of parameters, implying a mobility $\mu = 14000 \pm 4000$ cm s^{-1} Oe^{-1}. From the model it also follows that at the time when the velocity dq/dt reaches its maximum, it is equal to the stationary value v_s which corresponds with the field H_z felt by the wall at that moment, according to Eqs. (1) and (2). The information collected this way about v_s as a function of H_z is shown in the low-field part of the curves in Fig. 5.

For increasing fields H_{zo} the character of the wall motion increasingly deviates from the underlying nearly-exponential behaviour shown by the curves of Figs. 2.a - 4.a. In Figs. 2-4, c-d, the wall is seen to accelerate initially in the same way as for the lower fields, but then its velocity (the slope of the curves) drops sharply to a much lower and virtually field-independent value. The maximum velocity reached before the onset of non-linearity is, within experimental error, equal to the maximum velocity v_p observed in the "linear" low-field region. In Fig. 5 we have also shown the low velocity in the non-linear region, v_{nl}, for different values of the in-plane field H_{yo}. The peak velocity increases linearly with H_{yo}, as shown in Fig. 6; for low H_{yo}, it cannot be measured, since the cri-

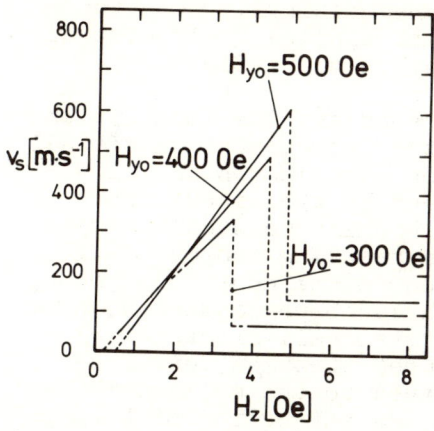
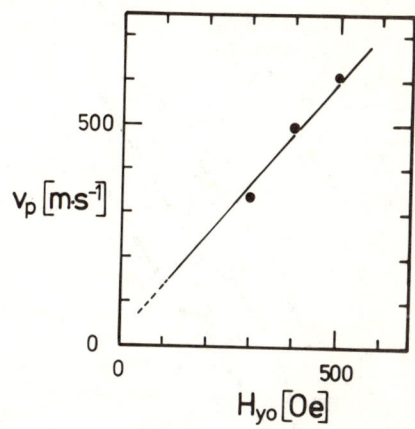

Fig.5. Stationary velocity as a function of drive field for different in-plane fields.

Fig.6. Peak velocity as a function of in-plane field.

tical field $H_p = v_p/\mu$ is not sufficiently large with respect to the coercive field, as discussed before. The "non-linear" mode velocity v_{nl}, measured at a constant applied field H_{zo} = 15 Oe, is plotted as a function of the in-plane field H_{yo} in Fig. 7. It is roughly linear with H_{yo} for values of $|H_{yo}| > 50$ Oe. The slight asymmetry in this curve is not understood at present and is the subject of further investigation.

Non-linear behaviour of this kind, with a peak in the stationary velocity vs field curve, has been predicted by Slonczewski[23]; similar phenomena have been observed recently by Malozemoff[18]. However, Figs. 2-4 show another feature not reported before: for applied fields high enough to cause non-linearities, the transient still indicates very high wall speeds[22] before converting to the "non-linear mode". A possible explanation for such behaviour will be discussed in the

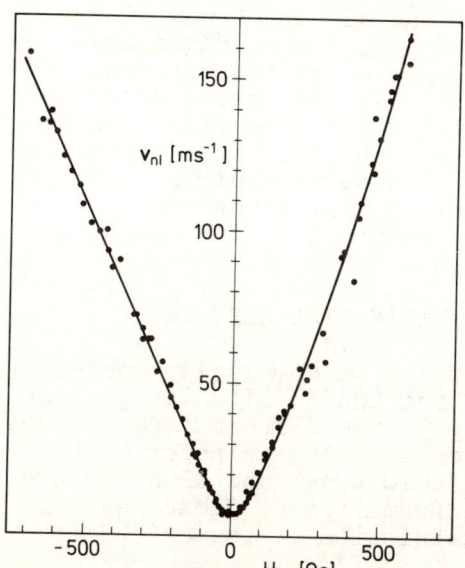

Fig.7. "Non-linear mode" velocity vs in-plane field.

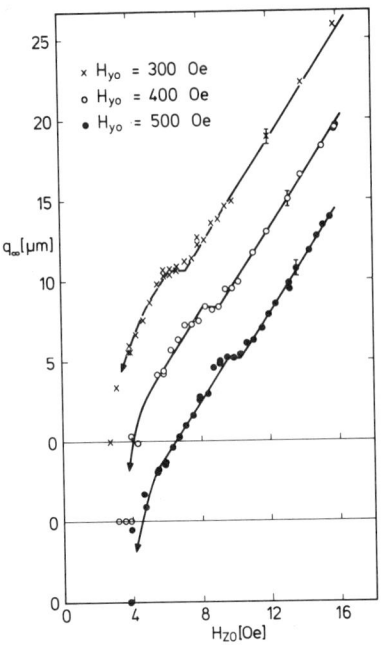

Fig.8. Total wall excursion measured optically as a function of applied field, for different in-plane fields.

next section: its implications for the interpretation of wall motion experiments are considered in the discussion.

Finally we want to report an interesting aspect of wall motion in the non-linear mode, which is observed when we plot the total wall excursion q_∞ measured at times very long compared to the wall motion time, as a function of the applied field. From Eq. (2), such a curve is expected to be linear with the applied field H_{zo}:

$$q_\infty = (H_{zo} - H_c)/H' \qquad (3)$$

where H_c is not necessarily equal to H_o and Eq. (3) is not valid for H_{zo} close to H_c. However, the experimental plots (Fig. 8) are seen to consist of two linear parts, shifted by about 1 Oe with respect to each other. The curves shown in Figs. 2.b-4.b correspond to the onset of the transition region, those in Figs. 2.c-4.c to the right-hand side of this region, indicating that the shift is correlated with the occurrence of non-linear behaviour. The slopes of the linear parts, seen most clearly for high H_{yo}, are indeed equal to $1/H'$, showing that Eq. (3) applies in both regions, but with a different value of H_c. We are inclined to interpret this shift as an increase in the dynamic coercive field: a wall in the non-linear mode stops moving when the field at the wall drops below a value 1 Oe higher than the corresponding field for the wall in the linear mode. A tentative explanation for this phenomenon will be proposed in the next section.

INTERPRETATION OF EXPERIMENTAL RESULTS

Most theoretical descriptions of damping-dominated domain wall motion start from the Landau-Lifshitz precession equation with the damping in the Gilbert form[24]. For a review of the theoretical development, we refer to recent articles by Hagedorn[19], Schlömann[25] and Slonczewski[1].

Here, we will follow the formalism of Slonczewski[1], because it allows a simple description of the effect of

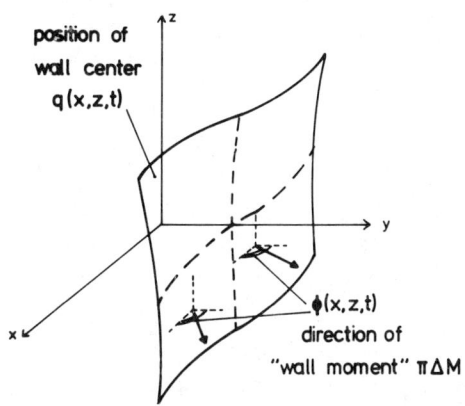

Fig.9. Geometry of domain wall.

fields, of internal or external origin, on the domain wall motion. The coordinate system employed is indicated in Fig. 9. The wall, on average normal to the y-direction, is assumed to have a spatially invariant magnetization structure, which can be completely characterized[1] by the local position $q(x,z,t)$ of its center on the y-axis, and by the direction of its "wall magnetic moment", integrated over the thickness, of magnitude $\pi\Delta M$. Here $\Delta = (A/K)^{\frac{1}{2}}$ is the usual wall thickness parameter, and the local moment direction makes an angle $\phi(x,z,t)$ with the x-axis in the x-y-plane.

From the Landau-Lifshitz equation follow relations for q and ϕ:

$$\dot{q} = (\gamma/2M)\delta\sigma/\delta\phi + \alpha\dot{\phi}\Delta, \qquad (4.a)$$

$$\dot{\phi} = - (\gamma/2M)\delta\sigma/\delta q - \alpha\dot{q}/\Delta, \qquad (4.b)$$

where the dots denote time derivatives, γ is the gyromagnetic ratio, σ is the domain wall (surface) energy and α is the Gilbert damping parameter. The usual functional derivatives are defined by $\delta\sigma/\delta q = \partial\sigma/\partial q - \nabla\cdot(\partial\sigma/\partial\nabla q)$, where the gradient operator $\nabla = (\partial/\partial x, \partial/\partial z)$ refers to two-dimensional spatial derivatives. The theory is limited to materials for which $2\pi M^2 \ll K$, and is valid in the limit that $|\nabla q| \ll 1$, $|\nabla\phi| \ll 1/\Delta$ and $M|H| \ll K$ for any field in the x-y-plane.

If the wall remains flat $(\nabla q=0)$, the only contribution to $\delta\sigma/\delta q$ in Eq. (4.b) comes from the external field H_z. Then for the stationary state $(\dot{\phi}=0)$ one has a uniform translation with

$$\dot{q} = v_s = \gamma\Delta H_z/\alpha. \qquad (5)$$

Only values of v_s which allow solutions of Eq. (4.a) are allowed: this determines a peak velocity v_p and a corresponding field $H_p = \alpha v_p/\gamma\Delta$ as mentioned in the previous section. Equation (5) is of the form (1), with $\mu=\gamma\Delta/\alpha$; the coercive field H_o is introduced ad hoc, it does not follow from the theory. The value of v_p depends upon the external or internal fields present, and thus on the geometry. For purely uniaxial bulk materials and no external fields, one finds $v_p = \gamma\Delta \cdot 2\pi M$, similar to other theories[19,25].

For thin films, still with external fields $H_{xo} = H_{yo} = 0$, Slonczewski[23] has developed a model in which the stray fields $H_{ys}(z)$ due to the surface poles induce a strong variation of ϕ with z. For a static wall, different equivalent structures $\phi(z)$ are possible. In a moving wall, $\phi(z)$ is close to one or another of these structures, except for a narrow region, a "Bloch line", where ϕ changes over. The motion is characterized by a shift in the z-direction of this Bloch line. Not all locations are allowed for the Bloch line; when it reaches the limit of its allowed region the peak velocity v_p is reached. For drive fields H_z higher than the field $H_p = \alpha v_p/\gamma\Delta$, Slonczewski has proposed[23] the idea of "instability dominated" motion. For most of the time the wall is in one of the possible modes for linear motion, described above; in such modes the average $<\phi(x,z)>$ increases continuously, accompanied by continuous changes in wall structure and velocity. When, at the velocity v_p, the limit of such a stable mode is reached, the structure changes irreversibly in a short time to one corresponding to a different, stable mode with the same value of $<\phi>$, but with a lower energy. The energy difference is transmitted to the entire magnetic system and hence to the lattice[23]. The wall motion is punctuated by sudden changes of velocity at these instability points. In lowest order the mean velocity turns out to be equal to that found by equating the power supplied by the drive field to the energy lost in the instabilities. This mean velocity is then identified with the stationary velocity in the non-linear mode, and turns out to be independent of H_z or α in this approximation[23].

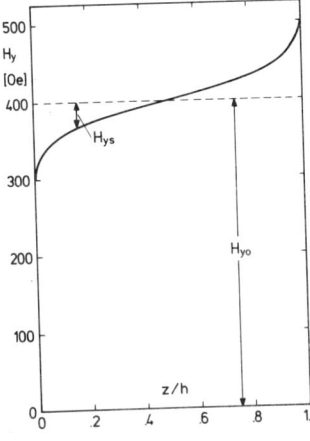

Let us now examine the consequences of an external in-plane field H_{yo}. The stray field $H_{ys}(z)$ due to the surface poles has the form $H_{ys} = 4M \ln(z/(h-z))$ far from the surface[23]. However, by integrating over the actual surface pole density $- M\tanh(y-q)/\Delta$ corresponding to the wall structure assumed, we find that H_{ys} levels off to a value $H_{ys}(h) = 4M \ln(2.27h/\Delta)$ at the surface, having a slope

Fig.10. In-plane field H_y as a function of z, when the external in-plane field $H_{yo} = 400$ Oe.

$dH_{ys}/dz = 2\pi M/\Delta$ there. For our material, $H_{ys}(h) \approx 100$ Oe, showing that H_{yo} will dominate the wall structure for large in-plane fields. The total field $H_y = H_{yo} + H_{ys}(z)$ is shown in Fig. 10 for $H_{yo} = 400$ Oe.

For a uniformly moving plane wall, invariant along the x-axis, one has $q(x,z,t) = q(t)$, $\phi(x,z,t) = \phi(z,t)$, and Eqs. (4) reduce to:

$$\dot{q} = 4\pi\gamma\Delta M \sin\phi\cos\phi - (2\gamma\Delta A/M)d^2\phi/dz^2 \quad (6.a)$$
$$\qquad - \tfrac{1}{2}\pi\gamma\Delta H_y \cos\phi + \alpha\dot{\phi}\Delta,$$

$$\dot{\phi} = \gamma H_z - \alpha\dot{q}/\Delta. \quad (6.b)$$

Here we will neglect the first term on the r.h.s. of Eq. (6.a), since $|H_y| \gg 8M$ for all z. If we assume that $|H_{ys}| \ll |H_{yo}|$, the spatial derivative in (6.a) may be dropped. Uniform, stationary motion with $\dot{\phi}=0$, $\dot{q}=v_s$ is possible, described by equation (5) and by

$$v_s = -\tfrac{1}{2}\pi\gamma\Delta H_{yo}\cos\phi. \quad (7)$$

If the linear regime extends to a critical angle ϕ_p, we expect a peak velocity and field given by

$$v_p = -\tfrac{1}{2}\pi\gamma\Delta H_{yo}\cos\phi_p, \quad H_p = \tfrac{1}{2}\pi\alpha H_{yo}\cos\phi_p. \quad (8.a)$$

For the static wall, the wall moment points along the y-axis, and $\phi=\tfrac{1}{2}\pi$. From (7) one expects that the linear region could extend to $\phi_p=\pi$ leading to

$$v_p = \tfrac{1}{2}\pi\gamma\Delta H_{yo}, \quad H_p = \tfrac{1}{2}\pi\alpha H_{yo}. \quad (8.b)$$

To describe the non-linear mode observed in the experiments, a possible explanation might be in terms of "instability-dominated" motion, like discussed above. One possible mode of motion has been described above in Eqs. (7/8). A second possible mode has $\phi=\tfrac{1}{2}\pi$ for $z \gg z_B$. At the position z_B a horizontal Bloch line occurs, where the angle ϕ rotates over 2π to $5\pi/2$, since in the strong H_y field only $\phi = (2k+\tfrac{1}{2})\pi$, k integer, is a stable position. When stationary, the Bloch line structure is described by Eq. (6.a) with $\dot{q}=0$, $\dot{\phi}=0$. Its excess energy W_B and its length parameter Λ are easily shown[23] to be:

$$W_B = 16A(\pi M|H_y|/K)^{\tfrac{1}{2}}, \quad \Lambda = (A/\pi M|H_y|)^{\tfrac{1}{2}}. \quad (9)$$

Substitution of parameter values shows that, for $H_y = 400$ Oe, $\Lambda = .07$ μm; since $d\phi/dz$ reaches a maximum value of $1/\Lambda$, we may be violating the condition $d\phi/dz \ll 1/\Delta$ mentioned before. In this model, there is a transition at the critical velocity v_p from the linear mode to the

Bloch-line mode. The Bloch line starts at a position z_B such that $\langle\phi\rangle = \phi_p = \frac{1}{2}\pi + 2\pi z_B/h$ remains continuous. Then it moves upwards, and in this mode the forward velocity of the wall is very low. When the Bloch line reaches the surface $(z_B=h)$ it annihilates. Then ϕ is $5\pi/2$ for all z, and a new cycle can start in the linear mode. Neglecting the small variation of the Bloch line energy with z_B, the mean velocity is

$$v_{nl} = \tfrac{1}{4}\gamma\Delta H_{yo}(1 - \sin\phi_p), \qquad (10.\text{a})$$

which for $\phi_p=\pi$ reduces to

$$v_{nl} = \tfrac{1}{4}\gamma\Delta H_{yo}. \qquad (10.\text{b})$$

We have analysed the q vs t curves in the linear region[22] by comparing them with calculated curves, obtained by numerical integration of Eqs. (6), starting from $q=0$ and $\phi=\frac{1}{2}\pi$ at $t=0$, and substituting (2) for the field with H_{zo} rising exponentially with a rise time of $8ns$. All curves, with different values of H_{zo}, H_{yo} and H', were fitted best by[22] $\alpha = .011 \pm .003$, leading to a mobility $\mu \approx 14000 \pm 4000$ cm s^{-1} Oe^{-1} according to (5). However, while Eq. (8.b) predicts $v_p \approx 240\ H_{yo}$, with v_p in cm s^{-1} and H_{yo} in Oe, we find only $v_p \approx 125\ H_{yo}$, implying (Eq. 8.a) a critical angle $\phi_p \approx \frac{1}{2}\pi + 0.56 = 90° + 32°$. A similar discrepancy is found in the non-linear mode; while Eq. (10.b) predicts $v_{nl} \approx 38\ H_{yo}$, the actual values give $v_{nl} \approx 24\ H_{yo}$, and according to Eq. (10.a) this corresponds to an angle $\phi_p \approx \frac{1}{2}\pi + 1.19 = 90° + 68°$. The discrepancies between these values for ϕ_p and the expected value $\phi_p=\pi$ may be partially explained by the fact that at the lower surface $z=0$ the field H_y is about 100 Oe lower due to the stray field H_{ys}, and that near the surface H_y varies rapidly and the derivative $d^2\phi/dz^2$ in Eq. (6.a) is probably not negligible. The region near $z=0$ is a critical one, and instabilities or Bloch lines are expected to originate there. The above seems a reasonable description, but some doubts remain, specially since on the time scale of our experiments one would expect to see something of the velocity oscillations predicted, which have a period of the order $2\pi/\gamma H_z$, i.e. 20 - 100 ns.

The observed increase in dynamic coercive force in the non-linear region (cf. Fig. 8) finds a reasonable explanation in such a picture of Bloch-line dominated motion. When the wall contains Bloch lines, which have strong stray fields caused by the local divergence of the magnetization, it can be expected to interact more strongly with material imperfections and consequently it would need a larger drive field to keep moving. A Bloch-line coercive force has also recently been invoked by Malozemoff[18] to explain some of his results on hard bubble motion.

DISCUSSION

The observation of non-linear domain wall behaviour has some implications for the interpretation of domain wall motion experiments. Many experimental methods, being essentially "time-of-flight" measurements, do not observe details of the motion. As long as no non-linearities occur, and if wall inertia has only a small influence, such methods yield reliable results; but when transient behaviour of the kind shown in Figs. 2.d-4.d occurs, an interpretation based on distance travelled over time gives a wrong picture. An example is given by the recent bubble collapse measurements of Malozemoff[18]. For bias fields far below the static collapse field large drive fields are needed, and he observes an apparently simple velocity saturation. However, when reducing the drive fields by raising the bias field, he observes several non-linearities, and for the lowest drive field a peak in the collapse time vs field curve is clearly resolved.

Measurements similar to ours, but on stripe domain patterns, have been reported[15,26,27]. Callen et al.[26], to explain their results, have suggested a non-linear wall motion equation, which implies a monotonous increase of v_s as a function of H_z up to a saturation value. However, the behaviour we observe (Fig. 5) shows that this assumption is not valid in our case. Argyle and Malozemoff[15] and Moody et al.[27] have observed wall oscillations in the strong gradient fields found in stripe domain patterns, applying relatively low in-plane fields parallel to the walls. In our case, the large in-plane fields strongly reduce the wall mass, and the much lower gradient field implies a lower stiffness; thus we do not observe such oscillations.

In conclusion we can say that evidence for non-linearity of domain wall motion in low-loss garnets has been presented; however, in the present state of knowledge no decisive evidence has been given to establish the nature of the wall motion after the onset of non-linearity.

Acknowledgements: the authors thank A.M.J. van der Heijden for his help in obtaining the experimental data, and U. Enz for stimulating discussions.

REFERENCES

1. J.C. Slonczewski, Int. J. Magn. **2**, 85 (1972).
2. H.J. Williams, W. Shockley and C. Kittel, Phys. Rev.

80, 1090 (1950).
3. A.H. Bobeck, I. Danylchuk, J.P. Remeika, L.G. van Uitert, and E.M. Walters, Proc. of the Int. Conf. on Ferrites (University of Tokyo Press, 1971), p. 361.
4. H.E. Callen and R.M. Josephs, J. Appl. Phys. 42, 1977 (1971).
5. J.W.F. Dorleijn and W.F. Druyvesteyn, Appl. Phys. 1, 167 (1973).
6. J.A. Cape, W.G. Hall and G.W. Lehman, Phys. Rev. Lett. 30, 801 (1973).
7. W.J. Tabor, A.H. Bobeck, G.P. Vella-Coleiro and A. Rosencwaig, Bell Syst. Tech. J. 51, 1427 (1972).
8. A.P. Malozemoff, Appl. Phys. L. 24, 149 (1972).
9. J.C. Slonczewski, A.P. Malozemoff and O. Voegeli, A.I.P. Conf. Proc. 10, 458 (1973).
10. W.J. Tabor, A.H. Bobeck, G.P. Vella-Coleiro and A. Rosencwaig, A.I.P. Conf. Proc. 10, 442 (1973).
11. G.P. Vella-Coleiro and W.J. Tabor, Appl. Phys. Lett. 21, 7 (1972).
12. F.C. Rossol and A.A. Thiele, J. Appl. Phys. 41, 1163 (1970).
13. F.B. Hagedorn, J. Appl. Phys. 41, 1161 (1970).
14. C. Kooy and U. Enz, Philips Res. Repts. 15, 7 (1960).
15. B.E. Argyle and A.P. Malozemoff, A.I.P. Conf. Proc. 10, 344 (1973).
16. B.E. Argyle, J.C. Slonczewski and A.F. Mayadas, A.I.P. Conf. Proc. 5, 175 (1972).
17. F.H. de Leeuw, 1973 Intermag paper 29.1, to be published in IEEE Trans. Magn., Dec. 1973.
18. A.P. Malozemoff, IBM report RC 4325, to appear in J. Appl. Phys.
19. F.B. Hagedorn, A.I.P. Conf. Proc. 5, 72 (1972).
20. J.A. Seitchik, W.D. Doyle and G.K. Goldberg, J. Appl. Phys. 42, 1272 (1971).
21. W. Döring, Z. Naturforschung 3a, 373 (1948).
22. F.H. de Leeuw, Int. Conf. on Magnetism, Moscow, 1973, paper 27m-W6; F.H. de Leeuw, to be published.
23. J.C. Slonczewski, J. Appl. Phys. 44, 1759 (1973).
24. L. Landau and L. Lifshitz, Phys. Z. Sowjet 8, 153 (1935); T.L. Gilbert, Phys. Rev. 100, 1243 (1955).
25. E. Schlömann, A.I.P. Conf. Proc. 5, 160 (1972); J. Appl. Phys. 43, 3834 (1972).
26. H. Callen, R.M. Josephs, J.A. Seitchik and B.F. Stein, Appl. Phys. Lett. 21, 366 (1972).
27. J.W. Moody, R.W. Shaw, R.M. Sandfort and R.L. Stermer, IEEE Trans. Magn. MAG 9, 377 (1973).

BUBBLE DYNAMICS IN AMORPHOUS MAGNETIC MATERIALS

M. H. Kryder and H. L. Hu
IBM Research Center, Yorktown Heights, N. Y. 10598

ABSTRACT

The velocity of magnetic bubble propagation in a magnetic field gradient in sputtered amorphous metallic films has been measured in several samples. The measurements show typical mobilities ranging from 500 cm/sec-Oe to 2000 cm/sec-Oe and velocities up to 10,000 cm/sec. With very high drive fields the bubble velocities become poorly reproducible and deflection of the bubble from the field gradient direction occurs suggesting a dynamic change in wall structure. Bubble domains with complex wall structures are also suggested by the observation of stable dumbbell shaped domains which rotate in a pulsed bias field. The eddy current contribution to the damping of wall motion in the films is approximately calculated and shown to be negligible.

INTRODUCTION

A large amount of experimental data on the dynamics of domain wall motion in materials supporting magnetic bubbles has been accumulated by numerous workers in recent years. Rossol et al.[1,2] used stroboscopic observations to measure domain wall mobility in orthoferrites and found mobilities ranging from 100 to 3000 cm/sec-Oe. Calhoun et al.[3] showed that the mobility may be a function of drive field amplitude by using Bobeck's method[4] of determining the speed at which bubble domains collapse. Vella-Coleiro[5] measured mobilities in epitaxial garnet films by measuring the bubble translation due to a pulsed gradient field and found mobilities ranging from 100 to 4500 cm/sec-Oe with velocities up to 3000 cm/sec. Malozemoff and Slonczewski[6] showed that different bubbles exhibit different mobilities and attributed this to varying numbers of "Bloch lines" like those observed by Grundy et al.[7]. Vella-Coleiro et al.[8] found different bubbles to deflect at different angles from the gradient and also attributed this to the "Bloch lines".
Recently it was reported that amorphous metallic films[9] support magnetic bubble domains. These films can have values of $Q = K_u/2\pi M_s^2$ greater than one and coercivities less than one percent of $4\pi M_s$. They are therefore potentially useful for bubble domain device applications. The dynamic properties of these materials are considered in this report. Data has been obtained which indicates that relatively high mobilities and velocities are achieved in amorphous magnetic films.

EXPERIMENTAL METHODS AND RESULTS

The circuit used to perform the bubble translation experiments is similar to that described by Vella-Coleiro and Tabor[10]. Conductor lines with configuration and dimensions shown in Fig. 1 are insulated by SiO_2 from the sample to be measured. The outer two lines spaced 50μm apart are used to apply a gradient field by pulsing equal currents down them in the direction indicated. Narrow (60-200 nsec) 12 nsec

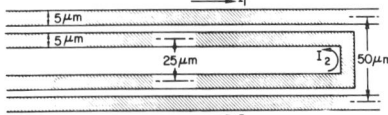

Fig. 1 Strip-conductor configuration used to translate magnetic bubble domains.

risetime pulses of current were used, and bubble position was measured before and after each pulse. Thus the bubble displacement due to the pulse was determined, and the average velocity was computed by dividing the displacement by the pulse width. Bubble-bubble interaction was eliminated by using only those bubbles which were ten or more diameters from their nearest neighbor. Attempts were made to compensate for the decrease in effective bias field on the propagating bubble by integrating the pulse current and applying it to the hairpin loop inside the outer two lines as Vella-Coleiro and Tabor[10] suggested. However, the displacement of a bubble during a given pulse varied within approximately ± 30% limits. This made it impossible to properly adjust the time constant of the current in the loop. Since the effect of the compensation achievable with this scatter seemed to be negligible in comparison to the scatter, it was not employed in obtaining the data in this report.

Data on bubble translation velocity versus the difference in bias

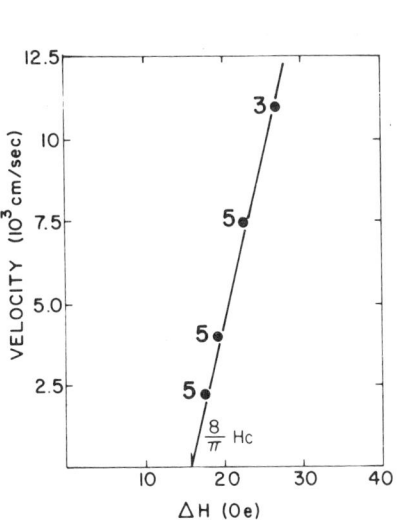

Fig. 2 A plot of translational velocity versus difference in bias field across the bubble diameter for a GdCo film with $4\pi M_s$=1100G, h=1.7μm, l=0.2μm. The number of measurements averaged for each point is indicated.

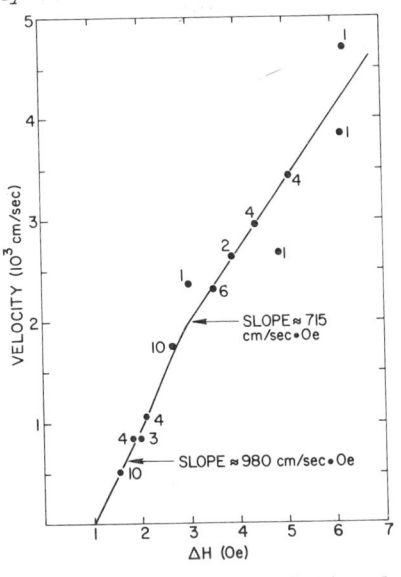

Fig. 3 A plot of translational velocity versus difference in bias field across the bubble diameter for a GdCoAu film with $4\pi M_s$=1000G, h=1.7μm, l=0.28μm. The number of measurements averaged for each point is indicated.

field across the bubble diameter is shown in Figs. 2 and 3 for two amorphous film samples. Each data point shown in the figures is an average of several measurements with the same bubble in the same area of the film. The number of measurements averaged for each point is indicated on the figures. The value of ΔH was taken as the product of the field gradient times the bubble diameter measured between pulses.

The data in Fig. 2 show that bubble velocities exceeding 10,000 cm/sec are observed in amorphous films. This velocity is significantly higher than those typically measured in garnets. It should be noted that in this film even at 10,000 cm/sec, there is no evidence of non-linearity. The data shows a linear relation of the predicted form[11]

$$V = \frac{\mu_w}{2} (\Delta H - \frac{8}{\pi} H_c) \qquad (1)$$

with wall mobility μ_w = 2000 cm/sec-Oe.

Data in Fig. 3 for a gold doped film show that similar mobilities may be obtained in films with quite different coercivities and compositions. The data of Fig. 3 show a dynamic coercivity less than one Oersted, while the data of Fig. 2 show a coercivity of 5.7 Oe. Either the mobility measured from the low drive region (1960 cm/sec-Oe) or the high drive region (1430 cm/sec-Oe) of Fig. 3 are comparable to the 2000 cm/sec-Oe measured from Fig. 2. In GdCo, GdCoAu, and GdCoCu samples, mobilities ranging from 500 to 2000 cm/sec-Oe and coercivities from a small fraction of an Oersted to several Oersteds have been measured. Data obtained thus far do not show a particular correlation of mobility or coercivity with doping element. No change in bubble behavior is found where the data of Fig. 3 change slope, however with ΔH greater than 5 Oe a change does occur. Above 5 Oe velocities in excess of 5000 cm/sec are measured, but very large scatter occurs. It is suggested that this scatter is due to dynamic changes in wall structure. This view is supported by the fact that in this field region the bubbles are occasionally observed to deflect from the field gradient direction. However, when such a deflecting bubble is later propagated at low drive, it again propagates at normal velocity in the gradient direction. This behavior is similar to that reported by Vella-Coleiro[5] in epitaxial garnets.

Additional evidence of complex wall structures in amorphous films is the observation of dumbbell shaped domains which are stable in a portion of the field region where normal bubble domains exist and which rotate in a pulsed bias field. Also, some bubbles formed by rapidly pulsing the hairpin loop of Fig. 1 consistently deflect at large angles from the pulsed field gradient direction. "Bloch lines" were observed by Herd and Chaudhari[12] in thin amorphous films and may be the cause of these phenomena as was suggested[6,8] in the case of garnets.

EDDY CURRENT DAMPING

Since the amorphous films are metallic, eddy currents are induced in them by the moving domain walls. A simple model was used to calculate the effects of eddy currents in a film. A straight, infinitely long domain wall of zero width was assumed to lie in an infinite conducting film of finite thickness h. The magnetization was assumed to have the constant y-component M where $x < 0$ and $-M$ where $x > 0$. A simple calculation following the approach of Williams et al.[13] leads to the mobility of the infinite domain wall in this configuration.

$$\mu_W = \frac{v}{H} = \frac{c^2}{32\pi \log 2 \, hM\sigma} \quad . \tag{2}$$

Typical values for films investigated are $M = 100$ emu, $h = 2 \times 10^{-4}$ cm, $1/\sigma = 2 \times 10^{-4}$ Ω-cm, yielding $\mu_W = 1.4 \times 10^5$ cm/sec-Oe.

Although the actual wall in a bubble domain has a finite width and is neither infinitely long nor straight, the calculation shows that the eddy currents are large where $-h < x < h$ and drop off rapidly outside this region. Since the wall width is significantly smaller than the film thickness h and since the bubble diameter is typically at least equal to the film thickness, the mobility given in Eq. 2 should be at least an order of magnitude approximation to the actual eddy current limited mobility of a bubble domain wall. For an actual cylindrical bubble domain it is expected that the eddy current damping is even less than that calculated for the straight wall since the contributions from the eddy currents of the leading and trailing segments of wall perpendicular to the direction of propagation tend to cancel except within the cylindrical domain itself. Because the measured mobilities are so much less than the mobility expected from eddy current damping alone, some other mechanism must be responsible for the major portion of the damping in the amorphous films.

ACKNOWLEDGMENTS

The authors thank Praveen Chaudhari and Larry Rosier for helpful discussions about eddy current damping and Jerry Cuomo and Dick Gambino for supplying the films used in this study.

REFERENCES

1. F. C. Rossol, J. Appl. Phys. 40, 1082 (1969).
2. F. C. Rossol and A. A. Thiele, J. Appl. Phys. 41, 1163 (1970).
3. B. A. Calhoun, E. A. Giess, and L. L. Rosier, Appl. Phys. Lett. 18, 287 (1971).
4. A. H. Bobeck et al., Proc. Int. Conf. Ferrites, July, 1970.
5. G. P. Vella-Coleiro, AIP Conf. Proc. 10, 424 (1973).
6. A. P. Malozemoff and J. C. Slonczewski, Phys. Rev. Lett. 29, 946 (1972).
7. P. J. Grundy et al., Phys. Status. Solidi (a) 9, 79 (1972).
8. G. P. Vella-Coleiro, A. Rosencwaig and W. J. Tabor, Phys. Rev. Lett. 29, 949 (1972).
9. P. Chaudhari, J. J. Cuomo, and R. J. Gambino, IBM J. Res. Develop., 17, 66 (1973).
10. G. P. Vella-Coleiro and W. J. Tabor, Appl. Phys. Lett. 21, 7 (1972).
11. A. A. Thiele, J. Appl. Phys. 41, 1139 (1970).
12. S. Herd and P. Chaudhari, IBM Research Report RC4315 (1973).
13. H. J. Williams, W. Shockley, and C. Kittel, Phys. Rev. 80, 1090 (1950).

DYNAMIC CONVERSION EFFECTS IN EPITAXIAL GARNET FILMS

G. P. Vella-Coleiro
Bell Laboratories, Murray Hill, N. J. 07974

ABSTRACT

Measurements of magnetic bubble velocity in a pulsed field gradient have shown that there exists a considerable scatter in the velocity of the bubble at any given value of drive field. This effect has been investigated as a function of the drive field amplitude in a high mobility $Lu_1Gd_2Al_{.6}Fe_{4.4}O_{12}$ film, and in a relatively low mobility $Sm_{.4}Y_{2.6}Ga_{1.2}Fe_{3.8}O_{12}$ film. At low drive fields, the scatter in the velocity is attributed to the effect of coercivity and other imperfections in the film. At high values of drive field, the scatter is believed to be the result of a dynamic conversion of the spin structure of the bubble wall into a less mobile one analogous to that of a hard bubble. Both ion implantation and an in-plane field of 100 Oe were found to have little effect on the velocity scatter.

Measurements of magnetic bubble mobility have recently been made in a number of rare earth garnet epitaxial films in an effort to formulate a composition suitable for use in bubble devices operating at 1 MHz bit rates. While several compositions with mobility greater than 1000 cm/sec Oe have been grown successfully,[1] a common characteristic is a large scatter in the bubble velocity at any given value of drive field. This scatter in the velocity is accompanied by a small probability that the path of the bubble makes a large angle with the direction of the field gradient, reminiscent of the motion of a hard bubble. These effects are believed to be due, at least in part, to a dynamic conversion of the spin configuration of the bubble wall to a less mobile spin structure analogous to that of a hard bubble. Hence they are collectively described as dynamic conversion effects. They were discussed briefly in Refs. 2 and 3; in this paper we present a more detailed account of the velocity scatter. A model for the dynamic conversion process is described by Hagedorn.[4]

The measurements described here were made on two types of material, a high mobility $Lu_1Gd_2Al_{.6}Fe_{4.4}O_{12}$ garnet (subsequently referred to as LuGd), and a relatively low mobility $Sm_{.4}Y_{2.6}Ga_{1.2}Fe_{3.8}O_{12}$ garnet (referred to as SmY). Somewhat different results were observed in these two materials, as discussed below. Measurements of bubble velocity were obtained by applying a pulsed field gradient to the bubble, using the technique described in Ref. 5. Although motion at a large angle to the field gradient was sometimes observed, the probability was not very large (~0.1 for LuGd, and much less for SmY). The data reported here comprise only those cases where the bubble motion was essentially parallel to the field gradient. It should be noted, however, that even when those pulses which result in large skew angles are ignored, the motion of the bubble resulting from the remaining pulses is not always exactly parallel to the direction of the field gradient, but shows a scatter of ~±10°. This has only a minor influence on the data reported here, and it is therefore ignored.

We discuss first the data on LuGd. Figure 1 shows the measurements of bubble velocity as a function of the drive field ΔH. At each value of ΔH, 10 measurements of the bubble velocity were made, and the total spread observed is shown by the vertical bars. It can be seen that at low drive fields considerable scatter in the velocity occurs. This can be ascribed to the effect of coercivity, pinning points, and other imperfections in the film. The importance of these imperfections is expected to diminish at high drive fields, whereas Fig. 1 shows that the scatter in the velocity actually increases with increasing drive field. Thus we ascribe the velocity scatter at high drive fields to a dynamic conversion process of the type described above.

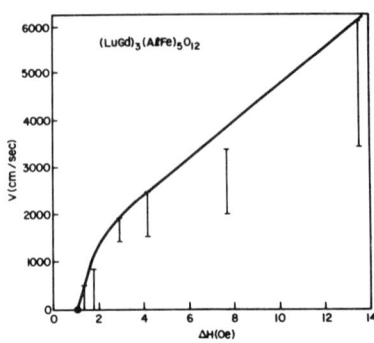

Fig. 1-Bubble velocity V versus field difference across the bubble, ΔH, in an epitaxial film of $Lu_1Gd_2Al_{.6}Fe_{4.4}O_{12}$. The vertical bars show the total spread observed among 10 successive measurements of velocity at each value of ΔH. The curve represents the maximum velocity that can be attained, and is analogous to the mobility curves reported previously (Ref. 3).

A more quantitative measure of the scatter in the velocity is shown in Fig. 2. Measurements of the bubble velocity at a fixed drive field of 7.9 Oe were made 100 times, and the results were plotted as a velocity probability distribution. A similar set of data to that discussed above was taken on a LuGd film implanted with 10^{14} Ne$^+$ ions/cm^2 at an energy of 200 keV. This implantation is very effective in suppressing the static occurrence of hard bubbles, yet, as can be seen from Figs. 3 and 4, it does not appear to have any significant effect on the velocity scatter. This result is not surprising when viewed in the light of Hagedorn's model[4] for the dynamic conversion process. In this model, the generation of vertical Bloch lines results from the motion of horizontal

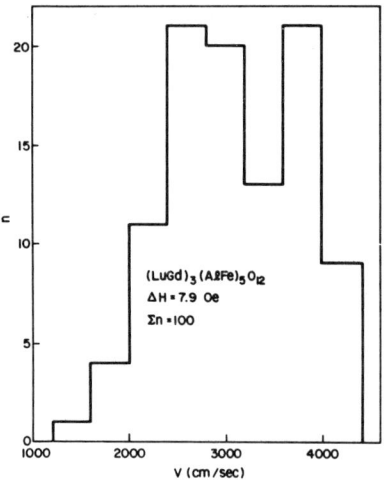

Fig. 2-Probability distribution of the bubble velocity in a LuGd film at a drive field $\Delta H = 7.9$ Oe. The value of n is the number of velocity measurements that fall within a 400 cm/sec interval.

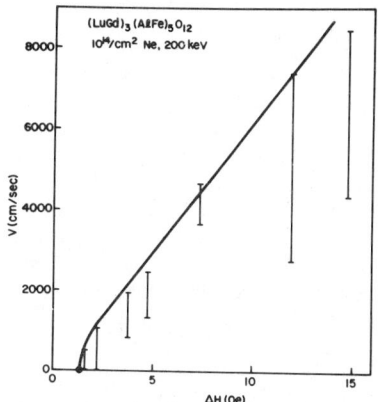
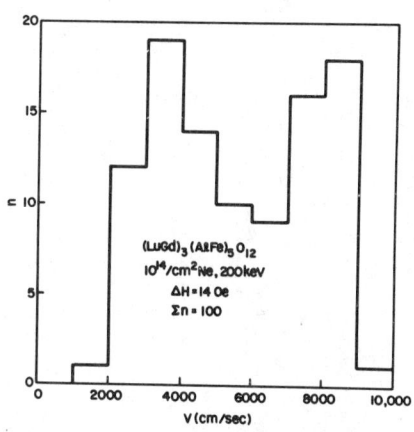

Fig. 3-Bubble velocity measurements in a LuGd film implanted with $10^{14}/cm^2$ Ne$^+$ ions at 200 keV.

Fig. 4-Probability distribution of the bubble velocity in a LuGd film implanted with $10^{14}/cm^2$ Ne$^+$ ions at 200 keV. The drive field $\Delta H = 14$ Oe.

Bloch lines from one surface of the film to the other. This horizontal Bloch line motion is a result of the interaction of the moving domain wall with the demagnetizing field at the film surfaces.[6] Since the level of ion implantation involved here does not materially affect the surface demagnetizing field, it cannot be expected to have much influence on the dynamic conversion process.

We have also investigated the effect of an in-plane field in the LuGd sample, and the results are shown in Fig. 5 for an in-plane field $H_t = 100$ Oe applied parallel to the drive field. Although the scatter in the velocity is not reduced by the in-plane field, an increase in the maximum velocity occurs at all drive fields. The increase in the velocity is quite small at low drive fields (the ratio of the velocities with and without the in-plane field is 1.04 at $\Delta H = 2$ Oe), but it becomes considerably larger at high drive fields (the ratio of the velocities is 1.6 at $\Delta H = 10$ Oe). Since the wall velocity is directly proportional to the wall width, it is tempting to attribute the increase in the velocity to the increase in wall width produced by the in-plane field. From the work of Johansen et al.[7] we obtain the following expression for the ratio of the wall widths with and without the in-plane field:

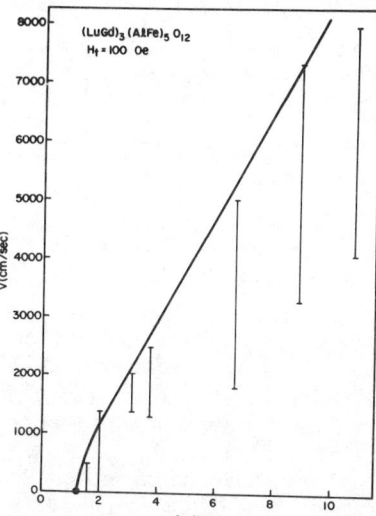

Fig. 5-Bubble velocity measurements in a LuGd film, with an in-plane field $H_t = 100$ Oe.

$$\frac{\ell_{wt}}{\ell_w} = \left(\frac{H_k}{H_k - H_t}\right)^{1/2} \tag{1}$$

where H_k is the equivalent anisotropy field (H_k = 1030 Oe for our sample). Hence, for H_t = 100 Oe, we find ℓ_{wt}/ℓ_w = 1.05. Although Eq. 1 is strictly applicable only to a straight wall, we do not expect the equivalent result for a bubble to differ greatly. Hence the increase in velocity at low drive fields can be understood in terms of the increased wall width, but the increase at high drive fields is too large to be explained in this way. Since the bubble shape is distorted by the in-plane field, becoming elongated in the field direction, we speculate that the distortion might increase when a large field gradient is applied, thus causing an increase in the velocity due to an increase in the field difference across the bubble.

Bubble velocity measurements were also made in a SmY film, and they are shown in Fig. 6. It is immediately evident that the scatter in the velocity is much smaller than it is in LuGd. This is borne out by the detailed measurements shown in Fig. 7. In an effort to

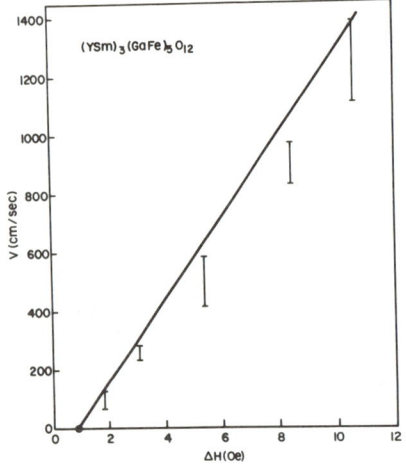

Fig. 6-Bubble velocity measurements in an epitaxial film of $Sm_{.4}Y_{2.6}Ga_{1.2}Fe_{3.8}O_{12}$.

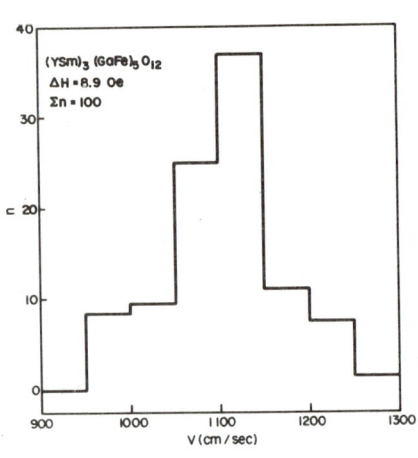

Fig. 7-Probability distribution of the bubble velocity in a SmY film at a drive field ΔH = 8.9 Oe.

determine which sample properties are responsible for this difference, we list the relevant parameters in Table I. It is seen that the film thickness h, saturation magnetization M, and wall energy σ are very similar for the two films. The main differences appear to be in the values of q ($\equiv H_k/4\pi M$) and the damping constant α. While a high q-value might be expected to

Table I: Comparison of Some Film Parameters for LuGd and SmY

	LuGd	SmY
α	0.023	0.20
q	5.5	8.8
h(μm)	6.3	5.6
$4\pi M$(G)	202	200
σ(erg/cm^2)	0.23	0.18

stabilize the spin structure of the bubble wall against the type of perturbation considered here, it is unlikely that the relatively small difference in q-value between these two samples can account for the markedly different behavior. A more significant difference appears to be in the value of α. Slonczewski[8] has derived the following inequalities for the wall velocity:

$$\frac{\alpha}{1+\alpha^2} \frac{\gamma \ell_w}{\pi} H \leq V \leq \frac{1}{\alpha} \frac{\gamma \ell_w}{\pi} H \qquad (2)$$

where γ is the gyromagnetic ratio. The quantity on the right hand side is the usual mobility relation, which represents the fastest rate at which the spins can be made to precess, and hence leads to an upper limit on the velocity. The expression on the left hand side is valid when the only torques acting on the spins are those due to the external magnetic field and damping. This therefore leads to the lower limit on the velocity, since any additional torques introduced into the spin system will cause the spins to precess faster and lead to a higher velocity. It is evident from Eq. 2 that when $\alpha \ll 1$, the range of allowable velocities is very large, and this range decreases with increasing α. The data reported here are in qualitative agreement with this, although the range of velocities observed is much smaller than that allowed by Eq. 2.

The data reported above have obvious implications for bubble devices utilizing the LuGd samples discussed here. Figures 1-4 clearly show that although high velocities can be attained with drive fields ~5 Oe, which are commonly encountered in bubble devices, the probability of lower velocities occurring is by no means negligible. Thus reliable device operation at frequencies which require velocities higher than the minimum value cannot be expected. This was confirmed by bubble propagation tests on the sample of Fig. 3 using a permalloy overlay circuit consisting of a 21-step T-X loop. The operating margins at 300 kHz were almost as large as the quasistatic margins, but a rapid drop in the margins occurred with increasing frequency.

The author thanks F. B. Hagedorn and W. J. Tabor for helpful discussions, S. L. Blank and B. S. Hewitt for the garnet films, and J. C. North for the ion implantation.

REFERENCES

1. J. E. Geusic, et al., to be presented at the 1973 Conf. on Magnetism and Magnetic Materials, Boston; J. W. Nielsen, et al. to be published in J. of Electronic Materials.
2. G. P. Vella-Coleiro, et al., Appl. Phys. Lett. 22, 324 (1973).
3. G. P. Vella-Coleiro, AIP Conf. Proc. 10, 424 (1973).
4. F. B. Hagedorn, to be presented at the 1973 Conf. on Magnetism and Magnetic Materials, Boston.
5. G. P. Vella-Coleiro, et al., Appl. Phys. Lett. 21, 7 (1972).
6. J. C. Slonczewski, J. Appl. Phys. 44, 1759 (1973).
7. T. R. Johansen, et al., J. Appl. Phys. 42, 1715 (1971).
8. J. C. Slonczewski, Int. J. Magnetism 2, 85 (1972).

A MODEL FOR DYNAMIC CONVERSION IN BUBBLE DOMAINS

F. B. Hagedorn
Bell Laboratories, Murray Hill, N.J. 07974

ABSTRACT

Erratic propagation behavior of magnetic bubble domains, as reported previously by Vella-Coleiro et al., has been attributed to dynamic conversion of a normal bubble domain wall into a relatively immobile but temporary state. A model for dynamic conversion is presented. This model is based on the nucleation and propagation of Bloch lines within the moving bubble domain wall. Material imperfections are postulated to play an important role in nucleating these Bloch lines, thereby accounting for the erratic behavior of the propagating bubbles.

INTRODUCTION

Dynamic conversion of a normal bubble domain wall into a temporary state that is relatively immobile has been recently reported.[1,2] This phenomenon was originally identified during bubble transport measurements, where a single bubble domain was repetitively translated back and forth over a distance of several μm by the application of a carefully controlled magnetic field gradient.[3] Erratic bubble propagation behavior was observed in this repetitive experiment, even though the same bubble was being subjected to a series of identical pulse sequences. Skew propagation (as seen in hard bubble dynamics[4]) was observed[1,2] to occur during some of the pulses, and the distance over which the bubble was translated during a fixed pulse length and amplitude was observed to vary by as much as a factor of 10 from one pulse sequence to the next.

QUALITATIVE MODEL FOR DYNAMIC CONVERSION

This model is based on nucleation and propagation of horizontal Bloch lines (HBL) within the bubble domain wall. For planar walls, Slonczewski[5] has shown that HBL can originate near one surface of the magnetic film and move through the domain wall to the other surface. While moving, such HBL can dissipate a large amount of energy.

One essential point of the present model is the hypothesis that the nucleation of such HBL is the result of interactions between the moving bubble wall and material inhomogeneities or imperfections, the size-scale of which is comparable with the domain wall thickness. Inspection of the equations of motion has not revealed an intrinsic instability for HBL nucleation during steady-state motion of the wall, so this imperfection-interaction mechanism has been postulated to account for a portion of the erratic behavior which is observed.

A second essential feature of the present model arises from the geometric differences between HBL motion in planar as opposed to cylindrical domain walls. Thiele's g-force formalism[6,7] provides a

223

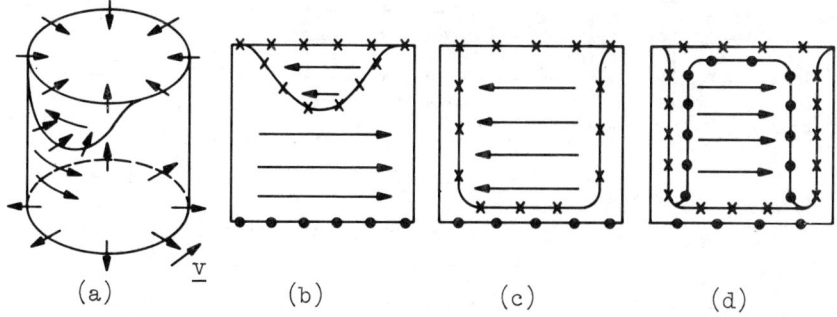

Fig. 1-Schematic representation of the magnetization at the center of a bubble domain wall during Bloch line growth.

convenient method of exploring this difference. Figure 1a, following Ref. 7, shows a moving bubble domain in which a HBL is propagating downward. The g-force, which is driving the HBL, vanishes where the wall tangents are parallel to the velocity direction. Consequently, the HBL is curved in the bubble domain wall. In Fig. 1b, the rear half of the curved domain wall is represented schematically in a planar sketch, again showing the curved HBL. Figure 1c shows the result of further HBL motion, after which a 2π HBL lies along the bottom surface of the wall in combination with a pair of <u>vertical</u> Bloch lines (VBL) through the thickness of the film. Figure 1d shows the spin configuration which results after a second HBL nucleates from Fig. 1c and then propagates to the top of the film. In Fig. 1d, there are two pairs of VBL, in addition to the 2π HBL dynamically trapped at the top and bottom film surfaces by the demagnetizing field gradient.

Similar nucleation and propagation of Bloch lines can occur independently in the front half of the bubble domain wall shown in Fig. 1a. Analysis of the g-forces[6,7] shows that all of the VBL are forced to the sides of the moving bubble wall. Each generated pair of VBL, as in Fig. 1c, contains one right-handed and one left-handed line. All right-handed lines end up on one side, while all left-handed lines are forced to the other side of the moving bubble wall. Dynamic stabilization of a statically unstable spin configuration thus results. The motion of this stabilized spin configuration is slowed down both by subsequently created HBL moving across the thickness of the film, as indicated schematically in Fig. 1, and by the existence of previously generated VBL. The latter create an added dissipation, which can be easily seen from Thiele's dissipation dyadic.[6,7] Consequently, the reduced velocity observed in dynamic conversion is accounted for by the combination of these two factors.

The third essential feature of this model is to postulate yet another result from the interaction between the moving bubble domain wall and imperfections near the magnetic film surfaces. Shown in Fig. 2a is a somewhat more detailed representation of Fig. 1d. Each line in Fig. 2 a represents a contour of constant direction for the magnetization. In Fig. 2b it is assumed that the lower 2π HBL shown in Fig. 1a has interacted with an imperfection and reached the film surface, that the VBL on the left are now attached to the surface in exactly the way that occurs in hard bubbles, and that the VBL on the

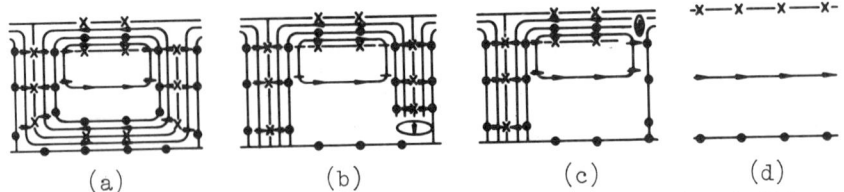

Fig. 2—Contours of the magnetization in a section of the bubble domain wall before and after severing the Bloch lines.

right terminate in a vortex which is somewhere away from the film surface. Surface tension will cause the vortex to be pulled toward the upper surface, as shown in Fig. 2c, after which there will be a net transverse force due to the unequal numbers of right-handed and left-handed VBL. This force has been calculated in detail to account for the dynamic properties of hard bubbles[6,8,9] and can explain the occasional skew propagation effects which are observed[1,2] in dynamic conversion. According to the model being presented herein, skew propagation is erratic because of the statistical nature of the interactions which lead to the severing of the Bloch line structure shown in Fig. 2. It is conjectured for this part of the model that the configuration of Fig. 2c may be dynamically stabilized but that the vortex is pulled around the bubble wall by surface tension as soon as motion terminates, thereby reverting to the initial normal configuration as shown in Fig. 2d.

QUANTITATIVE ASPECTS OF DYNAMIC CONVERSION

While a more complete quantitative description is given elsewhere,[10] it will be instructive to summarize some of these results here. It is shown in Ref. 10, for the case of a <u>planar</u> wall, that the wall velocity as a function of the driving magnetic field (H_d) would be expected to be as shown in Fig. 3. Below a critical velocity (v_p) defined in Ref. 5 to be

$$v_p = 24|\gamma| A/hK_u^{\frac{1}{2}}, \qquad (1)$$

Fig. 3 is linear. In Eq. (1), $|\gamma|$ is the gyromagnetic ratio of the electron, A is the magnetic exchange constant, h is the film thickness, and K_u is the uniaxial anisotropy constant of the material. Equation (1) defines the minimum velocity for propagating a HBL away from the film surface, once it has been nucleated, and this velocity will be attained in a given material when $H_d = H_p$, with

$$H_p = 24\alpha A^{\frac{1}{2}}/h \qquad (2)$$

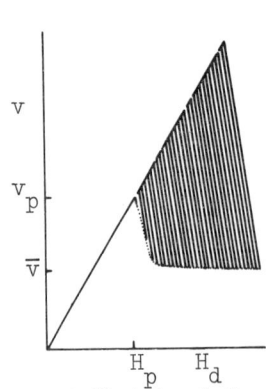

Fig. 3—Sketch of the wall velocity as a function of the driving field.

where α is the Gilbert damping parameter for the magnetic film.

For $H_d > H_p$, the wall motion becomes non-uniform in time, depending on the location of the moving HBL. The average wall velocity ($\bar{v}$) has been calculated[10] using the assumption that there is always one HBL moving through the film thickness, and $\bar{v}$ is shown in Fig. 3 as the lower bound of the cross-hatched region. In Ref. 10, it is shown from the equations of motion that $\bar{v} = 0.55\ v_p$.[11] If there are no HBL moving through the wall, then the dashed line (i.e., an extrapolation of the initial linear region) pertains. If a HBL is moving only part of the time during the motion of the wall, a time-average must be done; the result will fall somewhere in the cross-hatched region. Figure 3, therefore, pertains to a moving planar wall in which HBL may be nucleated sporadically. The observed[2] variations in the measured wall velocity can be accounted for in terms of Fig. 3, and this fact provides the motivation to consider the nucleation-from-imperfections hypothesis. In addition, Fig. 3 suggests the spread in measured velocity should increase with increasing drive field; the experimental results[2] are also consistent with this feature of the model.

For bubble domains, the curved wall complicates a corresponding analysis. However, v_p would appear to be a lower limit for the velocity required to propagate a HBL, once it has been nucleated. It is possible that a somewhat larger velocity could be required in order to achieve the "stretching" indicated schematically in Fig. 1. The effects of VBL have not been included in Fig. 3, though, and can be shown by using Thiele's[7] dissipation dyadic results and the observed[12] maximum Bloch line densities to be equivalent to an added dissipation of not more than a few times that due to the motion of a simple planar wall. The net result is that the average velocity of a bubble domain can appear below $\bar{v}$ in Fig. 3, as well as anywhere in the cross-hatched region.

DISCUSSION

Quantitative comparisons probably require one to take into account the excitation of the domain wall modes, as discussed by Thiele.[13] These modes represent additional damping and also are excited by interactions between the moving wall and imperfections in the magnetic film. Dynamic conversion and wall-mode excitation can and probably do coexist in a given bubble domain when it is moving, and a clear experimental separation of the two has not been demonstrated. However, erratic skew propagation would appear to be a unique sign of dynamic conversion, and the observed threshold value of about 1000 cm/sec as reported in Fig. 1 is in reasonable agreement with the value of 1100 cm/sec as calculated from Eq. (1) and the material parameters which pertain to Ref. 1.

Another aspect of the model which can be experimentally checked is related to the in-plane anisotropy dependence of Bloch line nucleation. Slonczewski's[5] model for the twisted wall structure makes it clear that a large anisotropy in the plane of the film, either from an externally applied magnetic field or from a material anisotropy, should make HBL nucleation much more difficult. Complete experimental verification of this point is not yet available, but the fact that dynamic conversion was never observed in orthoferrite supports this aspect of the model.

A final comment is that the model proposed herein leads one to suggest a new kind of film structure for bubble domains. With two similar magnetic films separated by a very thin nonmagnetic film, it is expected that the exchange coupling necessary to propagate the Bloch line will be interrupted but that a pair of bubbles in the two magnetic films will be strongly magnetostatically coupled and will appear approximately as one bubble. The wall structure will be dynamically stabilized as shown in Fig. 4, however. Only a single vertical Bloch line will appear on each side of the bubble, resulting in relatively little added damping. Experiments using such a multi-layered structure are presently in progress.

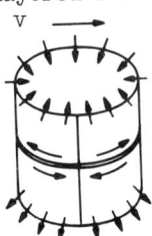

Fig.4-Dynamically stabilized configuration in a 3-layer structure which suppresses dynamic conversion.

ACKNOWLEDGMENTS

The author has benefited from discussions with A. A. Thiele, G. P. Vella-Coleiro, J. E. Geusic and A. H. Bobeck.

REFERENCES

1. G. P. Vella-Coleiro, F. B. Hagedorn, Y. S. Chen and S. L. Blank, Appl. Phys. Lett. 22, 324 (1973).
2. G. P. Vella-Coleiro, paper presented at 1973 M^3 Conference.
3. G. P. Vella-Coleiro and W. J. Tabor, Appl. Phys. Lett. 21, 7 (1972).
4. W. J. Tabor, A. H. Bobeck, G. P. Vella-Coleiro and A. Rosencwaig, Bell Syst. Tech. J. 51, 1427 (1972).
5. J. C. Slonczewski, J. Appl. Phys. 44, 1759 (1973).
6. A. A. Thiele, Phys. Rev. Lett. 30, 230 (1973).
7. A. A. Thiele, J. Appl. Phys., scheduled for December 1973.
8. J. C. Slonczewski, Phys. Rev. Lett. 29, 1679 (1972).
9. A. A. Thiele, F. B. Hagedorn and G. P. Vella-Coleiro, Phys. Rev. B 8, 241 (1973).
10. F. B. Hagedorn, J. Appl. Phys., to be published.
11. The velocity saturation effect shown in Fig. 3 was previously obtained in Ref. 5. In Ref. 5, however, the saturation velocity was calculated to be 0.3 v_p. Although the origin of this difference is discussed in Ref. 10, it is not important for the present discussion, which is predominantly qualitative.
12. D. H. Smith and A. A. Thiele, paper presented at 1973 M^3 Conference.
13. A. A. Thiele, Phys. Rev. B 7, 391 (1973).

THE INFLUENCE OF AN IN-PLANE FIELD ON BUBBLE DYNAMICS

R. M. Josephs and B. F. Stein
Sperry Univac, Blue Bell, Pa. 19422

ABSTRACT

The influence of an in-plane H_{in} on bubble dynamics, as measured by the bubble collapse technique, in an LPE film of $Y_{2.4}Eu_{.6}Fe_{3.8}Ga_{1.2}O_{12}$ is reported. The scatter in the collapse time (τ) data caused by a distribution of bubbles with varying wall structure was eliminated by taking data on bubbles generated by in-plane demagnetization. The saturation value of $1/\tau$ increases quadratically with H_{in} with the value at 90 Oe being 4 times the value at $H_{in} = 0$. The dependence of the saturation value of $1/\tau$ on bias field H_1 reveals an apparent increase in the saturation velocity as H_1 approaches the static collapse field.

INTRODUCTION

Recent measurements on stripe domains[1,2] and isolated straight walls[3] have demonstrated that the wall dynamics in bubble materials is affected by the application of an in-plane field (H_{in}). It is expected that the data obtained by the bubble collapse technique[4] should also be influenced by H_{in}. However, the frequently observed scatter in bubble collapse time (τ) data caused by a distribution of bubbles with varying wall structure can obscure this effect, particularly at low values of H_{in}. In this work, a technique for insuring that all the bubbles have the same wall structure is introduced. It will be shown that the results of bubble collapse measurements on such bubbles are in good agreement with the results reported for measurements on isolated straight walls.

EXPERIMENTAL DETAILS

The usual bubble collapse mobility apparatus,[5] consisting of a bias field (H_1) coil, a pulsed field coil (7 nsec risetime), and an adjustable sample holder, was located between the pole pieces of a 4" electromagnet which furnished the in-plane field (H_{in}) both for the sample demagnetization and also for the measurements of the wall dynamics. A HP-214A pulse generator and a Paravan 1500 delay line pulse generator were used for long and short pulse widths respectively. The data reported here were taken on a 200 mil square sample cut from the center of a 1" diameter $Y_{2.4}Eu_{.6}Ga_{1.2}Fe_{3.8}O_{12}$ (h = 11.3μ, $4\pi M_s$ = 139G, ℓ = 0.6μ, and H_K = 1060 Oe) film grown by liquid phase epitaxy on a [111] $Gd_3Ga_5O_{12}$ substrate.

All the data were taken on bubbles generated by in-plane demagnetization. The wall structure of such bubbles is shown in

Fig. 1. Within the experimental accuracy of 0.2 Oe, these bubbles have the same value of static collapse field (H_O) as do normal bubbles. [In bubble transfer measurements it was observed that the bubbles having the wall structure shown in Fig. 1 (b) propagate along the field gradient with no sidewise deflection.[6] This independently confirms that aspect of the model of Slonczewski et al[7] describing the dependence of the translational motion on the number of Bloch lines.] The bubble collapse measurements were made on a group of widely spaced bubbles. The τ data reported here were taken on the last bubble in such a group to collapse to avoid interaction effects. The measurements were frequently repeated and the data were reproducible to 10 nsec. Although all the bubbles were prepared with the same wall structure, in the course of the measurement, a given bubble would be subjected to a series of pulses until it eventually collapsed. It is possible that the wall structure might be permanently altered during these preliminary pulses.[8] For this reason, the bubbles were frequently subjected to such a series of pulses, but were not collapsed out. These bubbles were then quasi-statically collapsed by increasing H_1. Within the experimental accuracy of 0.2 Oe, no change was ever observed in H_O. Although this result does not exclude the possibility of a small change in the number of wall transitions, it does indicate that the wall structure was not changed to one consisting of a large number of transitions.

The alignment of H_{in} with respect to H_1 was ascertained by measuring H_O for different combinations of H_1 and H_{in} polarities. In this manner, a component of H_{in} along the direction of H_1 could be accounted for in the analysis. The variation of H_O with H_{in} is illustrated in Fig. 2 where the data have been corrected for a slight misalignment of H_{in} with respect to H_1. Over the region where the dynamic data were taken, $0 \leq H_{in} \leq 120$ Oe, H_O increased by $\sim 2\%$ due to a slight decrease in the material parameter.[9] In this range, the influence of the cubic anisotropy was ignored.

INFLUENCE OF IN-PLANE FIELD ON BUBBLE COLLAPSE TIME

The influence of H_{in} on τ is illustrated in Fig. 3. A number

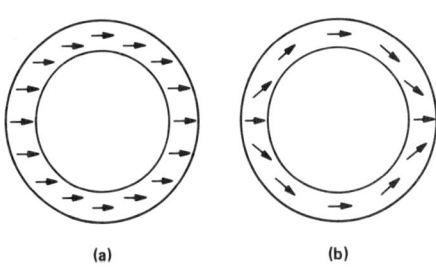

Fig. 1. Wall structure generated by in-plane demagnetization: (a) $H_{in} \geq 4\pi M_s$; (b) $H_{in} = 0$.

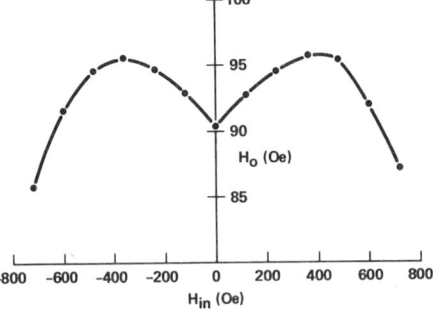

Fig. 2. Dependence of H_O on H_{in}.

of points in the low pulsed field region, all falling on the solid curve, have been omitted for clarity. Because of the slight variation of H_0 with H_{in} as seen in Fig. 2, H_1 was adjusted so that the quantity $(H_0 - H_1)$ was the same for all values of H_{in}; therefore, all the curves have the same $(1/\tau) = 0$ intercept. Although all the data shown were taken for the same polarity of H_{in}, a number of points were checked for the opposite sense of H_{in} with identical results.

The velocity (v), for a given value of H_1 and H_{in}, was obtained by multiplying $(1/\tau_1)$ by (r_+-r_-), the computed difference in stable and unstable radii.[4] Because of the slight increase in H_0 with increasing H_{in}, (r_+-r_-) varied from 2.02μ at $H_{in} = 0$ to 1.91μ at $H_{in} = 120$ Oe. This 5% variation in (r_+-r_-) was ignored and all the $1/\tau$ values in Fig. 3 were multiplied by 2μ to convert them to values of v. The most striking feature of the data is the effect of H_{in} in the saturation region. In Fig. 4 the saturation values of $1/\tau$ and v are plotted as a function of H_{in}. The solid curve, fit to the values of $1/\tau_{SAT}$ at $H_{in} = 0$ and 60 Oe, is

$$(1/\tau_{SAT}) = 4 + H_{in}^2/600 \tag{1}$$

where $1/\tau_{SAT}$ is in μsec^{-1} and H_{in} is in Oe. The saturation v in m/sec is then given by

$$v_{SAT} \approx 8 + 64 \, (H_{in}/4\pi M_s)^2 \tag{2}$$

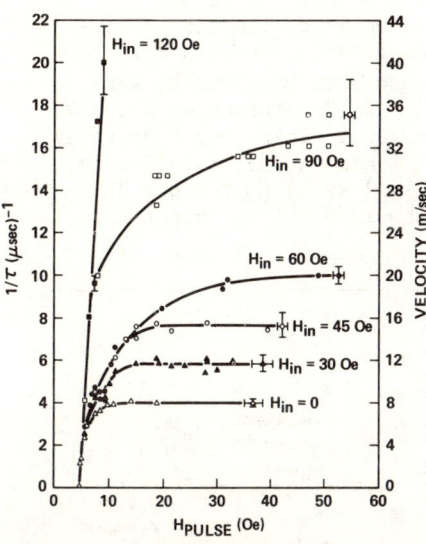

Fig. 3. Dependence of the reciprocal of the collapse time on pulsed field. An approximate velocity scale is plotted on the right hand ordinate.

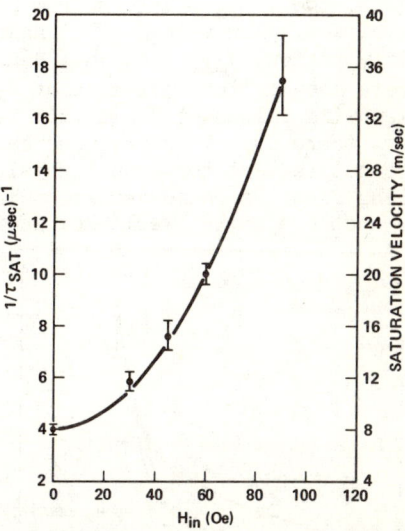

Fig. 4. Dependence of the reciprocal of the saturation collapse time on pulsed field. An approximate saturation velocity scale is plotted on the right hand ordinate.

Malozemoff has found that in the saturation region, bubbles having a small number of Bloch lines may, after repeating pulsing, change their state to the pure Bloch wall state.[8] Since the difference in v between the 2 Bloch line bubble and the zero Bloch line bubble is small, this effect would produce a negligible change on the v data shown in Figs. 3 and 4. De Leeuw has reported data on the effect of H_{in} on the v of a single straight wall.[3] For values of $(H_{in}/4\pi M_s)$ comparable to ours, he observed that

$$v_{SAT} \approx 7 + 46 \, (H_{in}/4\pi M_s)^2 \qquad (3)$$

This agrees favorably with our result given in Eq. (2). Since there are no published theoretical treatments for the effect of H_{in}, in the range of interest, on the wall dynamics, it is not possible to make a more detailed comparison between the two experimental results.

DEPENDENCE OF SATURATION COLLAPSE TIME ON BIAS FIELD

The absence of scatter in the τ data, due to the method of generating bubbles, led us to investigate the H_1 region in which reliable measurements of v_{SAT} could be made. The variation of $1/\tau_{SAT}$ with H_1 at $H_{in} = 0$ and 45 Oe is shown in Fig. 5. As H_1 approached H_0, τ became so short that limitations in the apparatus prevented us from taking meaningful data for $(H_0 - H_1) < 2$ Oe. In Fig. 6, the $1/\tau_{SAT}$ values from Fig. 5 have been converted to velocities. For values of H_1 near the center of the static stability region, v_{SAT} has an insignificant variation with H_1. The v data showing the influence of H_{in}, described previously, were taken with values of H_1 in this region. As H_1 approaches H_0, however, there is an apparent increase in v_{SAT}. This result probably does not reflect the actual physical situation, but, more likely, arises from the rapid variation in $(r_+ - r_-)$ as H_1 approaches H_0. Any subtle effects overlooked in the theory might be expected to be

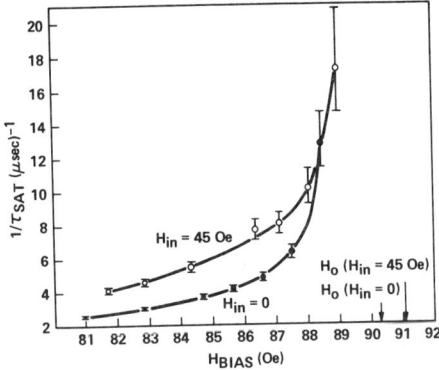

Fig. 5. Dependence of the reciprocal of the collapse time on bias field.

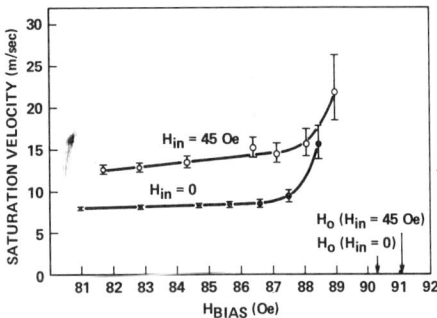

Fig. 6. Dependence of the saturation velocity on bias field.

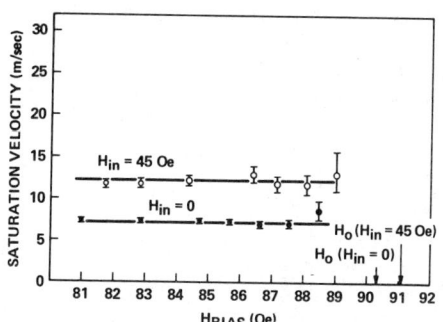

Fig. 7. Dependence of the saturation velocity on bias field for a 1.46% reduction in H_o.

most prominent in this region. The apparent dependence of v_{SAT} on H_1 can be eliminated by adjusting the respective values of H_o for $H_{in} = 0$ and 45 Oe. Because of the sensitivity of $(r_+ - r_-)$ to H_o, a reduction of merely 1.46% in the values of H_o produced the results shown in Fig. 7. Although this change in H_o is small, it is considerably greater than the experimental uncertainty in H_o which is 0.2%. The most obvious reason for the dependence of v_{SAT} on H_1 is that the approximation used for the demagnetizing field[4] overestimates $(r_+ - r_-)$ as H_1 approaches H_o. However, $(r_+ - r_-)$ was calculated using Thiele's theory[10] and similar behavior was observed. Changes in the wall energy due to the presence of the pair of Bloch lines[11], twisting of the wall structure by stray fields at the film surfaces,[12,13] and wall curvature[14] were also considered and found to have negligible influence.

It has been shown that in plane demagnetization provides a useful method for defining the wall structure. This technique enabled us to observe accurately the influence of H_{in} on bubble collapse dynamics. The similarity in the dependence of v_{SAT} on H_{in} as observed by bubble collapse and straight wall measurements demonstrates that the collapse technique provides a faithful representation of the domain wall motion in the drive field region where v is saturating.

REFERENCES

1. B. E. Argyle, A. P. Malozemoff, AIP Conf. Proc. 10, 344 (1972).
2. D. C. Fowlis, J. A. Copeland, AIP Conf. Proc. 10, 393 (1972).
3. F. H. De Leeuw, to be published in IEEE Trans. Mag.
4. H. Callen and R. M. Josephs, J. Appl. Phys. 72, 1977 (1971).
5. R. M. Josephs, AIP Conf. Proc. 10, 286 (1972).
6. Our bubble transfer results were corroborated by A. B. Smith.
7. J. C. Slonczewski, A. P. Malozemoff, and O. Voegeli, AIP Conf. Proc. 10, 458 (1972).
8. A. P. Malozemoff, to be published in J. Appl. Phys.
9. J. W. F. Dorleijn, W. F. Druyvesteyn, G. Bartels, and W. Tolksdorf, Philips Res. Repts. 28, 133 (1973).
10. A. A. Thiele, Bell System Tech J. 48, 3287 (1969).
11. H. Nishida, T. Koboyashi, and Y. Sugita, AIP Conf. Proc. 10, 493 (1972).
12. J. C. Slonczewski, J. Appl. Phys. 44, 1759 (1973).
13. E. Schlömann, Appl. Phys. Lett. 21, 227 (1972).
14. J. Kaczer and I. Tomas, Phys. Stat. Sol. 10, 619 (1972).

THE EFFECT OF A CONSTANT IN-PLANE MAGNETIC FIELD ON
MAGNETIC BUBBLE TRANSLATION IN $(YGdTm)_3(FeGa)_5O_{12}$*

D. C. Bullock
Central Research Laboratories, Texas Instruments Incorporated,
Dallas, Texas 75222

ABSTRACT

Bubble translational velocity studies on ion implanted epitaxial garnet films of $(YGdTm)_3(FeGa)_5O_{12}$ have demonstrated that several discrete deflection angles for a given bubble diameter can be obtained between the bubble velocity, $\vec{V}$, and the direction of maximum field gradient, $\vec{\nabla}H_m$, with one of the angles predominating. These angles are interpreted by a theory due to Slonczewski. These deflections can be suppressed by the application of a constant in-plane magnetic field of magnitude greater than or equal to 110 Oe which stabilizes the two Bloch line magnetic bubble. An increase in bubble velocity is also observed when the in-plane magnetic field is applied.

INTRODUCTION

Ion implantation has been used to suppress hard bubbles in epitaxial garnet films used for magnetic bubble domain devices.[1] It has been suggested that the ion implantation produces a capping layer near the surface of the film in which the magnetization lies in the plane. We have found that certain ion implants suppress hard bubbles in two compositions of $(YGdTm)_3(FeGa)_5O_{12}$ but that the bubbles that are stabilized move at an angle to $\vec{\nabla}H_m$. In order to induce motion parallel to $\vec{\nabla}H_m$ an in-plane magnetic field, $\vec{H}_p$, is required. This field produces an ellipticity in the bubble shape and presumably stabilizes the two Bloch line magnetic bubble which moves parallel to $\vec{\nabla}H_m$.[2] We have measured the deflection angle as a function of drive, ΔH, and bubble diameter, d. For a given d there is one deflection angle that predominates ($\theta \approx 33°$ for d = 4.8 µ) but by applying an in-plane magnetic field and then slowly reducing it to zero, bubbles with three other deflection angles (14°, 25° and 39°) have been observed for d = 4.8 µ. An attempt has been made to correlate the above set of deflection angles with the theory of Slonczewski.[2]

By applying an in-plane magnetic field greater than or equal to 110 Oe bubble motion parallel to $\vec{\nabla}H_m$ has been achieved together with an enhancement of the bubble velocity for a given $\Delta H = d\vec{\nabla}H_m$.

MAGNETIC BUBBLE DEFLECTIONS

$Y_{1.4}Gd_{0.6}Tm_{1.0}Fe_{4.2}Ga_{0.8}O_{12}$ and $Y_{1.0}Gd_{1.0}Tm_{1.0}Fe_{4.2}Ga_{0.8}O_{12}$ epitaxial garnet films on gadolinium gallium garnet substrates have been ion implanted with various ions (H^+, H_2^+ or Ne^+), energies (30 keV to 150 keV) and doses (1 x 10^{14}/cm² to 2 x 10^{16}/cm²). From bubble collapse measurements it has been observed that hard bubbles were suppressed for the implants given above.

Bubble translation studies were performed on about fifty films of the above compositions by applying a gradient field produced by two parallel conductors carrying equal currents as described by Vella-Coleiro.[3] In all but a few of the films the bubble velocity, $\vec{V}$, was rotated counterclockwise from the $-\vec{\nabla}H_m$ direction if the direction of observation is back along the direction of the bias field (left hand deflection). The few films which had no deflecting bubbles possessed bubbles which were elliptical in shape possibly due to substrate misorientation or an in-plane anisotropy. For garnet films of $Y_{2.6}Sm_{0.4}Fe_{3.8}Ga_{1.2}O_{12}$ and $Y_{1.0}Eu_{1.8}Yb_{0.2}Fe_{3.9}Al_{1.1}O_{12}$ which were also ion implanted this deflection behavior is not present, demonstrating that the effect is not a general behavior for all garnet film compositions.

The detailed experimental results given below were done on an epitaxial film of $Y_{1.4}Gd_{0.6}Tm_{1.0}Fe_{4.2}Ga_{0.8}O_{12}$ which was ion implanted with neon at 100 keV and $2 \times 10^{14}/cm^2$. The material parameters for the film were; $h = 5.5~\mu$, $\ell = 0.5~\mu$, $4\pi M_S = 229$ G, $\sigma_w = 0.21$ ergs/cm, $q = 4.4$ and $a_o = 12.375$ Å. The gyromagnetic ratio, γ, was estimated from garnet end members[4] to be approximately 1.64×10^7 Oe^{-1} sec^{-1} and the exchange constant, A, was assumed to be 3×10^{-7} ergs/cm. The bubble deflection angle, θ, was measured in a localized region of the film for several different in-plane orientations of this area relative to $\vec{\nabla}H_m$. No strong dependence on in-plane orientation was observed. For a fixed bubble diameter the deflection angle was measured as a function of the drive field, $\Delta H_\parallel = d\nabla H_m \cos\theta$. The bubble was moved one diameter per pulse. The velocity was determined when the initial position of the bubble was at the center line and then displaced one diameter away. The deflection angle was independent of drive as shown in Fig. 1 for $d = 4.8~\mu$. Twenty-five different bubbles with $d = 4.8~\mu$ were observed for $\Delta H = 3.84$ Oe and the mean value of θ was $33.5°$ with a standard deviation of $0.5°$. A small fraction of the

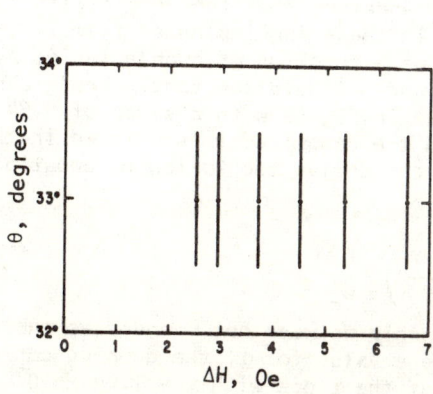

Fig. 1. Deflection angle, θ, vs ΔH for $d = 4.8~\mu$.

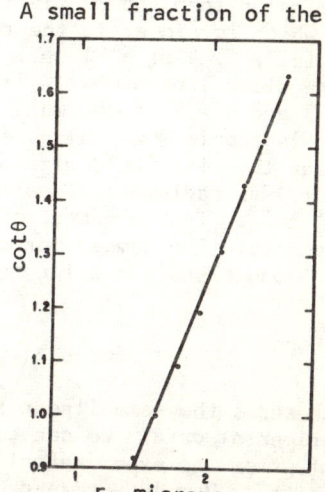

Fig. 2. Cot θ vs r measured by varying the bias field.

time the bubble would move straight along $\vec{\nabla}H_m$ or move at a substantially different angle than 33°. These events were omitted from the above average. At no time was a right hand deflecting bubble observed in any of the above ion implanted films.

In order to investigate bubbles having other characteristic angles it was found that these bubbles could sometimes be created by applying an in-plane magnetic field, H_p, of 40 Oe to the film and then slowly reducing it to zero while the vertical bias field remained constant. Three other deflection angles for d = 4.8 μ were found to be 14°, 25° and 39°. Once created these bubbles were stable enough to measure their velocity. Table II summarizes the deflection data for these four types of bubbles along with the revolution parameter, n'_r, calculated from Slonczewski's equation

$$n'_r = \frac{\gamma r^2 \nabla H_m \sin\theta}{2V}$$

and twice this value, n_r.

TABLE II

d(μ)	θ(°)	ΔH(Oe)	V(cm/sec)	n'_r	n_r
4.8	14	3.84	1828	0.50	1.00
4.8	25	3.84	1561	1.02	2.04
4.8	33.5	3.84	1465	1.42	2.84
4.8	39	3.84	1280	1.86	3.72

The above set of data show approximate half-integer values for the revolution number using the above formula which is unrealistic if we are to avoid divergences in the magnetization in the wall around the circumference of the bubble. If we omit the factor of two in the denominator then the revolution number, n_r, would be 1, 2.04, 2.84 and 3.72 which is closer to the required integer values and would correspond to 0, 2, 4 or 6 Bloch line bubbles since n = 2 (n_r - 1) where n is the Bloch line number. It is now apparent that the bubble with $n_r \approx 3$ and n = 4 is the most stable in these ion implanted films. For this bubble θ was also measured as a function of bubble radius by raising the bias field with H_p = 0 and $|\vec{V}|$ held constant. The graph of cot θ vs radius is linear as shown in Fig. 2 with a slope of 0.64 x 10^4 cm^{-1}. If the factor of two in the denominator is omitted in the above revolution number formula one can derive the following equation from Slonczewski's results to be

$$\cot\theta = \left(\frac{\alpha}{\Delta} + \frac{4H_c\gamma}{\pi V}\right)\frac{r}{n_r}$$

which shows the same linear relationship between cot θ and r as the experimental data. We can calculate a value for α, the damping constant, from the experimental value of the slope with γ = 1.64 x 10^7 sec^{-1} Oe^{-1}, V = 1454 cm/sec, $\Delta = \sqrt{A/K}$ = 5.7 x 10^{-6}cm, the stripe domain wall coercivity, H_c = 0.12 Oe and n_r = 3. The results give α = 0.10.

Since we are not at the initial slope region of the V vs ΔH curve this value of α doesn't necessarily equal the value calculated from the initial mobility but is the appropriate value for the above expression. With this value of α the following values of cot θ and θ can be calculated for $r = 2.4\,\mu$ and n_r assumed to be an integer.

TABLE III

n_r	cot θ_{cal}	θ_{cal}	θ_{exp}
1	4.519	12.5°	14°
2	2.288	23.6°	25°
3	1.535	33.1°	33.5°
4	1.165	40.6°	39°

The agreement between θ_{cal} and θ_{exp} is good for the $n_r = 1$, 2, and 4 cases and of course the $n_r = 3$ case is the value of n_r from which we calculated α. From the above results it is apparent that several different wall spin configurations can occur in ion implanted YGdTm with the most energetically favorable being the $n_r = 3$, $n = 4$ bubble. This implies that for YGdTm with the above material parameters the spin configuration in the cap is more complex than originally assumed and is considerably different than for the other two garnet compositions mentioned above. The physical origin of the different types of spin configuration is not known but may be related to the cubic anisotropy energies of the above films.

DEFLECTION SUPPRESSION AND VELOCITY ENHANCEMENT

In order to completely suppress the deflection effect the in-plane magnetic field was increased and motion parallel to $\vec{\nabla}H_m$ always occurred for $H_p \geq 110$ Oe. Between 50 and 110 Oe. there is a transition region where both straight and angular motion will occur. The particular orientation of $\vec{H}_p$ in the (111) plane was unimportant for suppressing the deflections. It is believed that the in-plane magnetic field magnetizes the capped region in a manner which favors the stabilization of the two Bloch line bubble ($n_r = 0$, $n = -2$).

The velocity of a bubble for a given ΔH was found to increase when the in-plane field was applied as shown in Fig. 3 a and 3b. deLeeuw has previously reported the increase in straight domain wall velocity in the presence of an in-plane magnetic field.[5] For the above bubble experiment data was taken for $\vec{H}_p$ parallel and perpendicular to $\vec{\nabla}H_m$. Since the bubbles become elliptical, 16% and 20% for $H_p = 125$ Oe and 215 Oe respectively, ΔH is now given by $d'\vec{\nabla}H_m$ where d' is the length of the elliptic axis parallel to $\vec{\nabla}H_m$. The above graphs show that the velocity for $\vec{H}_p \perp \vec{\nabla}H_m$ is larger than for $\vec{H}_p \parallel \vec{\nabla}H_m$. This difference may be due to the fact that higher mobility Bloch walls are perpendicular to $\vec{V}$ when $\vec{H}_p \perp \vec{\nabla}H_m$ while lower mobility Néel walls at the ends of the major axis of the ellipse are perpendicular to $\vec{V}$ for $\vec{H}_p \parallel \vec{\nabla}H_m$. It is known that for straight walls[6]

$$\mu_B/\mu_N = (1 + 1/q)^{1/2}$$

which is equal to 1.1 for the q = 4.4 film studied.

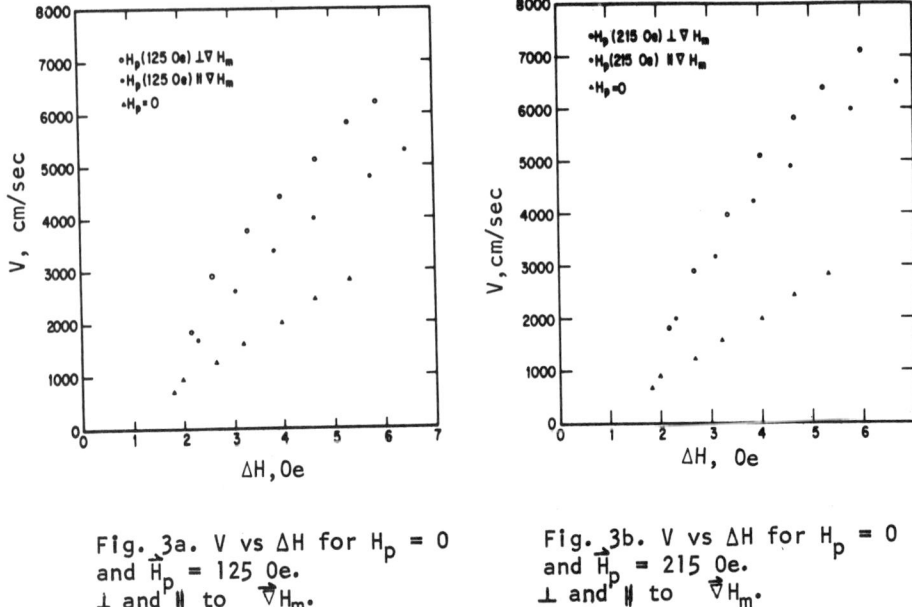

Fig. 3a. V vs ΔH for $H_p = 0$ and $\vec{H}_p$ = 125 Oe. ⊥ and ∥ to $\vec{\nabla} H_m$.

Fig. 3b. V vs ΔH for $H_p = 0$ and $\vec{H}_p$ = 215 Oe. ⊥ and ∥ to $\vec{\nabla} H_m$.

CONCLUSIONS

Bubble deflection modes have been suppressed in ion implanted YGdTm films and velocity enhancement effects have been achieved by the application of an in-plane magnetic field.

ACKNOWLEDGEMENTS

The author would like to thank L. G. Bailey, G. G. Sumner, and G. G. Rogers for supplying the films used in this work and F. G. West for many helpful discussions.

REFERENCES

1. R. Wolfe and J. C. North, B.S.T. J. **51**, 1426 (1972).
2. J. C. Slonczewski, A. P. Malozemoff and O. Voegeli, AIP Conference Proceedings No. 10, 458 (1972).
3. G. P. Vella-Coleiro and W. J. Tabor, Appl. Phys. Lett. **21**, 7 (1972).
4. W. H. vonAulock, "Handbook of Microwave Ferrite Materials," Academic Press, New York (1965).
5. F. H. deLeeuw, Intermag 1973, Washington, D. C. to be published.
6. E. M. Gyorgy and F. B. Hagedorn, J. Appl. Phys. **39**, 88 (1968).

*This work was partially supported by Air Force Avionics Laboratory Contract No. F33614-73-C-1029.

TEMPERATURE VARIATION OF MAGNETIC BUBBLE
GARNET FILM PARAMETERS

R. M. Sandfort, R. W. Shaw, and J. W. Moody
Monsanto Company, St. Louis, Mo. 63166

ABSTRACT

The dynamic response of domain walls to a step change in bias field both with and without an in-plane field has been measured from -10 to 60°C in $Sm_{0.4}Y_{2.6}Fe_{3.7}Ga_{1.3}O_{12}$, $Eu_{0.6}Y_{2.4}Fe_{3.7}Ga_{1.3}O_{12}$ and $Eu_{0.5}Y_{2.35}Yb_{0.15}Fe_{3.8}Ga_{1.2}O_{12}$ garnet films. In addition, characteristic length, saturation magnetization, domain wall energy density, anisotropy field, and coercivity have been measured up to the Neel transition. Domain wall mobility was found to increase by a factor of two over this temperature range as expected theoretically. Wall mobility also increased by a factor of two upon application of an in-plane field of 25 Oe. Over this temperature range the wall coercivity varied by an order of magnitude.

INTRODUCTION

The temperature variation of material parameters is important in bubble memory device design. A number of authors[1,2,3,4] have reported experimental work on the temperature characteristics of the static material parameters in a variety of garnet film compositions. However, little work [4,5] has been reported on the temperature behavior of domain wall mobility. In this paper we present data on the mobility of domain walls subjected to a step change in bias field measured from -10 to 60°C in several magnetic garnet films of interest for device application. The temperature dependence of the static material parameters is also presented and used to calculate theoretical wall mobilities for comparison with the measured values.

EXPERIMENTAL TECHNIQUES

The three garnet film compositions included in this study were adjusted to support 6μ bubbles at 24°C and to minimize room temperature lattice mismatch. The magneto-optical techniques used to measure characteristic length, ℓ, saturation magnetization, $4\pi M_s$, Neel temperature, T_N, and coercivity, H_c have been described in detail previously [6]. These measurements were carried out in a thermoelectric heating-cooling chamber similar to that described by Smith et al.[1] for temperatures in the range -10 to 70°C, and in a Leitz 350°C heating chamber from 24°C to the Neel transition. The anisotropy field, H_a was measured by the technique of Josephs [7]. The dynamic response of domain walls was measured by the Seitchik [8] method with the noise cancellation refinements suggested by Argyle and Malozemoff [9]. The total system risetime was about 7 ns so that mobilities as high as 1000 cm/sec/Oe could be measured when an overdamped exponential approximation was used to analyze the pulse response data.

RESULTS AND DISCUSSION

A typical pulse response for all three film compositions studied is shown in Fig. 1a. It is composed of a rapidly rising leading edge followed by a more slowly rising part which occasionally overshoots or oscillates with a very long period. The slowly varying portion of the response is variable in amplitude and time depending on the previous pulse and saturation history of the sample, the pulse amplitude, and temperature. On the other hand the leading edge is apparently insensitive to sample history and is sensitive only to temperature and, in the case of the EuY composition, to pulse amplitude. The effect of a moderate in-plane field of 25 Oe on the pulse response is to stabilize the amplitude and the time variability of the slow portion of the response while significantly decreasing the risetime of the leading edge of the response. In the EuYYb composition studied, the response became oscillatory as shown in Fig. 1b. Similar oscillatory behavior in the presence of an in-plane field has been observed by others [9,10].

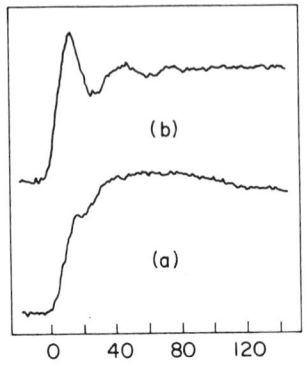

Fig. 1. Wall response of $Eu_{0.6}Y_{2.35}Yb_{0.15}Fe_{3.8}Ga_{1.2}O_{12}$ with the application of an 8 Oe step bias (a) $H_x = 0$, (b) $H_x = 25$ Oe.

We tentatively ascribe the fast leading edge of the response to normal Bloch walls which are essentially devoid of Bloch-Neel transitions and horizontal Bloch lines [11]. The slower response may be that of wall segments containing a variable but significant number of Bloch lines. In our analysis, therefore, only the time constant of the leading edge of the response is considered. The time constant data plus the measured stripe domain period P_0 and $4\pi M_s$ at a given temperature are used along with the susceptibility calculation technique of Shaw et al. [12] to obtain domain wall mobility as a function of temperature. The results are shown in Figs. 2,3, and 4. The dashed curve is a calculated mobility using the measured temperature dependence of $4\pi M_s$ and wall energy density, σ_w, assuming that mobility is related linearly to M_s/σ_w [13]. The mobility data of the more highly damped SmY garnet composition is well described by the theoretical mobility curve, while the data of the EuY and EuYYb garnets fit the theoretical curve reasonably well. In the case of the EuYYb garnet there is an apparent saturation of the mobility at a temperature around 25°C. The 25°C mobility value of 380 cm/sec/Oe at a field pulse amplitude of 8 Oe compares favorably with a wall

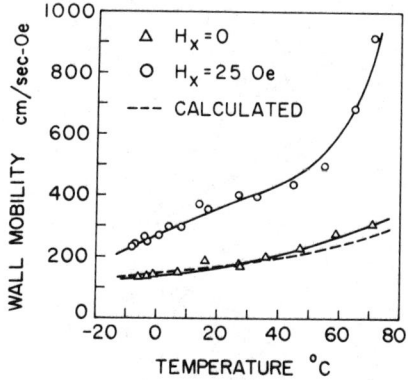

Fig. 2. Wall mobility of $Sm_{0.4}Y_{2.6}Fe_{3.2}Ga_{1.3}O_{12}$ versus temperature

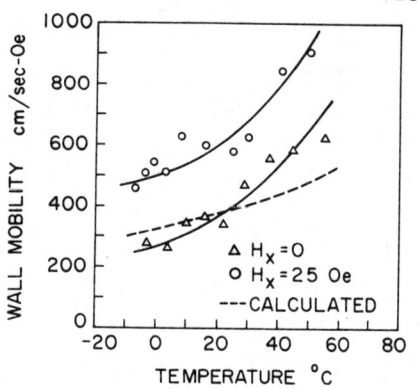

Fig. 3. Wall mobility of $Eu_{0.6}Y_{2.4}Fe_{3.7}Ga_{1.3}O_{12}$ versus temperature.

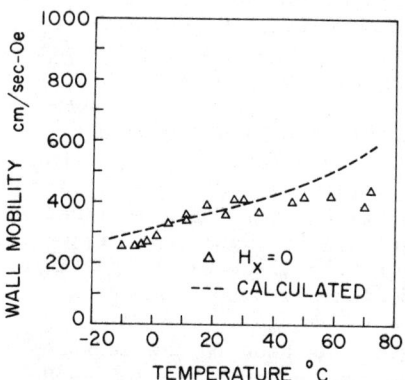

Fig. 4. Wall mobility of $Eu_{0.5}Y_{2.35}Yb_{0.15}Fe_{3.8}Ga_{1.2}O_{12}$ versus temperature

mobility of 320 cm/sec/Oe obtained from extrapolating the room temperature bubble velocity results of Bonner et al.[4] for EuYTm garnet films to a ΔH of 16 Oe.

The bubble mobility increases by a factor of two with the application of a device magnitude in-plane field of 25 Oe. In the case of SmY and EuY the wall response still resembled that of Fig. 1a (except near 70°C where the response of both compositions tended to be oscillatory) and could be analyzed in the same way as the $H_x = 0$ response curves. However the wall response of the EuYYb film shown in Fig. 1b was oscillatory over the whole temperature range. The dynamic response of this composition in the presence of an in-plane field was obviously a good deal faster than that of the other two compositions studied. However, lack of a suitable model for this oscillatory behavior prevented a complete analysis of these response curves from being made. If the response is analyzed in terms of the harmonic oscillator equation[10] a room temperature mobility of about 2600 cm/sec/Oe is obtained. On the other hand, based on the risetime of the room temperature response (6.4 ns corrected for system response) a mobility of about 1300 cm/sec/Oe is calculated.

The temperature dependence of the static material parameters for the three compositions included in this study are shown in Figs. 5, 6, and 7. The temperature behavior of all three samples are nearly identical, differing primarily in the absolute value of the parameters at a given temperature. The coercivity is seen to vary by an order of magnitude over the range -10 to 70°C in agreement

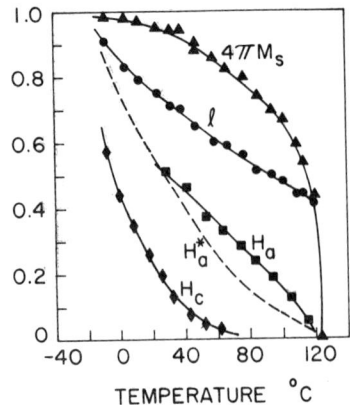

Fig. 5. Material parameter data for $Eu_{0.6}Y_{2.4}Fe_{3.7}Ga_{1.3}O_{12}$. Maximum ordinate values are 1 μ for ℓ, 160 G for $4\pi M_s$, 2 kOe for H_a and H_a^*, and 2 Oe for H_c.

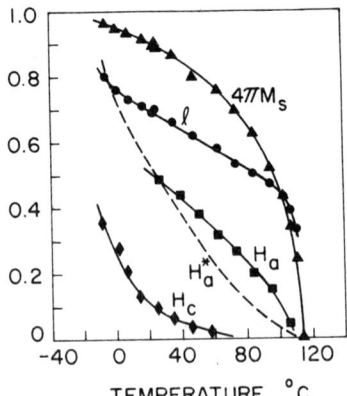

Fig. 6. Material parameter data for $Sm_{0.4}Y_{2.6}Fe_{3.7}Ga_{1.3}O_{12}$. Maximum ordinate values are 1 μ for ℓ, 170 G for $4\pi M_s$, 3 kOe for H_a and H_a^*, and 2 Oe for H_c.

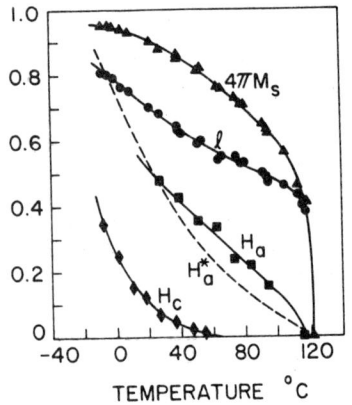

Fig. 7. Material parameter data for $Eu_{0.5}Y_{2.35}Yb_{0.15}Fe_{3.8}Ga_{1.2}O_{12}$. Maximum ordinate values are 1 μ for ℓ, 160 G for $4\pi M_s$, 2 kOe for H_a and H_a^*, and 2 Oe for H_c.

with the results of Shumate et al.[3]. This fact along with the approximate factor of two change in mobility over the same range is felt to be significant relative to the allowable temperature excursions of bubble memory devices.

In addition to the measured H_a versus temperature plot an H_a^* curve is shown dashed in these figures. This curve is derived from the σ_w data by assuming that the exchange coefficient, A, can be calculated from the energy density expression for a 180° Bloch wall in an infinite medium, $\sigma_w = 4(AK)^{1/2}$, and that A is constant over this temperature range. It is clear that this expression for wall energy density yields values for H_a significantly different from the measured H_a in thin films containing a striped array of domain walls. These data can also be used to calculate A versus temperature. In this case A is found to decrease with increasing temperature which, in turn leads to a nearly temperature independent μ_w. These results will be reported in detail elsewhere.

Temperature coefficients of ℓ, $4\pi M_s$ and H_a for the compositions studied are given in Table 1 for three different temperatures.

Table 1 Temperature Coefficient Data

Composition	h μm	T_N °C	T °C	$\frac{100}{\ell}\frac{\Delta\ell}{\Delta T}$ %/°C	$\frac{100}{M_s}\frac{\Delta M_s}{\Delta T}$ %/°C	$\frac{100}{H_a}\frac{\Delta H_a}{\Delta T}$ %/°C
$Sm_{0.4}Y_{2.6}Fe_{3.7}Ga_{1.3}O_{12}$	6.78	114	-10	-.58	-.45	-1.00
			25	-.40	-.24	-1.50
			80	-.80	-1.28	-4.00
$Eu_{0.5}Y_{2.35}Yb_{0.15}Fe_{3.8}Ga_{1.2}O_{12}$	6.81	121	-10	-.66	-.38	-1.60
			25	-.51	-.24	-1.90
			80	-.52	-.74	-3.00
$Eu_{0.6}Y_{2.4}Fe_{3.7}Ga_{1.3}O_{12}$	7.40	123	-10	-.64	-.18	-1.30
			25	-.58	-.16	-1.70
			80	-.59	-.66	-3.10

ACKNOWLEDGMENTS

The authors would like to express their appreciation to Mr. L. G. Hellwig and Mr. J. R. Scherr for their expert technical assistance.

REFERENCES

1. D. H. Smith and A. W. Anderson, AIP Conference Proceedings No. 5 120, (1972).
2. W. A. Bonner, J. E. Geusic, D. H. Smith, F. C. Rossol, L. G. Van Uitert, and G. P. Vella-Coleiro, J. Appl. Phys. 43, 3226, (1972).
3. P. W. Shumate Jr., D. H. Smith, and F. B. Hagedorn, J. Appl. Phys. 44, 449, (1973).
4. W. A. Bonner, J. E. Geusic, D. H. Smith, L. G. Van Uitert and G. P. Vella-Coleiro, Mat. Res. Bull., 8, 785, (1973).
5. G. P. Vella-Coleiro, AIP Conf. Proc. No. 10, 424, (1973).
6. R. W. Shaw, R. M. Sandfort, J. W. Moody, Magnetic Bubble Materials: Characterization Techniques Study Report ARPA order 1999. Contract No. DAAH01-720C-0490.
7. R. M. Josephs, AIP Conf. Proc. No. 10, 286, (1973).
8. J. A. Seitchik, W. D. Doyle, and G. K. Goldberg, J. Appl. Phys. 42, 1271, (1971).
9. B. E. Argyle and A. P. Malozemoff, AIP Conf. Proc. No. 10, 344, (1973).
10. J. W. Moody, R. W. Shaw, R. M. Sandfort, and R. L. Stermer, IEEE Trans. Magn. MAG-9 to be published.
11. J. C. Slonczewski, J. Appl. Phys. 44, 1759, (1973).
12. R. W. Shaw, D. E. Hill, R. M. Sandfort, and J. W. Moody, J. Appl. Phys. 44, 2346, (1973).
13. G. P. Vella-Coleiro, D. H. Smith, and L. G. Van Uitert, J. Appl. Phys. 43, 2428, (1972).

OBSERVATION OF STRAIGHT AND WAVELIKE
DOMAIN WALL MOTION IN BUBBLE FILMS
BY HIGH SPEED PHOTOGRAPHY

T. M. Morris[*] and A. P. Malozemoff
IBM Research Center, Yorktown Heights, New York 10598

ABSTRACT

The motion of domain walls in garnet bubble films has been studied by means of a high speed photographic technique with 10 nsec resolution. A straight, isolated stripe was stabilized in a DC bias field, and a strong reverse field pulse was applied. Initially the two walls remained straight as the stripe expanded. This wall motion proceeded with a saturation velocity which was generally higher than but within 50% of that determined from dynamic bubble collapse. After the stripe had widened to typically several times its initial size, a wavelike distortion with a characteristic wavelength developed on each side of the stripe. A simple magnetostatic model is shown to predict such a wavelike instability at the observed stripe width and wavelength.

INTRODUCTION

A straight isolated stripe domain in a magnetic bubble garnet film can be stabilized by a DC bias field applied perpendicular to the film. If the field is suddenly reduced, the stripe will expand toward a new equilibrium shape. We have observed this process directly with a high-speed photographic technique. In this paper we describe quantitative results on two aspects of the process: 1) The width and wavelength of distortions are measured and compared to a simple magnetostatic theory. 2) A saturation velocity for the straight wall motion is measured and compared to results of the dynamic bubble collapse technique.

EXPERIMENTAL TECHNIQUE

Experiments were done on five garnet films grown by liquid phase epitaxy and having the characteristics listed in the Table. In each case a straight stripe domain was formed by first saturating the film and then slowly lowering the bias field until a single stripe ran into the field of view of the polarizing microscope. The stripe could be moved to a given starting location by a pair of strip lines pulsed to give a gradient field. An effort was made to suppress Bloch lines by applying 100 Oe DC in-plane field parallel to the stripe during the stripe preparation. When the stripe had been appropriately positioned, the in-plane field was subsequently removed. During the pulse experiment to be described, no other stripes ever approached within 100μ of the original stripe.

In the experiment the net bias field was reduced by pulsing a

---

[*]Current address: California Institute of Technology, Pasadena, Calif.

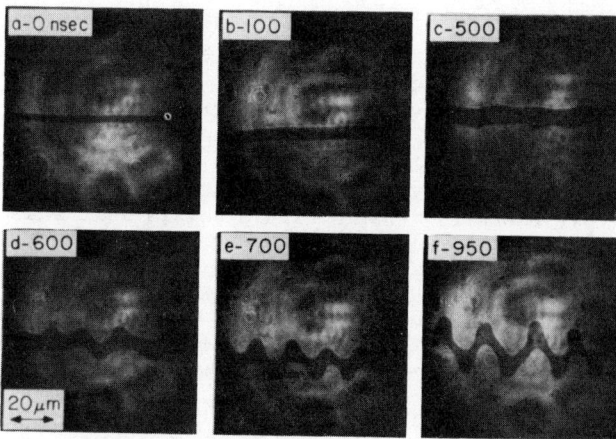

Fig. 1. High speed photographs on sample 5 of Table, with a 1 μsec, 45 Oe pulse and indicated time dealy nsec.

HP 214A pulse generator fed into a 1 mm i.d. pancake coil to create a magnetic field opposing the DC bias field. The domain was photographed in motion by a single 20kW yellow light pulse of <10 nsec duration, from a dye laser pumped by a nitrogen laser.[1,2] The light pulse could be triggered at any predetermined time delay after the onset of the field pulse. Delay jitter was <2 nsec. Between each pulse, the stripe was restored to its initial state as described above. Typical results taken on Type 57 Polaroid film are shown in Fig. 1.

MEASUREMENT AND INTERPRETATION OF WAVELIKE DISTORTIONS

Fig. 1 shows that initially the stripe expands with the two walls remaining straight (a and b). After the width of the stripe increases to several times its initial value, wavelike distortions develop independently on either side of the stripe (c). Then the stripe width decreases as the distortion amplitude simultaneously increases, causing the position of the waves on either side to correlate (d) so that the stripe as a whole develops a serpentine shape (e). Finally the stripe appears poised to expand laterally into the surrounding region (f).

Assuming constant wall energy and ignoring coercivity, which was typically <0.5 Oe, the equilibrium width w of a stripe domain in a film of thickness h and magnetization 4πM is determined by a balance between the applied bias field H and an internal demagnetizing field, such that[2,3]

$$H/4\pi M = \pi^{-1}\{2 \arctan (h/w) - (w/h) \ln [1+(h^2/w^2)]\} \qquad (1)$$

Fig. 2 is a plot of $H/4\pi M$ as a function of w/h; it also shows the measured static stripe widths for the five samples at the bias fields used in the pulse experiment. If the applied field is suddenly reduced, and as long as the walls remain straight, stripe expansion is driven by the imbalance between the resultant applied field and the demagnetizing field corresponding to the width attained. For a starting width w_i the initial excess demagnetizing field is indicated schematically in Fig. 2 by a downward arrow. As the stripe width increases this driving field decreases, and equilibrium should be attained at a width w_e where the excess demagnetizing field vanishes.

However, before such an equilibrium width is reached, wave like distortions are observed provided the drive pulse is sufficiently strong to expand the stripe beyond a critical width. The observed critical widths and characteristic wavelengths, are given in Figs. 3 and 4. Photographs such as Fig. 1c suggest that these distortions initially develop independently on each wall.

Now it has been shown on the basis of magnetostatic calculations[4,5] and verified experimentally[4] that a single domain wall stabilized in a gradient field will develop a sinusoidal instability of wavelength λ_c ($\equiv 2\pi h/Z_o$) when the magnitude of the gradient decreases below a critical value β_c. The values of β_c and λ_c are determined by[5]

$$\beta_c = (8M/h)[\ln(Z_o/2) + 0.5772 + K_o(Z_o) - (\pi \ell Z_o^2/4h)] \tag{2}$$

$$\ell/h = (2/\pi Z_o)[Z_o^{-1} - K_1(Z_o)] \tag{3}$$

where ℓ is the ℓ-parameter of the material[6] and K_o and K_1 are modified Bessel functions. In the case of the isolated stripe, each domain wall experiences an effective gradient field given by the derivative of Eq. (1):

$$\partial H/\partial w = -(4M/h)\ln[1+(h^2/w^2)] . \tag{4}$$

Since this gradient steadily decreases in magnitude as the stripe widens, the critical gradient for the onset of distortions will be reached at a critical width w_c. The predicted w_c and λ_c are shown in Figs. 3 and 4 as a function of ℓ/h; they are in good agreement with experiment.

MEASUREMENT OF WALL VELOCITY - COMPARISON WITH BUBBLE COLLAPSE

As long as the stripe width is less than the critical width for the onset of the wavelike distortions, the walls remain straight as in Fig. 1b. In this range an average velocity of the wall can be found directly from the width change of the stripe and the length of the pulse. For pulses less than 20 Oe the width change was too small to measure. Typical results for higher drives are shown in Fig. 5 where error bars indicate uncertainty in the width measurement and the leftward arrows indicate the calculated extent to which the effective drive field is reduced as the stripe expands (cf. Fig. 2). The velocities at drives of 47 and 67 Oe are approximately the same and indicate a velocity saturation. However at 25 Oe drive the velocity is much lower, presumably because the effective drive field has dropped almost to zero by the time of the photograph.

Saturation velocities at high drive fields were also measured by the dynamic bubble collapse technique[7,8] in the same area of the sample. Bubbles were first selected to contain no or small numbers of Bloch lines in a gradient pulse experiment.[8] Velocities, determined using the difference of theoretical starting and collapse diameters[7] and ignoring the relatively small bubble potential fields, are shown by crosses in Fig. 5 for one sample. In both experiments there is a small discrepancy between velocity results at different bias fields. Bubble collapse and straight wall saturation velocities

245

for the five samples are compared in the Table. For a given sample they differ at most by 50%, and they generally show similar trends from sample to sample. Considering the different nature of the two experiments, the agreement is encouraging.

Slonczewski[9] has proposed that the basic reason for velocity saturation is a breakdown in wall structure induced by nonuniform stray fields at the wall. Since the stray fields of stripes and bubbles of different sizes vary,[9] the precise value of saturation

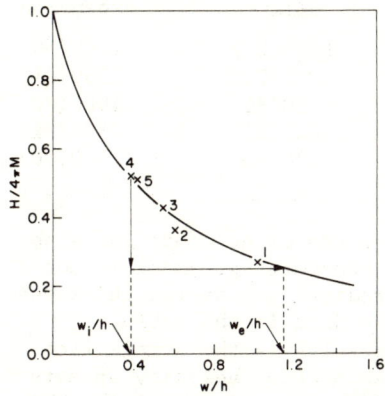

Fig. 2. DC bias field ÷ magnetization vs equilibrium width of an isolated stripe ÷ sample thickness from Eq. 1; also data for 5 samples of Table and schematic illustration of initial and equilibrium widths w_i and w_e in pulse experiment.

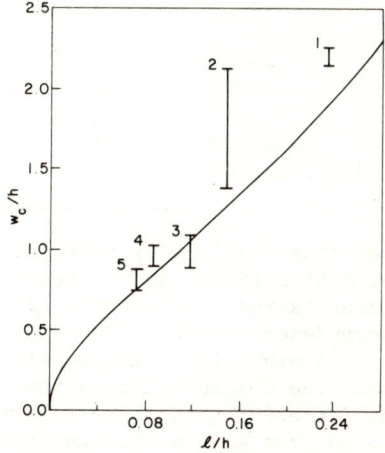

Fig. 3. Critical width for onset of distortion ÷ thickness vs ℓ-parameter ÷ thickness, from Eq. 2, 3, and 4; also data for 5 samples of Table.

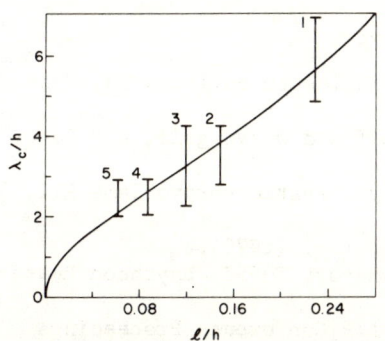

Fig. 4. Characteristic wavelength of distortions ÷ thickness vs ℓ-parameter ÷ thickness, from Eq. 3; also data for 5 samples of Table.

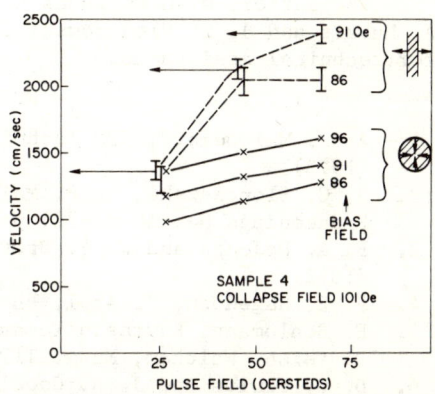

Fig. 5. Experimental velocities in sample 4 at different bias fields and pulse fields: from bubble collapse (x——x) and high speed photography (I---I).

TABLE

Parameters and Saturation Velocities for 5 Films
of Composition $Eu_{0.69}Tb_{0.01}Y_{2.3}Ga_{1.15}Fe_{3.85}O_{12}$

Sample	Thickness (μ)	$4\pi M$ (Oe)	ℓ (μ) (Fowlis-Copeland method[6])	Saturation velocity (cm/sec) 65 Oe drive, bias field in parenthesis	
				bubble collapse	stripe expansion
1	2.47	185	0.58	1270(50)	1260(50)
2	3.80	192	0.56	1050(71)	1360(71)
3	4.55	164	0.55	1380(80)	1980(70)
4	6.46	176	0.55	1400(91)	2390(91)
5	8.96	177	0.63	920(93)	1500(90)

velocity should vary somewhat between our two experiments and also with bias field, as we indeed observe. However, agreement is not quantitative; for example Slonczewski predicts an inverse thickness dependence[9] which is not supported by the data in the Table.

Another important aspect of these results is the simple fact that the domain boundary moving at the saturation velocity appears no different then the boundary at rest (cf. Figs. 1a and 1b). This means that within the resolution of our microscope, we can confirm experimentally some of the approximations on which domain wall theory for bubble materials is based. In particular we can eliminate such conceivable complications as diffuse boundary motion observed in permalloy,[10] or corrugated wall structures proposed in an earlier theory by Slonczewski,[11] or wall bowing through the thickness of the film as might arise from surface pinning or horizontal Bloch lines.[9]

The authors wish to thank E. A. Giess for the garnet films, A. Hubert and J. C. Slonczewski for helpful discussions, and L. Buszko for technical assistance.

REFERENCES

1. A. P. Malozemoff, IBM Technical Disclosure Bulletin 15, 2756 (1973).
2. J. C. Slonczewski, A. P. Malozemoff and O. Voegeli, AIP Conf. Proceedings No. 10 (1972) p. 458.
3. F. A. DeJonge and W. F. Druyvesteyn, Festkorperprobleme XII, 531 (1972).
4. F. B. Hagedorn, J. Appl. Phys. 41, 1161 (1970).
5. E. Schlomann, Raytheon Technical Report T-953, Raytheon Research Division, Waltham, Mass. (1973).
6. D. C. Fowlis and J. A. Copeland, AIP Conference Proceedings No. 5 (1971), p. 240.
7. H. Callen and R. M. Josephs, J. Appl. Phys. 42, 1977 (1971).
8. A. P. Malozemoff, J. Appl. Phys. 44, 5080 (1973).
9. J. C. Slonczewski, J. Appl. Phys. 44, 1759 (1973).
10. M. Kryder and F. B. Humphrey, J. Appl. Phys. 40, 1225 (1969).
11. J. C. Slonczewski, Intern. J. Magnetism 2, 85 (1972).

OBSERVATION OF DOMAIN-WALL RESONANCES BY A DIFFRACTION TECHNIQUE

G. R. Woolhouse and P. Chaudhari
IBM Research Center, Yorktown Heights, N. Y. 10598

ABSTRACT

A simple technique for observing domain-wall resonances in optically transparent magnetic films is described involving an understanding of the diffraction effects occurring when a magnetic sample behaves as an antiphase diffraction grating. Information about domain-wall mobility and mass has been obtained by studying resonances in garnet films.

INTRODUCTION

There have recently been several reports of the observation of domain-wall resonances.[1,2] In the first work, domain wall resonant frequencies were deduced from measurements of the r.f. susceptibility of garnet samples placed inside a solenoid. In the second work the intensity of a Faraday signal obtained in an optical microscope was monitored as a function of the field applied to a garnet sample. The geometry of the set-up made a complex electronic detection scheme necessary to separate the signal from the noise. In this paper we describe a simple technique for observing domain-wall resonances which only requires an understanding and use of the diffraction processes which arise when the light transmitted by a garnet sample between crossed polarizers is brought to a focus with a lens.

In a recent paper[3] the way in which a sample containing magnetic domains can behave as an antiphase diffraction grating was described. A simple theory was presented together with experimental results confirming the theory. The essential result is that when behaving as an antiphase grating in zero external magnetic field the even-order peaks of the diffraction pattern have zero (or at least minimum) intensity. In particular it is easily shown that:

$$I_o \propto (t_u - t_d)^2 \qquad (1)$$

where I_o is the intensity of the zero-order beam in the diffraction pattern and t_u, t_d are the widths of 'up' and 'down' magnetized domains respectively.

Thus if a sinusoidally varying 'z-field' (parallel or antiparallel to domain magnetizations) is applied to the sample the amplitude of the resulting domain-wall motion, X_o, is related to I_o by:

$$\Delta I_o \propto X_o^2 \qquad (2)$$

where ΔI_o is the change observed in I_o on application of the field. This is so provided the frequency of the domain-wall response is sufficiently great that the detector used to measure I_o senses the intensity averaged over a cycle of wall motion. Thus, monitoring I_o

as a function of applied field is a very convenient way of detecting domain-wall resonances.

EXPERIMENTAL TECHNIQUE

The equipment used is shown in Fig. 1. The analyzer was crossed with respect to the polarizer and the light focussed into the zero-order spot was monitored with the photo-multiplier tube. The optimum aperture diameter was about half the diameter of the first diffraction ring. The samples on which most measurements were made

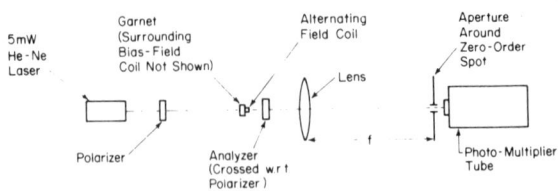

Fig. 1. Experimental arrangement used to record changes in the zero-order spot intensity.

were $Eu_{0.7}Y_{2.3}Ga_{1.15}Fe_{3.85}O_{12}$ epitaxial films grown on non-magnetic substrates and having 10μ and 6.8μ strip widths. A 15-turn 'pancake' coil of internal diameter 1.5 mms was cemented to the garnet and the laser beam directed through the coil. The electrical circuit used to deliver current to the coil is shown in Fig. 2. The field created by the coil at the garnet could be measured by comparing the D.C. response of the garnet to that obtained with a 12.5 cms diameter coil around the entire sample and holder. The current-field characteristics of this coil had been

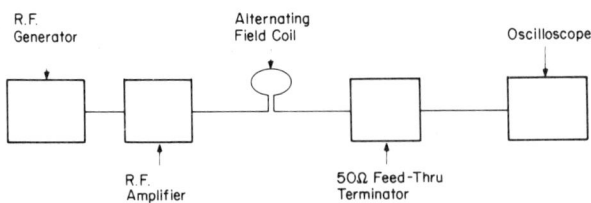

Fig. 2. Electrical circuit used to drive the alternating field coil.

previously determined with a Hall probe. Peak fields (H_p) up to 31 Oe at frequencies from D.C. to 50 Mc/s could be applied to the sample.

RESULTS

First of all a check was made on equation (2) by measuring ΔI_o as a function of H_p at a constant low frequency (10c/s). The results shown in Fig. 3 indicate that ΔI_o is indeed proportional to H_p^2 so that equation (2) is verified if one assumes that at this low frequency the domain wall motion is in phase with the applied field. This relationship was confirmed at several frequencies from D.C. to about 1Mc/s, using several garnets.

The response of the garnet of 10μ strip width up to 50 Mc/s is shown in Fig. 4 where ΔI_o is plotted against the frequency at which fields of different peak values (H_p) were applied. It is seen that at high fields double resonances are observed. At intermediate fields the resonance is not complete. This is because the maximum frequency at this value of H_p was not sufficient to permit its observation. In general the resonant frequency is a strong function of H_p. This has also

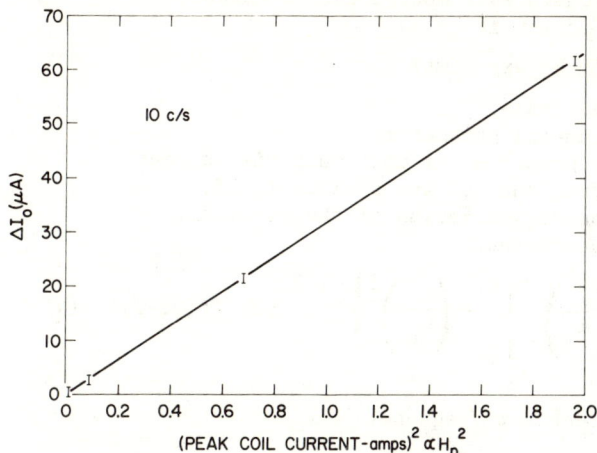

Fig. 3. ΔI_o plotted against H_p^2 at a constant low frequency (10c/s). See text for further explanation.

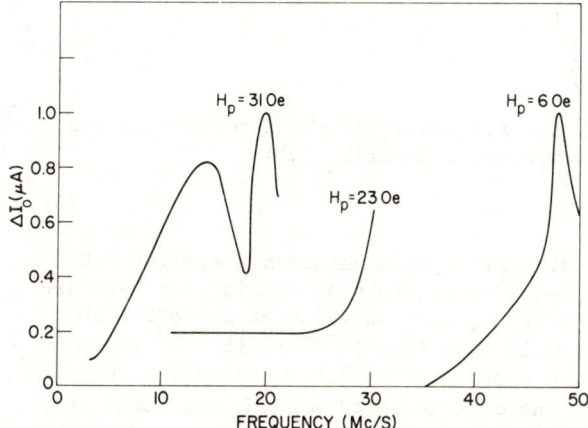

Fig. 4. Resonant frequencies observed in a garnet as a function of drive-field amplitude (H_p).

been observed by Argyle and Halperin.[4] Similar data were obtained using the 6.8μ strip width sample. Schlömann[5] has recently given a qualitative explanation for the observation of double peaks in terms of non-uniform distribution of wall mass density.

In general the magnetic structure was in the form of stripes but a bubble lattice could also be created in the area covered by the coil using a pulsing technique,[6] and the resonances of a bubble lattice obtained. They turned out to be almost identical to those of the stripes and so are not shown. In this experiment the diffraction pattern (observable on the aperture screen) was used to monitor the domain configuration. It was significant that after passing through a resonance the pattern changes from spot-type (characteristic of bubble lattice) to ring-type (characteristic of stripe domains). This was to be expected in view of the relative stabilities of these states. The use of diffraction in this way is discussed in more detail elsewhere.[7]

The results were checked for pick-up effects by making a run with just a coil and no sample and for substrate effects by making a run with the garnet saturated by the large coil. Both effects were negligible.

ANALYSIS OF RESULTS

Useful information could be obtained from the resonance data by

utilizing an analysis of domain wall motion due to Rossol.[8]
The equation of wall motion is:

$$m\ddot{X} + \beta\dot{X} + \alpha X = 2M\,H_p \exp(i\omega t) \tag{3}$$

where m is the domain wall mass,
       β is the viscous damping parameter,
       α is the restoring pressure per unit wall displacement,
       ω is the angular frequency of the applied field,
and· M is the saturation magnetization of the material.
The solution of (3) can be written:

$$X = \frac{2M\,H_p}{\alpha}\left\{\left[1-\left(\frac{\omega}{\omega_r}\right)^2\right]^2 + \left(\frac{\omega}{\omega_c}\right)^2\right\}^{-1/2} \exp[i(\omega t - \phi)] \tag{4}$$

where

$\omega_r = \left(\dfrac{\alpha}{m}\right)^{1/2}$ is the resonant frequency

$\omega_c = \dfrac{\alpha}{\beta}$ is the relaxation frequency

and φ is the phase lag of the wall motion behind the drive field.
Thus using (2) and (4) and the fact the $X^2 = XX^*$ we find that the ratio of intensity changes at resonance and at low frequency is:

$$\frac{(\Delta I_o)_{res}}{(\Delta I_o)_{l.f.}} = \left(\frac{\omega_c}{\omega_r}\right)^2 \tag{5}$$

Thus by measuring both the frequency and the intensity of the resonance we can get ω_c and hence the mobility (μ):

$$\mu = \omega_c \left(\frac{\Delta X}{\Delta h}\right) \tag{6}$$

Where ΔX/Δh is the domain wall displacement/unit applied D.C. field (obtainable from the B-H curve). This expression has been used by Seitchik et al.[9] For the 6.8μ garnet, a value of $\mu = 620 \pm 60$ cm/sec-Oe and values of the wall mass from 10^{-10} to 9×10^{-10} gm/cm^2 were found. These limits are approximately 0.8 m_o and 7.5 m_o where m_o is the Döring mass.[5,10] The dependence of m on H_p was found to be of the form: $m \propto H_p^y$ with y ≃ 1.4. It has been predicted[5,11] that m should be a function of H_p although the expected form of the dependence is not clear.

ACKNOWLEDGEMENTS

Thanks are due to Dr. E. A. Giess and his co-workers for preparing the garnet samples; to Drs. B. E. Argyle, A. P. Malozemoff and J. C. Slonczewski for useful discussions and to Dr. E. Schlömann for access to his results prior to publication. Dr. Argyle gave valuable assistance with the bubble generation technique. This work was supported in part by contract AFOSR F44620-71-C-0061.

REFERENCES

1. G. P. Vella-Coleiro, D. H. Smith, and L. G. Van Uitert, J.A.P. $\underline{43}$, 2428 (1972).
2. B. E. Argyle and A. P. Malozemoff, AIP Conf. Proceedings on Magnetism and Magnetic Materials, 344, (1972).
3. G. R. Woolhouse and P. Chaudhari, Phys. Stat. Sol.(a), $\underline{19}$, K3 (1973).
4. B. E. Argyle and A. Halperin, Intermag Conf. Proceedings, 238 (1973).
5. E. Schlömann, Raytheon Tech. Memo, T-959 (1973).
6. B. E. Argyle and P. Chaudhari, AIP Conf. Proceedings on Magnetism and Magnetic Materials, 403 (1972).
7. G. R. Woolhouse and P. Chaudhari (to be published).
8. F. C. Rossol, J.A.P. $\underline{40}$, 1082 (1969).
9. J. A. Seitchik, W. D. Doyle, and G. K. Goldberg, J.A.P. $\underline{42}$, 1272 (1971).
10. W. Döring, Z. Naturforsch, $\underline{3a}$, 373 (1948).
11. J. C. Slonczewski, J.A.P. $\underline{44}$, 1759 (1973).

NMR STUDY OF THE ELECTRON SPIN DENSITY NEAR IRON GROUP ATOMS IN Cu*

James B. Boyce **, Thomas J. Aton***,
and Charles P. Slichter
University of Illinois, Urbana, Illinois 61801

ABSTRACT

We have observed the nuclear resonances of Cu atoms which are near neighbors to 6 of the 3d transition element impurities in dilute Cu alloys, systems of interest in the Kondo problem. The resonances appear as satellites split from the strong resonance of the distant Cu atoms. Their splittings yield the magnitude and spatial shape of the conduction electron spin magnetization. We have seen one shell of neighbors to Ni, one to V, three to Co, five to Cr and Fe, and six to Mn. The satellite in CuNi was studied at liquid helium temperatures in fields from 6 to 60 kG and those in CuCo from 1.5 to 450 K and 6 to 63 kG. To date we have studied C̲u̲V from 8 to 63 kG at liquid helium temperatures, C̲u̲Cr and C̲u̲Mn from 7 to 15 kG at 300 K, and C̲u̲Fe from 7 to 61 kG and 77K to 330 K. The satellite positions are independent of concentration from about 500 ppm to 5000 ppm. The usual theoretical expressions for spin density are evaluated far from the impurity, where they depend only on the electrons at the Fermi surface. In contrast, to treat the near neighbors we have developed a d-wave phase shift analysis, in which conduction electrons of all energies are included to fit the experimental splittings in both the magnetic and non-magnetic cases.

INTRODUCTION

The detailed understanding of how an isolated magnetic atom behaves when dissolved in a non-magnetic metal, of the circumstances under which it possesses a permanent moment, has long been an area of interest. An important subclass of problems is concerned with the Kondo effect which may be characterized as the apparent change from the temperature independent susceptibility at low temperatures characteristic of an atom lacking a permanent moment to a near Curie's law behavior at high temperature characteristic of an atom possessing a permanent moment. Stimulated by the general interest in these dilute systems and in the Kondo effect, our colleagues and we have been studying dilute alloys of iron group atoms in copper.[1,2] In this paper we report on measurements of the spin polarization density, $\sigma(r)$, of the conduction electrons in the near vicinity of the impurity atoms. We report new results for V, Cr, Mn, and Fe. Previous results from our group have been reported on Co[2] and

Ni.[1] The measurements which we report in this paper were done at room temperature and liquid nitrogen temperature. Similar experiments to ours have been done by N. Karnezos and J. A. Gardner[3] on the system copper manganese. We appreciate their sending us reports prior to publication.

All nuclei in a metal experience a shift in their resonance frequencies due to the interaction with the conduction electrons. This is the well known Knight[4] shift. If, however, the conduction electron spin density is not uniform throughout the metal, some nuclei will experience different Knight shifts than others. This gives rise to a spectrum of resonance lines corresponding to the various inequivalent positions in the material. For example in the case of a dilute alloy of Co dissolved in Cu, the Cu nuclei which are near neighbors to Co atoms have a different Knight shift than the Cu nuclei far from Co atoms. Since the Co concentration is small, typically less than 0.5%, the resulting spectrum is a strong resonance (the main line) due to the Cu which are far from the Co and very weak satellite resonances due to the Cu which are near neighbors of Co atoms. Inhomogeneities in both the conduction electron spin density and charge density will contribute to a change in the Knight shift. We expect and find, however, that the spin density effect dominates for magnetic impurities.

EXPERIMENTAL METHOD

Some of the nuclear resonance apparatus used has been described in other papers.[1,2] In addition, for some of the data reported, we utilized a cross-coil Varian spectrometer through the courtesy of Professor T. J. Rowland. All of the data involved modulating the static field, lock-in detection, and averaging with a computer of average transients.

The sample preparation techniques are also described elsewhere.[1,2]

THEORY

In order to describe the spin polarization, $\sigma(r)$, we utilize the Anderson[5] model. We distinguish two cases: (1) magnetic atoms, and (2) non-magnetic atoms. In the Anderson model, the d-states of the free iron group atom are mixed with the conduction band states becoming broadened when the atom is dissolved in Cu. As a result of the Coulomb interaction, the orbital states with up spin repel those with down spin. If the repulsion is strong, the up and down states split spontaneously, leading to a permanent magnetic moment even without an applied magnetic field. In the simplest Anderson theory, either the spin up state lies above the spin down state in energy or the reverse is true. We call these two ion configurations "a" and "b", respectively. "a" and "b" are still degenerate, but they are split by a magnetic field. If the field lies in the up direction, state "a" is lower in energy than "b".

If the repulsion is weak, no spontaneous splitting occurs, but when a magnetic field splits the up and down spin states, the repulsion enhances the splitting, leading to an enhanced spin magnetic susceptibility.

(1) *Magnetic atoms.* A straightforward analysis shows that at position R (measured from the impurity) the changes in Knight shift $\Delta K(R)$ relative to the Knight shift of pure Cu is given by

$$\frac{\Delta K(R)}{K} = \frac{\chi}{\chi_s^e} \Sigma m_\sigma \int_0^{E_F} \rho_1(W_k) dW_k \frac{\left\{[n_2^2(kR) - j_2^2(kR)]\sin^2\delta_\sigma(k) - 2n_2(kR)j_2(kR)\sin\delta_\sigma(k)\cos\delta_\sigma(k)\right\}}{(\delta_+ - \delta_-)/2\pi} \quad (1)$$

where χ is the spin susceptibility of the impurity, χ_s^e is the spin susceptibility of the conduction electrons per unit volume, $\rho_1(W_k)$ the density of states of one spin orientation per unit volume. $\delta_\sigma(k)$ is the scattering phase shift for spin orientation σ for state "a". This result is similar to one utilized by Gardner and Flynn.[6] Their expression, derived for experiments on liquid metals, involves an average over lattice positions, but it is easily shown that their equations lead to Eq. (1). There is another term which we have omitted as unimportant for a strongly magnetic atom, but which should be included for an atom which is just barely magnetic.

(2) *Non-magnetic atoms.* When there is no splitting of the up and down spin d-wave resonances, one can show

$$\frac{\Delta K(R)}{K} = 5\left\{[n_2^2(k_F R) - j_2^2(k_F R)]\sin^2\delta(k_F) - 2n_2(k_F R)j_2(k_F R)\sin\delta(k_F)\cos\delta(k_F)\right\}$$

$$- 5\frac{\rho_d U}{1-\rho_d U} \int_0^{E_F} \frac{\rho_1(W_k)}{\rho_1(E_F)} dW_k \left\{[n_2^2(kR)-j_2^2(kR)]\frac{\partial}{\partial E_\sigma}\sin^2\delta(k) - 2n_2(kR)j_2(kR)\frac{\partial}{\partial E}\sin\delta(k)\cos\delta(k)\right\} \quad (2)$$

where ρ_d is the density of d-states per impurity atom at the Fermi surface arising from the impurity, U is the Coulomb repulsion term of the Anderson theory, and E_σ the energy of the resonant level. $1/(1-\rho_d U)$ is the well-known enhancement factor of susceptibility. This case is not treated by Gardner and Flynn.

As can be seen $\Delta K(R)/K$ depends on the location and width of the respective up and down spin states for both the magnetic and non-magnetic cases. Thus, in principle, sufficiently precise measurements of $\Delta K/K$ at enough non-equivalent neighbor sites would determine the width and location of the d-wave resonances.

EXPERIMENTAL RESULTS

The experimental results on $\Delta K/K$ to date are summarized in the Table. With the exception of our work on CuNi and CuCo we do not have experimental determinations of which neighbor shell goes with which $\Delta K/K$. We have studied CuV from 8 to 63 kG at liquid helium temperatures, CuCr and CuMn from 7 to 15 kG at 300 K, and CuFe from 7 to 61 kG and 77 K to 330 K. The satellite positions are independent of concentration from about 500 ppm to 5000 ppm.

Table I Experimental Values of $\Delta K/K$

Impurity	Temperature	Observed $\Delta K/K$
Fe	300 K	$-5.39 \pm .3$
		$-1.20 \pm .03$
		$-.36 \pm .02$
		$.28 \pm .03$
		$1.85 \pm .03$
Mn(a)	300 K	$-1.98 \pm .06$
		$-.53 \pm .08$
		$.34 \pm .03$
		$.58 \pm .09$
		$1.45 \pm .08$
Cr(b)	300 K	$-1.37 \pm .07$
	300 K	$.74 \pm .1$
	77 K	$1.52 \pm .09$
V	4.2 K	$-.66 \pm .03$

(a) An additional satellite was observed at $149 \pm 9G$ on the low field side of the Cu main line resonance at 8.7 kG and moves to $167 \pm 8G$ at 14 kG.
(b) Two additional satellites were seen on the low field side in CuCr. They are essentially field independent from 9 to 14 kG

one being at $57 \pm 6G$, the other at $129 \pm 9G$.

ACKNOWLEDGEMENT

We wish to thank Thomas Stakelon for his assistance in making samples, John Crues for help in taking data, and David Follstaed for providing some of the data on V. Special thanks are due to Professor Rowland for the use of his spectrometer.

REFERENCES

* This work was supported in part by the U.S. Atomic Energy Commission under Contract AT(11-1)-1198.

** IBM Postdoctoral Fellow, now at Xerox Research Laboratory, Palo Alto, California 94304.

*** IBM Predoctoral Fellow.

1. D. C. Lo, D. Lang, J. B. Boyce, and C. P. Slichter, Phys. Rev. B8, 973 (1973).
2. D. V. Lang, J. B. Boyce, D. C. Lo, and C. P. Slichter, **Phys. Rev. Letters** 29, 776 (1972). D. V. Lang, D. C. Lo, J. B. Boyce, and C. P. Slichter (to be published).
3. N. Karnezos and J. A. Gardner, AIP Conf. Proceedings 10, 801 (1973) report 3 of the 6 lines we report.
4. C. P. Slichter, <u>Principles of Magnetic Resonance</u> (Harper and Row, New York, 1963).
5. P. W. Anderson, Phys. Rev. 124, 41 (1961).
6. J. A. Gardner and C. P. Flynn, Phil. Mag. 15, 1233 (1967).

DETERMINATION OF MOMENT DISTRIBUTIONS AND HYPERFINE FIELDS AT AND SURROUNDING TRANSITION ELEMENT SOLUTE ATOMS IN FE

Mary Beth Stearns
Scientific Research Staff, Ford Motor Co., Dearborn, Michigan 48121

ABSTRACT

A new method has been developed which uses the hyperfine spectra of Fe atoms in dilute Fe alloys to obtain: (1) the moment on a transition metal solute atom, (2) the shape and magnitude of the Fe matrix moment perturbation and (3) the hyperfine field shifts at Fe atoms in the first four neighbor shells surrounding the solute atom. A dipolar-type moment perturbation varying as $1/r^3$ is found to fit the data best.

The moment distributions in dilute binary alloys have been mainly investigated thus far by elastic diffuse neutron scattering experiments.[1] Since hyperfine fields (hff) also originate from the moments, the hyperfine spectra contain information on the moment distributions. Thus the procedure developed here is to derive in detail the hff shifts at the Fe atoms in the near neighbor shells to a solute atom. Then using these shifts we calculate the Fe spectrum corresponding to a given alloy and compare this to the measured spectrum. With the assumptions discussed in the next paragraph this determines the moments of all the atoms in the alloys.

The solute atoms considered are small transition metal elements (no volume overlap) so the hff is due to a core polarization term, H_{cp}, and a conduction electron polarization (CEP) term, H_{ce}. In the usual way[2] we break up the H_{ce} term into a part due to the atom itself, H_s, and a part due to the surrounding atoms H_Σ. This enables us to combine the H_s and H_{cp} terms which only depend on the moment of the Fe atom itself. For pure Fe, $H_M = H_{cp} + H_s$ has been evaluated[2] to be -90.5 kG/μ_B, where μ_B is a Bohr magneton. The shifts due to the CEP through the first six neighbor shells have also been determined[2] for pure Fe. We assume that each atom in the dilute Fe alloys has a CEP oscillation around it with the same shape as in pure Fe but with a magnitude proportional to its moment. For very dilute alloys each solute atom is isolated and CEP shifts and H_M terms should be additive. Since we know that dipolar and non-saturation effects exist[2] we know that the method is less valid as the solute moments differ more from that of the Fe host. In general the additivity assumptions should be good for solute atoms whose moment and band structures are very close to that of Fe, e.g. Co and Ni. They are less valid for the higher transition series where the solute atoms might perturb the Fe matrix in a manner that can not be represented by the above assumptions. In practice we find that we can fit the spectra of FeRh and FePd fairly well but that an FePt spectrum cannot be fit at all satisfactorily.

The best dilute Fe alloy spectra taken to date are those of Budnick, et al[3] for Co, Ni, Rh, Pd and Pt. We shall use that data. Their alloys generally contained 1% of the solute atom. The neutron experiments[1] showed that for these solute atoms the moment on the solute atom was less than that of Fe and the surrounding Fe neighbors to the solute atom had increased moments over that in pure Fe. The procedure used was then as follows:

1) The Fe matrix moment perturbation was represented by $\Delta\mu_n (r_1/r_n)^m$. Where r_n is the distance to the nth neighbor shell, $\Delta\mu_n$ is the Fe moment perturbation in the nth neighbor shell and m gives the power of the radial variation.

2) The measured average saturation moment data gives an independent relation between the solute atom moment and the matrix moment perturbation. For example considering moment perturbations in the six nearest neighbor shells to a Co solute atom we find

$$\mu_{Co} = 3.26 - 19.6\ \Delta\mu_1 , \quad \text{for } m = 3 \quad (1)$$

where all moments are measured in μ_B.

3) Considering the detailed occupational configuration surrounding each Fe atom (i.e. the number and position of each atom with respect to the solute atom) we can write down the hff shifts for an Fe atom in the nth shell, H_{Fe}^{Nn}. For example for a $1/r^3$ perturbation and considering hff shifts caused by the six

nearest neighbor shells we find for the first two nearest neighbor Fe atoms to a Co atom.

$$\Delta H_{Fe}^{N_1} = -121.6 \, \Delta\mu_1 - 12.1(\mu_{Co}-\mu_{Fe}) \quad (2)$$

$$\Delta H_{Fe}^{N_2} = -106.2 \, \Delta\mu_1 - 2.7 \, (\mu_{Co}-\mu_{Fe}) \quad (3)$$

Thus for a given value of m the shifts contain only μ_{Co} and $\Delta\mu_1$ as parameters.

For a given value of m we thus derive a set of hff shifts as a function of μ_{Co}, eliminating $\Delta\mu_1$ by the equivalent of Eq. 1.

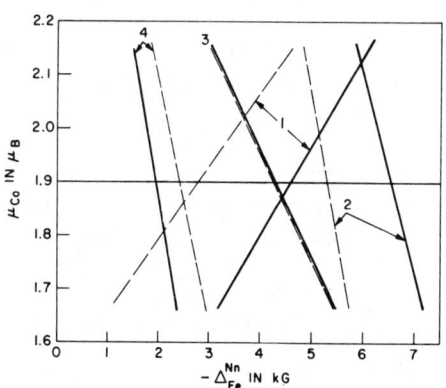

Fig. 1. Variation of the hff shifts as a function of the Co moment at Fe atoms in the first four neighbor shell surrounding a Co atom. The solid and dashed lines are for matrix moment perturbations which vary as $1/r^3$ and $1/r^2$ respectively.

Fig. 1 shows the values of ΔH_{Fe}^{Nn} for Fe atoms in the first four neighbor shells surrounding a Co atom as a function of μ_{Co} for $1/r^2$ (dashed) and $1/r^3$ (solid) matrix perturbations. We see that the N3 hff shift is insensitive to m but the N1 and N2 shifts are very sensitive to m. In Fig. 2 we show the spectra calculated for a 1% Co alloy for the best μ_{Co} value (1.9 μ_B in all cases) and for m = 2, 3 and 4 and compare it with the measured spectrum shown by the data points. We see that m = 3 clearly gives the best fit; the fits for the spectra for m = 2 or 4 are seen to be very poor. We can also see from Fig. 1 that the shifts are quite sensitive to the value of μ_{Co}. We thus find that $\mu_{Co} = 1.9 \pm 0.1 \, \mu_B$, $\Delta\mu_1 = 0.070 \pm 0.005 \, \mu_B$ and m = 3. A $1/r^3$ dipolar-type matrix moment perturbation is very satisfying since such a variation is considered to be most reasonable. The derived moments and hff shifts are listed in Table I. These hff shifts are

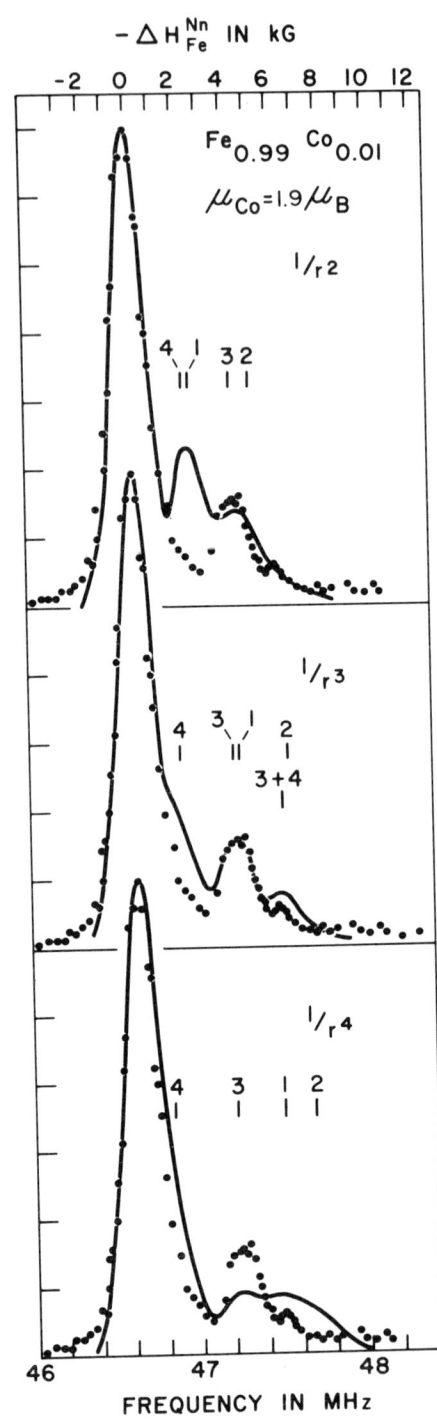

quite different from others[4] appearing in the literature. In previous treatments the shifts were constrained to the arbitrary condition that they decrease in value as the distance from the Co atom increased. There is no justification for such a constraint. In fact it is quite unreasonable since the basic CEP curve in pure Fe is known to oscillate in sign.

Fe alloys with Ni, Rh and Pd have been treated in a similar fashion. We found that the measured spectra could be fitted fairly well with the moment values and hff shifts listed in Table I.

Fig. 2. Calculated (solid line) and measured spectra (data points) for $Fe_{0.99} Co_{0.01}$. The data points are those of Budnick et al.[3] As discussed in the text the best fit is clearly for a $1/r^3$ matrix moment perturbation. The position of the hff shifts are indicated by the nearest neighbor shell number.

Table I. Moments and Hyperfine Field Shifts of Atoms in Dilute Fe Alloys

Solute	$\mu_{Sol}(\mu_B)$	$\Delta\mu_1(\mu_B)$	$\dfrac{\Delta H^{N_1}}{Fe}$	$\dfrac{\Delta H^{N_2}}{Fe}$	$\dfrac{\Delta H^{N_3}}{Fe}$	$\dfrac{\Delta H^{N_4}}{Fe}$
Co	1.9±0.1	0.070±0.005	-4.6±0.5	-6.5±0.3	-4.3±0.5	-1.9±0.2
Ni	1.4±0.1	0.09 ±0.005	-1.4±0.6	-7.7±0.3	-6.7±0.5	-2.8±0.2
Rh	1.1±0.2	0.11 ±0.01	-0.3±1.0	-9.1±0.6	-8.4±1.0	-3.5±0.4
Pd	0.7±0.2	0.07 ±0.01	+10.3±1.6	2.9±0.4	-7.0±1.0	-2.6±0.4

The moment values determined for Co and Ni with this method agree well with those determined by neutron experiments in Ref. 1. However, the moment values obtained here for Pd and Rh are not in good agreement with those obtained on dilute alloys in Ref. 1. On the other hand neutron experiments[5] on higher concentration ordered alloys gave a moment of μ_{Rh} = 1.0 ± 0.1 μ_B and μ_{Pd} = 0.4 ± 0.1 μ_B. These are believed to be the more reliable values since the experiments in Ref. 1 did not extend to large enough values of the scattering vector to give reliable solute moment values in all cases.

A variation of this method has also been applied to transition solute atoms to the left of Fe. It is found that the moment perturbations are due to the spin density oscillations of the itinerant d electrons. As predicted in Ref. 2 the first node moves inward as Z decreases due to the increase in number of itinerant d's with less nuclear charge. This confirms that the origin of ferromagnetism in Fe, Co and Ni is due to the coupling of the mainly localized d electrons (> 95%) through a small fraction (< 5%) of itinerant d electrons.

REFERENCES

1. M. F. Collins and G. G. Low, Proc. Phys. Soc. 86, 535 (1965).
2. M. B. Stearns, Phys. Rev. 147, 439 (1966); ibid 4B, 4069 and 4081 (1971); ibid 8B, Nov. (1973).
3. J. I. Budnick, T. J. Burch, S. Skalski and K. Raj, Phys. Rev. Letters 24, 511 (1970).
4. G. H. Stauss, Phys. Rev. 4B, 3106 (1971); G. K. Wertheim, D.N.E. Buchanan and J. H. Wernick, J. of Appl. Phys. 42, 1602 (1971).
5. Rh; G. Shirani, R. Nathans and C. W. Chen, Phys. Rev. 134, A1547 (1964): Pd; J. W. Cable, E. O. Wollan and W. C. Koehler, Phys. Rev. 138, A755 (1965).

SINGLE CRYSTAL STUDY OF HYPERFINE FIELDS IN Fe-3.2 at. pct. Mo*

A. Asano and L. H. Schwartz
Department of Materials Science, Northwestern University,
Evanston, Illinois 60201

ABSTRACT

Mössbauer spectra from single crystal Fe-3.2 at. pct. Mo were obtained with magnetization axes in the $(01\bar{1})$ plane along [100], [111] and intermediate directions. The complex spectra may be characterized by satellites whose appearance is greatly influenced by magnetization direction--one satellite with magnetization along [100], while two are clearly discernible with magnetization along [111]. A new deconvolution technique was used to eliminate thickness broadening. Analysis based on the pseudo dipole model of Cranshaw et al. yielded effective impurity induced dipole moments of -0.16 mm/sec and -0.14 mm/sec on the Fe atoms which are respectively first and second neighbors of Mo. The application of this analysis to the spectra for intermediate magnetization axes allows a test of the pseudo dipole model.

INTRODUCTION

Since the first detailed Mossbauer study of iron alloys by Wertheim et al.[1], extensive experimentation has continued in this field. In all of these experiments several unjustified assumptions have been employed in the data analysis. First, even though the effect is well known[2-4], most analyses were performed neglecting the distortion of the Mössbauer spectrum due to finite absorber thickness. Second, although reported nine years ago by Cranshaw et al.[5-7], the presence of an anisotropic component in the hyperfine field has been neglected by other workers. Finally most analyses assume that the atomic arrangement in the alloy is random, neglecting the effects of local atomic order. This paper describes the results of analysis of Mössbauer spectra from a single crystal of Fe-3.2 at. pct. Mo in which none of these erroneous assumptions were made. These results were reported briefly elsewhere.[8]

DECONVOLUTION-RECONVOLUTION METHOD

The analysis described here is intended to extract the absorption cross section $\sigma(E)$ from the measured Mössbauer absorption

---

*Supported by the Advanced Research Projects Agency of the Department of Defense and the National Science Foundation through the Northwestern University Materials Research Center.

spectrum which may be expressed as

$$P(V) = (1-f) + f \int_{-\infty}^{\infty} S(E-V)\exp[-\sigma(E)]dE \qquad (1)$$

where f = source Mössbauer fraction, V = Doppler velocity, and S(E-V) = normalized source line shape. This problem was solved by Ure and Flinn[9] who expressed their result in terms of the Fourier and inverse transformation operators F and F^{-1} as

$$\sigma(V) = -\ln\left\{F^{-1}\left(\frac{F[P(V)-1+f]}{f\,F[S(V)]}\right)\right\}. \qquad (2)$$

In practice a Gaussian filter was applied to the data to suppress the singularities introduced by the inherent noise in the raw data. As the effects of this Gaussian filter on the shape of the complex spectra dealt with here could not be assessed, an alternate approach was developed.

The absorption spectrum may be approximated first by a sum of "dummy" Lorentzian curves as

$$P(V) = 1.0 - \sum_i \frac{h_i \Gamma_i^2/4}{(V-V_i)^2 + \Gamma_i^2/4}. \qquad (3)$$

Note that while this is what the usual Lorentzian curve fit analyses do, no attempt is made here to give any physical meaning to the parameters, h_i, Γ_i, V_i, but simply to approximate the spectrum by Lorentzian curves. The Fourier transformation is then

$$\sigma(V) = \ln\left\{1 - \frac{1}{f}\sum_i \frac{\Gamma_i h_i}{(\Gamma_i - \Gamma_s)} \frac{(\Gamma_i - \Gamma_s)^2/4}{(V-V_i)^2 + (\Gamma_i - \Gamma_s)^2/4}\right\}, \qquad (4)$$

where Γ_s is the source line width. It was found that the result of Eq.(4) depends on the number of peaks assumed in the fit of Eq.(3). A reconvoluted or signal-averaged cross section $\sigma'(V)$ defined as

$$\sigma'(V) = \int S(E-V)\sigma(E)dE \qquad (5)$$

was insensitive to the mode of fitting Eq.(3) and could be analyzed by the usual curve fitting procedure, the effects of finite sample thickness now eliminated.

ANISOTROPIC INTERACTION

The effect of a substitutional impurity atom on surrounding iron atoms has been assumed to produce an isotropic change in magnetic hyperfine field ΔH, a change in isomer shift $\Delta\delta$, and a change in quadrupole splitting ΔQ, all referred herein to the resonance conditions for pure iron. ΔH, $\Delta\delta$, and ΔQ differ for each neighbor distance. ΔQ has always been assumed negligibly small. All studies except those of Cranshaw et al.[5-7] and Asano and Schwartz[8] have neglected the anisotropic perturbation of hyperfine field H_d due to the magnetic dipole created when an iron atom is replaced by an atom of different spin. Taking θ as the angle between the magnetization axis and the vector from the impurity atom to an iron atom, for a purely dipolar interaction, $H_d = D(3\cos^2\theta - 1)$, where D differs for each neighbor shell. Combining all effects, the resonance for a given iron atom is shifted relative to that of pure iron by an amount ΔP. For the lowest and highest energy resonances, denoted ΔP_- and $\Delta P'_+$,

$$\Delta P_\pm = \Delta\delta \pm \Delta H + (\Delta Q \mp D)(3\cos^2\theta - 1). \quad (6)$$

Cranshaw has shown that by varying the angle φ between the spin axis and the [100] in a single crystal sample, the satellite positions due to first and second neighbors could be identified. For example, with applied field H along [100], $\varphi = 0$, $\cos\theta = \pm 1/\sqrt{3}$ for all iron near neighbors to the impurity. For second neighbors, four have $H_d = -D$ while two have $H_d = 2D$ causing a splitting of this peak. For H along [111], $\varphi = 54°44'$, the first neighbor peak is split while that due to second neighbors is unsplit. Difference spectra between these two orientations clearly identify the locations of first and second neighbor resonances. Combining these results for both peaks represented by Eq.(6) allows the determination of $\Delta\delta$, ΔH, ΔQ and D.

EXPERIMENTAL RESULTS AND DISCUSSION

A single crystal disc of Fe-3.2 at. pct. Mo was cut with a surface normal of $[01\bar{1}]$, solution treated at 1050°C for one hour, quenched into water and reannealed at 540°C for 2½ hrs. followed by a chemical polish to approximately 1 mil thickness. The $[01\bar{1}]$ axis was aligned parallel to the γ-ray propagation with a magnetic field H of 2.7 KOe perpendicular to the γ-rays and at an angle φ from the [100]. The spectra for $\varphi = 0°$ ([100]), 11°, 20°, 28°, 45° and 54°44' ([111]) are shown in Fig. 1 as crosses. Sample thickness effects were eliminated as described above and the [111] spectrum and the difference spectrum between [100] and [111] simultaneously least squares fit. It was assumed that only first and second impurity neighbors produce measurable shifts described by Eq.(6) and that all resonance peaks are Lorentzian of the same half width. The parameters derived from these fits are shown in Table I.

Using the parameters of Table I and the variable angle φ, the solid lines shown in Fig. 1 were synthesized from Eq.(1). The remarkable success of this simulation confirms the dipole symmetry of the anisotropic H_d and the assumption that 1st and 2nd neighbor effects dominate.

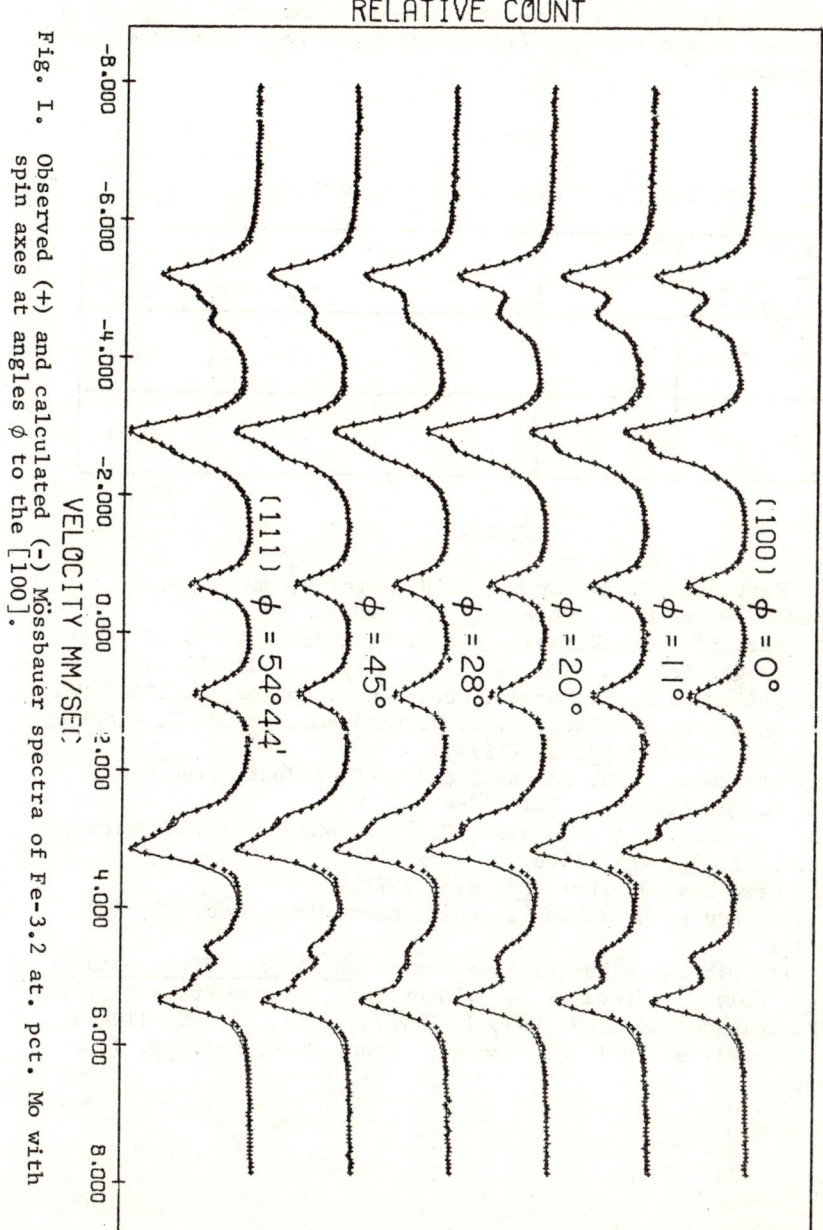

Fig. 1. Observed (+) and calculated (-) Mössbauer spectra of Fe-3.2 at. pct. Mo with spin axes at angles ϕ to the [100].

The magnitude of D is several times larger than that calculated using the known impurity site d-shell magnetic moments[10] suggesting a more complex origin for H_d. Details of these results will be published elsewhere.

TABLE I

FITTING PARAMETERS TO EQ.(6) IN MM/SEC

Parameters Neighbor Shell	ΔH	$\Delta \delta$	D	ΔQ
1st	-.58 (-11.0%)	-.11	-.16	<.01
2nd	-.28 (-5.4%)	-.07	-.14	<.01

REFERENCES

1. G. K. Wertheim, V. Jaccarino, J. H. Wernick, and D. N. E. Buchanan, Phys. Rev. Letters 12, 24 (1964).
2. W. C. Harper, C. W. Kimball, and A. T. Aldred, AIP Conf. Proc. No. 5 (Magnetism and Magnetic Materials), p. 533 (1971).
3. S. Margulies and J. R. Ehrman, Nucl. Inst. Method 12, 131 (1961).
4. R. E. Meads, B. M. Place, F. W. D. Woodhams, and R. C. Clark, Nucl. Inst. Method 98, 29 (1972).
5. T. E. Cranshaw, C. E. Johnson, and M. S. Ridout, Proc. Intl. Conf. on Magnetism, p. 141 (1964).
6. T. E. Cranshaw, C. E. Johnson, M. S. Ridout, and G. A. Murray, Phys. Lett. 21, 481 (1966).
7. T. E. Cranshaw, J. Phys. F2, 615 (1972).
8. A. Asano and L. H. Schwartz, Bull. Amer. Phys. Soc. 18, 113 (1972).
9. M. Celia Dibar Ure and P. A. Flinn in Mössbauer Effect Methodology, Vol. 7, edited by I. J. Gruverman, Plenum Press (1971).
10. M. F. Collins and G. G. Low, J. Phys., Paris, 25, 596 (1964); Data reanalyzed by I. A. Campbell, Proc. Phys. Soc. 89, 71 (1966).

CHARGE SCREENING AND THE MAGNETIC PROPERTIES OF THE Ni-Sb AND Ni-Ir ALLOY SYSTEMS*

J. Hudis, M. L. Perlman, and R. E. Watson
Brookhaven National Laboratory, Upton, NY 11973

ABSTRACT

Core and valence level soft x-ray photoemission results have been obtained and combined with Mössbauer isomer shift and magnetization data in a study of the Ni-Sb and Ni-Ir alloy systems. Analysis of the results shows that the reduction of the magnetic moment and the inferred filling of the Ni d bands are not concomitant with substantial change in the 3d electron count at Ni sites. The data also indicate that there is substantial electron depletion at Sb sites in the Ni-Sb system; no strong charging effects occur in the Ni-Ir system.

In this paper we report soft x-ray photoemission results for Ni-Ir and Ni-Sb alloys obtained in an effort to relate the charge disturbance associated with impurities in Ni to the well known magnetic effects.[1] Ir and Sb, which have quite different valence electron configurations, both reduce the magnetization of the Ni host. Mössbauer isomer shift data[2] is available for these alloys, allowing estimates to be made of changes in s-like behavior at the impurity sites upon alloying. In Friedel's model[3] as refined in recent calculations by Terakura and Kanamori[4] the alloying can be described crudely as a two step process: the perturbing potential from an Ir or Sb impurity is screened by the host valence electrons, resulting in a charge depletion on the host which is taken care of by filling the host d bands. In detail, the purpose of the present study is to estimate the extent of the screening and the character of the screening electrons.

For the alloy preparations Ni metal 99.999% purity (J. M. Specpure, United Mineral and Chemical Co., New York, N.Y.), Ir metal 99.9+% purity (Ventron Corp., Beverly, Mass.), and Sb metal 99.999+% purity (Cominco-American, Spokane, Washington) were used.

According to Bucher et al.[5] Ni and Ir form solid solutions across the entire range of composition and the Ni-Ir alloy specimens were made by arc melting proper weighed amounts of the constituents under Ti-purified Ar in a water cooled, non-wetting Cu hearth. To ensure uniformity in composition, the ingots were turned and remelted several times. The system Ni-Sb[6] is not as well behaved as Ni-Ir; here, in addition to the pure metals, the compound NiSb and a single phase, dilute Sb in Ni, were prepared. Weighed amounts of the constituents in the desired ratios were heated in silica tubes under pure helium for about ten days at 1050°C. As shown by x-ray diffraction the Ni-Sb compound thus produced was a single phase.

---
*Research performed under the auspices of the U.S. A.E.C.

The dilute solution, as first prepared, contained about 5 atom % Sb, perhaps enough to produce a second phase. Some of this Sb was vaporized away by briefly melting the specimen in the arc furnace. Then by grinding away the outer 0.5 to 1 mm of the resulting ingot, a region having composition in the desired range was exposed. Exact alloy compositions were determined by chemical analysis.

Specimens were cleaned first by filing under alcohol and then by Ar sputtering in a forechamber of the Varian IEE-15 photoelectron spectrometer. Surface conditions were monitored by observation of oxide-metal photoelectron peak intensity ratio and by oxygen 1s peak intensity.

The photoelectron lines measured originated from core levels of Ni, Ir, and Sb listed in Table I. Electron spectra were analyzed by fitting Gaussians to the peaks after background had been subtracted. No appreciable satellite structure was observed for these lines, and typical widths were 1.3 to 1.7 eV, FWHM. Spectrum analysis was straightforward except in the case of the dilute Sb in Ni; here care was taken that nearby intense Ni Auger lines were not confused with the Sb line. Binding energies in Table I are relative to zero at the Fermi level of each alloy. On this scale the Au $4f_{7/2}$ binding energy is 83.70 eV. Sb will be characterized in terms of s and p electron bands and Ir in terms of 5d and non-d "conduction" bands; and in the Table there are listed band count changes, Δn_{cond} for Ir and Δn_s for Sb, derived from existing Mössbauer isomer shift data[2] in the manner previously applied to Au and Sn.[7] These will be employed in the analysis of the level shift data.

There is a notably large shift, 1.4 eV, between the Sb 4f binding energy in pure Sb and that in the dilute alloy; this is the largest shift in metallic systems of which we are aware.

Valence band spectra were obtained for several of the samples; observable changes were seen but not in sufficient detail to be characterized here. This paper will concentrate on the implications of the core level shifts. The Ni photoelectron linewidths observed with the alloys are no greater than those of pure Ni, indicating that Ni atoms well away from impurities have chemical shifts similar to those of near neighbors. This is consistent with neutron diffraction results which show that the magnetic disturbance created by an impurity atom extends out from it to include fifty or more Ni atoms.[8] In what follows we assume a charge distrubance of similar extent; that is, that at the concentrations involved, all Ni atoms in each alloy are similarly perturbed.

If one neglects exchange and aspherical coulomb interactions, taking an atom from, say, a pure metal to an alloy causes a shift[7] in the one-electron energy of core level i

$$\Delta \varepsilon_i = \Delta[\sum_j n_j F^o(i,j)] - \delta F_{latt}(i). \quad (1)$$

Here n_j is the number of valence electrons, each normalized to the atomic site in question, and F^o is the coulomb interaction of each with the core level. The F^o's may change due to change in atomic volume, or wave function shape, upon alloying. The net electronic charge at the site in the alloy, i.e. a measure of the extent of

Table I. Core level binding energies in Ni-Ir and Ni-Sb alloys. Values are given relative to the Fermi level of each alloy. Conduction or s-band electron count changes inferred from Mössbauer isomer shift data are given in the last column.

Alloy	Level binding energies, $-E_B$, in eV			Δn_{cond}(Ir); Δn_s(Sb)
	Ni $2p_{3/2}$	Ir $4f_{7/2}$	Sb $3d_{5/2}$	
Ir		60.45(5)		
$Ni_{0.406}Ir_{0.594}$	852.38(5)	60.45(5)		+0.10
$Ni_{0.798}Ir_{0.202}$	852.47(5)	60.40(5)		+0.21
$Ni_{0.947}Ir_{0.053}$	852.39(5)	60.15(10)		+0.23
Ni	852.44(5)			
$Ni_{0.985}Sb_{0.015}$	852.37(5)		529.17(8)	−0.28
$Ni_{0.495}Sb_{0.505}$	852.40(8)		527.87(10)	−0.14
Sb			527.75(10)	

screening, is $\delta = \Sigma \Delta n_j$; and F_{latt} is the coulomb term associated with the charge, $-\delta$, outside the site in the rest of the lattice. In Eq. (1), $\Delta\varepsilon$ is the shift with respect to vacuum zero, if one neglects the barrier potential associated with the metal surface.[7] On the other hand, a photoelectron binding energy value, $-E_B$, from experiment is a measure of the position of ε_i with respect to the Fermi level ε_F, because the work function in the experiment is that of the spectrometer. Thus, the measured shift on alloying is

$$\Delta E_B = \Delta \varepsilon_i - \Delta \varepsilon_F, \qquad (2)$$

and relating a shift in E_B to valence electron changes upon alloying then requires knowledge of any shift of ε_F with respect to vacuum. We are not prepared to calculate these Fermi level shifts and, instead, make use of experimental work functions, a procedure which does not take into account any variation in surface barrier potential upon alloying. This and other problems associated with relating $\Delta\varepsilon_i$ to ΔE_B have been discussed elsewhere.[7] ΔE_B is expected to be rather small for atoms such as Ni, i.e. for transition metals which have a high density of d states at the Fermi level. In first approximation the core undergoes the same coulomb shift as the d bands, that is the shift in ε_i is "pinned" to the shift in the center of gravity, ε_d, of the d bands and in turn to the shift of ε_F. In Au alloys[7] $\Delta\varepsilon_d$ is roughly a factor of two greater than $\Delta\varepsilon_i$, presumably due to the fact that the d bands are not inert but play an active role in alloying. This behavior may be typical for transition metal alloys. The tendency towards pinning nevertheless makes it difficult to obtain numerical estimates of the quantities of interest at Ni sites, Δn_{3d} and Δn_{cond}, from the experimental data and Eqs. (1) and (2).

In the case of Sb it is appropriate to describe chemical effects in terms of s and p bands, employing Δn_s values from the isomer shift measurements to estimate the s band effects. Eq. (2) for the Sb 3d shifts may be rewritten

$$\Delta E_B = -\Delta\varepsilon_F(\text{Sb} \to \text{Ni}) - \Delta\varepsilon_F(\text{Ni} \to \text{alloy}) + \Delta n_s[F^o(\underline{s},3\underline{d}) - F^o(\underline{p},3\underline{d})]$$
$$+ \delta[F^o(\underline{p},3\underline{d}) - F_{latt}] + \Delta[\Sigma n_i F^o(\underline{i},3\underline{d})]_{vol} , \qquad (3)$$

where $\Delta\varepsilon_F$ is separated into two contributions: one reflecting the transition from Ni to the alloy, the same as that occurring at the Ni site, and the second a pure metal difference term, −0.7 eV, estimated from work function data.[9] The final term in Eq. (3) accounts for the fact that the Ni-Sb alloy sequence has a smaller average atomic volume than that derived from the atomic volumes in the pure metals.[10] We estimate this term to be +0.25 eV if Sb and Ni sites are compressed by the same volume percentage; it is much greater if the compression is primarily associated with Sb sites. Renormalized atom estimates[7,11] have been made of $F^o(\underline{s},3\underline{d})$ and $F^o(\underline{p},3\underline{d})$, yielding 13.7 and 12.1 eV, respectively, and if the charge disturbance, δ, is evenly distributed on the Ni lattice, $F_{latt} \simeq 3.7$ eV for $Ni_{0.985}Sb_{0.015}$. We then obtain for this alloy $\delta = -0.24 + 0.12 \Delta\varepsilon_F(\text{Ni} \to \text{alloy})$. This result indicates a substantial depletion of charge at Sb sites in the alloy, a result largely determined by ΔE_B and unlikely to be affected qualitatively by refinement in the estimates of the various terms of Eq. (3). Volume compression such as that occurring here may well be a good signature of charging. Eq. (3) is so conditioned that the magnitude of δ is not readily determinable for the 50-50 alloy, though Sb site depletion is indicated. The Ni d band filling indicated by the alloy magnetic behavior[1] implies $\delta \simeq -4$ if screening were turned off in the Terakura-Kanamori model; the fact that $\delta \simeq -1/4$ indicates that the screening is not quite complete.

Eq. (3) applies to Ir 4f shifts if the s and p band terms of Sb are replaced by Ir conduction and d terms. For $Ni_{1-x}Ir_x$, the alloy's atomic volume is $\simeq (1-x)V_{Ni} + xV_{Ir}$; hence the Ni and Ir volumes are probably independent of composition and the volume term of Eq. (3) is approximately zero. For Ir in the dilute $Ni_{0.947}Ir_{0.053}$ we obtain $\delta = 0.06 \pm 0.1$, essentially zero. A δ value of the same sign but still smaller is obtained for $Ni_{0.798}Ir_{0.202}$.

For any of these alloys the sign of δ given by Eq. (3) is the same as that of the Δn in Table I.[12] For Ni-Ir, δ is substantially smaller in magnitude than Δn_{cond}; we interpret this as being due to s-d compensation of the type encountered previously in Au alloys.[7]

The 2p binding energy shifts at Ni sites are essentially zero, and the Ni charge flow terms in the dilute alloys are small, since any charge shifted to or from Sb or Ir sites is shared by the larger relative number of Ni atoms. Therefore, one may expect from Eqs. (1) and (2)

$$\Delta n_{cond}\{F^o(\text{cond},2\underline{p}) - F^o(3\underline{d},2\underline{p})\} \simeq \Delta\varepsilon_F(\text{Ni} \to \text{alloy}). \qquad (4)$$

Specifically, including the small terms omitted from (4), we obtain for the $Ni_{0.985}Sb_{0.015}$ alloy

$$\Delta n_{cond} = -\{\Delta \varepsilon_F - 0.40\}/9.8. \qquad (5)$$

The value of Δn_{cond} provides evidence about the Ni screening. If the non-$\underline{d}$ conduction electrons are primarily responsible for the screening of the impurity sites, i.e. they spend less time on the Ni atoms, this screening taken with the concomitant $\underline{d}$ band filling would give the result $\Delta n_{cond} \simeq \Delta\mu$, the change in Bohr magneton count at a Ni site. A zero value for Δn_{cond} would imply that $\underline{d}$ band electrons are responsible for the screening. With the first supposition, Δn_{cond} is about -0.06 and $\Delta\varepsilon_F$ is about $+1.0$ eV for $Ni_{0.985}Sb_{0.015}$. For $Ni_{0.947}Ir_{0.053}$ $\Delta\varepsilon_F$ would be $\simeq 0.6$ eV. Although there is some uncertainty in these $\Delta\varepsilon_F$ values, they are unreasonably large in our view, and we take this to be evidence against the validity of the first supposition. It thus seems that the $\underline{d}$ band electrons are primarily responsible for impurity site screening.

1. E.g. J. Crangle and D. Parsons, Proc. Roy. Soc. (London) A255, 509 (1960) and J. Crangle and M. J. C. Martin, Phil. Mag. 4, 1006 (1959).
2. R. L. Mössbauer, M. Lengsfeld, W. von Lieres, W. Potzel, P. Teschner and F. E. Wagner, Z. Naturforsch. 26, 343 (1971) and H. Z. Dokuzoguz, L. H. Bowen, and H. H. Stadelmaier, J. Phys. Chem. Solids 31, 1565 (1970).
3. J. Friedel, Suppl. Nuovo Cimento 7, 287 (1958).
4. K. Terakura and J. Kanamori, Prog. Theoret. Phys. 46, 1007 (1971).
5. E. Bucher, W. F. Brinkman, J. P. Maita, and A. S. Cooper, Phys. Rev. B 1, 274 (1970).
6. M. Hansen, Constitution of Binary Alloys, 2nd ed. (McGraw-Hill Book Co., N. Y., 1958), p. 1037.
7. R. E. Watson, J. Hudis, and M. L. Perlman, Phys. Rev. B 4, 4139 (1971); R. M. Friedman, R. E. Watson, J. Hudis, and M. L. Perlman, Phys. Rev. B (2 publications in press).
8. E.g. J. B. Comly, T. M. Holden, and G. G. Low, J. Phys. C 1, 458 (1968).
9. D. E. Eastman, Phys. Rev. B 2, 1 (1970); J. C. Rivière, Solid State Surface Science 1, 179 (1969); G. A. Haas and R. E. Thomas in Measurement of Physical Properties, E. Passaglia, editor (Interscience, N. Y., 1972), p. 91; and M. Traum (private communication).
10. W. B. Pearson, Lattice Spacings and Structures of Metals and Alloys (Pergamon Press, London, 1958); see also Dokuzoguz et al. (ref. 2).
11. L. Hodges, R. E. Watson, and H. Ehrenreich, Phys. Rev. B 5, 3953 (1972).
12. The inferred direction of charge flow is to the atom evidently having the lower ε_F as indicated by the work functions.

The authors acknowledge J. J. Hurst, W. Kunnmann, and R. W. Stoenner for their assistance.

MAGNETIC IMPURITY STATES IN NiAℓ INTERMETALLIC COMPOUNDS*

J. R. Willhite, T. Yoshitomi, L. B. Welsh, and J. O. Brittain, Northwestern University, Evanston, Ill. 60201

ABSTRACT

The properties of the localized magnetic state formed by Fe impurities in stoichiometric NiAℓ have been studied from 50 to 2000ppm Fe. We have measured the low temperature resistivity, the magnetoresistance and the magnetization down to 1.3K and in magnetic fields to 25 kOe. The bulk properties are consistent with local moment behavior with an effective moment per Fe impurity of 4.9 μ_B and a concentration independent Curie-Weiss θ of 1.0K. Because of the long Aℓ^{27} spin lattice relaxation time ($T_1T=28.5$ sec K), the Fe impurity contributions to T_1 can be easily measured even for the lower Fe concentration samples. The T_1 results suggest that the radius of the nuclear spin diffusion barrier is considerably smaller in NiAℓ than in systems such as CuMn and LaAℓ$_2$ Gd.

INTRODUCTION

The intermetallic compound NiAℓ forms an ordered CsCℓ lattice over a wide range of composition, 45 to 60 at.% Ni. Previously reported low temperature resistivity,[1] magnetoresistance,[2,3] and magnetic susceptibility[4,5] (χ) measurements showed Kondo-like phenomena, i.e. a resistivity minimum, negative magnetoresistance, and Curie-Weiss χ. However, recent work[6] showed that the intrinsic properties of very pure stoichiometric NiAℓ are those of a simple Pauli paramagnet. This is consistent with the results of band structure calculations[7] which suggest the Ni d-bands are filled. In the MX compounds (X=Aℓ, Ga; M=transition metal) the replacement of an X by an M results in the formation of a nine atom M cluster. The concentration of these clusters depends on the amount of local atomic disorder in the compound. Of the more widely studied systems (NiAℓ, CoAℓ, FeAℓ) only in NiAℓ are these clusters nonmagnetic.[2,6,8] Thus in NiAℓ the behavior of magnetic impurity states in the MX systems can be systematically studied by the addition of transition metal impurities. NiAℓ is also a favorable compound in which to study the impurity induced NMR relaxation rate since the Aℓ^{27} Korringa relaxation rate[6,9] of $(T_1T)^{-1}=0.035$ sec^{-1}K^{-1} is even smaller than observed[10] in LaAℓ$_2$. In NiAℓ the Aℓ site has cubic symmetry so that the Aℓ^{27} NMR is not split by quadrupole interactions.

We find that the 3-d transition metal impurities which form magnetic moments in NiAℓ are Mn, Cr, and Fe. In this paper we report the effects of Fe impurities in the dilute impurity limit. Metallurgical problems which arise in low solubility systems such as CuFe are avoided because of the high solubility[11] of Fe in NiAℓ. The

---

*Supported by AFOSR, and by NSF through the NU Materials Research Center.

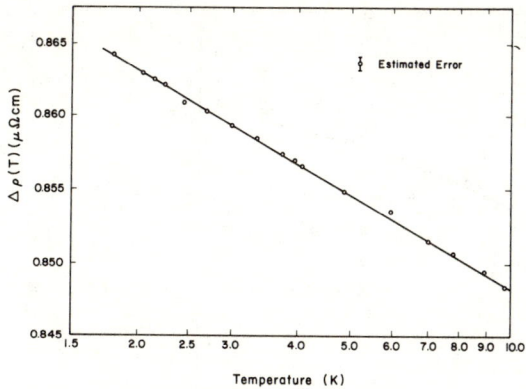

Fig. 1. Impurity contribution to the resistivity for a sample contain 210ppm Fe.

preparation of the $(NiAl)_{1-c}Fe_c$ samples and the experimental techniques used in this study have been described elsewhere.[6]

BULK PROPERTY RESULTS

Below 10K the resistivity of stoichiometric NiAl is temperature independent and for our best atomically ordered sample[6] is $\rho=0.597\mu\Omega$cm. The addition of Fe impurities produces a low temperature resistivity minimum. Fig. 1 shows the impurity contribution to the low temperature resistivity for a typical low concentration Fe sample. The impurity contribution to the resistivity is:[12]

$$\Delta\rho = (2\pi N(0)mc/zNe^2\hbar)J^2S(S+1)[1+4JN(0)\ln(kT/D)] \quad (1)$$

where $N(0)$ is the density of states at E_F/spin direction, zN the number of conduction electrons/unit volume, J is the s-d exchange constant, and D the conduction band cutoff. Using the value[7] of $N(0) = 0.46$ states/NiAl/ev, $d(\Delta\rho)/d(\ln T) = -9.38\times10^{-9}\Omega$cm from Fig. 1 and assuming S=2 consistent with the high temperature χ data presented below, we find that $J=-0.51$ev. This is smaller than the value[13] of $J \simeq -0.9$ev for Fe in Cu. The low temperature slope of $\Delta\rho$ vs $\ln T$ scales with concentration up to 1000ppm Fe. For higher concentrations a resistivity maximum appears above 1.8K.

The magnetization (M) and magnetic susceptibility (χ) of the Fe impurities in NiAl have been studied between 1.3 and 300K in magnetic fields to 25kOe. For those fields where M is linear in field the temperature dependence of χ^{-1} gives an effective moment per Fe impurity of $4.90\pm0.05\mu_B$ and a concentration independent θ of 1.0 ± 0.1K. As is shown in Fig. 2, for low concentration M is linear in c and follows a Brillouin function $B_S(x)$ with S=2 and $x=g\mu_BSH/k(T+\theta)$. The concentration independent of θ indicates that it is not due to impurity interactions but rather is related to Kondo-like behavior. This interpretation leads to a Kondo temperature of about 0.2K ($T_K=\theta/4.5$).[14] The field and temperature dependence of M and χ is similar to that of other local moment systems such as CuMn (S=2, $\theta=0.01$K) and CuFe (S=3/2; $\theta=29$K).

The impurity magnetoresistance for a typical dilute Fe concentration sample is shown in Fig. 3. The impurity contribution to the magnetoresistance is given by:[15]

$$\Delta\rho(H) = -(3\pi m/2E_Fe^2\mu_B^2\hbar)cvJ^2M^2[1+g\mu_B^2H/12kM(T+\theta)] \quad (2)$$

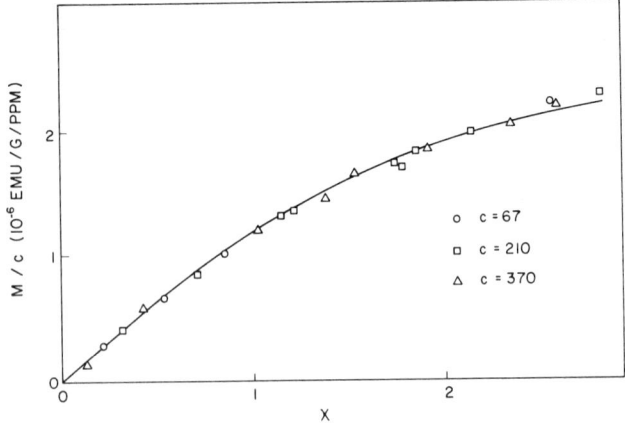

Fig. 2. Impurity M/c for several low Fe concentration samples. Data is for $1.3 \leq T \leq 4.2K$. The solid curve shown is $B_S(x)$; $x = g\mu_B SH/k(T+\theta)$; $S = 2$; $\theta = 1.0K$. The concentration is determined from $\rho(T)$.

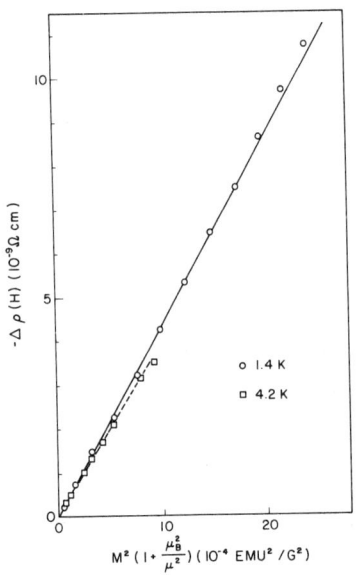

Fig. 3. $\Delta\rho(H)$ vs $M^2(1+\mu_B^2/\mu^2)$ for a 210ppm Fe sample where $\mu^2 = 12Mk(T+\theta)/gH$. Note that $\Delta\rho(H)$ is negative.

As shown in Fig. 3, the impurity magnetoresistance at 4.2 and 1.4K has the form given in Eq. 2. However the slopes differ slightly giving $|J| = 0.34$ev at 4.2K and $|J| = 0.35$ev at 1.4K. These values are in reasonable agreement with the value $J = -0.51$ev obtained from the temperature dependence of the resistivity.

The bulk properties presented above ($\rho(T)$, $M(H)$, $\rho(M)$) clearly show that Fe impurities in NiAℓ form local magnetic moments similar to those formed in systems such as CuMn and CuFe.

NMR RESULTS

The Fe impurity contribution to the measured Aℓ^{27} spin-lattice relaxation rate is linear in c for the low concentration samples. The field dependence of T_1^{-1} for a typical low Fe concentration sample is shown in Fig. 4 for 4.2 and 1.57K. The impurity induced excess relaxation rate is weakly temperature dependent between 4.2 and 1.57K and for low fields has

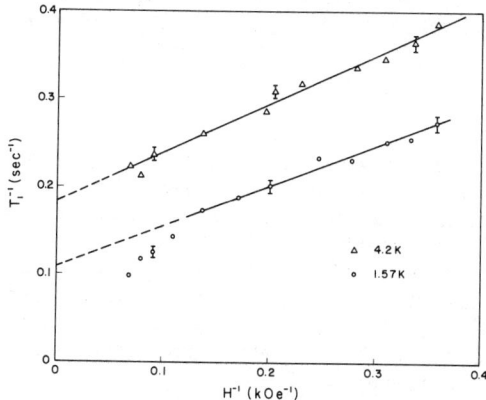

Fig. 4. Total NMR relaxation rate as a function of inverse magnetic field for a sample containing 210ppm Fe.

an H^{-1} dependence. The field and temperature dependence of the observed excess relaxation rate is similar to that observed in CuMn and CuFe.

Bernier and Alloul[16] have suggested that the low field H^{-1} dependence of the impurity induced relaxation rate in CuMn may be accounted for by the field dependence of a spin diffusion barrier. In such a model, nuclear spin diffusion is suppressed for n_b nuclei within a sphere of radius r_b about each magnetic impurity due to the magnetic field gradient at the Al sites resulting from RKKY interaction. Bernier and Alloul[16] found for CuMn that $n_b \propto \langle S_z \rangle$. If the induced relaxation rate at a distance r from an impurity is written as $T_1^{-1} = \zeta/r^6$, then[10,17]

$$T_1^{-1} = (16\pi^2/9)N^2 \zeta/n_b \quad (3)$$

where N is the density of lattice sites. The mechanism responsible for the excess relaxation rate is not well understood at this time, however the longitudinal dipolar flucuations in the impurity magnetization would seem to give the largest contribution and this mechanism has been used to describe the relaxation rates obtained in LaAl$_2$ Gd. The longitudinal dipolar mechanism gives for $H\gamma_n \tau_1 \ll 1$,[17]

$$\zeta = \frac{24\pi}{5} (\gamma_e \gamma_n \hbar)^2 S^2 \frac{dB(x)}{dx} \tau_1 \quad (4)$$

where γ_e and γ_n are the electron and nuclear gyromagnetic ratios, $x = g S \mu_B H/kT$, and τ_1 the impurity spin longitudinal relaxation time. For low fields the field dependence of T_1^{-1} is determined by that of n_b, which gives $T_1^{-1} \propto H^{-1}$. Since the Fe d-spin-conduction electron spin system is probably not bottlenecked and the observed excess relaxation rate is linear in concentration, τ_1 can be estimated from the Korringa rate of

$$\tau_1^{-1} \approx (4\pi/\hbar) J^2 N^2(0) kT. \quad (5)$$

Because of the difficulty of determining n_b, we can only make a rough comparison between the relaxation rate for the longitudinal dipolar mechanism and the experimentally determined relaxation rates. From the rf field dependence of the spin-echo amplitude and the fact that the recovery of the signal is exponential we estimate that at 4.2K, $n_b \leq 10^{-2}H$. Using this number, the slopes from Fig. 4, Eq. 3-5, and the previously determined value for J, we calculate that the longitudinal dipolar rate should be larger than the observed rate by a factor of ten. Thus it is not clear that this mechanism is correct for Fe in NiAℓ.

CONCLUSION

The low temperature resistance minimum, the negative magnetoresistance, and the impurity magnetization are consistent with the formation of a local magnetic moment for an Fe impurity in stoichiometric NiAℓ. These effects are consistent with an Fe impurity having an effective moment of $4.9\mu_B$ implying a spin of S=2 and a Kondo temperature of about 0.2K. The behavior of the Fe impurity induced excess Aℓ^{27} relaxation rate is similar to that of the excess rate seen in CuMn and suggest that the radius of the nuclear spin diffusion barrier is smaller than in CuMn or in LaAℓ_2 Gd.

REFERENCES

1) Y. Yamaguchi, T. Aoki and J. O. Brittain, J. Phys. Chem. Solids 31, 1325 (1970).
2) G. R. Caskey, J. M. Franz and D. J. Sellmyer, J. Phys. Chem. Solids 34, 1179 (1973).
3) Y. Yamaguchi and J. O. Brittain, Phys. Rev. Letters 21, 1447 (1969).
4) Von H. Hohl, Ann. Phys. (Paris) 19, 15 (1967).
5) M. B. Brodsky and J. O. Brittain, J. Appl. Phys. 40, 3615 (1969).
6) J. R. Willhite, L. B. Welsh, T. Yoshitomi, and J. O. Brittain, to be published.
7) J. W. D. Connolly and K. H. Johnson in NBS 3rd Mat'l. Res. Sym. on Elect. Density of States (Edited by J. H. Bennett) pg. 323, NBS (1971).
8) D. J. Sellmyer, G. R. Caskey and J. Franz, J. Phys. Chem. Solids 33, 561 (1972).
9) J. J. Spokes, C. H. Sowers, C. O. Van Ostenburg, and H. G. Hoeve, Phys. Rev. B1, No. 6, 2523 (15 March 1970).
10) M. R. McHenry, B. G. Silbernagel, and J. H. Wernick, Phys. Rev. B, 5, No. 8, 2958 (15 April 1972).
11) A. J. Bradley and A. Taylor, Proc. Roy. Soc. A156, 353 (1938).
12) J. Kondo in Solid State Physics (Edited by F. Seitz, D. Turnbull, and H. Ehrenreich) Vol. 23, pg. 220 (1969).
13) W. M. Star thesis University of Leiden (1971).
14) A. J. Heeger in Solid State Physics (Edited by F. Seitz, D. Turnbull, and H. Ehrenreich) Vol. 23, pg. 365 (1969).
15) M.-T. Beal-Monrod and R. A. Weiner, Phys. Rev. 170, No. 2, 552 (10 June 1968).
16) P. Bernier and H. Alloul, J. Phys. F: Metal Phys., 3, 869 (April 1973).
17) B. Giovannini, P. Pincus, G. Gladstone, and A. J. Heeger, J. Phys. (Paris) 32, C1-163 (1971).

CONDUCTION ELECTRON POLARIZATION IN FCC TRANSITION METALS

G. P. Huffman
U. S. Steel Corp., Research Laboratory
Monroeville, Pa. 15146

ABSTRACT

Mossbauer results are presented for the alloys NiAl, NiSn, CoAl, and CoSn doped with 0.3%Fe^{57} and for ordered FeNi$_3$ containing 1 and 2 a/o Sn. Values of the core and conduction electron polarization fields at the Fe nucleus are deduced and found to depend strongly on the moment decrement of an Fe atom with a non-magnetic nearest neighbor.

INTRODUCTION

It is often assumed that the conduction electron polarization (cep) in Ni and Co is negative at the nearest neighbor distance and of the same approximate size as that in Fe. However, there is little actual experimental data to support this view. Our work indicates that the cep contribution to the hyperfine field at Sn^{119} nuclei in fcc transition metals and alloys (Ni, Co, Co-Ni, FeNi, FeNi$_3$) is positive.[1] This is certainly the case in Ni where the Sn^{119} hyperfine field is positive and has exactly the temperature dependence expected for a field produced entirely by cep.[1,2] We report here the results of Fe^{57} Mossbauer studies aimed at clarifying the nature of cep in fcc transition metals.

EXPERIMENTAL TECHNIQUES

The alloys Ni$_{0.97}$Al$_{0.03}$, Co$_{0.97}$Al$_{0.03}$, Ni$_{0.98}$Sn$_{0.02}$, and Co$_{0.98}$Sn$_{0.02}$ (enriched to ~50% in Sn^{119}) doped with 0.3 atomic percent (a/o) Fe^{57} were prepared by arc-melting and thinned by mechanical and chemical methods to foils 1.0 to 1.5 mils thick for Mossbauer spectroscopy. Both Co alloys were given the required annealing treatments to produce pure fcc samples,[3] and a nearly pure hcp sample of the Co-Al alloy was produced by cold rolling. Two FeNi$_3$ alloys containing 0.94 and 1.89 a/o Sn, which had been previously studied,[1] were given an extensive annealing treatment (1 week at 480 C, 2 weeks at 450 C, and 4 weeks at 400 C) to improve the degree of long-range order. Sn^{119} spectra of these samples indicated a long-range order parameter of at least 0.9.[4] Sn^{119} spectra of the other Sn-bearing alloys were also obtained and showed no detectable amounts of any Sn-rich phases. X-ray, fluorescence and chemical analysis were performed on selected samples and indicated no significant divergences from the intended compositions and crystal structures. However, both x-ray and Sn^{119} Mossbauer spectra showed a small amount (~10%) of hcp material in the Co-Sn sample.

The Mossbauer spectrometer has been described elsewhere;[1] low temperature spectra were obtained using a continuous flow cryotip (Air Products Co.).

RESULTS

Spectra of several of the alloys obtained at 77 K are shown in Fig. 1. The satellite peaks are not well resolved as they are in Fe alloys;[5] nevertheless, good fits to all spectra were obtained on the assumption of three six-peak magnetic components, assumed to arise from Fe atoms with 0, 1 and 2 impurity (Al or Sn) nearest neighbors (nn). This analysis seems justified both by the simple appearance of the spectra and by previous NMR studies[6,7] which show resolved satellite peaks from different nn configurations but not from differing 2nd nn arrangements. The best fits were generally obtained by constraining the relative intensities of the three components to the values indicated by the binomial probability distribution for random fcc alloys. The Fe^{57} hyperfine fields at 77 K are given in Table I. Spectra were also obtained at room temperature and, for several of the alloys, at 4.2 K; these showed the expected variation of the hyperfine field with temperature[2] (close, but not identical to the temperature dependence of the magnetization).

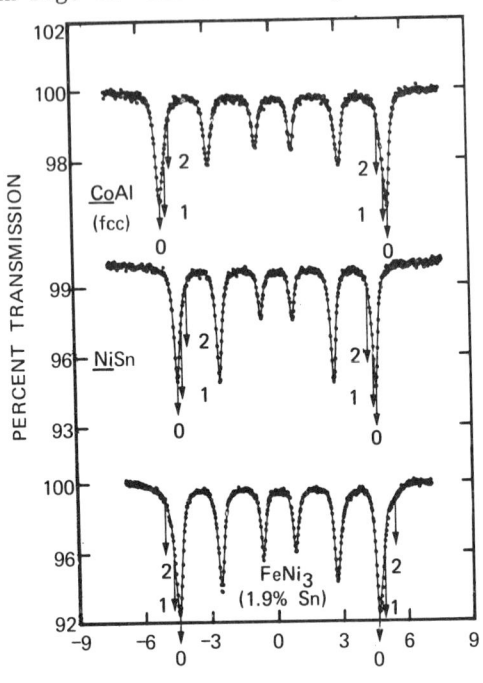

Fig. 1. Spectra at 77 K of several fcc alloys containing 0.3%Fe^{57} and of ordered $FeNi_3$ (1.9%Sn). The arrows denote the outermost peaks of the magnetic components arising from Fe atoms with 0, 1 and 2 impurity nn.

Assuming that the hyperfine field of an Fe atom is governed primarily by its own moment and by the total moment of its nn shell, we write

$$H_m = a\, \mu_{Fe}(m) + b\, M_{nn}(m) \tag{1}$$

where H_m and $\mu_{Fe}(m)$ are the hyperfine field and moment of an Fe atom with m Al or Sn nn, $M_{nn}(m)$ is the total moment of the nn shell containing m impurities, and a and b are the core polarization and 1st nn cep field per μ_B.

The neutron results of Comly et al[8] indicate that in Ni, a wide range of non-magnetic impurities (including Al and Sn) produce magnetic defects of differing amplitude but with approximately the same spatial shape. The defects are widespread, with significant moment losses produced out to about the 4th nn distance. Following the theoretical arguments of Comly et al,[8] we would expect a similar situation to prevail in Co, since its density of states is qualitatively similar to that of Ni. We put the moment decrement of a magnetic atom with an nth nn impurity equal to $\Delta\mu_1 f_n$, where $\Delta\mu_1$ is the decrement produced by a 1st nn impurity and $f_1 = 1.0$; $f_n(n>1)$ is simply the ratio of the moment decrement produced by an nth nn impurity to that produced by a 1st nn impurity. In Ni, these ratios are relatively insensitive to the type of non-magnetic impurity, and from Fig. 4 of Ref. 8, $f_2 \approx 0.52$, $f_3 \approx 0.31$ and $f_4 \approx 0.11$. To test the sensitivity of the results to the shape of the magnetic defect (since experimental data exists only for Ni alloys) we have also considered a more localized model for the moment decrement (Model II) for which $f_2 = 0.33$, $f_3 = 0.11$ and $f_4 = 0.04$.[9]

Table I. Hyperfine Fields (kG) at Fe Nuclei With $0(H_0)$, $1(H_1)$, and $2(H_2)$ Al or Sn nn (T = 77 K)

Alloy	H_0	H_1	H_2
Ni-Al	-278.4 ± 0.2	-270.8 ± 0.5	-260.8 ± 1.6
Ni-Sn	-279.8 ± 0.2	-270.3 ± 0.6	-256.6 ± 3.1
Co-Al(fcc)	-320.3 ± 0.2	-308.0 ± 0.5	-293.3 ± 1.9
Co-Al(hcp)	-320.6 ± 0.2	-306.1 ± 0.6	-290.0 ± 2.1
Co-Sn	-324.3 ± 0.2	-312.5 ± 0.6	-298.3 ± 3.6
FeNi$_3$(1.9 a/o Sn)	-281.9 ± 0.2	-294.5 ± 0.5	-321.2 ± 3.0

Neglecting more distant impurities, the average moment of an alloy M-X (M=Co or Ni, X=Al, Sn, etc.) at solute concentration C can be written as

$$<\mu(C)> = (1-C)\{\mu_M^0 - \Delta\mu_1^M C \sum_{n=1}^{4} Z_n f_n\} \quad (2)$$

where μ_M^0 is the moment per host atom at C = 0, $\Delta\mu_1^M$ is the host moment decrement produced by a 1st nn impurity and Z_n is the number of sites in the nth nn shell. Eq.(2) gives excellent agreement with the experimental μ vs. C curves[10] for both moment decrement models and yields the $\Delta\mu_1^M$ values shown in Table II. It is seen that Model II transfers a larger percentage of the total moment loss produced by a non-magnetic impurity to its nearest neighbor shell.

Table II. Values of $\Delta\mu_1^M$ (M=Ni or Co) for the Extended (Model I) and More Localized (Model II) Moment Deficit Models

	$\Delta\mu_1^M$ (μ_B) (M = Ni or Co)			
Model	Ni-Al	Ni-Sn+	Co-Al*	Co-Sn^x
I	.080	.141	.140	.216
II	.113	.199	.199	.374

+$\Delta\mu_1^{Ni}$ is assumed to be the same in FeNi$_3$(Sn) as in Ni-Sn.
*Negligible difference between fcc and hcp values.
^{x}Estimated from Ni-Al, Ni-Sn, and Co-Al μ vs. C data.

As a first approximation, we assume that the f_n values (but not the $\Delta\mu_1$'s) used for the Co and Ni host atoms are also appropriate for the Fe^{57} impurities in these hosts. In perfectly ordered $FeNi_3$ containing Sn, the nn shell of the Fe atoms would presumably be identical to that of Fe in the $\underline{Ni}$ Sn alloy; it seems plausible, therefore, to adopt the same treatment for this system.

Returning to Eq.(1), we can write the average moment values of Fe atoms in these alloys with 0 and 1 non-magnetic impurity nn as

$$\mu_{Fe}(0) \simeq 3.0 - \Delta\mu_1^{Fe} \, C \, \{6f_2 + 24f_3 + 12f_4\} \quad (3)$$

$$\mu_{Fe}(1) \simeq 3.0 - \Delta\mu_1^{Fe} \{1 + C [6f_2 + 24f_3 + 12f_4]\} \quad (4)$$

where $\Delta\mu_1^{Fe}$ is the moment decrement of an Fe atom produced by a 1st nn impurity. The total moments of nn shells containing 0 or 1 non-magnetic impurities are

$$M_{nn}(0) \simeq 12\mu_M^0 - \Delta\mu_1^M \, C \, \{84 + 48f_2 + 240f_3 + 132f_4\} \quad (5)$$

$$M_{nn}(1) \simeq 11\mu_M^0 - \Delta\mu_1^M [4 + 2f_2 + 4f_3 + f_4 + C \{77 + 44f_2 \quad (6)$$
$$+ 220f_3 + 121f_4\}]$$

where we have taken into account the surroundings of atoms in the nn shell out to the 4th nn distance.

Combining Eq.(1) with Eqs.(3)-(6), and inserting the results of Tables I and II, a and b can now be calculated, provided a value can be determined for $\Delta\mu_1^{Fe}$; however, essentially no experimental data are available for this parameter. In general, it would be surprising if $\Delta\mu_1^{Fe}$ were not $\gtrsim \Delta\mu_1^M$, and magnetization data for the two $FeNi_3(Sn)$ samples indicates a value for $\Delta\mu_1^{Fe} \sim 0.1$ to $0.3\mu_B$.[11] However, rather than assuming a particular value, we have instead evaluated the coefficients a and b as functions of $\Delta\mu_1^{Fe}$ and typical results found using the f_n values of Model I are shown in Fig. 2. Very similar results were found for the more localized form of the magnetic defect, Model II. For the Ni and Co alloys, b changes from negative to positive values at a rather low value of $\Delta\mu_1^{Fe}$, $\sim 0.1\mu_B$. It seems probable, therefore, that b, the nn cep field coefficient for Fe nuclei in those alloys, is positive. This conclusion depends, of course, on whether or not a model which considers only the core polarization and nn shell cep contributions to the hyperfine field is reasonable.

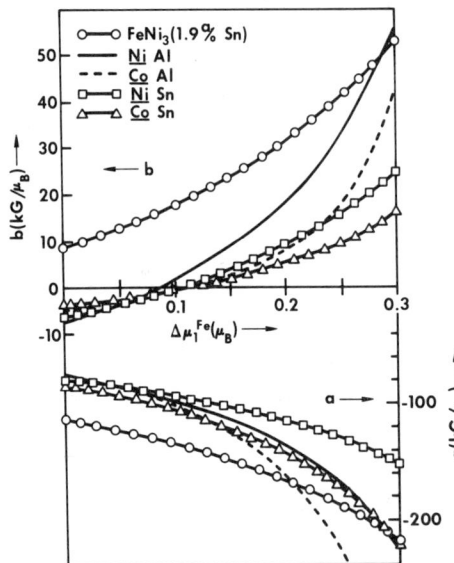

Fig. 2. Dependence of a and b on $\Delta\mu_1^{Fe}$. b(a) is shown by the upper (lower) set of curves and the left (right) hand ordinate. hcp and fcc $\underline{Co}Al$ give very similar results.

For ordered $FeNi_3$, b is

apparently positive even for $\Delta\mu_1^{Fe} = 0$. However, some fraction of the Fe atoms undoubtedly occupy Ni sites, even though the long range order is thought to be quite good in these alloys. Thus, the satellite peaks we have identified as arising from Fe atoms with 1 or 2 Sn nn in an otherwise pure Ni nn shell may instead be due to Fe atoms with one or more Fe nn. Both the observed positive sign of the Sn^{119} fields[1,4] and the absence of any satellite peaks on the low-field side of the main peaks tend to support our identification. Nevertheless, Mossbauer spectra of pure $FeNi_3$ given the same ordering treatment are required to confirm this result.

Similar results were obtained by applying this model to the NMR data of Kobayashi et al[6] on Co alloys. That is, b increases, and in some cases, changes from negative to positive on increasing $\Delta\mu_1^{Co}$ over the range allowable by the measured μ vs. C curves. In this case, however, the results are more sensitive to the assumed form of the magnetic defect. Thus, the extended defect model (Model I) gave values for b of about +5.7 and +10.0 kG/μ_B when applied to CoSi and CoSn, but a value of -4.5 kG/μ_B for CoAl; the more localized form of the magnetic defect (Model II), on the other hand, gives b ≈ +6.9 kG/μ_B for CoAl. These results indicate that the shape of the defect may be more dependent on the particular impurity causing it than has been previously supposed.

REFERENCES

1. G. P. Huffman, F. C. Schwerer and G. R. Dunmyre, J. Appl. Phys. 40, 1487 (1969). G. P. Huffman and G. R. Dunmyre, J. Appl. Phys. 41, 1323 (1970); J. Appl. Phys. 42, 1613 (1971); Proc. 1971 3M Conf., p. 544 (AIP, N. Y., 1972).
2. S. W. Lovessy and W. Marshall, Proc. Phys. Soc. 89, 613 (1966).
3. A. R. Troiano, Trans AIME 175, 728 (1948).
4. The Sn^{119} results will be reported separately.
5. See, for example, T. E. Cranshaw et al, Phys. Letters 21, 481 (1966) and references therein.
6. S. Kobayashi, K. Asayama and J. Itoh, J. Phys. Soc. Japan 21, 65 (1966).
7. T. J. Burch, J. I. Budnick and S. Skalski, Phys. Rev. Letters 22, 846 (1969).
8. J. B. Comly, T. M. Holden and G. G. Low, J. Phys. C 1, 458 (1968).
9. Model II is an approximate compromise between the results of reference (8) and those of Robbins, Claus, and Beck (Phys. Rev. Letters 22, 1307 (1969)) for Ni-Cu alloys.
10. J. Crangle, p. 51, <u>Electronic Structure and Alloy Chemistry of Transition Elements</u> (John Wiley, 1963).
11. These measurements were made by F. C. Schwerer of this Laboratory.

SPIN FLUCTUATIONS AND SPIN-SPIN INTERACTIONS IN AMORPHOUS METALLIC ALLOYS[*]

F. R. Szofran, J. W. Weymouth,
G. R. Gruzalski, and D. J. Sellmyer
Behlen Laboratory of Physics, University of Nebraska
Lincoln, Nebraska 68508

R. Ray and B. C. Giessen
Chemistry Department, Northeastern University
Boston, Massachusetts 02115

ABSTRACT

Electronic and magnetic properties were investigated in the amorphous metallic alloys $Zr_{40}Cu_{60-x}Fe_x$ for $0 \leq x \leq 12$. Evidence was obtained for spin fluctuation phenomena in the host alloy ($x = 0$) and for local moment behavior associated with Fe impurities. Resistance maxima are observed at higher Fe concentrations but, in general, the behavior is different from that exhibited by typical crystalline spin glasses.

INTRODUCTION

The development of techniques for rapidly cooling liquids has led to the discovery of a fairly large number of noncrystalline or amorphous structures. In addition to the possibility of technological applications for these materials, this work has brought new opportunities for studying the problem of electronic and magnetic states in disordered solids in general. Many of the amorphous materials obtained thus far were produced by splat cooling molten mixtures of transition metals and metalloids (P, Si, etc.).[1] However, Ray et al.[2] recently produced amorphous alloys in a system containing only transition metals, namely, Cu and Zr. The present work was motivated by the possibility of doping amorphous $Zr_{40}Cu_{60}$ alloys with small concentrations, x, of Fe, and thereby studying the influence of the disorder of the host on local-moment formation on the Fe atoms, the coupling of Fe moments as x increases, and the general nature of any low-temperature cooperative magnetic state.

EXPERIMENTAL PROCEDURE

Magnetic susceptibility χ at 10.5 kG was measured with a standard Faraday technique between 1.4 and 300 K. Electrical resistivity ρ was determined between 1.4 and 300 K by attaching current and potential leads to thin strips of the samples with Ag

[*]Research supported at the University of Nebraska by the National Science Foundation and at Northeastern University by the Office of Naval Research.

paint. The samples were produced by melting 100-250 mg of alloy in a modified arc-melting furnace equipped with a copper plunger; they were then quenched into foils of ~50 μ thickness by rapidly propelling the plunger onto the melted alloy. Cooling rates of 10^6-10^7 K/s were achieved. The nominal Fe concentrations were verified with x-ray fluorescence techniques and the amorphous nature of the alloys was confirmed by x-ray diffraction.

RESULTS AND DISCUSSION

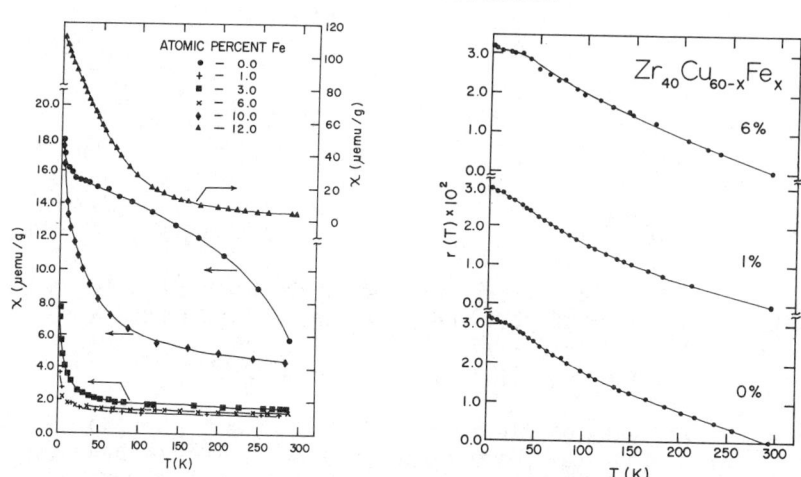

Fig. 1. $\chi(T)$ for $Zr_{40}Cu_{60-x}Fe_x$. Fig. 2. $r(T)$ for dilute alloys.

Figure 1 shows the general behavior of χ for selected samples in the field-cooled state. As a test of the effect of field cooling, measurements at 1.5 kG were made on the x = 12 sample cooled in zero field. No qualitative differences from Fig. 1 were observed in the low temperature behavior. The $Zr_{40}Cu_{60}$ host is rather strongly paramagnetic; its susceptibility is larger than that of Pd over the whole temperature range. χ for the Fe-doped alloys cannot be written simply as

$$\chi(x,T) = \chi_o + C(x)(T-T_o)^{-1} \qquad (1)$$

where χ_o is due to the host and the Curie-Weiss term is due to Fe local-moments. This follows since χ first decreases with x but then increases for the large x values. Also, the concentration dependence in general is not simple.

Resistivity data for several samples is shown in Fig. 2, where $r(T) \equiv [\rho(T)/\rho(296 \text{ K}) - 1]$. The absolute values of ρ for all the samples are $\simeq 350$ μΩcm. Except for the region below ~20 K, all of the samples appear to contain a T^2 term in $r(T)$ to about 50 K with a tendency towards a linear behavior at higher temperatures. $r(T)$ and $\chi(T)$ for the $Zr_{40}Cu_{60}$ sample are plotted vs. T^2 in Fig. 3

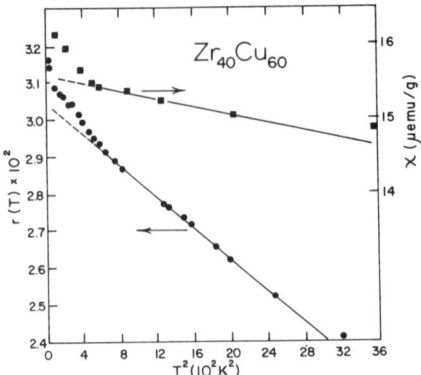

Fig. 3. χ and r vs. T^2 for the host alloy.

Fig. 4. $(\Delta\chi)^{-1}$ vs. T for $Zr_{40}Cu_{57}Fe_3$.

as a test of the low T behavior. If the upturn below 20 K is neglected, it is seen that $\rho(T)$ and $\chi(T)$ can be represented by

$$\rho(T)/\rho(0) = 1-(T/\theta)^2; \quad \chi(T)/\chi(0) = 1-(T/\theta')^2 \qquad (2)$$

for $T \lesssim 50$ K. The data of Fig. 3 give $\theta = 697$ K and $\theta' = 240$ K. $\rho(T)$ vs. T^2 plots for other samples (x = 1, 3, 4, 5) show behavior similar to Fig. 3, each with an upturn at low T and θ values between 700 and 840 K. The similarity of the anomaly below 20 K in the Fe-doped and undoped samples suggests that the upturn seen in the host alloy may be due to a small concentration of Fe impurities, perhaps present in the starting materials. In fact, x-ray fluoresence measurements did show the presence of ~0.2 at.% of Fe in the nominally pure host. $\rho(T)$ behavior very similar to that seen in our x = 0 sample has been reported for amorphous $(V_3Ni_{27}Pt_{70})_{75}P_{25}$ by Hasegawa.[3] In this latter alloy, also, the low-temperature upturn in $\rho(T)$ was attributed to scattering from local moments on V atoms--superimposed on a $1-(T/\theta)^2$ spin fluctuation background. If it is assumed, then, that Fe manifests itself as a local moment with a low Kondo temperature, the general behavior shown by $\rho(T)$ between 1.4 and 300 K (Fig. 2) must be intrinsic to the $Zr_{40}Cu_{60}$ host. A possible explanation is that there are localized spin fluctuations (LSF) arising from partially filled Zr d states which lead to the observed behavior. In fact, Zuckermann[4] has developed an LSF theory which predicts $\rho(T)$ behavior quite similar to that seen for the $Zr_{40}Cu_{60}$ sample over the whole temperature range. However, we are obviously not able to determine the phonon scattering contribution to ρ so a quantitative comparison with the LSF theory is impossible. In addition there appear to be no reliable LSF calculations which predict a $1-(T/\theta')^2$ dependence for the susceptibility.[5]

Despite the complex concentration dependence of χ seen in

Fig. 1, it seems clear that there is a local-moment-type contribution for the x = 3, 10, and 12 samples. Therefore, in order to obtain some information on the nature of the moments responsible for this contribution, we have analyzed the 3% Fe sample data in terms of a Curie-Weiss equation such as Eq. (1). A plot of $(\Delta\chi)^{-1} = (\chi-\chi_o)^{-1}$ vs. T is shown in Fig. 4, where χ_o = 1.55 µemu/g. Analysis of Fig. 4 gives T_o = -3.5 ± 1 K, a paramagnetic moment of μ_{eff} = 0.80 μ_B, and a spin value of S = 0.14 assuming g = 2. These small values of μ_{eff} and S, as well as the decrease of χ as x increases from 0, are both suggestive of an exchange-enhanced host with a strong <u>negative</u> polarization of neighboring Zr atoms by the Fe impurities. A similar suggestion has been made for antiferromagnetic coupling in Pd(Cr) alloys,[6] and for ferromagnetic coupling (positive polarization) in Pd(Fe) alloys. Among the possibilities for explaining the small value of $T_o \simeq$ -3.5 K are: (1) the spin clusters postulated above simply have a low Kondo temperature, or (2) if we are already in a regime where there are significant Fe-Fe interactions, it is possible that the theory of Kok and Anderson,[7] which predicts $T_o \simeq$ 0 for amorphous alloys, is operative.

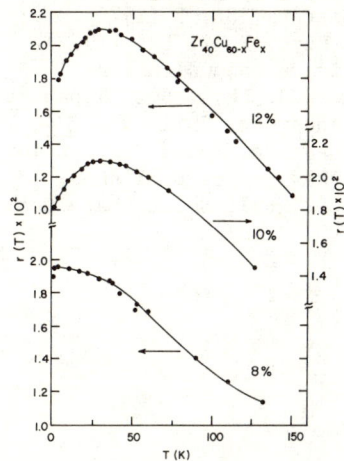

Fig. 5. r(T) for concentrated alloys.

Figure 5 shows the ρ(T) data below 150 K. There are resistivity maxima at $T_{max} \simeq$ 30 K for x = 10 and 12 and no maximum above 4 K for x = 8. It is remarkable that no anomalies such as maxima in χ are observed for these same samples in Fig. 1. This behavior is exactly the converse of that observed in amorphous $Cr_5Pd_{75}Si_{20}$ alloys where susceptibility maxima were seen at T_{max} but the resistivity showed no anomaly near this temperature.[8] The magnetic properties of both of these amorphous alloys are therefore quite distinct from a normal crystalline spin glass such as Cu(Mn) where maxima in ρ and χ normally are found together and $T_{max} \propto$ x. It would appear that a new model for the spin-spin interactions, probably not based on RKKY interactions, is required for understanding the magnetic state of these alloys at low temperatures.

It may be mentioned that room temperature Mössbauer studies of the present alloys yield quadrupole interaction doublets typical of many different local environments with similar electric field gradients;[9] analogous results were found in other amorphous Fe alloys above the Curie temperature.[10] It is clear also that further measurements including Mössbauer experiments below T_{max} and quench-rate studies will be useful for providing information on the effects of host short-range order, local environments, and spin clustering in these alloys.

We wish to thank R. Hasegawa and M. J. Zuckermann for helpful discussions and W. R. French for performing the x-ray fluorescence measurements.

REFERENCES

1. B. C. Giessen and C. N. J. Wagner, in Liquid Metals, S. Beer, ed., 1971.
2. R. Ray, B. C. Giessen, and N. J. Grant, Scripta Met. 2, 357 (1968).
3. R. Hasegawa, Phys. Letters 38A, 5 (1972).
4. M. J. Zuckermann, J. Phys. F. 2, L25 (1972).
5. M. J. Zuckermann, private communication.
6. See, e.g., H. Nagasawa, J. Phys. Soc. Japan 28, 1171 (1970).
7. W. C. Kok and P. W. Anderson, Phil. Mag. 24, 1141 (1971).
8. C. C. Tsuei and R. Hasegawa, Solid State Comm. 7, 1585 (1969).
9. B. C. Giessen and R. Ray, to be published.
10. T. E. Sharon and C. C. Tsuei, Solid State Comm. 9, 1923 (1971).

ELECTRON SPIN RESONANCE OF METALLIC Si:P WITH IRON, A VARIABLE ELECTRON CONCENTRATION ALLOY*

T. A. Kennedy, R. T. Longo† and J. H. Pifer
Rutgers University, New Brunswick, NJ 08903

ABSTRACT

An X band ESR study has been made of heavily-doped single crystal Si (3.-30. x 10^{18} P/cm^3) with 3 x 10^{16} Fe/cm^3. The CESR (g = 1.999) seen in samples without Fe is broadened by the addition of Fe. Above 20K this broadening is explained by spin-flip scattering through spin-orbit coupling to Fe. As the temperature decreases below 20K the line broadens further and g increases slightly. The low temperature excess linewidth increases linearly with electron concentration up to about 10^{19} P/cm^3. The data is analyzed assuming exchange coupling between the conduction electrons and the Fe d shell. In a strong coupling interpretation the g value of the Fe resonance is 2.00 instead of g = 2.07 observed for Fe^0 in insulating Si. This implies that the electron configuration or site of the Fe has changed. In a weak coupling interpretation the broadening is due to a distribution of g shifts first order in the exchange and proportional to the d-spin susceptibility. The relative merits of these two interpretations are discussed.

INTRODUCTION

In n-type semiconducting silicon, Fe^0 impurities occupy interstitial positions. The 3d^8 electron configuration produces a single ESR line located at g = 2.07.[1] When Si is doped with greater than 3 x 10^{18} P/cm^3, the material becomes metallic.[2] The exact nature of both the metallic transition and the electronic states in the resulting impurity band is not understood. At 2 x 10^{19} P/cm^3 the Fermi level rises into the Si conduction band. Above 3 x 10^{18} P/cm^3 a single narrow "free carrier" ESR line is observed at g= 1.9987. Below 20K the width is T-independent and increases linearly with n.[3]

Diffusing Fe into heavily doped Si makes it possible to study the interaction of a localized moment with conduction electrons of variable density. We describe here the results of such a study. In the following sections we present data on the T and n dependence of the ΔH and g of the ESR observed in (Si:P):Fe. We then discuss the data in the limits of strong and weak coupling between the d electrons and the conduction electrons. Although we understand some aspects of the data, no single interpretation explains all the feature observed.

---

* Work supported in part by the National Science Foundation.
† Present address: General Dynamics, Pomona, CA.

EXPERIMENT

In this section we describe the samples, the method of Fe diffusion, and the ESR spectra. Single crystal boules of Si:P were obtained from General Diode Corporation, Boston, MA. Flat samples (1x3x5 mm^3) were cut perpendicular to the <111> axis. The electron concentrations we quote were determined assuming $n = 1/R_H ec$ from room-temperature Hall effect measurements on other samples cut from the boules.

After the samples were etched with CP4, ESR spectra were taken at 77, 4.2 and 1.4K. Then the Fe diffusion was carried out with the sample and a section of pure Fe wire in an open quartz tube at 1250°C for 45 minutes.[1,4] Helium gas was passed through the tube during the diffusion. The sample was quenched as rapidly as possible in ethylene glycol. The entire process was kept as uniform as possible from sample to sample in an attempt to introduce a constant Fe concentration. After etching again the sample was mounted in a 9Ghz ESR cavity and cooled to 77K. We have studied and will present elsewhere the effects on the ESR spectrum of room temperature annealing of the Fe and of oxygen present in the samples.[5]

The spectrum of a (Si:P):Fe sample consists of a strong single absorption near the position of the Si:P CESR. The T-dependence of ΔH for four Fe-doped samples and sample 2.6E18 without Fe is shown in Fig. 1. For T > 20K, ΔH increases with T. Although there is a constant difference between ΔH for Fe-doped and pure samples, the T-dependence is the same for both. This increase in ΔH at 77K, $\delta H(77K)$, is nearly the same for different n. See Table I.

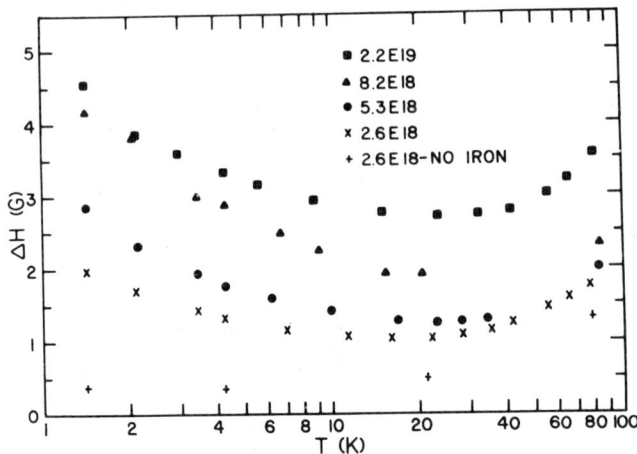

Figure 1. Temperature dependence of the linewidth for 4 (Si:P):Fe samples and one (Si:P) sample.

For T < 20K, ΔH in (Si:P):Fe is markedly different from ΔH in Si:P in magnitude and T-dependence. In all cases the Fe sample linewidths increase with decreasing T, whereas without Fe linewidths are independent of T. The excess ΔH at 1.4K, δH(1.4) in Table I, increases for samples below 10^{19} P/cm^3, but this trend does not continue for 2.2E19.

For T < 20K the g-values for (Si:P):Fe show small T-dependent changes from Si:P. For the 2.2E19 sample, Si:P and (Si:P):Fe both have g = 1.9986 (± 0.0001) at 77K. At 1.4K the Fe sample has g = 1.9991 while the pure sample has g = 1.9987.

For 2.2E19 the low T asymmetry of the line produced by the skin effect is close to that expected for very slowly diffusing electrons, but for lower n the asymmetry is considerably smaller and T-dependent.

Table I. ESR Linewidths

Linewidth(G) \ n	2.6E18	5.3E18	8.2E18	2.2E19
No Iron				
ΔH(77K)	1.31	1.51	1.72	3.05
ΔH(1.4K)	0.39	0.65	1.24	2.22
With Iron				
ΔH(77K)	1.77	1.98	2.32	3.57
ΔH(1.4K)	1.97	2.85	4.17	4.56
Excess				
δH(77K)	0.46	0.47	0.60	0.52
δH(1.4K)	1.58	2.20	2.93	2.34
Low T Excess Δ(δH) = δH(1.4K) - δH(77K)	1.12	1.73	2.33	1.82

DISCUSSION

The theory of the ESR of conduction electrons (s) exchange coupled to localized moments (d) is well understood.[7,8] The dynamics are found to depend upon the size of the d to s cross relaxation rate $1/T_{ds}$ relative to the s spin lattice relaxation rate $1/T_{sl}$ and the frequency splitting of the two resonances $\Delta g \omega_o/2$ where $\Delta g = g_d - g_s$, and ω_o is the angular frequency of the rf field. If $1/T_{ds}$ is much smaller than these rates, weak coupling prevails, with separate s and d resonances whose width and resonant frequencies are perturbed by the interaction. $1/T_{ds}$, the Korringa rate, is quadratic in $J\rho$ where J is the exchange coupling and ρ is the density of states at the Fermi level, while the g shift is linear in J and χ_d, the d spin susceptibility, and does not contain ρ. For weak exchange the g shift predominates.

If $1/T_{ds}$ is large, strong coupling prevails, and only a single composite s-d line is observed. The g-value is given by $g = (\chi_s g_s + \chi_d g_d)/(\chi_s + \chi_d)$ and ΔH is given by a similar weighted average of $1/T_{sl}$ and $1/T_{dl}$, the direct spin lattice relaxation rate of the local moment. For most alloys J cannot be calculated accurately enough to predict which limit will occur.

The important experimental facts from the (Si:P):Fe data for a particular n are the increase observed at all T in ΔH compared to the undoped ΔH, the T-dependence of ΔH for T < 20K, and the smallness of the g-shift from the Si:P CESR g-value. Any explanation must account for the observed dependences of these parameters on n.

For T > 20K, we attribute the T-independent excess linewidth to spin-flip scattering of conduction electrons from the Fe impurities through spin-orbit coupling. Assuming that Fe occupies one type of site with a particular cross-section for spin-flip scattering, the near constancy of $\delta H(77K)$ for different n (See Table I) implies that the Fe concentration is nearly constant in all our samples. This conclusion is important in the analysis of the low T results.

For T < 20K, we first discuss the data in the limit of weak coupling which would be most likely to apply to samples with low n. The requirement for weak coupling is $1/T_{ds} << \Delta g \omega_0 / 2$. For Fe^0 at 1.4K this means $J\rho << .02$. Estimating the density of states in the impurity band to be $\approx 10^{-2}$ (ev atom spin)$^{-1}$, weak coupling requires J<<2ev. This is probably satisfied, since in a metal J is typically .1 to 1 ev. Furthermore we do observe a weak resonance at g = 2.07 in sample 2.6E18 where the large skin depth offers the best chance of seeing the Fe resonance in a single crystal. Assuming weak coupling one can place a more stringent limit on J, since if $1/T_{ds} > 1/T_{sl}$, one would see an additional broadening of the CESR line that increases with T. We do see a broadening but it decreases with T. Thus we conclude that $1/T_{ds} < 1/T_{sl}$, or $J\rho < .002$ and J < .2ev.

The observed line broadening can be explained using weak coupling by assuming that the peaking of the s wave function near P ions leads to a J distribution P(J) for different Fe atoms. The low T broadening is then due to a distribution of g shifts produced by P(J). In the weak coupling limit a given value of J shifts the CESR line by an amount linear in J, and P(J) broadens the line proportionally to the width of P(J). The T^{-1} dependence of the g shift, which arises from the Curie behavior of χ_d, produces a broadening as T decreases. Note that the g shift and line broadening are intimately connected in this picture. For sample 2.6E18 we estimate the width of P(J) from $\Delta(\delta H)$ to be .1 ev. Numerical calculation of the lineshape and g shift for a Gaussian P(J) shows that it is possible to explain the T-dependence of the CESR in this picture.

We now consider the weak coupling explanation of the n dependence of the low T broadening. Table I gives the excess linewidth δH of the Fe doped samples over the undoped samples at 77K and 1.4K. δH (77K) is approximately constant for all n but $\delta H(1.4K)$ depends on n. If $\delta H(77K)$ is subtracted from $\delta H(1.4K)$ one obtains $\Delta(\delta H)$, the width that we attribute to P(J). We find $\Delta(\delta H)$ doubles as n increases from 2.6E18 to 8.2E18; then drops again for 2.2E19, where the Fermi level enters the conduction band. For constant Fe concentration the only factor that can produce this n dependence is a change in P(J). It is not clear that a distribution can be found that produces the observed broadening consistent with the very small g shift of the entire line and the observed lineshape. Assuming that the model of weak coupling including a distribution of J values proves to be correct, a fit to the observed lineshape for all n may provide information on the nature of the metallic state in Si:P.

Since our high n sample approaches metallic electron densities and has electronic properties characteristic of the Si conduction band,[2] one might expect strong coupling to be the proper limit as is the case for many iron-group impurities in metallic hosts. We therefore start with our 2.2E19 sample and discuss a strong and intermediate coupling model.

The 2.2E19 data has the lineshape expected for and has been fit to strong coupling, yielding g_d = 1.9995 and $1/\gamma T_{d1}$ = 6.3G. For n <2.2E19 the data show T and n dependent lineshapes.[8] These variations have been shown to occur for intermediate strength exchange where the s and d resonances can act somewhat independently and where the g-values are close enough that the two resonances are not resolved. Calculations show that the n <2.2E19 data may be fit with g_d = 2.000, $1/\gamma T_{d1} \sim 6G$, and $J\rho \sim .002$. Using $\rho = 10^{-2}$ (ev-atom-spin)$^{-1}$, $J\rho \sim .002$ implies that $J \sim 0.2ev$.

Thus our data is consistent with the picture of a local moment with g_d = 2.000±.001 and an exchange interaction that increases from intermediate to strong as n increases. Since g_d differs from the value of 2.07 observed for interstitial Fe^0, a second site or charge state for Fe is implied. Specific information about this site is lacking. This interpretation implies that an analysis of the high T spin-flip scattering must include the possibility of different Fe states with different scattering cross sections.

In this model the low T broadening arises from $1/\gamma T_{d1}$. Although low T broadening has been observed in other dilute alloy systems and attributed to $1/\gamma T_{d1}$,[10] the mechanism for direct local moment to lattice relaxation is not understood. It is interesting to speculate that the broadening attributed to $1/\gamma T_{d1}$ may be due to a distribution of local moment g values produced by inhomogeneities or nonequivalent local moment sites.

REFERENCES

1. G. W. Ludwig and H. H. Woodbury, Solid State Phys. 13, 223 (1962).
2. M. Alexander and D. F. Holcomb, Rev. Mod. Phys. 40, 815 (1968).
3. S. Maekawa and N. Kinoshita, J. Phys. Soc. Japan 20, 1447 (1965).
4. C. B. Collins and R. O. Carlson, Phys. Rev. 108, 1409 (1957).
5. W. Kaiser, Phys. Rev. 105, 1751 (1957).
6. J. H. Pifer (unpublished).
7. D. C. Langreth and J. W. Wilkens, Phys. Rev. B 6, 3189 (1972) and J. Dupraz, B. Giovannini, R. Orbach, J. D. Riley and J. Zitkova in Magnetic Resonance, edited by C. K. Coogan et al. (Plenum Press, New York, 1970), p. 197.
8. J. H. Pifer and R. T. Longo, Phys. Rev. B4, 379 (1971).
9. J. R. Asik, M. A. Ball, E. K. Cornell and C. P. Slichter in Magnetic Resonance, edited by C. K. Coogan et al. (Plenum Press, New York, 1970), p. 187.
10. P. Monod and S. Schultz, Phys. Rev. 173, 645 (1968).

ANISOTROPIC KONDO RESISTANCE IN Fe DOPED NbSe$_2$*

R. C. Morris[†], B. W. Young[†] and R. V. Coleman
University of Virginia, Charlottesville, Va. 22901

ABSTRACT

The resistivity and magnetoresistivity of NbSe$_2$ single crystals doped with iron have been measured in the temperature range 1-70 K. The crystals were grown with iron concentrations ranging from 0.25% to 5.0% and in zero magnetic field a strong resistance minimum is observed in the 10-15 K range followed by a relative maximum in the 1-5 K range. For magnetic fields perpendicular to the layers the increase in resistivity below the minimum is enhanced while for magnetic fields parallel to the layers the resistivity increase can be completely quenched. The detailed behavior is dependent on the iron concentration and has been studied in fields up to 151 kOe. Hysteresis in the temperature dependent resistivity has been observed and along with other effects suggests that some spin ordering may take place.

INTRODUCTION

The effect of dilute magnetic impurities on the electrical properties of metals has been studied in a number of alloy systems in recent years and the results have led to numerous theoretical papers on the subject. Reviews of the work can be found in several articles[1,2,3]. The majority of the systems experimentally investigated have been composed of noble metal or noble metal alloys doped with a dilute concentration of a transition metal element capable of forming a localized magnetic moment. Fe, Cr, and Mn appear to form local moments readily and have been studied extensively in the experimental work. One of the characteristic properties of such dilute magnetic alloys is a resistivity minimum observed at low temperatures, and referred to as the Kondo effect. Kondo[4] made the first breakthrough in understanding this behavior when he assumed that the localized magnetic moment of the impurity atom interacts with the conduction electrons via the s-d exchange interaction. This results in a spin-flip scattering of the conduction electrons which contributes a logarithmically divergent term to the resistivity as the temperature approaches zero and thus a resistivity minimum is observed. This paper reports on new observations of the Kondo effect in single crystals of the layer structure NbSe$_2$ doped with Fe.

---

*Work supported by the U. S. AEC and ONR.

[†]Present address: Florida State Univ., Tallahassee, Fla. 32306

EXPERIMENTAL PROCEDURE

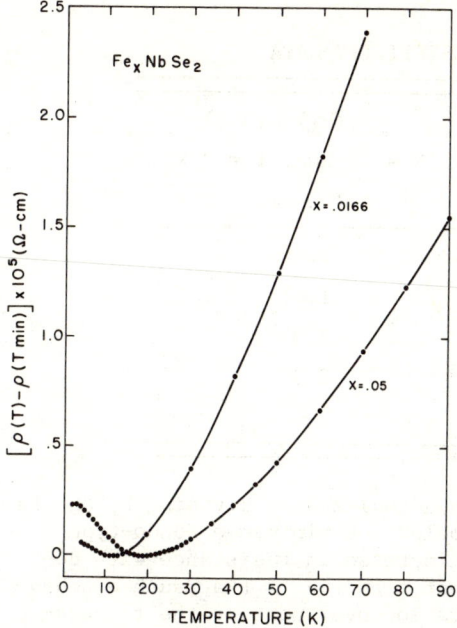

Fig. 1. Variation of the resistivity with temperature of Fe_xNbSe_2 for $x = .0166$ and $x = .05$.

Compounds of Fe_xNbSe_2 ($x = .0025, .0166, .0332, .05$) were prepared from the elements by reacting the appropriate mixtures in sealed quartz tubes. Single crystals of each compound were grown by the iodine vapor-transport reaction method.[5] Crystals for all Fe concentrations grow in a thin platelet geometry with shiny, mirrorlike surfaces and samples were prepared by slicing into bar-shaped specimens.[6]

$NbSe_2$ is a layer compound consisting of trigonal prismatic Se-Nb-Se sheets with a two layer packing sequence. In the weak binding gap between adjacent Se layers both tetrahedrally and octahedrally Se coordinated holes exist. It has been shown[7] that Fe enters the octahedral holes up to a maximum concentration of approximately $x = 1/3$. For our samples no direct check was made of the stochiometry with respect to either the position or the quantity of Fe in the single crystals, but a comparison of the superconducting transition temperature to that for Fe_xNbSe_2 powders has been made.[8] Our pure $NbSe_2$ crystals have transition temperatures of approximately 7.3 K with a width of 0.3 K. 0.25% Fe crystals had transition temperatures in the range 5.6 - 5.8 K with a width of about 0.5 K. For 0.5% Fe crystals the transition temperature was ~ 4.0 K with a width of 0.5 K. These numbers are in essential agreement with the values for powdered samples reported earlier.[8] Samples with higher Fe concentrations showed no signs of superconductivity down to 1.5 K. The indication is that the Fe concentration in the single crystals is essentially that of the starting powdered material.

RESULTS

Fig. 1 shows the temperature dependence of resistivity for a 1.66% and a 5% Fe doped sample. The measuring current was parallel to the layers for all results presented in this paper. At low temperatures a resistivity minimum is evident and it moves to higher temperature as the Fe concentration is increased. At temperatures above the minimum more concentrated samples show a less rapid increase in resistivity as expected. Table I tabulates T_{min} and ρ_{min} for the different Fe concentrations studied.

TABLE I Fe_xNbSe_2 RESISTIVITY DATA

x	T_{min} (K)	ρ_{min} ($\mu\Omega$-cm)	$\rho(H_\perp)-\rho(H_{//})$ [H = 72 kOe, T = 2 K] ($\mu\Omega$-cm)
.05	20	135.1	1.34
.0332	15	76.0	1.61
.0166	11	65.2	2.05
.005	9	55.0	2.32
.0025	-	11.2	-

In Fig. 2 the curves have been normalized by dividing by the Fe concentration and at temperatures below the minimum a Kondo-type theory would predict a logarithmic increase in resistance with decreasing temperature.[1] It is evident from Fig. 2 that such a behavior is followed by each Fe concentration over part of the temperature range. For the 0.5% Fe sample the low temperature data is limited by the onset of superconductivity. As the concentration is increased from 0.5% Fe to 3.32% Fe the maximum resistivity change per Fe atom decreases which is the expected behavior for Kondo scattering when interaction between the impurity Fe atoms commences at higher concentrations.[1] Such an interaction would also be consistent with the recent interpretation of the anomalous behavior of the transition temperature in $NbSe_2$ as a function of Fe concentration.[8] In addition for $Fe_{.26}NbSe_2$ antiferromagnetic ordering has been reported[7] at 120 K. However the 5% Fe samples do not continue this trend, a result which is unexplained. A Kondo effect at reasonably high concentrations has also been observed in La-Ce[9] and Y-Ce[10] alloys.

Further study of these effects has been made in transverse applied magnetic fields up to 151 kOe as shown in Figs. 3-5. The normalized resistance

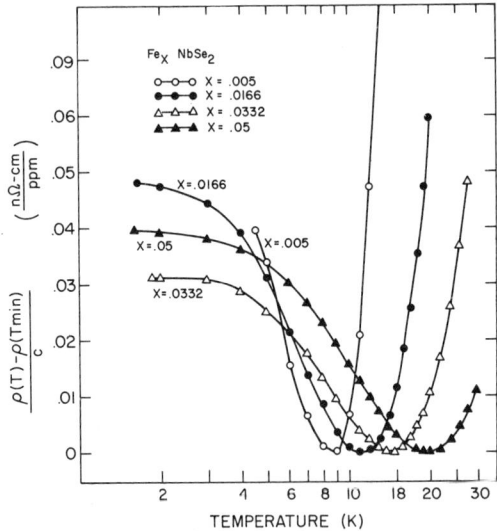

Fig. 2. Change in the resistivity for Fe_xNbSe_2 below the minimum divided by the impurity concentration in parts per million plotted versus log temperature.

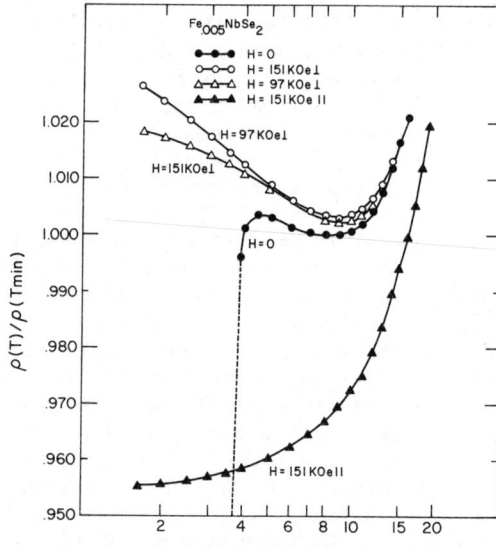

Fig. 3. Normalized resistivity for $Fe_{.005}NbSe_2$ versus log temperature. Data for magnetic fields parallel and perpendicular are shown.

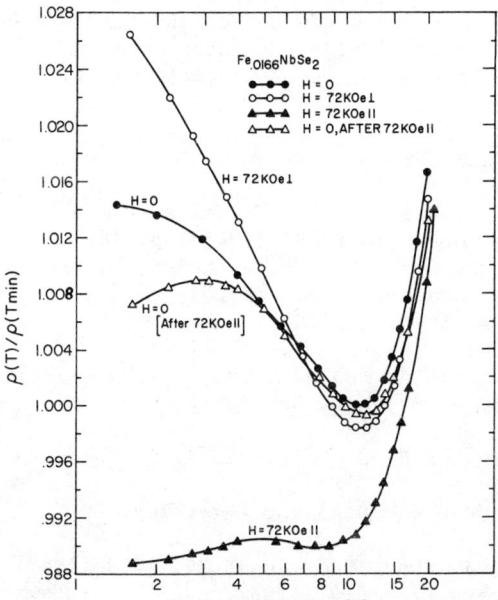

Fig. 4. Normalized resistivity and magnetoresistivity for $Fe_{.0166}NbSe_2$ versus log temperature.

$\rho(T)/\rho(T_{min})$ has been calculated using the resistance at the minimum measured in zero field. For field induced ordering a negative magnetoresistance is expected, since the magnetic field Zeeman splits the spin states of the impurity and makes it energetically more difficult for spin-flip scattering.[1] All doped samples show a negative magnetoresistance for magnetic fields parallel to the layers. (Negative magnetoresistance corresponds to a decrease in normalized resistance with field.) The relative size of the effect is greatest for 0.5% Fe concentration and least for 5% Fe concentration. The negative magnetoresistance appears to be much larger than in iron doped copper and does not appear to follow the form $-(\Delta\rho/\rho) \sim H^n$ observed for Kondo systems. Possible exceptions are the 0.5% Fe samples which follow H^n with $n = 1.19$.

When the field was oriented perpendicular to the layers of the crystal a positive magnetoresistance was observed at the lower temperatures for all samples. (An increase in normalized resistance with field.) This positive magnetoresistance has the effect of extending the temperature range over which an approach to a logarithmic temperature dependence is observed.

For the 0.5% Fe samples the positive magnetoresistance at the lowest temperatures peaks at around 97 kOe and afterwards decreases. The effect at high temperatures remains positive for the same fields. As Fig. 4

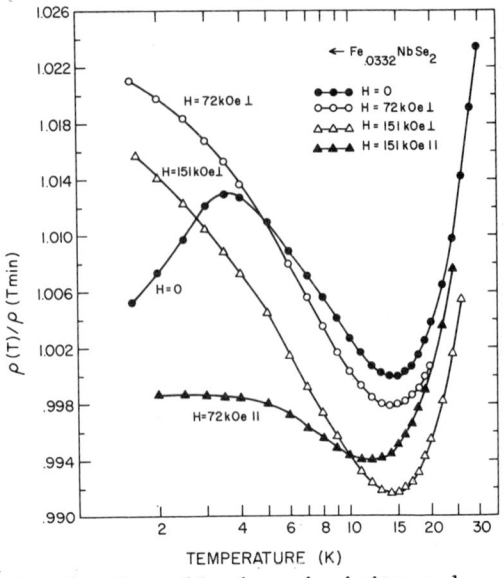

Fig. 5. Normalized resistivity and magnetoresistivity for $Fe_{.0332}NbSe_2$ versus log temperature.

shows a similar effect is observed in 1.66% Fe sample, however, here the magnetoresistance becomes negative at higher temperatures also. The behavior suggests that for magnetic fields perpendicular to the layers two competing processes exist, one giving positive magnetoresistance and the other negative with the negative part being the more concentration dependent. Table I gives a value for the total change in resistivity between parallel and perpendicular field orientations at 72 kOe and 2 K. The effect is largest for the lowest concentration.

As seen in Figs. 4 and 5 for the .0166 and .0332% alloys a decrease in resistivity is sometimes observed at 0-field at the lowest temperatures and is dependent on the application and removal of the magnetic field. This hysteresis effect is sample dependent and could reflect a saturation of the Kondo effect and a return to the normal resistive behavior.[11] It could also reflect a magnetic ordering of the Fe atoms or a clustering of Fe atoms particularly in the more concentrated alloys.

REFERENCES

1. J. Kondo, Solid State Physics, Vol. 23, Ed. F. Seitz, D. Turnbull, and H. Ehrenreich, Academic Press (1969), p. 184.
2. A. J. Heeger, Solid State Phys. Vol. 23, Ed. F. Seitz, D. Turnbull, and H. Ehrenreich, Academic Press (1969), p. 184.
3. K. Fischer, Phys. Status Solidi 46, 11 (1971).
4. J. Kondo, Progr. Theoret. Phys. (Kyoto) 32, 37 (1964).
5. R. Kershaw, M. Blasse, and A. Wold, Inorg. Chem. 6, 1599 (1967).
6. R. C. Morris, R. V. Coleman and Rajendra Bhandari, Phys. Rev. B5, 985 (1972).
7. J. M. Voorhoeve-van den Berg and R. C. Sherwood, J. of Phys. Chem. Solids 32, 167 (1970).
8. J. J. Hauser, M. Robbins, and F. J. Disalvo, Phys. Rev. B8, 1038 (1973).
9. T. Sugawara and S. Yoshida, J. Phys. Soc. Japan 24, 1399 (1968).
10. A. S. Edelstein, Phys. Letters 27A, 614 (1968).
11. P. Monod, Phys. Rev. Letters 19, 1113 (1967).

NMR AND SUSCEPTIBILITY STUDIES OF Pt-Rh ALLOYS[*]

H. T. Weaver and Rod K. Quinn
Sandia Laboratories, Albuquerque, New Mexico 87115

ABSTRACT

Low temperature ($T \approx 1 - 4°K$) measurements of the Knight shift and spin-lattice relaxation time of ^{103}Rh and ^{195}Pt in $Pt_{1-x}Rh_x$ ($0 \leq x \leq 1.0$) have been carried out using transient NMR methods. Magnetic susceptibilities of the alloys were also measured as a function of temperature. The ^{103}Rh Knight shift (K) and relaxation rate increase monotonically as x decreases, but the ^{195}Pt shift and the bulk susceptibility exhibit an extremum near $x = 0.2$. As in the related alloy Pd Rh, the Rh local susceptibility is larger than for Pt in the limit $x \rightarrow 0$ although the magnetization carried by the host (Pt) accounts for $\approx 70\%$ of the bulk susceptibility. The ratio of local susceptibilities, $\chi(Rh)/\chi(Pt)$, which is determined by the respective K, decreases with increasing x, with $\chi(Pt)$ becoming larger than $\chi(Rh)$ at $x \approx 0.3$. For $x \leq 0.2$ the ^{103}Rh Korringa product is constant with $\sim 3\%$ orbital shift indicated.

The Fermi energy for the elemental metals palladium and platinum is located near a peak in the density of d-states.[1,2] As a consequence, these metals possess large electronic specific heats and susceptibilities[3] which show strong variation with changes in the average electron-to-atom ratio produced by alloying and changes in the temperature. Qualitatively, the bulk properties of some Pd and Pt alloys can be accounted for using rigid band models, but microscopic measurements, such as NMR, always show intra-atomic effects. Generally, there is more interest in Pd due to its highly magnetic nature, but NMR measurements on Pd nuclei are severely limited due to quadrupole interactions. However, several NMR studies of Pd alloys have been carried out using the second constituent as the resonance probe. Platinum, on the other hand, exhibits many of the properties of Pd and possesses a convenient isotope for NMR experiments. We have carried out NMR and bulk susceptibility measurements on a series of $Pt_{1-x}Rh_x$ alloys that display anomalous magnetic behavior and which, in addition to their intrinsic interest, are of relevance to the Pd problem.

All NMR experiments were carried out at 10 MHz in the temperature range $1.2 - 4.0°K$ using a transient, phase-coherent spectrometer. Knight shifts (K) were determined by comparing the frequency-to-field ratios for maximum echo intensity to tabulated reference values.[4] Spin-echo line profiles were also generated. Spin-lattice relaxation

[*]Work was supported by U. S. Atomic Energy Commission.

times (T_1) were determined from echo recovery rates following saturation. Single pulse heating together with fast relaxation rates for ^{195}Pt prevented accurate measure of T_1 for platinum, but the longer ^{103}Rh T_1 values allowed their determination. Phase memory (T_2) plots were determined only for ^{195}Pt from which the nuclear-nuclear indirect exchange constant (J) was determined.[5] Bulk susceptibilities (χ_B) were measured using the Faraday technique over the temperature range 4-300°K. However, magnetic impurities masked the measurements at low temperatures so that an extrapolation from the 77-300°K range is used for the low temperature susceptibility.

Table I. Summary of NMR and susceptibility data for $Pt_{1-x}Rh_x$. The numbers in parentheses represent the error in the last quoted digit.

x	χ_B (10^{-4} emu/mole)	K(Rh) (%)	T_1T(Rh) (sec°K)	K(Pt) (%)	$J/2\pi$ (kHz)
0.00	2.04	-	-	-3.54	-
0.03	2.56(2)	-3.4 (2)	0.9(1)	-3.6(2)	5.2(7)
0.10	3.46(3)	-3.0 (2)	1.2(1)	-4.5(2)	4.2(7)
0.20	3.24(3)	-1.8 (2)	1.5(1)	-5.1(2)	4.7(7)
0.40	2.52(2)	-0.44(10)	2.6(2)	-3.1(2)	4.7(7)
0.60	1.68(1)	+0.22(1)	3.9(4)	-2.1(2)	4.5(7)
0.80	1.14(1)	+0.27(1)	5.3(5)	-1.3(2)	-
1.00	1.00(1)	+0.43	9.2	-	-

We have tabulated our results in Table I. In general, the NMR of ^{103}Rh and ^{195}Pt was characterized by wide ($\sim$ 1-2%) lines indicating the absence of rigid band behavior on the local level. This inhomogeneous broadening decouples the Pt spins allowing indirect exchange effects to be observed. The magnitude of the ^{195}Pt shift and χ_B exhibit maxima for $x \sim 0.1 - 0.2$, in contrast to the ^{103}Rh shift which decreases monotonically as x decreases. Thus, even as the Rh spins become relatively more localized, the Pt magnetization is mainly responsible for the bulk properties. The observation of maxima for $|K_{Pt}|$ and χ_B at different x is probably due to screening at the Pt sites.

Localization of magnetization at the Rh site is evident from the plot of K_{Rh} vs. χ_B shown in Fig. 1. The larger rate of increase for K_{Rh} than for χ_B suggest a larger local magnetization at the Rh sites. This is to be contrasted with the roughly linear K_{Pt} vs. χ_B plot. The ^{103}Rh data are very similar to data[6] obtained for $Pd_{1-x}Rh_x$ and suggest that the Pt resonance reflects qualitatively the behavior of Pd in this system.

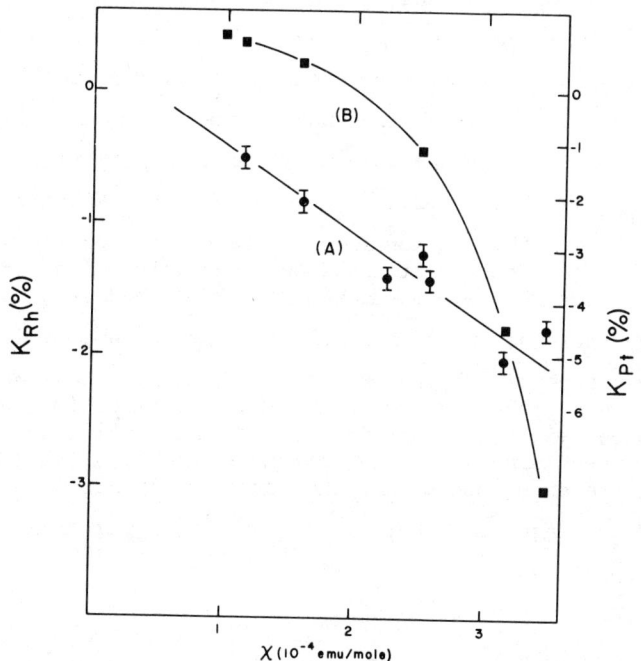

Fig. 1 Variation of ^{195}Pt(A) and ^{103}Rh(B) Knight shift with the bulk susceptibility for a series of $Pt_{1-x}Rh_x$ alloys.

The degree to which magnetization is localized near the Rh sites can be estimated using

$$\chi_B(x) = x\,\chi_{Rh}(x) = (1-x)\chi_{Pt}(x) \;, \qquad (1)$$

from which we find

$$\frac{\chi_{Rh}(0)}{\chi_B(0)} - 1 = \frac{d\ln\chi_B}{dx} - \frac{d\ln K_{Pt}}{dx} \;. \qquad (2)$$

In the low Rh concentration limit we have $\chi_B(0) \approx \chi_{Pt}(0)$ so that eq. (2) yields $\chi_{Rh}(0)/\chi_{Pt}(0) \approx 5.5$.

In principle this ratio can be obtained from the Knight shifts and d-hyperfine fields of ^{103}Rh and ^{195}Pt. The core polarization hyperfine field for ^{195}Pt is -2.36×10^6 G; however, there is some doubt as to the precise value of the hyperfine field ($A_{Rh}(d)$) for Rh.[6-8] If we use[7] $A_{Rh}(d) = -0.32 \times 10^6$ G (which is close to the value of $A_{Rh}(d) = -0.40 \times 10^6$ G derived from the initial slope of K_{Rh} in Fig. 1), we find $\chi_{Rh}(0)/\chi_{Pt}(0) = 6.1$, in fair agreement with the above value. In any case it is clear that at low x the Rh magnetization is much larger than for Pt. We also note that the smooth variation of the ratio $|K_{Rh}|/|K_{Pt}|$ suggest gradual development of this localization.

The large increase observed in $(T_1T)^{-1}$ for ^{103}Rh as x decreases is a direct consequence of the increase in the density of d-states. Quite likely the exchange enhanced core polarization interaction dominates, but this is uncertain since the orbital rate also tracks the density of states and is much larger than the core polarization rate in elemental Rh. However, the experimental Korringa produce ($K^2 T_1 T$) is the same for $x = 0.1$ and 0.03 suggesting that $(T_1T)^{-1}_{cp} \gg (T_1T)^{-1}_{orb}$ in this concentration region.

REFERENCES

1. A. C. Switendick, Ber. Bunsen Phys. Chem. **76**, 535 (1972).
2. F. M. Mueller, J. W. Garland, M. H. Cohen, and K. H. Bennemann, Ann. Phys. **67**, 19 (1971).
3. D. W. Budworth, F. E. Hoare, and J. Preston, Proc. Roy. Soc. (London) **A257**, 250 (1960).
4. We used the reference ratios $(\nu/H)_{Rh} = 0.1340$ KHz/sec and $(\nu/H)_{Pt} = 0.9089$ kHz/sec. The ^{195}Pt ratio was obtained by decreasing the Nuclear Data Tables result by 0.7% to account for orbital effects as discussed in A. M. Clogston, V. Jaccarino, and Y. Yafet, Phys. Rev. **134**, 650 (1964).
5. C. Froidevaux and M. Weger, Phys. Rev. Letters **12**, 123 (1964); H. Alloul and C. Froidevaux, Phys. Rev. **163**, 324 (1967).
6. A. Narath and H. T. Weaver, Phys. Rev. **B3**, 616 (1971).
7. J. A. Seitchik, A. C. Gossard, and V. Jaccarino, Phys. Rev. **136**, 1119 (1964).
8. G. N. Rao, E. Matthias, and D. A. Shirley, Phys. Rev. **184**, 325 (1969).

PARAMAGNON EFFECTS ASSOCIATED WITH Ni IMPURITIES IN A Pd_{90}-Pt_{10} MATRIX

C. F. Eagen[†] and A. I. Schindler
Naval Research Laboratory, Washington, D. C. 20375

ABSTRACT

The electrical resistivity from 2 to 50 K and the magnetic susceptibility at 2 K have been measured in Pd_{90}-Pt_{10}-Ni alloys with Ni concentrations of 0, 0.3, 0.6, 1, 2, and 5 at%. The Pd_{90}-Pt_{10} host has a magnetic susceptibility ~30% lower than pure Pd, and Ni additions result in a monotonically increasing susceptibility, with the critical concentration for the onset of ferromagnetism being ~5 at%. The resistivity of the Pd_{90}-Pt_{10} host is temperature independent below 5 K with no trace of the T^2 component associated with electron-paramagnon scattering. However, the difference in the resistivity between that of each ternary alloy and that of the Pd_{90}-Pt_{10} host does exhibit a low temperature T^2 dependence for Ni concentrations less than 2 at%. The Ni contribution to the resistivity agrees surprisingly well with the functional form of the Kaiser-Doniach[1] single impurity expression, with a corresponding value for the local spin fluctuation temperature, T_s, of ~80 K in the dilute Ni limit. Both the magnetic susceptibility and the coefficient of the T^2 term in the low temperature resistivity depend linearly on the Ni concentration for concentrations less than 1 at%, in agreement with the K-D model. However, T_s is found to fall rapidly as a function of the Ni concentration and is not constant in the dilute region as predicted by the K-D model.

[†]Present address: Scientific Research Staff, Ford Motor Company, Dearborn, Michigan 48121

[1]A. B. Kaiser and S. Doniach, Intern. J. Magnetism 1, 11 (1970).

EFFECT OF STRONG SPIN CORRELATIONS ON THE MAGNETIC SPECIFIC HEAT AND MAGNETIZATION OF DILUTE PdFe ALLOYS

N. C. Koon
Naval Research Laboratory, Washington, D. C. 20375

ABSTRACT

A simple model for the magnetic behavior of dilute PdFe alloys is proposed which assumes that the effective exchange interaction between iron impurities increases very rapidly as the separation decreases. When correlations between spins which are strongly coupled due to close proximity are taken into account, the model can explain both the anomalously low spin values obtained from measurements of the magnetic specific heat and the increase of the gS product deduced from magnetization and magnetoresistance measurements. Predictions of the model for the concentration dependence of the entropy per spin and the gS product are in qualitative agreement with experiment.

INTRODUCTION

Dilute Fe in Pd constitutes the classic giant magnetic moment system, with a total moment of 10-12 μ_B associated with each iron atom[1]. Neutron diffraction studies have shown that only about 3 μ_B of this moment is localized on the iron site, with the remainder being associated with a long range ($\gtrsim$ 10Å) polarization of the host Pd matrix[2]. This long range polarization of the host leads to a number of striking properties, including the fact that PdFe remains ferromagnetic down to very low iron concentrations--on the order of 0.1 at.% or less.

In spite of the fact that a rather large magnetic moment per iron atom in PdFe is well established, measurements of the magnetic specific heat[3,4] yield an entropy per spin corresponding to S = 1.1 ± 0.3, which is much lower than expected for a giant moment. Furthermore, it appears that the entropy per spin may be decreasing with concentration[3] at concentrations well below 2% Fe where the saturation magnetic moment per Fe atom begins to decrease[5]. Another interesting observation concerns the magnetization of dilute PdFe alloys in the paramagnetic region above T_c. Maley, et al[6] found in a Mössbauer study of extremely dilute Fe in Pd that the dynamics of an isolated Fe atom were well described by a Brillouin function with g = 2.95, S = 3.74 and a corresponding moment size of 11 μ_B. Manuel and McDougald[7] found in magnetization measurements above T_c on more concentrated samples, however, that the effective moment size (gS) increased with concentration to about 20 at 0.15 at.% Fe, extrapolating essentially to the result of Maley[6] at zero concentration. Qualitatively similar conclusions have been obtained from magnetoresistance data in the paramagnetic region[8,9].

Manuel and McDougald[7] suggested in their work that the increasing gS which they observed with concentration could be due to the fact that when two or more iron atoms were close enough to their polarization clouds to overlap, they effectively behaved as a "super" giant moment with a moment equal to the sum of several "isolated" giant moments. It is the purpose of this work to show how an extension of this basic idea can be used to explain both the magnetization and the specific heat data.

DISCUSSION

A question central to the discussion of local moment dynamics in dilute PdFe alloys is range dependence of the effective exchange energy $J(r)$ between two iron atoms in a Pd host. Kitchens and Trousdale[9] assumed that $J(r)$ had the form $J(r) = \exp(-r^2/\sigma^2)$, which exhibits a rather slow variation for small r. It is probably more reasonable to assume that $J(r)$ is proportional to the induced spin density, which neutron diffraction[2] and Pd NMR[10] measurements have shown is a rapidly varying function of distance from the iron atoms, even at very short distances.[10]

If $J(r)$ is a very rapidly varying function of r, then magnetization and specific heat results are relatively easy to understand qualitatively. The ordering temperature T_c will be determined primarily by the longer range part of the exchange, corresponding roughly to the average separation between iron atoms. Because of random location of the iron atoms, however, some will be close enough so that $J(r)$ is large compared with $k_B T_c$. Correlations between such spins will persist far above T_c so that in the vicinity of T_c and below they will behave dynamically like a single "super" giant moment of the type suggested by Manuel and McDougald.[7] The entropy associated with such correlations could only be accurately measured by extending the specific heat measurements to temperatures such that $k_B T \gg J(r)$ for all r which occur with significant probability. For practical reasons none of the specific heat measurements reported thus far have extended much above T_c.

To put the model in quantitative form we assume the experimental measurements to be made at temperatures of T_{max} and below, and at a separation r_o, $J(r_o) \simeq k_B T_{max}$. We also assume that $J(r)$ changes rapidly enough at r_o so that for all practical purposes if $r < r_o$, $J(r) \gg k_B T_{max}$, and if $r > r_o$, $J(r) \ll k_B T_{max}$. At $T = T_{max}$, therefore, we assume that any spins separated by less than r_o are completely correlated, and any separated by greater than r_o are completely uncorrelated unless "connected" by other spins separated by less than r_o. The entropy associated with the magnetic disordering which takes place between $T = 0$ and $T = T_{max}$ is then given by

$$S_m = \sum_n \frac{1}{n} P_n \ln(2nS + 1) \qquad (1)$$

In this expression P_n is the probability that a given spin will be a part of a "cluster" of n spins and S is the spin of a single iron spin and its associated polarization cloud. The total spin of a cluster is assumed to be nS, which should be approximately correct for alloys containing less than 2% Fe. The reduced "pseudo spin" S' which would be obtained from a measurement of the magnetic specific heat between T = 0 and T = T_{max} is then

$$S' = \left[\exp(S_m) - 1\right]/2 \qquad (2)$$

If we assume that all of the iron atoms are randomly located in a continuum, the probabilities P_n can be calculated in a straightforward way using Monte Carlo techniques. The spin S' obtained from such calculations is given in Fig. 1 for a number of different values of the spin S as a function of the parameter cr_o^3.

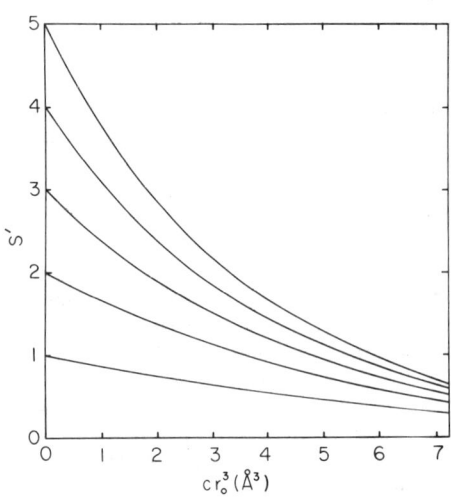

Fig. 1. Reduced spin S' as a function of cr_o^3 for several values of spin S.

It is easy to see from this figure that if $r_o \sim 10$Å, a reduction of S' to less than unity occurs for S = 5 at about 0.6 at.% Fe. Unfortunately, a direct comparison of the theoretical curves to the data of Veal and Rayne[3] is not strictly valid because for each of their samples the data only extended a short distance above T_c, and the high temperature "tail" was simply estimated. In the context of the present work it would be crucial to extend the measurements on all of the samples to the same temperature T_{max} greater than T_c for any of the samples. For comparison with the present work, therefore, it is probable that the lower concentration samples were assigned an entropy lower than they should actually have. If we arbitrarily assume that the entropy obtained by Veal and Rayne for their two high concentration samples is correct, then it is possible to fit them assuming S = 2.5 and r_o = 8.6Å. The resulting theoretical entropy vs c is given in Fig. 2 along with the data. It seems

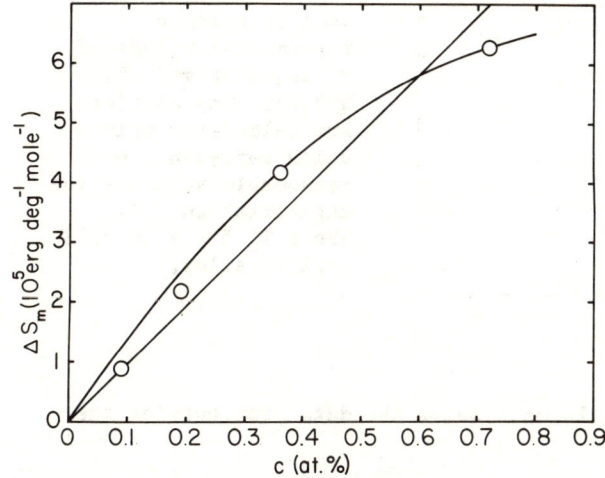

Fig. 2. Entropy as a function of concentration assuming S=2.5 and r_o =8.6Å. Circles are the data of Veal and Rayne and the straight line corresponds to S=1.1.

likely that improved data would raise both S and r_o in this fit. However, the spin S = 2.5 obtained is much more in accord with the giant moment size than the value of 1.1 which was obtained from the straight line fit to the data.

Comparison of the theory to magnetization data is somewhat more straightforward than specific heat. Maley, et al[6] found that isolated Fe atoms in Pd were well characterized magnetically by a Brillouin function with g = 2.95, S = 3.74. For a 0.23 at.% Fe sample we found[8] that in the paramagnetic region the data could be described to within a few percent by molecular field theory with g = 3.6, S = 5. To compare theory and experiment, we assume that the experimental magnetization for this sample as a function of Δ = 2 $\mu_B H/k_B T$ is given by $B_S(.5g\Delta)$ where S = 5 and g = 3.6. For the theoretical magnetization we assume that

$$\sigma(\Delta)/\sigma(0) = \sum_n P_n B_{nS}(.5g\Delta), \qquad (3)$$

where g = 2.95 and S = 3.74. In the dilute limit, therefore, the calculation agrees exactly with the results of Maley, et al. A comparison of the experimental magnetization data for the 0.23 at.% Fe sample and the theoretical curves for various assumed values of r_o is given in Fig. 3. It is interesting to note that, although none of the curves fit the data exactly, an r_o of 9.5Å seems to fit the data best at low values of Δ, where such a comparison should be most valid. The curve calculated using r_o = 0, which corresponds

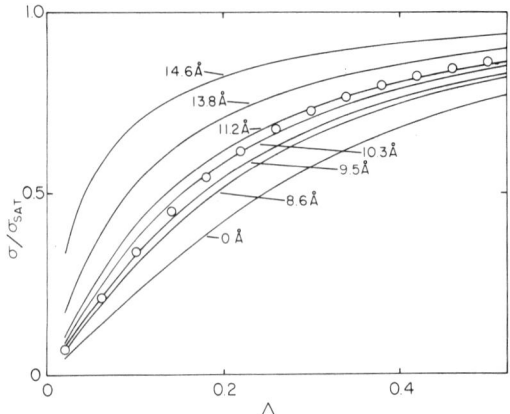

Fig. 3. Magnetization as a function of Δ for several values of r_o assuming g=2.95, S=3.74. The circles are calculated points which represent to reasonable accuracy the magnetization data above T_c for a Pd-0.23 at.% Fe alloy.

to infinite dilution, falls well below the data, far outside the estimated error.

Since increasing r_o is formally equivalent to increasing the concentration, it can be seen that the curves reflect qualitatively the increase in gS with c observed by Manuel and McDougald[7] and by Koon, Schindler, and Mills[8]. Both the magnetization and specific heat data, therefore, seem to be consistent in suggesting that in very dilute PdFe alloys there are very strong spin correlations which persist far above T_c for many of the alloys. The reasonable qualitative agreement of the present theory with both the magnetization and specific heat data indicates that such correlations are an understandable consequence of an exchange interaction between impurities which varies strongly with separation.

REFERENCES

1. R. M. Bozorth, P. A. Wolff, D. D. Davis, V. B. Compton, and J. H. Wernick, Phys. Rev. 122, 1157 (1961).
2. G. G. Low and T. M. Holden, Proc. Phys. Soc. (London) 89, 119 (1966).
3. B. W. Veal and J. A. Rayne, Phys. Rev. 135, A442 (1964).
4. G. J. Nieuwenhuys, B. M. Boerstoel, J. J. Zwart, H. D. Dockter, and G. J. van den Berg, Phsica 62. 278 (1972).
5. T. F. Smith, W. E. Gardner, and H. Montgomery, J. Phys. C, S370 (1970).
6. M. P. Maley, R. D. Taylor, and J. L. Thompson, J. Appl. Phys. 38, 1249 (1967).
7. A. J. Manuel and M. McDougald, J. Phys. C3, 147 (1970).
8. N. C. Koon, A. I. Schindler, and D. L. Mills, Phys. Rev. B6, 4241 (1972). N. C. Koon and A.I. Schindler, Phys. Rev., to be published.
9. T. A. Kitchens and W. L. Trousdale, Phys. Rev. 174, 606 (1968).
10. J. I. Budnick, private communication.

LOW FIELD MAGNETIC SUSCEPTIBILITY OF $(Pd_{1-x}Ag_x)_{.99}Fe_{.01}$*

J. I. Budnick
National Science Foundation, Washington, D.C. 20550

V. Cannella
Department of Physics, Wayne State University
Detroit, Michigan 48202

T. J. Burch
Department of Physics, Fordham University
Bronx, New York 10458

ABSTRACT

To investigate the effect of the reduction of the host matrix susceptibility on the details of the magnetic ordering in exchange enhanced dilute ferromagnetic alloys, we have examined the temperature dependence of the low-field magnetic susceptibility, $\chi(T)$, for $(Pd_{1-x}Ag_x)_{.99}Fe_{.01}$ alloys. $\chi(T)$ has been measured in the region of the magnetic ordering temperature, T_0, for alloys with x = .025, .010, .25, .33 and .50. In agreement with recent Mössbauer studies of these alloys we find a linear dependence of T_0 upon the host matrix susceptibility for $x \leq .25$. In sharp contrast to the enhanced ferromagnetic behavior found in $\chi(T)$ for $x \leq .25$, $\chi(T)$ shows a dramatic change in behavior for $x \geq .33$, where the host matrix susceptibility becomes small. For x = .50 we find a relatively sharp peak in $\chi(T)$, and a Curie-Weiss law above the ordering temperature. This behavior is similar to that found for $\chi(T)$ in "spin glass" alloys such as low concentration CuMn, AuMn, AuFe, etc.

INTRODUCTION

The ferromagnetic behavior of dilute concentrations of Fe in exchange enhanced hosts such as Pd has been extensively studied, both experimentally and theoretically.[1] The variation of the magnetic moment of Fe atoms as a function of electron concentration in different transition metal alloys was investigated in the classic work by Clogston and co-workers.[2] Since the enhanced susceptibility of Pd can be decreased and eventually suppressed by alloying with Ag,[3,4] a study of the effect of introducing Ag into PdFe alloys should reflect the dependence of the ferromagnetic interactions and ordering upon the host matrix susceptibility. A recent Mössbauer study of the $(Pd_{1-x}Ag_x)_{.99}Fe_{.01}$ alloy system

---

*Work supported in part by Air Force Office of Scientific Research under Grant AFOSR-71-2002.

by Levy and co-workers[1] has shown that for $x \leq .25$ the ferromagnetic Curie temperature of $(Pd_{1-x}Ag_x)_{.99}Fe_{.01}$ varies linearly with the host matrix susceptibility as is predicted by theoretical models.

In this paper we present a study of the temperature dependence of the low field (~ 5 gauss) magnetic susceptibility, $\chi(T)$, for $(Pd_{1-x}Ag_x)_{.99}Fe_{.01}$ alloys with $x = .025, .10, .25, .33$, and $.50$. The low field susceptibility is a useful measurement in studying magnetic transitions since critical effects are most faithfully seen in the limit of low fields, and even moderately high fields can shift, distort, or mask critical phenomena. A low field (~ 5 gauss) low frequency (50 Hz-155 Hz) a-c mutual inductance technique was used in these measurements. For these low frequencies, diamagnetic contributions due to eddy currents are negligible. Demagnetizing effects must be taken into account when measuring $\chi(T)$ for samples with large χ such as ferromagnets. If χ_r is the real χ and χ_m is the measured χ, then $\chi_r = \chi_m/(1 - \chi_m D)$ where D is the demagnetizing factor. For a bulk ferromagnet the measured low field $\chi(T)$ is expected to follow a Curie-Weiss law well above T_o, where demagnetizing effects are negligible. As the temperature is lowered to T_o, χ_r diverges and χ_m assumes the constant value $1/D$ below T_o since $1/\chi_m = 1/\chi_r + D$.

EXPERIMENTAL RESULTS

Plots of the measured low field $\chi(T)$ for $(Pd_{1-x}Ag_x)_{.99}Fe_{.01}$ with $x = .025, .10, .25, .33$ and $.50$ are shown in Fig. 1. The

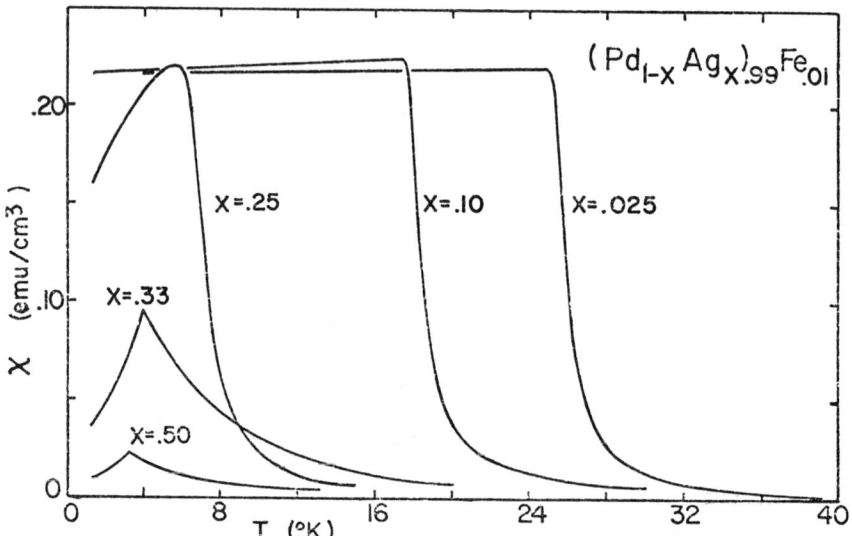

Fig. 1. Measured $\chi(T)$ for $(Pd_{1-x}Ag_x)_{.99}Fe_{.01}$ with $x = .025, .10, .25, .33$, and $.50$.

measured $\chi(T)$ for alloys with x = .025 and .10 clearly shows the behavior expected for a bulk ferromagnet: for temperatures below a critical divergence in χ_r in the measured $\chi(T)$ is nearly constant and equal in magnitude to 1/D where D is the demagnetizing factor of the sample. The sample with x = .25 shows a critical divergence in χ_r with the magnitude of χ_m equal to 1/D at a temperature just below the divergence, but $\chi(T)$ immediately decreases as T is lowered indicating the growing importance of interactions other than the ferromagnetic type that caused the critical divergence. Thus for samples with x $\leq$ .25 we have bulk ferromagnetism at some temperature. Well above the ferromagnetic Curie temperature when the susceptibility becomes small and demagnetizing effects are negligible $\chi(T)$ follows a Curie-Weiss law which gives paramagnetic Curie-Weiss values for θ slightly larger than T_o as is expected for ferromagnets. Values of the effective moment p_{eff} found from the paramagnetic data are discussed later.

Because of the rounding at the peak of the measured $\chi(T)$ for the ferromagnetic samples, some criterion had to be established for choosing the ordering temperature, T_o. Since by definition χ_r diverges at the ferromagnetic ordering temperature, it seems reasonable to choose T_o to be between the maximum in $\chi(T)$ and the maximum in $d\chi/dT$. The values given for T_o in this paper are halfway between χ_{max} and $(d\chi/dT)_{max}$.

$\chi(T)$ for samples with x = .33 and x = .50 shows a dramatic change from the ferromagnetic behavior found when x $\leq$ .25. $\chi(T)$ shows a relatively sharp peak at an ordering temperature T_o, then decreases to a non-zero intercept at T → 0. Above T_o, at higher temperatures were demagnetizing effects are negligible, $\chi(T)$ follows a Curie-Weiss law from which values of θ and p_{eff} were found. The behavior of $\chi(T)$ for alloys with x = .33 and .50 is very similar to the sharp peaks in $\chi(T)$ found in spin glass alloys,[5] especially for AuFe alloys with C $\simeq$ 12%, just below the ferromagnetic percolation limit.[6] For spin glass alloys this behavior in $\chi(T)$ is interpreted as due to random long range interactions of the RKKY type.

Values for the susceptibility of the Pd-Ag host matrix, χ_o, were interpolated from the work of other authors.[3] Table I shows these values for χ_o, and the values for T_o, θ, and p_{eff} found from our $\chi(T)$ measurements. The low value of θ for the x = .50 sample is consistent with values found for lower concentration spin glasses while the larger value for the x = .33 sample is similar to that for spin glasses with very large amounts of superparamagnetic clustering (this value of θ was found from $\chi(T)$ at temperatures well above θ and T_o). The very large values for p_{eff} in all of these samples seems to reflect a combination of the host matrix polarization expected for lower values of x, and superparamagnetic clustering which is also found to be very large near the percolation limit in AuFe alloys.[6] It is difficult at this point to analyze the values for p_{eff} further but it should be noted that low field $\chi(T)$ measurements seem more sensitive to superparamagnetic clustering than higher field measurements such as that of Clogston and co-workers[2]

TABLE I

x	χ_o (10^{-6} emu/g)	T_o (°K)	p_{eff}	θ (°K)
.025	6.3	25.3	12 ± 1	27 ± 1
.10	4.6	17.6	13 ± 1	20 ± 1
.25	2.3	6.5	16 ± 1	7.5 ± .5
.33	1.6	4.0	17 ± 1	8.5 ± .5
.50	.2	3.3	15 ± 1	1 ± .5

Table I. Values fo the host matrix susceptibility χ_o interpolated from reference 3, and values of T_o, p_{eff}, and θ found from $\chi(T)$ data, for $(Pd_{1-x}Ag_x)_{.99}Fe_{.01}$ alloys with various values of x.

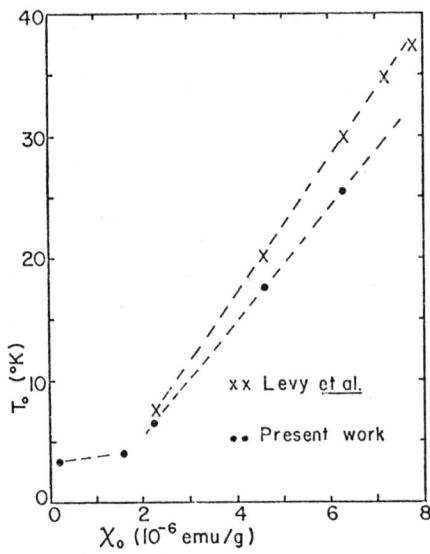

Fig. 2. T_o vs. the host matrix susceptibility χ_o, for $(Pd_{1-x}Ag_x)_{.99}Fe_{.01}$ alloys including the Mössbauer data of Levy and co-workers from reference 1.

for $(Pd_{.75}Ag_{.25})_{.99}Fe_{.01}$.

The plot of T_o vs. χ_o in Fig. 2 shows that T_o varies quite linearly with T_o for alloys with x ≤ .25. The values of T_o from the Mössbauer data really represent an upper limit rather than an accurate determination of T_o,[7] so the discrepancy between the sets of data is not significant. The change from ferromagnetic order to a type of antiferromagnetic order is seen in the change of the T_o dependence for samples with x = .33 and .50, but more data is necessary before this can be studied in detail.

CONCLUSIONS

The values of the ferromagnetic ordering tempreature determined from $\chi(T)$ for $(Pd_{1-x}Ag_x)_{.99}Fe_{.01}$ confirm that T_o varies linearly with the host matrix susceptibility χ_o for $x \leq .25$ as was concluded by Levy and co-workers. For larger values of x, however where χ_o becomes much smaller, the type of magnetic ordering gradually changes from ferromagnetic to the type of spin freezing behavior or antiferromagnetic order characteristic of random spin glass alloys. The values of p_{eff} determined from the paramagnetic Curie-Weiss data are quite large and suggest that a combination of host matrix polarization and superparamagnetic clustering is taking place. The type of freezing of spins or antiferromagnetism which occurs for the larger values of x is not well understood even in dilute spin glasses where no superparamagnetism occurs. The similarities, however, between this $(Pd_{1-x}Ag_x)_{.99}Fe_{.01}$ alloys system and the AuFe alloys system suggest that the continuous transition from the random type interactions of the spin glass regime to the ferromagnetic interactions of higher concentration regions should be a fruitful topic for future study.

REFERENCES

1. R.A. Levy, J.J. Burton, D.I. Paul, and J.I. Budnick, to be published in Phys. Rev. B.; R.A. Levy, Doctoral Thesis, Columbia Univ., 1973 (unpublished).
2. A.M. Clogston, B.T. Matthias, M. Peter, H.J. Williams, E. Corenzwit, and R.C. Sherwood, Phys. Rev. 125, 541 (1962).
3. F.E. Hoare, J.C. Matthews, and J.C. Walling, Proc. Roy. Soc. of London 216, 502 (1963); R. Doclo, S. Foner, and A. Narath, J. Appl. Phys. 40, 1206 (1969).
4. See for example, J.E. Noakes and A. Arrott, J. Appl. Phys. 35, 931 (1964).
5. See V. Cannella and J.A. Mydosh, this conference.
6. V. Cannella and J.A. Mydosh, Phys. Rev. B 6, 4220 (1972).
7. R.A. Levy, private communication.

THE LOW TEMPERATURE SPECIFIC HEAT OF
DILUTE RARE-EARTH IN TRANSITION METAL
ALLOYS: Sc(Gd), Pd(Dy), AND Pd(Gd)

L. L. Isaacs
Dept. of Chemistry, University of Washington,
Seattle, Wn. 98195

ABSTRACT

The magnetic solutes, Gadolinium and Dysprosium, strongly influence the magnetic and thermal properties of the exchange enhanced metals, Scandium and Palladium. The results of low temperature specific heat measurements on several series of dilute Gadolinium in Scandium, Dysprosium in Palladium and Gadolinium in Palladium alloys are presented. In Sc(Gd) alloys, an antiferromagnetic phase transition is observed. The results indicate that the interaction between the magnetic spins is of the Ruderman-Kittel-Kasuya-Yosida type. The specific heat of the Pd(Dy) alloys is dominated by crystal field splitting of the Dysprosium ground state and by spin-clustering effects. The specific heat of Pd(Gd) alloys shows both clustering and a ferromagnetic phase transition. The specific heat results on the three series correlate well with magnetization studies.

INTRODUCTION

The magnetic properties of dilute alloys of Scandium and Palladium with rare-earth elements have been investigated extensively.[1] The effects of solvent spin polarizations and phase transformations, antiferromagnetic and ferromagnetic, have been observed. In order to elucidate further the nature of the interactions between the magnetic spins (considered to be localized on the rare-earth atoms) and between the solvent and solute, specific heats were determined as a function of solute concentration.

EXPERIMENTAL RESULTS

The properties of Scandium and Palladium are extremely sensitive to impurities. To minimize extraneous impurity effects, the experimental alloy specimens were prepared by levitation melting and casting into a chilled copper mold. The melts were held in the liquid state for a prolonged time to ensure thorough mixing of the components, but no subsequent homogenizing treatments were applied. The specific heats were measured using a recirculating mode helium-three cryostat and the heat-pulse technique. The experimental results are presented in Figures 1, 2 and 3.

DISCUSSION

At low temperatures, the total specific heat of a metallic system is expected to follow the expression,

$$C_p(T) = \gamma T + \beta T^3 + \Delta C(T),$$

where γT and βT^3 represent the contributions of the conduction electrons and of the lattice to the total specific heat, respectively, and $\Delta C(T)$ is the contribution due to various interactions. Changes in the electronic specific heat of the alloys relative to that of the solvent also reflect the effect of interactions on the band structure of the solvent (enhancement effects).

Inspection of the C_p/T versus T^2 plots (not shown) for the dilute alloys (less than 0.5 at.% solute) indicates that the lattice specific heat may be assumed to be independent of the solute concentration and to be equal to that of the solvent itself. This assumption is further justified by noting that at high temperatures ($T > 10$ K), the specific

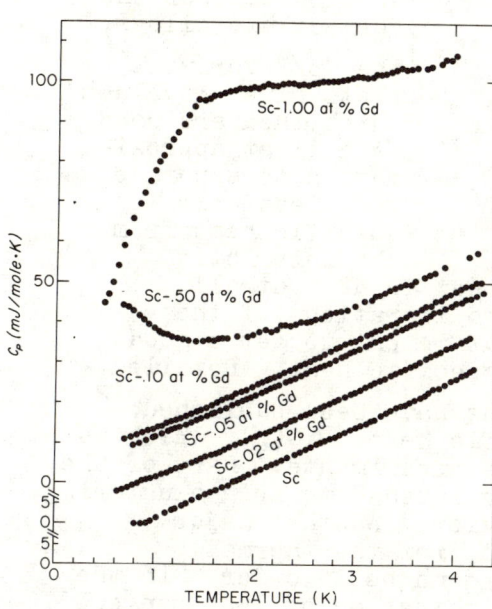

Fig. 1. Specific heat of Sc(Gd) alloys.

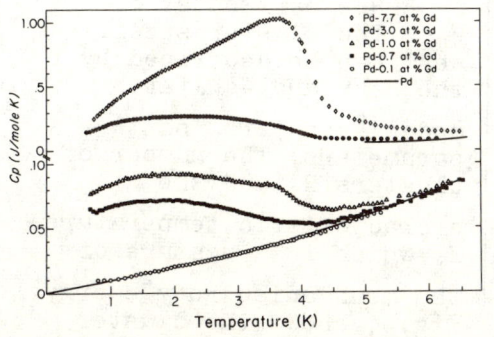

Fig. 2. Specific heat of Pd(Gd) alloys.

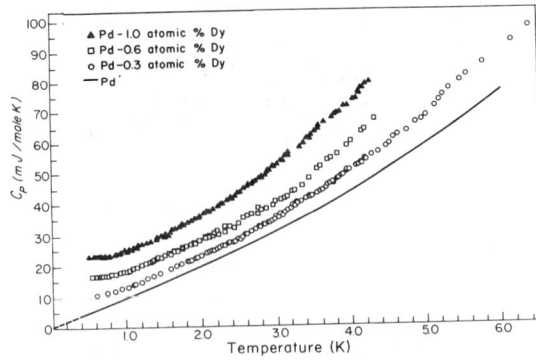

Fig. 3. Specific heat of Pd(Dy) alloys.

heat of Pd(Gd) alloys approaches that of pure Pd.

The extraction of the electronic specific heat from the data is somewhat more difficult. For the very dilute alloys, the C_p/T versus T^2 plots give γ values which then are used as a first approximation to separate the electronic specific heat from the interaction contribution to the total specific heat at all concentrations. Usually, two iterations of the separation procedures are sufficient to get a good estimate of the electronic specific heat. Detailed discussions of the procedures have been published.[2]

It is the excess specific heat, $\Delta C(T)$, which is of primary interest in these experiments. Part of the excess specific heat is associated with spin clustering, i.e. superparamagnetic behavior. Another major contribution to $\Delta C(T)$ comes from the thermal disordering of the long range magnetic order. In the Pd(Dy) alloys, there is no indication of long range magnetic order. However, a major contribution to the specific heat comes from a Schottky anomaly due to the partial removal of the ground state degeneracy of the Dy ion by the cubic crystalline field of the Pd matrix. This effect is confirmed by EPR experiments.[3]

The cluster specific heat can be described by the Hahn and Wohlfarth model[4] in terms of two parameters, the number of clusters N_A and the characteristic temperature T_A which is a measure of the mean anisotropy energy. Fig. 4 shows the cluster specific heat associated with the Pd(Dy) alloys. If T_A is of the order of

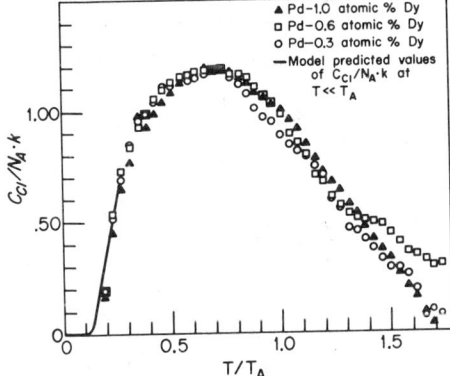

Fig. 4. Illustration of cluster specific heat.

10 K, then in the 1 to 4 K temperature range the cluster specific heat will appear to be a temperature independent constant. The dilute Sc(Gd) alloys show this type of behavior. The number of clusters per mole of alloy depends on the thermal history of the specimen and on the solute concentration.

In Pd(Gd) and Sc(Gd) alloys of sufficiently high concentration for the establishment of long range spin interactions, the rest of $\Delta C(T)$ is associated with magnetic phase transitions. The ferromagnetic transition in Pd(Gd) shows the typical temperature dependence associated with a cooperative phenomenon, slightly broadened by local compositional fluctuations. The more concentrated Sc(Gd) alloys undergo an antiferromagnetic phase transition. The results can be interpreted [5] as showing that the long range interaction between the magnetic spins is of the Ruderman-Kittel-Kasuya-Yosida type.

Thus, it appears that the onset of long range magnetic order is preceded and accompanied by superparamagnetic (cluster) behavior. Clustering, which leads to a decrease in the free energy of the system via a decrease in the entropy of the system, is promoted by the attractive interaction between the spins. However, the spins must be located within a certain critical volume for the interaction to be effective. Thus, the long range magnetic order is concentration dependent.

REFERENCES

1. J. Crangle, Phys. Rev. Letters $\underline{13}$, 569 (1964); F. Y. Fradin, J. W. Ross, L. L. Isaacs and D. J. Lam, Phys. Letters $\underline{28A}$, 276 (1968); L. L. Isaacs, D. J. Lam and F. Y. Fradin, J. Appl. Phys. $\underline{42}$, 1458 (1971); R. P. Guertin, H. C. Praddauda, S. Foner, E. J. McNiff Jr. and B. Barsoumian, Phys. Rev. $\underline{B7}$, 274 (1973).
2. L. L. Isaacs, Phys. Rev. $\underline{B8}$, in press; L. L. Isaacs, Phys. Rev. $\underline{B}$ in press.
3. R. A. Devine, J. M. Moret, J. Ortelli, D. Shaltiel, W. Zingg and M. Peter, Solid State Commun. $\underline{10}$, 575 (1972).
4. A. Hahn and E. P. Wohlfarth, Helv. Phys. Acta $\underline{41}$, 857 (1968).
5. J. Souletie and R. Tournier, J. of Low Temp. Phys. $\underline{1}$, 95 (1968).

THE ELECTRICAL AND MAGNETIC PROPERTIES OF GOLD-NICKEL ALLOYS

J. R. Clinton, E. H. Tyler, and H. L. Luo
Department of Applied Physics and Information Science
University of California, San Diego
La Jolla, California 92037

ABSTRACT

Completely random solid solutions have been obtained for the entire range of Au-Ni alloys by the splat-quenching technique. In an attempt to compare the Au-Ni system with the similar Cu-Ni system, electrical resistance and magnetization (for the ferromagnetic samples) measurements were made from below 4.2°K to room temperature. For corresponding compositions, magnetization and Curie temperatures are consistent in the two systems. Resistance measurements indicate that alloys with less than 35 at. % Ni appear normal, with slight paramagnetic effects. Alloys in the critical region (37-42 at. % Ni) display a temperature-independent resistance after an initial low-temperature decrease. Ferromagnetic samples (>45 at. % Ni) show a resistance maximum near the Curie temperature to 52 at. % Ni, while higher-Ni-content samples display normal ferromagnetic behavior: a decrease in slope at T_c. According to our data, the Au-Ni system is less complex than the Cu-Ni system. No evidence for the "giant magnetic clouds" seen in Cu-Ni was observed.

ALTERATION OF THE MAGNETIC PROPERTIES OF Au-Fe ALLOYS BY NEUTRON IRRADIATION[*]

R. J. Borg
Lawrence Livermore Laboratory, University of California
P.O. Box 808
Livermore, California 94550

ABSTRACT

Three alloys of Au-Fe containing nominally 13, 17, and 24 at.% Fe were irradiated to a total dose of $\sim 10^{19}$ neutrons/cm^2 (reactor spectrum neutrons). The magnetic hyperfine splitting was measured by means of the Mössbauer effect for both irradiated and unirradiated specimens. Only the 17% alloy demonstrated a significant change in the temperature dependence of the magnetic order. However, the change for this composition was extreme; the magnetic order commencing at ~128°K for the irradiated specimen as compared to ~65°K for the unirradiated. These results demonstrate that short range compositional order has a profound effect upon the magnetic properties and that the effect is a function of bulk composition. The indicated critical composition is in agreement with the predictions of Sato, et al.,[1] and also agrees with previously reported thermal annealing studies.[2]

[1] H. Sato, A. Arrot and R. Kikuchi, J. Phys. Chem. Sol. 10, 19 (1959).

[2] R. J. Borg, Phys. Rev. B, 5, 1035 (1972).

---

[*] Work done under the auspices of the U.S. Atomic Energy Commission.

MAGNETIC ORDER IN Ni_3Mn ALLOYS[*]

Carl E. Patton and Glenn L. Baker
Colorado State University, Fort Collins, Colorado 80521

ABSTRACT

Recent measurements reveal significant changes in the magnetic moment and superparamagnetic Curie point during the initial stages of atomic order in Ni_3Mn alloys. The magnetic moment at 8 kG and 300°K exhibits a small discontinuity at about 1.5 hours annealing time at 400°C. The superparamagnetic Curie temperature is also found to increase from 145°K for the disordered alloy to 250°K after an extremely brief (20 minute) anneal. These results show that significant atomic rearrangement is actually taking place rather quickly during the initial stages of ordering, contrary to the previously proposed model of homogeneous long range order development which saturates only after extremely long annealing times (1000 hrs or more).[1] The discontinuity at 1.5 hours is tentatively associated with the short range order saturation observed for anneals at higher temperatures, also at about 2 hours,[1] and the peak in the field-induced uniaxial anisotropy also observed at 2 hours annealing time.[2] Thus it may be possible to account for this anisotropy by preferential pairing of unlike nearest neighbors during short range order development.

---

[*] A full description of this work will be published elsewhere.
[1] M.J. Marcinkowski and N. Brown, J. Appl. Phys. 32, 375 (1961).
[2] C. E. Patton and S. Chikazumi, J. Phys. Rad. 32, 99 (1971).

MEASUREMENTS AND TRENDS OF
HYPERFINE FIELDS IN HEUSLER ALLOYS

C. C. M. Campbell and W. Leiper,
Dalhousie University, Halifax, N. S., Canada.

ABSTRACT

The hyperfine fields at the Sb sites in PtMnSb (+220 kOe), Co_2MnSb (+270 kOe) and CuMnSb ($\sim$0 kOe at 4.2K), were measured by means of the Mössbauer effect. The signs were determined using polarised γ-rays. Measurements were also made of the Sn hyperfine fields in the Sn-doped alloys (1% Sn^{119}) Pd_2MnSb ($\sim$+10kOe) and Ni_2MnSb (+53 kOe), and in the alloy Co_2TiSn (+76 kOe). The results for the Sn-doped alloys corroborate earlier evidence of marked local effects at the nonmagnetic sites. It is also seen that the above results, together with measurements by other workers at Cd (-147 kOe), In (-93 kOe) and Pb (+480 kOe) sites show the same general trend as in the case of dilute nonmagnetic impurities in Fe. Theoretical models which predict the experimental trends are available.

INTRODUCTION

Heusler alloys are a class of concentrated, ferromagnetic alloys of general composition A_2BC. The structure is designated $L2_1$, with the A atoms forming a simple cubic matrix and the B and C atoms occupying alternate body centres in the cubic structure. A typical member of this class of alloys (eg Pd_2MnSb) has the magnetic moment localised on Mn atoms at the B site.[1]

Both Heusler alloys and members of the related series ABC (Cl_b structure) have been the subject of extensive study in the last few years. In particular a number of attempts have been made to explain the hyperfine fields measured at A and C sites in Heusler alloys,[2,3] using theoretical models which are strictly applicable only to alloy systems with dilute magnetic impurities.

The purpose of the present study is to establish the existence of hyperfine field systematics at non-magnetic sites in the Heusler alloys, and to emphasize the effect of local screening on the hyperfine field value.[4]

EXPERIMENTAL

The alloys measured in the present work were prepared by heating the constituents in a r. f. furnace in an argon atomosphere, and annealing for 4 days at 500°C. X-ray diffraction experiments indicated the Heusler-like structure (either $L2_1$ or Cl_b) in each case. In the case of Co_2MnSb, Webster[1] reported the presence of an impurity phase containing precipitated Co. The main phase was of composition

Co_xMnSb with $x=1.75$[1(a)] or $1.35 \leq x \leq 1.65$[1(b)], and was of the $L2_1$ Heusler structure. An x-ray photograph of our Co alloy indicated an $L2_1$ structure and impurity lines were not apparent. The Mössbauer spectra shown in Figures 1 to 4 were all taken at 77°K. The experimental spectra were computer-fitted to Lorentzian shapes, using a least-squares procedure.

The weak, central lines in the Co_2 Ti Sn spectrum are due to the hyperfine field (-21 kOe) of an impurity phase, probably Co Sn^5.

The hyperfine field measured at the Sn site in the doped alloy Pd_2 Mn Sb was found to vary between about 0 kOe and +28 kOe depending on the heat treatment.

The measurement on Cu Mn Sb was made at 4.2°K, below the Neel temperature of the alloy (55°K).[6]

DISCUSSION

The zero hyperfine field found at the Sb site in Cu Mn Sb supports neutron diffraction evidence[7] that this alloy is an antiferromagnet with successive Mn ion moments along the cube edge oriented at 180° to each other.

Table 1 lists results of measurements of hyperfine fields at A and C sites in a number of Heusler alloys, and at dilute impurity sites in Fe. Two main conclusions can be drawn. (1) In the Heusler alloy systems, the nature of the ion at the site is the dominant factor in determining the magnitude and sign of the hyperfine field. Antimony fields, for example, are large and positive. At Sn sites the fields tend to be smaller, and mainly positive. This is most noticeable in Cu_2 Mn In where the In hyperfine field has been measured as -93 kOe[3] whereas the Sn field in Sn-doped Cu_2 Mn In (1% Sn^{119}) is +200 kOe[4(a)] (2) The same trend exists for the variation in hyperfine field across the sp series Cd, In, Sn and Sb in both the Heusler alloys and in Fe. In both cases there is a positive increase in the value of the hyperfine field across the sp series, although the position of the crossover point from positive to negative is different.

Two theoretical models have been proposed,[8,9] both incorporating the effects of local screening and both predicting the trends displayed by the experimental measurements.

Blandin and Campbell incorporate the local nature of the non-magnetic ion, taken to be at the origin, by extending the RKKY approach to include the s-wave phase shifts δ_o produced by a spherical, spin-independent potential centred at the origin. The phase shift is calculated from the Friedel sum rule, assuming that $\delta_o = \delta_1$ for the s-p conduction band. The effect of the magnetic ions, at distance Ro from the origin is represented by an additional delta-function-type spherical, spin-dependent potential at $r=R_o$. The effective hyperfine field at the non-magnetic sites is found to have the form, in a first-order approximation $\cos(2k_FR + 2\delta_o^F)$ where δ_o^F is the phase-shift at the Fermi surface. The variation in hyperfine field across an sp series is obtained from the change in the value of δ_o^F.

Geldart and Jena[9] have constructed a model, based on earlier work by Daniel and Friedel[10], in which the non-magnetic ion is considered to be an impurity in a ferromagnetic background (the

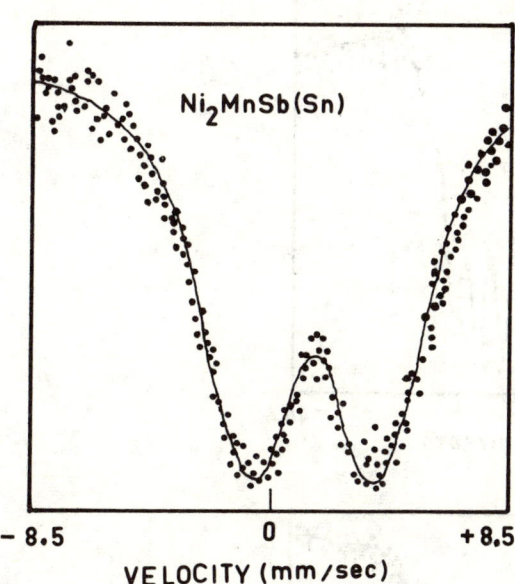

Fig. 1. $Ni_2MnSb(Sn)$ at 77K

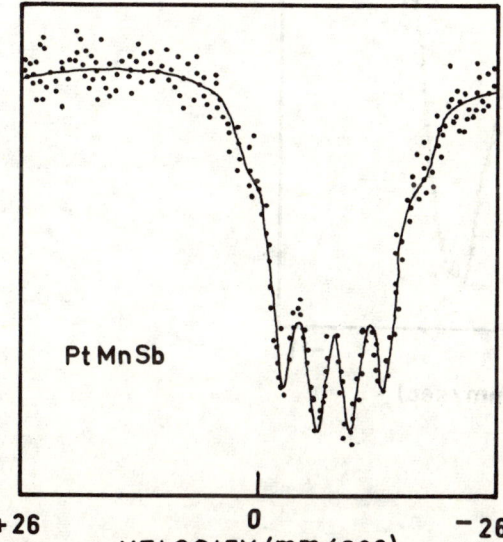

Fig. 2. PtMnSb at 77K

Atom	Host	H kOe
Cu	Cu_2MnIn	-200^3
	Cu_2MnSn	-175^3
	Cu_2MnAl	-214^3
	Fe	-213^5
Cd	$Cu_2MnIn(Cd)$	-147^{13}
	Fe	-348^5
In	Cu_2MnIn	-93^3
	Fe	-290^5
Sn	Ni_2MnSn	$+87^{14}$
	Cu_2MnSn	$+200^{14}$
	Pd_2MnSn	-35^{14}
	PtMnSn	$+20^{19}$
	Co_2MnSn	$+105^{15}$
	$Cu_2MnIn(Sn)$	$+200^4$
	$Pd_2MnSb(Sn)$	$\sim 10^{15}$
	$Ni_2MnSb(Sn)$	$+53^{15}$
	Fe	-81^5
Sb	PtMnSb	$+220^{15}$
	Co_2MnSb	$+270^{15}$
	Ni_2MnSb	$+300^{16}$
	Pd_2MnSb	$+579^{16}$
	CuMnSb	$\sim 0^{15}$
	Fe	$+230^5$
Al	Cu_2MnAl	$\pm 68^{17}$
	Fe	-55^5
Pb	$Cu_2MnBi(Pb)$	$+450^{18}$
	Fe	$+262^5$

Table 1

Experimental results for Hyperfine fields, H, at non-magnetic sites in Heusler alloys and iron.

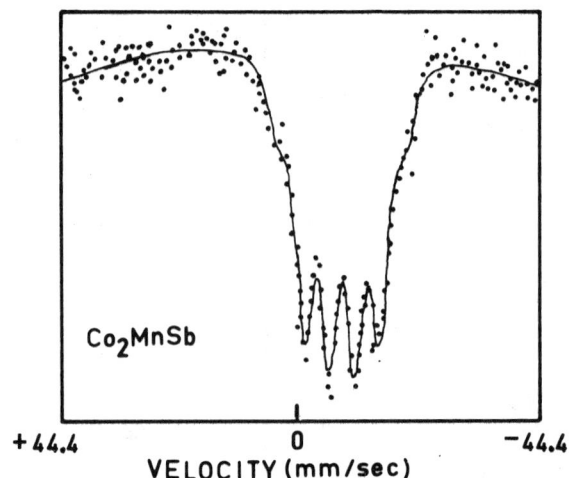

Fig. 3. Co$_2$MnSb at 77K

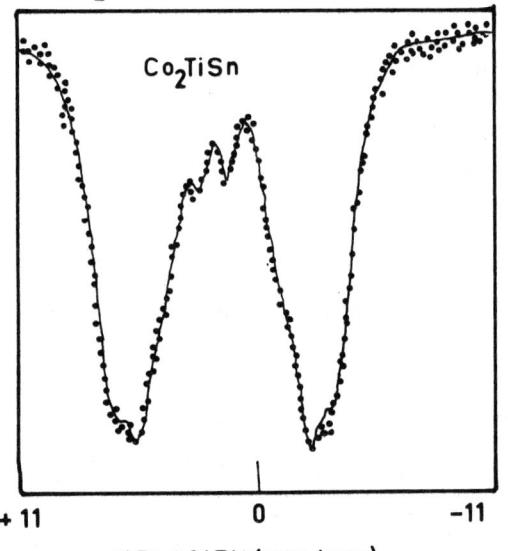

Fig. 4. Co$_2$TiSn at 77K

spin-split conduction band). The polarised conduction electrons are scattered from a spin-dependent square well potential at the non-magnetic site. Two contributions to the scattering potential at the impurity sites are considered; a spin-independent part due to the local screening requirement and a spin-dependent component due to the suppression of the s-d exchange interaction at the impurity site. The sign of the hyperfine field depends on the ratio of these two components and we also believe that the sign change with temperature of the hyperfine field at dilute Sn sites in Co^{11} can be understood in terms of this model.

Both theories predict the absence of a hyperfine field at the Sb sites of antiferromagnetic CuMnSb but a comparison of the Sn site field in Co_2TiSn and Co_2MnSn tends to favour the second approach. It increases from +76 kOe to +105.2 kOe^{12}, a small increase considering the large increase in magnetization due to the presence of the Mn moments in Co_2MnSn, showing that the effect of the ferromagnetic background on the hyperfine field is of second order.

However, as both models described above predict the general experimental trend, from different physical assumptions, future theoretical work should be based on more detailed calculation involving band structure effects.

REFERENCES

1. (a) P. J. Webster, (thesis) Sheffield University, (1968).
   (b) P. J. Webster, J. Phys. Chem. Solids 32, 1221 (1971).
2. B. Caroli and A. Blandin, J. Phys. Chem. Solids, 27, 503 (1966).
3. T. Shinohara, Jour. of Phys. Soc. of Japan, 27, 1127 (1969).
4. (a) W. Leiper and C.C.M. Campbell, Möss. Conf. Proc. (Israel), (August, 1972).
   (b) L. J. Swartzendruber and B. J. Evans, A.I.P. Conf. Proc. on Magnetism and Magnetic Materials, 10, 1369 (1972).
5. E. Matthias and D. A. Shirley, "Hyperfine Structure and Nuclear Radiations", p.981, (North Holland Publishing Co.) (1968).
6. K. Endo, Jour. of Phys. Soc. of Japan, 29, 643 (1970).
7. R. H. Forster and G. B. Johnston, J. Phys. Chem. Solids, 29, 855 (1968).
8. A. Blandin and I. A. Campbell, Phys. Rev. Lett., 31, 51 (1973).
9. D.J.W. Geldart and P. Jena (private communication).
10. E. Daniel and J. Friedel, J. Phys. Chem. Solids, 24, 1601 (1963). (The applicability of the Daniel-Friedel model to Heusler alloy systems was first suggested by L. J. Swartzendruber and B. J. Evans, A.I.P. Conf. Proc. on Magnetism and Magnetic Materials, 5, 539 (1971)).
11. T. E. Cranshaw, J. Appl. Phys. 40, 1481 (1969).
12. J. M. Williams, J. Phys. C, 1, 473 (1968).
13. W. Walus (reported as private communication in ref. 8).
14. D.J.W. Geldart, C.C.M. Campbell, P. J. Pothier and W. Leiper, Can. J. Phys. 50, 206 (1972).
15. (Present Work.)
16. L. J. Swartzendruber and B. J. Evans, Phys. Lett. 38A, 511 (1972).
17. N. R. Sharpe, J. M. Titman and G. G. Wood, J. Phys. C. Sol. St. Phys. 3, 560 (1970).
18. F. Zawislak (private communication).
19. C.C.M. Campbell and W. Leiper, C.A.P. Conf. Proc. (Edmonton), (1972).

ANALYSIS OF MÖSSBAUER ^{57}Fe ABSORPTION IN AN F.C.C. Fe-Co-V ALLOY

G. Bambakidis and J. P. Cusick
NASA, Lewis Research Center, Cleveland, Ohio 44135

ABSTRACT

Mössbauer ^{57}Fe absorption spectra of superparamagnetic F.C.C. Fe$_{.34}$ Co$_{.52}$ V$_{.14}$ (Vicalloy II) have been analysed using a cluster model and magnetic resonance relaxation theory. A least-squares fit to the spectrum at 4.2 K gives H_{hyp} = 175. kG, substantially greater than the value 55 kG inferred just from the observed broadening of the single-line spectrum. From this broadening an estimate of 40 kG is obtained for the width of the distribution in H_{hyp} at 4.2 K arising from compositional fluctuations and small-cluster effects. The model can be applied to unresolved or poorly resolved Mössbauer spectra in superparamagnetic multi-component single-phase alloys of arbitrary composition.

I. INTRODUCTION

In the investigation of weakly magnetic systems using the Mössbauer effect, unresolved or poorly resolved spectra are frequently encountered.[1-6] The coexistence of several non-instrumental broadening mechanisms can make the interpretation of such spectra difficult. In particular Wickman has shown how magnetic resonance relaxation theory can be applied in the analysis of Mössbauer paramagnetic hyperfine structure arising from spin relaxation effects.[2] Lundquist et al have applied this theory to superparamagnetic Ni particles, using a simplified model for the spin relaxation.[3] In this paper we apply the theory to F.C.C. Fe$_{.34}$ Co$_{.52}$ V$_{.14}$ (Vicalloy II). Allowance is made for the possibility of more than one superparamagnetic cluster size and corresponding cluster relaxation time. Although the effect of compositional fluctuations are not included explicitly in the model, an estimate of the width of the distribution of hyperfine fields is obtained from a consideration of the residual broadening at low temperatures.

II. CLUSTER MODEL FOR F.C.C. VICALLOY

We wish to describe a disordered, single-phase superparamagnetic alloy in terms of a cluster model. Such models have been used in discussing the magnetization of many superparamagnetic binary alloys.[7] Thus at a given temperature T, we assume that the alloy consists of superparamagnetic regions separated by paramagnetic regions. This fine-grained magnetic inhomogeneity is presumed to arise from compositional fluctuations. We assume that each super-paramagnetic "cluster" has a magnetic ordering temperature Θ, with T < Θ, which results in a non-vanishing net moment $\vec{\mu}$, the direction of which changes in a random manner with a characteristic spin-flip time τ. This quantity depends on the cluster volume Ω through the bulk anisotropy energy E_a; for zero applied field we have

$$\tau = \tau_0 \exp(E_a/k_B T) \quad , \qquad (1)$$

where the pre-exponential factor τ_0 is, for $T \ll \Theta$, of the order of the inverse of the Larmor precession frequency $\omega_L(H_c) = \gamma H_c$ of the electronic spin in the anisotropy field H_c.[8,9] If we assume that the first-order anisotropy constant K_1 gives the major contribution to E_a and that the [111] direction is the easy direction in F.C.C. Vicalloy, then $E_a = |K_1|\Omega/12$. The Larmor frequency for this case ($K_1 < 0$) is given by $4\gamma|K_1|/3(\mu/\Omega)$,[10] so we take

$$\tau_0 = \tfrac{3}{4} \, \frac{\mu}{\gamma |K_1| \Omega} \quad . \qquad (2)$$

In general we expect there to be a distribution of cluster sizes. The exchange interaction between clusters, although weak, will be sufficient to cause existing clusters to coalesce and new ones to form, as the temperature is lowered. Hence the distribution will be rather sensitive to temperature. Previous work on F.C.C. Vicalloy II indicates that this alloy is superparamagnetic down to 4.2 K.[1] The Mössbauer data show a slightly broadened single-line spectrum at 300 K which broadens with decreasing temperature without development of resolvable hyperfine splitting. The magnetization data was analysed in Ref. 1 assuming only one size of cluster at given T, and indicated the presence of a large number of small clusters at low T. The data has therefore been reanalyzed for $T \leqslant 150$ K assuming two different cluster sizes at each T, one "large" and one "small". If we treat the clusters as non-interacting, then the sample magnetization is

$$M(H,T) = \mu_L N_L L(\mu_L H/k_B T) + \mu_S N_S L(\mu_S H/k_B T) + \chi_p(T) \cdot H, \qquad (3)$$

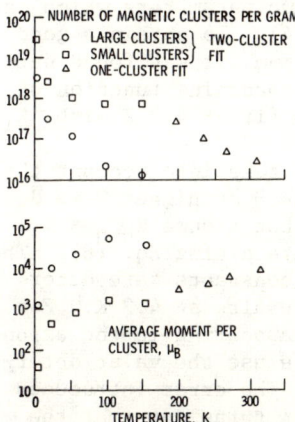

Figure 1. - Cluster parameters vs. temperature for F.C.C. Vicalloy II, determined from magnetization data of Ref. 1 using a two-cluster fit for T≤150 K.

for $T \leqslant 150$ K, where $N_{L,S}$ and $\mu_{L,S}$ are the number per unit volume and moment of the large and small clusters, respectively, and $\chi_p(T)$ is the susceptibility of the paramagnetic regions. This results in a substantially better fit to the data, and the variation of the cluster parameters with T is shown in Fig. 1. The points for $T > 150$ K were obtained previously using the single-cluster fit. As in Ref. 1, the parameters of Fig. 1 can be used to obtain an average moment $<\mu_a>$ per magnetic atom, shown in Fig. 2.

The Mössbauer ^{57}Fe absorption spectrum $I(\omega)$ at a given temperature for zero applied field can be similarly written as

$$I(\omega) = C\left[f_L I_0(\tau_L,\omega) + f_S I_0(\tau_S,\omega) + f_p I_0(\tau_p,\omega)\right], \qquad (4)$$

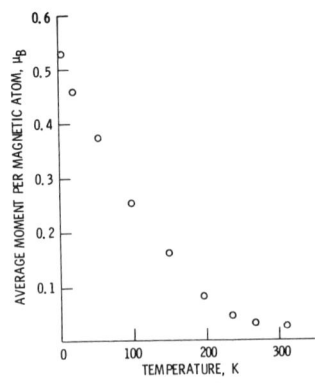

Figure 2. - Average moment per magnetic atom vs. temperature obtained from the parameters of fig 1.

where $f_L = \Omega_L N_L$, $f_S = \Omega_S N_S$ and $f_p = 1 - (f_L + f_S)$ are the fractions of the sample consisting of large clusters, small clusters and paramagnetic regions, respectively. The parameter C fixes the scale of the absorption at each T and was determined from the experimental data by fitting the integrated area under $I(\omega)$ to the observed total sample absorption. The expression for $I_o(\tau,\omega)$ has been derived by others using magnetic resonance relaxation theory based upon the modified Bloch equations.[2,11] It depends implicitly on the hyperfine field H_{hyp} and the isomer shift v_o. In this paper no attempt was made to analyze the magnitude of v_o or its variation with T. The unknown parameters entering Eq. (4) are therefore the cluster volumes, the relaxation times and the hyperfine field. The cluster volumes are not directly determinable from the data. At sufficiently low T their values can be inferred by assuming that the clusters have a common magnetization and that the decrease in $\langle \mu_a \rangle$ with T in Fig. 2 is brought about by an identical decrease in $(f_L + f_S)$ from a value of unity at T = 0. Thus taking

$$f_L(T) + f_S(T) = \langle \mu_a(T) \rangle / \langle \mu_a(0) \rangle , \quad (5)$$

$$\mu_L / \Omega_L = \mu_S / \Omega_S , \quad (6)$$

enables us to obtain Ω_L and Ω_S and hence f_L, f_S and f_p. From Eq. (6), the value of τ_o is the same for the two cluster sizes. The paramagnetic relaxation time τ_p is taken to be τ_o. The quantity $|K_1|$ is not known, but together with H_{hyp} gives a set of two parameters which can be fitted to the spectrum at low T using Eq. (4). To obtain a good fit in the neighborhood of the absorption maximum, it was found necessary to convolute the data with a Gaussian broadening function of full width 2δ. The results of a least-squares fit at 4.2 K with $|K_1|$, H_{hyp} and δ as parameters is shown in Fig. 3.

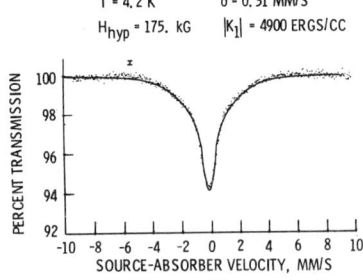

Figure 3. - Mössbauer ^{57}Fe spectrum at 4.2 K. The calculated curve is a least-squares fit using a relaxation model with the parameters indicated.

In order to take into account the effects of finite Θ at higher T we do not use Eq. (5) but assume $H_{hyp} \propto \mu/\Omega \propto \tau_o$, while retaining Eq. (6). The proportionality constants were determined from the results at 4.2 K. For the value of K_1 appearing in the exponent in Eq. (1) we use the value determined at 4.2 K. The error introduced is small since it turns out that the Boltzmann factor is nearly unity at all temperatures investigated.

III. RESULTS AND CONCLUSIONS

The parameters resulting from a least-squares fit to the spectra at various T are shown in Table I. Included is an estimate of the cluster diameter D obtained assuming spherical clusters. The values of N and μ for the clusters were determined for each T by linear interpolation between the data of Fig. 1. The value of $|K_1|$ is $4.9 \cdot 10^3$ ergs/cc, indicating a weak magnetic anisotropy. In fact the exponent

Table I. Values of H, τ_o, f_p, 2δ and D obtained from a least-squares fit to the data

T (K)	H_{hyp} (kG)	$\tau_o (10^{-8}$ s)	f_p	2δ (mm/s)	D_L (Å)	D_S (Å)
4.2	175.	.30	.032	.62	38.	12.
25.	151.	.26	.056	.46	84.	30.
75.	111.	.19	.131	.18	134.	43.
150.	69.	.12	.246	.22	162.	52.
300.	55.	.10	.832	.20	105.	

in Eq. (1) is so small that τ never exceeds τ_o by more than five percent. Therefore, for this system at least, it appears that including more than one cluster size makes very little difference in the analysis of the Mössbauer data. On the other hand the variation of H_{hyp} with T should resemble a Brillouin function if the clusters are characterized by a unique Curie temperature Θ. The plot of H_{hyp} versus T given in Fig. 4 shows that this is not the case. We therefore regard our H_{hyp} as some average over a distribution of hyperfine fields.

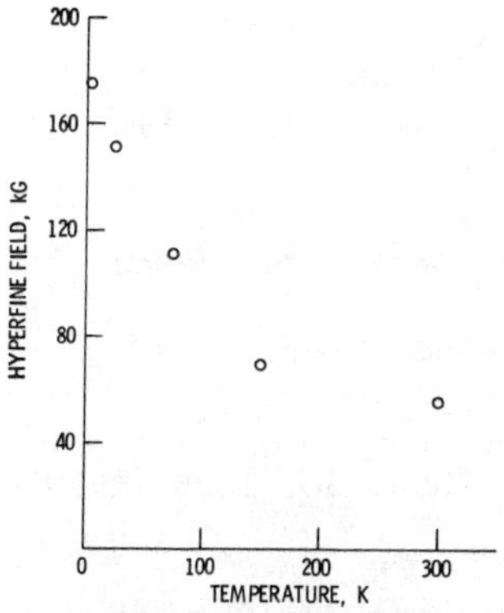

Figure 4. - Hyperfine field vs. temperature.

Such a distribution can arise for small clusters from variations in composition (and hence in Θ) from cluster to cluster and also from the dependence of Θ on cluster volume (and cluster shape). The critical cluster diameter for the onset of these size effects is of order 100 Å or less, which is the case here. (If the cluster size turned out to be greater than a few hundred Ångströms our model would actually be inconsistent, since multiple-domain formation would occur.) These small-cluster effects have been neglected in our model. Eq. (6), for example, essentially assumes that Θ is independent of cluster size. However an estimate of the width of the distribution of H_{hyp} at low T can be obtained in the following way. We assume that the constancy of 2δ at ~20 mm/s for T ⩾ 75 K represents an essentially temperature-independent collision-induced broadening arising from the finite sample thickness; such a broadening corresponds to about three detector channel widths and seems to us to be a reasonable estimate of the instrumental broadening. The additional broadening at 25 K and 4.2 K we attribute to a distribution of H_{hyp}. Eq. (4) can then be used to determine the increase in H_{hyp} required to make up the difference. The result is about 20 kG. This can be interpreted as the half-width of a distribution having a mean value of 175 kG. Thus H_{hyp} = (175. ± 20.) kG, which is substantially larger than the value of 55 kG inferred just from the observed line broadening.

The distribution in H_{hyp} presumably exists at higher T, but the fall-off with T of the mean H_{hyp} and the corresponding relaxation time combine to give an intrinsically narrower Mössbauer line, and this is why we do not consider the effect of the distribution of H_{hyp} on the observed line width at higher T.

Linear extrapolation of Fig. 2 to zero average moment gives 510 K as the point where superparamagnetic behavior disappears. Fig. 4 indicates that there are relatively few clusters large enough really to have a Curie point this high.

The magnetic coupling of the iron in the clusters is not known, but may be ferrimagnetic or antiferromagnetic. It is known that metallic Fe in an F.C.C. matrix tends to be antiferromagnetic. Furthermore the small average moment of the system is difficult to account for if all the magnetic atoms couple ferromagnetically, but can be explained rather simply if it is assumed that the Fe makes no net contribution to the moment while that contributed by the Co is reduced by the V by an amount roughly equal to that in an F.C.C. Co-V alloy having the same V/Co atom ratio.

REFERENCES

1. J. P. Cusick, G. Bambakidis and L. C. Becker, "Amorphous Magnetism," H. O. Hooper and A. M. de Graaf, Eds. (Plenum Press, N.Y., 1973), p. 291.
2. H. H. Wickman, "Mössbauer Effect Methodology," V. 2, I. J. Gruverman, Ed. (Plenum Press, N.Y., 1966), p. 39.
3. R. H. Lindquist, G. Constabaris, W. Kundig and A. M. Portis, J. Appl. Phys. 39, 1001 (1968).
4. U. Gonser, H. Wiedersich and R. W. Grant, ibid, p. 1004.
5. Y. Nakamura, M. Shiga and N. Shikazono, J. Phys. Soc. Japan 19, 1177 (1964).
6. C. E. Johnson, M. S. Ridout and T. E. Cranshaw, Proc. Phys. Soc. 81, 1079 (1963).
7. P. A. Beck, Metall. Trans. 2, 2015 (1971).
8. C. P. Bean and J. D. Livingston, J. Appl. Phys. 30, 120S (1959).
9. W. F. Brown, Jr., ibid, p. 1305.
10. J. O. Artman, Phys. Rev. 105, 62 (1957).
11. H. S. Gutowsky, D. W. McCall and C. P. Slichter, J. Chem Phys. 21, 279 (1953).

Section 8.   Phase Transitions   329

MAGNETO-OPTICAL STUDIES OF METAMAGNETIC PHASE TRANSITIONS IN $FeCl_2$

E. Yi Chen, J. F. Dillon, Jr., and H. J. Guggenheim
Bell Laboratories
Murray Hill, New Jersey   07974

ABSTRACT

In the course of magneto-optical studies of the metamagnet $FeCl_2$ with $\vec{H}$ along [0001], we have encountered unexpected hysteresis phenomena on traversing the mixed phase region. The basic experiment consisted of measuring the net Faraday rotation of monochromatic light passing through a macroscopic area in the center of a much larger sheet as the magnetic field was slowly swept (10^2 Oe/sec) at constant temperature. This provided a sensitive measure of the magnetization. Below the 21.15K tricritical point, antiferromagnetic and paramagnetic phases may coexist over a finite range of applied field. Especially below 4.2K, hysteresis is seen along the low field boundary of the mixed phase region. The variation of the hysteresis with temperature suggests that the process is thermally activated.

INTRODUCTION

We have been stimulated to use optical techniques in examining the metamagnetic phase transition in $FeCl_2$ following their revealing use in the case of dysprosium aluminum garnet (DAG),[1] and the pioneering experimental work on $FeCl_2$ by Jacobs and Lawrence.[2] The anhydrous dichloride of iron has a hexagonal layer structure with Fe^{++} ions in layers perpendicular to the six-fold axis. Their spins are confined to the hexagonal axis. These spins order in zero field below T_N = 23.55K. Within a single layer, the moments are all parallel, and antiparallel to the moments of both adjacent layers. The magnetic properties of $FeCl_2$ have been extensively investigated both experimentally and theoretically by many earlier workers.[2,3] It has been shown that the antiferromagnetic-paramagnetic (A-P) transition is of first order at temperatures below T_t = 21.15K,[4] the tricritical point. Between T_t and T_N this transition is second order. Jacobs and Lawrence[2] have carried out steady-field magnetization measurements over a wide range of temperature and field. They also studied metamagnetic hysteresis at 4.2K with pulse fields. In the present work we report optical Faraday rotation - applied field measurements which give a measure of the magnetization versus applied field. A considerable hysteresis was observed between rotation curves taken in increasing and decreasing magnetic field at temperatures below 4.2K. Such experiments as these should be valuable in clarifying details of the metamagnetic transition and its hysteresis.

EXPERIMENTAL TECHNIQUES

The $FeCl_2$ crystals were prepared by reacting high purity iron metal sponge with Cl_2 at about 950C. This was done by placing a quartz boat containing the iron in a reaction tube. The tube was positioned in a resistance furnace mounted on a traveling zone apparatus. With a short hot zone traversing the boat at 5 mm/hr, 20-40μ thick $FeCl_2$ samples were produced from which our samples were carefully selected.

Thin (0001) sheets with dimensions roughly equal to 25μ×3mm× 4mm were used in all the measurements. The variation of magnetic rotation versus field was made as shown in Fig. 1. Monochromatic linearly polarized light illuminated a 1.2mm diameter spot on the sample. The sample was immersed in cryogenic fluid (H_2 or He). The specimen was subjected to an external magnetic field parallel to the propagation direction and the [0001] crystal axis. The light then encountered a rotating Glan-Thomson prism driven by a synchronous motor with velocity ω. The magneto-optical rotation was obtained from the phase of the 2ω signal at the detector relative to a signal obtained from the rotating analyzer. The output of the phase computer is proportional to rotation. Field strength was detected by means of a Hall-probe.

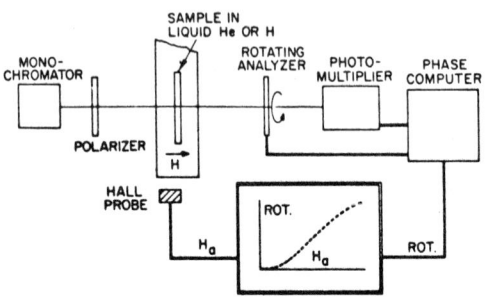

Fig. 1 Apparatus used to plot $\rho(H_a)$ automatically.

ROTATION MEASUREMENTS

Magneto-optical rotation curves for various temperatures are shown in Fig. 2. For temperatures below the tricritical point, there exists a straight line portion of the rotation - field plot identifiable as the first order phase transition region. The ends of the straight line are the boundary points of the mixed phase region on the H_a-T plane. Fields greater than 16 kOe were not available to us and so for temperatures below about 17K we were unable to pass through the mixed phase into the paramagnetic phase. At .35μ the rotation measured at 19.05K at the coexistence field for the antiferromagnetic and paramagnetic phases are respectively: $\rho(anti) = 3.6 \times 10^3$ deg/cm and $\rho(para) = 26.0 \times 10^3$ deg/cm. Clearly, in a single phase region the rotation serves as an excellent measure of magnetization as it varies with field and temperature. It can also be shown that in a mixed phase regime the rotation can be a good measure of net magnetization if the inhomogeneities are so small that most of the light samples several regions. Specifically we may write

$$\rho = R(\lambda)M(H,T) \qquad (1)$$

where ρ = specific rotation in deg/cm, and M = magnetization in emu/cm^3. The validity of this approximation will be discussed elsewhere. We quote a preliminary value of $R(.35\mu)$ = 63 deg cm^2/emu. This was measured at 19.0K and contains a value of the magnetization as measured by R. J. Birgeneau et al.[4] The shape of these rotation curves agrees quite well with the magnetization curves given by Jacobs and Lawrence.[2] Furthermore, the plots are consistent with the phase diagram deduced by Birgeneau et al. from neutron studies.

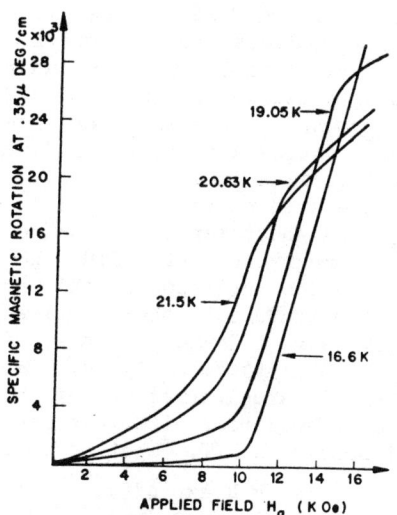

Fig. 2 Plots of $\rho(H_a)$ at several temperatures.

In the course of the rotation measurements we have encountered a diffraction effect in the mixed state region when using thick samples (e.g. 215µ). This scattering by A-P domains is similar to the domain diffraction seen earlier in a number of compounds with strong magneto-optical effects.[5] A few hundred oersteds inside the mixed phase at 1.5K for this 215µ thick sample, we deduced a domain spacing of 12µ from the size of the diffraction rings observed with .6238µ light. The diffraction effect is so pronounced that it is essential to use only very thin samples for rotation measurements in the mixed phase.

METAMAGNETIC HYSTERESIS

Below 4.2K we have observed a hysteresis phenomenon on entering the mixed state which we believe to be closely related to the "metamagnetic hysteresis" studied by Jacobs and Lawrence[2] with pulsed fields. Though the fields available limited us to the low field boundary of the mixed phase, we confidently believe similar effects will be observed at the high field edge. When as in Fig. 3 the applied field increases linearly with time, the rotation does not change significantly until we reach the edge of the mixed phase near 11.0 kOe. Drawing on the insight gained from microscope observations of the metamagnet DAG,[1] we believe that nuclei of the paramagnetic phase then appear and grow in response to the applied field. If H_a changed infinitely slowly, $\rho(H_a\infty)$ would be a straight line from one edge of the mixed phase region to the other with a slope determined by the demagnetizing factor of the sample. This we might call the equilibrium curve. However, for finite sweep rates the magnetization may be unable to follow this curve. We suggest below that the wall motion at these low temperatures is activated and is relatively slow. The wall area of the nuclei is at first small, and this limits $\dot{M}$. For a time M lags the equilibrium value for H_a. As the area of moving metamagnetic wall

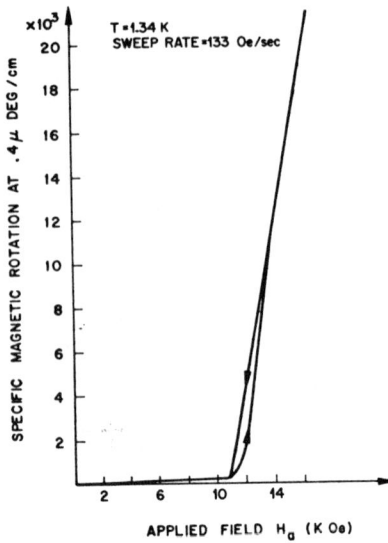

Fig. 3 Plots of $\rho(H_a\uparrow)$ and $\rho(H_a\downarrow)$.

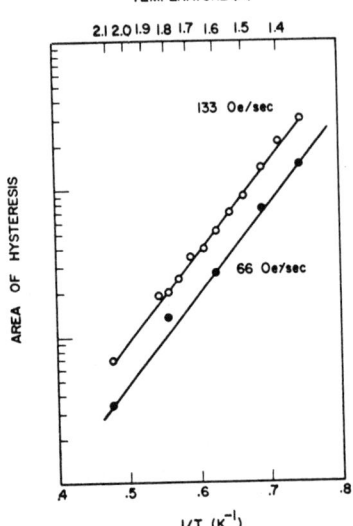

Fig. 4 Data on the area of the metamagnetic hysteresis at several temperatures plotted to show the logarithmic dependence on 1/T.

increases, the total M can easily follow H_a. We find that $\rho(H_a\downarrow)$ going down follows the straight line $\rho(H_a\infty)$ right to the phase boundary. This is possible because coming from the center of the mixed phase region, there is a complicated domain structure with sufficient wall area.

The area enclosed by $\rho(H_a\uparrow)$ and $\rho(H_a\downarrow)$ may be taken as a measure of the metamagnetic hysteresis. We have examined the variation of this area with temperature in the relatively narrow range 1.34-2.1K. Early results are plotted in Fig. 4. The hysteresis area varies by a factor of 50. Furthermore its logarithm is linear with 1/T. That is to say

$$\text{Area} \propto \dot{H} \exp(\delta/T), \quad (2)$$

where $d\ln(\text{Area})/d(1/T) = \delta = 14.2K$. The factor $\dot{H}$ in the first expression has been tentatively inserted on the slight evidence that a single factor 2 change in dH_a/dt has halved the area. Note that this dramatic change in hysteresis takes place over a temperature range in which such properties as sublattice magnetization and anisotropy of this 23.55K antiferromagnet do not vary perceptibly. We are inclined to interpret the logarithmic dependence on 1/T as indicating that the progress of the metamagnetic phase transition by (A-P) domain walls is thermally activated. One would not expect a thick A-P wall to see energy wells in a lattice. Thus activation of wall motion seems to imply that the A-P walls are of thickness comparable to lattice constant. The obvious analogy here is to thermally activated motion of similarly thin walls in ferroelectric crystals.[6]

An alternative mechanism for (2) has been suggested by A. Rosencwaig and L. R. Walker, namely that the 14.2K may simply correspond to the activation energy of the nucleation process. It should be possible to determine by experiment whether the activated process is in the nucleation or the wall motion.

The mechanism underlying this hysteresis may be different from that responsible for the effect seen by Jacobs and Lawrence.[2] Note that their sweep rates were perhaps 10^7 Oe/sec whereas ours are only 10^2 Oe/sec. With further experiments it may be possible to resolve the question as to whether the nucleation-wall motion mechanisms mentioned above are pertinent to the high speed magnetization in pulse field experiments.

It should be noted that the phenomenon described here is distinct from the "I-S Hysteresis" recently reported in another Ising antiferromagnet, DAG. That process pertains to the different magnetization curves obtained with a variety of different magnetic past histories. No trace of it was seen in $FeCl_2$.

CONCLUSION

We have reported preliminary magneto-optical measurements of the metamagnetic phase transition in the simple compound $FeCl_2$. The techniques give sensitive information on the transition and we believe it will be of great value in studies close to the tricritical point. Measurements of the hysteresis observed on entering the mixed phase at various temperatures suggest that a thermally activated process is involved. This may lie in the nucleation itself or in the motion of the A-P walls. If the wall motion is thermally activated, it implies that the A-P walls are very thin, perhaps comparable to a lattice constant. Direct examination by optical microscope should greatly facilitate the interpretation of the nucleation and wall motion processes.

We are pleased to acknowledge a series of helpful discussions with R. J. Birgeneau, with L. R. Walker and with R. C. Miller.

REFERENCES

1. J. F. Dillon, Jr., E. Yi Chen, and W. P. Wolf, Proceedings of the International Conference on Magnetism, Moscow, 22-28 August, 1973 (In Press).
2. I. S. Jacobs and P. E. Lawrence, Phys. Rev. 164, 866-878 (1967).
3. R. J. Birgeneau, W. B. Yelon, E. Cohen, and J. Makovsky, Phys. Rev. B 5, 2607-2615 (1972). Many references to earlier work are given in this and in Ref. 2.
4. Birgeneau, Shirane, Blume and Koehler (to be published).
5. J. F. Dillon, Jr., and J. P. Remeika, J. Appl. Phys. 34, 637-640 (1963); J. C. Suits, J. Appl. Phys. 38, 1500 (1967) and 39, 570 (1968).
6. R. C. Miller and G. Weinreich, Phys. Rev. 117, 1460-1466 (1960).

MICROSCOPE STUDIES OF THE MIXED PHASE REGION OF THE ISING ANTIFERROMAGNET DYSPROSIUM ALUMINUM GARNET (DAG)

J. F. Dillon, Jr. and E. Yi Chen
Bell Laboratories
Murray Hill, New Jersey 07974

W. P. Wolf*
Yale University
New Haven, Connecticut 06520

ABSTRACT

The magnetic phase diagram of DAG has been of interest recently because it appeared to possess a tricritical point in a readily accessible region of field and temperature[1] (T_t=1.66K, H_i^t=3.35 kOe). The first order region below T_t is of special interest since the paramagnetic and antiferromagnetic phases coexist over a range of applied field by virtue of demagnetizing effects. It is thus possible to study the transition in some detail. Recently we have reported optical studies of DAG in this region.[2] These comprised measurements of magneto-optical rotation as a function of field and temperature as well as microscopy of the mixed phase. Here we present further results of the microscope studies. Our samples were thin (111) sheets (25-200μm thick) with the field along [111]. The two phases are distinguishable by virtue of their different Faraday rotations (e.g. at 1.30K: ρ(para)=5500°/cm, ρ(anti)=400°/cm for H_{int}=3.7 kOe). On entering the mixed phase from the high or low field boundaries, nuclei of the new phase appear, then rapidly grow to extend all the way through the specimen. Soon this structure apparently breaks up into separate needles of the new phase, which repel each other. A complicated evolutionary process develops over a long period of time, greater than 1000 seconds, and a final state is reached in which the two phases are very finely intermixed on a scale of less than 3μm. In the center of the coexistence region the two phases seem disposed in a sponge-like fashion. As in the more familiar growth of crystals from a liquid the details of the magnetization structure seen at any point in field and temperature depend greatly on the way in which that point was reached.

A detailed report of this work will be published elsewhere.

*Supported in part by AROD.

1. D. P. Landau, B. E. Keen, B. Schneider, W. P. Wolf, Phys. Rev. B 3, 2310 (1971). This paper contains many references to earlier work.
2. J. F. Dillon, Jr., E. Yi Chen, W. P. Wolf, Proceedings of International Conference on Magnetism, Moscow, September 22-28, 1973, in press.

MAGNETIC PHASE BOUNDARIES AND SPIN WAVE RENORMALIZATION IN TWO ANTIFERROMAGNETIC COMPOUNDS[+]

John E. Rives, Som N. Bhatia, and V. Benedict
University of Georgia, Athens, GA 30602

ABSTRACT

Differential magnetic susceptibility measurements on $MnCl_2 \cdot 4H_2O$ and $CoCl_2 \cdot 6H_2O$ have been used to map out complete H-T phase diagrams for these two antiferromagnetic compounds down to $T/T_N \simeq 0.15$. The phase boundaries bordering the paramagnetic state near absolute zero with the external field along the preferred direction, as well as along perpendicular directions are well represented by the relation $H_c = H_c(0)(1 - C_2 T^{5/2} - C_3 T^{7/2})$. It has been shown by Anderson and Callen that this phase boundary should reflect the temperature dependence of the spin-wave renormalization. They find a dominate $T^{3/2}$ dependence. Their argument is based on the fact that the two $T^{3/2}$ terms which cancel in the Dyson ferromagnetic formalism, in fact add in this case, due to the repulsive interaction among antiferromagnetic spin-waves. The data follow a predominately $T^{5/2}$ dependence with a small $T^{7/2}$ correction. No evidence for a $T^{3/2}$ dependence is found. A further discussion of this point will be presented.

INTRODUCTION

The magnetothermal properties of simple two-sublattice antiferromagnets with relatively small single-ion anisotropy have been the subject of considerable interest for many years. It is well known that such materials can exist in three distinct phases. With an external magnetic field applied along the preferred direction of spin alignment the antiferromagnetic(AF) phase is stable at low fields. At a certain critical field the spin system undergoes a first order transition to the spin-flop(SF) phase, and finally, at still higher fields there is a transition to the paramagnetic(P) phase. For systems with sufficiently small exchange energy it is possible to investigate the entire phase diagram with reasonable laboratory magnets.

We have used isothermal differential susceptibility measurements to study the phase transitions in $MnCl_2 \cdot 4H_2O$ and $CoCl_2 \cdot 6H_2O$. In this report we will restrict our attention to the phase boundaries bordering the paramagnetic phase near absolute zero.

THEORY

The statistical mechanics of a simple uniaxial Heisenberg antiferromagnet has been analyzed by a Green-function technique by Ander-

[+]Research sponsored in part by the National Science Foundation.

son and Callen[1], and by standard spin-wave theory by Feder and Pytte[2]. The temperature dependence of the renormalized spin-wave energies was determined by including spin-wave interactions, in a manner similar to that of Dyson[3] for the ferromagnet. The effective spin-wave interaction in the antiferromagnet is repulsive rather than attractive as in the ferromagnet. Thus the dynamical correction inverts the sign of the dynamical-kinematical term and doubles the effect of the pure kinematical term. This yields a dominate $T^{3/2}$ term in the renomalization factor for antiferromagnets.

In the paramagnetic phase the spin configuration is identical to that of a ferromagnet. However, due to the negative exchange interaction, the spectrum is inverted with respect to the ferromagnetic case. The minimum in the spectrum in a simple cubic structure is expected to occur at the corner of the Brillouin zone. If the field is reduced in the paramagnetic phase a transition to the spin-flop phase occurs when the minimum in the spectrum reaches zero energy. Since the spin-wave energies are renormalized by the temperature dependent renormalization factor, the critical field for the SF to P transition should reflect the temperature dependence of the spin-wave renormalization. Following Anderson and Callen we expect a temperature dependence given by $H_c = H_c(0)(1-C_1 T^{3/2} - C_2 T^{5/2})$. Feder and Pytte obtain an identical result, but do not extend their work to include the $T^{5/2}$ term.

CRYSTAL STRUCTURE AND SPIN ORIENTATION

$MnCl_2 \cdot 4H_2O$ has a monoclinic crystal structure with the angle $\beta = 99°6'$. The preferred direction for spin alignment below $T_N = 1.62$ K has been determined by neutron diffraction[4] to be between the crystallographic c-axis and the c^* direction ($\perp$ to ab-plane), $2.8° \pm 1.4°$ from the c^* direction. $CoCl_2 \cdot 6H_2O$, which orders at 2.29 K is also monoclinic with the angle $\beta = 122°20'$. The preferred direction coincides with the crystallographic c-axis[5]. Samples of both compounds were shaped into spheres for these measurement.

RESULTS AND DISCUSSION

The differential susceptibility $\chi = dM/dH$, is measured as a function of magnetic field at constant temperature for a number of temperatures between about 0.3 K and T_N. Data for $MnCl_2 \cdot 4H_2O$ at T=0.297 K are shown in Fig. 1. The data are essentially the same as those obtained earlier by one of us (JER)[6] on unshaped samples. At the AF to SF transition χ rises abruptly to a value of 1/N, where $N = 4\pi/3$ is the demagnetizing factor for a sphere. The susceptibility remains constant over a range of external field equal to $N\Delta M$, where ΔM is the increase in magnetization at the first order transition. It then drops to a value approximately equal to $\chi_\perp$, and remains almost constant throughout the SF phase. At the SF to P transition there is a discontinuous change in slope of χ, after which χ decreases rapidly toward zero. The critical fields for both transitions are easily determined by direct examination of the data.

From data taken at a series of temperatures the entire H-T phase

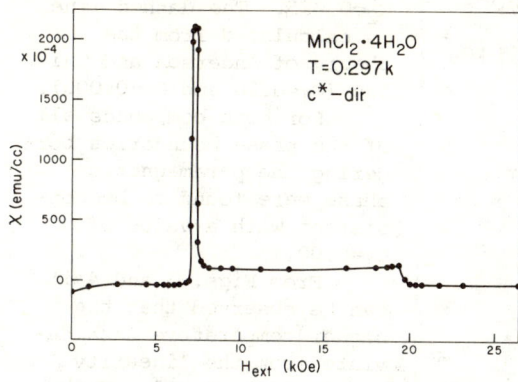

Fig. 1 Differential susceptibility of $MnCl_2 \cdot 4H_2O$ at T=0.297 K.

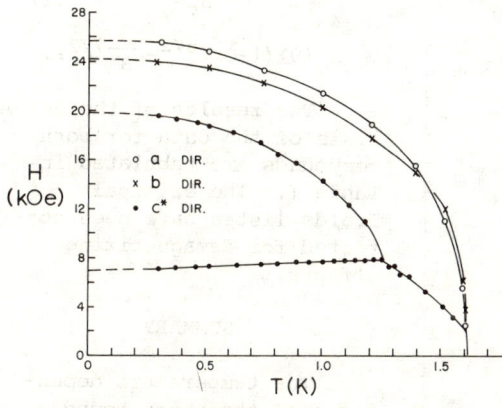

Fig. 2 Magnetic phase diagram for $MnCl_2 \cdot 4H_2O$.

diagram can be determined. Figure 2 shows the resulting phase boundaries for $MnCl_2 \cdot 4H_2O$ for the external field along the preferred direction, as well as along the two perpendicular directions. In a perpendicular direction only a single AF to P transition exists. Although the AF to SF and the AF to P transitions, with the field along the preferred direction, are also of interest, we will restrict our discussion to the transitions bordering the paramagnetic phase near T=0 K.

In order to compare the data with the above relation for H_c it is convenient to rewrite this equation in the form

$$\Delta H_c / H_c(0) T^{3/2} =$$
$$(H_c(0) - H_c)/H_c(0) T^{3/2} = \quad (1)$$
$$C_1 + C_2 T,$$

where the left hand side of the equation is now a linear function of temperature. The data for the SF to P transition in $MnCl_2 \cdot 4H_2O$ are presented in the form of Eq. 1 in Fig. 3.

The solid line represents the best fit to the data which yields values of the coefficients $C_1=0.00$ and $C_2=0.265$. The dashed line is the behavior expected from the theory of Anderson and Callen[1] for this compound ($C_1=0.43$ and $C_2=0.07$). The data for the phase boundaries in the perpendicular directions are also consistent with $C_1=0.00$, but with a somewhat smaller value of C_2. The total results are tabulated below.

The H-T phase diagram for $CoCl_2 \cdot 6H_2O$ is similar to that for $MnCl_2 \cdot 4H_2O$, but with the SF to P transition occuring at an external field of 41.6 kOe. at T=0.

The data for the SF to P transition in $CoCl_2 \cdot 6H_2O$ is presented in Fig. 4. The solid line again represents the best fit to the data

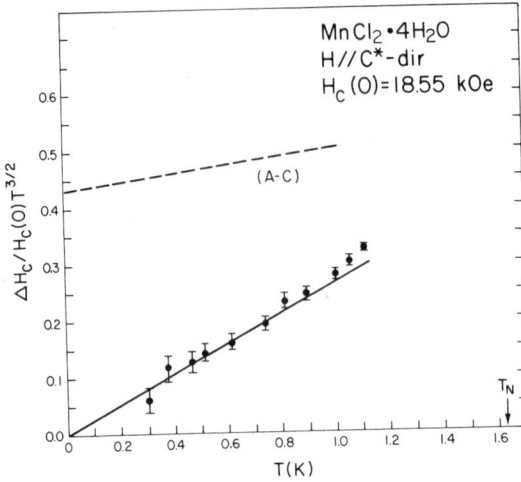

Fig. 3 Temperature dependence of the SF to P transition in $MnCl_2 \cdot 4H_2O$.

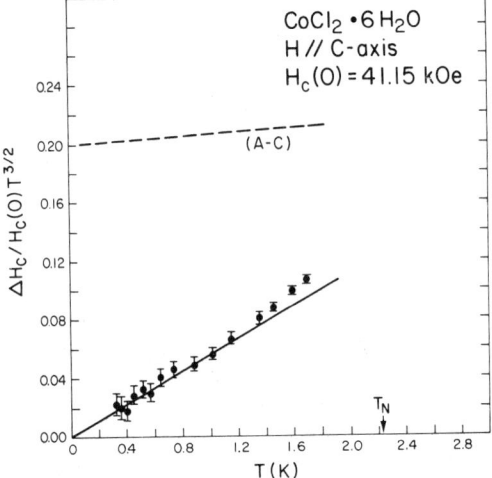

Fig. 4 Temperature dependence of the SF to P transition in $CoCl_2 \cdot 6H_2O$.

which yields values of the coefficients $C_1=0.000$ and $C_2=0.055$. The dashed line is calculated from the theory of Anderson and Callen ($C_1=0.20$ and $C_2=0.006$).

For both compounds all of the phase boundaries bordering the paramagnetic phase were found to be consistent with a value of $C_1=0.00$.

From Figs. 3 and 4 it can be observed that the higher temperature data deviate from the linearity predicted by Eq. 1. With $C_1=0.00$ further analysis of the data indicated that it was possible to obtain a good fit to the empirical relation

$$H_c = H_c(0)(1-C_2T^{5/2}-C_3T^{7/2}). \quad (2)$$

The results of the analysis of the data for both compounds are tabulated in Table I. The critical fields listed have been corrected for demagnetizing effects.

SUMMARY

The temperature dependence of the phase boundaries bordering the paramagnetic phase of $MnCl_2 \cdot 4H_2O$ and $CoCl_2 \cdot 6H_2O$ are found to be inconsistent with the theories of Anderson and Callen[1] and Feder and Pytte[2]. The method of treating the spin-wave interactions which leads to the renormalization factor appears to be theoretically sound. It has been suggested that the argument that the critical field should directly reflect the temperature dependence of the renormalization factor might break down in the present cases if for any reason the minimum in the spectrum occurs at some point other than at the corner of the Brillouin zone. The spectra of these compounds have not been experimentally determined to date, so that further speculation as to the absence of the $T^{3/2}$ term is difficult.

TABLE I. Comparison of theoretical and experimental critical field parameters for $H_c=H_c(0)(1-C_1T^{3/2}-C_2T^{5/2}-C_3T^{7/2})$.

		$H_c(0)$ kOe	C_1	C_2	C_3
		$MnCl_2 \cdot 4H_2O$			
Th.	A-C		0.43	0.07	
	F-P		0.43		
Exp.					
c*-dir.		18.55±0.05	0.00±0.02	0.240±0.02	0.04 ±0.01
b -axis		22.95±0.05	0.00±0.015	0.160±0.015	0.008±0.004
a-axis		24.55±0.05	0.00±0.015	0.180±0.01	0.010±0.008
		$CoCl_2 \cdot 6H_2O$			
Th.	A-C		0.20	0.006	
	F-P		0.20		
Exp.					
c-axis		41.15±0.05	0.000±0.004	0.050±0.006	0.007±0.001
b-axis		43.20±0.05	0.000±0.004	0.048±0.005	0.005±0.001

REFERENCES

1. F. Burr Anderson and Herbert B. Callen, Phys. Rev. **136**, A1068 (1964).
2. J. Feder and E. Pytte, Phys. Rev. **168**, 640 (1968).
3. F. J. Dyson, Phys. Rev. **102**, 1217, 1230 (1956).
4. R. F. Altman, S. Spooner, J. E. Rives and D. P. Landau, Winter Meeting of the Amer. Crystallographic Soc., 1971, Columbia, SC
5. R. Kleinberg, J. Chem. Phys. **33**, 2660 (1971).
6. John E. Rives, Phys. Rev. **162**, 491 (1967).

MAGNETIC AND STRUCTURAL TRANSITIONS IN NdS, DyS and ErS

L. J. Tao, J. B. Torrance, and F. Holtzberg
IBM Research Center, Yorktown Heights, New York 10598

ABSTRACT

Magnetic susceptibility measurements on single crystals of the metallic compounds NdS, DyS, and ErS indicate an antiferromagnetic transition at 11, 34 and 7.3°K, respectively. Below T_N the magnetization is anisotropic. For example, both NdS and DyS have a maximum χ along <u>one</u> of the original [100] axes and a minimum along another. Note that the symmetry corresponding to this χ is lower than the cubic symmetry of the crystal at room temperature, implying that a crystallographic distortion has occurred. A tetragonal distortion is able to describe the χ result. The behavior of ErS is more complicated. For a small field ($H \lesssim 250$ G), the anisotropy of χ is similar to but much smaller than that of DyS and NdS. For intermediate fields $\chi = M/H$ is no longer independent of field, but for fields larger than 2kG, χ becomes stable again. Rotating about a [001] axis, χ_{max} is along the [100] and [010] directions and χ_{min} is along the [110] and [1$\bar{1}$0]. These results can be interpreted as due to a small tetragonal distortion whose direction can be rotated by a large field.

The types of distortions for different Rare-Earth crystals have been discussed recently using a tunneling model[1]. With two simple assumptions about the crystal field parameters, the directions of the minimum crystal field potential can be determined. There are either six potential minima in the [100] directions for negative β_J or eight minima in the [111] directions for positive β_J, where β_J is the reduced matrix element. These minima states are composed of low-lying states of the Rare-Earth ions, and a Jahn-Teller type distortion is expected if these states have appropriate degeneracy.

Assuming the strains of the tetragonal and trigonal distortion are comparable, the theory predicts a tetragonal distortion for negative β_J with [100] minima and trigonal for positive β_J with [111] minima. Since Nd^{3+} and Dy^{3+} have negative β_J, and tetragonal distortions are observed, the tunnelling model is adequate. Er^{3+} has positive β_J, however, yet a small tetragonal distortion is observed. This seems to indicate that the strain of a tetragonal distortion is larger than that of a trigonal. A detailed report will be published elsewhere.

1. G. T. Trammell, Phys. Rev. <u>131</u>, 932 (1963). E. Pytte, and K. W. H. Stevens, Phys. Rev. Letters <u>27</u>, 862 (1971).

MAGNETIC STRUCTURE OF FeI_2 AND PHASE TRANSITIONS IN HIGH MAGNETIC FIELDS PARALLEL TO THE SPIN DIRECTION

J. Gélard, A.R. Fert and P. Carrara,
Laboratoire de Physique des Solides
INSA, Av. de Rangueil, 31077 TOULOUSE (France).

ABSTRACT

We have determined the magnetic structure [1] of FeI_2 by neutron diffraction experiments at 4,2 K (T_N = 9,3 K). This structure is identical to $MnBr_2$ structure, but the spins are directed along the c axis.

This magnetic structure and an estimate of the principal exchange coupling parameters lead to an original behavior [2] in a magnetic field parallel to c axis. At low temperature, two successive first order transitions are observed at 46 and 120 kOe and saturation is reached for 140 kOe.

An experimental study of phase transitions of FeI_2 by magnetization measurements in high static (150 kOe) and pulsed (250 kOe) magnetic fields is presented. An original phase diagram is given.

This compound is different from $FeCl_2$ and $FeBr_2$ (previously studied) which are two sublattice antiferromagnets and have a typically metamagnetic behavior.

(1) J. Gélard, A.R. Fert, P. Mériel, Y. Allain, Sol. State Comm. (to be published Dec. 73).

(2) A.R. Fert, J. Gélard, P. Carrara, Sol. State Comm. (to be published Oct. 73).

THE ROLE OF HARMONICS IN THE FIRST ORDER ANTIFERROMAGNETIC TO PARAMAGNETIC TRANSITION IN CHROMIUM

C. Y. Young
Physics Department, Northeastern University,
Boston, Massachusetts 02115
and
Institute of Physics, National Tsing Hua University,
Taiwan, Republic of China

J. B. Sokoloff
Physics Department, Northeastern University,
Boston, Massachusetts 02115

ABSTRACT

Physical reasons are given for why many simple two band models of chromium, even those with imperfect nesting, fail to give a first order transition. It is then shown that a three band model, for which the spin density wave can have harmonics can give a first order transition. Since we find that when $T_N = 0$ the transition is always second order, it seems likely that this is the mechanism which gives a first order transition in the Kimball-Falicov model. Effects of alloying on the order of the transition are also studied.

The existence of a first order antiferromagnetic to paramagnetic transition in chromium[1] is a phenomenon in the itinerant theory of magnetism which has defied explanation for many years. Kimball and Falicov[2] proposed a model band structure for chromium which gave a first order transition, but it is not known which feature of their model makes the transition first order. Rice,[3] and Malaspinas and Rice[4] have shown by model calculations that the first order transition most likely cannot be explained by a simple two band Hartree-Fock theory for the parameters relevant to chromium, even when magneto-elastic coupling is included.

In the two band model, the problem is to find under what conditions the "gap equation,"

$$1 = \frac{U}{N} \sum_{\vec{k}} \frac{f(E_1(\vec{k})) - f(E_2(\vec{k}))}{\left[\left(\frac{\varepsilon_a(\vec{k}) - \varepsilon_b(\vec{k}+\vec{Q})}{2}\right)^2 + g^2\right]^{\frac{1}{2}}}, \quad (1)$$

has no solutions for small g but has solutions for larger g, for sufficiently low temperatures. Here g is the gap parameter, $E_{1,2}(\vec{k})$ are the energies of lower and upper bands in the antiferromagnetic state, U is the interaction, $f(E)$ is the Fermi function, $\varepsilon_{a,b}(\vec{k})$ are the energies of the two paramagnetic state bands which are mixed to form the antiferromagnetic state, $\vec{Q}$ is the wave vector of the spin density wave, and N is the number of atoms in the crystal. One way that equation (1) can have solutions for larger g but no solutions near g = 0 is if the Fermi level never intersects the surface $\epsilon_a(\vec{k}) = \epsilon_b(\vec{k} + \vec{Q})$, because then when g becomes very small, the right hand side of equation (1) becomes completely independent of g. At this point any further increase in temperature will reduce the right hand side below 1 and decreasing g further will not keep this from happening. This happens in Rice's "Fermi spheres of unequal radius" model[3] if $\vec{Q}$ is held fixed and there is no reservoir. Also the model proposed by Mattis and Langer[5] to explain a first order metal-insulator transition has a similar gap equation, and their way of doing the next nearest neighbor tight binding approximation has the feature that the Fermi energy never falls on the $\epsilon_a(\vec{k}) = \epsilon_b(\vec{k} + \vec{Q})$ surface. On the other hand Liu's model,[6] which has the Fermi surface intersect this surface, always gives a second order phase transition. So, it is clear what features a model should have to give a first order transition; there should be few points of intersection of the two Fermi surface sections. In such a case, however, the probability of having a stable antiferromagnetic state is very small.

One feature of real chromium not included in the two band model is the existence of harmonics. To study the effects of harmonics on the order of the transition, we consider a three band model like that studied by Penn and Falicov.[7] In our model, there is a band whose Fermi surface surrounds the origin in $\vec{k}$-space denoted by the symbol a and two bands with Fermi surfaces of similar shape centered around points near $\vec{k} = \pm\vec{Q}$, where $\vec{Q}$ is the wave vector of the spin density wave state (so as to allow for imperfect nesting, they are not centered at exactly $\vec{k} = \pm\vec{Q}$.) In the Hartree-Fock approximation

$$G^{-1}(\vec{k},i\omega_\nu) = \begin{pmatrix} (i\omega_\nu - \epsilon_a(\vec{k}))1 & -g_{ab_-} & -g_{ab_+} \\ -g^+_{ab_-} & (i\omega_\nu - \epsilon_{b_-}(\vec{k}))1 & -g_{b_-b_+} \\ -g^+_{ab_+} & -g^+_{b_-b_+} & (i\omega_\nu - \epsilon_{b_+}(\vec{k}))1 \end{pmatrix} \quad (2)$$

where 1 is the unit 2 × 2 matrix, g is a 2 × 2 matrix in the spin indices, and the one electron energies are denoted by the notation $\epsilon_{b_+}(\vec{k}) = \epsilon_b(\vec{k} + \vec{Q})$ and $\epsilon_{b_-}(\vec{k}) = \epsilon_b(\vec{k} - \vec{Q})$. The notation of Abrikosov, et al. is used.[8] G is found to have b_-b_+ elements which are diagonal in the spin index, implying a harmonic of the spin density wave with wave vector $2\vec{Q}$ which is a charge density wave. The self-consistency conditions on the g's are

$$g_{ab_-} = 1 \; \Sigma_{\nu'\vec{k}'} [U_{ab_+b_+a} Tr_\sigma G_{b_+a}(\vec{k}',i\omega_{\nu'}) + U_{ab_-ab_-} Tr_\sigma G_{ab_-}(\vec{k}',i\omega_{\nu'})]$$

$$- \Sigma_{\nu'\vec{k}'} [v_{ab_+b_-a} G_{b_+a}(\vec{k}',i\omega_{\nu'}) + v_{aa,b_-b_-} G_{ab_-}(\vec{k}',i\omega_{\nu'})], \quad (3a)$$

$$g_{ab_+} = 1 \; \Sigma_{\nu'\vec{k}'} [U_{ab_-b_-a} Tr_\sigma G_{b_-a}(\vec{k}',i\omega_{\nu'}) + U_{ab_+ab_+} Tr_\sigma G_{ab_+}(\vec{k}',i\omega_{\nu'})]$$

$$- \Sigma_{\nu'\vec{k}'} [v_{aa,b_+b_+} G_{ab_+}(\vec{k}',i\omega_{\nu'}) + v_{ab_-b_+a} G_{b_-a}(\vec{k}',i\omega_{\nu'})], \quad (3b)$$

$$\delta = g_{b_-b_+} = 1 \; \Sigma_{\nu'\vec{k}'} [U_{b_-b_+b_-b_+} Tr_\sigma G_{b_-b_+} - v_{b_-b_+b_+b_-} G_{b_-b_+}(\vec{k}',i\omega_{\nu'})],$$

$$\quad (3c)$$

where $v(\vec{r},\vec{r}')$ and $U(\vec{r},\vec{r}')$ are respectively the screened and unscreened interactions used by Halperin and Rice[9] and

$$U_{iji'j'}(\vec{k},\vec{k}') = \int d^3r d^3r' \varphi^*_{i\vec{k}}(\vec{r})\varphi_{j\vec{k}}(\vec{r}) U(\vec{r},\vec{r}') \varphi_{i'\vec{k}'}(\vec{r}') \varphi^*_{j'\vec{k}'}(\vec{r}'),$$

and similarly for v. Tr_σ signifies taking the trace over spin indices. When we invoke inversion and time reversal symmetries we have only two parameters, δ and g, given by

$$\delta = \frac{v'}{\beta} \frac{1}{N} \Sigma_{\nu \vec{k}} \frac{(i\omega_\nu - \epsilon_a(\vec{k}))\delta + g^2}{DET}, \quad (4a)$$

$$g = \frac{v}{\beta} \frac{1}{N} \Sigma_{\nu \vec{k}} g \frac{i\omega_\nu - \epsilon_{b_-}(\vec{k}) + \delta}{DET}, \quad (4b)$$

where DET is the determinant of the 3 x 3 matrix of equation (2) treating each element as a c-number instead of a 2 x 2 matrix. [Then, the determinant is a c-number times the unit 2 x 2 matrix (since the square of a Pauli matrix is a unit matrix). DET is this c-number.]

$$v' = v_{b_-b_-b_+b_+} - 2U_{b_-b_+b_-b_+},$$

$$v = v_{aab_-b_-} + v_{ab_+b_-a},$$

evaluated at relevant values of $\vec{k}$ and $\vec{k}'$. To get an idea of what is happening, we expand equations (4a) and (4b) to second order in g and first order in δ and substitute for δ in (4b) using (4a). Using Fedders and Martin[10] like energy bands, we obtain the gap equation

$$1 = \rho(\epsilon_f)v \int_0^\infty \frac{d\xi}{\xi} \tanh \frac{\xi}{2kT} - \rho(\epsilon_f)v \frac{7}{8}\zeta(3) \frac{g^2}{(\pi k T_N)^2} + g^2 vv'A^2 \quad (5a)$$

where $\quad A = \frac{1}{N} \Sigma_{\vec{k}} \frac{2}{\epsilon_{b_+}(\vec{k}) - \epsilon_{b_-}(\vec{k})} \frac{f(\epsilon_a(\vec{k})) - f(\epsilon_{b_+}(\vec{k}))}{\epsilon_{b_+}(\vec{k}) - \epsilon_a(\vec{k})}$,

and where ζ is the Rieman zeta function and $\rho(\epsilon_f)$ is the density of states at the Fermi energy. Clearly if $vv'A^2 > \rho(\epsilon_f)v \times \frac{7}{8}\zeta(3) \frac{g^2}{(\pi k T_N)^2}$ the transition is first order. One immediate result is that when T_N is zero, the transition is always second order. This is consistent with what Kimball and Falicov found.[2]

We have performed a calculation based on the following model energy bands

$$\epsilon_a(k) = 5(k - 0.45) , \quad (6a)$$

$$\epsilon_{b_+}(k) = -5(k - 0.45) , \quad (6b)$$

$$\epsilon_{b_-}(k) = -5(k - 0.55) , \quad (6c)$$

where all wave vectors are in units of π/a, where a is the lattice constant, and all energies are in eV. The parameters were chosen so as to agree with band structure calculations[11] and optical data.[12] If we choose parameters v and v' to be 1.075 eV and 1.98 eV respectively we obtain a first order transition with the observed discontinuity[1] in the gap g at the Neel temperature. The results are shown in Fig. 1.

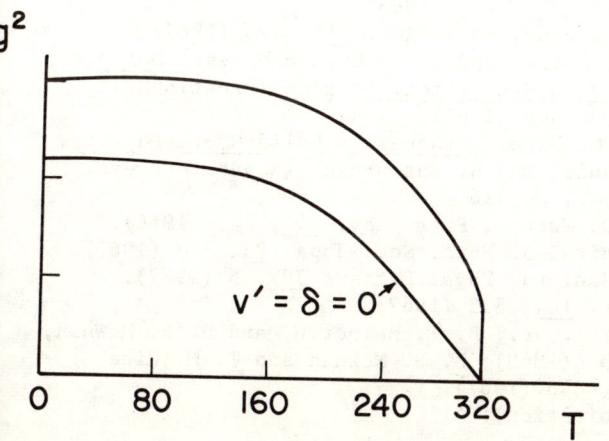

Figure 1. The square of the gap (g^2) versus temperature. Temperature is in degrees, and the gap is in arbitrary units. Also plotted is g^2 for v' equal to zero.

Effects of alloying on the order of the transition in our model have been studied by straightforward application of the methods of Zittartz[13] to our model. We find that for Γ (the Born approximation paramagnetic state lifetime broadening of the one electron energies) much less than kT_N, if the bandwidth were smaller by a factor of 5 or 6, the transition would be made second order by adding 2% vanadium impurities. If the transition were not made second order by impurities (as appears to be the case), a possible check to see if our theory really explains chromium would be to look for a first order transition in alloys of chromium. The transition should be first order in alloys with incommensurate spin density waves, but in alloys with a commensurate spin density wave, our theory predicts a second order transition since there are no harmonics. Another test would be to observe the order of the transition in pure chromium as a function of pressure since pressure lowers T_N[14] and our model predicts a second order transition when the Neel temperature is lowered. Another experimental check would be to look for harmonics. We expect that δ, and hence the amplitude of the harmonic, should scale as g^2 (from Eqs. 4a and 5) and hence have the temperature dependence of g^2. Since the harmonic is a charge density wave, we might be able to observe it by electron diffraction or by observing the resulting lattice distortion by neutron diffraction. Rough estimates of the amplitude of the lattice distortion wave accompanying the charge density wave, based on chapter 2 of reference 8, show that it is slightly less than 1% of a lattice constant. This small a distortion should be observable by neutron diffraction which can see a distortion of 0.3% of a lattice constant.[15]

1. A. Arrott, S. A. Werner, and H. Kendrick, Phys. Rev. Letters 14, 1022 (1965).
2. J. C. Kimball and L. M. Falicov, Phys. Rev. Letters 20, 1164 (1968); J. C. Kimball, Phys. Rev. 183, 533 (1969).
3. T. M. Rice, Phys. Rev. B2, 3619 (1970).
4. A. Malaspinas and T. M. Rice, Phys. Kondens. Materie 13, 193 (1971).
5. D. C. Mattis and W. D. Langer, Phys. Rev. Letters 25, 376 (1970).
6. S. H. Liu, Phys. Letters 27A, 493 (1968).
7. L. M. Falicov and D. R. Penn, Phys. Rev. 158, 476 (1967).
8. A. A. Abrikosov, L. P. Gorkov and I. E. Dzyoloshinski, Methods of Quantum Field Theory in Statistical Physics (Prentis Hall, Englewood Cliffs, New Jersey, 1963).
9. B. I. Halperin and T. M. Rice, Advances in Solid State Physics, ed. F. Seitz, D. Turnbull, and H. Ehrenreich (Academic Press, New York, 1968), Vol. 21, p. 116.
10. P. A. Fedders and P. C. Martin, Phys. Rev. 143, 245 (1966).
11. S. Asano and J. Yamashita, J. Phys. Soc. Japan 23, 714 (1967).
12. M. A. Lind and J. L. Sanford, Phys. Letters 39A, 5 (1972).
13. J. Zittartz, Phys. Rev. 164, 575 (1967).
14. T. M. Rice, A. S. Barker, Jr., B. I. Halperin, and D. B. McWhan, J. Appl. Phys. 40, 1337 (1969); D. B. McWhan and T. M. Rice, Phys. Rev. Letters 19, 846 (1967).
15. J. Axe (private communication).

THE PROXIMITY EFFECT FOR VERY
WEAK ITINERANT FERRO AND
ANTI FERRO MAGNETS

M. Kiwi
Universidad de Chile

M. J. Zuckermann
McGill University, Montreal

ABSTRACT

The integral equation for the spatial variation of the order parameter in a very weak itinerant ferromagnet is reduced to a Landau equation using Werthamer's method and is applied to ferromagnetic colloidal suspensions. It is further shown that the integral equation is identical in form for very weak itinerant antiferromagnets. The antiferromagnetic proximity effect is discussed.

The linear integral equation for the spatial variation of the magnetisation for very weak itinerant ferromagnets[1] is given in the molecular field approximation for a one-band Hubbard model by :

$$\Delta(r) = I(r) \int dr' \, \chi_T (r-r';0) \, \Delta(r') \qquad (1a)$$

where $\Delta(r) = I(r) [n_\uparrow(r) - n_\downarrow(r)]$ (1b)

$n_\sigma(r)$ is the number of electrons of spin σ at point r, $\chi_T(r;0)$ is the magnetic susceptibility function of the conduction band and $I(r)$ is the Hubbard interaction constant at point r.

Use of Werthamer's analysis[2] in conjunction with (1) yields the following Landau equations for $\Delta(r)$ at temperature T:

in a <u>ferromagnet</u>; $\xi_o^2 \, \nabla_r^2 \, \Delta(r) + [1-t^2] \, \Delta(r) = 0$ (2a)

and in a <u>paramagnet</u> ; $\xi_o^2 \, \nabla_r^2 \, \Delta(r) - [t^2+x^2] \, \Delta(r) = 0$ (2b)

where:
$t = T/T_{c1}$; $\xi_o = \lambda T_F/T_{c1}$; $x^2 = [(1-N(0)I)T_F^2/N(0)IT_{c1}^2]$ (3)

N(0) is the density of states of conduction electrons at the Fermi level, I is the Hubbard interaction constant for the bulk paramagnet, and λ is defined below.

Equation (2) was used in a recent communication[1] to predict an observable proximity effect for very weak itinerant ferromagnets. The proximity sample was taken to be a thin ferromagnetic film of thickness d_f in good electrical contact with a thick film of an enhanced paramagnetic metal. It was shown that the proximity sample is paramagnetic for d_f less than a critical thickness d_f^o and ferromagnetic for $d_f > d_f^o$. Here $d_f^o = \lambda T_F/T_{c1}$ when λ describes the spatial extent of the polarization of the conduction band, T_F is a characteristic temperature of the conduction band and T_{c1} is the Curie temperature of the bulk ferromagnet. The band structure is assumed to be the same on both sides of the proximity sample. For a proximity sample of Zr Zn$_2$ coated on Pd, $d_f^o \sim 100$ Å using the parameters of reference 3.

In this communication we investigate the following systems (2).

A. Colloidal ferromagnetic suspensions

Equation (2) is applied to a suspension of ferromagnetic colloidal particles in a matrix composed of an enhanced paramagnet or another ferromagnet. The colloidal particles are assumed to be spherical and far apart. It is therefore sufficient to solve the problem of a single colloidal particle of radius a_f in an infinite matrix[5]. The use of equation (2) in conjunction with the theory of the spatial variation of the order parameter due to Silvert and Singh[4] and the Werthamer boundary conditions yields the following equations for the Curie temperature T_c:

(Ai) <u>for ferromagnetic particles in a parametic matrix</u>

$$Z_f = (1-t_c^2)^{-1/2} \left\{ \pi - \tan^{-1}[(1-t_c^2)/(t_c^2+x^2)]^{1/2} \right\} \qquad (4a)$$

(Aii) <u>for ferromagnetic particles in a ferromagnetic matrix</u>

$$Z_f = (1-t_c^2)^{-1/2} \left\{ \pi - \tan^{-1}[(1-t_c^2)/(t^2-t_F^2)]^{1/2} \right\} \qquad (4b)$$

Here $t_c = T_c/T_{c1}$, $t_F = T_{c2}/T_{c1} < 1$ and T_{c2} is the Curie temperature of the ferromagnetic matrix. Z_f is a non-dimensional parameter which is related to the radius a_f of the ferromagnetic colloidal particles as follows:

$$Z_f = a_f/\xi_o \qquad (5)$$

where ξ_o is given by equation (3)

(4a) shows that the suspension is paramagnetic in case (Ai) if the colloidal radius a_f is less than a critical radius $a_f^o = \xi_o [\pi - \tan^{-1}(1/x)]$. Note that $a_f^o = \pi \xi_o$ when $I = 0$ and $a_f^o = \pi \xi_o/2$ when $IN(0) = 1$.

(4b) shows that the Curie temperature of the matrix in case (Aii) is unchanged by the colloidal particles for $a_f < a_f^o$ where $a_f^o = (1-t_F^2)^{-1/2} (\pi \xi_o/2)$. Note that a_f^o increases slowly as T_{c1} approaches T_{c2} e.g. for $t_F = 0.9$, $a_f^o = 2.29 \, (\pi \xi_o/2)$. The colloidal suspension differs from the proximity sample in that there is always a non-zero critical distance.

B. <u>Proximity Effect in very weak itinerant antiferromagnets</u>

We have extended the above analysis for ferromagnetic systems to the case of a proximity sample consisting of a thin film (thickness d_A) of a very weak itinerant antiferromagnet ($Cr_{0.8}V_{0.2}B_2$) in contact with a thick film of a paramagnetic metal having a similar band structure ($Mo\, B_2$) We have used the two band model of Jerome[5] for antiferromagnetic and nearly antiferromagnetic systems to analyse this case.

It can be shown that the linear integral equation for the antiferromagnetic case has the same form as (1) except that $\Delta_Q(r)$ of a linear spin density wave (LSDW) of wave vector Q and $\chi_T(r-r';0)$ is to be interpreted as the Fourier transform of the interband susceptibility $\chi_T(Q+q,0)$ The resulting Landau equation is given by:

$$a_T(Q) \nabla^2 \Delta_Q(r) + [(1-I'(r)\chi_T(Q,0))/I'(r)] \Delta_Q(r) = 0 \quad (4)$$

$a_T(Q)$ is the coefficient of the q^2 term in the expansion of $\chi_T(Q+q,0)$ and $I(r)$ is the interband interaction. Use of (4) in conjunction with the Werthamer boundary conditions yields an equation for the Neel temperature T_N' of the proximity sample, provided that the LSDW is assumed to attenuate exponentially in the paramagnetic film. The equation is:

$$Z_N = (1-t_N)^{-1/2} \tan^{-1} [\alpha \, (t_N+x^2)^{1/2} / (1-t_N)^{1/2}] \quad (6)$$

where Z_N, t_N and X are defined as follow:

$$t_N = T_N'/T_{N1}, \quad x^2 = \delta_p [1-N(0)I|\ln \delta_p|] (\alpha IN(0) T_{N1}')^{-1} \quad (7)$$

$$\alpha = (\delta_p/\delta_A)^{1/2}, \quad Z_N = kd_A (\delta_A T_N')^{1/2}. \tag{8}$$

δ_p and δ_A are parameters describing the mismatch of the Fermi surfaces in the paramagnetic and antiferromagnetic films respectively in Jerome's model[5] ($\delta_{A,p} \ll 1$). k is related to the Fermi momentum, T_N is the Neel temperature of the bulk antiferromagnet and I is the interband interaction in the paramagnet. It can be seen from (6) that the proximity sample is paramagnetic for thickness $d_A < d_A^o$, where the critical thickness $d_A^o = \hbar (\delta_p T_N' k^2)^{-1/2}$, and that a stable LSDW state exists for $d_A > d_A^o$, the Neel temperature being given by (6). This proximity effect in antiferromagnets should be observable using resonance experiments provided d_A^o is macroscopic.

We wish to acknowledge stimulating conversations with Dr. B. Chornik and to thank Dr. R. Harris for a critical reading of the manuscript.

REFERENCES

1. M.J. Zuckermann, Solid State Comm., 12, 745 (1973).
2. N.R. Werthamer, Phys. Rev., 132, 2440 (1963).
3. E.P. Wohlfarth, J. Applied Phys., 39, 1061 (1968).
4. W. Silvert and A. Singh, Phys. Rev. Let. 28, 222, (1972)
5. D. Jerome, Solid State Comm. 8, 1793 (1970).

LATTICE DISORDER AND MAGNETIC PHASE TRANSITIONS

B. A. Huberman
Xerox Palo Alto Research Center, Palo Alto, California 94304

ABSTRACT

A general model describing the role of lattice disorder on magnetic phase transitions is presented. It takes into account interactions between defects as well as the fact that the magnetic state of a defect may be different than that of a lattice site.[1] Depending on the values of the parameters we obtain the following behavior with increasing temperature: i) a first order magnetic phase transition accompanied by melting or, ii) a second order magnetic phase transition with a negligible amount of defects. The phase diagrams are calculated and applications to the ferromagnet MnBi are discussed.

(1) B. A. Huberman, Phys. Rev. Letters **31**, 1251 (1973)

Section 9. Magnetic Order and Structure

MAGNETICALLY INDUCED LATTICE DISTORTIONS IN THE NEPTUNIUM MONOPNICTIDES*

M. H. Mueller, G. H. Lander, H. W. Knott, and J. F. Reddy
Argonne National Laboratory, Argonne, Illinois 60439

ABSTRACT

The lattice parameters of the neptunium monopnictides (NaCl crystal structure) have been measured between 5 and 300°K by X-ray diffraction. NpN becomes ferromagnetic at 87°K. At T_C a rhombohedral distortion, which indicates that the <1$\bar{1}$1> is the easy axis, is observed. At 5°K, the rhombohedral angle is 60.46 $\pm$ 0.02°. NpP exhibits a tetragonal distortion at 74°K, at which temperature the antiferromagnetic 3+, 3- structure becomes commensurate with the lattice. At 5°K c/a = 0.9958 $\pm$ 2. NpAs becomes tetragonal at T_N = 175°K when the material has a 4+, 4- structure. However, NpAs returns to being a cubic at 142°K at which temperature the magnetic structure is the simple + - (type I) configuration. The 'return' to cubic is first order; the volume of the unit cell expands by 0.23%. NpSb, which orders antiferromagnetically at 207°K with the type I structure remains cubic in the ordered regime.

INTRODUCTION

The onset of magnetic order in a crystal lattice may lead to a reduction in symmetry of that lattice. For example, if a magnetic vector parallel to the [001] axis is introduced into a cubic lattice the symmetry is reduced from cubic to tetragonal. The ordering process is caused by exchange interactions, which are frequently anisotropic and depend on the interatomic distances. Another important interaction is electrostatic in nature, and occurs when neighboring atoms have aspherical electron-cloud distributions. This electrostatic interaction is especially important when the spin-orbit coupling is strong and L is a good quantum number. The magnetization density is then effectively tied to the direction of the magnetic moment. The magnetostriction of the rare-earth materials is primarily a consequence of this electrostatic interaction. From a macroscopic point of view, the energy of these interactions is frequently minimized if the lattice distorts at the ordering temperature. The magnetic and crystallographic symmetry are then compatible. Examples of this process are the transition-metal oxides[1] and the rare-earth monopnictides.[2] As part of a general study that includes magnetization, Mössbauer, resistivity, and neutron-diffraction measurements on the neptunium monopnictides,[3,4] we have measured the lattice parameters of these compounds (NaCl structure) between 5 and 300°K.

---

*Performed under the auspices of the U.S. Atomic Energy Commission.

EXPERIMENTAL

The sample preparation[3] and experimental technique[5,6] have been described previously. Polycrystalline samples have been used. We analyze the X-ray diffraction profiles with a least-squares process that fits a calculated X-ray profile to the experimental intensities. A simple Cauchy function is used to simulate the diffraction profiles,[7]

$$y_{ij} = \frac{m \, Lp \, S}{\Delta} \bigg/ \left[1 + \left(\frac{2\theta_j - 2\theta_i}{\Delta/2}\right)^2\right],$$

where y_{ij} is the intensity from a reflection hkl_j at the position $2\theta_i$, m is the multiplicity of the reflection, Lp is the Lorentz-polarization factor, S is the scale factor, $2\theta_j$ is the calculated central position of the diffraction peak, and Δ is the full width at half maximum (FWHM) of the diffraction profile. At the position $2\theta_i$ the total intensity is $\sum y_{ij}$, where the sum is over all reflections and all incident wavelengths (i.e. the α_1 and α_2 components). The parameters S, Δ, and the constant background level are obtained by measuring the diffraction profiles above the ordering temperature. The FWHM is a function of the diffractometer and of the sample homogeneity and strain. In most cases, the FWHM does not vary when the lattice distorts. The least-squares process rapidly converges and gives an excellent fit ($\chi^2 \sim 1$) between the calculated and observed diffraction profiles. This analysis gives accurate values for the c/a ratio, which is the main interest in the experiment. To obtain absolute values of c and a we have used a computer program,[8] which requires a series of diffraction peak positions as input.

NpN

Neptunium nitride becomes ferromagnetic at T_C = 87°K. At the same temperature the unit cell distorts from cubic to rhombohedral. Above T_C a rhombohedral unit cell with an angle α of 60° may be used to describe the structure. The trigonal axis of this unit cell is the [111] axis of the cubic unit cell. When the distortion at T_C occurs the [111] axis is either compressed or stretched, the rhombohedral angle becoming greater than or less than 60°, respectively. A convenient description of this distortion is to define a length 'c' as a distance along the unique trigonal axis and 'a' as a distance in the plane perpendicular to 'c' such that 'c'/'a' = 1.00 in the cubic phase. This definition is especially useful in comparing the magnitude of trigonal and tetragonal distortions. The relationship between the change in the rhombohedral angle $\Delta\alpha$ and the 'c'/'a' ratio is given by $\Delta\alpha = -8/\sqrt{27} \, (c-a)/a$ radians. In Fig. 1, we present the variation of these equivalent lattice parameters in the trigonal phase, as well as the variation of the cube edge in the paramagnetic phase.

The rhombohedral distortion in NpN indicates that the easy direction of the magnetic moments is either in the (111) plane or perpendicular to it. This information cannot be obtained from neutron-diffraction experiments on polycrystalline ferromagnets. At 5°K the value of $(c-a)/a = -52 \times 10^{-4}$ where c and a are defined above.

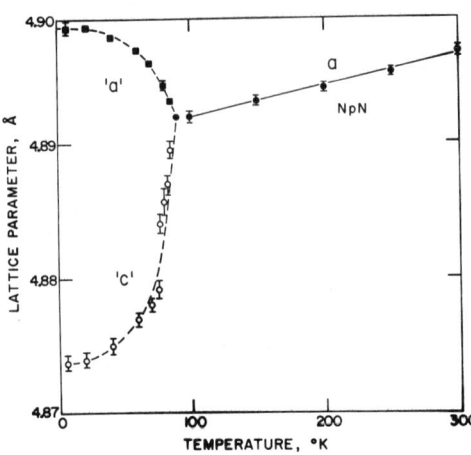

Fig. 1. Variation with temperature of the equivalent lattice parameters in NpN.

NpP

The variation with temperature of the lattice parameters of NpP have been reported.[6] Briefly stated, NpP orders antiferromagnetically at 130°K. The (001) ferromagnetic sheets are stacked along the c axis with a sinusoidal variation of the moment from sheet to sheet.[4] The repeat distance is 2.78 ± 0.08 unit cells. This magnetic structure has tetragonal symmetry, however, the cubic lattice parameter shows no discontinuity at T_N (see Fig. 2). At 74°K another transition occurs and the magnetic cell becomes commensurate with 3.0 chemical unit cells. At the same temperature, the X-ray experiments observe a large tetragonal distortion. At 5°K $c/a = 0.9958$, i.e. $(c-a)/a = -(42 \pm 2) \times 10^{-4}$.

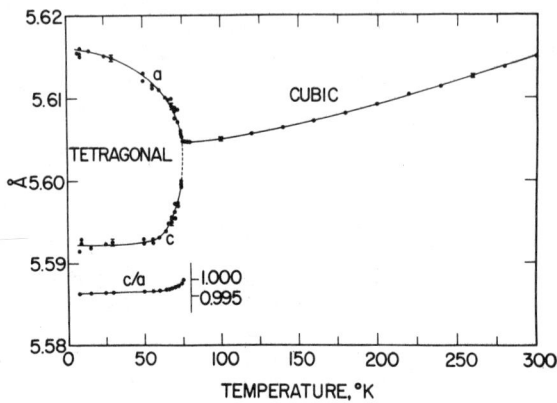

Fig. 2. Variation with temperature of the lattice parameter of NpP.

NpAs

NpAs becomes antiferromagnetic at $T_N = 175$°K. The high-temperature magnetic structure of NpAs is very similar to that described for NpP in that it consists of a stacking of ferromagnetic sheets in a 4+, 4− sequence along the c axis. A tetragonal distortion is observed by X-rays at T_N, as shown in Fig. 3. At 143°K $(c-a)/a = -(8 \pm 2) \times 10^{-4}$. Neutron experiments on NpAs at temperatures below ~ 145°K indicate that the magnetic structure is type I, in which the ferromagnetic sheets are

355

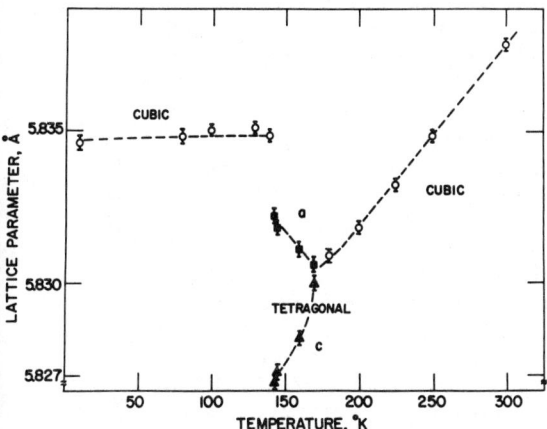

Fig. 3. Variation with temperature of the lattice parameter of NpAs.

stacked in the simple + − sequence. The transition from the 4+, 4− to the + − sequence appears from the neutron experiments to be spread over several degrees, but in the X-ray experiments we observe a first-order transition in which the unit cell distorts from tetragonal (143°K) to cubic (142°K). We are unaware of any other material that transforms from tetragonal to cubic in the ordered regime. Not surprisingly, considerable strain is observed in the lattice just below 142°K, but the symmetry is clearly cubic $|(c-a)/a| \leq 3 \times 10^{-4}$. On further cooling, the lattice parameter remains unchanged (see Fig. 3).

NpSb

NpSb becomes antiferromagnetic at 207°K. The magnetic structure at all temperatures below T_N is the type I. Neptunium antimonide proved difficult to prepare and the X-ray diffraction profiles were very broad. The sample also contained a small amount of Np_3Sb_4. Because of these difficulties precise lattice parameters could not be obtained. No evidence for a crystallographic distortion was found in NpSb. The limit of our sensitivity is such that $|(c-a)/a| \leq 15 \times 10^{-4}$ at 5°K.

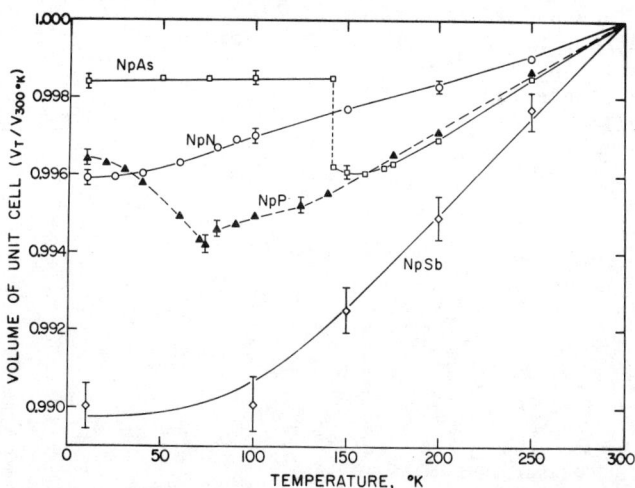

Fig. 4. Temperature dependence of the volume of the unit cell (relative to 300°K) for the neptunium monopnictides.

DISCUSSION

The volume changes relative to room temperature are shown in Fig. 4 as a function of temperature for the four compounds. At high temperatures the straight-line behavior reflects the regular thermal expansion of the lattice and, as expected, this value increases with increasing lattice parameter. The coefficients of thermal expansion (5.5 for NpN, 9.6 for NpP, 10.3 for NpAs, and 17.0 for NpSb; all $\pm$ 1.0 and in units of $10^{-6}/K°$) are comparable to those of the uranium monopnictides. At lower temperatures the volume contraction in NpN ($T_C = 87°K$) and NpSb ($T_N = 207°K$) appear normal, indicating that the transitions are not first order. Recall that NpN distorts rhombohedrally at T_C, but, within our experimental uncertainty, the volume of the unit cell varies smoothly through this transition. Similarly, second-order transitions occur at T_N in NpP (130°K) and NpAs (175°K). However, at lower temperatures both NpP and NpAs exhibit first-order transitions in which a discontinuity in the unit-cell volume is observed. These transitions are associated with changes in the magnetic structures. Large anomalies are also observed in the electrical resistivities of NpP and NpAs.[9]

In conclusion, the X-ray measurements provide a wealth of new and valuable information on the neptunium monopnictides. Indeed a complete understanding of these materials will clearly have to account for the crystallographic symmetry and the unexpected distortions that occur in the ordered state.

REFERENCES

1. W. L. Roth, Phys. Rev. 110, 1333 (1958).
2. F. Lévy, Phys. Kondens Materie, 10, 85 (1969).
3. D. J. Lam, B. D. Dunlap, A. T. Aldred, and I. Nowik, Amer. Inst. Physics, Conf. Proc. 10, 83 (1973).
4. G. H. Lander et al., ibid, 10, 88 (1973).
5. J. A. C. Marples, J. Phys. Chem. Solids, 31, 2431 (1970).
6. M. H. Mueller, G. H. Lander, H. W. Knott, and J. F. Reddy, Phys. Lett. 44A, 249 (1973).
7. E. R. Pike, Acta Cryst. 12, 87 (1959).
8. M. H. Mueller, L. Heaton, and K. T. Miller, Acta Cryst. 13, 828 (1960).
9. Ref. 4 and A. R. Harvey, private communication.

MAGNETIC PROPERTIES OF UCu_5 AND UNi_5*

M. B. Brodsky
Argonne National Laboratory, Argonne, Ill. 60439[†],
and Imperial College, London, S. W. 7, U.K.

N. J. Bridger
Atomic Energy Research Establishment,
Harwell, Didcot, U.K.

ABSTRACT

The magnetic susceptibilities and electrical resistivities of isostructural UCu_5 and UNi_5 have been measured between 2° and 300°K. Whereas UCu_5 is antiferromagnetic at $T_N = 16 \pm 1$°K, and $p_{eff} = 2.3\ \mu_B$, UNi_5 has a nearly temperature independent susceptibility. Apparently, non-stoichiometry of the samples causes some uncertainties in the results. The occurrence of long-range magnetic order in UCu_5 is explained by the large U - U distance which prevents 5f-level broadening due to direct f-f overlap. The lack of magnetic order in UNi_5 may be due to d-f hybridization which causes sufficient 5f-level broadening and no magnetic order.

INTRODUCTION

Hill[1] has shown that the lack of magnetic order or superconductivity in actinide systems may be directly related to the inter-actinide spacing. Thus, in uranium compounds, no magnetic order or superconductivity is found for U - U distances less than about 3.5 Å. The intermetallic compounds UCu_5 and UNi_5 form in the UNi_5-type structure which is derived from the $MgCu_2$-cubic Laves phase by the substitution of one Mg-type atom by an additional Cu-type atom.[2] As a result, the U - U distances are 4.97 Å and 4.81 Å in UCu_5 and UNi_5, respectively. These may be compared to a U - U distance of 3.1 Å for the hexagonal Laves phase compound UNi_2. And, in keeping with the observation of Hill, UNi_2 does

---

*Work performed under the support of the U.S. Atomic Energy Commission, U.K. Science Research Council, and U.K. Atomic Energy Authority.
[†]Permanent address.

not order magnetically.[3]

RESULTS

Samples were made by arc-melting stoichiometric amounts, followed by anneals at 1210°K. One set of each compound was analyzed microscopically, by fluorescence analysis, and X-ray diffraction, and found to be single phased, and of the desired structure. Although UNi_5 is reported to form congruently,[4] and the relatively constant values of the resistivity at room temperature, ρ_{300}, bear this out, some deviations from stoichiometry are seen. Conversely, UCu_5 is formed peritectically,[4] and the results presented below indicate difficulty with the preparation of the stoichiometric compound. Susceptibility measurements were made by the Faraday method and standard, 4-probe, dc resistivity methods were used.

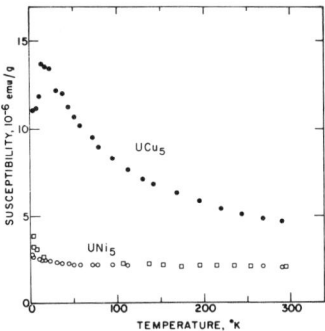

Fig. 1. Magnetic susceptibility of UCu_5 (●); as-cast UNi_5 (□); and annealed UNi_5 (○).

Fig. 1 shows the susceptibility data for both compounds. The UCu_5 sample had been annealed at 1210°K for five days and both as-cast and annealed-sample data are shown for UNi_5. The results for both UNi_5 samples are nearly the same in their susceptibility.

The maximum for UCu_5 is likely due to an antiferromagnetic transition at 16 ± 1°K. The uncertainty in T_N is due to a rounded maximum. This is likely due to non-perfect stoichiometry leading to a range of ordering temperatures. In the paramagnetic region, down to 23°K, the data fit the expression $\chi = 1.4 + C/(T + 78)$ in 10^{-6} emu/g. The Curie-Weiss constant, C, is equivalent to an effective moment of 2.28 μ_B/formula unit. In the as-cast sample, the results are $\chi_o = 1.5$; $\theta = -75°K$; and C is equivalent to 2.33 μ_B/F.U.

The temperature dependence of the UNi_5 susceptibility nearly follows a modified Curie-Weiss law above 10°K (of the form given for paramagnetic UCu_5) with $\chi_o = 2.15 \times 10^{-6}$ emu/g, $\theta \simeq 0°K$, and $p_{eff} = 0.17$ μ_B/F.U. for as-cast UNi_5 and 0.14 μ_B/F.U. for the annealed sample. Below 10°K, the susceptibility is less temperature dependent than is predicted by the Curie-Weiss law. The electrical resistivities of the annealed UNi_5 and both as-cast and annealed UCu_5 are shown in Fig. 2. The data for as-cast UNi_5 are essentially the same as those shown in the figure except for a lower resistivity ratio, $\rho_{300}/\rho_4 = 4.1$ as opposed to 8.5. Because of the shape of the ρ-T curves for UCu_5, resistivity ratios do not have too much meaning, but it is obvious that annealing has made the curve simpler, i.e., lower values for ρ_{300} and ρ_4; a positive $d\rho/dT$ above T_N; and a much smaller magnitude for the size of the "bump" immediately below T_N. Intersection of the flat portion just above T_N with the small region of negative $d\rho/dT$ gives $T_N = 15.55 \pm 0.10°K$.

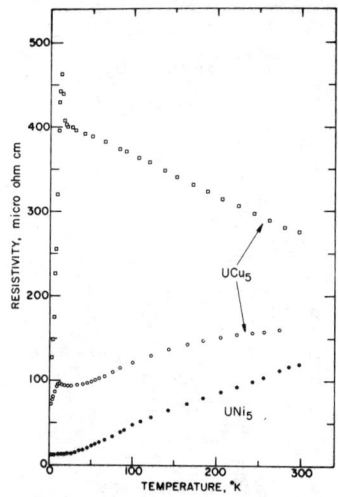

Fig. 2. Resistivity of annealed UNi_5 (●); as-cast UCu_5 (□); and annealed UCu_5 (○).

In the temperature region below the maximum in the ρ-T curve, the resistivity does not follow a simple power law in the temperature. In an expression of the form $\rho - \rho_o = AT^n$, $\underline{n}$ varies from 1.3 to 2.5. Although $\underline{n}$ does not equal two rigorously below T_N, it seems close enough to attribute it to the spin-disorder scattering below T_N.[5] The resistivity data for UNi_5 follow a law of the form $\rho - \rho_o = AT^3$ up to about 40°K.

DISCUSSION

As discussed in the Introduction, the compound UCu_5 seems to fit Hills' criterion.[1] That is, the inter-uranium distance,

4.97 Å is well in excess of the minimum needed for long range magnetic order, viz. 3.5 Å. There seems little doubt that the susceptibility maximum at 16°K is due to an antiferromagnetic transition, as seems borne out by the large decrease in the resistivity, beginning several degrees below T_N. The effective moment, 2.3 μ_B, is in fair agreement with the value 2.6 calculated for the $5f^4$ configuration.[6]

The nature of the resistivity "bump" is not clear. If it is due to additional Brillouin zone scattering,[7] then it should be seen more strongly in the purer compound, or in this case the annealed sample. However, if the defect structure stabilizes a nested Fermi surface, then the larger rise in ρ below T_N for the poorer sample could be explained. It should be noted that in antiferromagnetic $PuPt_3$ similar behavior was seen. In the purest samples there is no rise in ρ for $T < T_N$, but in two-phased samples there can be a large rise in ρ below T_N before the resistivity begins to drop again.[8]

At first glance, the ρ-T data for UCu_5 above T_N look as though they might follow spin fluctuation behavior, i.e., $T^2 \to T^1$ dependence. However, the data follow a much higher power law than T^2, perhaps T^6 up to 35°K. It would seem necessary to clarify the question of the resistivity "bump" and non-stoichiometry before any attempt is made to conclude the presence of spin fluctuation effects above T_N, as in $PuRh_3$[9] or UGa_2.[10]

The situation in UNi_5 is less clear. The smaller temperature dependent susceptibility (and smaller p_{eff}/F.U.) for the annealed sample indicate that the temperature dependence is due primarily to displaced atoms, perhaps clusters of Ni atoms. As in the case of UCu_5, a quick glance at the ρ-T curve could be indicative of localized spin fluctuations as have been shown to exist in many actinide intermetallic compounds.[11] However, the power law dependence below the linear ρ-T which begins at 150°K is $\rho-\rho_o = AT^3$ up to 40°K. There is, however, a region which seems to depend as $T^{1.84}$ from 50 - 100°K, which could be a $T^{2.0}$ dependence modified by other effects. This would lead to a spin fluctuation temperature of about 600°K, much too high to explain the temperature dependent susceptibility below 300°K. As in the case of UCu_5, this type of question requires much better samples before the various effects could be assigned to non-stoichiometric impurity effects, pure compound effects, etc. Thus far, attempts to make better samples have not been successful.

The lack of magnetic order or magnetic moments in UNi_5, while they exist in UCu_5 is obviously not a result of f-f overlap since the U-U distance is 4.81 Å. It is more likely that f-d hybridization causes sufficient 5-f level broadening to prevent the formation of magnetic moments in UNi_5. Although there is no direct evidence for f-d hybridization in UNi_5, there is evidence for the lack of hybridization in the antiferromagnetic UCu_5. This is obtained from the coppery color of the compound, which is presumably due to the filled 3d band lying just below the Fermi level, as in pure copper.

ACKNOWLEDGEMENTS

The authors wish to thank Prof. B. R. Coles for his hospitality and encouragement and the assistance of Drs. H. E. N. Stone and A. R. Harvey in various aspects of the work.

REFERENCES

1. H. H. Hill, Nucl. Met. 17, Part I, 2 (1970).
2. N. C. Baenziger, R. E. Rundle, A. E. Snow, and A. S. Wilson, Acta. Cryst. 3, 34 (1950).
3. Y. Hamaguchi, S. Komura, N. Kunitomi, and M. Sakamoto, J. Phys. Soc. Japan Suppl. B-III, 17, 46 (1962).
4. M. Hansen, "Constitution of Binary Alloys" (McGraw-Hill, N. Y., 1958), pp. 647 and 1054.
5. P. G. de Gennes and J. Friedel, J. Phys. Chem. Solids 4, 71 (1958).
6. S. K. Chan and D. J. Lam, Nucl. Met. 17, Part I, 219 (1970).
7. A. L. Trego and A. R. Mackintosh, Phys. Rev. 166, 495 (1968).
8. M. B. Brodsky and N. J. Nellis, unpublished work.
9. A. R. Harvey, M. B. Brodsky, and W. J. Nellis, Phys. Rev. B. 7, 4137 (1973).
10. K. H. J. Buschow and H. J. van Daal, "Magnetism and Magnetic Materials - 1971", (A. I. P., N. Y., 1972), 17th Annual Conference (C. D. Graham and J. J. Rhyne, eds.) Part 2, pp. 1464 - 14777.
11. M. B. Brodsky, Phys. Rev., in press.

MAGNETIC PROPERTIES OF Gd-RICH Gd-Sm ALLOYS*

Sigurds Arajs, D. L. Adour, E. E. Anderson, T. F. DeYoung
and K. V. Rao
Department of Physics, Clarkson College of Technology
Potsdam, New York 13676

ABSTRACT

Ferromagnetic Curie temperatures (T_c) of Gd alloys containing 5, 10, 20, 25, 30, and 35 at.% Sm have been determined using the $d\rho/dT$ method, where ρ is the electrical resistivity. It has been found that this technique is consistent with the kink-point method and conventional magnetization vs internal field extrapolations. The effect of Sm on T_c of Gd is nonlinear, the initial decrease (up to about 5 at.% Sm) being only 0.4 K/at.Sm. The decrease in the mass magnetization of the ferromagnetic alloys due to the addition of Sm in Gd is qualitatively similar to the T_c vs concentration dependence.

INTRODUCTION

It is well-known that the interesting magnetic structures of the rare earth metals are formed by the spin and orbital angular momenta of the approximately-localized 4 f electrons. Although the detailed arrangements of the magnetic moments of most of the rare earth metals have been established for some time, the structure of Sm has been determined only recently.[1] According to the magnetic susceptibility and electrical resistivity studies,[2] polycrystalline Sm exhibits two magnetic transitions: one at about 14 K and another at 106 K. Koehler and Moon,[1] using neutron-diffraction studies, have determined that the low temperature transition is associated with the ordering of the magnetic moment on the hexagonal sites, while the high temperature one is due to the ordering on the cubic sites. Because of the existence of these unique magnetic structures in pure Sm and due to the fact that practically nothing is known about the magnetic properties of binary Sm alloys, we have initiated extensive investigations of the physical properties of Sm alloys with other rare earth elements. In this paper we report the results of magnetic and transport property measurements of Gd-rich Gd-Sm alloys.

SAMPLES AND APPARATUS

The polycrystalline samples of Gd-Sm used in this investigation were prepared by the method of arc-melting using the facilities described elsewhere.[3] The stocks of pure (distilled) samarium and gadolinium (both of 99.9% purity) were purchased from Research

---

* Work supported by the Office of Naval Research under grant number N0014-70-A-0311-0001-05.

Chemicals. From the homogenized (at 1150 K for 170 hours) single-phase ingots weighing about 20 g, samples were prepared in the form of spheres (dia. ~ 3 mm) and parallelepipeds (~25 mm x 4 mm x 4 mm). The spheres were used for the measurements of the magnetic moments by means of the PAR vibrating sample magnetometer associated with the Magnion 15-in. electromagnet.[4] Various temperatures between 77 and 300 K were obtained and controlled using an Andonian Dewar specifically designed for the vibrating sample magnetometer and PAR Cryogenetic Temperature Controller (Model 152). The parallelepipeds were used for the electrical resistivity studies. These measurements were carried out between 4.2 and 310 K by means of the equipment described earlier.[5]

EXPERIMENTAL RESULTS

One of the physical quantities which characterizes a binary magnetic system is the composition dependence of some particular magnetic transition. Gd is the simplest of all the magnetically-ordered rare earth metals,[6] possessing the Curie temperature $T_c = 293.3 \pm 0.1$ K.[7] When Sm is added to Gd, it can be expected that T_c would decrease. Experimentally, we have found that this, indeed, is the case. From our studies of the electrical resistivity (ρ) and the behavior of the magnetization (σ) at low applied fields, we have found that either the method of the first derivative of the electrical resistivity ($d\rho/dT$) or the kink-point method[8,9] is a good technique for the determination of T_c of Gd alloys. Figure 1 shows

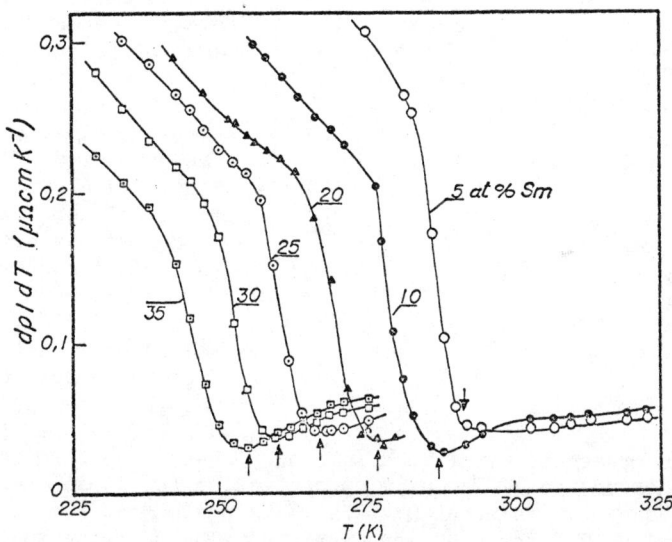

Fig. 1. Temperature derivatives of the electrical resistivity of Gd-Sm alloys in the neighborhood of their Curie temperatures.

the plots of dρ/dT vs T, calculated by means of a computer, for Gd-Sm alloys containing 5, 10, 20, 25, 30, and 35 at.% Sm in the neighborhood of their respective Curie temperatures. The exact location of T_c is shown by arrows. Figure 1 clearly reveals that the temperature variation of ρ around T_c is very similar in all Gd-rich alloys containing various amounts of Sm. It appears that T_c can be determined reliably within at least a few degrees from the dρ/dT vs T plots by finding the temperature where dρ/dT exhibits a minimum, which becomes more pronounced in alloys of higher Sm concentration. In the 5 at.% Sm alloy the minimum disappears and we chose T_c at the point where the slope of dρ/dT changes abruptly (see Fig. 1). This method of the determination of T_c is supported by the kink-point and conventional magnetization studies. A typical plot of the mass magnetization (σ) as a function of T in low applied fields is shown in Fig. 2, where the data are presented for the spherical sample containing 5 at.% Gd. Although the transitions from the nonuniform to the uniform magnetization (the kink-point) are not as sharp in alloys as in pure substances, the values of T_c still can be clearly determined from such plots. In general, the determination of T_c from the kink-point data are in good agreement with the resistivity data as mentioned above. Furthermore, we also have examined the conventional σ vs T plot associated with low (a few hundred Oe) internal fields. The values of T_c estimated from these curves, (although a less accurate technique) are also in reasonable greeement with the dρ/dT and kink-point determinations.

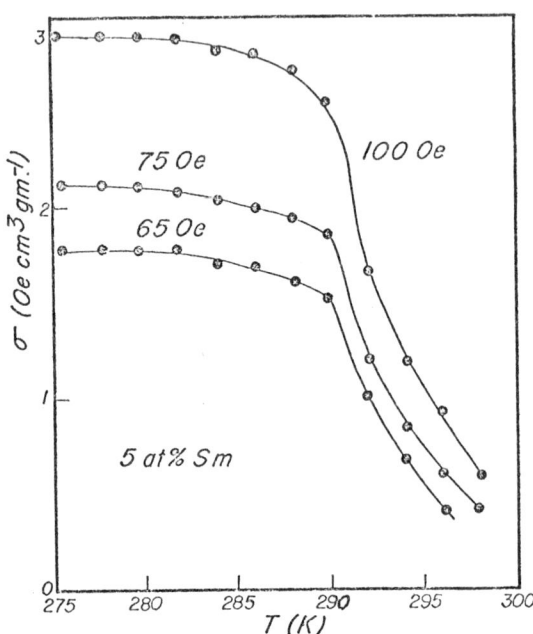

Fig. 2. Kink-point plot for Gd 5 at.% Sm alloy.

The dependence of T_c on Sm concentration (c) is shown in Fig. 3. The decrease in T_c is about 0.4°K/at.% Sm up to 5 at.%Sm; for larger concentration, T_c decreases more rapidly. The decrease in the mass magnetization at constant temperature (for example, 250 K) and constant internal field (∼10 KOe) as a function of c behaves (not shown in this paper) in a similar fashion up to at least 20 at.% Sm level. It should be remarked that the effects of Sm on Gd are quite different from those of, for instance, Sc and Y. In particular, Nigh et al.,[10] have found that additions of Sc to Gd greater than 25 at.% very rapidly eliminates the ferromagnetic phase giving rise to an antiferromagnetic state. Whether such behavior also takes place in the central portion of the Gd-Sm system is presently under investigation.

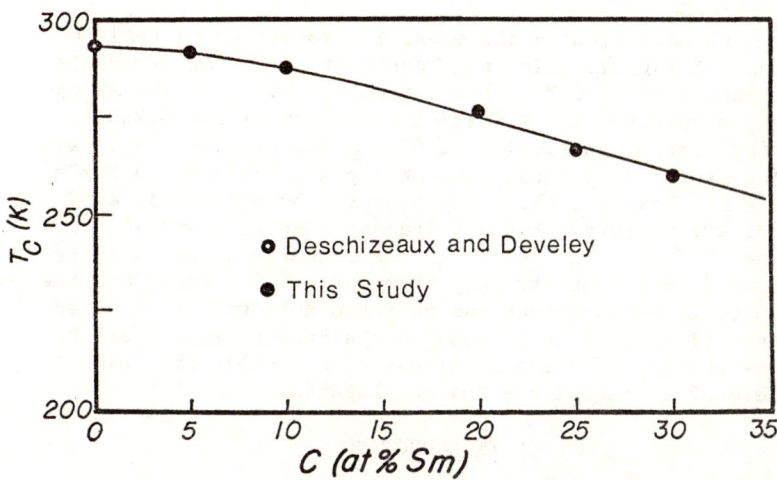

Fig. 3. Curie temperature of Gd-Rich Gd-Sm alloys as a function of Sm concentration.

ACKNOWLEDGMENTS

The authors are thankful to C. D. Levermore and J. R. Kelly for their assistance with computer calculations.

REFERENCES

1. W. C. Kohler and R. M. Moon, Phys. Rev. Letters 29, 1468 (1972).
2. S. Arajs and G. R. Dunmyre, Z. Naturforsch 21a, 1856 (1966).
3. S. Arajs and G. P. Wray, J. Phys. E 2, 518 (1969).
4. S. Arajs, B. L. Tehan, E. E. Anderson, and A. A. Stelmach, phys. stat. sol. 41, 639 (1970).
5. S. Arajs, Canad. J. Phys. 47, 1005 (1969).
6. C. D. Graham, Jr., J. Phys. Soc. Japan 17, 1310 (1962).
7. M. N. Deschizeaux and G. Develey, J. Physique 32, 319 (1971).
8. M. Rayl and P. J. Wojtowicz, Phys. Letters 28A, 142 (1969).
9. E. E. Anderson, S. Arajs, A. A. Stelmach, B. L. Tehan, and Y. D. Yao, Phys. Letters 36A, 173 (1971).
10. H. E. Nigh, S. Legvold, F. M. Spedding, and B. J. Beaudry, J. Chem. Phys. 41, 3799 (1964).

MAGNETIC PROPERTIES OF $NpAl_3$*

A. T. Aldred, B. D. Dunlap, and D. J. Lam
Argonne National Laboratory, Argonne, Illinois 60439

ABSTRACT

We have studied the magnetic properties of $NpAl_3$ by means of magnetization and Mössbauer-effect measurements between 4 and 300°K. This compound, which has the $AuCu_3$-type structure with a lattice parameter of a = 4.260 Å, orders ferromagnetically at 62.5 ± 0.5°K. The magnetization cannot be saturated by an applied field of 13.5 kOe at 4°K. However, the ^{237}Np magnetic hyperfine field of 2630 kOe measured at 4.2°K indicates an Np magnetic moment of ~ 1.2 μ_B. Although the magnetic susceptibility above T_c deviates strongly from Curie-Weiss behavior, the temperature dependence can be reasonably well reproduced theoretically if it is assumed that the neptunium ion has a +4 charge ($5f^3$) and is situated in a cubic crystal field of 12 negatively charged ligands.

INTRODUCTION

Magnetic measurements on actinide compounds with the $AuCu_3$-type structure have been confined primarily to uranium[1,2] and plutonium[3] compounds. These results have been interpreted in terms of either crystal-field theory[4] or a localized spin-fluctuation model.[2,5,6] In the present work, we have measured the magnetization of $NpAl_3$ ($AuCu_3$-type structure) between 4 and 300°K and determined the hyperfine field at 4.2°K. Our results are consistent with crystal field theory.

EXPERIMENTAL

The sample was prepared by arc-melting the requisite amounts of high-purity neptunium and aluminum into a 500-mg button that was heat treated at 1000°C for 4 hr and furnace cooled. The lattice parameter was determined to be 4.260 ± 0.001 Å, in good agreement with the value of 4.262 Å given by Dwight,[7] and the x-ray pattern showed no evidence of any second phase. The sample was powdered, and $\sim$250 mg was sealed into an aluminum capsule for the Mössbauer measurements. An additional $\sim$100 mg was separately encapsulated for the magnetization experiments.

The hyperfine spectra were obtained at 4.2°K by means of the 59.6 keV Mössbauer resonance of ^{237}Np with a source of ^{241}Am in α-Am metal. The magnetization experiments utilized a conventional force technique with applied fields from 0.5 to 13.5 kOe. An approximate correction for the demagnetizing field of the specimen

---
*Work performed under the auspices of the U. S. Atomic Energy Commission.

was made, although this was small (~150 Oe). Measurements were made at 4°K intervals, except near the ferromagnetic transition temperature where the increments were 0.5°K.

RESULTS

Some typical magnetization data at low temperatures are shown in Fig. 1. It is evident from the large field dependence at 4°K that the magnetocrystalline anisotropy of this material is large, and it is not possible to obtain magnetic saturation with a field of 13.5 kOe. The maximum measured magnetization of 7.57 emu/g corresponds to a magnetic moment of 0.43 μ_B/mole. Attempted extrapolation of the field dependence to $H = \infty$ to obtain a saturation moment yields poor fits when conventional H^{-1} or H^{-2} terms are used. Inclusion of a linear susceptibility term improves the fit but much better agreement with the data is obtained when a term linear in H is combined with a slowly saturating term such as $H^{-0.05}$. This fit gives a moment of ~1.3 μ_B/mole and a susceptibility of ~6.7 x 10^{-5} emu/g, with an RMS error of 0.011 emu/g. However, such an extrapolation, with no theoretical justification, should be approached cautiously. A more accurate determination of the saturation moment can be obtained from the hyperfine field, inasmuch as the ratio of Np hyperfine field to magnetic moment has been well established in other Np compounds with cubic symmetry.[8] A value of 2630 kOe was determined at 4.2°K, and this corresponds to a moment of ~1.2 μ_B/Np atom.

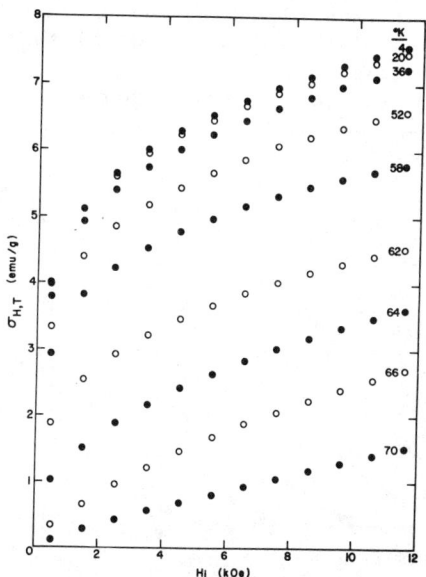

Fig. 1. Specific magnetization $\sigma_{H,T}$ of NpAl$_3$ as a function of internal field H_i at various temperatures.

The susceptibility well above the ferromagnetic transition temperature was determined by fitting the magnetization $\sigma_{H,T}$ to an equation of the form

$$\sigma_{H,T} = \sigma_o + \chi H_i \qquad (1)$$

in the range from 3.5 to 13.5 kOe. The data from 0.5 to 3.5 kOe systematically fell below the line given by Eq. (1), and this fact, together with the nonzero value of σ_o, is consistent with the presence of a small, easily saturated, magnetic component. The value of σ_0 showed a small linear temperature dependence, varying from 0.0472 (±2) emu/g at 300°K to 0.0479 (±3) emu/g at 100°K. The origin of σ_o is not known at present. Below 90°K, the $\sigma_{H,T}$ versus H_i plots showed curvature characteristic of the approach to a ferromagnetic transition (Fig. 1). When the data were replotted as σ^2 versus H/σ in the usual manner, the lines were strongly curved because of the magnetic component. Trial values of σ_o as a function of temperature were therefore calculated by extrapolation of the σ_o versus T values above 90°K, and the data were replotted as $(\sigma - \sigma_o)^2$ versus $H/(\sigma - \sigma_o)$. The linear temperature dependence of σ_o was adjusted until the best linear set of $(\sigma - \sigma_o)^2$ versus $H/(\sigma - \sigma_o)$ curves was obtained. Some typical corrected data of this type are shown in Fig. 2. The curvature at low fields, which remains evident near the transition temperature, indicates that σ_o

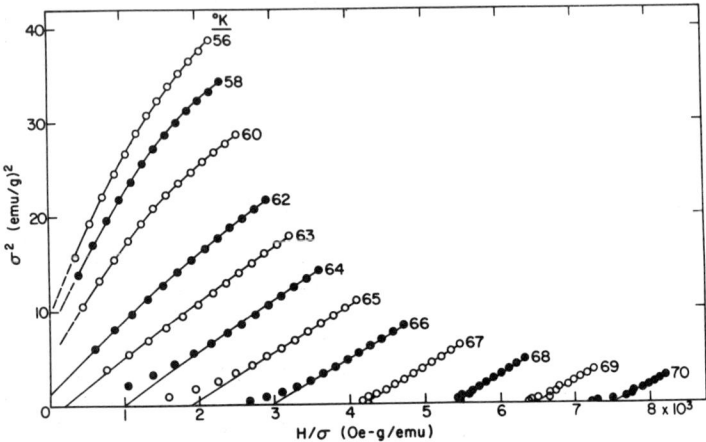

Fig. 2. Plots of σ^2 vs H/σ for NpAl$_3$ at various temperatures in the vicinity of the ferromagnetic ordering temperature.

increases more rapidly with a decrease in temperature than the linear extrapolation. The values of $(H/\sigma)_{H \to 0}$ vary smoothly with temperature and extrapolate to zero to give a Curie temperature of 62.5 ± 0.5°K. The σ^2 versus H/σ plots become curved in the opposite direction immediately below the transition temperature, which is consistent with the presence of magnetocrystalline anisotropy, as indicated by the low-temperature data.

A plot of the reciprocal molar susceptibility versus temperature is shown in Fig. 3. The data are strongly curved over the entire range and do not follow the Curie-Weiss law. The various lines through the data are based on theoretical calculations, and these will be discussed in the next section.

DISCUSSION

The temperature dependence of the susceptibility has not yet been analyzed within the framework of the localized-spin-fluctuation model, therefore, we will consider our results solely in terms of crystal-field theory. To evaluate the temperature-dependent

magnetic properties of $NpAl_3$ in terms of the crystal-field model, we
need to know the ionicity of the neptunium ion in the compound, the
effective charge of the ligands, the crystal-field parameters, and
the molecular-field constants. The molecular-field constants can be
estimated from the magnetic-ordering temperature in ferromagnetic
compounds. The ionicity of the magnetic ion and the effective charge
of the ligands are not known in conducting compounds, consequently,
a large amount of numerical computation is required to determine
a configuration that will uniquely yield the experimental results.
However, the $5f^3$ configuration is the only one that gives an ordered
moment in reasonable agreement with that deduced from the magnetic
hyperfine field ($\sim$1.2 μ_B/Np). Given the configuration, there are two
choices for the effective charge of the ligands. If the effective
charge is positive, the ground state of the Np ion is a quadruplet
Γ_8 with the doublet Γ_6 excited state 2711 cm^{-1} above the ground
state; the ordered moment is 1.2 μ_B/Np. Inasmuch as the excited
state is far from the Γ_8 ground state, the reciprocal susceptibility
versus temperature curve is almost linear. When the effective charge
of the ligands is negative, the Γ_6 doublet is the ground state with
the Γ_8 excited state only 10 cm^{-1} above the ground state. The ordered magnetic moment in this case is 1.35 μ_B/Np, and the reciprocal
susceptibility versus temperature curve has substantial curvature.
Several calculated reciprocal susceptibility curves are shown in Fig.
3. The numerical values of the fourth-order ($A_4<r^4>$) and the sixth-
order ($A_6<r^6>$) crystal-field parameters are indicated with each
curve. The experimental results (circles) show a somewhat larger
curvature than the theoretical lines. Until we have more experimental data available to constrain the fit, there is no justification for
trying to determine precise values of the parameters. In particular
we need accurate values for the ordered moment and its temperature

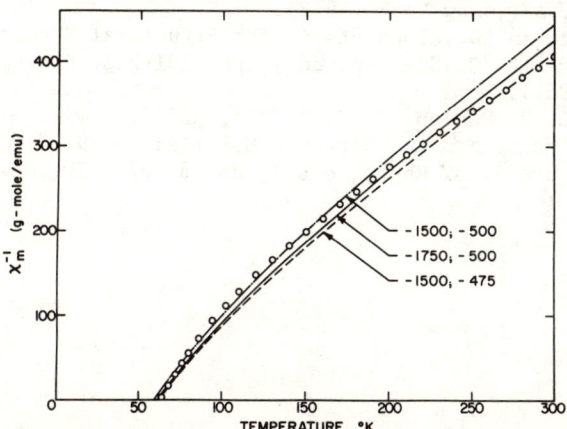

Fig. 3. Reciprocal molar susceptibility as a function of temperature for $NpAl_3$. The numerical values (in units of cm^{-1})
for each curve are the crystal-field parameters $A_4<r^4>$
and $A_6<r^6>$, respectively.

dependence in the ferromagnetic region. We hope to rectify this situation by neutron-diffraction measurements.

The experimental results obtained so far, if interpreted in terms of a crystal-field model, suggest that the configuration of the Np ion is $5f^3$ (Np^{+4}), and the effective charge of the ligands is negative. Inasmuch as the degeneracy between the Γ_6 and Γ_8 ground states differs by a factor of two, depending on whether the effective charge is positive or negative, it would be interesting to determine the magnetic entropy change through the ferromagnetic transition by specific-heat measurements.

ACKNOWLEDGMENTS

We would like to thank A. W. Mitchell and S. D. Smith for able experimental assistance.

REFERENCES

1. A. Misiuk, J. Mulak, and A. Czopnik, Bull. Acad. Polon. Sci., ser. sci. chim. 20, 891 (1972).
2. K. H. J. Buschow and H. J. Van Daal, AIP Conf. Proc. #5, "Magnetism and Magnetic Materials - 1971" (C. D. Graham, Jr. and J. J. Rhyne, eds.), pp. 1464-1477, AIP, New York (1972).
3. A. R. Harvey, M. B. Brodsky, and W. J. Nellis, Phys. Rev., to be published.
4. J. Mulak and A. Misiuk, Bull. Acad. Polon. Sci., ser. sci. chim. 19, 207 (1971).
5. N. Rivier and M. J. Zuckerman, Phys. Rev. Letters 21, 904 (1968).
6. S. Doniach, AIP Conf. Proc. #5, "Magnetism and Magnetic Materials - 1971" (C. D. Graham, Jr. and J. J. Rhyne, eds.), pp. 549-557, AIP, New York (1972).
7. A. E. Dwight, in "Developments in the Structural Chemistry of Alloy Phases" (B. C. Giessen, ed.), pp. 181-226, Plenum Press, New York (1969).
8. D. J. Lam, B. D. Dunlap, A. T. Aldred, and I. Nowik, AIP Conf. Proc. #10, "Magnetism and Magnetic Materials - 1972" (C. D. Graham, Jr. and J. J. Rhyne, eds.), pp. 83-87, AIP, New York (1973).

A STUDY OF IRON IMPURITIES IN THE LINEAR MAGNETIC SYSTEMS TMMC AND CsNiF$_3$*

P. A. Montano
University of California, Santa Barbara, Calif. 93106

ABSTRACT

Mössbauer measurements have been carried out on TMMC (^{57}Co) and CsNiF$_3$ (^{57}Co). In TMMC (^{57}Co) the Mössbauer spectrum has been measured down to 0.3 K. A strong temperature dependence of the quadrupole splitting below 4.2 K has been observed. Below the three-dimensional transition point no magnetic hyperfine splitting was observed. This indicates either a cancellation of the Fermi contact field by the orbital internal field or a reduction of the sublattice magnetization by fast spin fluctuations as expected in one-dimensional systems. Mössbauer measurement were also taken in CsNiF$_3$ (57Co) down to 1.4 K (T_N = 2.61 K). In CsNiF$_3$ the Ni ions are ferromagnetically coupled along the c-axis. A magnetic hyperfine splitting in the ^{57}Fe nucleus was observed well above T_N indicating the presence of very strong spin correlations along the chains for nonsubstitutional ^{57}Fe in CsNiF$_3$. When the ^{57}Fe is substitutional there was no h.f.magnetic splitting observed down to 1.3 K. Higher concentration of iron up to 10% in CsNiF$_3$ tends to break the linear chain structure of CsNiF$_3$.

INTRODUCTION

Recently, considerable interest has been shown in the study of the magnetic properties of nearly one-dimensional magnetic systems.[1-4] The best known example of a linear-chain antiferromagnet is $(CH_3)_4$ NMn Cl_3 (TMMC)[1-4] and of a ferromagnetic chain is CsNiF$_3$.[5] In order to study the behavior of Fe impurities in TMMC and CsNiF$_3$ a small amount of radioactive ^{57}Co was introduced into the material. This nucleus is transformed to ^{57}Fe by electron capture. Then the iron nuclei were studied with the help of the Mössbauer effect.

PREPARATION OF THE SAMPLES

It is not possible to introduce ^{57}Co into TMMC by a simple diffusion technique due to the decomposition of the sample at temperatures higher than room temperature (r.t.). It was necessary to grow single crystals of TMMC from a solution containing ^{57}Co. Due

*Supported by the National Science Foundation

to the strong γ-ray absorption of the crystal it was necessary to prepare a powdered source. The source was mounted on a metallic holder and sealed with a Mylar window to avoid decomposition of the sample when it is exposed to the atmosphere.

Single crystals of $CsNiF_3$ were grown in an HF atmosphere by mixing and melting stoichiometric amounts of CsF and NiF_2. ^{57}Co was deposited on the surface of the sample and diffused during a few minutes below the melting point in an HF atmosphere and then rapidly quenched to r.t.. The sample was checked by X-ray and the powder pattern of $CsNiF_3$ was observed.[6] If one tries to diffuse for longer times a partial decomposition of the sample takes place.

A source of $CsNiF_3$ was prepared also by melting $CsNiF_3$ with ^{57}Co and growing a single crystal (in an HF atm). The sample showed a definite phase separation after this process. At the top there was a more yellow phase that gave a different X-ray pattern than $CsNiF_3$ (This phase has not been completely analyzed but it seems to be Cs rich and slightly hygroscopic, probably $Cs_4Ni_3F_{10}$).[6] At the bottom of the sample the needle-like crystals of $CsNiF_3$ were present.

In order to understand more the behavior of the Fe impurities in $CsNiF_3$, absorbers were prepared by sintering $CsNiF_3$ and $CsFeF_3$ in the correct percentages (Samples with 5% and 10% $CsFeF_3$ were prepared).

EXPERIMENTAL RESULTS AND DISCUSSION

TMMC (^{57}Co): In TMMC there are chains of Mn ions coupled antiferromagnetically along the c-axis. The Mn^{2+} ions are in the center of Cℓ octahedra which have a trigonal distortion. The cobalt can go substitutionally in the Mn site. At r.t. it is reasonable to assume that the crystal field at the Mn site is axial symmetric, but below 127 K[7] there is a crystallographic transition to a monoclinic structure.

The presence of divalent and trivalent Fe following the decay of ^{57}Co in TMMC was observed in Mossbauer spectra. Higher charged states are produced in the source during the Auger cascade initiated by the K capture decay of ^{57}Co. There are several examples of Mössbauer spectra where Fe^{3+} has been observed following the decay of ^{57}Co in an insulator.[8] Fig. 1 shows a typical Mössbauer spectra of the TMMC source with a $K_4Fe(CN)_6 \cdot 3H_2O$ absorber with 0.25 mg/cm^2 of ^{57}Fe (90% ^{57}Fe). Zero velocity is with respect to a metallic iron absorber.

The Fe^{2+} shows a large Q.S. and its I.S. is consistent with a high spin electronic configuration. The Fe^{3+} spectra could be fit best by assuming only one line. It was observed that this line

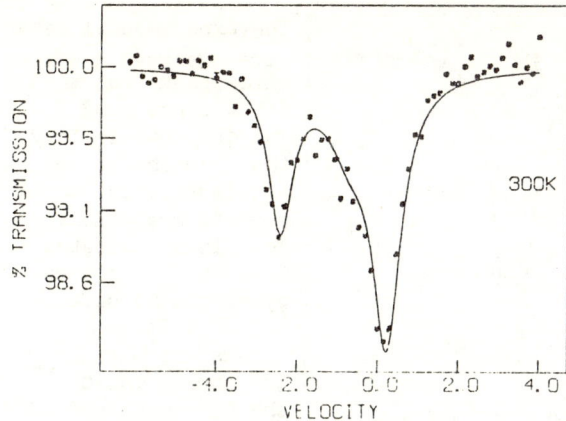

broadened considerably at low temperatures. The linewidth of the Fe^{2+} increases only slightly at low temperature. There is a marked increase in the Q.S. below $4.2°K$ (see Table I).

Fig. 1. TMMC (^{57}Co) at r.t. velocity in mm/sec.

Table I   Q. S. of Fe^{2+} in TMMC

	300K	78K	4.2K	1.3K	0.3K
ΔE_Q [mm/sec]	2.62±0.05	2.69±0.05	2.93±0.05	3.18±0.05	3.2±0.1

This increase is probably associated with the onset of short range order. The magnetic interaction can alter the electric field gradient at the Fe nucleus at low temperatures.[9] Measurements were carried out below the transition point ($T_N = 0.85°K$) at T = 0.3K.

No appreciable change in linewidth and Q.S. was observed at 0.3K and there was no magnetic hyperfine splitting. The absence of magnetic splitting indicates either a cancellation of the internal field at the nucleus by opposite Fermi contact and orbital contributions or a very small magnetic moment strongly reduced by the zero point motion.[10,11] Further information on this point might be obtained from spectra taken in an external field.

$CsNiF_3$ (^{57}Co): It was observed that the Mössbauer spectra are different depending on the way the radioactive nuclei ^{57}Co were introduced into the lattice. In Table II the values of the Mössbauer parameters at r.t. are given.

Table II   I.S. and Linewidth at R.T. of $CsNiF_3$ (^{57}Co)

	Sample I		Sample II	
	Fe^+	Fe^{3+}	Fe^{2+}	Fe^{3+}
I.S. [mm/sec]	-1.84±0.04	-0.37±0.02	-0.70±0.02	-0.05±0.02
Γ [mm/sec]	0.45±0.06	0.73±0.04	0.68±0.04	0.47±0.04

Sample I was prepared by diffusion of the ^{57}Co. In this sample Fe^{2+} was not observed. The main charge state observed was Fe^{3+}. There was a small peak at a high velocity (Fig. 2).

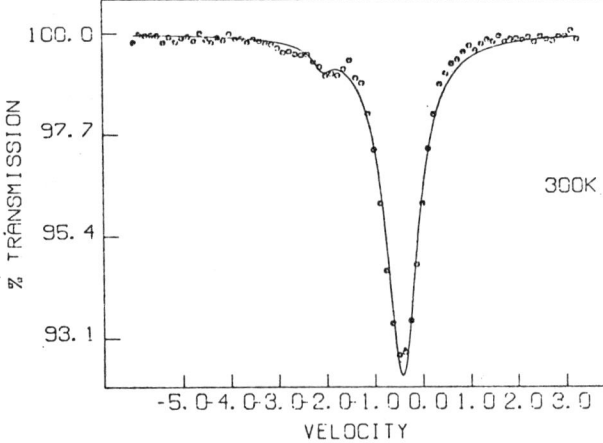

Fig. 2. CsNiF$_3$(^{57}Co) Sample I Velocity in mm/sec.

The experimental data were analyzed assuming three peaks but the fit was not good. The only possibility that remains is to attribute this small peak to monovalent Fe. This indicates that the Fe is occupying the Cs site. The I.S. is consistent with a $3d^7$ configuration.[12] The Fe^{3+} observed is probably either in place of the Cs or interstitial. The Fe^{3+} line is very broad at r.t. This may be produced by the presence of charge compensating vacancies around the impurity and/or by magnetic relaxation effects.[13] When the temperature is lowered the Fe^{3+} line width increases and near 4.2°K it begins to show hyperfine magnetic splitting. Below 4.2°K the magnetic splitting is characterized by a constant field of around 576 kG (characteristic of Fe^{3+}) and strong temperature dependence of the intensities.

In Sample II the ^{57}Co was introduced by melting together with CsNiF$_3$ and growing single crystals. The phase that was identified as CsNiF$_3$ shows two peaks with I.S. of -0.05 and -0.70 mm/sec with respect to Fe(Fig. 3). In order to know if the lines are independent one uses a simple technique. The source was run with a polarized Fe absorber. If there is a quadrupole splitting in the source one line corresponds to the transition $\pm \frac{3}{2} \to \pm \frac{1}{2}$ ($\Delta M = \pm 1$) and the other to $\pm \frac{1}{2} \to \pm \frac{1}{2}$ ($\Delta M = \pm 1, 0$). When running against a Fe polarized absorber a marked asymmetry between the $\pm \frac{3}{2} \to \pm \frac{1}{2}$ and $\pm \frac{1}{2} \to \pm \frac{1}{2}$ transitions should be present. That was not the case, so the two lines are independent.[15] The lines are broad and the linewidth is practically temperature in-

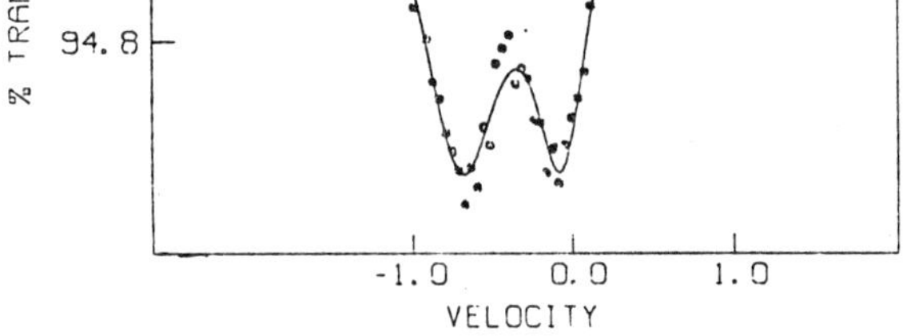

Fig. 3. CsNiF$_3$ (^{57}Co) Sample II

dependent. (This is probably associated with small unresolved Q.S.) The line with the largest I.S. with respect to Fe can be attributed to Fe^{2+}. The Fe^{2+} seems to be in an unusual electronic configuration for being in a noncubic fluorine compound. One can think that the iron is in a very strong crystal field that produces a low spin configuration. Down to low temperatures but above $T_N(2.61K)$ there is no indication of magnetic splitting either in the Fe^{2+} or the Fe^{3+}.

When impurities of Fe were introduced with percentages of 4% to 10%, a break of the linear chain was observed. A Mössbauer spectrum very similar to that of $CsFeF_3$ was obtained.[10] This indicates that Fe has the tendency to break the chain and produce a new crystalline structure more like $CsMnF_3$ (built up of monomers and dimers of Fe octahedrally coordinated with the fluorine).

In summary, in $CsNiF_3$ the presence of Fe^{2+} impurities substitutionally for Ni^{2+} break the linear chain. When the Fe^{3+} was not replacing Ni, a magnetic hfs was observed at low temperatures indicating the presence of relaxation effects. The Mössbauer spectrum of Fe^{2+} impurities in TMMC shows a very strong temperature dependent Q.S. at low temperature and no magnetic hfs below T_N was detected.

REFERENCES

1. T. Smith and S. A. Friedberg, Phys. Rev. <u>176</u>, 660 (1968).
2. R. Dingle, M. E. Lines and S. L. Holt, Phys. Rev. <u>187</u>, 643(1969).
3. R. J. Birgeneau, R. Dingle, M. T. Hutchings, G. Shirane and S. L. Holt, Phys. Rev. Lett. <u>26</u>, 718 (1971).
4. M. T. Hutchings, G. Shirane, R. J. Birgeneau and S. L. Holt, Phys. Rev. <u>B5</u>, 1999 (1972).
5. M. Steiner, W. Krüger and D. Babel, Solid State Comm. <u>9</u>, 227 (1971).
6. D. Babel, Zeit. für Anorg. und Allg. Ch. <u>369</u>, 117 (1969).
7. P. S. Peercy, B. Morosin and G. A. Samara, to be published Phys. Rev.
8. G. K. Wertheim, H. J. Guggenheim and D.N.E. Buchanan, J. of Chem. Phys. <u>51</u>, 1931 (1969).
9. P. A. Montano, E. Cohen, H. Shechter and J. Makovsky, Phys. Rev. <u>B7</u>, 1180 (1973).
10. D. E. Cox and V. J. Minkiewicz, Phys. Rev. <u>B4</u>, 2209 (1971).
11. P. A. Montano, E. Cohen and H. Shechter, Phys. Rev. <u>B6</u>, 1053 (1972).
12. H. Micklitz and P. H. Barrett, Phys. Rev. Letters <u>28</u>, 1547 (1972).
13. H. H. Wickman and G. K. Wertheim, Chemical Applications of Mössbauer Spectroscopy, edited by V. I. Goldinskii and R. H. Herber, Academic Press (1968).
14. M. Steiner and B. Dorner, Solid State Commun. <u>12</u>, 537 (1973).
15. P. A. Montano, in preparation.
16. M. Eibschutz, L. Holmes, H. J. Guggenheim and H. J. Levinstein, J. Appl. Phys. <u>40</u>, 1312 (1969).

SPECIFIC HEAT OF THE MAGNETIC CHAIN TMMC BELOW 1.2 K *

H. W. White and K. H. Lee
Physics Dept., Univ. of Mo., Columbia, MO 65201

J. Trainor and D. C. McCollum
Physics Dept., Univ. of Ca., Riverside, CA 92502

S. L. Holt
Chemistry Dept., Univ. of Wy., Laramie, WY 82070

ABSTRACT

Specific heat measurements are reported for single crystal $(CH_3)_4NMnCl_3$ (TMMC) from 0.55 to 1.2 K in zero applied magnetic field. A peak in the magnetic contribution to the specific heat was found at 0.850 ± 0.005 K which could be attributed to a 1d to 3d transition. The peak height was 360 mJ/mole-K. The ratio of the interchain to intrachain exchange constants is estimated to be 8×10^{-5} via a Green's function calculation using a Néel temperature of 0.85 K and a intrachain exchange constant of 6.6 K.

INTRODUCTION

Specific heat measurements on the linear magnetic chain $(CH_3)_4NMnCl_3$ (TMMC) below 1.2 K were undertaken to investigate the location and nature of the transition from 1d to 3d ordering. Magnetic susceptibility,[1,2] neutron diffraction,[3] and proton magnetic resonance[4] measurements indicated the transition to be between 0.8 to 0.9 K.

EXPERIMENTAL PROCEDURE

The measurements were made in a He^3 cryostat which has been well described by Taylor,[5] using a discontinuous heating method. Data were taken from 0.55 to 1.2 K, a temperature range in which the magnetic transition occurred. The calibration of the germanium thermometer was against the vapor pressure of He^3, which was measured with an MKS Baratron Capacitance Manometer.[6] Procedures were checked by measuring the specific heat of 99.999% pure copper, the result being within 1% of accepted values.[7] The sample (TMMC II) was a single crystal grown from a saturated solution using seeding techniques and had a mass of about 6 grams. The sample was analyzed and found to contain the following weight percent impurities: 0% Co, 0.0006% Ni, 0.0034% Fe, and 0.0007% Cu. The addenda heat capacity amounted to about 0.6% of the total at 1 K, which was measured separately using a similar thermometer and heater assembly.

*Supported in part by Research Corporation, University of Missouri Research Council, AFOSR, and NSF.

EXPERIMENTAL RESULTS

It was assumed that the total specific heat below 1.2 K was composed of a magnetic contribution C_M and a lattice contribution C_L. Below 1.2 K we let $C_L = 11.3\,T^3$ mJ/mole-K since specific heat results in the region 1.4 to 2.5 K could be fit very well by the expression $C = (133\,T + 11.3\,T^3)$ mJ/mole-K, where the first and second terms could represent the magnetic and lattice contributions, respectively.[8] To this approximation, the lattice contribution at 1 K was about 6% of the total specific heat. The total and magnetic specific heats are listed in Table I.

A peak of height 360 mJ/mole-K in the magnetic specific heat was found, which was located at 0.850 ± 0.005 K. Figure 1 is a plot of the magnetic specific heat versus temperature illustrating the λ-shaped peak.

Table I Values for the total specific heat C and the magnetic specific heat C_M for TMMC, sample II, between 0.55 and 1.2 K. C and C_M are expressed in units of mJ/mole-K and T in units of degrees Kelvin.

T	C	C_M	T	C	C_M	T	C	C_M
0.569	119	116	0.789	237	231	0.954	189	179
0.593	120	118	0.792	236	230	0.971	187	177
0.602	122	119	0.800	246	240	0.977	186	175
0.605	121	119	0.803	255	249	0.989	188	177
0.618	126	124	0.816	272	265	0.996	183	172
0.637	132	129	0.820	281	275	1.01	187	175
0.639	132	129	0.825	304	298	1.01	187	175
0.643	134	131	0.831	311	305	1.03	187	175
0.669	145	142	0.836	318	311	1.03	189	177
0.669	144	141	0.843	344	337	1.06	188	175
0.681	150	146	0.850	361	354	1.06	190	177
0.700	159	155	0.858	306	299	1.07	188	174
0.706	164	160	0.863	281	274	1.09	193	178
0.716	170	166	0.884	206	198	1.10	191	176
0.722	174	170	0.884	214	206	1.11	195	180
0.731	176	171	0.890	201	193	1.12	195	179
0.734	182	177	0.893	201	193	1.13	196	180
0.745	193	188	0.922	192	183	1.16	200	182
0.761	199	194	0.928	190	181	1.18	202	184
0.764	208	203	0.929	190	181	1.19	202	183
0.766	213	209	0.929	190	181	1.20	204	184
0.764	208	202	0.939	190	180	1.20	206	187

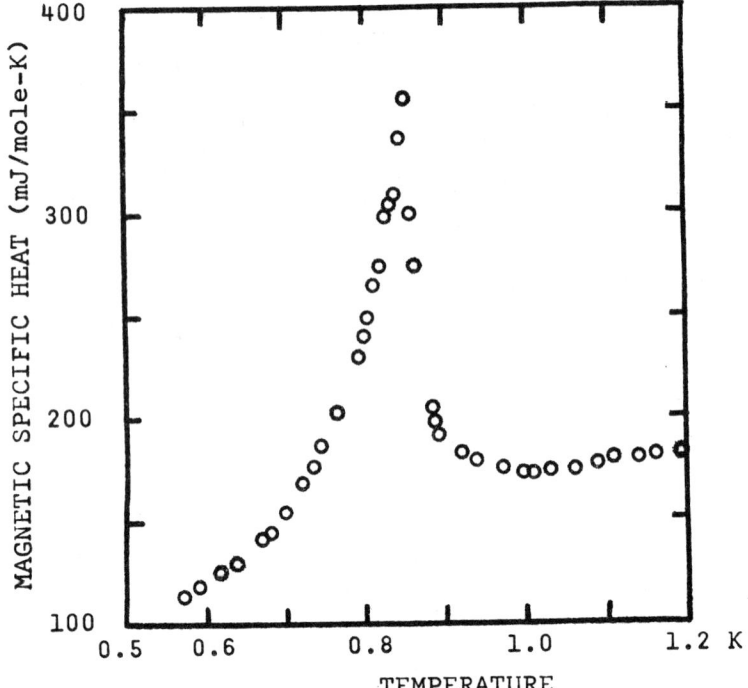

Figure 1. Magnetic specific heat of TMMC, sample II, between 0.55 and 1.2 K.

DISCUSSION

Since the proton magnetic resonance data of Dupas and Renard[4] provide convincing evidence that a transition to a 3d ordering occurs in TMMC near 0.85 K, we conclude that the small, sharp peak in the specific heat at 0.850 ± 0.005 K locates the onset of 3d ordering rather than being due to a hyperfine interaction. For our calculations we have assumed that the 3d ordering is precipitated by exchange forces, even though Dingle, Lines, and Holt[1] have concluded that it is entirely possible that the long range order is precipitated by interchain dipole-dipole interactions.

Assuming a Néel temperature $T_N = 0.85$ K, we proceeded to correlate the location of T_N to the ratio of the interchain to intrachain exchange constants by use of a simplified model identical to that used by Oguchi[10] for $Cu(NH_3)_4SO_4 \cdot H_2O$; namely, a <u>three</u> dimensional Heisenberg antiferromagnet with tetragonal structure and an interchain exchange constant J' which is smaller than the intrachain exchange constant J. In the Green's function approximation, he obtains an expression for T_N given by

$$\frac{4S(S+1)J}{3k_B T_N} = \frac{1}{\pi^3} \int\!\!\!\int\!\!\!\int_0^\pi \frac{dxdydz}{(1-\cos x) + \eta[(1-\cos y) + (1-\cos z)]}, \quad (1)$$

where S is the spin quantum number, k_B is Boltzmann's constant, η is the ratio J'/J. It is to be noted that, in the limit of J' → J, Eq. (1) gives the usual T_N value for a simple antiferromagnet (J' = J).[9] However, Eq. (1) is different from the result of Dingle, Lines, and Holt,[1] which is a calculation for a one dimensional antiferromagnet (J' = 0) with the result giving a one dimensional integral on the right hand side of Eq. (1).

With J = 6.6 K,[3] S = 5/2, and T_N = 0.85 K, Eq. (1) gives

$$\eta = J'/J \simeq 8 \times 10^{-5} \text{ K} \quad (2)$$

or, equivalently,

$$J' \simeq 5 \times 10^{-4} \text{ K}. \quad (3)$$

It is also to be noted that

$$4J'S \simeq 5 \times 10^{-3} \text{ K} \quad (4)$$

obtained here is comparable to the quantity

$$g\mu_B H_E \simeq 4.1 \times 10^{-3} \text{ K} \quad (5)$$

given by Dingle, Lines, and Holt,[1] where H_E is an interchain effective field.

REFERENCES

1. R. Dingle, M. E. Lines, and S. L. Holt, Phys. Rev. 187, 643 (1969).
2. L. R. Walker, R. E. Dietz, K. Andres, and S. Darack, Solid State Comm. 11, 593 (1972).
3. M. A. Hutchings, G. Shirane, R. J. Birgeneau, and S. L. Holt, Phys. Rev. B 5, 1999 (1972).
4. C. Dupas and J-P. Renard, Phys. Letters 43A, 119 (1973).
5. W. A. Taylor, Ph.D. Thesis, University of California, Riverside (unpublished)
6. MKS Instruments, Inc., Burlington, MA 01803
7. D. W. Osborne, H. E. Flotow, and F. Schreiner, Rev. Sci. Instr. 38, 159 (1967).
8. H. W. White, J. M. Milan, K. H. Lee, and S. L. Holt (to be published).
9. M. E. Lines, Phys. Rev. 135, A1336 (1964).
10. T. Oguchi, Phys. Rev. 133, A1098 (1964).

PHASE TRANSITIONS IN $FeCl_2 \cdot 2H_2O$ IN EXTERNAL MAGNETIC FIELDS: MÖSSBAUER SPECTROSCOPY

L. D. Kandel, M. A. Weber[*] and R. B. Frankel
Francis Bitter National Magnet Laboratory[**]
Massachusetts Institute of Technology
Cambridge, Massachusetts 02139

C. R. Abeledo
Observatorio Nacional de Fisica
Cosmica, San Miguel, Buenos Aires, Argentina

ABSTRACT

Monoclinic $FeCl_2 \cdot 2H_2O$ orders antiferromagnetically at $T_N = 23$ K, and the magnetic structure consists of two sublattices of $-FeCl_2-$chains lying along the c-axis. The coupling along the chains is ferromagnetic with weak antiferromagnetic coupling between chains. Application of an external magnetic field along the easy axis (a) induces phase transitions at $H_1 = 39$ kOe and at $H_2 = 46$ kOe.
We report the observation of the three phases using the Mössbauer effect in a single crystal of $FeCl_2 \cdot 2H_2O$ cut parallel to the c-axis and placed at 32° to the γ ray beam and magnetic field H_o so that H_o was parallel to the easy axis a. The results may be summarized as follows: (1) For $H_0 < H_1$ we observe two superposed spectra with equal intensities due to the external field adding and subtracting to the hyperfine field in the spin down and spin up sublattices respectively; (2) For $H_1 < H_o < H_2$ the relative intensities of the spin up to spin down spectra are roughly 3:1. (3) For $H_2 < H_o$, only one spectrum is observed. These observations are consistant with the antiferromagnetic →ferrimagnetic →paramagnetic model of Narath[1] from susceptibility measurements.

[*] Supported by the Organization of American States.
[**] Supported by the National Science Foundation.
[1] A. Narath, Phys. Rev. A139, 1221 (1965).

MÖSSBAUER STUDY OF ORIENTED YbCrO$_3$ POWDER

G. R. Davidson* and B. D. Dunlap*
Argonne National Laboratory, Argonne, Illinois 60439

M. Eibschütz and L. G. van Uitert
Bell Telephone Laboratories, Murray Hill, New Jersey 07974

ABSTRACT

The ^{170}Yb Mössbauer effect has been used to study the local susceptibility α_{ij} of Yb^{3+} ions in the orthorhombic weak ferromagnet YbCrO$_3$ at 4.2°K. By applying an external field $\vec{H}$ to a loosely packed powder absorber, YbCrO$_3$ crystallites were oriented so that each had its weak ferromagnetic a-axis parallel to $\vec{H}$. Absorption spectra then yielded the direction z of the EFG principal axis: z lies in the a-b plane at an angle of 25 $\pm$ 1° to the a-axis. For H $\leq$ 6 kOe, the component of the magnetic hyperfine field parallel to z increases in magnitude linearly with H: $|H_z^{hf}| = (56 \pm 12) + (40 \pm 6)$ H, with both H_z^{hf} and H in kOe. This linear dependence is interpreted in terms of an effective-field model. By combining our results with published single crystal magnetization data, we estimate the off-diagonal component α_{ab} of the Yb^{3+} susceptibility to be $\pm$ (15 $\pm$ 13) x 10^{-6} μ_B/Oe. This value is shown to be consistent with expectations for a ground state that is predominantly a $|\pm 7/2\rangle$ Kramers' doublet. Results of measurements in zero external field at 4.2 and 77.4°K are also reported.

INTRODUCTION

The weak ferromagnet YbCrO$_3$ is an orthorhombically distorted perovskite (space group Pbnm) having four formula units per unit cell.[1] Below T_N = 118°K, the Cr^{3+} spins are ordered nearly antiferromagnetically along the c-axis with a small ferromagnetic component along the a-axis. A net opposing magnetization is induced on the Yb^{3+} sublattice.[2]

Recently, Shtrikman et al.[2] have made a <u>bulk</u> measurement of the magnetic susceptibility χ_i (i = a,b,c) of <u>single</u> crystals of YbCrO$_3$. By subtraction of the Cr^{3+} contribution,[3] the diagonal components of the Yb^{3+} ionic susceptibility α_{ij} may be evaluated.

In this paper we report a direct <u>local</u> measurement of the Yb^{3+} magnetic moment by means of the ^{170}Yb Mössbauer effect. Powder absorbers of YbCrO$_3$ were studied in small external magnetic fields. By combining local and bulk data, we determine the off-diagonal susceptibility component α_{ab}. In addition, we have obtained

---
*Work performed under the auspices of the U. S. Atomic Energy Commission.

information on the EFG tensor acting on the ^{170}Yb nuclei.

EXPERIMENT AND RESULTS

Absorbers were prepared by powdering single crystals grown from a flux. The 84.3 keV 0→2 transition of ^{170}Yb was studied using a source of ^{170}Tm in TmAl$_2$ and a standard transmission geometry.

Spectra in zero external field at 77.4 and 4.2°K were similar in appearance and characteristic of nearly pure quadrupole splitting [See Fig. 1(a)]. The 77.4°K spectrum could be fit with magnetic hyperfine field $\vec{H}^{hf}$ = 0, quadrupole coupling constant e^2qQ/h = 1860 ± 30 MHz and asymmetry parameter η = 0.16 ± 0.05. At 4.2°K good fits were obtained with e^2qQ/h = 2060 ± 30 MHz, η values of 0 to 0.2, and $|H_z^{hf}|$ = 60 ± 10 kOe, where H_z^{hf} denotes the component of $\vec{H}_{hf}$ along the principal axis z of the EFG tensor. The component perpendicular to z could not be determined because of its small size. (This component has only a second order effect on the energy levels of the I = 2 state.) Linewidths (~ 2.8 mm/sec) were comparable to those obtained with a single-line YbAl$_2$ absorber, indicating fast relaxation ($T_R \ll 10^{-9}$ sec) and equivalent hyperfine parameters at all four Yb^{3+} sites.

At 4.2°K measurements were also made in external magnetic fields H = 1 to 6 kOe. The absorber consisted of a <u>loose</u> powder of crystallites 3 to 6 mils in diameter. The external field oriented the weak ferromagnetic a-axis of each crystallite parallel to the γ-ray direction enabling us to obtain information normally available only in single crystal experiments.

Good fits to these spectra [Fig. 1(b)] could be obtained by assuming η = 0 and again ignoring the component of $\vec{H}^{hf}$ perpendicular to z. In this approximation the relative line intensities are simple functions[4] of $\cos^2\theta$, where θ is the angle between the γ-ray direction and the z direction. Fits to our spectra yielded an average value for θ of ± 25 ± 1° (or ± 155 ± 1°) with the value of θ independent of H. Since the Yb^{3+} ions lie at sites of mirror planes perpendicular to the c-axis, z must be along c or in the a-b plane. Our result indicates the latter is true. The values obtained for $|H_z^{hf}|$ in these fits were found to increase linearly with H: $|H_z^{hf}|$ = (56 ± 12) + (40 ± 6)H with both H_z^{hf} and $\bar{H}$ in kOe (see Fig. 2). These uncertainties take into account both the scatter of the

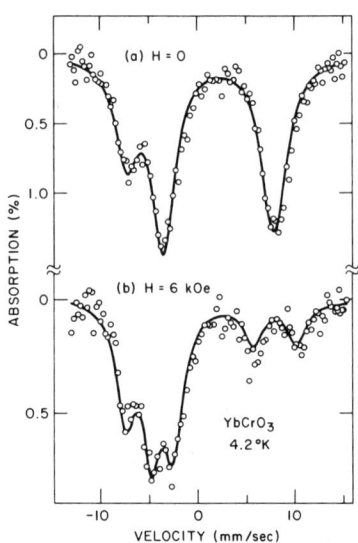

Fig. 1. Mössbauer spectra of ^{170}Yb in absorber of loosely packed YbCrO$_3$ crystallites at 4.2°K (a) for external field H = 0 and (b) with H = 6 kOe parallel to γ-ray direction.

data points and estimated systematic errors due to our fitting approximations.

ANALYSIS

The Yb^{3+} hyperfine field is proportional to its magnetic moment $\vec{\mu}$: a field of 4.2 MOe corresponds to a moment of $4\,\mu_B$.[5] Thus, our external field measurements indicate that μ_z has a linear dependence on H:

$$|\mu_z| = c_0 + c_1 H . \quad (1)$$

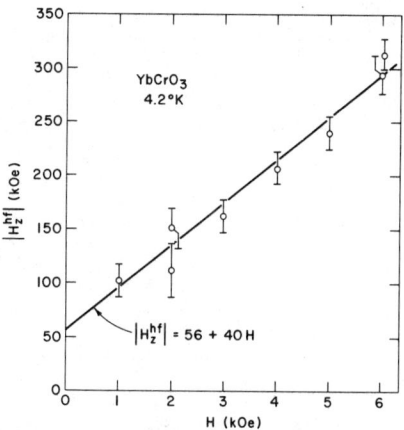

Fig. 2. Dependence of hyperfine field H_z^{hf} (where z is the EFG principal axis) on external field H parallel to the a-axis.

This result may be understood on the basis of a simple effective-field model.

Let us label the four Yb^{3+} sites 5,6,7, and 8 in accordance with the notation of Bertaut.[6] Symmetry considerations[6,7] require that at each site both $\vec{\mu}$ and the effective field $\vec{h}$ acting on it lie in the a-b plane. Also prescribed are the relative orientations of $\vec{\mu}$, $\vec{h}$, and z (indicated in Fig. 3). These conditions, plus the fact that α_{ij} must be symmetric,[8] enable us to relate $\vec{\mu}$ to an external field $\vec{H}$ parallel to the a-axis by means of five constants (α_{aa}, α_{bb}, α_{ab}, h_a, and h_b) as follows:

$$\begin{pmatrix} \mu_a \\ \mu_b \end{pmatrix} = \begin{pmatrix} \alpha_{aa} & \pm\alpha_{ab} \\ \pm\alpha_{ab} & \alpha_{bb} \end{pmatrix} \begin{pmatrix} h_a + H \\ \pm h_b \end{pmatrix} . \quad (2)$$

Also,

$$\mu_z = \mu_a \cos\theta \pm \mu_b \sin\theta . \quad (3)$$

The + signs apply to sites 5 and 6 and the – signs to sites 7 and 8. We assume that the external field is small enough not to affect the values of α_{ij} or $\vec{h}$. Substitution from Eq. (2) into Eq. (3) yields for all sites Eq. (1) with

$$c_0 = (\alpha_{aa} h_a + \alpha_{ab} h_b)\cos\theta + (\alpha_{ab} h_a + \alpha_{bb} h_b)\sin\theta . \quad (4)$$

and

$$c_1 = \alpha_{aa}\cos\theta + \alpha_{ab}\sin\theta . \quad (5)$$

Three additional equations relating our five unknowns involve the bulk magnetization data[2] at 4.2°K:

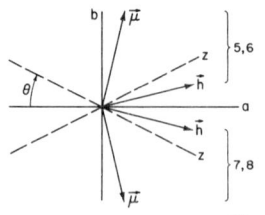

Fig. 3. Relationship among orientations of moments $\vec{\mu}$, effective fields $\vec{h}$, and EFG principal axes z at the four Yb^{3+} sites. The directions at sites 7 and 8 are obtained from those at 5 and 6 by 180° rotation about the a-axis.

$$\alpha_{aa} = (\chi_a^{Yb}/N_o\mu_B) = 48.9 \times 10^{-6} \, \mu_B/Oe; \tag{6}$$

$$\alpha_{bb} = (\chi_b^{Yb}/N_o\mu_B) = 6.4 \times 10^{-6} \, \mu_B/Oe; \tag{7}$$

and

$$(\alpha_{aa}h_a + \alpha_{ab}h_b) = M_a^{Yb}(0)/N_o\mu_B = 0.124 \, \mu_B. \tag{8}$$

In these expressions N_o is Avogadro's number, and χ_i^{Yb} and $M_a^{Yb}(0)$ are the Yb^{3+} contributions to the susceptibility and to the spontaneous weak ferromagnetic moment.

Simultaneous solution of Eqs. (5) and (6) yields $\alpha_{ab} = \pm (15 \pm 13)$ or $\pm (194 \pm 15)$ in units of $10^{-6} \, \mu_B/Oe$. The multiplicity of solutions arises from our uncertainty as to the sign of μ_z and the quadrant of z. Diagonalization of the resulting α_{ij} tensors gives the following principal values (in units of $10^{-6} \, \mu_B/Oe$ and referred to susceptibility principal axes a',b'): $\alpha_{a'a'} = 54 \pm 6$ and $\alpha_{b'b'} = 1 \pm 6$ (for $\alpha_{ab} = \pm 15 \pm 13$); $\alpha_{a'a'} = 223 \pm 11$ and $\alpha_{b'b'} = -168 \mp 11$ (for $\alpha_{ab} = \pm 194 \pm 15$). The solution resulting from $\alpha_{ab} = \pm 194$ is not physically permissible since a paramagnetic ion cannot have a negative principal component of susceptibility.

In principle, our formalism allows us to determine the directions of the principal axes of α_{ij}, the values of h_a and h_b, and the magnitudes and directions of the moments and effective fields at all Yb^{3+} sites. Unfortunately, the large uncertainty of α_{ab} precludes us from obtaining any useful information about these quantities. For example, using $\alpha_{ab} = \pm (15 \pm 13)$ and straightforward error propagation, we obtain $h_a = -17 \pm 400$ kOe.

DISCUSSION

The major result of our analysis of the field dependence of H_z^{hf} is establishing that α_{ab} is considerably smaller than α_{aa}. This enables us to determine the principal values $\alpha_{a'a'}$ and $\alpha_{b'b'}$ with reasonable precision.

The high anisotropy of α_{ij} in its principal axis system is significant. Such a result is consistent with expectations for a Kramers' doublet $|\pm J_{z'}\rangle$ having $J_{z'} > 1/2$ (for which only one principal component of α_{ij} would differ from zero). According to Ref. 2, χ_a^{Yb} has a Curie-Weiss dependence on temperature $\chi_a^{Yb} = C/(T - \theta_a)$, where $\theta_a = -12°K$. If we assume a $|J_{z'} = \pm 7/2\rangle$ ground state and a similar Curie-Weiss dependence for $\alpha_{z'z'}$, we would have

$$\alpha_{z'z'} = g_J^2 J_z^2 \mu_B/k(T - \theta_a) = 66 \times 10^{-6} \mu_B/\text{Oe} \quad .$$

This is close to our experimental result, indicating that the ground state of Yb^{3+} in $YbCrO_3$ is predominantly a $|\pm 7/2\rangle$ Kramers' doublet.

Such a doublet would yield an ionic quadrupole coupling $e^2qQ/h = 3500$ MHz.[9] This is nearly twice the value observed at 4.2°K, indicating a substantial opposing lattice contribution to the EFG.

ACKNOWLEDGMENTS

We are grateful to D. J. Lam for helpful discussions.

REFERENCES

1. S. Quezel-Ambrunaz and M. Mareschal, Bull. Soc. Franc. Miner. Crist. 86, 204 (1963).
2. S. Shtrikman, B. M. Wanklyn, and I. Yaeger, Intern. J. Magnetism 1, 327 (1971).
3. V. M. Judin and A. B. Sherman, Sol. St. Commun. 4, 661 (1966).
4. B. D. Dunlap, in Mössbauer Effect Data Index-1970 (Plenum Press New York, 1971), p. 25.
5. G. M. Kalvius, G. K. Shenoy, and B. D. Dunlap, in Les Elements des Terres Rares (Centre national de la recherche scientifique, Paris, 1970), p. 477.
6. E. F. Bertaut, in Magnetism, Vol. III, ed. G. T. Rado and H. Suhl, (Academic Press, New York, 1963), p. 149.
7. D. L. Wood, L. M. Holmes, and J. P. Remeika, Phys. Rev. 185, 689 (1969).
8. J. F. Nye, Physical Properties of Crystals (Oxford University Press, London, 1960), p. 67.
9. This estimate is based on the assumptions $\langle r^{-3}\rangle_{4f} = 13$ au, $Q = 2.14$ b, and Sternheimer shielding parameter $R = 0.2$.

MAGNETIC ORDERING AND HYPERFINE PARAMETERS IN THE LINEAR CHAIN COMPOUND RbFeBr$_3$

M. Eibschütz and G. R. Davidson*
Bell Laboratories, Murray Hill, N.J. 07974

D. E. Cox**
Brookhaven National Laboratory, Upton, N.Y. 11973

ABSTRACT

Neutron diffraction and Mössbauer effect (ME) measurements on RbFeBr$_3$ have revealed a low-temperature magnetically ordered state ($T_N \approx 5.5°K$) with almost zero hyperfine magnetic field at the Fe^{57} nucleus. Neutron diffraction data taken at 150°K show that the compound has the CsNiCl$_3$ structure with space group P6$_3$/mmc, in which the Fe^{2+} ions are arranged in linear chains parallel to the hexagonal c-axis. At 20°K superlattice reflections are present which are consistent with a crystallographic transformation to the lower symmetry space group P6$_3$cm. The new unit cell edges are $a' = a\sqrt{3}$, $c' = c$ in terms of the P6$_3$/mmc cell, and the Fe^{2+} ions now occupy two sets of nonequivalent octahedral sites in the ratio 2:1.

Below T_N the Fe^{2+} magnetic moments are coupled antiferromagnetically along the chains and lie in the basal plane. Two models give a good fit to the intensity data. The first is a triangular arrangement in which the Fe^{2+} moments all equal 2.2 μ_B at 1.7°K, and the second a collinear antiferromagnetic structure in which one-third of the Fe^{2+} ions have zero moment, and two-thirds a moment of 2.7 μ_B at 1.7°K. The ME absorption spectra between 1.54 and 300°K show two resonance absorption lines due to the electric field gradient at the iron nucleus. At 1.54°K both sites have the same values for the quadrupole splitting ($|e^2qQ/2| = 1.75 \pm 0.01$ mm/sec), the isomer shift (IS = 1.20 ± 0.01 mm/sec relative to iron) and the hyperfine field ($H_{hf} = 0 \pm 10$ kOe).

RbFeBr$_3$ is one of a class of isostructural compounds ABX$_3$ which have been extensively studied recently as a result of their quasi one-dimensional magnetic properties.[1] In these compounds there are chains of octahedrally coordinated 3d-transition-metal ions along the c-axis. The one-dimensional character arises primarily because these (FeBr$_3$)$^{-1}$ chains are widely spaced in the basal plane: the interchain separation is about 7 Å, more than twice the 3 Å distance between Fe^{2+} ions in the chains. We report here the results of neutron diffraction and Mössbauer effect studies which

---
*Present address: Argonne Nat. Lab., Argonne, Ill. 60439.
**Work supported by U.S. Atomic Energy Commission.

were undertaken to clarify the nature of the magnetic ordering in RbFeBr$_3$.

Neutron powder data were collected at several temperatures with a beam of 1.03 Å wavelength. At 150°K (Fig. 1, top pattern) all the peaks can be indexed on a hexagonal cell with a = 7.38 Å, c = 6.28 Å. Least squares refinement based on the CsNiCl$_3$ structure with P6$_3$/mmc symmetry yielded an excellent fit between observed and calculated intensities with a Br parameter of 0.1643 and an overall temperature factor of 1.3 Å^2.

At 20°K (Fig. 1, middle pattern) additional peaks were observed which could all be indexed on a hexagonal cell tripled with respect to the P6$_3$/mmc cell (a' = a$\sqrt{3}$, c' = c). Although this enlarged cell is characteristic of the triangular magnetic structure observed for many compounds of this type,[2,3] the intensities were quite

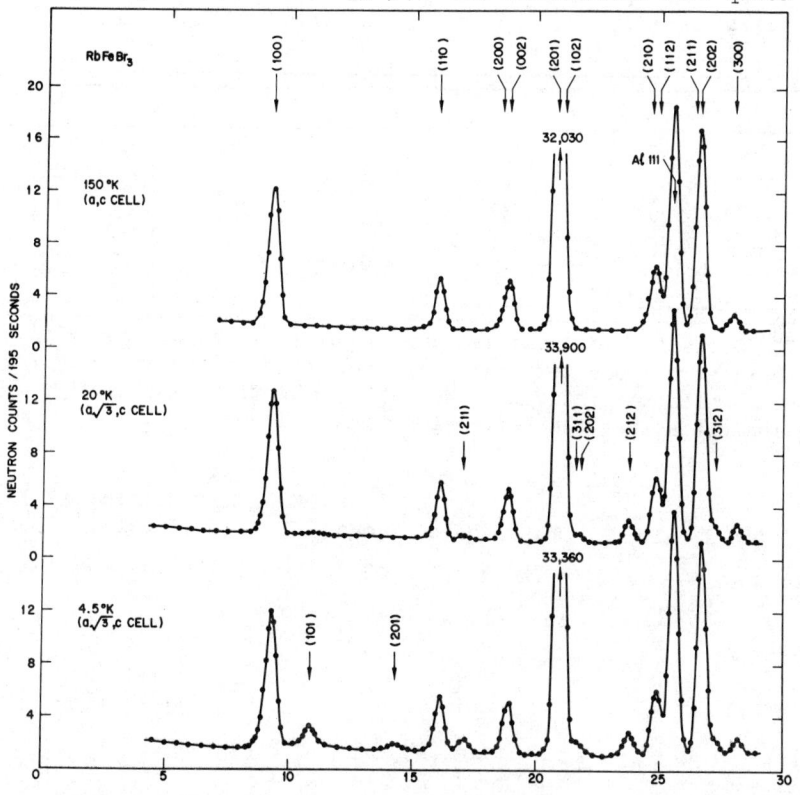

Fig. 1 Powder neutron data for RbFeBr$_3$ at 150°K, 20°K and 4.5°K. At 150°K (top pattern) the peaks can be indexed on a P6$_3$/mmc cell. At 20°K (middle pattern) there are additional nuclear peaks which can be indexed on an a$\sqrt{3}$, c hexagonal cell. At 4.5°K (bottom pattern) there are magnetic peaks which can also be indexed on the a$\sqrt{3}$, c cell.

different from those expected on this basis and much more suggestive of a crystal structure transition. By a trial-and-error process, it was found that good agreement was obtained with a structure based on the space group $P6_3cm$ and the parameters listed in Table I. Observed and calculated intensities for the first few peaks are shown in Table II. The temperature dependence of this phase was not studied in detail, but a few rough measurements were made on the strongest of the extra reflections, and extrapolation of these indicated a transition temperature around 120°K. It is to be noted that there are two crystallographically distinct kinds of Fe site, and the distortion is such that two-thirds of the Fe ions are shifted about 0.5 Å out of the plane of the others. However, there is a similar shift in the positions of the Br octahedra so that the

Table I Structure parameters for $RbFeBr_3$ at 20°K. Space group $P6_3cm$, cell edges $a' = a\sqrt{3}$, $c' = c$ in terms of 150°K cell.

Atom	Position	x	y	z
Fe(I)	2(a)	0	0	0
Fe(II)	4(b)	1/3	2/3	-0.076
Rb	6(c)	0.333	0	0.232
Br(I)	6(c)	-0.170	0	0.250
Br(II)	12(d)	0.492	0.829	0.192

Table II Comparison of observed and calculated neutron intensities for $RbFeBr_3$ at various temperatures: $hk\ell$ and $h'k'\ell'$ refer to $P6_3/mmc$ and $P6_3cm$ cells respectively. Magnetic reflections are labeled "M". A dash indicates that the reflection is not allowed.

		150°K		20°K		4.5°K Triangular	Collinear	
$hk\ell$	$h'k'\ell'$	I calc	I obs	I calc	I obs	I calc	I calc	I obs
-	100	-	-	1	<2	1	1	<2
100 / 001	110 / 001(M)	283 / - } 283	288	292 / - } 292	290	284 / 0 } 284	283 / - } 283	275
-- / -	200 / 101(M)	- / - } -	-	0 / - } 0	<2	0 / 27 } 27	0 / 30 } 30	32
101	111 / 111(M)	1 / - } 1	<3	2 / - } 2	<2	2 / 1 } 3	2 / - } 2	<2
-	210 / 201(M)	- / - } -	-	1 / - } 1	<3	1 / 12 } 13	1 / 12 } 13	10
110	300	96	88	96	96	93	93	89
- / -	211 / 211(M)	- / - } -	-	7 / 1 } 7	7	7 / 15 } 22	7 / 14 } 21	20
Weighted R Factor		0.042		0.038		0.073	0.071	

local environments remain quite similar, and the structure can be viewed mainly as a puckering of the octahedral framework.

Finally, at 4.5°K still more peaks are present, which can also be indexed on the enlarged cell (Fig. 1, bottom pattern). The intensities of these are quite consistent with a triangular magnetic structure (Fig. 2a and Table I). In these calculations, the moments of the two kinds of Fe were assumed to be equal and directed within the basal plane. This gives a value of 1.6 ± 0.2 μ_B per Fe. However, a collinear model in which one-third of the moments are zero (Fig. 2b) is equally consistent with the data, with a moment of 2.0 ± 0.2 μ_B. This is similar to the structure proposed for the high temperature magnetic phase of $CsCoBr_3$.[4] At 1.7°K, the moment for this model is about 2.7 μ_B. Above 4.5°K, the magnetic scattering rapidly becomes more diffuse, and the Bragg component disappears around 5.5°K, which may therefore be taken as the three-dimensional ordering temperature.[5]

The ^{57}Fe Mössbauer (ME) absorption spectra were taken using a constant acceleration spectrometer and a ^{57}Co in Pd source. The ME absorption spectra between 1.54°K and 300°K show two resonance absorption lines due to the electric field gradient at the iron nucleus. Hyperfine parameters obtained at selected temperature, are summarized in Table III. Both crystallographic sites have the same value for the quadrupole splitting (QS) and isomer shift (IS), (see Fig. 3 and Table III).

The magnitudes of the QS and IS are indicative of covalency in the Fe^{2+} bonding. The low temperature IS corresponds to a $3d^6 4s^{0.25}$ configuration according to the IS calibration of Danon.[6] Comparison of the 4.2°K QS with splitting observed in other trigonally distorted octahedral compounds,[7] yields a radial parameter $\langle r^{-3} \rangle_{3d}$ ≈1.9 a.u., indicative of substantial d-electron delocalization with respect to the free ion ($\langle r^{-3} \rangle_{3d}$ = 5.1 a.u.).

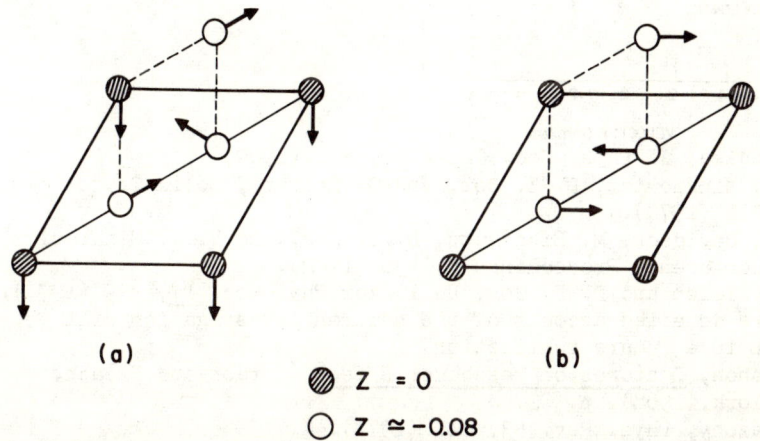

Fig. 2 Two possible magnetic structures for $RbFeBr_3$ consistent with the neutron data. The absolute orientation of the moments within the basal plane cannot be determined. Adjacent moments along the c axis are coupled antiparallel.

Below the three dimensional magnetic ordering temperature, $T_N \approx 5.5°K$, the spectra continue to show only two absorption lines due to the QS (Fig. 3). The change in line width upon going below the transition is less than the uncertainty of the line width. Based on this uncertainty we estimate a hyperfine field H_{hf} of 0 ± 10 kOe. A small magnetic hyperfine field can arise from a cancellation among the various components contributing to H_{hf}.[8,9] Thus, the low value observed is not inconsistent with the magnetic moments measured in the neutron diffraction experiments.

Table III Hyperfine parameters at four selected temperatures.

T (°K)	Q.S.[a] (mm/sec)	I.S.[b] (mm/sec)	H_{hf}[c] (kOe)
296	1.27±0.01	1.04±0.01	
70	1.58±0.01	1.19±0.01	
4.2	1.74±0.01	1.20±0.01	0±10
1.54	1.74±0.01	1.20±0.01	0±10

[a.] Q.S. = quadrupole splitting $\frac{e^2qQ}{2}$; $\eta = 0$ by symmetry; only the absolute value has been determined.

[b.] Isomer shift relative to iron foil at 296°K.

[c.] Hyperfine field.

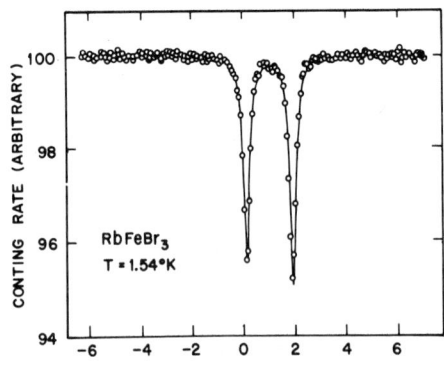

Mössbauer absorption spectra of $RbFeBr_3$ at 1.54°K.

1. N. Achiwa, J. Phys. Soc. Japan <u>27</u>, 561 (1969).
2. V. J. Minkiewicz, D. E. Cox, and G. Shirane, Solid State Comm. <u>8</u>, 1001 (1970).
3. G. R. Davidson, M. Eibschütz, D. E. Cox, and V. J. Minkiewicz, AIP Conference Proceedings <u>5</u>, 436 (1972).
4. W. B. Yelon and D. E. Cox, Bull.Amer.Phys.Soc. <u>17</u>, 339 (1971).
5. A more detailed account of the neutron investigation will be given in a future publication.
6. J. Danon, <u>Lectures on Mössbauer Effect</u> (Gordon and Breach, New York, 1968), p. 92.
7. Y. Hazony, Phys. Rev. <u>B3</u>, 711 (1971).
8. K. Ono, A. Ito and T. Fujita, J. Phys. Soc. Japan <u>19</u>, 2119(1964).
9. D. J. Simkin, Phys. Rev. <u>177</u>, 1008 (1969).

MAGNETIC STUDIES OF A CANTED ANTIFERROMAGNETIC: $MnBr_3(CH_3)_3NH \cdot 2H_2O$[†]

P.R. Newman, J.A. Cowen and R.D. Spence
Michigan State University,
East Lansing, Michigan 48824

ABSTRACT

The low temperature NMR and applied field susceptibility and magnetization are reported. The experimental data suggests a 4-sublattice canted antiferromagnet described by the magnetic space group $P_{2s}2_1/m$. Below $T_N=1.5K$ the system undergoes a magnetic phase transition in an applied field, which appears to be metamagnetic.

Above T_N, there is a discontinuity in the slope of the plot of magnetization versus applied field. This behavior appears to be caused by short range spin correlations which result from a one dimensional antiferromagnetic exchange.

INTRODUCTION

The class of compounds $M[(CH_3)_3NH]X_3 \cdot 2H_2O$ (where X is chlorine or bromine and M is a transition metal) exhibit interesting low-dimentional magnetic behavior.[1,2,3] X-ray studies show the existence of chain-like structures of the form $M\genfrac{}{}{0pt}{}{x}{x}M\genfrac{}{}{0pt}{}{x}{x}M$. Magnetic measurements made by D.B. Losee et al. and Stirrat and Cowen, show evidence of short-range low-dimensional magnetic correlations above the ordering temperature. NMR and magnetization studies by Spence[4] on the compound $Co[(CH_3)_3NH]Cl_3 \cdot 2H_2O$ show the ordered state to be a weak canted ferromagnet, which exhibits a metamagnetic transition in an applied field.

The specific heat and near-zero-field susceptibility of $Mn[(CH_3)_3NH]Br_3 \cdot 2H_2O$ have been measured by Carlin et al.[5] The specific heat exhibits a broad peak which may be associated with one-dimensional magnetic behavior. An ordering temperature of 1.56K is inferred from a small sharp peak that occurs in the specific heat. At T_N, the susceptibility has a large spike in the b-direction. The a and c direction susceptibilities exhibit no such spike. At the lowest experimental temperature, the c-axis susceptibility has the smallest value. The chemical cell is monoclinic with a=8.45, b=7.65 and c=8.54 angstroms and a monoclinic angle of 91°56'. There are two molecular units in the unit cell.

[†] Supported by the National Science Foundation

Fig.1. Resonance diagram for protons. The rotation is in the b-c plane.

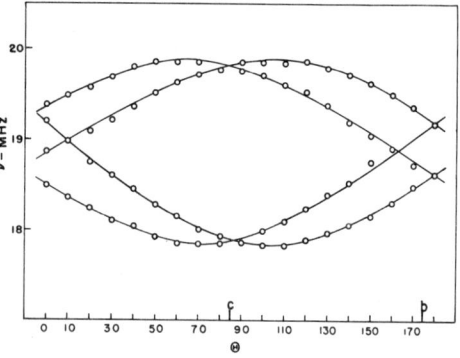

This paper will discuss the NMR and magnetization measurements and the magnetic space group and a sublattice model which are inferred from the experimental data.

EXPERIMENTAL

A) <u>NMR</u> 12 distinct lines have been observed in zero applied field. Analysis of line widths, splitting factors and temperature dependences indicates that the lower 9 lines are proton and the upper 3 are bromine resonances. The nine proton lines occur at frequencies ranging from 1.4 MHz to 18.7 MHz, while the bromine lines are found from 23-27 MHz. The remaining bromine lines are presumed to be above 30 MHz which was the highest experimental frequency.

In the presence of a small applied field (∼300 Oe), the rotation diagram of each of the proton lines shows 4 distinct directions (see Fig 1) for each local field magnitude.

B) <u>Susceptibility</u> In addition to the previously mentioned near-zero field data, the susceptibility was also measured in an applied field. The susceptibility along the b-axis exhibits a sharp peak in an applied field at temperatures of T_N and below. The position of the peak is dependent on field and temperature. A plot (see Fig 2) of the critical field and temperature values for this peak indicates the presence of at least 3 distinct magnetic states. No such peak was observed in the other axes in fields up to 16kOe.

C) <u>Magnetization</u> As the temperature is lowered from 4.2K to T_N, one observes a linear increase in magnetization with small increasing applied fields. The slope dM/dH, is consistent with the near-zero field susceptibility. However, at a field of several hundred oersteds in the b-direction, a discontinuity in the slope occurs. If one extrapolates the magnetization curve for fields beyond this point back to zero applied field, one finds a non-zero magnetization whose magnitude increases as the temperature is lowered toward T_N. There is no discontinuity in the other directions.

At temperatures below T_N, the magnetization in the b-direction

Fig.2. Magnet phase diagram derived from susceptibility. The applied field is parallel to the b-axis.

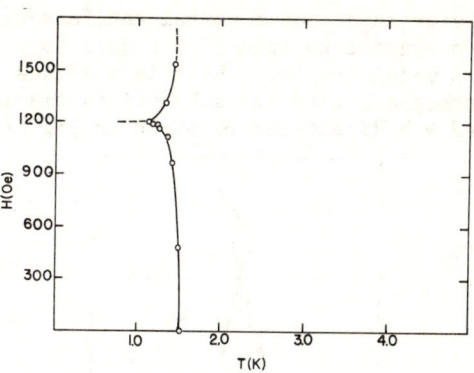

shows a sudden (90 Oe wide) increase which occurs at 1.2 kOe. At higher fields, the magnetization continues to rise smoothly, although not linearly, with increasing applied field. There is no evidence of saturation in fields up to 15 kOe. The magnetization in the other directions show only smooth increases. There is no magnetization in zero applied field in any direction.

DISCUSSION

It appears from the external morphology and x-ray extinctions, that the chemical space group is $P2_1/m$, with 2-fold axis parallel to b. The magnetic space group may be formed from this group, by replacing some of the elements with anti-elements. If one considers the monoclinic Heesh point groups for axial vectors, which result from 2/m, one can show that if all the protons in the unit cell occupy general symmetry positions, there should be 7 distinct proton local field magnitudes. However, since there are 2 NH protons and 4 operations in the point group, the NH protons must occupy special positions. There are 18 CH_3 protons, which indicates that at least 2 of these protons must also occupy special positions. These conditions raise the total number of local field magnitudes to 9. Further, by noting the number of operations in the Heesh groups, and the number of local field magnitudes and directions, it can be shown[6] that the magnetic point group must contain both inversion and anti-translation or neither. There are 3 members of the Heesh group which satisfy these requirements: 2/m', 2'/m and 2/m1'.

The magnetization data shows there is no net moment in zero applied field. This fact, along with the absence of any asymetry in the NMR rotation data at low fields indicates that the magnetic structure exhibits no net moment. The direction of sublattice magnetization as inferred from the near zero field susceptibility is approximately along the c-axis. The observed magnetic transition occurs not with the applied field along c, but rather along b or approximately perpendicular to the sublattice magnetization. This behavior rules out a normal antiferromagnet with an associated antiferromagnetic to spin-flop transition.

A model suggested by this evidence is a 4 sublattice canted antiferromagnet, with the sublattices nearly along c but canted in the + and - b directions as shown in Fig 3A.

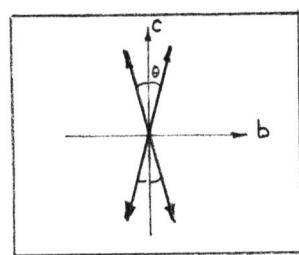 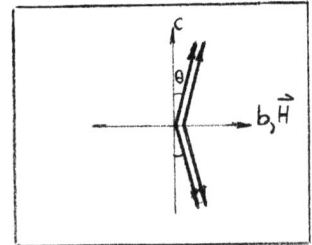

Fig.3A. Model for sublattice magnetization in zero field.

Fig.3B. Sublattice model after having undergone a magnetic phase transition.

If the manganese atoms are placed in the mirror plane and on the 2-fold axis, as required by the groups $P2_1/m'$ and $P2_1'/m$, one can show that the spins must either lie in the plane or perpendicular to it. The canting implied by our model is not allowed by these groups. In the remaining group $P_{2s}2_1/m$ (where s is a direction perpendicular to b) the canting is allowed if the manganese atoms are placed at the inversion centers.

We postulate that the transition which occurs in an applied field of approximately 1.2 kOe is metamagnetic and produces the spin arrangement shown in Fig 3B. The canting angle θ, may be calculated from the model by assuming the total magnet moment induced by this transition is:

$$\Delta M = Ng\mu_B S \sin \theta.$$

Using the experimental magnetization data, the latter expression yields an angle of approximately 4°.

The bulk magnetization at temperatures above T_N shows the onset of a net moment with an applied field of several hundred oersteds in the b-direction. The fact that this behavior is observed in the same temperature region as the broad peak in the specific heat indicates one-dimensional antiferromagnetic chains with a net canting in the b-direction. In addition, we believe there is some crystalline anisotropy, whose direction varies from site to site, which causes a net canting in the direction of the chains. In zero field, neither direction along the chain is energetically favored, and there are equal numbers of spins canted in each direction, resulting in no net moment. An applied field destroys this symmetry and a net moment results. The magnitude of this moment increases as T approaches T_N and the correlations become larger.

At T approximately equal to T_N, a second weaker anti-ferromagnetic exchange between chains couples adjacent chains. It is this second exchange which produces real three-dimensional ordering.

CONCLUSIONS

In summation it appears that the ordered magnetic ground state of $Mn[(CH_3)_3NH]Br_3 \cdot 2H_2O$ is a 4-sublattice canted anti-ferromagnet. The direction of the sublattice magnetization is nearly along c with a 4° canting in the + and - b directions. The fact that 9 distinct proton local field magnitudes were experimentally observed indicates that the magnetic space group belongs to the family of $P2_1/m$. From the experimental evidence, it appears that the magnetic space group is $P_{2s}2_1/m$.

The observed magnetic transition is metamagnetic and results in a spin configuration having a net moment. Finally, the induced moment recorded above T_N in an applied field of a few hundred oersteds is attributed to short range one dimensional spin correlations.

REFERENCES

1. D.B. Losee, J.M. McElearney, A. Siegel, R.L. Carlin, A.A. Khan, J.P. Roux, and W.J. James, Phys Rev <u>B6</u>, 4342 (1972)
2. D.B. Losee, J.N. McElearney, G.E. Shankle, R.L. Carlin, P.J. Cresswell and W.T. Robinson; to appear in Phys Rev B
3. C.R. Stirrat, S. Dudzinski, A.H. Owens and J.A. Cowen, unpublished
4. R.D. Spence. Proceedings of International Conference on Magnetism, Moscow USSR, Aug 1973, to be published
5. S. Merchant, J.N. McElearney, G.E. Shankle and R.L. Carlin, private communication
6. W.J. M De Jonge, J.G. Rensen, C.W. Swüste and R.D. Spence, Physica Vol 58 No 4, 544, (1972)

NEUTRON MICROSCOPY OF SPIN DENSITY WAVE DOMAINS IN CHROMIUM[*]

J. B. Davidson
Oak Ridge National Laboratory, Oak Ridge, Tenn. 37830

S. A. Werner
Ford Motor Co., Dearborn, Mich. 48121 and
Univ. of Michigan, Ann Arbor, Mich. 48105

A. S. Arrott
Simon Fraser University, Burnaby, B.C., Canada

ABSTRACT

The ORNL neutron-counting camera has been used to study structures in two Cr crystals. One crystal cut from an arc melted ingot showed two large domains which differ in the direction of the wave vector of the transverse linear spin density wave. The spatial variation of the mean polarization direction was analysed for one of these domains and found to vary continuously across the crystal. Presumably the mean direction follows the local strain field. The other crystal grown from vapor and presumably rather less influenced by strain effects shows that the two components of the polarization coexist over most of the specimen in regions small compared to the resolution of the experiment, ~ 0.5 mm.

INTRODUCTION

Despite numerous studies of the antiferromagnetism in pure chromium, there are still uncertainties concerning the exact nature of the spin configuration. These uncertainties are primarily on the scale of distances large compared to the atomic spacing, and concern the factors influencing the selection among the three cubic axes for the direction of the fundamental wave vector $\vec{Q}$ of the spin density wave and of its associated polarization $\hat{\eta}$. It is experimentally known that an applied magnetic field or stress can affect the choice of the direction of $\vec{Q}$. It is presumed that internal strains are also important. In addition to the effects of an applied magnetic field on the spin direction in the transversely polarized spin density wave (TSDW) state, stable between 122K to 312K, one must consider the role of strain (internal or externally applied), crystalline anisotropy, the influence of imperfections, and what is possibly most important, the role of the entropy coming from the thermal excitations of this TSDW.

The scale of distances over which changes in the direction of $\vec{Q}$ and $\hat{\eta}$ occurs can be determined experimentally. The recent work of Ando and Hosoya has shown that these distances can be comparable to

---

[*]Experimental work done at Oak Ridge National Laboratory, which is sponsored by U.S. Atomic Energy Commission, under contract with the Union Carbide Corp.

the size of crystals and hence observable by methods of neutron photography.[1] This paper describes a method of neutron microscopy developed in recent years at the Oak Ridge National Laboratory,[2] and shows its application to the study of chromium.

ORNL "FLY'S EYE" NEUTRON CAMERA

This wide-angle, position-sensing counting system for thermal neutrons has as its main components a scintillation screen of Li^6F-ZnS, fiber optics, an image intensifier, a TV camera and a digital memory. The localized light flash at the scintillator is intensified and focused on the camera tube. The light is converted to electrons at the photocathode, and then to localized positive charges on a thin KCℓ target. Thus, the neutron event is stored temporarily until it is erased and converted to a voltage pulse by a low velocity electron beam scanned over the target. The center of the pulse is located and stored in a memory location and seen simultaneously as a flash on the TV monitor. The coordinates of the events are obtained by tracking the position of the scanning beam with horizontal and vertical time-to-amplitude converters which generate x and y address pulses for storing in a 2-parameter-analyzer. The camera resolution is determined mainly by the spreading of the scintillations in the 0.5mm thick phosphor. By storing the centers of the events the resolution is improved over photographic or TV methods in which the scintillations are integrated directly.

The resolution of the system for images on the scintillation screen is 150μm horizontal and 500μm vertical. To take advantage of this resolution, it is necessary to consider the neutron optics; the angular divergence of the beam (in the horizontal plane) diffracted from a single crystal is greatest at the parallel position. The work described here was, however, carried out near the parallel position in order to gain intensity while giving up some of the resolution capability of the detector system. For a beam collimated by the crystal and a slit of comparable size (1.2 × 0.7 cm) placed 180 cm in front of the crystal, and with the detector set back 20 cm to make room for an electromagnet, the resolution of the overall system was 400μm horizontal and 600μm vertical.

RESULTS

We have taken pictures of the magnetic reflections of two Cr crystals. One of these, Cr #1, is the same crystal used in previous work by two of us.[3] It was prepared by vapor deposition in the decomposition of chromium iodide. It has the shape of two truncated pyramids placed base-to-base and of volume 83 mm.^{3} The second crystal,[4] Cr #2, was cut from an arc melted ingot and is in the shape of a slab (1.2 × 1.0 × 0.2 cm). These crystals were chosen because they represent two extremes: Cr #1 appears to be relatively free of internal strains and imperfections, while Cr #2 does not.

Fig. 1a shows the intensity for scattering from the (200) nuclear reflection of Cr #1 as a function of position on the detector in an isometric plot. The structure indicates that the center of this crystal is either hollow or that it is a misoriented crystallite. Fig. 1b is the same plot, but when the screen is 38 cm away from the crystal rather than 3 cm as in fig. 1a. Fig. 2a shows an isometric plot for the magnetic reflection (1, ε, 0) in 0 magnetic field, and fig. 2b shows this same reflection in 10kG. The shape of intensity distribution in these two plots is similar. Results for 0 field and 16 kG are compared in fig. 8. The crystal is in a single-Q state.

Fig. 3 shows the (200) nuclear reflection for Cr #2. Fig. 4 shows (on a scale with the horizontal axis expanded) the magnetic reflection (1-ε, 0, 0) with the detector set 4 cm back from the crystal. Fig. 5 gives an isometric plot of the intensities of fig. 4. This intensity comes from domains with $\vec{Q}_1$ along the x-axis, and does not distinguish the direction of the polarization. The x-axis is horizontal, in the plane of the slab, and also the direction in which the crystal was field-cooled in 60kG. (The desired single-Q state was not produced.) Fig. 6 shows the magnetic reflection (1, ε, 0) with the crystal angular setting almost the same (rotated 2°) as for fig. 5. This reflection corresponds to domains with $\vec{Q}_2$ ⊥ to the plane of the slab (i.e. along the y-axis), and shows intensity only for the vertical component of polarization, i.e. $\vec{\eta}$ ⊥ to both $\vec{Q}_1$ and $\vec{Q}_2$. As we had previously determined that $\vec{Q}_3$ is not present in this crystal, we would have expected fig. 6 to show a rather even intensity over the region missing in fig. 4. The fact that this distribution is not flat indicates that the mean direction of the polarization is changing across the region of the $\vec{Q}_2$ domain, with $\vec{\eta}$ being primarily ⊥ to $\vec{Q}_1$ and $\vec{Q}_2$ in a vertical stripe parallel to this polarization. To check that the polarization is primarily parallel to the x-axis in the region to either side of this stripe, it is sufficient to rotate the crystal 90° about $\vec{Q}_2$ and examine the (0, ε, 1) reflection. The result of this experiment is shown in fig. 7. To interpret this, it is necessary to remember what is vertical in fig. 6 is horizontal in fig. 7. A reasonable interpretation of these figures is that there is a continuous variation of the polarization across the $\vec{Q}_2$ domain.

These results are consistent with the thermal activation model.[5] This model envisions the excitations of the linear SDW as rapid fluctuations of the direction of the polarization in the plane ⊥ to $\vec{Q}$ on a scale short in comparison to the resolution of these measurements. To the extent that the curves in fig. 8 are similar, one can conclude that the two components of polarization coexist on the scale of the resolution of this experiment. To the extent to which they differ one presumes that internal strain fields favor one component of the polarization over the other in different parts of the crystal. The apparent difference in curve width may be due to surface pinning of the polarization.

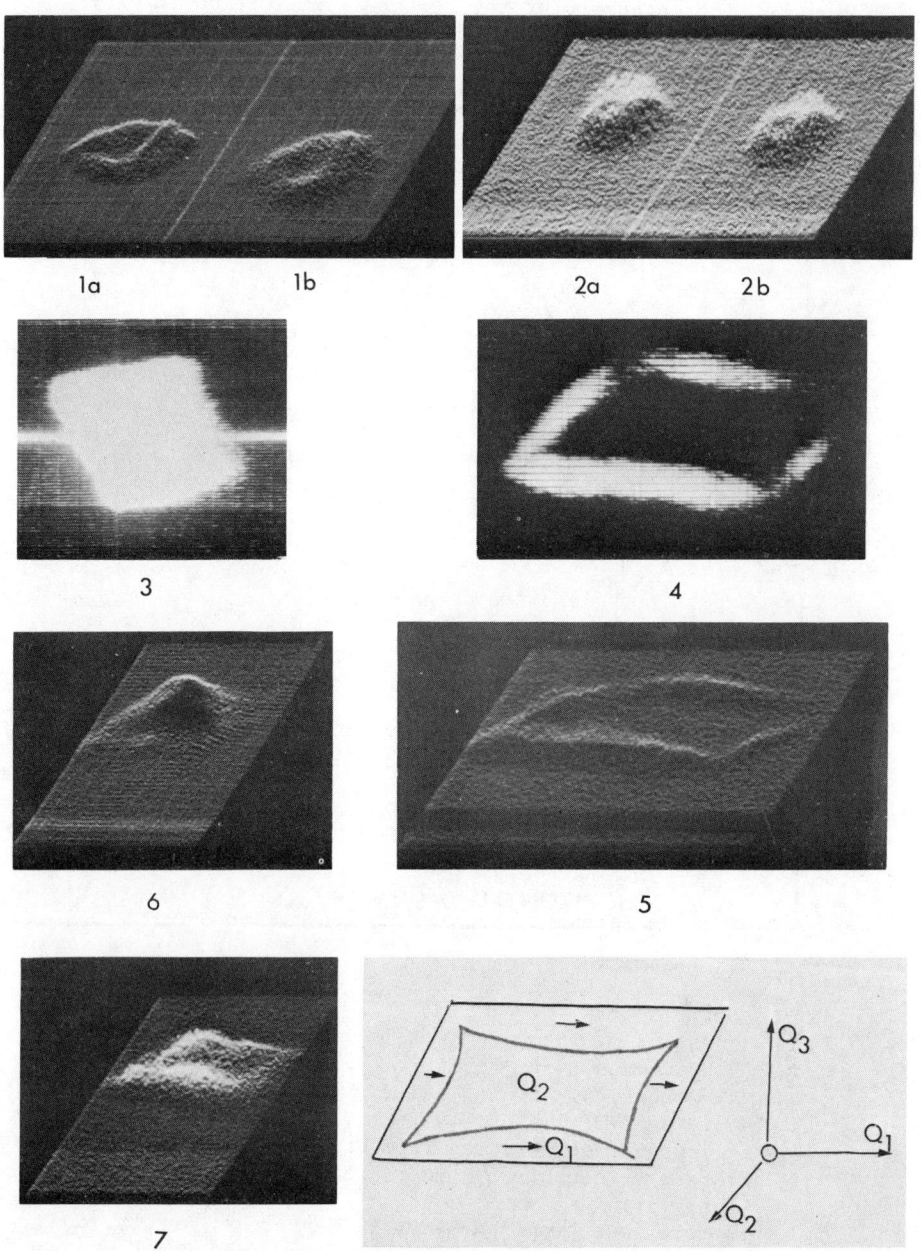

Fig. 1-7 Neutron microscopy of Cr. See text.

Fig. 8. The intensity of the $(1, \epsilon, 0)$ reflection in Cr #1, integrated along a vertical slice for 0 kG, and 16 kG,. The 0 kG data is scaled up by the factor 1.72 which corresponds to the increased intensity from the entire sample due to polarization rotation as measured previously in a field of 16 kG at room temperature.[4]

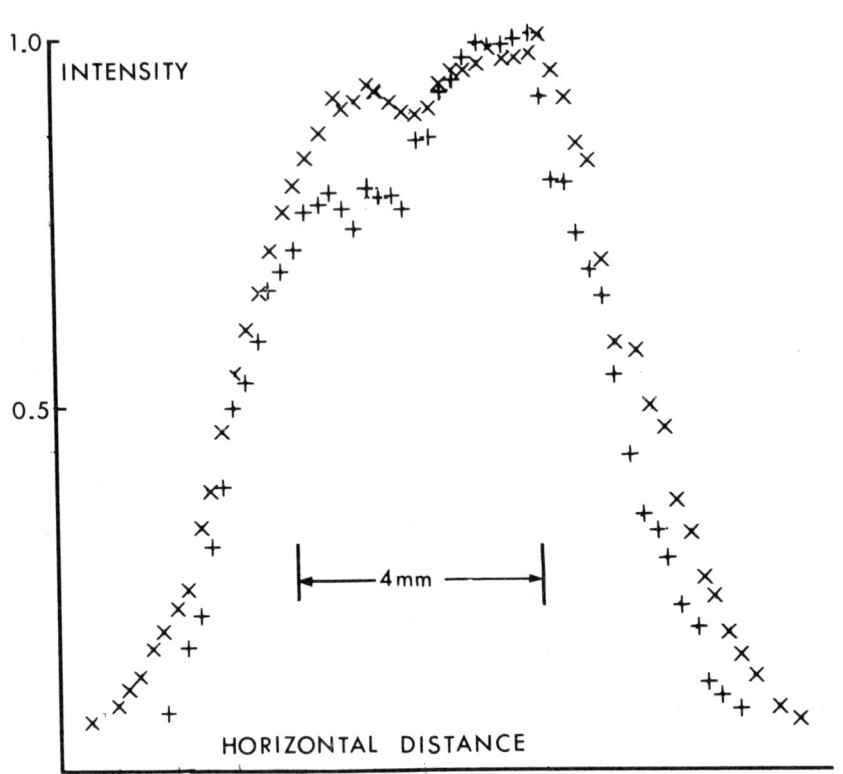

REFERENCES

1. M. Ando and S. Hosoya, Phys. Rev. Lett. 29, 281 (1972).
2. J. B. Davidson, Acta Cryst., A25, S66 (1969), and to be published.
3. S. A. Werner, A. Arrott, and M. Atoji, J. Appl. Phys. 39, 671 (1968), and 40, 1447 (1969).
4. This crystal was borrowed from Dr. R. L. Thomas, Wayne State University. See D. F. Snider and R L. Thomas, Phys. Rev., B3, 1091 (1971).
5. This model was proposed by S. A. Werner, A. Arrott, and H. Kendrick, Phys. Rev. 155, 528 (1967) and has been used by a number of authors, see for example M. O. Steinitz, F. Fawcett, C. E. Burleson, J. A. Schaefer, L. O. Frishman, and J. A. Marcus, Phys. Rev. B5, 3675 (1972), and B. C. Munday and R. Street, J. Phys. F1, 498 (1971).

EFFECTS OF PRESSURE ON THE MAGNETIC ORDERING IN Cr-Fe[*]

L. R. Edwards and I. J. Fritz
Sandia Laboratories, Albuquerque, New Mexico 87115

ABSTRACT

The effect of hydrostatic pressure (up to 3.3 kbar) on the magnetic transitions of a Cr-2.8 at.% Fe <u>single crystal</u> alloy has been determined via electrical resistance and acoustic measurements. These are the first ultrasonic measurements on these alloys and the combination of the two techniques allows a clear identification of the phase boundaries. Our results show that only the paramagnetic to commensurate-antiferromagnetic transition occurs at 1 bar and that the incommensurate-antiferromagnetic phase does not exist below ~ 0.5 kbar. These observations are compared to previous studies on alloys containing between 2 and 4 at.% Fe.

INTRODUCTION

The effect of pressure on the itinerant antiferromagnetic transitions in Cr-rich alloys containing a variety of transition metals has been of recent interest.[1,2] Typically there are three distinct magnetic phases that can be observed in these types of alloys: (1) paramagnetic (P), (2) incommensurate antiferromagnetic (I), and (3) commensurate antiferromagnetic (C). In the case of the Cr-Fe alloys at a pressure of 1 bar, early neutron diffraction measurements indicated that for alloys containing between 2 and 4 at.% Fe the ordering sequence is P-I-C with decreasing temperature.[3] Arrott <u>et al</u>,[4] however, suggested from their neutron diffraction data on a Cr-2.3 at.% Fe crystal that the appearance of the I-phase may be due to concentration inhomogeneities (i.e. there may be regions in the crystal with low Fe concentrations). Nityananda <u>et al</u>[1] have studied via the electrical resistance the pressure dependence of the magnetic transitions in a polycrystalline Cr-3 at.% Fe alloy and deduced that the P-I and I-C phase boundaries meet at a triple point in the negative pressure region. These results show that the I-phase exists at 1 bar; however, the transition temperatures in some cases are not clearly defined. First, the abrupt increase in resistance which is associated with the I-C transition progressively loses sharpness with increasing pressure making the I-C transition difficult to identify, and second, the Néel temperature (T_N, P-I transition) is identified by a broad resistance versus temperature minimum. In view of these results, and since the magnetic properties of Cr are very sensitive to dilute concentrations of transition metals and to the effect of pressure, it appears that well defined alloys are needed for a study of the magnetic phase transitions.

[*]This work was supported by the U. S. Atomic Energy Commission.

The purpose of this work was to determine the magnetic transition temperatures and their pressure dependences for a carefully prepared Cr-Fe single crystal within the composition range 2 to 4 at.% Fe. Here we report the results of electrical resistance, acoustic velocity, and acoustic attenuation measurements on a single crystal Cr-2.8 at.% Fe alloy. The combination of these techniques gives a clear indication of the phase boundaries. A triple point is found in the positive pressure plane for this alloy.

EXPERIMENTAL RESULTS

The single crystal used in this study was grown by an arc-zone melting technique.[5] Both of the constituent elements contained less than 50 ppm of other transition metals and the composition of the crystal was determined by atomic absorption spectroscopy to be 2.8 at.% Fe. The homogeneity of the crystal was determined from an electron microprobe analysis to be ± 0.15 at.% Fe over a distance of several millimeters. Both the resistance and acoustic samples were cut from the same single crystal grain.

Both measurements were made in a high pressure helium gas vessel as described by Hammons.[6] The resistance measurements were made by the standard 4-probe technique with the current parallel to the <100> crystallographic direction. For the acoustic velocity and attenuation measurements the standard pulse echo technique was used at a frequency of 5 MHz. Only the longitudinal mode propagating along the <100> direction was studied.

The magnetic transitions for the Cr-2.8 at.% Fe single crystal as determined by resistance measurements are shown in Fig. 1. The resistivity at 20°C and 1 bar is ~ 32 $\mu\Omega$-cm. For pressures in excess of 0.8 kbar two magnetic transitions are clearly visible. With decreasing temperature it is observed that there is a sharp continuous increase in resistance followed by a discontinuous jump in resistance. At the higher pressures, where the two transitions are well separated in temperature, a "hump" in the resistance is found between the transitions. This "hump" is due to magnetic superzone scattering and is characteristic of a P-I transition. The temperature region where the resistance exhibits a sharp continuous increase defines the Néel temperature (P-I transition). The second transition, which is characterized by a discontinuous jump in the resistance, appears as a superimposition over the magnetic superzone resistance anomaly. Neutron diffraction

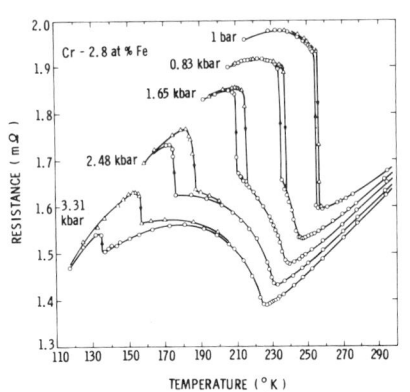

Fig. 1. The magnetic transitions as determined by resistance measurements. The arrows indicate either heating or cooling.

measurements[3,4] have established for alloys containing more than 2 at.% Fe that the low temperature phase has a C structure. It can then be concluded that the discontinuous jump in resistance defines the I-C transition. The 1 bar resistance isobar appears to be different from the higher pressure isobars in that only one transition is observed. The discussion of this result will be deferred until later.

In Fig. 2 the relative acoustic transit time as a function of temperature for various pressures is shown. The transit time was measured over a sample of length 1.50 cm, and at 20°C and 1 bar the acoustic velocity was measured to be 7.0 mm/μsec. The acoustic data clearly show that there is only one transition at 1 bar, while at elevated pressures there are two transitions. For pressures greater than 0.8 kbar a "cusp" type anomaly is observed and is similar to the anomaly found in pure Cr at the P-I transition.[7] This "cusp" corresponds to the sharp continuous increase in resistance in Fig. 1. The discontinuous jump in transit time, which follows the "cusp" anomaly at a lower temperature corresponds to the I-C transition. At the higher pressures the I-C transition was difficult to observe reliably because the signal was severely attenuated. We attribute this to domain scattering effects.[4,8] Acoustic attenuation measurements were also made over the same temperature and pressure ranges as the velocity measurements. Both of the acoustic measurements gave essentially identical transition temperatures.

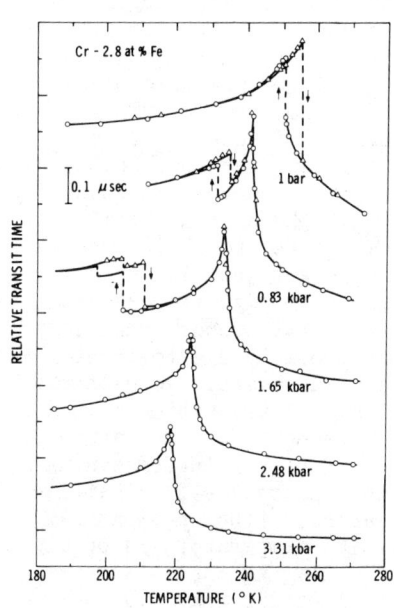

Fig. 2. The magnetic transitions as determined by acoustic transit time measurements. The arrows indicate either heating or cooling.

DISCUSSION

The pressure-temperature magnetic transition phase diagram for the Cr-2.8 at.% Fe alloy, as obtained from the acoustic and resistance measurements, is shown in Fig. 3. Of the two measurements the "cusp" anomaly in the acoustic transit time gives the best definition of the P-I transition temperature. On identical isobars the temperature at which the acoustic "cusp" occurs corresponds to the temperature where an inflection point is observed in the resistance data. The inflection points occur ~ 5°K below the minima in the resistance curves. The I-C transition temperatures as defined by the resistance anomalies are used in the construction of this phase diagram. The resistance measurements yield a better definition of this transition, especially at the higher pressures (cf. Figs. 1 and 2).

The phase diagram as depicted in Fig. 3 has several significant features. The P-I and I-C phase boundaries converge and intersect at ~ 0.5 kbar and ~ 245°K. These results show for an alloy containing 2.8 at.% Fe that a triple point among the P, I, and C phases exists in the positive pressure plane. The transition that occurs in the resistance and acoustic transit time at 1 bar must then be the P-C transition. However, it should be noted that there is a significant increase in the acoustic transit time as this transition is approached from high temperature (cf. the 1 bar isobar in Fig. 2). This is probably due to significant fluctuations associated with the impending P-I transition. At 1 bar the P-C transition temperature is only ~ 6°K higher in temperature than the apparent P-I transition temperature. (The apparent P-I transition temperature is obtained by extrapolating the P-I transition to 1 bar in Fig. 3.)

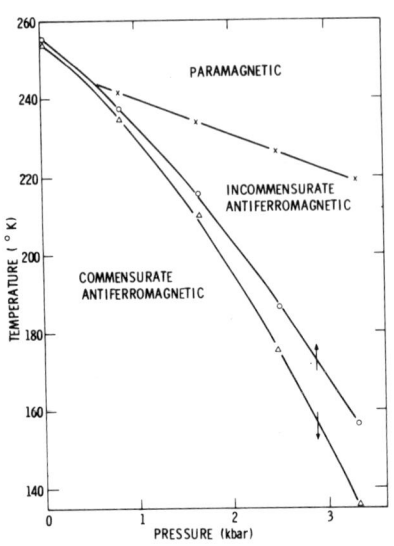

Fig. 3. The pressure-temperature magnetic phase diagram for Cr-2.8 at.% Fe. The arrows indicate whether the transition temperature was determined by heating or cooling.

It is evident that the magnetic transitions for this alloy are sensitive to pressure, especially the I-C transition. The pressure sensitivity of the P-I transition is -9.3°K/kbar, while for the I-C transition (cooling cycle) the pressure sensitivity increases (in magnitude) from -27°K/kbar at 0.83 kbar to -55°K/kbar at 3.3 kbar. These pressure coefficients are about an order of magnitude greater than those observed for itinerant ferromagnets such as the Invar alloys. If the I-C transition temperature (cooling cycling) is extrapolated to 0°K, then it is found that the C-phase would be inaccessible for pressures in excess of 5.5 kbar.

All transitions were observed to exhibit a temperature hysteresis. That of the P-I transition was very small (< 1°K). The I-C transition had the most notable hysteresis effect, which increased with increasing pressure. This further supports the contention[4,9] that all three transitions are first order. It should be noted that in previous resistance studies[1,2] on polycrystalline alloys no hysteresis effects were observed.

The results of this study show that the I-phase does not exist at 1 bar for an alloy containing 2.8 at.% Fe. Earlier work[1,4] has indicated that the I-phase exists at 1 bar for alloys containing up to 4 at.% Fe and that the C-phase exists at a lower temperature than the I-phase for alloys containing between 2 and 4 at.% Fe. Arrott et al,[4] however, have speculated that the appearance of the I-phase in their study on Cr-2.3 at.% Fe may be due to concentration inhomogeneities. Since our crystal was quite homogenous, it is our contention that the I-phase does not exist at 1 bar for an Fe concentration of 2.8 at.% and most probably for Fe concentrations greater than 2.8 at.%. The precise Fe concentration above which the I-phase does not exist at 1 bar still needs to be determined by further measurements on pure homogenous single crystals containing less than 2.8 at.% Fe.

The authors are especially grateful to Mr. F. A. Schmidt of the Ames Laboratory, Iowa State University, for the preparation of the single crystal alloy. Acknowledgment is also due to Mr. W. Chambers and Mr. B. Seely for sample analysis, and to Mr. J. D. Pierce and Mr. R. Tyler for their expert technical assistance.

REFERENCES

1. R. Nityanada, A. S. Reshamwala, and A. Jayaraman, Phys. Rev. Lett. 28, 1136 (1972).
2. Y. Syono and Y. Ishikawa, Phys. Rev. Lett. 19, 747 (1967).
3. Y. Ishikawa, S. Hoshino, and Y. Endoh, J. Phys. Soc. Jap. 22, 1221 (1967).
4. A. Arrott, S. A. Werner, and H. Kendrick, Phys. Rev. 153, 624 (1967).
5. O. N. Carlson, F. A. Schmidt and W. M. Paulson, Trans. Am. Soc. Metals 57, 356 (1964).
6. B. E. Hammons, Rev. Sci. Instr. 42, 1889 (1971).
7. E. J. O'Brien and J. Franklin, J. Appl. Phys. 37, 2809 (1966).
8. D. I. Bolef and J. de Klerk, Phys. Rev. 129, 1063 (1963).
9. T. M. Rice, A. Jayaraman, and D. B. McWhan, J. Phys. (Paris) Colloq. 32, C1-39 (1971).

GENERALIZED NEUTRON POLARIZATION ANALYSIS

F. Mezei
Institut Laue-Langevin, 38042 Grenoble, France

ABSTRACT

Until recently it was not realized that neutron beam polarization could be controlled and analysed in any direction independently of the actual magnetic field direction. We describe a simple, well-understood method of producing a beam polarization P of any direction at the sample and of determining the polarization vector P' of the scattered or transmitted beam, in contrast to classical methods where one can keep control only of the components P_z and P'_z, the z axis being the magnetic guide field direction at the sample.

Considering the fields of application of this generalized polarized neutron technique, its use in magnetic structure studies in non-collinear, single domain antiferromagnetic crystals is discussed in more detail. Furthermore it is pointed out that the method could be used in domain topology and anisotropy investigations on ferromagnetic (crystalline or not) thin films using a transmission method which has a typical sensitivity of about 1000 Å in film thickness.

INTRODUCTION

The suggestions described in this paper are the immediate outcome of a simple new experimental technique in polarized neutron beam work: the use of neutron polarization not parallel to the magnetic field direction. Historically such polarizations were used first by Rekvelt[1] and Forte[2], in a rather ad-hoc way, however. A systematic, easily controllable method was suggested by the author[3] some time ago, and it has now been experimentally verified in detail. In the present paper we briefly describe the technique of this generalized neutron polarization analysis and then go on to the discussion of possible applications. The purpose of the present paper is mainly to call attention to these and similar possibilities.

THE CONCEPT OF GENERALIZED POLARIZATION ANALYSIS

Polarized neutron beam work means in the classical sense the control and retention of only the polarization component parallel to the magnetic field at any point along the neutron path. In this approach a polarized beam is regarded as characterized by the probabilities of the occupation of the two Zeeman states (parallel and antiparallel to the field). In reality, however, between the polarizer and analyser we always have a coherent superposition of the two Zeeman states, i.e. a well defined, general polarization direction for each neutron. It is easy to show (see e.g. Slichter[4]) that this polarization direction evolves according to the classical equation of motion. This means that if the magnetic field configuration fulfills the adiabatic guidance condition (field direction varying sufficiently slowly), we have a constant polarization component P_z parallel to the field, and another one precessing around the field. If the adiabatic condition is not fulfilled, P_z and the angle between $\vec{P}$ and $\vec{H}$ changes. Thus neutrons starting from the same point and having the same direction, velocity and polarization will have the same polarization direction after passing through any magnetic field configuration. The only thing that may happen is that this polarization does not stay parallel to the field. According to this concept a neutron beam can only be depolarized by having different polarization directions for neutrons with different paths and velocities, but probabilistic Zeeman transitions never occur when traversing any constant or time-dependent field. (If the field is time-dependent, the final polarization will also be time-dependent.)

Thus a real depolarization of a neutron beam can be caused only by either of the following:

i) Different spin turns in the non-adiabatic field regions for different neutron velocities.

ii) Dephasing of the precessing polarization components during passage through the adiabatic field regions due to
    a) differences in neutron velocities
    b) field inhomogeneities across the beam
    c) traverse time differences due to beam divergence

In fact, points i, ii/b and ii/c can easily be made negligible in most cases, while ii/a is either negligible or can be cancelled out by the use of the spin-echo compensation method as described in Ref. 3. So for the following it is assumed that general polarization directions may exist and be retained, and we now have to describe how to produce and determine them.

Basically, any nonadiabatic region in the magnetic guide field produces a change in the angle between $\vec{P}$ and $\vec{H}$. The method we have found the most clear and simple is described here. Others are given in Refs. 1 and 3.

We put a rectangular coil across the neutron beam in the presence of the uniform guide field $\vec{H}_o$ (Fig. 1). Thus inside the coil the field is $\vec{H}_o + \vec{H}_1 = \vec{H}$, while it stays practically unchanged outside if the coil is long enough. The change of the field from $\vec{H}_o$ to $\vec{H}$ is so quick for thermal neutrons crossing the windings that there is no time for spin direction changes. So the neutron spins do not change upon entering the coil. Inside the coil they perform a precessory motion on a conical surface around $\vec{H}$, for a period of time determined by the width of the coil and the neutron velocity, and leave the coil again without further change. Adjusting the value and direction of H_o, the right value of $\vec{H}$ can be set to produce a net spin turn into any desired direction (e.g. as shown on the right-hand side of Fig. 1, where $\vec{P}$ was taken originally to coincide with $\vec{H}_o$ and it is turned into $\vec{P}'$). Using such a coil device after the polarizer, we can produce a beam polarized in a given direction and, inversely, putting a coil before the analyser, we can perform the polarization analysis referring to another general direction, namely to that which is turned into the H_0 direction by the second coil.

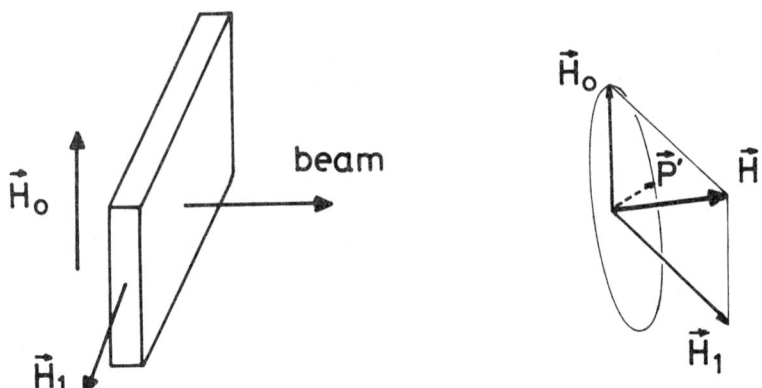

Fig. 1. The coil device for turning neutron polarization relative to the field direction and a scheme of its action.

GENERALIZED POLARIZATION ANALYSIS IN STRUCTURE STUDIES

The first application suggested is for the study of antiferromagnetic crystals. The reason why only antiferromagnets can be dealt with is that in ferro or ferrimagnets the polarization direction would precess very rapidly inside the sample, due to the magnetization of the sample contributing to the field value $\vec{B}$. (Let us recall a typical figure: for 1 Å neutrons in 10 Oe field, one Larmor precession takes 15 cm).

Such studies could be made on any polarized neutron diffractometer, with or without polarization analysis. During the measurement the sample region should be kept in a low magnetic guide field (typically 10 Oe), which allows the control of the Larmor precessions between the spin-turn coils and the sample. The sample itself should be much smaller than the Larmor precession distance in the given field.

The interest of such generalized diffraction studies can be indicated as follows[5]. In the most general case the magnetic structure of a monodomain crystal can be described by a magnetization distribution $\vec{M}(r)$, which contains spin and orbital contributions and is not necessarily collinear. The neutron magnetic scattering amplitude of the crystal for a given scattering vector $\vec{k}$ is given by

$$\vec{Q}_\perp(\vec{k}) = -\frac{1}{2\mu_B} \int d\vec{r} \, [\vec{k} \times (\vec{M}(\vec{k}) \times \vec{k})] e^{i\vec{k}\vec{r}} \qquad (1)$$

Note that $\vec{Q}$ is vector perpendicular to $\vec{k}$. Denoting the nuclear amplitude by the scalar $N(\vec{k})$, the elastic scattering (Bragg) cross section and the polarization $\vec{P}'$ of the scattered beam is given by

$$\frac{d\sigma}{d\tau} \propto \vec{Q}_\perp^* \vec{Q}_\perp + N^*N + \vec{P}(N^*\vec{Q}_\perp + \vec{Q}_\perp^*N) + i\vec{P}(\vec{Q}_\perp^* \times \vec{Q}_\perp) \qquad (2)$$

and

$$\vec{P}'\frac{d\sigma}{d\tau} \propto N^*\vec{Q}_\perp + \vec{Q}_\perp^*N - i\vec{Q}_\perp^* \times \vec{Q}_\perp + \vec{P}(N^*N - \vec{Q}_\perp^*\vec{Q}_\perp) +$$
$$+ \vec{Q}_\perp^*(\vec{Q}_\perp\vec{P}) + (\vec{Q}_\perp^*\vec{P})\vec{Q}_\perp + i\vec{P} \times (\vec{Q}_\perp^*N - N^*\vec{Q}_\perp) \qquad (3)$$

It can be seen that the determination of the quantities N, N^*, $\vec{Q}$, $\vec{Q}^*$ (8 parameters) for a given reflection is not straightforward. The best that experiments can give is the determination of all terms in Eqs. (2) and (3). For example, the determination of the third term in Eq. (3) involves the rotation of $\vec{P}$ in the plane perpendicular to $\vec{k}$ in which $\vec{Q}_\perp^*$ and $\vec{Q}_\perp$ lie. This could also be done by the classical technique, i.e. either rotating $\vec{H}$ in this plane or rotating $\vec{Q}_\perp$ and $\vec{Q}_\perp$ by rotating the crystal around the $\vec{k}$ vector[6]. In this case the generalized polarization technique represents only an alternative solution which might be more easily realized. Consider, however, the last term in Eq. (3), which occurs in non-centrosymmetrical structures and which gives a scattered beam polarization component perpendicular to the incoming polarization, which can be directly detected only by the present method.

FERROMAGNETIC DOMAIN TOPOGRAPHY

Making use of the simple fact that the real magnetic field determining the Larmor precessions of a polarized neutron beam in a medium is $\vec{B} = \vec{H} + 4\pi\vec{M}$, information can be obtained about $\vec{M}$ by observing the polarization of the transmitted beam. Basically this means a generalization of the classical beam depolarization studies[7].

Consider a polarized beam transmitted through a thin ferro (ferri) magnetic sample and having a cross section small enough so that it traverses the sample within a single domain. The effect of the sample on the beam polarization will be very similar to that of our spin turn coil described in Fig. 1 since the surface wall in typical ferromagnets is sufficiently sharp. Analysing the change in spin direction for different incoming polarization directions, one can infer the direction of $\vec{B}$ in the domain. For example, if $\vec{B}$ happens to be parallel to the polarization of the incoming beam, nothing happens to the polarization during the traverse. Making a scan with the beam across the sample, a domain map could be constructed using one or more incoming polarizations. Concerning the sensitivity of the method, let us consider a special case: non-perfect alignment of $\vec{M}$ to the external field $\vec{H}_o$ in an anisotropic material. If we denote the angle between $\vec{M}$ and $\vec{H}_o$ by ν, and assume that $\vec{M} \gg \vec{H}_o$, the maximum observable deviation of polarization direction $\vec{P}$ during the transmission is ν. In the classical technique this could be observed as a maximum depolarization corresponding to $1 - \cos 2\nu \approx 2\nu^2$. On the other hand, the corresponding precessing component of the polarization is 2ν. So small deviations are much more easily detected in the precessing component than in the P_z component.

The typical sensitivity figures of this method for $\vec{M}$ of the order of 10,000 Gauss and 5 Å neutrons are: 1,000 Å film thickness in domain topography and .5° for at least .01 mm thick sample in magnetization misalignment studies.

REFERENCES

1. M. Th. Rekveldt, J. de Phys. Suppl. 32, C1-579 (1971).
2. Private communication.
3. F. Mezei, Z. für Phys. 255, 146 (1972).
4. Ch. P. Slichter, Principles of Magnetic Resonance (Harper & Row, 1963).
5. For more details see W. Marshall and S. W. Lovesey, Theory of Thermal Neutron Scattering (Oxford, 1971).
6. J. P. Brown, Lecture at a Harwell School on Neutron Physics, 1972.
7. G. E. Bacon, Neutron Diffraction (Oxford, 1962) p. 355.

NUCLEAR POLARIZATION DEPENDENCE OF COHERENT NEUTRON SCATTERING FROM HOLMIUM

G. R. Little and R. A. Erickson
The Ohio State University, Columbus, Ohio 43210

ABSTRACT

Neutron diffraction measurements with unpolarized neutrons have been made on a single crystal of Ho in the temperature range 4.2 K to 0.4 K with the aim of studying the change in coherent scattering produced by the h.f.s. induced polarization of the Ho nuclear spins. It was found that the magnetic satellite reflections decrease linearly with nuclear polarization, but that no effect could be seen on the ordinary hexagonal reflections. The incoherent scattering amplitude was found to be $b_{inc} = -0.086 \pm .008 \times 10^{-12}$ cm. The magnetic structure reported by Koehler et.al.[1] was confirmed except for the spin bunching parameter δ. We find $\delta = 8.3°$ where Koehler finds $\delta = 5.8°$. Since $\delta = 0°$ represents complete bunching, our measurement suggests that the magnetic anisotropy does not affect the magnetic structure as much as Koehler suggests. Using the Ho coherent scattering amplitude $b_{coh} = 0.85 \times 10^{-12}$ cm we find that the two compound nucleus scattering amplitudes are $b_+ = +0.70 \pm .04 \times 10^{-12}$ cm and $b_- = +1.04 \pm .04 \times 10^{-12}$ cm.

[1] W. C. Koehler, J. W. Cable, M. K. Wilkinson, and E. O. Wollan, Phys. Rev. 151, 414 (1966).

INTRODUCTION

In this paper, a neutron diffraction study of single crystal holmium is presented. The primary aim of this investigation is the utilization of the nuclear polarization dependence of the coherent scattering to determine the sign and magnitude of the Ho-165 incoherent scattering amplitude. Some measurements of the magnetic structure parameters have also been made and are included here.

The magnitude of the incoherent scattering amplitude is traditionally determined from the total and coherent nuclear scattering cross sections (σ_s and σ_c). In holmium, however, the total scattering cross section has not been accurately measured due to the large thermal neutron capture and paramagnetic scattering cross sections. Koehler, Wollan and Wilkinson[1] deduced a value of 9.1 barn for σ_c and estimated a value of ~13 barn for σ_s yielding $b_{inc} = \pm .28 \times 10^{-12}$ cm. We note that the sign of b_{inc} can not be determined with this technique. Schermer[2] used the nuclear polarization dependence of the transmission of polarized neutrons to determine the spin dependence of the holmium absorption cross section. Assuming a 1/v dependence to the absorption he attempted to deduce the spin dependence of the scattering as well. He concludes the $\sigma_s \approx \sigma_c$ and, based on his results, we calculate $b_{inc} = -.03 \times 10^{-12}$ cm. Both estimates are unreliable due to the large absorption cross section in holmium. We eliminate this problem by investigating the spin dependence of the coherent scattering.

The general theory of the scattering of thermal neutrons from magnetically ordered systems with nuclear polarization has been given by Schermer and Blume[3]. In applying their formulae to holmium, we find that no new reflections appear, but rather that nuclear polarization dependent terms appear in conjunction with the magnetic terms such that the magnetic portion of the square of the structure factor F_m^2 can be written

$$F_m^2 = F_{0m}^2 \left(1 + \frac{b_{inc} \cdot I \cdot f_N}{p(K)} + \frac{b_{inc}^2 \cdot I^2 \cdot f_N^2}{4 \cdot p(K)^2 \cdot q_{eff}^2} \right) \quad (1)$$

where f_N is the nuclear polarization, $p(K)$ is the magnetic scattering amplitude and q_{eff}^2 is the magnetic structure parameter appropriate to the reflection and is normalized to unity. That the f_N dependence follows the magnetic scattering can be seen from the fact that the nuclear spins align with the ionic moments and hence obey the same periodicities.

The holmium single crystal used in this study was kindly supplied by W. C. Koehler of ORNL and is the one referred to as Ho-B in his study[4]. The crystal was encased in a tightly fitting aluminum mount which was in turn attached to the He-3 pot of a He-3 cryostat. Since the alignment of holmium nuclear spins produces a large specific heat anomaly, special care was taken to insure good thermal contact.

The He-3 cryostat used in this study was constructed especially for use in neutron diffraction studies. One principle feature is its small size in that the entire unit is only two feet long with a maximum diameter of three inches and a weight of less than ten pounds. The cryostat can therefore be easily oriented, eliminating the need for either an internal goniometer or a complex system of external supports. A second feature of the cryostat is the inclusion of 0.085 mole of holmium metal (in addition to the crystal) to the He-3 stage. This holmium provides an excellent thermal ballast and the holmium internal energy is used as the thermometric standard in the calibration of thermometers. Detailed descriptions of the cryostat and the thermometer calibration procedure are to be published elsewhere.

Standard diffraction data were obtained for reflections in the (h0l) set at room temperature, 77 K, and 4.2 K. Magnetic intensities were placed on an absolute scale by normalizing to the normal hexagonal reflections. Debye-Waller corrections were made using the characteristic temperatures 183 K, 192 K and 194 K appropriate at 300 K, 77 K and 4.2 K[5] respectively. Extinction-absorption corrections were calculated using the numerical technique outlined by Hamilton[6]. The mosaic spread parameter was chosen to be twelve minutes of arc and the linear absorption coefficient was calculated to be 1.63 cm^{-1} at room temperature decreasing to 1.32 cm^{-1} at 4.2 K as the magnetic moments order.

The $(002)^{\pm}$, $(101)^{\pm}$, $(201)^{+}$, $(300)^{-}$ and $(401)^{\pm}$ satellite reflections and the (002), (101) and (401) central reflections were selected for study to 0.4 K. Due to the nature of the cryostat, constant temperature integrated intensity measurements could not be obtained below 1.5 K. In addition, resolution problems on several reflections made integrated intensities suspect. It was therefore decided to use the maximum peak intensity as the measure of the reflection. The validity of this technique was verified by measuring the $(002)^{-}$ reflection between 4.2 K and 0.4 K using both techniques.

RESULTS

The magnetic structures observed at 77 K and 4.2 K agree with those reported by Koehler et.al.[4] We find the interlayer turn angle to be $41°$ at 77K and $30.3\pm.3$ at 4.2 K. The effective basal component of the holmium moment is found to be $7.04\pm.07$ at 77 K and $9.5\pm.2$ at 4.2 K. Due to resolution problems and limited data, we are unable to give a value for the ferromagnetic component at 4.2 K. The magnetic form factor was obtained at 77 K and 4.2 K for several reflections and is shown in Figure 1 along with the theoretical curve calculated using the theory of Trammell[7] and wavefunctions of Blume, Freeman and Watson[8]. We find good agreement with the theory for $K/4\pi < .5 A^{-1}$. The deviations at the $(401)^{\pm}$ reflections may be due to resolution problems.

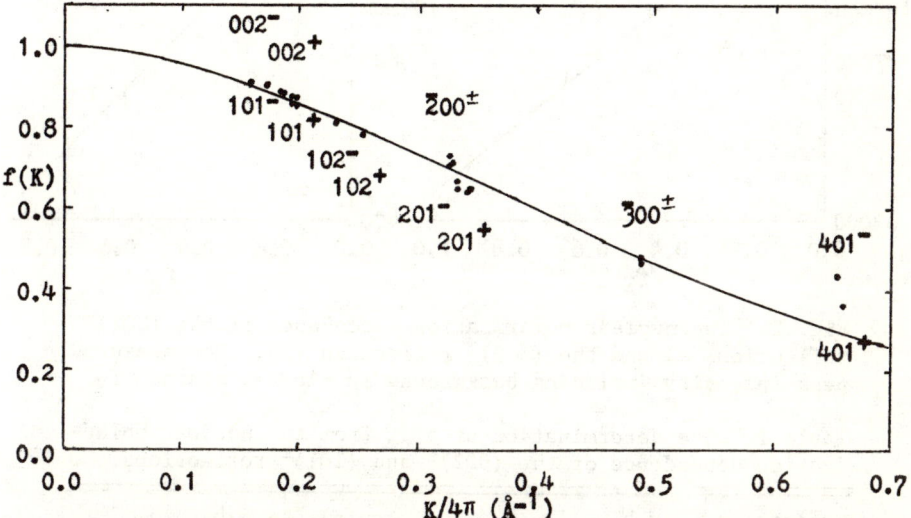

Fig. 1. The holmium magnetic form factor. The solid curve represents the theoretical form factor.

We also observe apparent satellites of the forbidden (001) reflection, indicating the perturbation to the helical structure as reported by Koehler et. al.[4] However, we find these reflections to be weaker by a factor of two relative to the $(002)^{\pm}$ reflections.

Based on our results, we calculate the spin bunching parameter to be $\delta=8.3°$ as opposed to Koehler's value of $5.8°$. Checks were made for the presence of simultaneous reflection with negative results. We note that this change in δ produces only 1-2% changes in the predicted intensities of the other reflections.

All satellite reflections were observed to decrease in intensity as temperature was lowered below 4.2 K, the amount of decrease ranging from 6% to 25%. The polarization dependence of the central reflections and the spin incoherent scattering was too small to be accurately measured. The nuclear polarization was calculated as a function of temperature using the hyperfine parameters of Lounasmaa[9] and the intensity data was plotted against $|f_N|$. (Figure 2) yielding the result that the intensities decrease linearly with $|f_N|$. Least squares fits of the data were made and the results are presented in Tables I and II. The factors χ are corrections calculated to account for the nuclear polarization dependence of the extinction-absorption factors introduced through their dependence on F_m^2. The large corrections for the $(002)^{\pm}$ satellites are due to the substantial extinction present in these reflections.

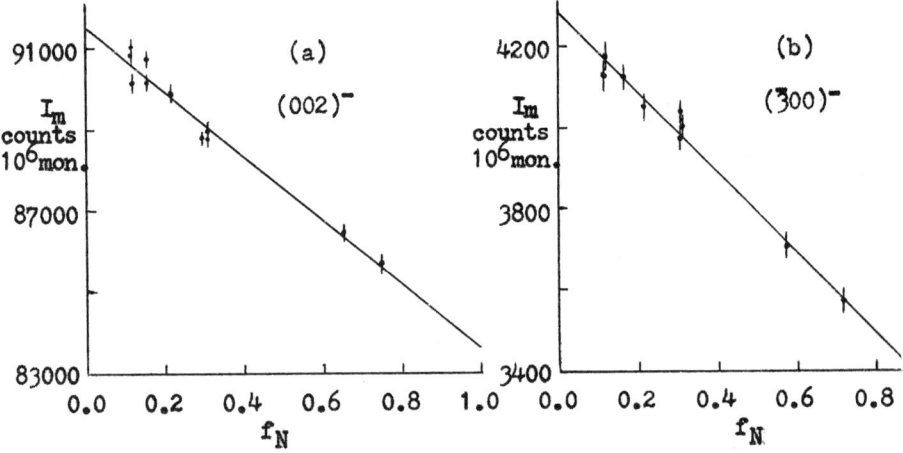

Fig. 2. The nuclear polarization dependence of the $(002)^-$ reflection (a) and the $(300)^-$ reflection (b). The measured peak intensity including background is plotted against f_N.

Table I. The determination of b_{inc} from the nuclear polarization dependence of the $(002)^{\pm}$ and $(101)^{\pm}$ reflections.

Reflection	$\left|\dfrac{\chi \cdot b_{inc} \cdot I}{p(K)}\right|$	χ	$p(K)$ $(10^{-12}\,cm)$	$|b_{inc}|$ $(10^{-12}\,cm)$
$(002)^-$	0.090 ± 0.009	0.75	2.36	0.081 ± 0.008
$(002)^+$	$.103 \pm .010$	.80	2.30	$.084 \pm .008$
$(101)^-$	$.113 \pm .011$	.94	2.34	$.080 \pm .008$
$(101)^+$	$.141 \mp .014$	.94	2.30	$.098 \pm .010$

Table II. The determination of the holmium magnetic form factor from the nuclear polarization dependence of the coherent scattering.

Reflection	$\left|\dfrac{b_{inc} \cdot I}{p(K)}\right|$	f(K)	f(K) (4.2 K)	f(K) (theory)
$(201)^+$	0.17 ± 0.02	0.65 ± 0.08	0.65	0.66
$(\bar{3}00)^-$	$.26 \pm .02$	$.44 \pm .04$	.48	.48
$(401)^-$	$.22 \pm .07$	$.52 \pm .16$	.44	.31
$(401)^+$	$.38 \pm .11$	$.30 \pm .09$	.36	.31

The $(002)^\pm$ and $(101)^\pm$ reflections were used to extract $|b_{inc}|$ because the magnetic amplitude p(K) is most accurately known for these reflections. We find $|b_{inc}|=0.086\pm.008\times10^{-12}$ cm. The sign of b_{inc} is obtained from the fact that all satellite reflections were observed to decrease with increasing nuclear polarization. Under consistent assignment of the directions of the vector quantities involved (the magnetic vector q and the nuclear spin I^{10}), the signs of f_N and p(K) are identical here and b_{inc} must be negative. We then have $b_{inc} = -0.086\pm.008\times10^{-12}$ cm. The scattering amplitudes b_+ and b_- associated with the compound nucleus states of total angular momentum $I+\tfrac{1}{2}$ and $I-\tfrac{1}{2}$ are readily calculated to be $b_+ = +0.70 \pm.04\times10^{-12}$ cm and $b_- = +1.04\pm.04\times10^{-12}$ cm using the coherent amplitudes of $b_{coh} = +0.85\pm.02\times10^{-12}$ cm^I. The incoherent scattering cross section is 0.37 barn yielding the value 9.5 barn for the total nuclear scattering cross section. We note that the recent work of Herpin and Meriel[11] is in exact agreement with our results.

As a final check on the value of b_{inc}, the polarization dependence of the remaining reflections and the selected value of b_{inc} were used to determine the magnetic form factor for these reflections. The resulting values are compared with those obtained using a standard technique and those calculated from the theory in Table 2. The agreement within error limits supports our value of b_{inc}.

REFERENCES

1. W. C. Koehler, E. O. Wollan and M. K. Wilkinson, Phys. Rev. 110, 37 (1957).
2. R. I. Schermer, Phys. Rev. 136, B1285 (1964).
3. R. I. Schermer and M. Blume, Phys. Rev. 166, 554 (1968).
4. W. C. Koehler, J. W. Cable, M. K. Wilkinson and E. O. Wollan, Phys. Rev. 151, 414 (1966).
5. M. Rosen, Phys. Rev. 174, 504 (1968).
6. W. C. Hamilton, Acta. Cryst. 10, 180 (1957).
7. G. T. Trammell, Phys. Rev. 92, 1387 (1953).
8. M. Blume, A. J. Freeman and R. E. Watson, J. Chem. Phys. 37, 1245 (1962).
9. O. V. Lounasmaa, Phys. Rev. 128, 1136 (1962).
10. V. L. Sailor, R. I. Schermer, F. J. Shore, C. A. Reynolds, H. Marshak and Hans Potsma, Phys. Rev. 127, 1124 (1962).
11. A. Herpin and P. Meriel, J. Physique 34, 423 (1973)

A NEUTRON DIFFRACTION STUDY OF ANTIFERROMAGNETIC CoO WITH NUCLEAR POLARIZATION FROM THE HFS INTERACTION IN THE REGION 0.35-4.2K

D. A. Goer and R. A. Erickson
The Ohio State University, Columbus, Ohio 43210

ABSTRACT

Using neutron diffraction techniques, a single crystal of antiferromagnetic CoO has been studied in the temperature range 0.35-4.2K. In this temperature region the hyperfine interaction produces a partial polarization of the cobalt nuclei which results in an observed temperature dependence of the Bragg intensities of the magnetic peaks. This temperature dependence enables both the sign and magnitude of the product of the nuclear incoherent scattering amplitude, b_i, and the hyperfine coupling constant, A, to be determined. If the sign of A is known, it is then possible to unambiguously determine the correct spin-dependent scattering amplitudes, (b_+, b_-). For cobalt, b_i, is known and is negative, and thus, A was determined to be: $A = +.0121 \pm .001$K. The temperature dependence of the intensity also permits a simultaneous and independent determination of the form factor, f, and the magnetic parameter, q^2 in contrast to the standard form factor analysis technique which determines only the absolute value of f and depends on the proposed magnetic structure through q^2. For CoO q^2 was measured directly and it was demonstrated that only two of the five proposed magnetic structures for CoO are viable: Van Laar's colinear model A and Van Laar's multispin model. It was not possible to decide between these two models since they both have the same effective q^2 due to the presence of spin domains.

INTRODUCTION

When thermal neutrons are scattered from a crystal containing unpolarized nuclei of spin I, there exists a well known ambiguity in the determination of the appropriate spin dependent scattering amplitudes b_+ and b_- that are associated with the compound nuclear states $I+\frac{1}{2}$ and $I-\frac{1}{2}$ respectively. In a magnetically ordered crystal the hyperfine interaction (HFS) can produce a partial alignment of the nuclei and give rise to a measureable temperature dependence in the magnetic (Bragg) reflections. An analysis of the temperature dependent magnetic peaks permits a determination of the product, $A(b_+ - b_-)$, both in magnitude and algebraic sign, where A is the hyperfine coupling constant. In addition, the temperature dependence of the magnetic scattering permits a simultaneous and independent determination of the magnetic form factor and magnetic structure factor. This paper reports on the application of this technique to antiferromagnetic CoO.

THEORY

If $A \ll kT$, the nuclear polarization from the HFS can be well approximated by,

$$f_N = -(I+1) AS/3kT \qquad (1)$$

Where S is the effective spin of the magnetic electrons, and it is assumed that T is far below the Néel temperature of the antiferromagnetic crystal. The integrated intensity of neutron scattering*in the magnetic peaks is:

$$P(T) = P_o (1 + \alpha/fT + \alpha^2/4f^2T^2q^2) \qquad (2)$$

where P_o is the intensity for unpolarized (high temperature) nuclei, f is the magnetic form factor, q^2 is the magnetic structure parameter ($\vec{q} = \hat{K} \times \hat{K} \times \hat{S}$, for scattering vector $\vec{K}$), and

$$\alpha = 2I(I+1)(b_+ - b_-) A/3kp(2I+1) \qquad (3)$$

where p is the magnetic scattering constant, $e^2|\gamma|/mc^2$. The intensity of the purely nuclear peaks is expected to show no temperature dependence due to the nuclear (antiferromagnetic) polarization.

Equation (2) indicates that the temperature dependence of the neutron scattering from the antiferromagnetic peaks consists of two terms: 1) a term linear in reciprocal temperature and form factor, arising from the interference between magnetic scattering and nuclear spin dependent scattering and; 2) a term quadratic in reciprocal temperature and form factor, due entirely to the nuclear spin dependent scattering. As written, the linear term is independent of q, since P_o is linear in q^2. Insofar as the two temperature dependent terms can be separately measured, so too can the <u>absolute</u> form factor be determined independently from the magnetic structure parameter.

EXPERIMENTAL

The CoO crystal supplied by Dr. Y. Nakazumi, was cut to cubical shape of side 6mm. The same crystal was used in the previous studies of CoO reported from this laboratory[1]. The crystal was mounted in a small orienting, four-stage He^3 cryostat which permitted fixed temperature diffraction measurement at 80K, 20K, 4.2→1.4K and 0.35K. Diffraction data in the interval 0.35 → 1.4K was obtained during the slow warm-up (1-2mK/min) after exhaustion of the He^3. Temperatures below 1.4K were deduced from resistance thermometers calibrated against the internal energy of holmium metal, using the technique reported by Culbert and Sungaila[2]. Further details of the cryostat and thermometer calibration will be published elsewhere.

A standard form factor analysis of CoO was done using 80K data and the results are given in Fig. 1 along with a comparison with the calculation of Mahendra and Khan.[3]

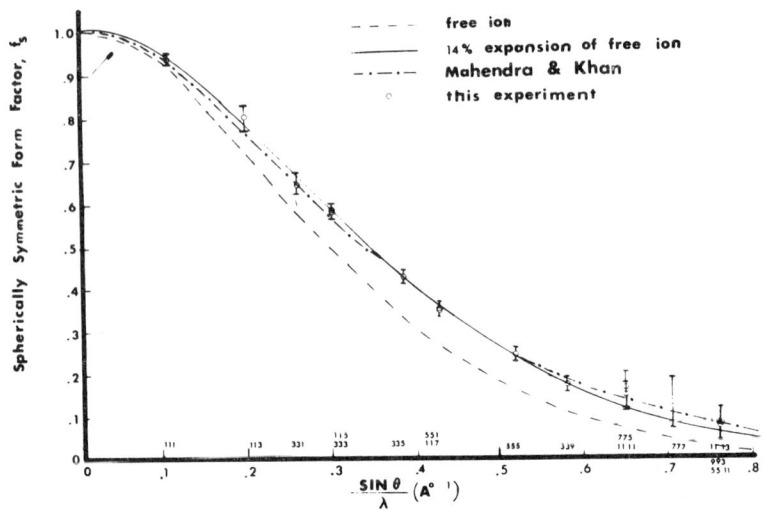

FIG. 1. CoO Form Factor

The temperature dependence of five magnetic and three nuclear peaks was measured to 0.35K. As expected, the nuclear peaks { (222), (444) and (004) } showed no temperature dependence. The temperature dependence of two of the magnetic peaks, (111) and (555) is demonstrated in Fig. 2 and Fig. 3.

The constant α (Eq 2) is not well known for CoO, but can be found from the linear T^{-1} dependence of a peak of known form factor. The (111) peak was used to this purpose, since this form factor has virtually no assymetry, is well-known, and is large enough to make negligible the term quadratic in T^{-1}. A weighted least square fit of the temperature dependent (111) intensity, corrected for secondary extinction effects, gives $\alpha = +0.0406 \pm .0018$K, using $f_{111} = 0.92 \pm .01$. Using Schermer's[4] value, $b_+ - b_- = -1.38 \pm .05 \times 10^{-12}$cm, gives A = $+0.0121 \pm .0010$K for the hyperfine coupling parameter. With α known, the form factor and magnetic structure parameter (q^2) of the other magnetic peaks can be deduced, as given in Table I. Also shown in Table I are the integrated intensity (80K) form factor (with Van Laar structure) and q^2eff based upon the magnetic structure models proposed by Van Laar[5] Li[6] and Roth.

In the earlier work[1] on CoO form factor, we chose the Van Laar multispin model because it allowed for a more realistic form factor. From the present measurements we conclude, independently, that the form factor and q^2eff are best accounted for with either the Van Laar multispin model or a colinear model A with $\gamma = 27.4°$.

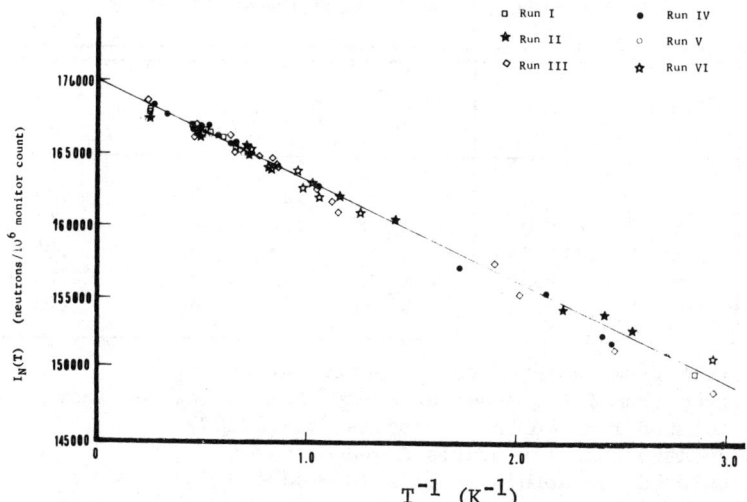

FIG. 2. Temperature Dependence of CoO (111).

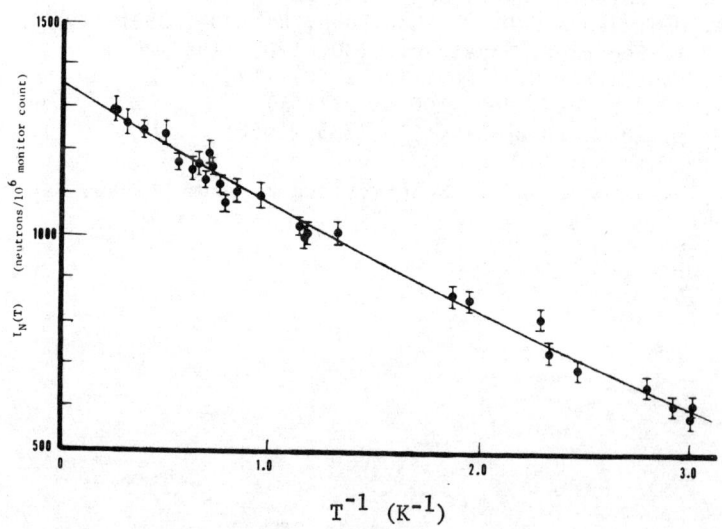

FIG. 3. Temperature Dependence of CoO (555)

Table I Form Factors and q^2_{eff} from Temperature Dependent Intensities, with model Comparisons

hkl	Form Factor A	Form Factor B	q^2_{eff} A	q^2_{eff} C	q^2_{eff} D	q^2_{eff} E
111		.92±.01		.98	.67	.84
113	.80±.08	.80±.03	.32±.07	.36	.65	.45
333	.53±.01	.54±.01		.98	.67	.84
555	.19±.02	.22±.01	1.08±.20	.98	.67	.84
339	.14±.20	.17±.02	.44±.18	.36	.65	.45

A- Derived from temperature dependent intensity
B- Derived from integrated intensity, 80K, using Van Laar's models.
C- Calculated for Van Laar's models ($\gamma = 27.4°$)
D- Calculated for Li's models A and B ($\gamma = 0°$)
E- Calculated for Roth's model A ($\gamma = 11.5°$)

REFERENCES

1. D. C. Khan and R. A. Erickson, Phys. Rev B 1, 2243 (1970)
   D. C. Khan and R. A. Erickson, J. Phys. and Chem of Solids 29 2087 (1968)
2. H. V. Culbert and Z. Sungaila, Cryogenics 8, 386 (1968)
3. A. Mahendea and D. C. Khan Phys. Rev. B4, 3901 (1971)
4. R. I. Schermer, Phys. Rev. 130, 1907 (1963)
5. B. Van Laar, Phys. Rev. 138 A 584 (1965)
6. Y. Y. Li, Phys. Rev. 100 627 (1955)
7. W. L. Roth, Phys. Rev. 110 1333 (1958)

*See, for example, R. I. Schermer and M. Blume, Phys. Rev. 166, 554 (1968)

NEUTRON DIFFRACTION STUDY OF THE Mn-Pd SYSTEM

G. Kádár, E. Krén and L. Pál
Central Research Institute for Physics,
H-1525 Budapest, P.O.B. 49, Hungary

ABSTRACT

Crystallographic and magnetic structural data, obtained from neutron diffraction studies on powdered samples in the Mn-Pd system between 20 and 52 at% Mn, are reported. Structures of the $MnPd_3$, $MnPd_2$, Mn_3Pd_5 and MnPd phases are presented and their interrelations are discussed.

INTRODUCTION

In the past few years a large number of papers has dealt with the crystallographic and magnetic structures of the intermetallic phases occurring in the Mn-Pd system. In order to clarify the structural problems, a systematic neutron diffraction study was performed on 18 polycrystalline samples uniformly distributed in the 20-52 at% Mn concentration range. Some results of this study have been published in a few short communications. In the present paper the crystallographic and magnetic structural data for the $MnPd_3$, $MnPd_2$, Mn_3Pd_5 and MnPd phases are summarized and the relation between the observed structures are discussed.

THE $MnPd_3$ AND MnPd PHASES

The intermetallic compound $MnPd_3$ has a one-dimensional long-period superlattice based on the Cu_3Au type order with periodic antiphase domains[1]. By x-ray diffraction on single crystal the crystal structure was refined[2] in the space group I4mm. The special values of the parameters showed the actual space group to be I4/mmm[3]. The existence of the long-period structure was observed up to 36 at% Mn in samples quenched from 800° C and the periodicity 2M, where M is the number of the Cu_3Au type cells per domain, was found to be 4 from 23 to 30 at% Mn increasing to about 6 at higher Mn concentrations[4-6]. Above 25 at% Mn concentration the excess Mn atoms occupy preferentially the 4(e) Pd sites leading in quenched samples to a continuous transition to the CuAu-I type MnPd phase which is stable from 36 to 52 at% Mn[3,4].

The magnetic structure of the $MnPd_3$ phase was shown to be collinear antiferromagnetic with the magnetic moments being below 25 at% Mn parallel and above this concentration perpendicular to the tetragonal axis[7]. At stoichiometric composition the dependence of the direction of the magnetic moments on the atomic long-range order was observed[8]. In the 4(c) and 4(e) positions the existence of ordered moments on the Pd atoms coupled antiparallel to those of the nearest neighbour Mn atoms was established.

The magnetic structure of the MnPd phase is also collinear antiferromagnetic with magnetic moments perpendicular to the tetragonal axis. The magnetic moment on the Pd atoms is assumed to be zero[9].

THE $MnPd_2$ AND Mn_3Pd_5 PHASES

After lengthy annealing below 480° C two new phases, $MnPd_2$ and Mn_3Pd_5 appear in the 32-36 and 36-41 at% Mn concentration range, respectively.

The crystal structure of the $MnPd_2$ phase belongs to the orthorhombic space group Pnma, the unit cell contains 12 atoms[10]. The accuracy of the powder measurements is insufficient for establishing a unique set of values for the six atomic position parameters. The magnetic structure is collinear antiferromagnetic with magnetic moments parallel to axis b. No magnetic moment on the Pd atoms was observed.

The Mn_3Pd_5 phase has an orthorhombic crystal structure of the space group Cmmm[11]. In the unit cell there are two non-equivalent Mn and three non-equivalent Pd positions. A small atomic disorder was observed even after very long annealings. The magnetic structure is collinear ferrimagnetic due to the antiparallel alignment of the moments having different values at the two non-equivalent sites. A small magnetic moment on the Pd atoms, coupled antiparallel to the resultant Mn moment, can also be inferred. The magnetic moments point in the direction of axis c.

DISCUSSION OF THE STRUCTURES

The crystal and magnetic structures of the $MnPd_3$, $MnPd_2$, Mn_3Pd_5 and MnPd phases are illustrated in Fig. 1. In the Mn_3Pd_5 structure the axes y and z are interchanged for an easier comparison. The relative directions of the magnetic moments are indicated by plus and minus signs. For clarity, the magnetic moments on the Pd sites are omitted. Some crystallographic and magnetic structural data are listed in Table I.

As seen in Fig. 1, the observed crystal structures

Table I Crystal and magnetic structure data for the $MnPd_3$, $MnPd_2$, Mn_3Pd_5 and MnPd phases.

	$MnPd_3$		$MnPd_2$	Mn_3Pd_5	MnPd
	25%Mn	32%Mn	32%Mn	38%Mn	50%Mn
a_x (Å)	3.87	4.02	3.84	4.036	4.07
a_y (Å)	3.87	4.02	4.03	3.044	4.07
a_z (Å)	3.87	3.70	3.84	3.640	3.58
V (Å³)	57.96	59.79	59.42	59.41	59.30
μ_{Mn} (μ_B)	4.1	4.1	4.0	4.2; 3.4	4.4
μ_{Pd} (μ_B)	uncertain	0.2	0	0.15	0
φ_y (°)	90	uncertain	0	0	uncertain
φ_z (°)	0	90	90	90	90
T_N (°K)	220	225	415	540	820

can be built up from a face-centred quasi-cubic basic unit whose parameters a_x, a_y, a_z and volume V are given in Table I. The basic unit is cubic at 25 at% Mn and it is gradually distorted with increasing concentrations. It should be noted that in Fig. 1 the repetition period of the structure for $MnPd_2$ consists in both the x and z directions of six basic units, but the true unit cell shown by thick solid lines in Fig. 1 is composed of only three basic units.

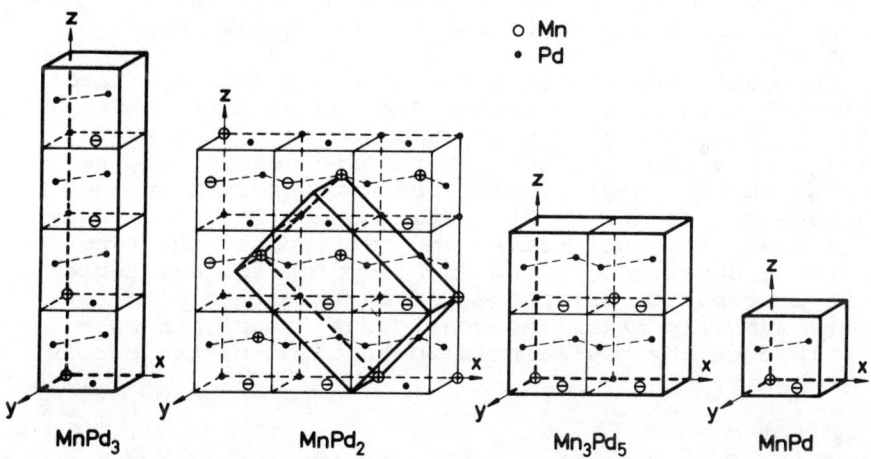

Fig. 1 Structures in the Mn-Pd system. The relative directions of the magnetic moments are indicated by plus and minus signs. The magnetic moments on the Pd sites are omitted.

In $MnPd_3$, Mn_3Pd_5 and MnPd every second atomic layer perpendicular to the z-axis contains only Pd atoms while the other layers have Mn and Pd atoms in an ordered arrangement. On the other hand, in $MnPd_2$ all layers are equivalent containing Pd and Mn atoms in a ratio 2 to 1. Due to this special order, the distortion of the basic unit is different from that in the other structures.

The common feature of the observed magnetic structures is that the layers perpendicular to the y-axis are ferromagnetic and the successive layers are coupled antiparallel to one another. The dominant antiferromagnetic interaction responsible for the magnetic order is the nearest neighbour interaction in $MnPd_2$, Mn_3Pd_5, MnPd and the third neighbour interaction in $MnPd_3$. The interaction along the edges of the basic unit is ferromagnetic. The ferrimagnetic structure of Mn_3Pd_5 follows from the fact that the successive layers perpendicular to the y-axis contain different numbers of Mn atoms and that the magnetic moments on the two nonequivalent Mn sites are different due to their unlike environments.

In $MnPd_3$ and Mn_3Pd_5 ordered moments on the Pd atoms could be established at the sites where the resultant exchange field is nonzero. At the other sites the moments are supposed to be zero.

The angle φ_z of the magnetic moments to the z-axis is $90°$ except below 25 at% Mn. The powder diffraction measurements did not allow the determination of φ_y in structures with tetragonal symmetry. However, the value of $\varphi_y = 0°$ measured for orthorhombic $MnPd_2$ and Mn_3Pd_5 is probably valid for the tetragonal $MnPd_3$ and MnPd, too.

The estimated signs of the Mn-Mn interactions are consistent with those obtained for the similar Mn-Pt system[12]. However, the strong dependence of the exchange interactions on the interatomic separation, as observed in the Mn-Pt system, was not observed in the Mn-Pd system. In the Mn-Pt system no continuous transition between the cubic Cu_3Au and tetragonal CuAu type phases was observed. In the Mn-Pd system the existence of the long-period antiphase domain structure with tetragonal symmetry makes the continuous transition possible through the preferential occupation of the excess Mn atoms.

ACKNOWLEDGEMENTS

The authors are indebted to Drs. G. Konczos and J. Paitz for preparing the samples, to Mrs. É. Zsoldos for x-ray investigations and to Mrs. K. Zámbó for the chemical analysis.

REFERENCES

1. J. W. Cable, E. O. Wollan, W. C. Koehler and H. R. Child, Phys. Rev. 128, 2118 (1962).
2. H. Iwasaki, K. Okamura and S. Ogawa, J. Phys. Soc. Japan 31, 497 (1971).
3. E. Krén, G. Kádár and M. Márton, Solid State Comm. 10, 1195 (1972).
4. E. Krén, G. Kádár, L. Pál, É. Zsoldos, M. Barberon and R. Fruchart, J. Phys. 32, C1-980 (1971).
5. H. Sato and R. S. Toth, Phys. Rev. 139, A1581 (1965).
6. J. Gjönnes and A. Olsen, Phys. Stat. Sol. (a) 17, 71 (1973).
7. E. Krén and G. Kádár, Phys. Letters 29A, 340 (1969).
8. E. Krén, G. Kádár and L. Pál, J. Appl. Phys. 41, 941 (1970).
9. E. Krén and J. Solyom, Phys. Letters 22, 273 (1966).
10. G. Kádár, E. Krén and M. Márton, J. Phys. Chem. Solids 33, 212 (1972).
11. G. Kádár and E. Krén, Solid State Comm. 11, 933 (1972).
12. E. Krén, G. Kádár, L. Pál, J. Sólyom, P. Szabó and T. Tarnóczi, Phys. Rev. 171, 574 (1968).

MAGNETIC FORM FACTOR OF PALLADIUM AT 4.2 K[*]

J. W. Cable and E. O. Wollan
Oak Ridge National Laboratory, Oak Ridge, Tennessee 37831

G. P. Felcher, T. O. Brun, and S. P. Hornfeldt
Argonne National Laboratory, Argonne, Illinois 60439

ABSTRACT

The magnetic form factor of palladium was measured by a polarized neutron experiment. The measurements were made on a single crystal in a magnetic field of 57 kOe along the (110) axis. The form factor was measured at the position of the first ten Bragg reflections of the [110] zone at 4.2 K.

Two specimens were used that were characterized by magnetization measurements. One of the samples was very pure, while the second contained 44 parts per million of iron. At the rated conditions, the first sample had a net magnetic moment of 0.0075 μ_B/atom, the second 0.0080 μ_B/atom. The form factor of the two samples, once corrected for the minute contribution of iron, were found to be essentially identical.

The measurements indicated that the form factor of palladium drops sharply, going to zero at $\sin\theta/\lambda \approx 0.5$ Å^{-1}. By Fourier transformation of the data, magnetization density maps were obtained. No areas of negative polarization were found outside of the experimental error. The measured form factor was also compared with 4d form factor of Pd^{2+} calculated by the Hartree-Fock method.[1] The comparison was made after correcting the data for the diamagnetism of the closed shell and allowing for 10% orbital contribution. The experimental form factor appears to be slightly more expanded than the calculated one. The asphericity of the induced magnetization is similar to that of nickel; in terms of population of d-levels 85 ± 4% of the magnetic states have t_{2g} character.

1. A. J. Freeman, private communication.

[*]Work supported under the auspices of the USAEC.

ANTIFERROMAGNETIC IRON IN SMALL γ-PHASE Fe-Ni CRYSTALS*

J. M. Crowell and J. C. Walker
The Johns Hopkins University, Baltimore, Maryland 21218

ABSTRACT

Recent investigations of the magnetic properties of face-centered cubic (fcc) nickel-iron alloys suggest that they are antiferromagnetic in the range of low nickel concentration. The study of antiferromagnetism in these alloys is hindered by the fact that they undergo a transformation to a body-centered cubic structure at temperatures above any possible antiferromagnetic ordering temperature. In the body-centered phases the alloys are universally ferromagnetic. This change in crystal structure can, however, be suppressed if the alloys have the form of fine particles. We have used the Mössbauer effect and a 33 kilogauss superconducting solenoid to show that antiferromagnetism exists in these low nickel concentration fcc Ni-Fe alloys and to determine that the Neél temperature ranges from 24°K to 30°K for alloys with nickel concentration from 18 to 28 atomic percent nickel.

Powdered alloys ranging in nickel concentration from 18 to 32 atomic percent nickel were synthesized by reduction of nickel-iron oxalates. X-ray diffraction showed that the fcc phase was retained in all cases. At low temperatures fcc Ni-Fe alloys are ferromagnetic if the nickel concentration is greater than 29 atomic percent and apparently antiferromagnetic below this concentration. All our samples exhibit a mixture of ferromagnetic and antiferromagnetic iron sites which is attributable to a Gaussian distribution of concentration which bridges this critical value.

Recent investigations of the magnetic properties of face-centered-cubic nickel-iron alloys suggest that in the range of low nickel concentration they are antiferromagnetic. However, the study of antiferromagnetism in these metals is hindered by the fact that at low temperatures they undergo a martensitic transformation to body-centered-cubic. In this state they are universally ferromagnetic. Cech and Turnbull[1] and Kachi et al.[2] demonstrated that the martensitic transformation could be suppressed in fine particles. A number of investigations[3-7] of thermodynamic and magnetic

---

*Work supported by National Science Foundation.

properties of such powdered iron-nickel alloys indicate the possibility of antiferromagnetism. In particular, Asano[5,7] observed a broadening of the paramagnetic Mössbauer spectrum of the powdered alloys at a temperature of 4.2°K, and attributed this to an antiferromagnetic phase whose ordering temperature was between 77°K and 4.2°K. No direct evidence of antiferromagnetism was offered. In the work reported here, powdered alloys ranging from 18 to 32 at.% Ni were placed in a magnetic field of 30 kOe. The resulting Mössbauer spectra cannot be explained by any assumption of paramagnetism, ferromagnetism or superparamagnetic relaxation, whereas a simple model of randomly oriented antiferromagnetic particles fits the data quite well.

Alloys samples of 18, 20, 22, 25, 28, 29 and 32 at.% Ni were synthesized by a method similar to that of Kachi et al.[2] Solutions of $NiSO_4$ and $FeSO_4$ were mixed in the ratio x:1-x by volume, and the oxalate $Ni_xFe_{1-x}C_2O_4$ was precipitated by adding oxalic acid. The filtered precipitate was then reduced in hydrogen at 350°C for ten hours resulting in the alloy Ni_xFe_{1-x}. Alloys prepared at this temperature are a mixture of fcc and bcc phases. The samples were annealed at 650°C for 30 minutes to bring the mixture to a single phase. The alloys thus obtained were in the form of granules of 1-10µ diameter which themselves were clusters of .1-1µ particles.[2] X-ray diffraction analysis revealed that the fcc phase was retained in all samples even after they had been cooled to liquid helium temperature. (Samples of 10, 12 and 15 at.% Ni showed mixed fcc and bcc phases.)

The Mössbauer spectra of the fcc alloys showed a mixture of ferromagnetic and antiferromagnetic states. Fig. 1(a) and (b) shows a typical spectrum. The central peak in the spectrum comes from the antiferromagnetic portion of the alloy whose hyperfine field is so small that the six lines of the normal hyperfine spectrum cannot be clearly resolved. The spectrum of the ferromagnetic portion of the alloy, which has a much larger hyperfine field, is identified by the usual six-line pattern. The outer lines are easily discernible in the figure. The innermost lines of the ferromagnetic spectrum overlap the central peak.

The coexistence and the relative abundance of the two magnetic states are accurately described by the model of Kachi and Asano.[8] The alloys prepared in the manner indicated above are not homogeneous. The nickel concentration is described by a Gaussian distribution; i.e., the fraction of the alloy with nickel concentration between x and x+dx is given by,

$$P_\alpha(x)dx = \frac{1}{\sqrt{2\pi}\,\sigma} e^{-(x-\alpha)^2/2\sigma^2} dx$$

where α is the mean nickel concentration and σ is the width of the distribution to be determined empirically. According to a model by R. J. Weiss[9] an ideally uniform Ni-Fe alloy would be ferromagnetic if the nickel concentration is greater than 29 at.%. Below this concentration the alloys are assumed to be antiferromagnetic. In the

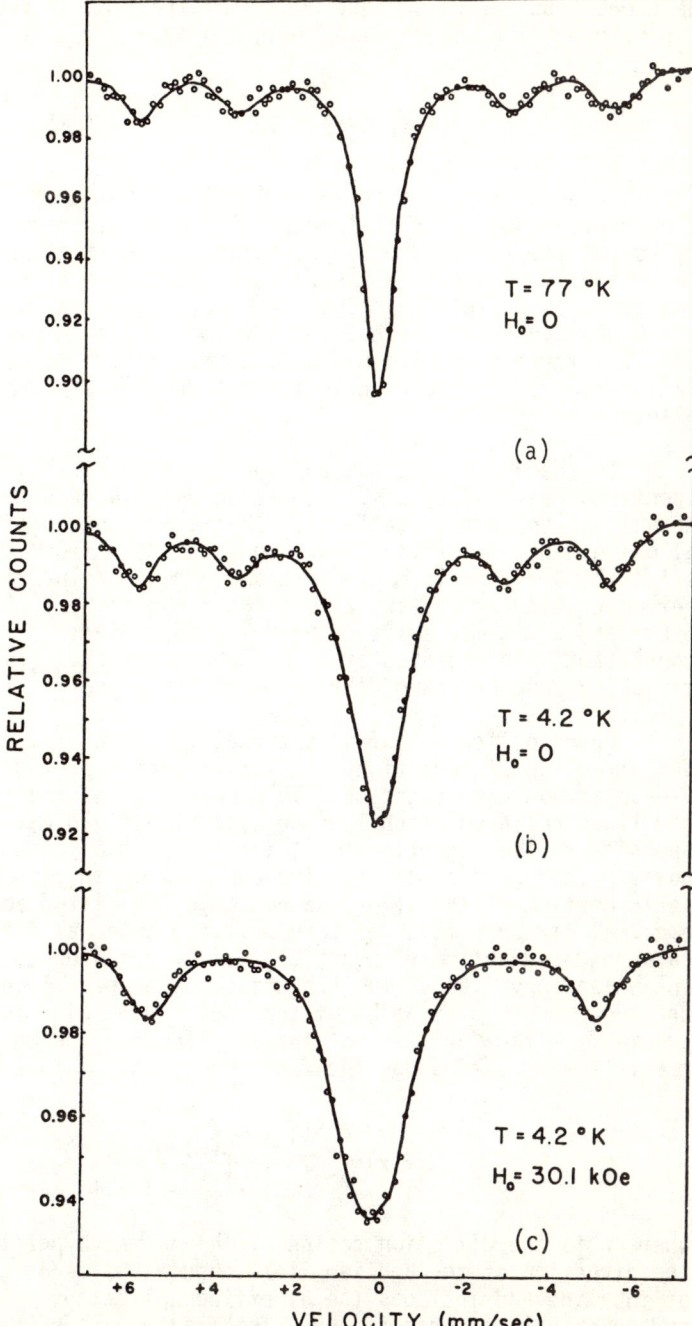

Fig. 1. Effect of an External Field Upon 22 at-% Ni-Fe Alloy

heterogeneous samples used here, the fraction of iron in the ferromagnetic portion of the alloy is given by

$$\frac{1}{1-\alpha} \int_{.29}^{1} (1-x)P_\alpha(x)dx \cong \frac{1}{1-\alpha} \int_{.29}^{\infty} (1-x)P_\alpha(x)dx$$

(So long as $\sigma \ll 1$, the infinite limit on x introduces no appreciable error.) Comparison of spectral intensities (total area under peaks) of the ferromagnetic and antiferromagnetic parts of the Mössbauer spectra shows that the relative abundance of the two magnetic phases is accurately described by a Gaussian distribution with $\sigma \approx 0.06$. This is in excellent agreement with Asano[7] who found that a distribution with $\sigma \approx 0.06$ describes the magnetic moment, lattice parameters, and Curie temperatures of similarly prepared alloys.

In the absence of the applied magnetic field, the Mössbauer spectra of each alloy sample could be described as the sum of two normal six-Lorentzian patterns. The ferromagnetic part of the alloys exhibit hyperfine fields on the order of 340 kOe and isomer shifts of 0.29 ± 0.05 mm/sec (relative to Fe^{57} in Cr). The antiferromagnetic part of the alloys has a hyperfine field of 27 ± 5 kOe and an isomer shift of 0.18 ± 0.05 mm/sec. These values are consistent with those obtained by Johnson et al.[10] for ferromagnetic alloys and by Asano[5] for the antiferromagnetic alloys.

Upon application of an external magnetic field parallel to the direction of radiation, the ferromagnetic part of the spectra behave as would be expected. The separation of the outer and inner lines collapse slightly (the hyperfine field being directed opposite to the magnetization), and the middle lines disappear. However, the part of the spectra due to iron in the antiferromagnetic portion of the alloy can no longer be fitted accurately by a normal six-Lorentzian pattern. If the external field applied to a powdered antiferromagnetic is small compared to the exchange and anisotropy fields, the sublattices are only slightly canted, and the random orientation of the individual spins is not appreciably affected. Each of the six lines in the Mössbauer spectra is then described by the function

$$f(x) = \int_{-1}^{1} \frac{D(\Theta)du}{(\frac{x-x_o(H)}{\Gamma})^2 + 1}$$

where u is the direction cosine of the normal hyperfine field to the direction of the applied field, and $x_o(H)$ denotes the center of the line and includes the hyperfine splitting, isomer shift, and quadrupole effects if any. The fractional depth of the line, $D(\Theta)$, is determined by the Clebsch-Gordon coefficients for the levels involved and the angular distribution for magnetic dipole radiation. It is given by

$$D(\Theta) = \begin{cases} 3D_o(1+\cos^2\Theta) & \text{for } m= \tfrac{3}{2} \to \tfrac{1}{2} \text{ and } m= -\tfrac{3}{2} \to -\tfrac{1}{2} \\ D_o(1+\cos^2\Theta) & \text{for } m= \tfrac{1}{2} \to -\tfrac{1}{2} \text{ and } m= -\tfrac{1}{2} \to +\tfrac{1}{2} \\ 4D_o \sin^2\Theta & \text{for } \Delta m=0. \end{cases}$$

Such a model fits the observed spectra quite well as shown in Fig. 1(c). Moreover, when the magnitudes of the normal hyperfine field and the applied field were left as free parameters in the fitting routine, the computed values of these variables for the best fit agree very well with the values of the hyperfine field measured without an applied field, and with the known value of the applied field. If the central peak were due to iron in any state other than the antiferromagnetic, the resulting Mössbauer spectrum would exhibit a normal hyperfine structure when an external field is applied. It is therefore clear that the central peak in the Mössbauer spectra of the powdered alloys is indeed the result of iron in an antiferromagnetic lattice. By observing the width of the central peak as a function of temperature, the Néel temperatures of the alloys were found to range from 24°K to 30°K in alloys of 18 to 28 at.% Ni.

These results can be compared with those of Williamson and Gonser[11] recently reported at the International Conference on Magnetism in Moscow. They have investigated fcc iron precipitates in copper using very similar techniques. In this case Néel temperatures varied greatly with particle size. The largest values reported were about 67°K. In the present case, both particle size and the nickel-iron exchange interaction have the effect of lowering the Néel temperature.

REFERENCES

1. R. E. Cech and D. Turnbull, Journal of Metals 8, 124 (1965).
2. S. Kachi, Y. Bando, and S. Higuchi, Jap. J. Appl. Phys. 1, 307 (1962).
3. Y. Nakamura and M. Shiga, J. Phys. Soc. Japan 19, 1177 (1964).
4. S. Kachi, H. Asano, and N. Nakanishi, J. Phys. Soc. Japan 25, 285 (1968).
5. H. Asano, J. Phys. Soc. Japan 25, 286 (1968).
6. Y. Nakamura, Y. Takeda, and M. Shiga, J. Phys. Soc. Japan 25, 287 (1968).
7. H. Asano, J. Phys. Soc. Japan 27, 542 (1969).
8. S. Kachi and H. Asano, J. Phys. Soc. Japan 27, 536 (1969).
9. R. J. Weiss, Proc. Phys. Soc. 82, 281 (1963).
10. C. E. Johnson, M. S. Risout, and T. E. Cranshaw, Proc. Phys. Soc. 81, 1079 (1963).
11. D. L. Williamson and U. Gonser, Private Communication.

VARIATION WITH CHEMICAL ORDER OF SPIN COUPLING IN NiPt

C. W. Chen, J. D. Greiner, and R. W. Buttry
Ames Laboratory USAEC
Iowa State University, Ames, Iowa 50010

ABSTRACT

The effect of chemical order on the coupling mode of spins in NiPt was studied by measurements of magnetic susceptibility, magnetization, and electrical resistance. Ferromagnetic and antiferromagnetic couplings were detected in the disordered and ordered states, respectively. These results confirm a previous report by Watanabe and Miyahara. The present study, however, obtained a lower (98 versus 136°K) Curie temperature and a different mean atomic moment in the ferromagnetic structure. Irradiation of the ordered alloy by fast neutrons tends to destroy the antiferromagnetism, which eventually disappears when the irradiation fluence reaches 4×10^{18} neutrons (E⩾1 MeV) per cm^2. Irradiation of the disordered alloy, on the other hand, causes peculiarities in the temperature dependence of susceptibility. Evidence for the existence of superparamagnetism in the irradiated samples is described. The complex magnetic behavior of NiPt resembles that of $FePt_3$ strikingly.

INTRODUCTION

In a previous paper[1], Watanabe and Miyahara (W&M) reported that alloys of the isomorphous Ni-Pt system near the equiatomic composition exhibit either ferromagnetism or antiferromagnetism, depending upon the state of chemical order. For instance, in the 53.3 atomic % Ni-Pt alloy, the disordered state, presumably produced by quenching, showed ferromagnetic behavior in the temperature dependence of reciprocal susceptibility, with a paramagnetic Curie temperature (T_c) of 136 ± 2°K and a mean saturation moment of 0.06 Bohr magneton (μ_B) per atom. The ordered state of the same alloy, on the other hand, displayed antiferromagnetic behavior, with an unknown, probably low (∼3°K), Néel temperature (θ_N).

Should these observations prove to be true, it would seem reasonable to expect the Ni-Pt alloys around the 50 at.% to show a mixed and varying magnetic behavior corresponding to the degree of long- and short-range order. As an example, we might anticipate the existence of both ferromagnetic and antiferromagnetic spin couplings in the same specimen that is originally ordered and subsequently partially disordered by irradiation with energetic particles or plastic deformation.

The purpose of the present study is twofold: to confirm and further characterize the dual magnetic behavior of the Ni-Pt alloys and to explore the possible coexistence of ferromagnetic and antiferromagnetic spin couplings in the partially ordered samples of these alloys. This paper will deal only with the equiatomic alloy NiPt, which obviously represents a simpler case from the viewpoint of atom distribution than other neighboring alloys.

EXPERIMENTAL PROCEDURE

An ingot of NiPt was prepared from high-purity (>99.99%) Ni and Pt by arc melting. To attain composition homogeneity, the ingot was remelted several times on alternate sides and homogenized at 1450°C in a dry hydrogen atmosphere for 48 hrs. Emission spectroscopic analysis detected no trace of magnetic impurities such as iron and cobalt in the ingot. The ordered state was obtained by annealing the specimen at 580°C for 60 hrs, whereas the disordered state was produced by quenching the sample from 1200°C to an ice-water bath. X-ray diffraction patterns showed no evidence of chemical order in the quenched sample and a face-centered tetragonal crystal structure (the $L1_0$-, or CuAuI-type superlattice) with superlattice reflections in the annealed specimen. The irradiation experiment was performed in a combined cryostat-flux converter system[2] permanently installed inside the Ames Laboratory Research Reactor. The system was capable of maintaining the sample temperature at 80°K during the irradiation and shielding out thermal neutrons. The flux of fast neutrons with energy greater than 1 MeV was estimated at 2.1×10^{12} n/cm^2/sec and the fluence attained the experiment was monitored to be 4.3×10^{18} n/cm^2.

Magnetic susceptibility was measured by a modified Faraday method[3] in the temperature range 50-298°K. A vibrating sample magnetometer equipped with a liquid helium metal dewar was used to determine magnetization under a field of maximum strength of 20 kOe. Electrical resistance measurements were made on wire specimens in a cryostat using a 4-point probe and a Rubican 6-dial potentiometer.

RESULTS AND DISCUSSION

Experimental data obtained from susceptibility measurements are summarized in Figs. 1 and 2. Fig. 1 plots the reciprocal of susceptibility corrected for the presence of the diamagnetic lead container against temperature. Data for the annealed specimen (3s) fall upon a straight line, which, by extrapolation, intercepts the temperature axis at -176 ± 5°K. The emergence of such a negative paramagnetic Néel temperature (T_N) thus confirms the presence of antiferromagnetism in the ordered NiPt. Neutron irradiation of the ordered alloy tends to raise the susceptibility. Consequently, the best-fit lines for two ordered-and-irradiated samples (3x and 3b) are shifted to the right with higher T_N values. No reason can be given, however, for the divergent results for these two samples, because they received identical heat treatments and the same dose of irradiation. On the same ground, the 0°K intercept for specimen 3b is considered coinci-

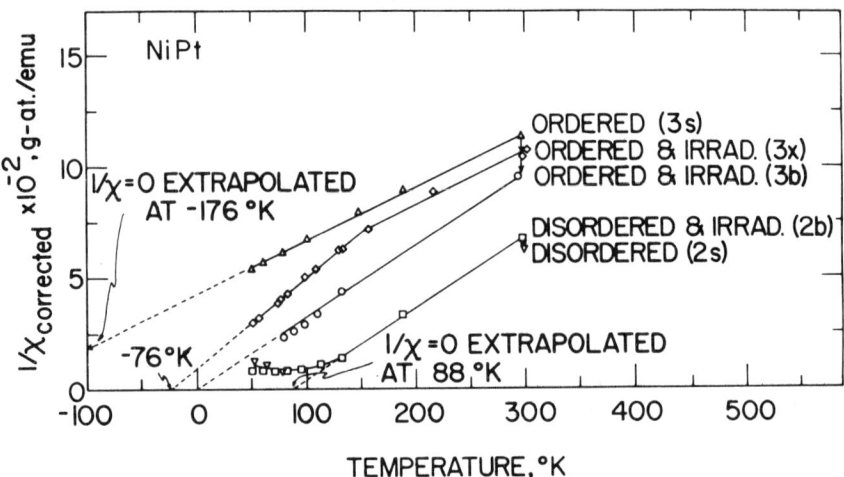

Fig. 1. Temperature dependence of the reciprocal susceptibility observed for NiPt in the ordered and disordered states with or without the effect of neutron irradiation.

dental and should not be taken as evidence for the existence of typical paramagnetism in this specimen.

Susceptibility data for the disordered samples show the following features: (1) Irradiation exerted a decreasing effect on χ at room temperature as opposed to the increasing effect in the ordered state. Such opposite effects of irradiation are self-consistent, of course.

(2) At temperatures below 78°K, and thus in the ferromagnetic state, the effect of irradiation on χ is hardly seen.

(3) In the same low temperature range, susceptibility becomes essentially independent of temperature. Similar $1/\chi$ vs T curves to that shown for specimens 2s and 2b in Fig. 1 were observed in a series of face-centered cubic Pr-Th alloys by Bucher et al.[4], who ascribed the temperature-independent part at low temperatures to the van Vleck-type paramagnetism. It is not known and the authors seriously doubt that this type of paramagnetism prevails in the disordered NiPt.

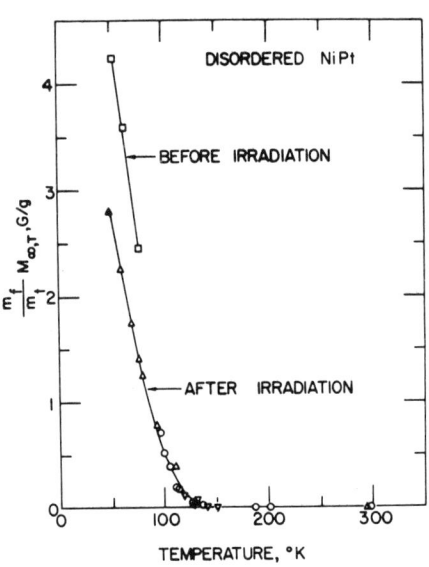

Fig. 2. Comparison between saturation magnetization of disordered NiPt before and after irradiation.

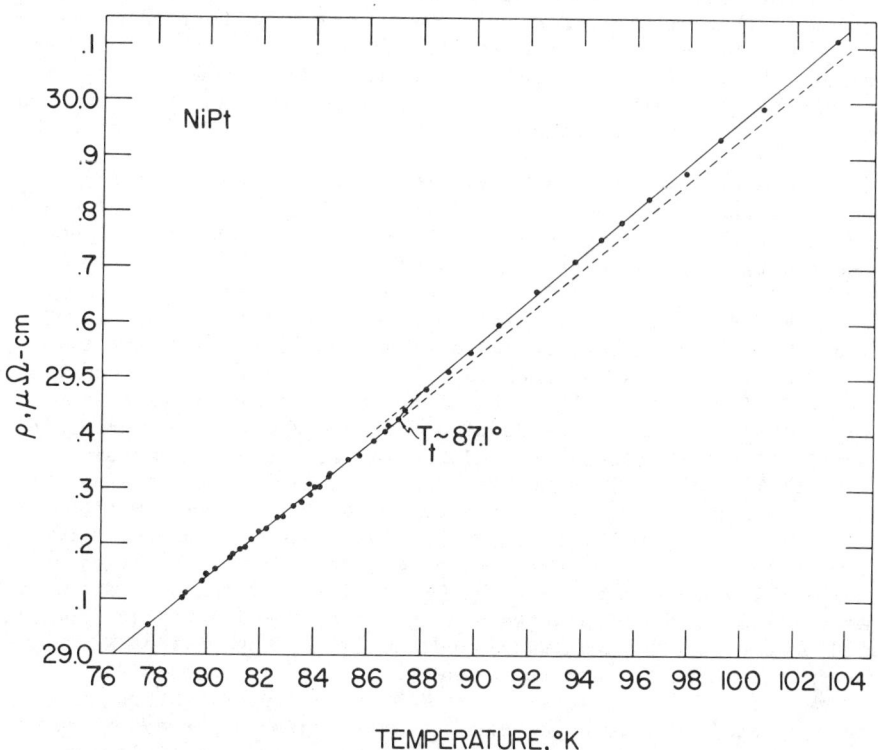

Fig. 3. A resistivity anomaly was detected at the transition temperature of 87 ± 2°K in the disordered NiPt.

Extrapolation of the upper part of the $1/\chi$ vs T plot for the disordered samples to zero $1/\chi$ gives a paramagnetic Curie temperature of 88 ± 3°K. This temperature is much lower than T_c = 136 ± 2°K deduced for the 53.3 at.% Ni-Pt alloy by W&M. Also, our deduced value for T_N differs considerably from that given by W&M. These differences have prompted us to conduct resistance measurements on a quenched NiPt wire in the temperature range 4.2-298°K. An anomaly was detected at 87 ± 2°K in the resistivity vs temperature plot shown in Fig. 3. Thus according to the resistivity data, the disordered NiPt alloy undergoes a magnetic transition at 87°K, which is very close to the paramagnetic Curie temperature of 88°K.

To further explore the effect of irradiation on magnetization of the disordered alloy, saturation values ($M_{\infty,T}$) were deduced at the infinite field from susceptibility data. By applying a multiplying factor m_f/m_t, where m_f is the mass associated with the field-dependent term in the Faraday force equation[3] and m_t is the total mass, the quantity $(m_f/m_t) M_\infty T$ gives the saturation magnetization of the NiPt specimen. Fig. 2 plots the deduced values for this quantity against temperature. Clearly there is a considerable decrease in saturation caused by neutron irradiation. The decrease amounts to 37% at 52°K and 38% at 76°K.

Results of magnetization measurements yield valuable information in three respects. First, we used the relation, $a(T-\theta_c)M + bTM^3 = H$,[5] to determine the Curie temperature of disordered NiPt, where a and b are constants. In the plot of M^2 vs H/M, a linear relationship emerges at low fields up to 8 kOe at each temperature. Hence the residual value of M^2 at zero field can be deduced reliably by extrapolation. Then by plotting the residual values of M^2 vs T, the temperature at which $M^2_{H=0}$ vanishes marks the Curie point. The deduced value of θ_c for NiPt is 98 ± 4°K. Note this transition temperature is higher by $\sim 10°$ than those deduced from other data; however, this value of θ_c has not been corrected for the effect of demagnetization.

Another respect of magnetization of the disordered NiPt is the mean saturation moment, which we deduced, from the M vs $1/H$ plots and M_s vs T plot, to be 0.25 μ_B per atom. This value is in poor agreement with the 0.06 μ_B value reported for the 53.3 at.% Ni-Pt by W&M.

The third respect deals with the possible existence of superparamagnetism in the irradiated samples. For simplicity, we concentrated on the ordered and irradiated samples in the present work and chose two criteria for superparamagnetism: (1) no hysteresis and (2) superposition of data points in the M/M_s vs H/T plot. Strong evidence was obtained for the absence of hysteresis in the cyclic measurements of magnetization. Evidence for the second criterion at fields up to 10 kOe is also quite good, but less conclusive at higher fields because of an appreciable spread of the data points.

In summary, the magnetic behavior of the equiatomic alloy NiPt is complex, varying and strikingly resembles that of FePt$_3$[6]. Both alloys show antiferromagnetism in the ordered state, ferromagnetism in the disordered state, and superparamagnetism in the ordered-and-irradiated state[7]. What makes the varying magnetic behavior of NiPt even more remarkable is that the onset of antiferromagnetism in a metallic system of Ni is rarely seen, nor would it be predicted from the Bethe-Slater criterion for the sign of exchange integral in the Heisenberg theory.

REFERENCES

1. M. Wanatabe and S. Miyahara, J. Phys. Soc. Japan 23, 451 (1967).
2. C. W. Chen, R. G. Stuss, and J. Crudele, J. Phys. E2, 746 (1969).
3. E. C. Stoner, Magnetism and Atomic Structure, p. 39 (1926).
4. E. Bucher et al., Helv. Phys. Acta 41, 723 (1968).
5. J. S. Kouvel and M. E. Fisher, Phys. Rev. 136, A1626 (1964).
6. P. W. Bellarby and J. Crangle, J. Appl. Phys. 39, 463 (1968).
7. C. R. Piercy, J. Phys. Soc. Japan 18 Suppl. III, 169 (1963).

TEMPERATURE DEPENDENCE OF THE Co HYPERFINE FIELD IN $GdCo_2$ and ITS RELATION TO MAGNETIC STRUCTURE

I.Wang and T.J.Burch
Department of Physics, Fordham University
Bronx, N.Y. 10458

J.I.Budnick
National Science Foundation
Washington, D.C. 20550

J.J.Murphy
Physics Department, Iona College
New Rochelle, N.Y. 10801

J.A.Cannon
Department of Physics, Pace University
Pleasantville, N.Y. 10570

ABSTRACT

To understand more fully the magnetic structure of $GdCo_2$ we have investigated the field and temperature dependence of the Co hyperfine field in this cubic-Laves phase ferrimagnet. The hyperfine field at Co nuclei is 61.3 kOe at 1.4 K and the sign of this field is positive with respect to the sample magnetization. Below 200 K the Co resonance is basically a single line. Above this temperature, near which a pronounced inflection appears in the saturation magnetization vs. temperature curve,[1] the Co spectrum becomes complex. Above 260 K it is composed of two closely spaced but resolved lines. The temperature dependence of the sublattice magnetizations are analyzed in the light of the bulk magnetization behavior and this hyperfine field data. The significance of the spectral structure and the temperature dependence of the Co sublattice magnetization are explored within the context of existing models of the magnetic structure.

1. J.Crangle and J.W.Ross, Proc. Intern.Conf. on Magnetism (Nottingham 1964) p.240.

LOW TEMPERATURE STUDY OF THE $Co_{1-x}Fe_xSi$ ALLOYS*

P. H. Barrett, P. A. Montano, and Z. Shanfield
University of California, Santa Barbara, Calif. 93106

ABSTRACT

A study of the internal magnetic field at the Co nucleus in $Co_{1-x}Fe_xSi$ alloys was carried out at low temperatures using nuclear orientation techniques. Radioactive ^{57}Co was diffused into the samples and the anisotropy of the 14.4 KeV γ-ray was observed at temperatures in the region of 40 mK. In the ferromagnetic region (x = 0.2 to 0.8) no detectable nuclear orientation was observed, thus placing an upper limit to the internal magnetic field at the ^{57}Co nucleus of 50 kG. We have also carried out Mössbauer measurements at low temperature in the above samples. It was observed that the field at the ^{57}Fe nucleus is very small, around 20 kG. These results seem to indicate that the ferromagnetism in these alloys is itinerant. This is in agreement with the spin echo measurements of Kawarazaki, et al.[1] Our measurements have been extended also to the paramagnetic region (x below 0.2).

INTRODUCTION

There have been several investigations [1,2,3] of the magnetic properties of the intermetallics FeSi, CoSi and their alloys. A ferromagnetic phase exists in the concentration range 0.2 < x < 0.8 of $Co_{1-x}Fe_xSi$.[2] For x below 0.2 the system is paramagnetic with an effective magnetic moment of 3.08 μ_B per Fe atom. Mössbauer, as well as NMR techniques, have been applied to the study of the different phases. It was found that the magnetic moments in the ferromagnetic phase are very small at the Fe^2 and Co^1 sites (of the order 0.1 μ_B). In the paramagnetic region (x < 0.2) there remains the possibility of a localized magnetic moment at the Co site.

It is the purpose of this work to study the magnetic behavior of ^{57}Co and ^{57}Fe nuclei in $Co_{1-x}Fe_xSi$ (x < 0.8) at low temperature by using Mössbauer and nuclear orientation techniques.

EXPERIMENTAL

Samples of $Co_{1-x}Fe_xSi$ were prepared by melting stoichiometric

*Supported by the National Science Foundation

amounts of high purity Co, Si and Fe in an Ar/H_2 atmosphere. Next, the samples were remelted and annealed at 900°C for 12h in an inert atmosphere. The homogeneity and composition of the samples were checked with an electron microprobe and lattice constants were calculated from X-ray diffraction patterns. The value of the lattice constants for different concentrations are given in Table I.

Table I Lattice Parameters

	CoSi	7% Fe	13% Fe	50% Fe	70% Fe
a[Å]	4.433	4.437	4.442	4.4585	4.4697

Our values are slightly different from those previously reported.[4] In the preparation of samples with concentrations of Fe below 10%, we observed that up to 15% of the material was separated into different phases. Only after very carefully quenching and annealing was it possible to obtain homogeneous samples. We conclude that the preparation of low concentration of Fe in CoSi requires careful checks on the homogeneity and composition of the resultant samples. The CoSi and $Co_{0.5}Fe_{0.5}Si$ samples were from single crystals. Radioactive ^{57}Co was diffused into the samples at 1150°C for 15 min and rapidly quenched to room temperature in an H_2 atmosphere. It was observed that when ^{57}Co was diffused at temperatures below 1000° C for several hours some clustering was always present. The appearance of the Mössbauer spectra of iron in metallic cobalt was taken as evidence of the clustering of cobalt atoms.

A He^3/He^4 dilution refrigerator was used to reach temperatures below 100 mK. The samples were mounted on a pure copper holder connected to the mixing chamber. In each run four samples were mounted on the holder, one of them has a $^{57}Co/Fe$ foil that was used as an absolute temperature thermometer and another was CoSi as a counting rate standard.

A gear system was used to rotate a shutter that exposed the detector to one sample at a time. A proportional counter was used to detect the 14-keV gamma ray. An external magnetic field of 1.5 kG was used to align the spins. The γ-rays were detected at an angle of π/2 with respect to the magnetic field. The intensity is given by[5]

$$I(\frac{\pi}{2}) = \sum_k B_k U_k F_k P_k(\frac{\pi}{2}) \qquad (1)$$

where B_k, the orientation parameter, is a measure of the population distribution among the substates of the initial spin I_o (for ^{57}Co $I_o = \frac{7}{2}$). F_k depends on the angular momentum properties of the observed transition; U_k is a function of the angular momenta of all transitions preceding that observed; P_k are the Legendre polynomials. For the observed transition only the first two terms are needed.

$$I(\tfrac{\pi}{2}) = 1 + B_2(\beta)\, U_2\, F_2\, P_2(\tfrac{\pi}{2}) \qquad (2)$$

where $\beta = g_N \mu_N H_z / k_B T$. For $\beta \ll 1$ and a Hamiltonian $\mathcal{H} = g_N \mu_N H_z I_z$,

$$B_2 = \frac{\beta^2}{6}\sqrt{\frac{1}{5} I_o(I_o+1)(2I_o-1)(2I_o+3)}\ . \qquad (3)$$

Since all the parameters except the field at the nucleus are known the above formula can be used to calculate H_z. In every measurement the counting rate was normalized to that of CoSi so that electronic drifts of the system could be accounted for. CoSi, being diamagnetic, shows no orientation.

RESULTS AND DISCUSSION

In our measurements we found no magnetic splitting in the Mössbauer spectrum of $Co_{1-x}Fe_xSi$ (x = 0.07, 0.13) below 100 mK even when a small magnetic field was applied. The spectra show only a small quadrupole splitting that is essentially temperature independent.

Nuclear orientation was used to detect the presence of a localized magnetic moment at the Co site. We observed an internal magnetic field of about 400 kG at the Co nucleus in the paramagnetic region, indicating the presence of a localized magnetic moment at the Co site. Table II summarizes all the values obtained from our measurements.

Table II Nuclear Orientation and Mössbauer Internal Magnetic Fields at the Fe and Co Sites (T = 37 mK)

	Fe	Co	Co[1 Fe-nn]	Co[2 Fe-nn]
7% Fe	0	440 ± 70kG	440 ± 70kG	380 ± 60kG
13% Fe	0			
50% Fe	~20kG	0 ± 50kG	-	-
70% Fe	~20kG	0 ± 50kG	-	-

The experimental results in the paramagnetic region were analyzed by assuming (a) one effective magnetic field for all sites with Fe nearest neighbors, and (b) different fields for sites with one and with two Fe nearest neighbors. The differences between these analyses are not statistically significant.

Measurements were extended to include the ferromagnetic region. The internal field measured at the iron site using the Mössbauer effect agrees with the value found by Wertheim, et al.[2] This small field observed for the concentrations $x = 0.5$ and 0.7 at the Fe site and the small field measured by Kawarazaki, et al.[1] at the Co site indicate that the ferromagnetism is probably produced by itinerant electrons or that there is an approximate cancellation of the core polarization, orbital, dipolar, and transferred hyperfine fields.

The magnetic behavior in the paramagnetic region ($x < 0.2$) can be understood by the presence of a localized magnetic moment at the Co site when there are one or more Fe nearest neighbors as has been suggested by Kawarazaki, et al.[1]

ACKNOWLEDGMENTS

We wish to acknowledge the helpful discussions with V. Jaccarino, the assistance in the early phases of this experiment of H. Micklitz, and the very helpful crystallographic measurements of W. Wise.

REFERENCES

1. S. Kawarazaki, H. Yasuoka and Y. Nakamura, Magnetism and Magnetic Materials, 1972 (18th Annual Conference - Denver) p.p. 1632, edited by C. D. Graham, Jr. and J. J. Rhyne, American Institute of Physics, New York (1973).
2. G. K. Wertheim, J. H. Wernick and D. N. B. Buchanan, J. Appl. Phys. 37, 333 (1966).
3. H. J. Williams, J. H. Wernick, R. C. Sherwood and G. K. Wertheim, J. Appl. Phys. 37, 1256 (1966).
4. S. Asanabe, D. Shinoda and Y. Sasaki, Phys. Rev. 134, A774 (1964).
5. R. J. Blin-Stoyle and M. A. Grace, Handbuch der Physik, Band XLII, p.p. 555, edited by S. Flügge, Springer Verlag, Berlin, Göttingen, Heidelberg (1957).

MAGNETIC PROPERTIES OF UC-UN SOLID SOLUTIONS*

G. H. Lander, D. J. Lam, J. F. Reddy, and M. H. Mueller
Argonne National Laboratory, Argonne, Illinois 60439

ABSTRACT

Both UC and UN are semi-metallic materials with the NaCl crystal structure and lattice parameters 4.96 and 4.89Å, respectively. UC is nonmagnetic with a susceptibility that is temperature independent, whereas UN becomes antiferromagnetic at 53°K with ordering of the first kind. Solid solutions of the form $UC_{1-x}N_x$ exist for $0 \leq x \leq 1$. Neutron-diffraction experiments have been performed on five samples with $0.85 \leq x \leq 0.95$. Over this region of x both T_N and the ordered magnetic moment per U atom decrease linearly from T_N = 53°K and 0.75 μ_B for UN (x=1) to T_N = 30°K and 0.35 μ_B for x = 0.88. The composition x = 0.86 shows no ordering in the neutron experiments. The absence of magnetic order for x < 0.88 has been confirmed by magnetization measurements. In the range of x examined the effective paramagnetic moment is 3.25 ± 0.10 μ_B/U atom, close to the value of ~ 3.1 μ_B for UN.

INTRODUCTION

A number of solid solutions of uranium compounds with the NaCl crystal structure have been examined during the last few years, for example, UAs-US.[1] In these systems, both end members are magnetic, the pnictides (N,P,As) being antiferromagnetic and the chalcogenides (S,Se) being ferromagnetic. Neutron-diffraction studies of intermediate compositions in these systems have shown that long-range magnetic structures exist. Such magnetic structures suggest that the exchange interactions are long range, but the precise nature of the exchange interaction remains obscure. In principle, studies of systems in which the magnetic interactions are diluted (rather than changed from antiferro- to ferromagnetic) should be easier to interpret. Magnetization studies on UP-ThP[2] and US-ThS[3] compositions have shown that the critical concentration of thorium (i.e. that composition at which no magnetic ordering is observed is ~ 40%).

A potentially similar system is that of UC-UN. Uranium carbide is nonmagnetic with a temperature-independent susceptibility. This temperature-independent susceptibility has been explained in terms of both a band model,[4] and a localized $5f^2$ configuration in which the crystal-field ground state is the Γ_1 singlet.[5] Uranium nitride is antiferromagnetic at 53°K with ordering of the first kind in which ferromagnetic (001) sheets of uranium moments are stacked in a + − sequence along the [001] axis. The magnetic moments within any sheet

---

*Performed under the auspices of the U.S. Atomic Energy Commission.

are parallel to the [001] axis.[6] The effective paramagnetic moment, derived from the slope of the $1/\chi$ versus T curve at high temperature is ~ 3.2 μ_B/U atom. This is close to both the free ion $5\underline{f}^2$ and $5\underline{f}^3$ values of 3.58 μ_B and 3.62 μ_B, respectively. Surprisingly, therefore, the ordered moment at 5°K is only 0.75 μ_B/U atom. This low value of the ordered moment in UN poses the major problem in trying to understand the electronic structure of this material. The magnetic behavior of UN has been discussed (a) in terms of a $5\underline{f}^3$ configuration[5] with J-mixing that results in a lowering of the magnetic moment from the free-ion value, (b) in terms of a $5\underline{f}^2$ configuration with a singlet-ground state (i.e., as in UC) but with a large exchange field that induces a magnetic moment at low temperature,[7] and (c) in terms of an itinerant-band model[8] which assumes a hybridization of the $\underline{d}$ and $\underline{f}$ electrons.

EXPERIMENTAL

The uranium carbide was obtained from high-purity fused rods. The uranium nitride was made by a gas reaction of uranium powder and NH_3. An initial batch of stoichiometric $UC_{0.5}N_{0.5}$ was prepared by mixing the correct amounts of the constituent compounds, pressing a pellet and firing at 1600°C for 4 hours. It was necessary to crush the pellet and repeat, firing three times in all, to obtain a single phase material. Other compositions were produced from the $UC_{0.5}N_{0.5}$ by adding the necessary amount of UN, pressing a pellet, and repeating the firing procedure. The lattice parameters as determined by X-ray diffraction are given in Table I and are in good agreement with Anselin et al.[9] No evidence for additional phases was observed in either the X-ray or neutron experiments.

Table I Properties of UC-UN Solid Solutions

Composition	a_o (Å)	T_N (°K)	μ_B/U atom
UN	4.892	53 ± 2	0.75 ± 0.05
$UC_{.05} N_{.95}$	4.896	40	0.65
$UC_{.075} N_{.925}$	4.898	36	0.54
$UC_{.10} N_{.90}$	4.900	32	0.45
$UC_{.12} N_{.88}$	4.901	30	0.35
$UC_{.14} N_{.86}$	4.904		
$UC_{.50} N_{.50}$	4.932		
UC	4.960		

Neutron-diffraction experiments on the samples listed in Table I were performed at the CP-5 Research Reactor. The variation of the ordered magnetic moment with composition is shown in Fig. 1. The magnetic structure in all ordered samples is the type-I arrangement (i.e., as in UN). The Néel temperatures determined by neutron diffraction are given in Table I. The neutron experiments are insensitive to ordered moments less than 0.25 μ_B (indicated by the dashed line in Fig. 1). We have therefore, performed magnetization experiments on the samples $x = 0.88$ and $x = 0.86$ to test whether the critical composition x_c is less than the simple extrapolation of the straight line, which gives $x_c \sim 0.80$, in Fig. 1. The variation of $1/\chi$ versus T for the two samples is given in Fig. 2. From the high-temperature portion of the curve we obtain an effective paramagnetic moment of 3.2 ± 0.1 μ_B/U atom for both samples. The value for UN is 3.2 μ_B/U atom. The paramagnetic Curie temperature (i.e., the intercept when $1/\chi = 0$) is $\theta = -420°K$ for both samples. This is different from the value of $\theta = -320°K$ quoted for UN. However, the determination of this constant is very sensitive to the extrapolation; to obtain an accurate value, the susceptibility at temperatures greater than 300°K

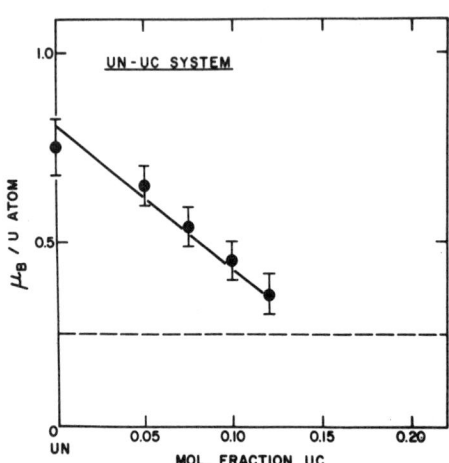

Fig. 1. Variation of the ordered magnetic moment of 5°K as a function of composition

must be measured. The low-temperature behavior of the susceptibility is quite different for the two samples and is illustrated in the insert. The neutron experiments indicate that the Néel temperature of the $x = 0.88$ sample is $30 \pm 2°K$. At this temperature, we observe a minimum in $1/\chi$ indicating a paramagnetic to antiferromagnetic transition. For $x = 0.86$, on the other hand, no minimum is observed in $1/\chi$ and we conclude that this composition does not order magnetically. No evidence for short or long-range magnetic order was observed in the neutron experiments with the $x = 0.86$ sample.

DISCUSSION

de Novion and Costa[10] have measured the magnetic susceptibility, specific heat, and electrical resistivity of eleven samples in the UC-UN system and obtained a critical concentration $x_c \sim 0.90$. Similarly, Ohmichi and coworkers have reported magnetic susceptibility[11] electrical resistivity, and thermoelectric power[12] of six samples and obtained a critical concentration $x_c = 0.40$. Although agreement between the authors exists on measurements of other properties, the

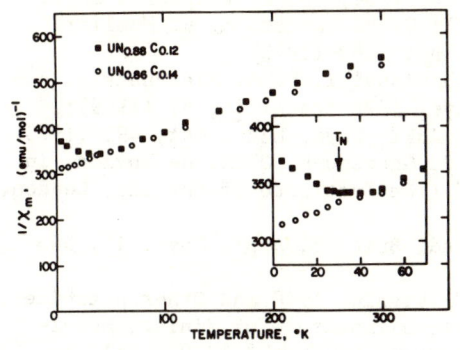

Fig. 2. Inverse molar susceptibility as a function of temperature for the two samples near the critical concentration.

major difference in x_c, which is the most important magnetic property of the system, remains unexplained. This difference has in fact been the subject of a series of controversial exchanges in the literature.[13]

Our measurements give a value of $x_c = 0.86 \pm 0.02$ and are therefore in agreement with de Novion and Costa.[10] The rapid quenching of the magnetic moment in UN with carbon substitution is particularly surprising considering that x_c for the UP-ThP and US-ThS systems is ~ 0.40. In these latter systems the uranium atoms are replaced by nonmagnetic thorium atoms, whereas in the UC-UN system the uranium atoms are left unaltered. For low values of x in the $Th_{1-x}U_xP$ and $Th_{1-x}U_xS$ systems the uranium atoms behave as magnetic impurities in a nonmagnetic matrix and exhibit an effective paramagnetic moment per uranium atom in dilute alloys, e.g., 2% UP in ThP. Magnetic ordering occurs when the concentration of uranium is sufficient to sustain the exchange interactions. However, the situation in $UC_{1-x}N_x$ is quite different, and cannot be treated as a magnetic impurity problem. Not only does the ordered moment vanish with a small carbon concentration ($x_c=0.86$) but no effective paramagnetic moment is observed for x <u>less than</u> 0.40.[10] On the other hand, if the $5f^2$ configuration is correct for both UN and UC then the effective moment (as measured in the range 100-300°K) should disappear for values of x just less than x_c (i.e., <u>greater than</u> x ~ 0.40). This is because the exchange in the molecular-field theory is very small for $x < x_c$ (since no ordering is observed) and a temperature-independent susceptibility similar to that in UC should be observed at low temperature. Alternatively, if the ground state of UN is $5f^3$ and UC is $5f^2$, the magnetic properties of the system depend strongly on the nearest-neighbor environment around the uranium atom; this might explain the rapid quenching of the ordering as the carbon concentration is increased.

A more quantitative interpretation of the results on the UC-UN system must, in our opinion, await a more detailed understanding of the electronic properties of both UC and UN.

REFERENCES

1. J. Leciejewicz, A. Murasik, R. Troc, and T. Palewski, Phys. Status Solidi 46, 391 (1971); G. H. Lander, M. H. Mueller, and J. F. Reddy, Phys. Rev. B6, 1880 (1972).
2. V. I. Chechernikov, T. M. Shavishuili, V. A. Pletyushkin, and V. K. Slovyanskikh, Sov. Phys. JETP trans. 28, 81 (1969); H. Adachi, S. Imoto, and T. Kuki, Phys. Lett. 44A, 491 (1973).
3. J. Danan, J. P. Marcon, J. P. Gatesoupe, C. H. de Novion, in "Rare Earths and Actinides" (The Institute of Physics, London, (1971) p. 176.
4. L. F. Bates and P. B. Unstead, Brit. J. Appl. Phys. 15, 543 (1964).
5. S. K. Chan and D. J. Lam, "Plutonium 1970 and Other Actinides," edited by W. N. Miner (American Institute of Mining, Metallurgical and Petroleum Engineers, New York, 1970), Vol. 1, p. 219.
6. N. A. Curry, Proc. Phys. Soc. (London), 86, 1193 (1965).
7. J. Grunzweig-Genossar, M. Kuznietz, and F. Friedman, Phys. Rev. 173, 562 (1968).
8. H. L. Davis in Ref. 5, p. 209.
9. F. Anselin, G. Dean, R. Lorenzelli, and R. Pascard; "Proc. Symp. Carbides in Nuclear Energy," edited by L. E. Russell (MacMillan and Company, London, 1964) Vol. I p. 133.
10. C. H. de Novion and P. Costa, J. de Physique 33, 257 (1972).
11. T. Ohmichi, S. Nasu, and T. Kikuchi, J. Nucl. Sci. and Tech. (Japan) 9, 11 (1972).
12. T. Ohmichi, T. Kikuchi, and S. Nasu, J. Nucl. Sci. and Tech. (Japan) 9, 77 (1972).
13. M. Kuznietz, J. Nucl. Sci. and Tech. (Japan) 8, 51 (1971); H. Adachi and S. Imoto, ibid 8, 53 (1971); T. Ohmichi and S. Nasu, ibid 8, 54 (1971); T. Ohmichi and T. Saito, ibid 8, 314 (1971).

Section 10. Physics and Chemistry of Transition Metal Compounds   447

DEMAGNETIZATION OF RARE EARTH IONS IN METALS DUE TO VALENCE FLUCTUATIONS*

M. B. Maple and D. Wohlleben
Institute for Pure and Applied Physical Sciences
University of California, San Diego
La Jolla, California 92037

ABSTRACT

Magnetic-nonmagnetic transtions of rare earth ions in metals are reviewed for both dilute and concentrated systems. Whereas <u>magnetic</u> behavior of a rare earth ion is associated with <u>integral</u> occupation of the 4f electron shell, <u>nonmagnetic</u> behavior is linked with <u>nonintegral</u> occupation. Integral occupation (valence) implies a single 4f configuration with infinite lifetime, while nonintegral occupation is a consequence of temporal valence fluctuations between two adjacent configurations. These temporal fluctuations result in nonmagnetic behavior below a characteristic temperature $T_f \sim h/k_B \tau$ because of the finite lifetime τ of the Zeeman levels.

INTRODUCTION

It is the purpose of this review to point out that, the traditional view not withstanding, rare earth (hereafter RE) ions provide a most interesting field of research on the problem of local moment formation in metals. Recent experiments show clearly that 4f electrons, which normally appear to be localized, definitely can delocalize, i.e., they can mix strongly with the conduction electrons, whenever two valence states of the local 4f shell are nearly degenerate. When this mixing occurs, the time averaged occupation of the 4f shell becomes nonintegral, and simultaneously the ion loses its magnetic moment at low temperatures. This finding, interesting on its own right, casts new light on local moment formation on transition metal impurities as well.

Experimentally, magnetic-nonmagnetic transitions are easier to induce and less complicated to interpret in RE metals than in transition metals. First, because of the narrower energy levels of concern, it is sometimes possible to follow magnetic-nonmagnetic transitions over the entire range of interest by application of pressure or variation of concentration in a binary alloy matrix. Secondly, in the magnetic limit the Curie-Weiss moment unequivocally identifies a Hund's rule integral valence ground state. In the case of transition metal impurities, this fact is obscured by partial quenching of the orbital moment and by a rather

---

*Supported by Air Force Grant No. AF-AFOSR-71-2073.

large conduction electron polarization. Thirdly, since there is no direct overlap of 4f shells on neighboring cells, experiments can be performed on concentrated RE systems without the complications inherent in band formation of the shells of interest.

In the next two sections, we present experimental results concerning magnetic-nonmagnetic transitions and valence changes of RE ions in metals for both dilute and concentrated cases. The last section attempts to interpret the results qualitatively within the context of an ionic model due to Hirst.[1]

MAGNETIC-NONMAGNETIC TRANSITIONS OF DILUTE RARE EARTH IMPURITIES IN METALLIC MATRICES

Magnetic-nonmagnetic transitions of certain RE impurities, notably Ce and Yb, dissolved in various metallic matrices have received considerable attention in recent years. Induced by means of the application of pressure, or the variation of the composition of a binary alloy matrix, the demagnetization of RE solutes is readily discernible through profound changes in the normal and superconducting state properties of the matrix-impurity system. Since the first and in many respects the most interesting manifestations of RE solute demagnetization have been observed in the superconducting state, we emphasize this aspect of the problem in this section concerning dilute RE systems. Some of these experiments led to the work on concentrated RE systems which is taken up in the following section.

Pressure-Induced Magnetic-Nonmagnetic Transitions

The first observations of a pressure-induced magnetic-nonmagnetic transition of a RE impurity in a metallic matrix were reported for the superconducting matrix-impurity systems $(La, Ce)_3In$[2] and $LaCe$.[2] The transition was inferred from the variation of the superconducting transition temperature T_c with pressure. The most extensive investigation was made on the $LaCe$ system which we briefly consider here. More detailed reviews appear elsewhere.[4]

In an fcc La host at zero pressure, Ce impurities are trivalent and carry well-defined local moments as evidenced by measurements of the magnetic susceptibility as a function of temperature. The temperature dependence of the magnetic susceptibility further suggests that a cubic crystal field splits the Ce^{3+} $J = 5/2$ multiplet into an excited state quartet and a ground state doublet with a splitting $\sim 100°K$.[4] Thus, at superconducting temperatures conduction electrons apparently exchange scatter from a doublet ground state with an effective spin $\underset{\sim}{S}$ of $1/2$. The exchange scattering proceeds via the exchange Hamiltonian

$$\mathcal{H}_{int} = -2 \mathcal{J} \underset{\sim}{S} \cdot \underset{\sim}{s} \tag{1}$$

where $\mathcal{J}$ characterizes the strength and sign of the interaction and $\underset{\sim}{s}$ is the conduction electron spin density at the impurity site. Low temperature electrical resistivity measurements on $LaCe$ alloys in the normal state reveal a minimum indicative of the Kondo effect,[5] which implies that the conduction electron-impurity spin exchange coupling is antiferromagnetic ($\mathcal{J} < 0$). The temperature dependence of the impurity contribution to the resistivity $\Delta\rho$,[6] and other properties as well,[4] indicate that the Kondo temperature T_K is low — much smaller than the La host T_c. In the range of Ce impurity concentrations where interaction effects are negligible, the superconducting properties can be rather well accounted for by <u>temperature dependent pair breaking</u> theories such as one by Müller-Hartmann and Zittartz [7] (hereafter MHZ).

The primary superconductivity data documenting the magnetic-nonmagnetic transition of Ce impurities in the <u>La</u>Ce system under pressure are shown in Fig. 1. With increasing Ce concentration, a minimum in T_c as a function of pressure P develops near 15 kbar which becomes more pronounced the higher the Ce concentration. The initial depression of T_c with Ce impurity concentration n, $\Delta T_c/n$, markedly increases with pressure up to a maximum near 15 kbar and thereafter decreases to a value which, above ~ 100 kbar, is more than an order of magnitude smaller than the maximum depression. Isobars of T_c as a function of n in the inset of Fig. 1 show that the shape of the T_c vs. n curves evolves with increasing pressure from nearly linear depressions with slight negative curvature to depressions with strong positive curvature. This latter behavior is very similar to that of the <u>nonmagnetic</u> ThCe system for which T_c exhibits a nearly exponential decrease with n at zero pressure.[8] The nonmagnetic behavior of ThCe at superconducting temperatures is well established by the BCS behavior of the specific heat jump ΔC as a function of T_c[8]: The <u>pair weakening</u> effect of short-lived local moments (nonmagnetic impurities) results in a reduced specific heat jump $\Delta C/\Delta C_0$ vs. reduced transition temperature T_c/T_{c_0} which conforms to the BCS law of corresponding states (i.e., $\Delta C/\Delta C_0 = T_c/T_{c_0}$ where ΔC_0 and T_{c_0} are respectively ΔC and T_c of the matrix), whereas the <u>pair breaking</u> effect of long-lived local moments (magnetic impurities) is manifested as deviations of

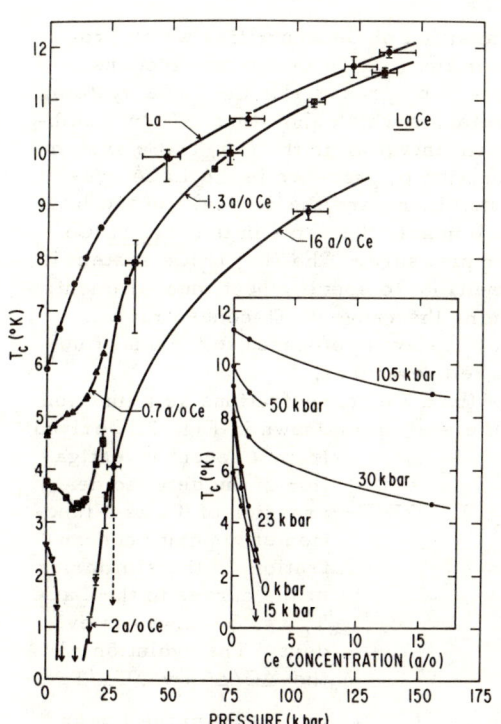

Fig. 1. T_c vs. pressure for the <u>La</u>Ce system. Isobars of T_c vs. Ce concentration are shown in the inset (from Ref. 3).

$\Delta C/\Delta C_0$ vs. T_c/T_{c_0} from the BCS law of corresponding states.[4] Unfortunately, experimental difficulties prevent direct confirmation of the nonmagnetic character of La̲Ce alloys at pressures $\geqslant$ 100 kbar by specific heat measurements.

Finally, we remark that Ce demagnetization under pressure has also been studied by resistivity measurements in La̲Ce[9] and Y̲Ce[10] alloys. In the La̲Ce system, it was found that the slope of the Ce impurity contribution to the resistivity $\Delta\rho$ vs. ln T also exhibits a maximum near the same pressure as the maximum in the depression of T_c.

Magnetic-Nonmagnetic Transitions Induced by Variation of the Composition of a Binary Alloy Matrix

The magnetic-nonmagnetic transition of Ce impurities which proceeds with increasing concentration in the (La, Th)Ce system appears to be the analogue of the transition induced by pressure in the La̲Ce system. This is suggested by recent measurements which show that $\Delta T_c/n$ exhibits a maximum as a function of Th concentration in the (La, Th)Ce system similar to that which occurs as a function of pressure in the La̲Ce system.[11] Moreover, the depth of the maximum and the decrease of the lattice parameter at which it occurs are nearly the same in the (La, Th)Ce system as in the La̲Ce system under pressure. The (La, Th)Ce system is especially interesting since it is amenable to specific heat measurements which provide a means for determining the range of Th concentrations over which the Ce moments remain long-lived compared to thermal fluctuation lifetimes at superconducting temperatures.

The variation of T_c at various fixed Ce concentrations as a function of Th concentration for the (La, Th)Ce system is shown in Fig. 2. Derived from a recent investigation of the detailed depression of T_c as a function of impurity concentration,[12] the similarity of the curves to the La̲Ce T_c vs. P curves is evident. The evolution of the $\Delta C/\Delta C_0$ vs. T_c/T_{c_0} curves from the linear BCS law of corresponding states, a measure of pair breaking, was also found to exhibit a maximum as a function of Th concentration.[13] This is illustrated in a plot of the initial slope

$$\left. \frac{d(\Delta C/\Delta C_0)}{d(T_c/T_{c_0})} \right]_{T_c = T_{c_0}} \text{vs. Th}$$

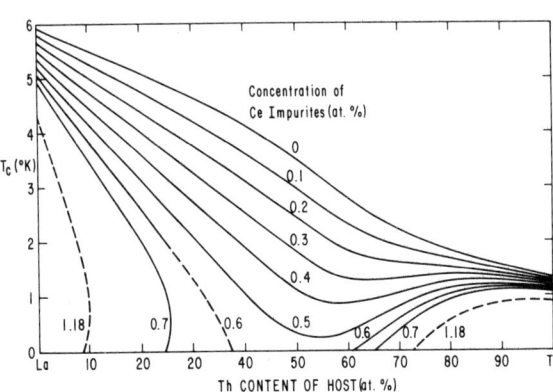

Fig. 2. Superconducting transition temperature vs. Th concentration at various fixed Ce concentrations for (La, Th)Ce (from Ref. 12).

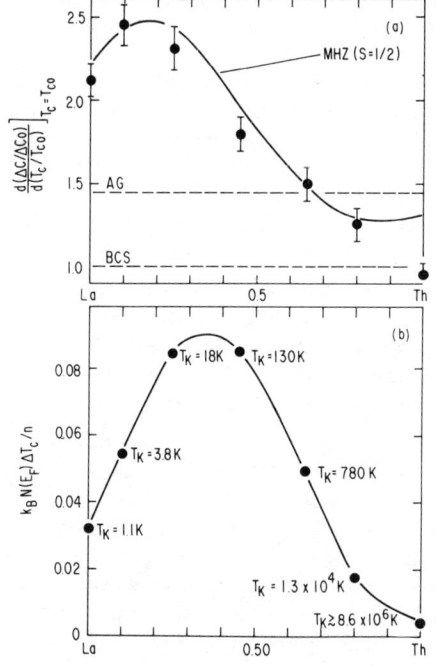

Fig. 3. (a) Initial slope
$\overline{\dfrac{d(\Delta C/\Delta C_o)}{d(T_c/T_{c_o})}}\bigg]_{T_c = T_{c_o}}$ vs. Th composition for the system (La, Th)Ce.

(b) $k_B N_b(E_F) \Delta T_c/n$ vs. Th composition for the system (La, Th)Ce (from Ref. 13).

concentration in Fig. 3(a).

In order to determine the range of Th concentrations over which the Ce moments remain long-lived, a self-consistent analysis which incorporates the temperature dependent pair breaking theory of MHZ [7] was applied to the specific heat data. First, the maximum in the experimental curve of the quantity $k_B N_b(E_F) \Delta T_c/n$ [$N_b(E_F)$ is the bare density of states at the Fermi level] vs. Th concentration shown in Fig. 3(b) was scaled to the maximum in the theoretical MHZ curve of $k_B N_b(E_F) \Delta T_c/n$ vs. T_K/T_{c_o}. Values of T_K indicated in Fig. 3(b) were obtained by matching the experimental points to the MHZ curve and then used to construct the theoretical MHZ curve for $\overline{\dfrac{d(\Delta C/\Delta C_o)}{d(T_c/T_{c_o})}}\bigg]_{T_c = T_{c_o}}$ vs. Th composition in Fig. 3(a) (without scaling). It can be seen that the MHZ theory and the specific heat data are in good agreement to Th concentrations ~70 at.%, but they definitely diverge at higher Th concentrations—the MHZ theoretical curve approaching the value given by the temperature independent pair breaking theory of Abrikosov and Gor'kov (hereafter AG) [14] in the limit $T_K/T_{c_o} \to \infty$, and the experimental data converging to the BCS value realized at 100 at.% Th. Although the maximum in both the theoretical and experimental curves near 15 at.% Th is especially striking, it should be pointed out that the agreement to such high Th concentrations may be somewhat fortuitous since the superconducting properties during the magnetic-nonmagnetic transition must go smoothly from pair breaking to pair weakening behavior as the solute spin lifetime passes from long-lived to short-lived. Clearly, a theory is needed which takes into consideration the variation of the solute spin lifetime and reduces to the MHZ theory in the long-lived (magnetic) limit and to a theory, such as one due to Kaiser, [15] in the short-lived (nonmagnetic) limit. The experiments reviewed here lay the foundation for such a theory and emphasize the necessity of taking into consideration the solute spin lifetime. In the following section which concerns magnetic-nonmagnetic transitions in concentrated RE systems, the lifetime of the RE local moments is again of central importance.

The variation of the solute spin lifetime during the magnetic-nonmagnetic transition can also be seen in the detailed curves of T_c vs. n in the (La, Th)Ce system which evolve with increasing Th concentration from re-entrant curves with negative curvature at low Th concentration to nearly exponential curves with positive curvature at high Th concentration. Characteristic of matrix-impurity systems with long-lived solute spins and low Kondo temperatures $T_K < T_{c_o}$, re-entrant T_c vs. n curves were predicted by MHZ and observed only once before, in the system (La, Ce)Al$_2$.[16] (The re-entrant behavior is also evident in the iso-Ce impurity concentration curves of T_c vs. Th concentration in Fig. 2). Characteristic of matrix-impurity systems with short-lived solute spins and localized spin fluctuation temperatures $T_f \gg T_{c_o}$, nearly exponential T_c vs. n curves were predicted by Kaiser for nonmagnetic solutes and documented in detail for ThCe and a number of other systems as well.

The behavior of normal state properties such as the magnetic susceptibility, electrical resistivity, thermoelectric power, etc., during magnetic-nonmagnetic transitions is also of interest. Planned or in progress on the (La, Th)Ce system, these measurements will be of great value in comparing the features of 4f shell demagnetization in dilute RE systems with those in concentrated RE systems. Some interesting work in this direction has been done on the system (Au, Ag)Yb which has been studied by measurements of electrical resistivity[17] and magnetic susceptibility.[18]

MAGNETIC-NONMAGNETIC TRANSITIONS IN CONCENTRATED RARE EARTH SYSTEMS

The magnetic-nonmagnetic transitions in the LaCe system as a function of pressure and in the (La, Th)Ce system as a function of Th concentration are consistent with a continuous change of valence of the Ce ion from three in the magnetic limit towards four in the nonmagnetic limit. Thus the magnetic moment indicates a valence of three in LaCe while the lattice constant vs. concentration in ThCe indicates a valence of 3.25 on the Th rich end. Unfortunately direct experimental verification of the valence change is generally difficult to obtain in dilute alloys.

On the other hand, in certain concentrated RE systems, abrupt or continuous valence changes as a function of pressure or concentration are well established. They manifest themselves as large changes of the lattice constant without change of crystal structure and can occur in systems with Ce, Sm, Eu, Tm and Yb constituents; i.e., ions which can have valence states other than three in metals. By comparison of lattice constants in a series of compounds it is possible to determine the valence within about 10% absolute, and better relative, accuracy. In some cases, the valence has also been found from the Mössbauer isomer shift and from the position of the soft x-ray absorption edge.

Valence transitions in concentrated RE systems do not proceed directly from one to the next integral valence phase. Instead, invariably an intermediate, nonintegral valence phase is found to be stable over a considerable portion of the P-T or concentration-T phase diagram. The integral and nonintegral valence phases are separated from each other by first or second order phase boundaries.

While such an ambivalent RE metal shows normal <u>magnetic</u> behavior (Hund's rule moments and magnetic ordering) in the <u>integral</u> valence phases, it is invariably <u>nonmagnetic</u> (intermediate high temperature Curie constant, temperature independent low temperature susceptibility, absence of magnetic order) in the <u>nonintegral</u> valence phase. For a recent review of this topic, see reference 19.

As examples we consider the pressure-induced valence transitions which occur in Ce metal and in the compound SmS.

Ce Metal

In the γ phase at room temperature and zero pressure Ce metal exhibits nearly integral valence of 3.08 and a well-defined local moment according to magnetic susceptibility measurements. In the β phase which can be stabilized at low temperature and also possesses nearly integral valence and a well-defined magnetic moment, it orders antiferromagnetically at 12.5°K.[20] When pressure is applied to γ-Ce (fcc) at room temperature, one observes a first order transition near 7.6 kbar to the α phase (fcc) with a volume decrease of ~12%.[21] At the transformation pressure, lattice parameter measurements indicate that α-Ce has a valence of 3.7[22]; 0.7 electrons per cell are promoted into the conduction band while 0.3 electrons remain, in time average, in the 4f shell. Thus α-Ce is a nonintegral valence phase. Figure 4 shows the magnetic susceptibility of Ce metal at room temperature as a function of pressure.[23] After a slow decrease in the γ-phase, the susceptibility drops abruptly at the hysteretic γ-α phase boundary, and decreases slowly in the α phase. The inset of Fig. 4 shows that the susceptibility of the α phase drops linearly with pressure between 8 and 18 kbar and extrapolates at 50 kbar, the α-α' phase boundary at room temperature, to a value close to that of the quadrivalent metals Th and Hf. Quadrivalence of α'-Ce is also suggested by the lattice constant[21] and by the existence of superconductivity in this phase.[24] The α phase

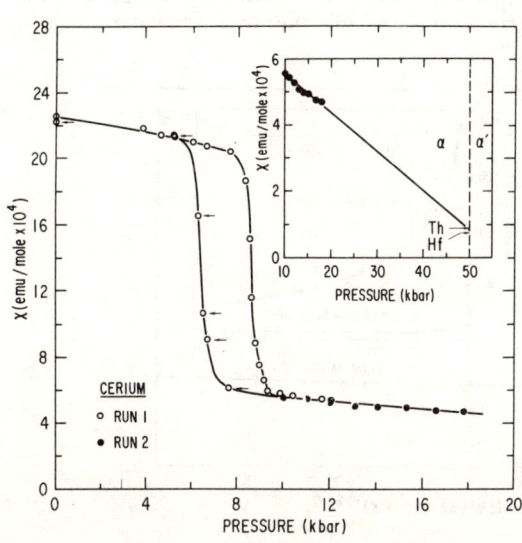

Fig. 4. Pressure dependence of the magnetic susceptibility of Ce metal at room temperature (from Ref. 23).

is nonmagnetic. This is indicated in Fig. 5 by the relatively weak temperature dependence of the susceptibility and by the absence of any indication of magnetic order[23] which contrasts sharply with the Curie-Weiss behavior and the antiferromagnetic transition at 12.5°K in the magnetic β phase. It was recently demonstrated[25] that the low temperature rise of the susceptibility of α-Ce is due to inclusions of the β phase. Near T = 0, the intrinsic susceptibility of α-Ce is given by a straight line extrapolation of the behavior between 100 and 300°K.

The absolute value of the susceptibility of α-Ce at 10 kbar is about four times larger than in normal quadrivalent metals like Th and Hf; the excess is obviously due to the remaining 0.3 4f electrons which, however, no longer carry a moment: If one localized 4f electron existed on every third ion all of the time, χ in α-Ce should show a Curie-Weiss law (with one third of the Curie constant of β-Ce) and magnetic order, like a concentrated magnetic alloy.

The electronic specific heat coefficient γ of α-Ce is much larger than in four valent metals (~ 10 mJ/mole°K^2), again suggesting that there are contributions from 4f electrons.[26]

SmS

More interesting examples of demagnetization of RE ions in metallic environments are those for which the occupation of the 4f shell is greater than unity, since the absence of magnetism in the intermediate valence phase of such systems cannot be trivially due to the absence of electrons or holes in the 4f shell, as a skeptic might like to believe in the case of Ce or Yb. Such a case is provided by the compound SmS in its high pressure collapsed metallic phase.

At zero pressure, SmS, SmSe and SmTe are semiconducting and the Sm ions are divalent.[27] Conductivity occurs via thermal activation of

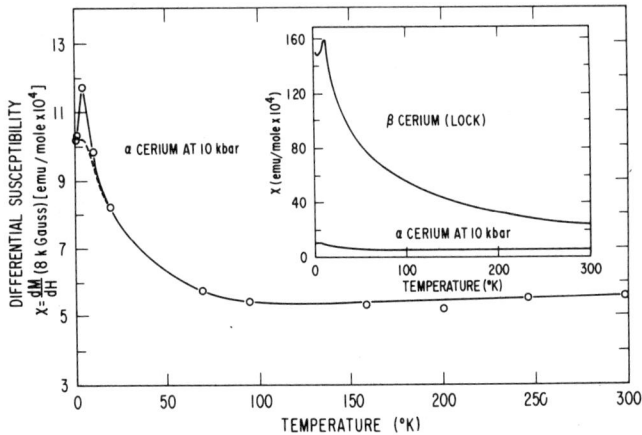

Fig. 5. Differential susceptibility of α-Ce at 8 kG and 10 kbar vs. temperature. The inset emphasizes the weak temperature dependence of the nonmagnetic α phase by a comparison with the β phase (from Ref. 23).

electrons from the Sm 4f shells into the conduction band across relatively small energy gaps of 0.20, 0.46, and 0.63 eV, respectively. The Sm 4f shells contain 6 electrons and the compounds exhibit ionic Van Vleck paramagnetism with a nonmagnetic ground state.[27]

Jayaraman et al.[28] discovered that these materials undergo semiconductor-metal transitions under pressure. The transition is discontinuous in SmS and continuous in SmSe and SmTe, and is in each case accompanied by large volume contractions without change of structure as in the γ-α transition of Ce. In the case of SmS the lattice constant[28] indicates a valence of 2.7 in the metallic state just after the transition. The transition in SmS was of particular interest since it occurred at a rather modest pressure of 6.5 kbar and was thereby accessible to accurate magnetic susceptibility studies under pressure.

The magnetic susceptibility is shown in Fig. 6[29] as a function of pressure at room temperature. At first it increases slowly, then drops abruptly at the semiconductor-metal transition at 6.5 kbar, and thereafter decreases slowly. It shows the same hysteresis as the resistivity measurements of Jayaraman et al.[30] The susceptibility as a function of temperature in the collapsed high pressure metallic phase is shown in Fig. 7(a).[29] It exhibits weak, but definite temperature dependence below $\sim 200°K$ and saturates to a constant value below $\sim 40°K$ with no sign of magnetic order. The constant low temperature susceptibility is in sharp contrast to what would be expected if the transition had proceeded directly to the trivalent $4f^5$ configuration. In the latter configuration the ground state must be at least a doublet, which results in a low temperature divergence or magnetic order, as shown in Fig. 7b for two normal $4f^5$ compounds ($SmPd_3$ and Sm_2In_3). Note that at high temperature, the susceptibility of SmS is intermediate between normal $4f^6$ and $4f^5$ behavior. The heat capacity of SmS in the high pressure collapsed phase has recently been measured and the electronic specific heat coefficient was found to be extremely large (145 mJ/mole°K^2).[30]

SmB_6

The compound SmB_6 at atmospheric pressure exhibits the same magnetic anomaly as SmS in the collapsed high pressure phase[31] (Fig. 7b). The valence was found to be 2.7 from lattice constant, Mössbauer isomer shift[32] and soft x-ray absorption[33] measurements. The valence is completely temperature independent.[32] The Mössbauer and soft x-ray data show a very interesting apparent contradiction: while the Mössbauer data exhibit a single line centered at valence 2.7, the x-ray spectrum has two lines,

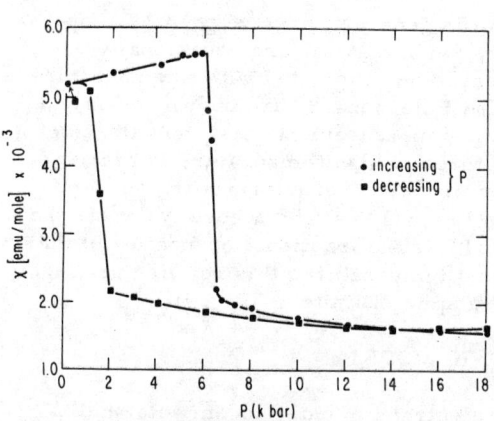

Fig. 6. Magnetic susceptibility of SmS vs. pressure at room temperature (from Ref. 29).

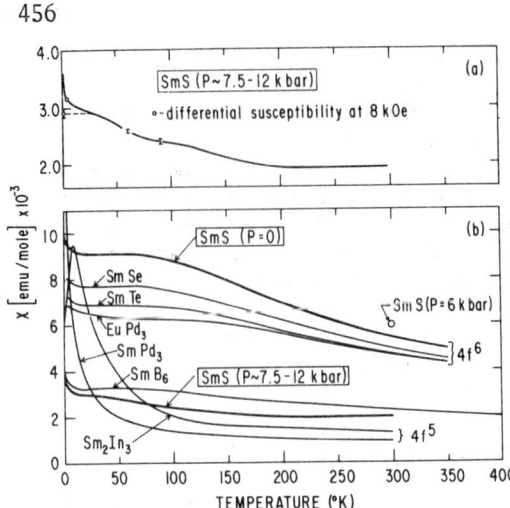

Fig. 7 (a). Temperature dependence of the susceptibility of SmS in the collapsed high pressure phase.
(b) Comparison of $\chi(T)$ for ions in $4f^6$ and $4f^5$ configuration with "collapsed" SmS and SmB_6 (from Ref. 29).

one at valence 2 with weight 0.3 and one at valence 3 with weight 0.7. This point will be discussed below.

Nonmagnetic RE Compounds at Atmospheric Pressure

The observation of a link between nonmagnetic behavior and nonintegral 4f shell occupation in Ce metal, SmS, and TmTe[34] in their collapsed high pressure metallic phases, and in SmB_6 at zero pressure led us to study the temperature dependence of the susceptibility of metallic RE compounds which according to lattice constants in the literature seemed to possess nonintegral valence at atmospheric pressure. Without exception they were found to be nonmagnetic in the sense that there was no sign of magnetic order nor a low temperature divergence (except for small low temperature impurity "tails" which differ in magnitude from one sample to the next and can be suppressed partially or entirely by careful preparation). Some examples of Ce and Yb compounds are shown in Figs. 8 and 9.[35] Other examples have been studied elsewhere (see reference 19).

While nonmagnetic Ce compounds generally have very little temperature dependence of the susceptibility, wide maxima are occasionally observed in Yb compounds. They seem to be associated with a temperature dependent valence. Thus Iandelli and Palenzona[36] have shown recently from lattice constant and susceptibility measurements that the valence of $YbAl_3$ goes continuously from near three at high temperature to near two at 70°K. In their temperature interval they could interpret the susceptibility maximum quantitatively by $\chi(T) = \varepsilon(T) \cdot \chi(f^{13})$ where $\chi(f^{13})$ is the susceptibility of trivalent Yb and $\varepsilon(T)$ is the fractional occupation of the trivalent configuration. At the lowest temperatures this relation no longer holds since there is no divergence in spite of finite ε (Fig. 9).

THE IONIC MODEL

When atoms form metals the electrons in outer incomplete shells delocalize to varying degrees because of nearest neighbor overlap and form bands, while electrons in inner noble gas shells remain localized

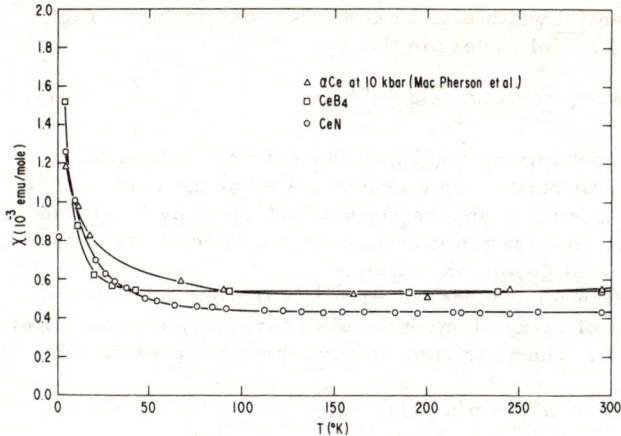

Fig. 8. Susceptibility vs. temperature for two nonmagnetic Ce compounds with nonintegral valence (from Ref. 35).

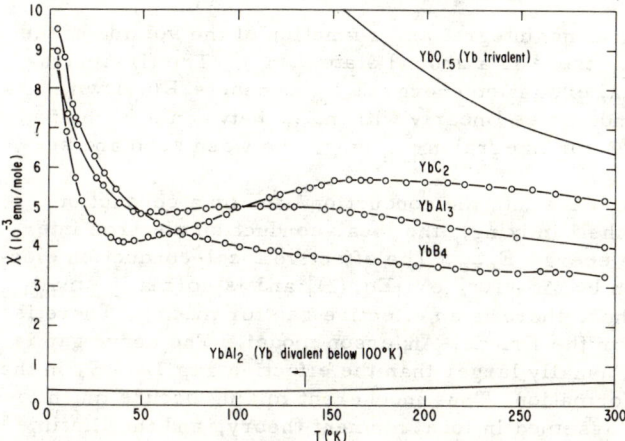

Fig. 9. Susceptibility vs. temperature for several nonmagnetic Yb compounds with nonintegral valence (from Ref. 35).

because there is no significant overlap and because strong inner shell Hund's rule correlations block band formation completely. Complete localization implies integral occupation and infinite intrinsic lifetime of the ground state of each individual noble gas shell, i.e., very sharp local many electron levels.

The Hund's rule magnetic moments of most rare earth metals indicate integral occupation (valence) of the 4f shell. There seems to be no 4f band formation even though these shells are only partially filled. Clearly the very small direct overlap between nearest neighbors is insufficient to break up the Hund's rule correlations. Under normal conditions, the much larger overlap between 4f and conduction electrons does not seem to be able to do this either. Although it does produce a small charge oscillation and spin polarization in the conduction electron gas and indirect RKKY interaction between neighboring shells, it generally fails to shorten the lifetime of the local state of the shell: The depression of the superconducting transition temperature due to most RE impurities follows the AG theory which assumes infinite lifetime of the local spin; the susceptibility of dilute RE impurities follows a Curie law down to the lowest measuring temperatures ($T < 0.1\,°K$), which implies an intrinsic width of the Hund's rule Zeeman levels of less than 10^{-5} eV; and EPR linewidths of dilute RE impurities can be as narrow as a few tens of gauss, indicative of shell lifetimes of more than 10^{-8} sec. This is surprising. Real mixing of local and conduction electrons should shorten the lifetime to nearly 10^{-13} sec since the <u>potential</u> width of a local 4f state due to incoherent mixing is, in the usual Friedel Anderson theory:

$$\Delta = \langle V_{k\ell}^2 \rangle \, \rho(E_\ell) \approx 10^{-2} \text{ eV} \qquad (2)$$

($V_{k\ell}$ is the matrix element mixing local and conduction electron states and $\rho(E_\ell)$ is the density of conduction electron states at the energy of the local level E_ℓ). Thus, mixing is apparently blocked again by local Hund's rule correlations, and it appears justified to assign a special stability to the <u>integral</u> valence state of 4f ions in metals.

The basic idea of the ionic model of Hirst[1] is that each local shell has a discrete spectrum of many electron levels characterized by a different integral occupation n. The spectrum, in zero-order approximation, is given by

$$E(n) = n V_0 + n(n-1) F_0/2 \qquad (?)$$

where V_0 is the nuclear potential and F_0 the spectroscopic term from the Coulomb repulsion of the local electrons. This can be rewritten as

$$E(n) = (n - n_{min})^2 F_0/2 + \text{const} \qquad (4)$$

where n_{min} is in general nonintegral and a function of the volume of the cell. For a given n_{min} there is a ground state $E(n_0)$. The first intra-ionic interconfigurational excitation energy E_{exc} connects $E(n_0)$ with $E(n_0+1)$ or $E(n_0-1)$ and varies linearly with n_{min} between 0 for half integral n_{min} and $F_0/2$ for integral n_{min} (e.g., between zero and several eV for RE ions).

If a local electron leaks into the conduction band or a conduction electron hops onto the shell (mixing), the local-conduction electron interaction has to supply the energy E_{exc}. The effective local-conduction electron interaction is given by $\Delta \sim 10^{-2}$ eV [Eq. (2)] and is normally <u>small</u> compared to E_{exc}. Thus, there is an effective gap for mixing. There is no such gap for mixing in the Friedel-Anderson model. The above gap is also <u>different</u> from and usually larger than the effective gap $E_F - E_\ell$ in the Schrieffer-Wolff transformation. Thus incoherent mixing occurs much less frequently than usually assumed in local moment theory, and the lifetime of the shell is enhanced with respect to h/Δ.

Let us apply the Hirst model to a dilute RE impurity. We assume first that n_{min} is near a half integer; i.e., $E_{exc} \lesssim \Delta$. In this case local and conduction electrons can mix freely since $E(n_0)$ and $E(n_0 \pm 1)$ lie within Δ. The lifetime τ of the shell in any of the two accessible configurations is then of order of the maximum possible value h/Δ and according to the uncertainty principle, the Zeeman levels of both configurations are broadened to the maximum possible width Δ if emission and readsorption of conduction electrons by the local shell is <u>incoherent</u>, i.e., occurs at random phase in time. The Zeeman precession in the external magnetic field will be interrupted for each configuration as if there were thermal motion at a temperature $T_f \lesssim \Delta/k_B$, the spin fluctuation temperature. This gives a clear physical reason for the spin fluctuation theories postulated several years ago. The static susceptibility contributed by each shell as function of temperature will be approximately

$$\chi(T) \approx \frac{\epsilon(n_o)[\mu(n_o)]^2 + [1 - \epsilon(n_o)][\mu(n_o \pm 1)]^2}{3k_B(T + T_f)} \tag{5}$$

where $\epsilon(n_o)$ is the fraction of time the shell spends in the configuration with occupation n_o, and $\mu(n_o)$ and $\mu(n_o \pm 1)$ are the Hund's rule effective moments in the two adjacent configurations with occupation n_o and $n_o \pm 1$. At high temperature ($T \gg T_f$) the susceptibility exhibits a Curie-Weiss law with a Curie constant intermediate between that of the neighboring (resonating) configurations, while the normal low temperature divergence is cut off by the fluctuation at T_f. The inflection point of $\chi(T)$ is thus a rough measure of the fluctuation temperature T_f which, in the case of high T_f, can also be extracted from the low temperature susceptibility if ϵ can be measured independently — for instance, by the Mössbauer isomer shift.

The quantities ϵ and T_f are not independent of each other. While one expects $\epsilon \sim 1/2$ and a fluctuation frequency $w_f \lesssim \Delta/h$ when $E_{exc} \ll \Delta$, one anticipates that the lower energy configuration is relatively stable, i.e., $\epsilon \to 1$, when $E_{exc} \gg \Delta$. In the latter case, transitions to the upper configuration occur very rarely, i.e., $w_f \ll \Delta/h$. If $1 - \epsilon < 10^{-2}$, it may be difficult to distinguish between a stable and an unstable shell from the Curie constant alone. It is then necessary to lower the temperature through T_f to establish the absence of a divergence. The ground state of the shell is most stable when n_{min} is an integer, i.e., when E_{exc} has its maximum value. Because of the very strong dependence of the lifetime $\tau = 2\pi/w_f$ on Δ/E_{exc}, the criterion for the transition from magnetism to nonmagnetism becomes $E_{exc} \gtrless \Delta$ in the Hirst model.

Note that the temperature dependence of χ in Eq. (5) is qualitatively very similar to that of a magnetic impurity exhibiting the Kondo effect, especially if $\epsilon \gg 1 - \epsilon$. The Kondo effect may, in general, be associated with an incipient valence instability. This seems certain in the case for LaCe.

While the excitation energy E_{exc} depends on the volume through n_{min} (linearly in first approximation), Δ does not. Thus application of pressure changes E_{exc} continuously and may cause a crossover of the energy levels associated with the two configurations. If $E_{exc} > \Delta$ decreases, ϵ and the lifetime τ of the moment decrease drastically. This provides a mechanism for the passage of the Ce impurity moment lifetime from long-lived to short-lived with increasing pressure in the LaCe system or increasing Th concentration in the (La, Th)Ce system discussed before. The higher energy $4f^0$ level may be driven towards the ground state $4f^1$ level by the volume decrease which results upon application of pressure or addition of Th to the La matrix. For example, in the (La, Th)Ce system, ϵ apparently decreases from unity at La to 0.75 at Th where T_f is $\geq 10^2 °K$.

Although the Hirst model was designed for dilute impurities, the same basic ideas apply in concentrated RE systems, because the 4f shells do not overlap to form bands but merely interact with the conduction electrons via $V_{k\ell}$ and with the crystal field. There are, however, two important new aspects. First the RKKY interaction between neighboring shells must be taken into account. It follows that while in the dilute case a static z-component of the magnetic moment operator cannot exist at $T = 0$ if there is the slightest instability of the shell, in the concentrated case such a component may exist if the RKKY interaction energy $k_B T_M$ is larger than $k_B T_f$. Whether $k_B T_M \lessgtr k_B T_f$ depends on E_{exc}, on the concentration and on the nature of $V_{k\ell}$. It is expected that at given concentration and E_{exc}, $k_B T_M / \Delta$ is largest near Eu and Gd and relatively small at both ends of the lanthanide series, since $k_B T_M$ is roughly proportional to the square of the spin. If $k_B T_M > k_B T_f$, a magnetically ordered state may be stable at the lowest temperatures (the energy $k_B T_M$ may enforce <u>coherent</u> hopping between shells). If $k_B T_M < k_B T_f$, no static z-component may exist at $T = 0$. In that case we speak of a nonmagnetic RE metal, since it has qualitatively the temperature dependence of the susceptibility in Eq. (5); i.e., the same as a nonmagnetic transition metal like Pd or Sc. Examples of nonmagnetic RE metals include α-Ce and SmS above 8 kbar, TmTe above 30 kbar, and SmB_6 and $YbAl_2$ at atmospheric pressure.

Secondly, during the valence transition, roughly one electron per cell must go into the conduction band, which causes an upward shift of the Fermi level of the order of several eV and comparable to E_{exc}. This kinetic energy change must be included in the energy balance to determine the equilibrium configuration of the entire system. If $E_{exc} \gg \Delta$, i.e. $\epsilon \approx 1$, E_{exc} will depend on volume in a different manner than if $E_{exc} \ll \Delta (\epsilon \approx 1/2)$. In the former case E_{exc} will change while ϵ and the position of the Fermi level will not. In the latter case ϵ will change by order unity; E_{exc} will not change by more than a fraction of Δ and the center of gravity of $E(n_o)$ and $E(n_o \pm 1)$ will track the Fermi level which increases by the order of eV. During this process, the volume will decrease strongly because of the order of one electron is transferred into the conduction band, increasing the metallic bond. The result is a strongly nonlinear PV relationship delineating first or second order phase boundaries between integral and nonintegral valence states.

From Eq. (5) one finds $T_f \approx 400°K$ for α-Ce, assuming $\epsilon(4f^1) = 0.3$. This equation is of course not applicable to the $4f^5$ and $4f^6$ shells which exhibit Van Vleck anomalies in the ionic susceptibility. Assuming that the multiplet splitting is not seriously affected by the valence fluctuation, one may write in these cases

$$\chi(T) = \begin{cases} \epsilon(n_o) \chi(n_o, T) + (1 - \epsilon(n_o)) \chi(n_o \pm 1, T) & (T > T_f) \\ \epsilon(n_o) \chi(n_o, T_f) + (1 - \epsilon(n_o)) \chi(n_o \pm 1, T_f) & (T \ll T_f) \end{cases} \quad (6)$$

Inspection of Fig. 7(b) shows that the susceptibility is indeed intermediate between $4f^6$ (e.g., $Sm^{2+}S$) and $4f^5$ (e.g., $SmPd_3$) for the intermediate valence cases SmB_6 and metallic SmS at high temperatures. From the susceptibility at helium temperature one extracts $T_f \approx (200 \pm 50)°K$ for both compounds, giving a shell lifetime of about 2×10^{-13} sec. This time is intermediate between the measuring time of the Mössbauer experiment[32] ($\tau \approx 3 \times 10^{-8}$ sec) and that of the soft x-ray experiment[33] ($\tau < 10^{-16}$ sec) on SmB_6. The latter measurement will therefore be done in one or the other valence state, resulting in one line at valence two and one at valence three with relative weight 0.3 and 0.7, respectively. On the other hand, the Mössbauer measurement is very slow compared to the lifetime and should therefore show a single line at intermediate isomer shift corresponding to valence 2.7. The measurement showed a line which was centered at valence 2.7 but was too broad to distinguish between a single line and a two line spectrum. Recently the compound $EuCu_2Si_2$ was studied by the Mössbauer effect and found to have a temperature dependent intermediate valence.[37] The spectrum was clearly a single line at intermediate isomer shift.

REFERENCES

1. L. L. Hirst, Phys. Kondens. Mat. 11, 255 (1970).
2. M. B. Maple and K. S. Kim, Phys. Rev. Lett. 23, 118 (1969).
3. M. B. Maple, J. Wittig and K. S. Kim, Phys. Rev. Lett. 23, 1375 (1969).
4. See, for example, M. B. Maple, in "MAGNETISM: A Treatise on Modern Theory and Materials," (edited by H. Suhl), Chapter 10, Academic Press (1973); and in AIP Conference Proceedings (No. 4) "Superconductivity in d- and f-Band Metals," (edited by D. H. Douglass), pp. 175-203 (1972).
5. T. Sugawara and H. Eguchi, J. Phys. Soc. Japan 21, 725 (1966).
6. J. J. Wollan and D. K. Finnemore, Phys. Rev. B4, 2996 (1971).
7. E. Müller-Hartmann and J. Zittartz, Z. Phys. 234, 58 (1970), Phys. Rev. Lett. 26, 428 (1971), and Solid State Commun. 11, 401 (1972).
8. J. G. Huber and M. B. Maple, J. Low Temp. Phys. 3, 537 (1970).
9. K. S. Kim and M. B. Maple, Phys. Rev. B2, 4696 (1970); W. Gey and E. Umlauf, Z. Phys. 242, 241 (1971).
10. M. B. Maple and J. Wittig, Solid State Commun. 9, 1611 (1971); M. Dietrich, W. Gey and E. Umlauf, Solid State Commun. 11, 655 (1972).
11. S. Ortega, M. Roth, C. Rizzuto and M. B. Maple, Solid State Commun. 13, 5 (1973).
12. J. G. Huber, W. A. Fertig and M. B. Maple, submitted to Solid State Comm.
13. C. A. Luengo, J. G. Huber, M. B. Maple and M. Roth, submitted to Phys. Rev. Lett.
14. A. A. Abrikosov and L. P. Gor'kov, Zh. Eksp. Teor. Fiz. 39, 1871 (1960); Soviet Phys. JETP 12, 1243 (1961).
15. A. B. Kaiser, J. Phys. C3, 409 (1970).
16. G. Riblet and K. Winzer, Solid State Commun. 9, 1663 (1971); M. B. Maple, W. A. Fertig, A. C. Mota, L. E. DeLong, D. Wohlleben and R. Fitzgerald, Solid State Commun. 11, 829 (1972).
17. J. Boes, A. J. Van Dam and J. Bijvoet, Phys. Lett. 28A, 101 (1968).
18. V. Allali, P. Donzé and A. Treyvaud, Solid State Commun. 7, 1241 (1969).
19. D. Wohlleben and B. R. Coles, in "MAGNETISM: A Treatise on Modern Theory and Materials" (edited by H. Suhl), Ch. 1, Academic Press (1973).
20. J. M. Lock, Proc. Phys. Soc. London:Sect. B70, 566 (1957).
21. A. W. Lawson and T. Y. Tang, Phys. Rev. 76, 301 (1949).
22. E. Franceschi and G. L. Olcese, Phys. Rev. Lett. 22, 1299 (1969).
23. M. R. MacPherson, G. E. Everett, D. Wohlleben and M. B. Maple, Phys. Rev. Lett. 26, 20 (1971).
24. J. Wittig, Phys. Rev. Lett. 21, 1250 (1968).
25. A. J. T. Grimberg, C. J. Schinkel and A. P. L. M. Zandee, Solid State Commun. 11, 1579 (1972).
26. N. E. Phillips, J. C. Ho and T. F. Smith, Phys. Lett. 27A, 49 (1968).
27. E. Bucher, V. Narayanamurti and A. Jayaraman, J. Appl. Phys. 42, 1741 (1971).

28. A. Jayaraman, V. Narayanamurti, E. Bucher and R. G. Maines, Phys. Rev. Lett. 25, 1430 (1970).
29. M. B. Maple and D. Wohlleben, Phys. Rev. Lett. 27, 511 (1971).
30. S. D. Bader, N. E. Phillips and D. B. McWhan, Phys. Rev. B7, 4686 (1969).
31. A. Menth, E. Buehler and T. H. Geballe, Phys. Rev. Lett. 22, 295 (1969).
32. R. L. Cohen, M. Eibschütz and K. W. West, Phys. Rev. Lett. 24, 383 (1970), and J. Appl. Phys. 41, 898 (1970).
33. E. E. Vainshtein, S. M. Blokhin and Yu. B. Paderno, Sov. Phys. Solid State 6, 2318 (1965).
34. D. Wohlleben, J. G. Huber and M. B. Maple, AIP Conf. Proc. Vol. 5 (H. C. Wolfe, ed.), p. 1478 (1972).
35. B. Sales, D. Wohlleben and M. B. Maple, to be published.
36. A. Iandelli and A. Palenzona, J. Less Com. Met. 29, 293 (1972).
37. E. R. Bauminger, D. Froindlich, I. Nowik, S. Ofer, I. Felner and I. Mayer, Phys. Rev. Lett. 30, 1053 (1973).

CORRELATION BETWEEN LATTICE CONSTANT AND MAGNETIC MOMENT IN 3d TRANSITION METAL ALLOYS

M. Shiga

Department of Metal Science and Technology
Kyoto University, Kyoto, Japan

ABSTRACT

An empirical relation between the lattice constant and the magnetic moment was found in binary solid solutions of 3d transition metals expressed as $a(x) = a_A (1-x) + a_B x + C \langle|\mu|\rangle$, where x is the atomic fraction, a_A, a_B and C are parameters, $\langle|\mu|\rangle$ the average magnitude of atomic magnetic moments. It is shown that the equation holds for all possible combinations of 3d transition metals which form solid solutions over a considerably wide range of concentrations. The analysis of lattice constants at high temperatures leads to the conclusion that localized atomic moments are retained above Tc in most 3d ferromagnetic alloys. The thermal expansion anomaly observed in the Invar alloy ($Fe_{65}Ni_{35}$) is explained as the result of collapse of localized moments above Tc. The physical meanings of the parameters are discussed in terms of the atomic size. It is shown that expansion of the atomic size is caused by the formation of the localized moment.

INTRODUCTION

It is well known that 3d electrons, which are responsible for the magnetic moment of transition elements, partake in the cohesion of transition metals. Therefore, a correlation between the magnetic moment and the atomic distance may be expected. Little attention has been paid to this point so far. In this paper, an empirical relation between the lattice constant and the atomic magnetic moment of transition metal alloys is proposed. We shall show that the analysis of the lattice constant versus concentration curve (L-C curve) can provide information about the atomic magnetic moments in alloys and can serve as an experimental method to detect the existence of localized moments above the Curie temperature.

EMPIRICAL RELATION

The lattice constant of binary solid solutions varies linearly with atomic concentration in the first approximation (Vegard's law). When applied to metalic solid solutions, deviations from the law invariably appear. These deviations are caused by many factors. In the transition metal alloys, magnetic properties may be one of the important factors[1]. We have found an empirical relation between the lattice constant and the atomic moment[2]. The relation is given by a simple equation,

$$a(x) = a_A (1-x) + a_B x + C <|\mu|> \qquad (1)$$

where $a(x)$ is the lattice constant of the solid solution $A_{1-x}B_x$, a_A, a_B and C are adjustable parameters and $<|\mu|>$ the average magnitude of the atomic magnetic moments, which can be estimated according to the following rules. (i) For simple ferromagnetic alloys, $<|\mu|>$ may be equated with μ_{so}, the spontaneous atomic moment at 0 K. (ii) In some ferromagnetic alloys, which usually exhibit an exchange anisotropy and/or a rotational hysteresis loss, a part of atomic moments aligns antiferromagnetically to the bulk magnetization. In this case, $<|\mu|> > \mu_{so}$. Modern experimental techniques such as neutron scattering and NMR may be helpful to estimate $<|\mu|>$.

(iii) It seems that $<|\mu|>$ can be equated with the sublattice moment for antiferromagnetic alloys. However, the spin structure of antiferromagnetic disordered alloys is not so simple as that of antiferromagnetic insulators. They are occasionally lacking in a long range spin order and should rather be described as "spin glasses". Therefore, the sublattice moment determined by neutron diffraction is not always equal to $<|\mu|>$. As for the case (ii), the estimation of $<|\mu|>$ could be done with the help of neutron scattering or some other experiments. (iv) $<|\mu|> = 0$ for nonmagnetic alloys, namely Pauli paramagnetic alloys or alloys having no localized magnetic moments.

In any case, it is desirable to use the data at T=0 K for both lattice constants and $<|\mu|>$. Nevertheless, we are mostly concerned with lattice constants at room temperature (RT) because of a lack of data at T=0 K. Even in this case, better agreement is obtained by using the atomic moments at T=0 K for $<|\mu|>$. However, if the alloy exhibits a thermal expansion anomaly below RT, we have to use the data at T=0 K for both $a(x)$ and $<|\mu|>$. The effect of temperature on the lattice constant will be discussed later. We first demonstrate how well equation (1) holds for all possible combinations of 3d transition metals which make a solid solution over considerably wide range of concentrations.

ALLOYS OF FERROMAGNETIC METALS

bcc $Fe_{1-x}Co_x$: Fig. 1 shows the observed and the calculated lattice constants of bcc Fe-Co. As can be seen in the figure, the agreement is excellent. Moreover, it is worth noting a change of the lattice constant in the vicinity x=0.5 by atomic ordering. Generally speaking, the formation of a superlattice causes a shrinkage of the lattice. On the contrary, the lattice constant increases by atomic ordering in this case. This exceptional behavior may be explained as the result of increase of the magnetization due to atomic ordering.

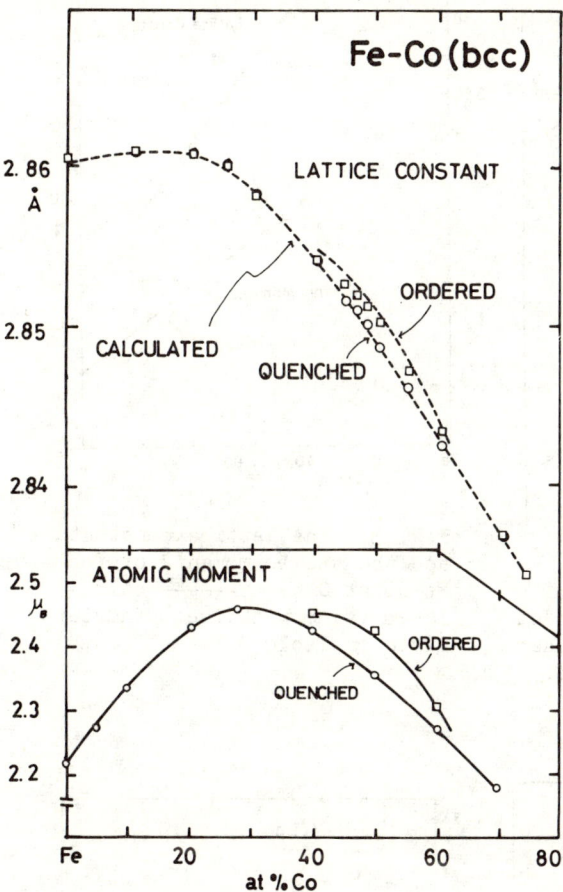

Fig. 1. The lattice constant of bcc Fe-Co at RT (Ref. 1, p.505). ------ represents calculated values with the parameters given in Table 1. $<|\mu|> = \mu_{so}3$ are also given.

fcc $Co_{1-x}Ni_x$: (Fig. 2) Both the lattice constant and the atomic moment vary linearly with concentration. This means that Eq. (1) holds well in this alloy. In this case, however, the parameters cannot be determined uniquely. It should be noted that the linear appearance of the L-C curve may be regarded as the evidence to show that contributions to the deviation from Vegard's law other than the magnetic one are negligible.

fcc $Fe_{1-x}Ni_x$: Since the alloys with $0.3 < x < 0.4$ have the

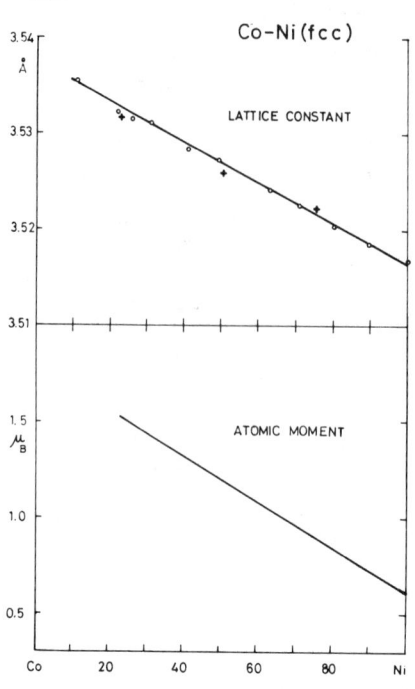

Fig. 2. The lattice constant at RT (Ref. 1, p.517) and the atomic moment at 0 K of fcc Co-Ni.

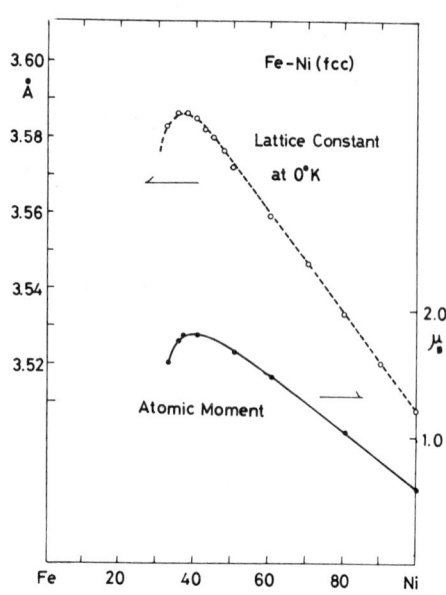

Fig. 3. The lattice constant[4] and the atomic moment[5] of fcc Fe-Ni at 0 K. ------ calculated value with the parameters given in Table 1.

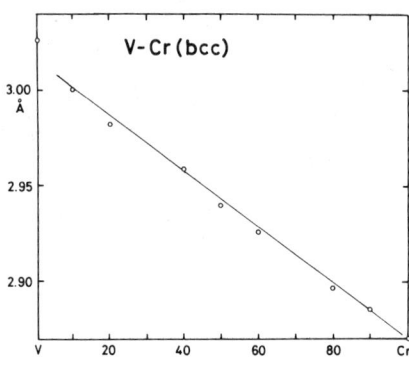

Fig. 4. The lattice constant at RT (Ref. 1, p.567) of bcc V-Cr.

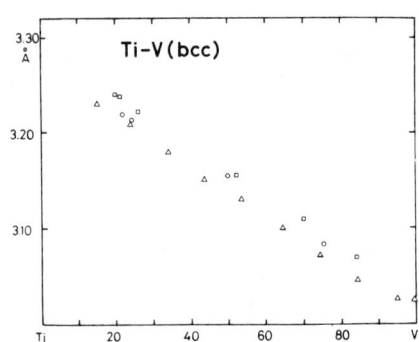

Fig. 5. The lattice constant of bcc Ti-V (Ref. 1, p. 875)

thermal expansion anomaly known as the Invar anomaly, we have to use the lattice constant at 0 K. Fortunately, we can estimate it from the available data on the thermal expansion. The lattice constant at 0 K thus obtained is in good agreement with the calculated values as seen in Fig. 3. The deviation of the lattice constant from a straight line near the Invar region ($0.3 < x < 0.4$) can be explained as the result of the deviation of the magnetization from the Slater-Pauling curve.

ALLOYS OF NONFERROMAGNETIC METALS

bcc $V_{1-x}Cr_x$ and bcc $Ti_{1-x}V_x$: (Figs. 4 and 5) In these systems, one may assume $<|\mu|> = 0$, since they are Pauli paramagnetic over all concentrations. Therefore, the lattice constant is expected to obey Vegard's law. In fact, the observed lattice constant varies linearly with concentration within the scatter of the data points.

bcc $V_{1-x}Mn_x$ (Fig. 6): In the vanadium rich region, the lattice constant decreases linearly with x. This linear appearance breaks at about x=0.5 and the L-C curve deviates upwards for $x > 0.5$, which means a relative expansion of the lattice. On the other hand, the temperature dependence of the magnetic susceptibility is Pauli paramagnetic in the vanadium rich region. There are some experimental results indicating the formation of magnetic moments in the manganese rich region[6] corresponding with the change of the slope in the L-C curve. From these observations, we may consider that $<|\mu|> = 0$ for $x < 0.5$ and $<|\mu|> \neq 0$ for $x > 0.5$. The change of the slope in the L-C curve can thus be explained on the basis of Eq. (1).

fcc $Cu_{1-x}Mn_x$ (Fig. 7): Although intensive investigations have been done, the magnetic and the electronic structures are still not clear except at the two sides of the system. At the copper rich end, it is believed that Mn atoms have well defined localized moment with $S = 5/2$ or $\mu_{Mn} = 5 \mu_B$. It has been revealed that Mn rich Mn-Cu alloys are antiferromagnetic with sublattice moment of about $2 \mu_B$ [7]. On the other hand, the lattice constant increases linearly with x in Cu rich region up to x= 0.4 and then the slope of the L-C curve becomes smaller and finally the lattice constant decreases with increasing x. Such a complicated behavior of the L-C curve might be explained by Eq. (1) as the result of the decrease of μ_{Mn} with increasing x.

ALLOYS OF FERROMAGNETIC AND NONFERROMAGNETIC METALS

bcc $Fe_{1-x}V$ and fcc $Ni_{1-x}Cu_x$ (Figs. 8 and 9) : Fairly good

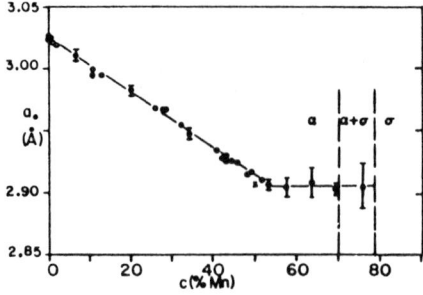

Fig. 6. The lattice constant of bcc V-Mn[6].

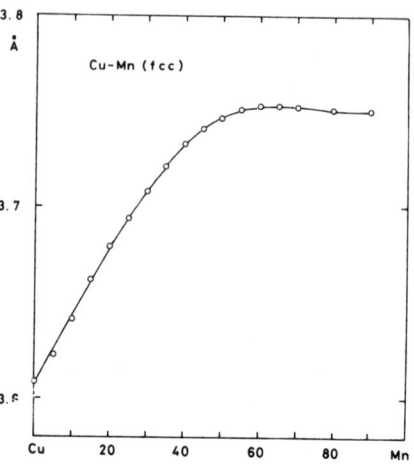

Fig. 7. The lattice constant of fcc Cu-Mn at RT (Ref. 1, p. 588).

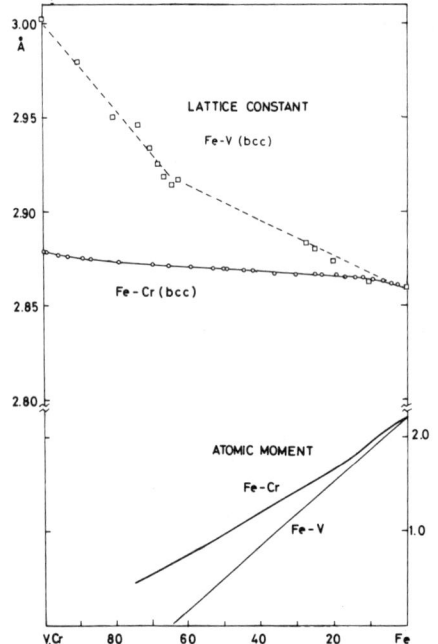

Fig. 8. The lattice constant of bcc Fe-V and bcc Fe-Cr at RT (Ref. 1, p.663 and p.533) ---- calculated value with the parameters given in Table 1.

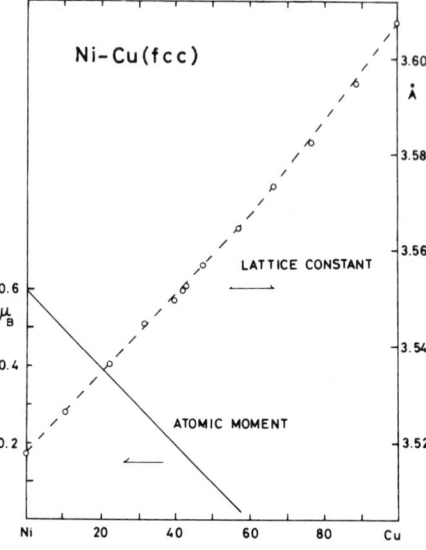

Fig. 9. The lattice constant at RT and the atomic moment at 0 K of fcc Ni-Cu. ---- culculated value with the parameters given in Table 1.

agreement is obtained in these system with $<|\mu|> = \mu_{so}$. This indicates the absence of localized moments in the paramagnetic phase.

fcc $Mn_{1-x}Co_x$ (Fig. 10) : The equation holds fairly well between the lattice constant at RT and μ_{so}. A small discrepancy is observed at the concentrations where ferromagnetism disappears ($0.6 < x < 0.8$). The origin of this discrepancy may be explained as follows. It is considered that some of the atomic moments are antiferromagnetically aligned at the critical concentrations because an exchange anisotropy has been observed. Consequently, $<|\mu|>$ should be larger than the bulk magnetization, i.e. $<|\mu|> > \mu_{so}$. The agreement in the nonferromagnetic region ($x < 0.5$) with $<|\mu|> = 0$, indicates absence of atomic moments in this concentration range. In fact, the magnetic susceptibility is Pauli paramagnetic. However, it was reported that the alloys with $x < 0.65$ becomes antiferromagnetic at low temperatures[10]. In order to discuss the effect of the antiferromagnetism on the lattice constant, the data at $T = 0$ are necessary.

bcc $Cr_{1-x}Fe_x$ (Fig. 8) : The magnetization versus concentration curve of this system is similar to that of Fe-V, while the appearance of the L-C curve is quite different. The equation does not hold if one takes simply $<|\mu|> = \mu_{so}$. In this system, however, it is believed that an iron atom has a localized atomic moment of about 2 μ_B[11] in the nonferromagnetic region. Assuming $\mu_{Fe} = 2\mu_B$ and $\mu_{Cr} = 0$ over all concentrations, we may write $<|\mu|> = 2 \cdot x$. Then, it is expected that $<|\mu|>$ and consequently the lattice constant varies linearly with concentration in rough agreement with experimental

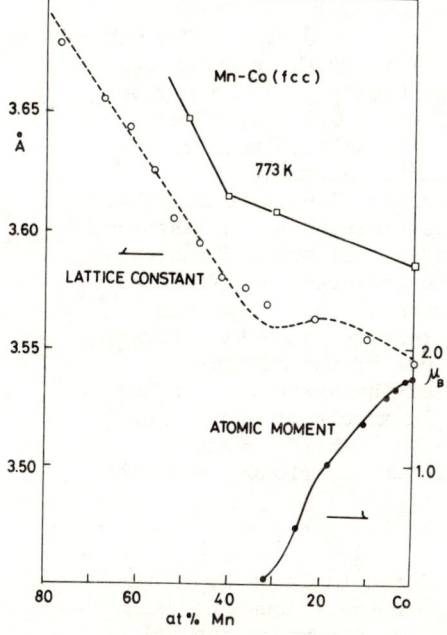

Fig. 10. The lattice constant at RT (o) (Ref. 1, p.510) and at 773 K (□)[8] of fcc Co-Mn. ---- calculated value with the parameters given in Table 1.

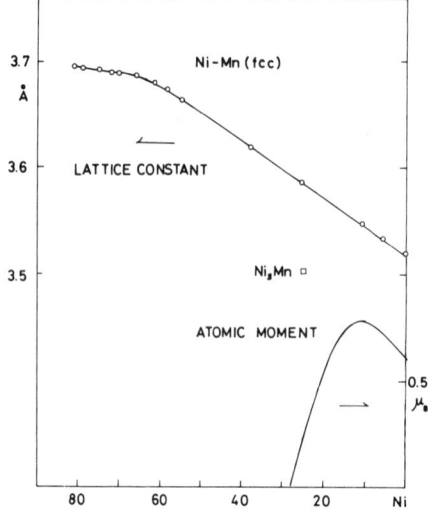

Fig. 11. The lattice constant at RT and atomic moment at 0 K of disordered Ni-Mn. □ represents the atomic moment of ordered Ni_3Mn[12].

observations. Strictly speaking, the L-C curve deviates slightly from a straight line at two sides of the system. Atomic moments on Cr atoms could be the cause of this behavior.

fcc $Ni_{1-x}Mn_x$ (Fig. 11):
The equation does not hold if $<|\mu|>$ is equated with μ_{so}. This apparent disagreement may be explained on the same basis as above, that is, the magnitude of atomic moments on each atom would remain unchanged in the concentration range where the L-C curve is linear ($0 < x < 0.6$). The sharp decrease of the spontaneous magnetization in the region $0.1 < x < 0.3$ might be attributable to antiferromagnetic aligment of surviving localized moments. It should be noted that no difference is observed in the lattice constants between the atomically ordered and disordered states at the composition of Ni_3Mn despite the remarkable difference of the magnetization. The antiferromagnetic spin coupling is supported from recent experimental results such as neutron scattering[13] and NMR[14]. The magnitude of the Mn moment differs slightly between the two sides of the system. In the Ni rich region, Mn atoms have an atomic moment of about 3 μ_B[12] and in the Mn rich region about 2 μ_B[15]. The decrease of the slope of the L-C curve for $x > 0.6$ might be attributable to the decrease of the Mn moment.

The parameters thus obtained are listed in Table 1. One should note that the parameter C has a roughly constant value of about 0.03 A/ μ_B. It is interesting to compare the values of a_A or a_B which are determined for different systems with the same structure. For example, the three values of a_{Fe} which are determined for bcc Fe-Co, bcc Fe-V and bcc Fe-Al are 2.788 A, 2.775 A and 2.788 A, respectively. It is worth noting that these three values are very close. This indicates a_A (or a_B) has a definite physical meaning.

Table I Values of parameters

system A - B	lattice structure	a_A (Å)	a_B (Å)	C (Å/μ_B)	temperature
Fe-Co	bcc	2.788	2.757	0.0324	RT
Fe-V	bcc	2.775	3.000	0.0386	RT
Fe-Al	bcc	2.788	3.030	0.0355	RT
Fe-Ni	fcc	3.551	3.489	0.0317	0K
Ni-Cu	fcc	3.506	3.607	0.0191	RT
Co-Mn	fcc	3.464	3.750	0.045	RT

LATTICE CONSTANT AT HIGH TEMPERATURES

It is interesting to see whether Eq. (1) holds or not at high temperatures or above the Curie temperature T_c. However, we encounter a difficulty in estimating $<|\mu|>$. According to the simple itinerant picture, it is possible that $<|\mu|> = 0$ above T_c, since the polarization of 3d bands and accordingly that of the Wannier functions at each atomic sites should vanish. On the other hand, on the basis of the localized spin model, $<|\mu|>$ would remain finite even above T_c. Bearing this in mind, we can expect to get information about the validity of different models from the appearance of the L-C curves at high temperatures.

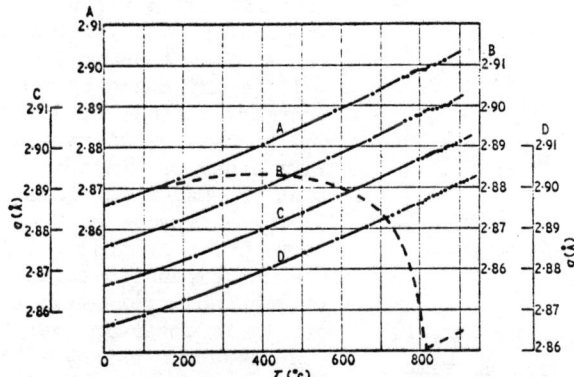

Fig. 12. Lattice constant a plotted against temperature for bcc Fe-Co. Co content: A 3% B 6%, C 9%, D 12%. ----- represents the expected change of lattice constant on the assumption that the magnetic term decreases with temperature.

Firstly, let consider the lattice constant of bcc Fe-Co alloys. According to our analysis at a low temperature, the magnetic term $C<|\mu|>$ is so large that a notable thermal expansion anomaly should be observed around T_c if $<|\mu|>$ decreased with the decreasing bulk magnetization as illustrated in Fig. 12. On the contrary, such an anomaly has not been observed[16]. From this fact, it was previously considered that the deviation from Vegard's law could not be ascribed to a magnetic origin. However, noting that the agreement at room temperature is too good to be accidental, we should rather conclude that $<|\mu|>$ remains constant over a wide temperature range including T_c and that thermal demagnetization is caused by fluctuations of localized atomic moments. One can describe such a situation in terms of Wannier functions which are still polarized above T_c within a certain time interval. In this sense, we may say that the localized spin picture is a better discription of this phenomenon. Similar behavior is observed in the L-C curve of fcc Co-Mn alloys at a high temperature, namely 773 K, as seen in Fig. 9.

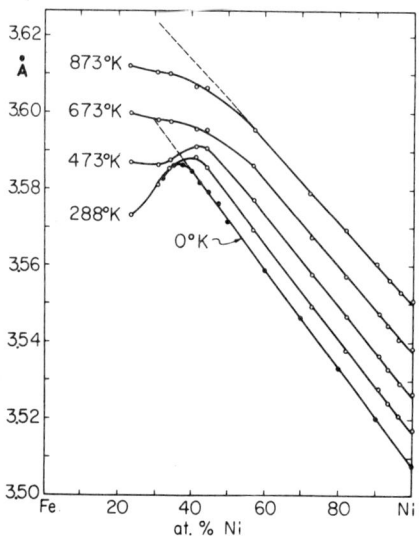

Fig. 13. The lattice constant of fcc Fe-Ni at various temperaures.

Particular attention should be paid to the L-C curve of fcc $Fe_{1-x}Ni_x$ alloys in connection with the Invar problem. As seen in Fig. 13, the lattice constant at 870 K, where the alloys are paramagnetic over all concentrations, varies linearly with x for $x>0.6$ and the increase of its value from that at 0 K may well be explained by the ordinary thermal expansion due to lattice vibrations. This behavior indicates that the magnetic term $C<|\mu|>$ remains unchanged above Tc. Actually, no thermal expansion anomaly attributable to a magnetic origin is observed over this concentration range. For $0.4<x<0.6$, the lattice constant at 870 K deviates downwards from a straight line, indicating the reduction of the magnetic term.

At 0K, however, the lattice constant and μ_{so} are still increasing linearly with increasing iron composition over this range. This behavior may be explained as follows. Both iron and nickel atoms have full moments, namely, μ_{Fe} = 2.8 μ_B and μ_{Ni} = 0.6 μ_B at 0 K for $0.4 < x < 1.0$. Therefore, $<|\mu|>$, which may be written as $<|\mu|> = 2.8 \cdot (1-x) + 0.6 \cdot x$, is linear in x. Above Tc, they still have full moments for $0.6 < x < 1.0$ but for $x < 0.6$, the magnitude of the atomic moments is reduced by raising temperature, which results in the thermal expansion anomaly observed in this region. Thus, the anomalous thermal expansion of the Invar alloy could be explained as the result of the reduction of the magnitude of the atomic moments[17]. This is in agreement with the result of neutron scattering[18].

ATOMIC SIZE MODEL

The change of the lattice constant in a solid solution $A_{1-x}B_x$ can be described in terms of the mean atomic size d expressed as

$$F\ a(x) = d = d_A \cdot (1-x) + d_B \cdot x, \qquad (2)$$

where d_A and d_B are atomic sizes of pure elements and F is a structure factor which depends on the lattice structure. F = 1 for simple cubic, F = $\sqrt{2}/2$ for fcc and F = $\sqrt{3}/2$ for bcc. We will show that Eq. (1) can be derived from extension of this idea. The procudure will give a certain physical meaning to the parameters a_A, a_B and C in Eq. (1). One should bear in mind, however, that the concept of atomic size does not necessarily have physical reality but may simply be a convenient representation of the atomic distance. We assume that constituent atoms take different states with the atomic moment $\mu_{A/B}^i$ and the atomic size $d_{A/B}^i$. The lattice constant and the mean atomic moment can be given by

$$a(x) = (1-x) \Sigma\ p_A^i(x)\ d_A^i + x \Sigma\ p_B^i(x)\ d_B^i \qquad (3)$$

$$<|\mu|> = (1-x) \Sigma\ p_A^i(x)\ \mu_A^i + x \Sigma\ p_B^i(x)\ \mu_B^i, \qquad (4)$$

where $p_{A/B}^i(x)$ is the probability of finding an A/B atom in the i th state. Eq. (1) can be derived from these equations if the following conditions are filfulled.

<u>Case I</u>: If $d_{A/B}^i = d_{A/B}^o + k\ \mu_{A/B}^i$, a(x) can be given by

$$F a(x) = (1-x) \{\Sigma p_A^i(x)\} d_A^o + x \{\Sigma p_B^i(x)\} d_B^o$$

$$+ k \{ (1-x) \Sigma p_A^i(x) \mu_A^i + x \Sigma p_B^i(x) \mu_B^i \}$$

$$= (1-x) d_A^o + x d_B^o + k <|\mu|> \quad (5)$$

Comparing with Eq. (1), we get $a_A = d_A^o / F$, $a_B = d_B^o / F$ and $C = k/F$

Case II. (Pseudo Ternary Alloy Model)   If the element A has two states and B has an only one state, $a(x)$ and $<|\mu|>$ are given by

$$F a(x) = (1-x) \{p_A^I(x) d_A^I + p_A^{II}(x) d_A^{II}\} + x d_B^o \quad (6)$$

$$<|\mu|> = (1-x) \{p_A^I(x) \mu_A^I + p_A^{II}(x) \mu_A^{II}\} + x \mu_B^o \quad (7)$$

Noting $p_A^I(x) + p_A^{II}(x) = 1$, we can eliminate $p_A^I(x)$ and $p_A^{II}(x)$ from Eqs. (6) and (7). Then we get Eq. (1) with

$$a_A = \{d_A^{II} - K \mu_A^{II}\} / F \quad , \quad a_B = \{d_B^o - K \mu_B^o\} / F$$

and $C = K/F$, where $K = \{d_A^{II} - d_A^I\}/\{\mu_A^{II} - \mu_A^I\}$.

One should note that, in this case, the atomic size of indivisual atoms is not necessarily a linear function of the atomic moment.   Some other experimental information is necessary to determine which case is realized in actual alloys.

It is possible to estimate $d_{A/B}^i$ for some special cases. For example, in bcc $Fe_{1-x}Co_x$ alloys, it has been revealed by neutron scattering that the individual atomic moments remain constant over the concentration range $x > 0.5$, where both the lattice constant and the magnetic moment vary linearly with concentration[21].   We may assume that each element takes only one state with $\mu_{Fe} = 3 \mu_B$ and $\mu_{Co} = 1.8 \mu_B$ over this range.   By extrapolating the linear region $0.5 < x < 0.7$ to pure iron, we can estimate the atomic size of bcc Fe in the state of $\mu_{Fe} = 3 \mu_B$. $d_{Fe} = 2.492$ A is thus obtained. Pure iron may be another state with $\mu_{Fe} = 2.2 \mu_B$ and $d_{Fe} = 2.477$ A.   Since iron atoms are thought to be nonmagnetic in V rich Fe-V, we can estimate the atomic size of nonmagnetic iron .   In this case, evidently $d^o = \sqrt{3}/2 \, a_{Fe}$, where a_{Fe} is the parameter of Eq. (1) for bcc Fe-V system.   The atomic sizes thus obtained are listed in Table 2 with other examples estimated in a similar way.

Table II Atomic sizes of some elements in different atomic states.

element and structure	atomic moment	atomic size (A)	expansion (%)	system
bcc Fe	0	2.403	0	Fe-V
	0	2.408	0	Fe-Al
	2.2	2.477	2.88	pure Fe
	3.0	2.492	3.53	Fe-Co
fcc Fe	0	2.521	0	Fe-Mn
	2.8	2.574	2.1	Fe-Ni
fcc Co	0	2.464	0	Co-Mn
	1.8	2.507	2.3	pure Co
fcc Mn	0	2.59	0	after Weiss19
	2.4	2.69	2.8	"
	4.5	3.82	6.4	"

The expansion of the atomic size with the atomic moment is plotted in Fig. 14. It should be emphasized that the rate of the expansion is almost same for different elements. It appears that the expansion is a linear function of the atomic moment, corresponding to the Case I. However, the number of data points is not adequate to make a definite conclusion.

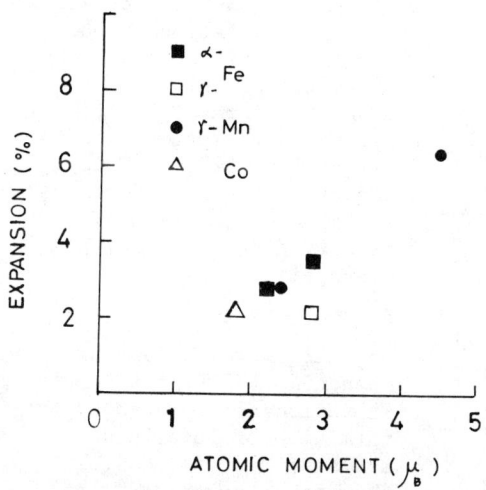

Fig. 14. Expansion of atomic size with atomic moment. The atomic sizes of γ - Mn in different states are estimated by Weiss[22].

DISCUSSION

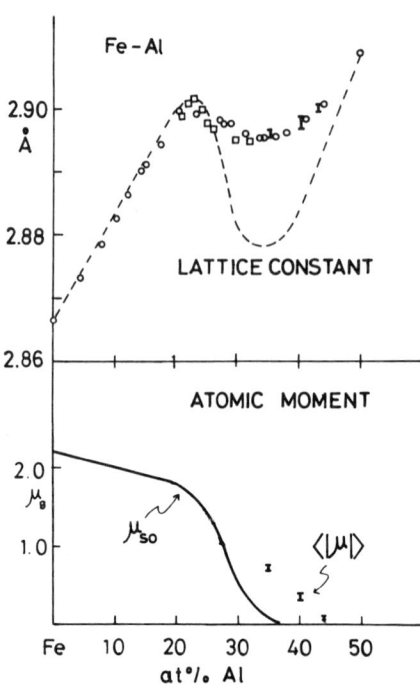

Fig. 15. The lattice constant of bcc Fe-Al at RT (Ref. 1, p.344). ------ calculated values with the parameters given in Table 1, where μ_{so} is used for $<|\mu|>$. Bars represent the calculated lattice constant with $<|\mu|>$ which are estimated from the distribution of the internal magnetic fields acting on Fe nuclei (bars in the lower figure).

So far, we have referred only to binary solid solutions of 3d metals with cubic structure. The equation may be more widely applicable, for example, one of the constituents is not a 3d metal. Nice agreement is obtained in bcc Fe-Al alloys as seen in Fig. 15. We may apply the relation to pseudo-binary alloys. The peculiar appearance of the L-C curve of cubic Laves compounds such as $Y(Fe_{1-x}Co_x)_2$[20] may be explained by adding a magnetic term. For alloys or metallic compounds with non-cubic structure, a similar relation might be found between the atomic volume and the magnetic moment.

Presumably, the magnetic term $C<|\mu|>$ is attributable to volume magneto-striction. However, theoretical[21] and experimental[22] investigations predict quadratic dependence of the spontaneous volume magneto-striction on the magnetic moment, while the magnetic term of Eq. (1) is linear with respect to the magnetic moment. This contradiction may be removed by invoking the pseudo-ternary alloy model with an assumption
$$d_A^i = d_A^o + k\,(\mu_A^i)^2.$$
Or there might be another mechanism to cause a volume expansion accompanied by the formation of localized moments. It is possible that the attractive force between magnetized atoms is weaker than that between nonmagnetic atoms. Theoretical investigation on this subject are desirable.

ACKNOWLEDGEMENTS

The author would like to thank Professor Y. Nakamura for his advice and encouragement and Professor E. Fawcett for reading the manuscript.

REFERENCES

1. W. B. Pearson, A Handbook of Lattice Spacings and Structure of Metals and Alloys (Pergamon Press, N.Y., 1958).
2. M. Shiga, Solid State Commun. 10, 1233 (1972).
3. D. L. Bardos, J. appl. Phys. 40, 1371 (1969).
4. M. Hayase, M. Shiga and Y. Nakamura, J. Phys. Soc. Japan 34, 925 (1973).
5. J. Crangle and G. C. Hallam, Proc. Roy. Soc. A272, 119 (1963).
6. E. von Meerwall and D. S. Schreiber, Phys. Rev. B 3, 1 (1971).
7. G. E. Bacon, I. W. Rummur, J. H. Smith and R. Street, Proc. Roy. Soc. 241, 223 (1957).
8. M. Matsui, Private Communication.
9. J. S. Kouvel, J. Phys. Chem. Solid 16, 107 (1960).
10. M. Matsui, T. Ido, K. Sato and K. Adachi, J. Phys. Soc. Japan 28, 791 (1970).
11. Y. Ishikawa, R. Touenier and J. Flilippi, J. Phys. Chem. Solid 26, 1727 (1965).
12. C. G. Shull and M. K. Wilkinson, Phys. Rev. 97, 304 (1955).
13. J. W. Cable and H. R. Child, J. de Phys. 32 C1-67 (1971).
14. R. L. Streever, Phys. Rev. 173, 591 (1968).
15. P. Wells and J. H. Smith, J. de Phys. 32, C1-70 (1971).
16. H. Stuart and N. Ridley, J. Phys. D 2, 485 (1969).
17. M. Shiga, Trans. IEEE on Magnetics 8, 666 (1972).
18. M. F. Collins, Proc. Phys. Soc. 86, 974 (1965).
19. R. J. Weiss, Phil. Mag. 26, 261 (1972).
20. J. T. Christpher, A. R. Piercy and K. N. R. Taylor, J. Less. Common Metals 17, 59 (1969).
21. R. H. Donaldson, Phys. Rev. 157, 366 (1961).
22. W. F. Schlosser, Internat. J. Magnetism 2, 167 (1972).

EFFECTS OF LATTICE PRESSURE ON Sm VALENCE STATES IN MONOSULFIDE SOLID SOLUTIONS

F. Holtzberg
IBM Research Center, Yorktown Heights, N. Y. 10598

ABSTRACT

A study of the effect of lattice compression of small trivalent rare earth sulfides on SmS has been carried out for the $Sm_{1-x}La_xS$ and $Sm_{1-x}Y_xS$ solid solution systems and critical regions of the $SmS_{1-x}As_x$ and $Sm_{1-x}Gd_xS$ systems. $Sm_{1-x}La_xS$ forms a continuous series of solid solutions whereas $Sm_{1-x}Y_xS$ phase separates at $x \sim 0.22$. The lattice pressure reduces the pressure induced phase transition observed in pure SmS from 6.5 k bar to atmospheric pressure. As a consequence of the negative slope in the temperature pressure diagram, lowering the temperature results in an unusual lattice expansion of the collapsed phase. This is interpreted on the basis of an electronic transition from a more trivalent to a more divalent valence state of Sm.

In critical concentration regions of their collapsed phases both $SmS_{1-x}As_x$ and $Sm_{1-x}Gd_xS$ explode on cooling. We prove that the explosive transition, which is electronically initiated, is actually the consequence of phase separation.

INTRODUCTION

Rare earth monosulfides have the simple NaCl structure and are relatively stable compounds with melting points in the vicinity of 2400°C. The structure is common to two well defined valence states of ionic members of the series. Most of the rare earths are found to form the smaller trivalent ion in their compounds with sulfur. Characteristically these phases are antiferromagnetic,[1] metallic, gold colored, brittle, and show [100] cleavage. We have measured resistivities as low as $10^{-5} \Omega$ cm for the metallic compounds. Photoemission measurements[2] indicate that the 4f levels are well below the valence band. Europium, samarium and ytterbium which can be trivalent in many compounds, are divalent in their monosulfides and are essentially black in bulk crystal form. EuS with a half filled 4f shell is ferromagnetic and we have measured resistivities as high as 10^{6} Ω cm on this material depending on stoichiometry. SmS with a $4f^6 - {^7F_0}$ ground state is generally an n-type semiconductor and exhibits temperature independent Van Vleck paramagnetism at low temperature. YbS is non-magnetic, the cation having a filled $4f^{14}$ electronic configuration. Photoemission studies done on EuS and SmS indicate that the 4f level for these divalent compounds lies between the valence and conduction bands.[2]

The existence of compounds having two distinct valence states with the same structure (with approximately 5% difference in lattice constant) provides a means of systematically varying conduction electron concentration and electronic structure by the formation of their solid solutions. Since the dominant interaction between spins in the rare earth semiconductors is an indirect exchange via the polarization of real or virtual electrons, one observes extremely large effects on physical properties (e.g. giant negative magnetoresistance,[3] increase in Curie temperature,[4] exchange enhanced susceptibility[5]) in these systems, depending upon the number of electrons donated by the trivalent metallic to the divalent semiconducting compounds.

Recently Jayaraman et al[6] reported that SmS undergoes a pressure induced phase transformation at room temperature and 6.5k bar. The semiconducting black SmS is converted to a collapsed isostructural phase which is a shiny gold color, resembling the trivalent rare earth monosulfides. They have also determined that the boundary between the black and gold phases in the temperature pressure diagram has a negative slope.

SmS, because of the proximity of the 4f level to the conduction band (~ 0.06 eV)[7], appears to lie on the borderline between the di- and trivalent rare earth monosulfides. Maple and Wohlleben[8] measured the magnetic susceptibility of the collapsed phase and found that it was non-magnetic. They interpret this result on the basis of a model which involves a resonance between the two valence states.

During a study of the effect of conduction electron concentration on Sm^{+2}-Sm^{+2} and Sm^{+2}-Eu^{+2} (the europium being an impurity) exchange interactions in the $Sm_{1-x}La_xS$ system[5] we observed an anomoly in the dependence of lattice constant on concentration. The data showed a minimum in lattice parameter at about x=0.5 which appeared to be related to the pressure induced phase transition. It was this clue to the effects of lattice pressure which led to the extensive study of samarium sulfide-trivalent rare earth sulfide solid solution systems reported here.

EXPERIMENTAL

The materials preparation of rare earth solid solution systems presents many difficulties, the principal one being the achievement of homogeneity. We have, therefore, developed a procedure which has been designed to minimize concentration gradients in the single crystals used for physical measurements.

All materials are handled in a helium purged dry box to prevent contamination of the readily oxidizable rare earth metals. The end members of the system are synthesized by reaction of sulfur vapor with the rare earth metal in a sealed, evacuated two chamber quartz tube. The temperature of this reaction is not permitted to exceed 600°C since above this temperature the sample begins to react with the quartz. The inhomogeneous product of this synthesis is pressed into pellets and heated in evacuated electron beam welded tungsten crucibles to 1600°C for several hours to homogenize the sample material. These charges are pulverized, analyzed and used to form the various compositions for crystal growth. Single crystals are grown in 3/8 x 2" electron beam welded tungsten crucibles which have been previously outgassed. The crucibles are heated to 2400°C in a vacuum chamber using r.f. heating and a calibrated L & N automatic pyrometer for temperature measurement. Crystal growth is achieved by slowly cooling the sample to 1800°C. It is annealed at this temperature for 16 hours and then either quenched or slowly cooled to room temperature. Single crystals can often be cleaved with the full diameter of the ingot.

Electron microprobe analysis is used to establish chemical composition and also to aid in the selection of homogeneous regions which are cleaved from the crystal for measurement.

Room temperature lattice constant data were obtained on the same samples used for physical measurements. The x-ray data were collected with a Guinier focussing camera using Cu K_α radiation and an internal Si standard. X-ray measurements at low temperatures were obtained on a GE-XRD-5 goniometer with Eulerian geometry and Ni filtered Cu radiation. Cooling was achieved by adapting a cold stream designed for an oscillation camera^9 to the diffractometer. The temperature was monitored with a copper-constantan thermocouple which had six inches of its length in the cold stream and the junction about 1/8" below the exit of the dewar tube. The crystals (0.1 mm) were supported on glass fibers at approximately 1/8" below the thermocouple junction. The highest angle reflection (in most cases the 640) was measured at various temperatures to 77°K and again on return to room temperature. Both the $K_{\alpha 1}$ and $K_{\alpha 2}$ were used for lattice constant data.

RESULTS AND DISCUSSION

The variation of lattice parameter with composition, x, is shown in Fig. 1 for the system $Sm_{1-x}La_xS$. The data indicate that semiconducting SmS forms a continuous series of solid solutions with metallic LaS. For 0<x<0.4 the lattice parameter of SmS is decreased by the

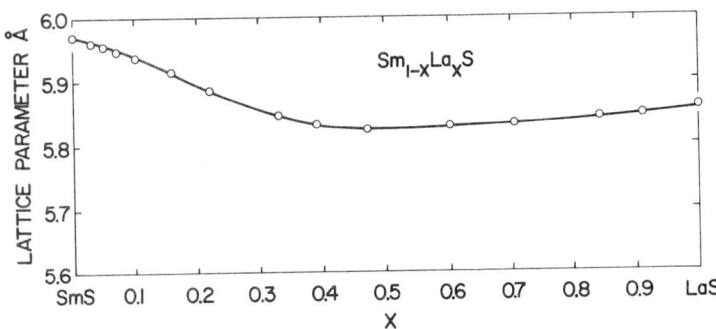

Fig. 1. Lattice parameters vs composition, x, in the solid solution system $Sm_{1-x}La_xS$.

addition of the smaller La^{+++} ion, more than expected from Vegard's law. There is also a small decrease in lattice constant in the La rich region. If this were a simple solid solution between divalent SmS and LaS, introduction of low concentrations of the larger Sm^{++}

into the LaS lattice would increase the lattice parameter (assuming hard spheres). We assumed, therefore, that the decrease is the result of the formation of a small fraction of Sm^{+++}, due to lattice pressure exerted by the La^{+++}. Comparison of the data in Fig. 1 with the lattice constant of pure SmS in the divalent state, $a_o=5.97Å$, and the collapsed phase, $a_o=5.70Å$ indicates that the relatively large La^{+++} ion restricts the compressive effect to a minimum value of 5.824Å. Since LaS ($a_o=5.855Å$) has a lattice parameter only slightly smaller than SmS, it seemed reasonable to attempt to form solid solutions with a smaller ion and drive the SmS completely to the trivalent state. In order not to complicate magnetic susceptibility measurements, non-magnetic YS ($a_o=5.49Å$) was used and the $Sm_{1-x}Y_xS$ system was studied.

The dependence of lattice constant on composition, x, for this system is shown in Fig. 2. The sample at x=0.16 is black and the one

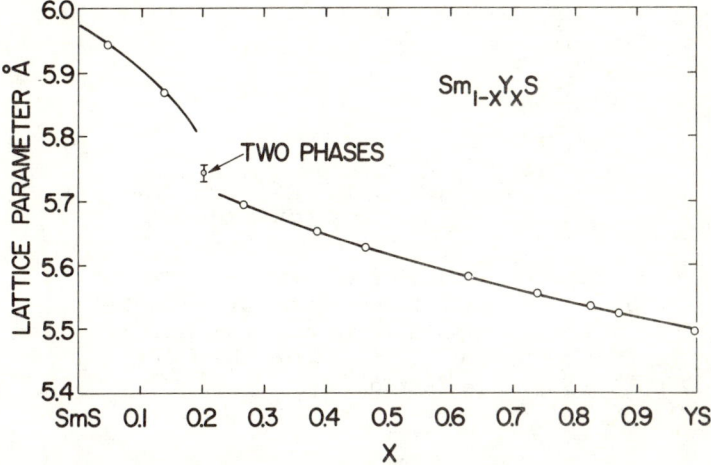

Fig. 2. Lattice parameters vs composition, x, in the solid solution system $Sm_{1-x}Y_xS$. The extent of the two phase region is not completely established.

at x=0.28 is bright yellow gold. The lattice parameter plotted for x≃0.20 is actually an average value of broad diffraction lines defining a two phased region at the discontinuity between the two regions of the diagram. The sample at this concentration contains both gold and black phases. The black portion cannot be powdered for x-ray analysis because the additional lattice pressure has reduced the total pressure required for transformation of SmS to essentially zero. Assuming the negative slope[6] of the temperature pressure phase boundary curve for pure SmS is retained in the solid solution system (the gold phase being stable at higher temperatures and pressures) one would expect that cooling the gold colored sample near the phase boundary would transform it to the divalent phase. The sample at 72 at. % Sm changes from a gold to black color on cooling in liquid nitrogen and the change is completely reversible.

If this color change reflects a change from a more trivalent to a more divalent phase, the sample would have to show a most unusual expansion on cooling. The temperature dependence of the variation of lattice parameter of a single crystal with composition $Sm_{.72}Y_{.28}S$ is shown in Fig. 3. The lattice parameter expands from its room temperature value of 5.706Å to 5.765Å at 104°K and remains constant at least to 77°K. The crystal contracts reversibly upon warming to room temperature. A similar plot is shown in Fig. 4 for a sample richer in Y. The room temperature lattice constant in this case is smaller, expansion begins at a lower temperature, and is expected to reach a lower maximum value because of the increased concentration of the smaller Y^{+++} ion.

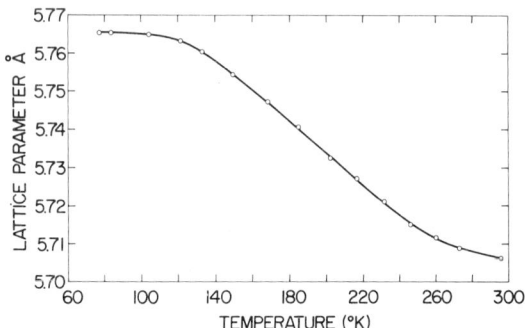

Fig. 3. Variation of lattice parameter with temperature for a single crystal with the composition $Sm_{0.72}Y_{0.28}S$ from the (640) reflection.

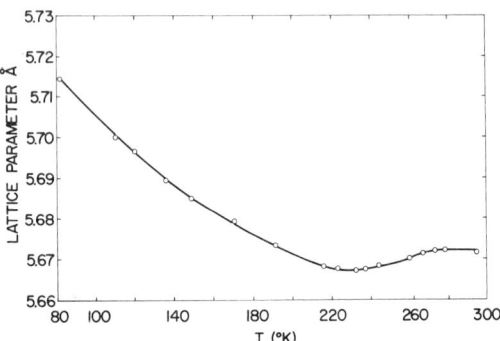

Fig. 4. Variation of lattice parameter with temperature for a single crystal with composition $Sm_{0.66}Y_{0.34}S$ from the (640) reflection.

Figure 5 is a plot of the susceptibility vs temperature of samples representing three regions of the phase diagram $Sm_{1-x}Y_xS$. In the composition range $0 \leq x < 0.20$ the samples are nearly divalent as shown in Fig. 6 and the susceptibility data is similar to that for pure SmS.[10] Well beyond the lattice parameter discontinuity (see Fig. 2) at, for example x=0.40 the susceptibility is similar to that of the mixed valence state described by Maple and Wohlleben.[8] For x=0.28 the room temperature susceptibility has the same value as the sample at x=0.40 but with decreasing temperature the susceptibility becomes more divalent-like following the temperature dependence of the lattice parameter (Fig. 3) for the same concentration.

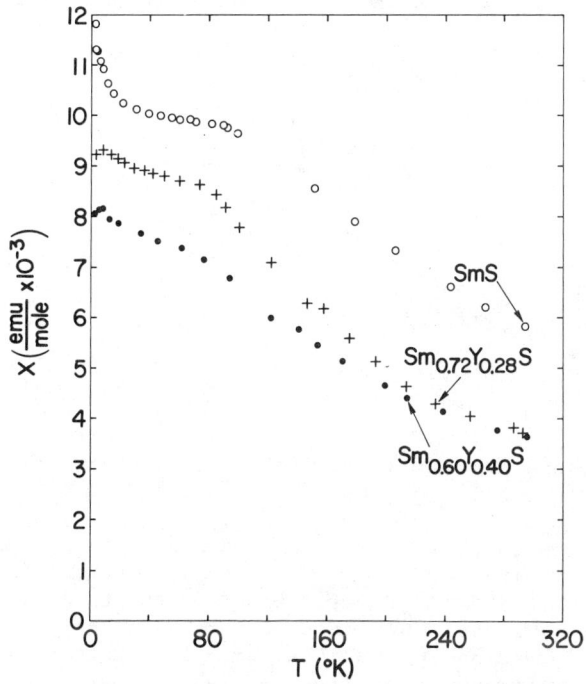

Fig. 5. Molar susceptibility vs temperature for two samples in the $Sm_{1-x}Y_xS$ solid solution system compared with SmS. Low temperature impurity contribution subtracted out for Y samples.

Changes in lattice parameter and susceptibility with composition and temperature, all suggest changes in the valence of Sm. If we assume that the non-linear behavior of lattice parameters in Figs. 1 and 2 is entirely due to changes in Sm valence, then we can use the deviations from linearity as a quantitative estimate of the Sm valence. The reference valence states can be developed as follows; assuming that Nd and Gd are purely trivalent in their monosulfides, an estimate of the lattice constant of SmS, with Sm^{+++}, can be obtained by interpolation. A value of a_o=5.62Å, so derived, would be the expected lattice constant of the collapsed phase of SmS if it were purely trivalent. To establish a reference state of purely divalent Sm in SmS one can use the difference between Eu^{+++} in the monosulfide estimated as above, and the value for divalent EuS. This difference is 0.38Å. On this basis the monosulfide of Sm^{++} would have a lattice constant $a_o = 6.00$Å. We further assume that both La and Y remain trivalent throughout the solid solution systems. The Sm valence for a specific composition, x, in the systems $Sm_{1-x}Y_xS$ and $Sm_{1-x}La_xS$ can be estimated from the deviation of the lattice constant a_x from a straight line joining the trivalent end member lattice constants.[11] The results of this analysis (Fig. 6) indicate that the Sm ion in SmS has non-integral values for all compositions in both systems. In addition the data for both systems show a break in valence as a function of composition.

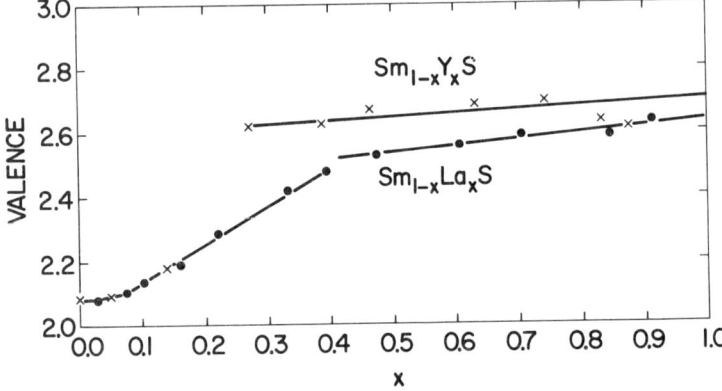

Fig. 6. Valence of Sm ion obtained from lattice parameter data for the systems $Sm_{1-x}La_xS$ and $Sm_{1-x}Y_xS$.

The changes in slope in the $Sm_{1-x}La_xS$ data is not as pronounced as the discontinuity in the $Sm_{1-x}Y_xS$ data in Fig. 6. It is not clear from these results that the slope change in the La data represents the same kind of separation between a more divalent-like phase and the collapsed phase observed for the $Sm_{1-x}Y_xS$ system. We, therefore, measured the temperature dependence of lattice parameter for the two samples in the $Sm_{1-x}La_xS$ system bracketing the abrupt change in slope (Fig. 6). The results of these measurements are plotted in Fig. 7. The top curve exhibits the normal contraction upon cooling of the more divalent-like phase. The expansion of the lattice parameter with decreasing temperature in the bottom curve, demonstrates that this composition is in the partially collapsed phase at room temperature.

The curves in Fig. 6 are based on the assumption that the lattice constant truly reflects the Sm valence. If the large La^{+++} ion restricts the contraction of the lattice (as mentioned earlier), then the actual valence in the collapsed region could be higher than that shown in Fig. 6.

A recent report by Darnell et al presented evidence of the formation of solid solutions between the isostructural SmS and SmAs.[12] This information was intriguing since Sm is purely trivalent in the

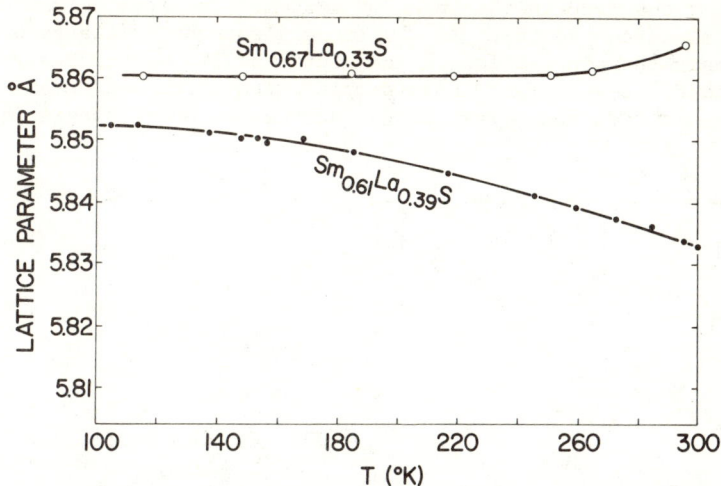

Fig. 7. Temperature dependence of lattice parameters for samples change in compositions in the $Sm_{1-x}La_xS$ system bracketing the slope change in Fig. 6.

arsenide. Furthermore their lattice constant data (obtained on sintered samples) showed a larger discontinuity with composition than observed in the $Sm_{1-x}Y_xS$ system. Although both SmS and SmAs are black, we found, that in $SmS_{1-x}As_x$ for 0.05<x<0.15 the single crystals had a range of color from black to gold. Selecting some of the gold colored crystals, we again cooled the material in liquid nitrogen expecting a change toward the divalent black color. In this case the crystals shattered on cooling leaving a black powder, which turned to gold on warming to room temperature.

The explosive transformation is the consequence of subsolidus phase separation. This can be deduced from the magnetic data in Fig. 8. As the sample is cooled the susceptibility rises abruptly at about 150°K. At this temperature the gold phase separates into a divalent and a more trivalent-like phase and the susceptibility rises due to conversion of the more trivalent-like Sm to Sm^{++}. The increase in susceptibility below 25°K is caused by impurities. With increasing temperature there is hysteresis but we also find that the final susceptibility at room temperature is slightly higher than its original value. The displacement comes from the presence of the separated divalent phase. A new curve is generated which reproduces with temperature cycling and represents the susceptibility of the two phases. Room temperature microscopic examination of an exploded sample, showed that at this concentration, the crystal separated into a predominantly gold phase and a very small quantity of the black phase. It is interesting to note that the first explosive transition and subsequent cooling cycles have the abrupt increase in susceptibility at the same temperature ∼150°K. The sharp change in susceptibility and associated hysteresis indicate that in this case the transition is first order differing significantly from the continuous transition observed in the $Sm_{1-x}Y_xS$ system. The limited data we have obtained on the $SmS_{1-x}As_x$ system suggest the existance of two phases near the composition region where the explosive transformation is observed. Considerably more sample preparation and detailed analysis are required before we can discuss valence changes in this system.

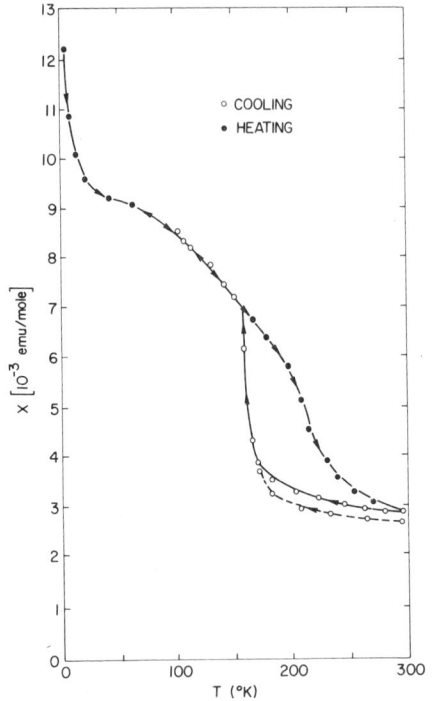

Fig. 8. Susceptibility, χ, in emu/mole for $SmS_{0.95}As_{0.05}$ as a function of temperature. Dotted line shows first cycle through explosive transformation. Solid line shows hysteresis in susceptibility.

Independently Jayaraman, Bucher, Dernier and Longinotti (JBDL) have reported on a "Temperature-Induced Explosive First Order Electronic Phase Transition in Gd-Doped SmS."[13] The paper states that in the system $Sm_{1-x}Gd_xS$ for a limited concentration range, $0.16<x<0.22$, a gold metallic phase explosively converts upon cooling to a black powder and on warming becomes gold again. The transformation is termed reversible. Concentration dependence of the lattice constant is given for room temperature and 4.2°K. The room temperature data is similar to that for $Sm_{1-x}Y_xS$ (Fig. 2), the discontinuity in $Sm_{1-x}Gd_xS$ being at x=0.15. The lattice parameter of GdS (a_o=5.563Å) is larger than that of YS. The change in lattice parameter at the discontinuity is from a_o=5.875 for the divalent phase to a_o=5.68Å for the gold metallic phase. At 4.2 there is a small decrease in lattice parameter from the room temperature value in the range $0 \leq x \leq 0.15$, but in the range $0.16 \leq x < 0.22$, the explosive region, the lattice parameter expands to a value 5.82Å and is found to be independent of Gd concentration over this 6 at. % interval.

The model JBDL invoke to explain the explosive transformation describes the 4f level as a very narrow band lying above but very close to the Fermi level in the 5d conduction band. With lowering temperature the 4f level lowers and overlaps the Fermi level, resulting in localization of a fraction of the conduction electrons in the 4f band. This then results in a lowering of the Sm valence causing a lattice expansion which in turn pushes the Fermi level further in the 4f band, resulting in further localization. This then leads to an instability and hence a catastrophic phase transition.

This model may account for an electronic transition responsible for lattice expansion but it is not an adequate explanation of the explosive nature of the transformation. Why would a uniformly expanding lattice explosively disintegrate into a powder? How can an explosive transformation be reversible? To clarify these points we prepared single crystals in the system $Sm_{1-x}Gd_xS$ which microprobe analysis showed to have the composition $Sm_{0.84}Gd_{0.16}S$. The sample, cooled from its melting point to room temperature was homogeneous within the resolution of the microprobe analysis and had a lattice constant $a_o = 5.683 \pm 0.001$Å in agreement with the value reported for this composition. A 2x2x2mm single crystal was cooled in an evacuated sealed glass tube to liquid nitrogen temperature. As the sample cooled one corner of the crystal began to spew forth a black powder. As the temperature continued to fall the process became more violent until the crystal completely disintegrated. The powder, an intense black color slowly changed back to gold as the sample was heated again to room temperature. The color change was reversible upon temperature cycling from then on without explosion. Since the resulting particle size averaged about 50μ, a powder pattern was obtained without additional grinding. The room temperature x-ray pattern showed two clearly resolved x-ray patterns, with slightly broadened lines. These were readily indexed on the basis of two fcc lattices with $a_1 = 5.681$Å and $a_2 = 5.841$Å.

Microscopic examination at 500X revealed separate gold and black colored crystallites with some fragments retaining both phases as shown in Fig. 9.

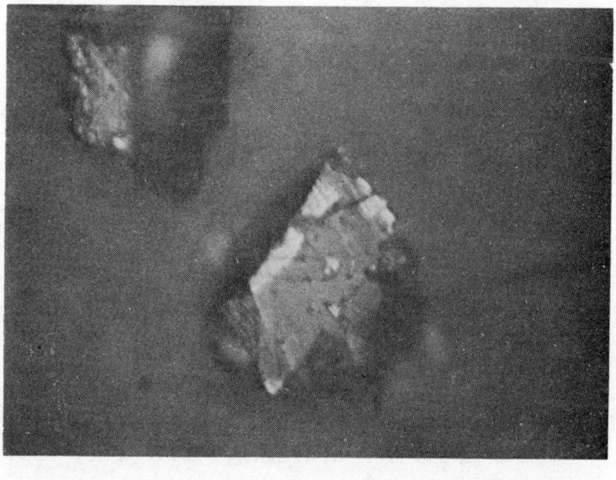

Fig. 9. Micrograph demonstrating existence of two phases after explosive transformation magnefication 500X.

Since we have shown that the crystals separate explosively into two phases on cooling it is possible to deduce a plausible mechanism for the sample mestability.

We know that crystals grown from the melt by slow cooling almost always have concentration gradients. In this case the sample is cooled from 2400°C to room temperature and depending on the thermal history there will generally be concentration fluctuations in the crystals and most likely regions of strain. If the black phase contracts and the gold phase expands on cooling then a very large differential stress will develop. As in the $SmS_{1-x}As_x$ system, the $Sm_{1-x}Gd_xS$ phase transition in the critical concentration range is first order and the large lattice expansion probably occurs over a very narrow temperature interval. In the presence of concentration fluctuations the abrupt lattice expansion would create differential stresses and an explosive phase separation. After the explosive transition we have separated small gold crystals and find that they become black upon cooling but suffer no further deteroration with repeated temperature cycling.

The explosive transformation is an interesting consequence of the change in valence state of Sm. We would, however, like to emphasize the changes occurring under equilibrium conditions in these systems. We demonstrate that by appropriate selection of trivalent ions one can systematically vary the electronic structure by the formation of solid solutions of trivalent ions in SmS. This has been accomplished by taking advantage of the peculiarities of the rare earth monochalcogenides which provide us with a divalent semiconductor and trivalent metal having identical crystal structure. We have discovered intriguing regions of solid solution systems in which additional lattice pressure of the smaller trivalent rare earth ions shift electronic transitions to atmospheric pressure and any selected temperature. Results of analysis of lattice parameter and susceptibility data strongly indicate the existance of nonintegral Sm valence at atmospheric pressure throughout the solid solution systems.

ACKNOWLEDGEMENTS

The author wishes to thank P. Lockwood, H. Lilienthal and C. Guerci for technical assistance; C. Aliotta, F. Cardone, J. Kuptsis for microprobe measurements; L. J. Tao for the use of his magnetic data; F. Jellinek, L. J. Tao, J. B. Torrance and R. Tournier for many helpful discussions. The author is indebted to T. Penney for his critical evaluation of the manuscript.

REFERENCES

1. L. J. Tao, J. B. Torrance, and F. Holtzberg, (this conference).
2. D. E. Eastman, F. Holtzberg, J. Freeouf and M. Erbudak, (this conference).
3. S. von Molnar and S. Methfessel, J. Appl. Phys. **38**, 959 (1967).
4. T. R. McGuire and F. Holtzberg, AIP Conf. Proc. #5 Magnetism and Magnetic Materials, (1972), p. 855.

5. F. Mehran, J. B. Torrance and F. Holtzberg, Phys. Rev. B $\underline{8}$, 1268 (1973).
6. A. Jayaraman, V. Narayanamurti, E. Bucher and R. G. Maines, Phys. Rev. Letters $\underline{25}$, 1430 (1970).
7. E. Kaldis and P. Wachter, Solid State Communications $\underline{11}$, 907 (1972).
8. M. B. Maple and D. Wohlleben, Phys. Rev. Letters $\underline{27}$, 511 (1971).
9. B. Post, R. Schwartz and I. Fankuchen, Rev. Sci. Inst. $\underline{22}$, 218 (1951).
10. The system is actually somewhat more complicated in this region because the increased electron concentration increases the susceptibility before the discontinuity at $x \approx 0.20$. See Ref 5 for effect in $Sm_{1-x}La_xS$.
11. This method of estimating valence was described to us by R. Tournier.
12. A. J. Darnell, N. C. Miller, R. C. Saunders and R. G. Breckenridge, ARPA Contract DAHC 15-70-C-0145, Final Report, March 1973.
13. A. Jayaraman, E. Bucher, P. D. Dernier and L. D. Longinotti, Phys. Rev. Letters $\underline{31}$, 700 (1973).

MAGNETIC PROPERTIES OF SOME SIMPLE METALLIC AND SEMIMETALLIC COMPOUNDS OF f-ELECTRON METALS WITH MAIN GROUP ELEMENTS

B. Staliński
Institute for Low Temperature and Structure Research,
Polish Academy of Sciences, Wroclaw.

ABSTRACT

This paper reviews experimental work, which was performed recently at the above-mentioned Institute, on the magnetic properties of rare earth and uranium compounds with main group elements. The results of low-temperature magnetic measurements carried out for some Ce, Pr, Nd, Sm, Eu, Gd, and Yb compounds with In, Tl, Sn, and Pb ($AuCu_3$ or $CdMg_3$ type of structure) as well as for uranium compounds with main group elements ($AuCu_3$ type of structure) are reported. Also, magnetic data for uranium compounds of the form UXY, $U_2N_2X(Y)$, and U_3Y_5 as well as for UX-UY solid solutions (where X -element of the Vth, and Y- of the VIth group) are included.

RARE EARTH COMPOUNDS

Only recently the magnetic properties of the rare earth intermetallic compounds with In, Tl, Sn, and Pb, of the composition REX_3 and RE_3X and with either the cubic $AuCu_3$ or the hexagonal $CdMg_3$ type of structure, have received considerable attention. The first extensive work on this subject was published by Tsuchida and Wallace in 1965.[1] Others followed in the years 1969-1973.[2-8] It has been found that most of these compounds exhibit antiferromagnetic order at low temperatures, and, more or less clearly, the influence of the crystal field. Moreover, there is now much interest in the magnetic behavior of systems with a crystal-field only singlet ground state like Pr_3Tl [9-12] where due to the competitive action of the exchange and crystal field interactions, one can find the examples of induced-moment ferromagnets.

I. EuX_3 AND GdX_3 COMPOUNDS (X = In, Tl, Sn, Pb)

From among all these compounds showing transitions to an antiferromagnetic state at low temperatures, $EuPb_3$, $GdPb_3$, $GdIn_3$,[2] and $GdSn_3$ [1] have been investigated previously at low magnetic fields. Our results reported in part last year[13] are consistent with these findings except in the case of $GdSn_3$ for which no transition has been established previously.[1] The Néel points are given in Table I.

Table I. Néel Points for EuX$_3$ and GdX$_3$ Compounds

Compound	T_N (K)		Compound	T_N (K)	
	H=6.5 kOe	H=70 kOe		H=6.5 kOe	H=70 kOe
EuSn$_3$	36.5	43.5	GdSn$_3$	16.5	13
EuPb$_3$	20	25	GdPb$_3$	17	14
EuIn$_3$	10	4.5	GdIn$_3$	45.5	48
EuTl$_3$	6.5	5.5	GdTl$_3$	13	12

One can see that for most of the compounds these transition temperatures shift towards lower temperatures with increasing magnetic field. This behavior is typical for most antiferromagnets that show second order transitions. Due to relatively large shifts, it was possible to obtain a magnetic phase diagram for EuIn$_3$ (Fig. 1). Anomalous behavior was found in the case of EuSn$_3$ and EuPb$_3$ where the transition temperatures increase markedly with increasing magnetic field. This in turn may indicate a first order transition which, according to the theory of Bean and Rodbell,[14] may take place in the case of strong volume dependence of the exchange interaction. Above the Néel points the magnetic susceptibilities follow the Curie-Weiss law with effective magnetic moments slightly lower than the expected one (7.94 BM), and θ<0 except in the case of EuIn$_3$ for which θ = +10K.

Preliminary Mössbauer effect studies of europium compounds[15] confirmed the data recently obtained for EuPb$_3$ by Loewenhaupt[5] and gave for EuIn$_3$ the effective field H_{hfs} (0 K) = -230 kOe, and the isomer shift against Eu$_2$O$_3$ of -11.8 mm/s. These values are very similar to those obtained for EuPb$_3$.

For all of the compounds under consideration, a slight nonlinearity in the magnetic field dependence of the magnetization (connected with spin flopping) has been observed below 20-30 kOe at 4.2 K (Fig. 2). The effect is most distinct in the case of EuIn$_3$ and GdIn$_3$. Moreover, the difficulty in approaching the saturation even at 80 kOe may indicate a large predominance of the exchange over the magnetocrystalline energy.

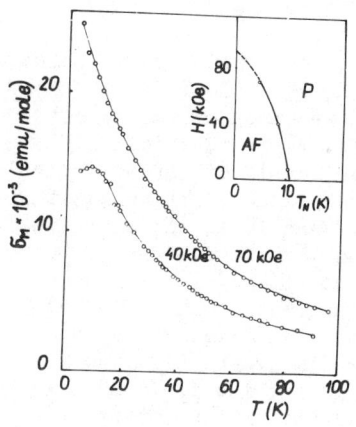

Fig. 1. σ$_M$ vs. T and magnetic phase diagram for EuIn$_3$.

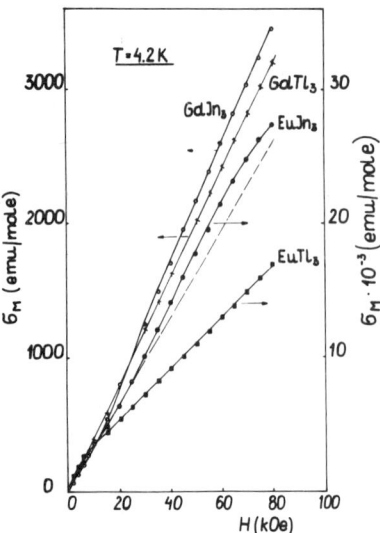

Fig. 2. σ_M vs. T for EuX_3 and GdX_3 compounds (X = In, Tl)

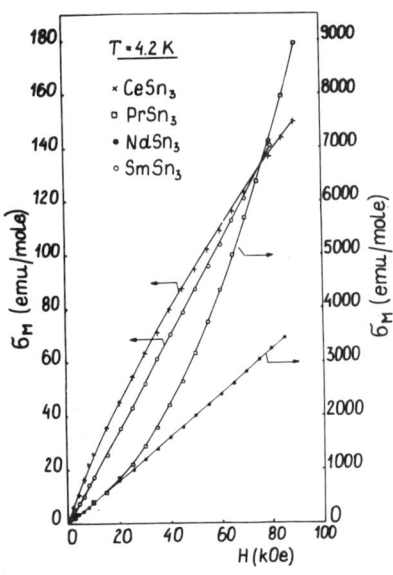

Fig. 3. σ_M vs. T for $RESn_3$ compounds (RE = Ce, Pr, Nd, Sm)

II. $RESn_3$ COMPOUNDS (RE = Ce, Pr, Nd, Sm)

All of these compounds except for $CeSn_3$ are known to be antiferromagnets[1,16] with low values of the Néel points. The magnetic susceptibility measurements performed in fields up to 80 kOe (Fig. 3) have confirmed the metamagnetic transition in the case of $PrSn_3$, which was recently observed by Lethuillier et al.[7]

Because of the anomalous physical properties of $CeSn_3$ in comparison to other isostructural CeX_3 compounds (X = Pb, In, Tl, Pd), the measurements of magnetic properties as well as of electrical resistivity on monocrystalline samples of $CeSn_3$ have been performed [16,17]. It has been found that the magnetic susceptibility (Fig. 4) follows the Curie-Weiss law at temperatures T<3 K, and T>220 K with θ = -6.8 K, μ_{eff} = 0.41 BM, and θ = -218 K, μ_{eff} = 2.65 BM, respectively.

In the temperature range between 12 and 26 K, $\chi_M \sim (1-\alpha T^2)$. The temperature dependence of the magnetization (Fig. 4) in the field of 80 kOe is slightly different from that in weak fields. It is constant between 4.2 and 12 K, and exhibits an additional maximum at about 80 K. Moreover, nonlinear field dependence of the magnetization at 4.2 K has been established (Fig. 3); $\sigma_M(H) = \chi_M(0)H (1+\beta H^\alpha)$, where $\beta<0$, $\alpha \neq 2$, and $\chi_M(0)$ is the low field susceptibility at 0 K. According to Wohlfarth's findings[18] such a dependence with $\alpha = 2$ may be expected in the case of a highly variable density of states near the Fermi level.

The electrical resistivity of $CeSn_3$ monocrystals in zero magnetic field is proportional to T^2 between

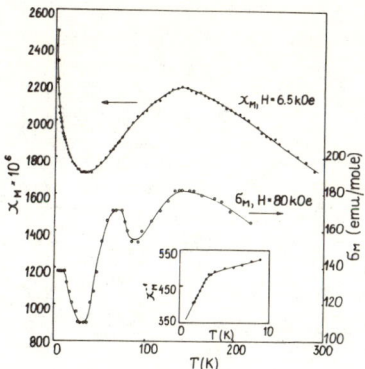

Fig. 4. χ_M and σ_M vs. T for CeSn$_3$.

2 and 12 K, and to $T^{2.7}$ above 12 K (up to about 50 K). One can see that the temperature ranges for which the relations $\chi_M \sim (1 - \alpha T^2)$ and $\rho \sim T^2$ hold are different. Thus it is expected that the localized-spin-fluctuations mechanism[19,20] is not responsible for the observed behavior. For this reason a simple model of the band structure qualitatively explaining the above observations as well as the observed achievement of saturation[21] by the transverse magnetoresistance of CeSn$_3$ in a magnetic field of 30 kOe has been proposed.[17]

III. YbX$_3$ COMPOUNDS (X = In, Tl, Sn, Pb)

The magnetic properties of this group of compounds as well as of the metallic ytterbium used for their synthesis have been investigated in magnetic fields up to 80-90 kOe.[22] The existence of a thermal hysteresis of the magnetic susceptibility of ytterbium, recently discovered by Bucher et al.[23] and connected with fcc-hcp transformation, has been confirmed.

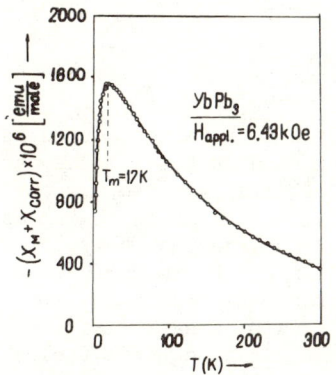

Fig. 5. $-\chi_M$ vs. T for YbPb$_3$.

Anomalous temperature dependent and strong diamagnetism with a minimum in the magnetic susceptibility at 17 K (Fig. 5) has been found in the case of YbPb$_3$. Moreover, this compound displays weak de Haas-Van Alphen oscillations in weak magnetic fields, the occurrence of which may be expected for the Fermi surface lying in the neighborhood of the Brillouin zone boundaries. The remaining compounds (except YbIn$_3$) were found to be paramagnetic at low temperatures and diamagnetic at higher temperatures. Their magnetizations at 4.2 K approach saturation in the fields of 80-90 kOe (Fig. 6) pointing to the presence of a few percent of Yb^{3+} ions.

IV. Pr$_3$In

It has been found[24] that Pr$_3$In, like Pr$_3$Tl[9-12], is an induced-moment ferromagnet with a saturation moment of 1.4 BM. However, the Curie temperature of 56.5 $\pm$ 0.5 K is much higher than for Pr$_3$Tl. Above 80 K the susceptibility follows the Curie-Weiss law with an effective magnetic moment of 3.4 BM and θ = 22 K.

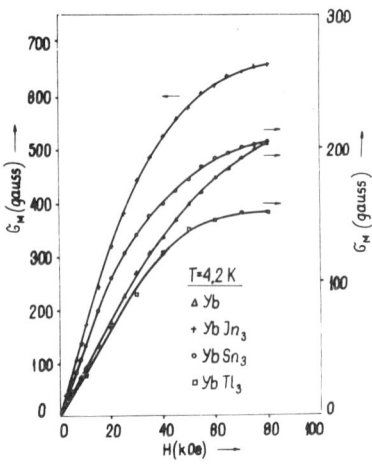
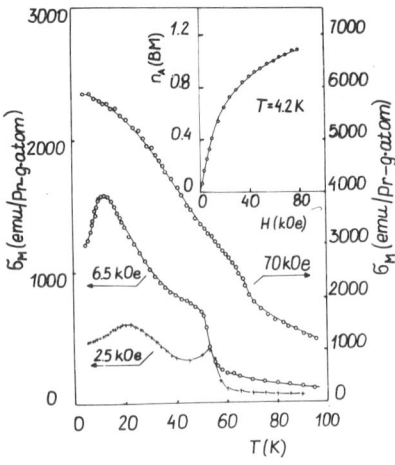

Fig. 6. σ_M vs. H for YbX$_3$ compounds (X = In, Tl, Sn)

Fig. 7. σ_M vs. T, and n_A vs. H for Pr$_3$In

Figure 7 presents the temperature dependence of the magnetization as well as the field dependence of the magnetization at 4.2 K. One can see that the first curve has unusual form. It displays two maxima, one at T $\leq$ 20 K, and the other at about 54 K. Both maxima vanish at the magnetic field strengths of ca. 30 and 3 kOe, respectively. There is some evidence for a much stronger anisotropy in Pr$_3$In than in Pr$_3$Tl. A complete explanation of the experimental data necessitates, however, the use of single crystals, and the performance of neutron diffraction experiments in a magnetic field.

It should be mentioned here that some preliminary results concerning the magnetic properties of Pr$_3$In have been reported recently by Hutchens et al.[8]

URANIUM COMPOUNDS

Another interesting group of compounds is that containing uranium and main group elements. The investigation of these materials was initiated by Trzebiatowski and expanded in our Institute (see e.g.[25-27]). Since the discovery of ferromagnetism in uranium hydride[28] in the early fifties, about 40 simple uranium compounds have been found to be magnetically ordered in the low temperature region. In this review only the recent, mostly unpublished data concerning the magnetic properties of some simple binary and ternary compounds as well as of the solid solutions will be presented in some detail.

I. THE AuCu$_3$-TYPE COMPOUNDS[29-31]

The results of magnetic measurements for this group of compounds are presented in Table II and Fig. 8. These are in fair

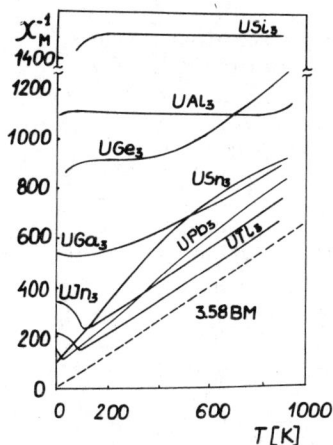

Fig. 8. χ_M^{-1} vs. T for UX_3 compounds.

agreement with the results of Buschow[32,33]. One can see that these compounds can be divided into two groups. The compounds of the first group exhibit temperature independent paramagnetism in a broad temperature range (e.g. USi_3, UAl_3, and UGe_3) or in a limited temperature range (e.g. UGa_3 and USn_3). To the second group belong UIn_3, UTl_3, and UPb_3 which become magnetically ordered in the low temperature region in agreement with the neutron diffraction experiments.[34,35] For UPb_3 the uranium magnetic moments are aligned oppositely in adjacent (001) sheets while the magnetic unit cell of both UIn_3 and UTl_3 is doubled in all three directions. Here the magnetic moments are aligned oppositely in adjacent (111) ferromagnetic planes.[35]

The properties of AuCu-type compounds can be explained in terms of the point charge model by assuming the initial ground term of uranium to be pure $^3H_4$. The negative ligand charge model gives then either the doublet Γ_3 or the triplet Γ_5 as the ground state depending on the ratio of the crystal field parameters B_6/B_4. On the other hand, the positive ligand charge model gives the ground level singlet Γ_1[29] which is not supported satisfactorily by experiment.

Table II. Magnetic Properties of UX_3 Compounds

Compound	U-U distance (Å)	Temp. indep. $\chi_M \cdot 10^6$	T_N (K)
USi_3	4.04	660	
UAl_3	4.29	1 080	
UGe_3	4.20	1 100	
UGa_3	4.26	1 950	
USn_3	4.63	9 500	
UIn_3	4.61		95-105
UTl_3	4.67		80-90
UPb_3	4.70		32

The magnetic properties of the first group mentioned above can be explained in terms of the nonmagnetic doublet Γ_3 as the ground state. However, the fall of the reciprocal susceptibility for the first three compounds in Table II (USi_3, UAl_3, and UGe_3) is not yet understood. In the case of the second group, the antiferromagnetic ordering observed in the low temperature region and the fulfillment of the Curie-Weiss law at higher temperatures suggest that the triplet Γ_5 ground state might be assigned to uranium combined with the heavy ligands. However, the positive ligand charge model cannot be excluded here (see e.g.[36]).

II. URANIUM MONOPNICTIDES AND THEIR SOLID SOLUTIONS WITH MONOCHALCOGENIDES

The magnetic properties of the antiferromagnetic monopnictides UX, the simple ferromagnetic monochalcogenides UY, and their solid solutions have received considerable attention in recent times.

The magnetic and electric properties of UP and UAs and their solid solutions, as well as the magnetic properties of UX-UY alloys, were investigated recently. It should be mentioned here that previous magnetic measurements made by Gulick and Moulton for UP only[37] have revealed, in addition to the normal susceptibility peak corresponding to the Néel point, a plateau in the low temperature region and a second, broad peak in the susceptibility at about 27 K. According to the work done in our Laboratory[38] both UP and UAs exhibit additional peaks in the temperature dependence of the susceptibility at 23 and 64 K, respectively (Fig. 9). Moreover, it should be mentioned that for UAs another less pronounced anomaly at 25 K has been also observed.

In UP the peak at 23 K is probably connected with a magnetic moment jump alone,[39] while in UAs, in addition to a magnetic moment jump,[40] the change of the magnetic structure from type I to type IA should be considered.[41] Moreover, the additional susceptibility maximum for UP is accompanied by a sharp increase in the electrical resistivity. For UAs a hump-backed maximum in the resistivity at 64 K is observed.[42] The above behavior stimulated the magnetic[38] and neutron diffraction investigations[43] of UP-UAs solid solutions. As shown in Fig. 10 the temperature of the transition: type I ⇌ type IA gradually decreases with increasing UP content and reaches 23 K for $UP_{0.75}As_{0.25}$. Between 55 and 75 mol. % of UP two phases of the type I and type IA structure were found together. Moreover, for all of the samples containing less than 75 mol.% of UP, the existence of a susceptibility anomaly has been observed at about 25K. This is similar to the case of pure UAs.

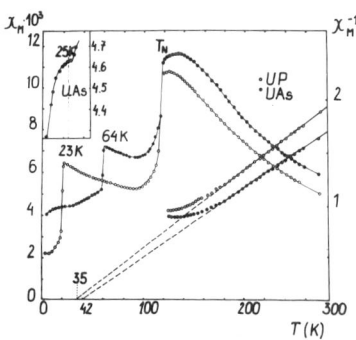

Fig. 9. χ_M vs. T for UP and UAs.

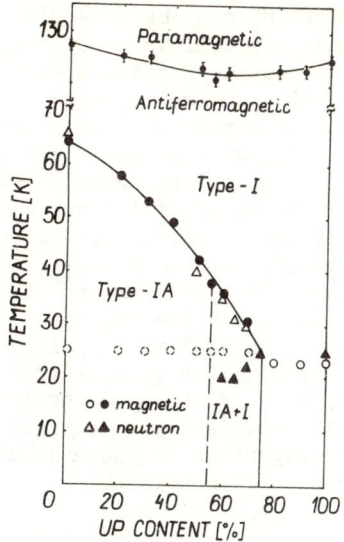

Fig. 10 Magnetic phase diagram for UP-UAs system

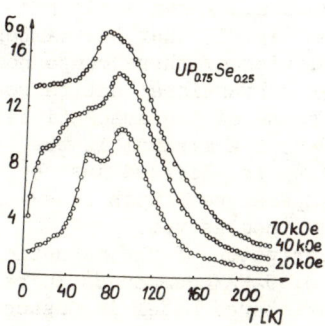

Fig. 11. σ_g vs. T for $UP_{0.75}Se_{0.25}$.

On the basis of the above results, it seems that the peaks observed for UP at 23 K and for UAs at 64 K have different origins (see e.g. [44,45]).

Some other problems arise in the case of the solid solutions involving uranium monopnictides and monochalcogenides (see e.g. [27,46,47]). The results of magnetic investigations of UAs-US and UP-USe systems in magnetic fields up to 80 kOe are as follows: [48,49]
In the $UAs_{1-x}S_x$ system the samples with $x \leq 0.15$ appear to be antiferromagnetic, while those with $x > 0.4$ are ferromagnetic with a peculiar temperature dependence of the magnetization. This latter feature is probably due to the high anisotropy observed in many ferromagnetic uranium compounds. The samples with $x = 0.25$ and 0.20 exhibit antiferromagnetic order in magnetic fields below 3 and 20 kOe, respectively, and become ferromagnetic at higher field strengths.

In the $UP_{1-x}Se_x$ system ferromagnetism is observed for $x \geq 0.3$, and antiferromagnetism for $x \leq 0.2$. For the samples with $0.25 \geq x \geq 0.15$, the temperature dependence of the magnetization seems to provide some evidence for the existence of two magnetic phase transitions in this system: the first one of the order-disorder type at about 100 K (see Fig. 11), and the second probably of the order-order type at about 50 K. The ordered phase of $UP_{0.75}Se_{0.25}$, which exists in the low temperature region transforms to a ferromagnetic one in a magnetic field of about 40 kOe. On the other hand, a preliminary neutron diffraction study[50] in zero magnetic field shows that the antiferromagnetic IA type of structure appearing at about 100 K exists in the range down to 4.2 K. Moreover, it was found that the second low-temperature maximum is connected with the moment jump without any change of magnetic structure.

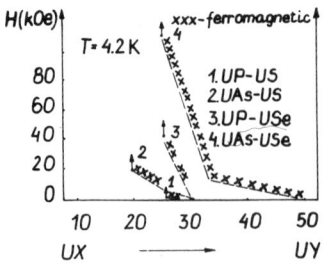

Fig. 12. Simplified phase diagrams for UX-UY systems (the data for UP-US system according to Ref. 51).

In Fig. 12 the simplified phase diagrams for UX-UY systems at 4.2 K are presented. It follows from it that the selenium presence increases the stability of the nonferromagnetic phases. Some attempts to explain the different stability of magnetic phases in terms of the RKKY interaction[26,49,52] as well as in terms of a more advanced approach, have been presented by Grunzweig-Genossar and Cahn.[53] At the same time it has been shown[54] that the correction due to the magnetic fields as compared with the magnitude of the RKKY interaction without the magnetic field is of the order of 10^{-8}, and therefore negligibly small. These models, however, have serious deficiencies in accounting for the magnetic properties of these alloys. Thus this problem needs some further examination from both the experimental and theoretical viewpoints.

III. $U_2N_2X(Y)$ TYPE COMPOUNDS

Uranium forms a series of ternary hexagonal or tetragonal $U_2N_2X(Y)$ compounds with the Vth and VIth groups of elements.

The hexagonal U_2N_2As, U_2N_2P, U_2N_2S and U_2N_2Se are antiferromagnetic below 415, 370, 242, and 233 K, respectively.[55] The first two compounds have the largest Néel temperatures of all known antiferromagnetic uranium compounds. One can see from Fig. 13 that the temperature dependence of the reciprocal susceptibilities of these compounds in the ordered state is rather nontypical for conventional antiferromagnets. All of these compounds show an increase of the susceptibility at the lowest temperatures. Moreover, for U_2N_2As a minimum in the susceptibility is observed just below the Néel temperature. Such behavior is not understood as yet.

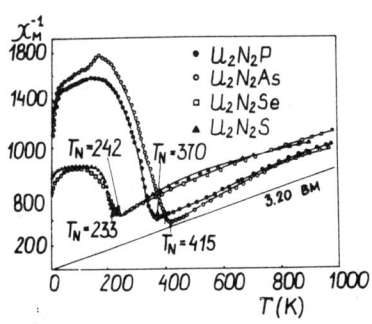

Fig. 13. χ_M^{-1} vs. T for U_2N_2P, U_2N_2As, U_2N_2S, and U_2N_2Te.

The description of the magnetic properties of U_2N_2X compounds in terms of the point charge model is unsuccessful because the real sequence of the crystal field levels is very sensitive even to a small variation in the nitrogen atom positions in coordination polyhedra, and the positions are not known with reasonable accuracy at the present time.

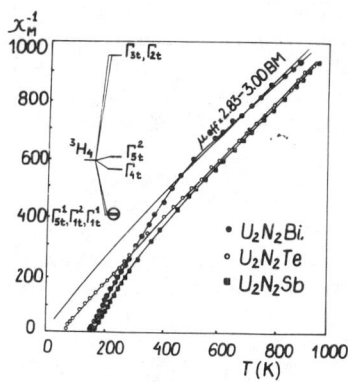

Fig. 14. χ_M^{-1} vs. T for U_2N_2Sb, U_2N_2Bi, and U_2N_2Te.

Contrary to the hexagonal compounds, the tetragonal U_2N_2Sb, U_2N_2Bi, and U_2N_2Te become ferromagnetic below 166, 160, and 75 K, respectively. The temperature dependence of the reciprocal susceptibilities (Fig. 14) exhibit considerable curvature in the lower temperature region and become almost linear at higher temperatures. Some explanation of such behavior can be obtained by applying the point charge model with a doublet Γ_{5t} as a ground state. Further investigation of this class of compounds is now in progress.

IV. UXY TYPE COMPOUNDS

Contrary to the antiferromagnetic tetragonal UOY compounds (see e.g. 25,27), the other ternaries were found to be ferromagnetic in accordance with the theoretical predictions.56,57 Their magnetic properties are listed in Table III.

Table III. Magnetic Properties of UXY Type Compounds

Compound	Crystal structure type	$\mu_{eff.}$ (BM)	Curie point (K)	Ref.
UPS	PbFCl	3.1	120	58
UPSe	PbFCl	3.3	105	59
UPTe	UGeTe	3.2	85	59
UAsS	PbFCl	3.3	124	60, 61
UAsSe	PbFCl	3.4	113	60, 61
UAsTe	UGeTe	3.3	66	60, 61
USbSe	PbFCl	3.3	125	59
USbTe	PbFCl	3.4	129	59

The temperature dependence of the reciprocal magnetic susceptibility of UsbTe (Fig. 15) is typical of all of the UXY compounds considered. It has curvature at low temperatures and becomes linear in the higher temperature region. The value of the effective magnetic

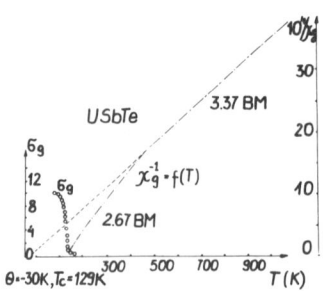

Fig. 15. χ_g^{-1} and σ_g at 7.8 kOe vs. T for USbTe

moment estimated from the linear part of the χ_M^{-1} vs. T plot (3.1 - 3.4 BM) may be obtained from the crystal field level scheme with two very closely spaced singlets Γ_{t2} and Γ_{t3} as the lowest levels.[35] Moreover, it follows then that the perpendicular susceptibility should be one order of magnitude smaller than the parallel one. This was confirmed by experiment.[62]

At low temperatures single crystals of UAsS and UAsSe exhibit an abrupt increase in the magnetization along the c-axis at some critical field, above which the magnetization is practically constant. The hysteresis loop is almost rectangular, possibly due to the fact that the ratio of the magnetocrystalline anisotropy energy to the exchange energy is so great that very narrow domain walls (of the thickness of a few interatomic distances) formed under such conditions remain frozen until the applied field exceeds some critical value.[61]

Neutron diffraction experiments showed that UAsSe and USbSe are uniaxial ferromagnets with magnetic moments along the c-axis.[63]

V. U_3Y_5 TYPE URANIUM CHALCOGENIDES[64]

Figure 16 shows the temperature dependence of the experimental and calculated susceptibilities for tetragonal U_3S_5 as well as for orthorhombic U_3S_5 and U_3Se_5. The calculated curves have been obtained for the temperatures above 50 K on the basis of the point charge model by assuming that all of the uranium ions are in the U^{4+} configuration. Although the stoichiometry of this class of compounds made the $U_2^{3+}U^{4+}Y_5$ model very tempting, the susceptibilities calculated for this valence scheme (considering the doublet for the U^{3+} ions, and the two singlets for the U^{4+} ion as the lowest levels, and neglecting all temperature independent terms) were much lower than the experimental ones.

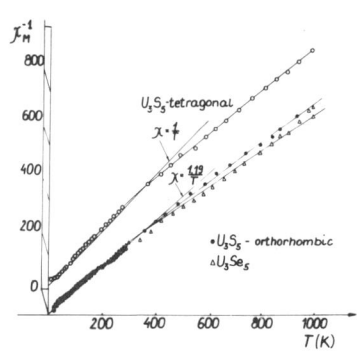

Fig. 16. χ_M^{-1} vs. T for U_3Y_5 compounds (Y = S, Se)

At low temperatures all of the U_3Y_5 compounds exhibit anomalies in the temperature dependence of the magnetization which seem to be connected with the transitions to magnetic phases with some ferromagnetic contribution. The magnetization maxima were observed at low magnetic fields at 10, 12, and 16 K for tetragonal U_3S_5, orthorhombic U_3S_5 and U_3Se_5, respectively, and disappeared in higher magnetic fields. However, without neutron diffraction experiments one cannot determine the character of these transitions unambiguously.

Finally it should be pointed out that although the magnetic properties of many uranium compounds in the paramagnetic region can be more or less satisfactory explained in terms of the point charge model, one has to be aware that this model offers only an approximate description of reality.

Some of the results reported here were obtained in close cooperation with the International Laboratory of High Magnetic Fields and Low Temperatures in Wroclaw (Poland).

The author is greatly indebted to Dr. W. Suski and Mr. A. Czopnik for their kind help in preparing the manuscript.

REFERENCES

1. T. Tsuchida, W. E. Wallace, J. Chem. Phys. 43, 3811 (1965).
2. K. H. J. Buschow, H. W. de Wijn, A. M. Van Diepen, J. Chem. Phys. 50, 137 (1969).
3. R. D. Hutchens, W. E. Wallace, J. Solid State Chem. 3, 564 (1971).
4. G. K. Shenoy, B. D. Dunlap, G. M. Kalvius, A. M. Toxen, R. J. Gambino, J. Appl. Phys. 41, 1317 (1970).
5. M. Loewenhaupt, Z. Physik 258, 209 (1973).
6. P. Lethuillier, G. Quezel-Ambrunaz, A. Percheron, Solid State Commun. 12, 105 (1973).
7. P. Lethuillier, J. Pierre, G. Fillion, B. Barbara, Phys. Stat. Sol. (a) 15, 613 (1973).
8. R. D. Hutchens, W. E. Wallace, N. Nereson, 10th Rare Earth Res. Conf. Carefree, Arizona, April 30 - May 3, 1973.
9. E. Bucher, J. P. Maita, A. S. Cooper, Phys. Rev. B6, 2709 (1972).
10. K. Andres, E. Bucher, S. Darack, J. P. Maita, Phys. Rev. B6, 2716 (1972).
11. R. J. Birgeneau, J. Als-Nielsen, E. Bucher, Phys. Rev. B6, 2724 (1972).
12. B. R. Cooper, Phys. Rev. B6, 2730 (1972).
13. B. Stalinski, A. Czopnik, N. Iliew, T. Mydlarz, Conf. on Rare Earth Metals, Alloys, and Compounds, Moscow, Sept. 12-17 (1972).
14. C. P. Bean, D. S. Rodbell, Phys. Rev. 126, 104 (1962).
15. K. Tomala et al. (priv. commun.).
16. B. Staliński, Z. Kletowski, Z. Henkie, Phys. Stat. Sol. (a) 19, K165 (1973).
17. B. Staliński, A. Czopnik (to be published).
18. E. P. Wohlfarth, Phys. Letters 22, 280 (1966).
19. N. Rivier, M. J. Zuckermann, Phys. Rev. Letters 21, 904 (1968).
20. A. B. Kaiser, S. Doniach, Int. J. Magn. 1, 11 (1970).
21. Z. Kletowski (priv. commun.).
22. B. Staliński, A. Czopnik, N. Iliew, T. Mydlarz, Int. Conf. on Magnetism, Moscow, August 22-28, 1973.

23. E. Bucher, P. H. Schmidt, A. Jayaraman, K. Andres, J. P. Maita, N. Nassau, P. D. Dernier, Phys. Rev. B2, 3911 (1970).
24. B. Staliński, A. Czopnik, N. Iliew, T. Mydlarz, Phys. Stat. Sol. (a) 19, K161 (1973).
25. W. Trzebiatowski, Int. Kong. Magnetismums, Dresden 1966. VEB Deutscher Verlag f. Grundstoffindustrie, Leipzig 1967, p. 88.
26. J. Grunzweig-Genossar, M. Kuznietz, F. Friedman, Phys. Rev. 173, 562 (1968).
27. D. J. Lam, A. T. Aldred in "Electronic Structure and Related Properties of Actinides," Eds. J. B. Darby and A. J. Freeman, (Academic Press, Inc., NY to be published).
28. W. Trzebiatowski, A. Sliwa, B. Staliński, Roczniki Chem. 26, 110 (1952), 28, 12 (1954).
29. J. Mulak, A. Misiuk, Bull. Acad. Polon. Sci., sér. sci. chim. 19, 207 (1971).
30. A. Misiuk, J. Mulak, A. Czopnik, Bull. Acad. Polon. Sci., sér. sci. chim. 20, 459 (1972).
31. A. Misiuk, J. Mulak, A. Czopnik, Bull. Acad. Polon. Sci., sér. sci. chim. 20, 891 (1972).
32. K. H. J. Buschow, H. J. Van Daal, Magnetism and Magnet. Mater. AIP Conf. Proc. No. 5, 1464 (1971).
33. K. H. J. Buschow, J. Less-Common Metals 31, 165 (1973).
34. J. Leciejewicz, A. Misiuk, Phys. Stat. Sol. (a) 13, K79 (1972).
35. A. Murasik, J. Leciejewicz, S. Ligenza, A. Misiuk, Phys. Stat. Sol. (in press).
36. E. Bucher, J. P. Maita, Solid State Commun. 13, 215 (1973).
37. J. M. Gulick, W. G. Moulton, Phys. Letters A 35, 429 (1971).
38. R. Troć (to be published).
39. N. A. Curry, Proc. Phys. Soc. London 89, 427 (1966).
40. G. H. Lander, M. H. Mueller, J. P. Reddy, Phys. Rev. B6, 1880 (1972).
41. J. Leciejewicz, A. Murasik, R. Troć, Phys. Stat. Sol. 30, 157 (1968).
42. R. Troć, Z. Kletowski (to be published).
43. J. Leciejewicz, A. Murasik, T. Palewski, Phys. Stat. Sol. (b) 46, K67 (1971).
44. C. Long, Y. L. Wang, Phys. Rev. B3, 1656 (1971).
45. P. Erdös, J. M. Robinson, 18th Ann. Conf. Magnetism, Magnet. Mater., Denver 1972 (in press).
46. G. H. Lander, M. H. Mueller in "Electronic Structure and Related Properties of the Actinides," Eds. J. B. Darby and A. J. Freeman (Academic Press, Inc., NY, to be published).
47. T. Palewski, W. Suski, T. Mydlarz, Int. J. Magn. 3, 269 (1972). K. P. Belov, A. S. Dmitrievski, R. Z. Levitin, T. Palewski, Yu, F. Popov, W. Suski, Zh. Eksp. Teor. Fiz. Letters 17, 81 (1973).
48. W. Suski, T. Palewski, T. Mydlarz, Int. J. Magn. (in press).

49. W. Suski, T. Palewski, T. Mydlarz, H. Reizer-Netter, Int. Conf. on Magnetism, Moscow, August 22-28, 1973.
50. J. Leciejewicz, S. Ligenza, A. Murasik, T. Palewski, R. Troć (priv. commun.).
51. J. Crangle, M. Kuznietz, G. H. Lander, Y. Baskin, J. Phys. C$\underline{2}$, 925 (1969).
52. H. Adachi, S. Imoto, J. Phys. Chem. Solids $\underline{34}$, 1537 (1973).
53. J. Grunzweig-Genossar, J. W. Cahn, Int. J. Magn. (to be published).
54. A. Kolodziejczak, Acta Phys. Polon. A $\underline{41}$, 51 (1972).
55. R. Troć, Z. Zolnierek, Int. Conf. on Magnetism, Moscow, August 22-28, 1973.
56. A. Murasik, J. Niemiec, Bull. Acad. Polon. Sci., sér. sci. chim. $\underline{13}$, 291 (1965).
57. J. Przystawa, W. Suski, Phys. Stat. Sol. $\underline{20}$, 451 (1967).
58. A. Zygmunt, A. Czopnik, Phys. Stat. Sol. (a) $\underline{18}$, 731 (1973).
59. A. Zygmunt (to be published).
60. A. Zygmunt, M. Duczmal, Phys. Stat. Sol. (a) $\underline{9}$, 659 (1972).
61. C. Bazan, A. Zygmunt, Phys. Stat. Sol. (a) $\underline{12}$, 649 (1972).
62. K. P. Belov, A. S. Dmitrievski, A. Zygmunt, R. Z. Lewitin, W. Trzebiatowski, Zh. Eksp. Teor. Fiz. $\underline{64}$, 582 (1973).
63. J. Leciejewicz, A. Zygmunt, Phys. Stat. Sol. (a) $\underline{13}$, 657 (1972).
64. W. Suski, H. Reizer-Netter (to be published).

STUDIES ON RARE EARTH COBALTITES, $La_{1-x}Sr_xCoO_3$ AND RELATED SYSTEMS

C. N. R. Rao
Department of Chemistry
Indian Institute of Technology, Kanpur 208016, India

V. G. Bhide
National Physical Laboratory
New Delhi 110012, India

ABSTRACT

In rare earth cobaltites, the cobalt ions are present mainly in the diamagnetic low-spin Co^{III} state at low temperatures. The Co^{III} ions transform to high-spin Co^{3+} ions with increase in temperature. At higher temperatures, there is electron-transfer from Co^{3+} to Co^{III} ions producing intermediate states. We see spin-state transitions in these cobaltites in the range 250-870 K. At very high temperatures, the cobaltites show evidence for localized-itinerant electron transitions. In $La_{1-x}Sr_xCoO_3$, there is onset of ferromagnetism at $x > 0.125$, at which point there is a structural discontinuity and electrons become itinerant. The composition with $x = 0.5$ is metallic and $T_c = 230$ K. The ferromagnetic component in $La_{1-x}Sr_xCoO_3$ increases with x in the range 0.125-0.50. In $LaCo_{1-x}Fe_xO_3$, the iron ions are always present in the high-spin state and the d-electrons get more localized with increase in x. Effect of substitution by ions like Th^{4+} and Ni^{3+} in $LaCoO_3$ have also been examined. Catalytic properties of rare earth cobaltites appear to be related to the spin state equilibria.

Transition metal oxides with perovskite structure exhibit metallic conductivity and Pauli paramagnetism if the spin of the transition metal ion $S \leq 1/2$. The outer d-electrons in such systems are well described by band theory.[1-3] Typical examples of oxides exhibiting collective d-electron behavior are $LaTiO_3$ and $LaNiO_3$ (S = 1/2). If $S \geq 2$, however, these oxides show an atomic moment which is described by crystal field theory; thus, d-electrons in oxides like $LaMnO_3$ and and $LaFeO_3$ generally exhibit localized behavior.[4] In oxides where there is coexistence of the low- and the high-spin states, it becomes possible to study their properties in terms of the localized versus collective behavior of d-electrons. Such a coexistence of high- and low-spin states is found in $LaCoO_3$ and Raccah and Goodenough[5] have shown that this oxide undergoes a first-order transition from a localized-electron state to a collective-electron state. Extensive studies in our laboratories[6] employing Mössbauer spectroscopy and other measurements have enabled us to understand the nature of spin-state equilibria responsible for the interesting magnetic and electrical properties of $LaCoO_3$. We have carried out detailed investigations of the spin-state equilibria and the associated magnetic and electron transport properties in several rare earth cobaltites besides $LaCoO_3$, employing Mössbauer spectroscopy and other measurements, and we shall review these results briefly. We have also focussed our attention on the so-called low spin-high spin transitions[7] exhibited by rare earth cobaltites and we shall also also discuss this aspect.

We have recently investigated the system, $La_{1-x}Sr_xCoO_3$, employing Mössbauer spectroscopy and other measurements. In this system, for every Sr^{2+} ion introduced we create a tetravalent cobalt ion. Since $Co^{3+}-O^{2-}-Co^{4+}$ interaction is ferromagnetic,[8-10] the $La_{1-x}Sr_xCoO_3$ system is expected to show interesting electrical and magnetic properties. Jonker and Van Santen[8] as well as Raccah and Goodenough[11] have carried out studies of these properties. Paramagnetic susceptibility of these oxides could be explained by the Curie-Weiss law, with Θ changing from a negative value for low Sr^{2+} content ($x < 0.1$) to a positive value for higher x. Consequently, these compounds exhibit ferromagnetism in the range $0.1 < x < 0.5$. Further, the composition, $La_{0.5}Sr_{0.5}CoO_3$ is metallic. The simultaneous observation of itinerant electron behavior and ferromagnetism in these oxides makes them very novel indeed.

Besides $La_{1-x}Sr_xCoO_3$, we have studied $LaCo_{1-x}Fe_xO_3$ and similar systems. Since rare earth cobaltites are reported to be good catalysts for the oxidation of CO,[12] we have examined the catalytic properties of the cobaltites in relation to spin-state equilibria of cobalt ions.

RESULTS AND DISCUSSION

<u>Rare Earth Cobaltites</u> : Magnetic susceptibility data on $LaCoO_3$, $PrCoO_3$, $NdCoO_3$, and $HoCoO_3$ distinctly show three regions in the $1/\chi_g$-T curves: (i) a low temperature region where $1/\chi_g$ is essentially linear with temperature giving a lower effective magnetic moment; (ii) an intermediate temperature region where $1/\chi_g$ is independent of temperature, and (iii) a high temperature region where $1/\chi_g$ is again linear with temperature, but gives a higher effective magnetic moment. It appears that the cobalt ions exist predominantly in the low-spin $(t_{2g}^6 e_g^0)$ Co^{III} state ($^1A_{1g}$) at lower temperatures. As the

temperature is increased, the diamagnetic Co^{III} ions are transformed progressively into paramagnetic Co^{3+} ions ($t_{2g}^4 e_g^2$). The energy of Co^{III} state is lower than that of the Co^{3+} state ($^5T_{2g}$); the energy difference between the two states is around 0.05 eV. The plateau in the susceptibility curve in the intermediate temperature region in $LaCoO_3$ has been identified with a variation of the population of the low versus high spin ions and establishment of short range order.[5] Evidence for such ordering is found by the observation of phase transitions in DTA curves as well as in other measurements.[5,6] After the establishment of short range order, $LaCoO_3$ undergoes a symmetry change from $R\bar{3}c$ to $R\bar{3}$. A similar situation is likely to exist in the other cobaltites as well; if so, we would expect to find ordering transitions in the corresponding temperature ranges. $GdCoO_3$ does not show the plateau region; however, there is a change in the slope of the susceptibility curve around 550 K where there could be an ordering transition.

DTA curves of $LaCoO_3$, $NdCoO_3$ and $GdCoO_3$ show the presence of small endothermic transitions in the region 500-700 K, possibly due to the ordering transitions referred to earlier. In addition, the DTA curves show endothermic transitions in the 1000-1200 K range due to localized-itinerant electron transitions similar to that described in the case of $LaCoO_3$.[5,6]

The spin-state equilibria in cobaltites can be understood by an examination of the variation of the magnetic susceptibility with temperature. Ordinarily, $\chi_g T$ of substances does not vary with temperature. However, when the proportion of the paramagnetic species varies with temperature, $\chi_g T$ would not remain constant, but would obey the relation (1) where n and m represent the proportions of the paramagnetic (high-spin) and the diamagnetic (low-spin) ions respectively:

$$\chi_g T = \frac{N^2 \mu^2}{3R} \cdot \frac{n}{n+m} \quad (1)$$

Thus, the ratio Co^{3+}/Co^{III} should keep increasing with temperature. Plots of $\chi_g T$ against temperature for $LaCoO_3$, $NdCoO_3$ and $GdCoO_3$ show that $\chi_g T$ increases with temperature as expected, although not in a uniform manner.

Mössbauer spectra obtained by using the cobaltites as sources show two resonances, one centered close to 0.0 mm/sec and another around 0.5 mm/sec.[6] Based on isomer shift systematics, we can safely assign the high energy resonance to Fe^{3+} and the low energy resonance to Fe^{III} arising out of electron - capture decay of Co^{3+} and Co^{III} respectively.[6] If the assignment of the two resonance is correct, the relative intensities of the two states should comply with equation 1. However, a plot of the relative proportion of Fe^{3+} (as measured by the intensity of the high energy resonance) shows that the $[Fe^{3+}]$ increases up to a temperature and then decreases, eventually becoming zero at a high temperature. The temperatures at which $[Fe^{3+}]$ reaches zero concentration correspond to the localized-itinerant electron transitions.

We see that the $\chi_g T$ data and Mössbauer data show similar behavior only in the low-temperature region. This apparent discrepancy between the susceptibility and the Mössbauer data can be understood if we propose that at high temperatures, there is electron-transfer Co^{3+} to Co^{III} giving rise to divalent $Co^{II}(t_{2g}^6 e_g^1)$ and tetravelent $Co^{4+}(t_{2g}^3 e_g^2)$ which are paramagnetic. The corresponding states of iron, Fe^{II} and Fe^{4+}, would however, have isomer shifts near the zero velocity position just as Fe^{III}. Such electron-transfer via orbitals of e_g symmetry is compatible with $\Delta_{cac}^{\sigma} > \Delta_{cac}^{\pi}$. We do not see in the Mössbauer spectra, evidence for the formation of Co^{2+} and Co^{IV} proposed by Raccah and Goodenough[7] in the case of $LaCoO_3$; Fe^{2+} produced by Co^{2+} would have given a very large isomer shift of the order of 1.0 mm/sec which we have not detected. We, therefore, conclude that at low temperatures, the cobaltites contain mainly diamagnetic Co^{III} ions which are partly transformed to paramagnetic Co^{3+} ions up to a particular temperature. Above this temperature, an e_g electron is transferred from Co^{3+} to Co^{III} to form Co^{4+} and Co^{II} ion pairs. At very high temperatures, the concentration of Co^{3+} decreases below the limit of detection. Our explanation for the difference between Mössbauer and susceptibility data finds support from our studies on $HoCoO_3$. In this heavy rare earth cobaltite, the proportions of low and high spin states obtained by the two methods is identical. This shows that where there is no such correspondence, electron-transfer between Co^{3+} and Co^{III} may occur. Accordingly, the resistivity of $HoCoO_3$ is much higher than that of other cobaltites. In $HoCoO_3$, the relative proportions of Fe^{3+} and Fe^{III} remain 1:1 beyond a certain temperature.

The temperature at which Co^{3+} (or Fe^{3+}) reaches a maximum (200 K in $LaCoO_3$ and $NdCoO_3$ and 400 K in $GdCoO_3$) is lower than the ordering temperatures evidenced in magnetic susceptibility and Debye-Waller factor measurements or in DTA curves. Apparently, the ordering of the different spin-states of cabalt takes place after the electron-transfer from Co^{3+} to Co^{III}. However, there seems to be some relation between the ordering temperatures seen in the various measurements and the temperatures at which the relative proportion of Fe^{3+} becomes unity. In $HoCoO_3$, where there is an exact correspondence between magnetic susceptibility and Mössbauer data with respect to the proportions of Co^{3+} and Co^{III}, a transition is seen when the two spin states get equally occupied (1:1).

As mentioned earlier Fe^{3+} eventually becomes zero at 1210 K in $LaCoO_3$ corresponding to the localized → collective electron transition. At this temperature, $LaCoO_3$ shows a significant thermal anomaly ($\Delta H \geqslant 1$ kcal mol^{-1}) and the material becomes metallic beyond this transition. $NdCoO_3$ and $GdCoO_3$ show thermal anomalies around 1000 K at which temperature $[Fe^{3+}]$ becomes zero; the ΔH of these transitions (as shown by DTA) is, however, considerably smaller than in $LaCoO_3$. Debye-Waller factor data also clearly indicate these transitions. It appears that the high temperature transitions in these cobaltites also correspond to the localized-collective electron transition although the nature of the transitions may be different. It is relevant to note at this point that after the high-temperature transition, the cobaltites exhibit only a single resonance (corresponding to the lower energy resonance assigned to Co^{III}, Co^{4+} or Co^{II}). Under these conditions, there would be no localized e_g electrons, but σ^* bands characteristic/itinerant systems. A possible description of the two cobalt sites at very high temperatures would be, $(t_{2g}^6 \sigma^{*5})_a$ $(t_2^5 \sigma^{*1})_{\frac{1}{2}-a}$ and $(t_{2g}^4 \sigma^{*2-\xi})_a (t_2^5 \sigma^{*1})_{\frac{1}{2}-a}$ where the $t_{2g}^5 \sigma^{*1}$ corresponds to the intermediate Co^{iii} state with $t_{2g}^5 e_g^1$ configuration.

In the case of $LaCoO_3$, $NdCoO_3$ and $GdCoO_3$, the Debye-Waller factor increases with temperature significantly before the ordering transition peak in the DTA curves, indicating either short range ordering or a temperature region of larger atomic vibrations. In view of the susceptibility data, this transition indicated by Debye-Waller factor may be taken to indicate establishment of short range order. These phase transitions can be conveniently studied by an examination of the temperature variation of the Lamb-Mössbauer factor (area under resonance). A plot of the area under the resonance against temperature clearly brings out these transitions. The Lamb-Mössbauer results agree well with the DTA and x-ray data. It is significant that high-temperature transitions (1000-1200 K) are preceded by a decrease in the Lamb-Mössbauer factor probably due to the occurrence of the large ionic vibrations similar to those discussed by Raccah and Goodenough[5] in the case of $LaCoO_3$. Above the transition temperature, the Lamb-Mössbauer factor increases indicating the establishment of long-range order. There appears to be no doubt that these transitions are electronic in origin particularly because of the absence of any structural changes in the transition region. These transitions support the hypothesis of Goodenough that the crystal field (localized) and band limits of d-electrons are distinct thermodynamic states. Accordingly, the transitions between the two states can be first or second order in rare earth cobaltites.

Spin-State Transitions in Cobaltites : Bari and Sivardiere[7] have discussed models for low-spin-high-spin transitions in transition metal compounds. The two-sublattice-spin-structure case considered by them shows a variety of behaviors and this model has been applied to $LaCoO_3$. In this model the low-spin ground state first goes to a two-sublattice structure which then goes to the high-spin state. If the first transformation is first order, the second one can be first or second order; if the first one is second order, the second one can only be second order. In the case of $GdCoO_3$, Bari and Sivardiere[7] suggest a progressive increase of high-spin population and decreasing level separation with increasing temperature. Since little information is available on spin-state transitions, we considered it worthwhile to characterize them in the case of rare earth cobaltites based on our studies discussed in the previous section.

Detailed studies carried out by us on rare-earth cobaltites show that the transitions in most of the cases are associated with ΔH values which are much less than one kcal mol^{-1}. The transitions are not accompanied by any change volume, but may be associated with a change in symmetry ($R\bar{3}c$ to $R\bar{3}$ as in $LaCoO_3$). Ordering sets in only after Co^{III} is transformed to Co^{3+} to a fair measure. All the Co^{III} ions do not continuously transform to Co^{3+} as suggested by Bari.[7] After considerable transformation takes place, other spin states (Co^{II}, Co^{4+}, etc.) are produced in all the cobaltites except $HoCoO_3$ where Co^{3+} / Co^{III} stabilizes at one. Even in the other cobaltites, transitions are found when the relative population of $[Co^{3+}]$ is near unity. The transitions in most of the cobaltites appear to be thermodynamically second order, although it is difficult to strictly classify them. We are presently characterizing these transitions by a variety of techniques.

$La_{1-x}Sr_xCoO_3$: $La_{1-x}Sr_xCoO_3$ compounds possess a rhombohedral structure with the rhombohedral distortion decreasing with increase in x; when x = 0.5, the oxide is cubic. These oxides show discontinuities in their lattice parameters around x = 0.125. In the DTA curves, $La_{1-x}Sr_xCoO_3$ compounds do not show the ordering transition seen in $LaCoO_3$ around 680 K beyond x = 0.05; beyond this composition neither do they show the plateau in the $1/\chi$-T curve as in $LaCoO_3$. The endothermic transition in $LaCoO_3$ corresponding to the localized-itinerant electron transition around 1200 K is not seen in

$La_{1-x}Sr_xCoO_3$ for compositions with $x > 0.2$. In fact, beyond $x = 0.3$, these oxides become metallic as indicated by resistivity and Seebeck coefficient data.

Mössbauer spectra of $La_{1-x}Sr_xCoO_3$ ($x \leq 0.125$) at room temperature show two resonances similar to $LaCoO_3$, the lower velocity resonance being slightly broader due to the superposition of resonance due to tetravalent cobalt and Co^{III}. For $0.125 < x \leq 0.50$, only a single peak located around 0.25 mm/sec is noticed, this location being close to that in $La_{1-x}Sr_xFeO_3$ ($x > 0.4$) reported by Shimony and Knudsen.[13] This observation implies that the time-averaged electronic configuration of all the cobalt ions is the same. That is, the 3d hole (tetravalent cobalt) around the Sr^{2+} impurity is mobile providing evidence for the formation of the impurity band.

Mössbauer spectra of $La_{1-x}Sr_xCoO_3$ ($x \leq 0.125$) at 78 K were similar to the room temperature spectra. However, for compounds with $0.125 < x \leq 0.50$, a magnetically split spectrum was observed against the background of a single line spectrum. As x increases, the intensity of the magnetically split spectrum increases at the cost of the single line spectrum. Further, the magnetically split spectrum is seen for compositions with the positive paramagnetic Curie temperatures. Only those samples ($x > 0.125$) which give a single line spectrum at room temperature exhibit the magnetically split spectrum at 78 K. That is, the formation of the impurity band (itinerant electron behavior) is connected with magnetic ordering.

Temperature variation of the h.f. field as well as the area under the resonance gave T_c of $La_{0.5}Sr_{0.5}CoO_3$ to be 230 K. The h.f. field is practically constant (320 ± 20 kOe) for all the compositions in the range $x = 0.2-0.5$. The variation of the normalized h.f. field reduced spontaneous magnetization (from the data of Watanabe[9]) and the Brillouin function for $S = 3/2$ are all similar ruling out the possibility of biquadratic exchange.[14] We find the value of β near the Curie temperature to be 0.361, a value similar to that in orthoferrites, ruling out the validity of molecular field theory.

It is most interesting that for samples with $x > 0.125$ showing magnetic ordering, the hyperfine field (320 ± 20 kOe) and the isomer shift (0.25 ± 0.04 mm/sec) are intermediate between those for high spin Fe^{3+} and tetravalent iron. Above T_c also the isomer shift corresponds to a similar intermediate state.

Goodenough[10] has modified Anderson's theory[15] and pointed out that if octahedral-site magnetic cations are on opposite sides of a common anion, the interaction is ferromagnetic provided one cation has completely empty e_g orbitals and the other has half-filled e_g orbitals. He postulates the existence of Co^{3+} and low-spin Co^{IV} to explain ferromagnetism in $La_{1-x}Sr_xCoO_3$. According to Goodenough, Co^{3+} transforms to Co^{2+} on the transfer of a 2p electron from oxygen while the Co^{IV} would transform to Co^{III}. The observed Mössbauer spectra, however, do not entirely support this mechanism, since we would expect to see more than one six-finger pattern below T_c whatever be the relative values of the relaxation time and Larmour precession time. According to Zener's double exchange mechanism,[16] there would be simultaneous transfer of an electron from the oxygen 2p orbital to Co^{IV} and from the Co^{3+} to the 2p orbital of oxygen, the resonance energy between the configurations, $Co^{3+} - O^{2-} - Co^{4+}$ $Co^{4+} - O^{2-} - Co^{3+}$ being large if spins are parallel. The observation of one six-finger pattern below T_c may be taken to indicate that the above transformation is very rapid. The resonance energy would be maximum if the tetravalent cobalt had the intermediate spin configuration, Co^{iv} ($t_{2g}^4 e_g^1$) and this configuration would permit ferromagnetism by Anderson-Goodenough as well as by Zener's mechanism.

The ferromagnetic component in $La_{1-x}Sr_xCoO_3$ ($0.125 < x \leq 0.50$) increases with x. This is because, the Sr^{2+}-rich region is associated with ferromagnetic clusters. Apparently, the ferromagnetic particle size distribution curve shifts towards lower particle size as Sr^{2+} content decreases. Variation of the paramagnetic-ferromagnetic ratio with temperature also supports this conclusion. We also find that ferromagnetism cannot be observed below x = 0.1. It is at this composition that the paramagnetic Curie temperature changes sign. Around this composition, we also see the structural discontinuities. Another consequence of the chemical inhomogeneities separating the La^{3+}-rich paramagnetic regions from Sr^{2+}-rich ferromagnetic regions is the difference between the paramagnetic Curie temperature and T_c. Further, La^{3+}-rich and Sr^{2+}-rich regions would be associated with localized and itinerant electrons respectively, making the material 'electronically inhomogeneous' by virtue of coexistence of the localized and itinerant electrons.[11] We are presently investigating the $La_{1-x}Sr_xCoO_3$ system by neutron diffraction.

Other Systems: We have investigated $La_{1-x}Th_xCoO_3$, $LaCo_{1-x}Fe_xO_3$ and $LaCo_{1-x}Ni_xO_3$ systems. Addition of Th^{4+} to $LaCoO_3$ renders it progressively an n-type material. Trivalent iron is always present in the high spin state in $LaCo_{1-x}Fe_xO_3$ and the resistivity of this system increases with x; up to x = 0.1, the general features of $LaCoO_3$ are seen. Addition of nickel to $LaCoO_3$ makes it progressively itinerant; thus, the resistivity decreases with increase in x in this system. We have recently initiated studies on $Sm_{1-x}Ca_xCoO_3$ and $Gd_{1-x}Ca_xCoO_3$ to see if ferromagnetism is exhibited in these compounds.

Catalytic Properties : We have examined the catalytic activity of various rare earth cobaltites and their solid solutions for the oxidation of CO. We find that $NdCoO_3$ and $HoCoO_3$ are much better catalysts than $LaCoO_3$. The catalytic activity seems to be related to the spin state populations in the cobaltites. In $HoCoO_3$ and $NdCoO_3$, the relative proportion of high spin Co^{3+} reaches unity at fairly low temperatures.

ACKNOWLEDGEMENTS

We are thankful to our coworkers, Messrs D. S. Rajoria, G. Rama Rao and Om Prakash, and Dr. P. Ganguly for their active collaboration.

REFERENCES

1. J. B. Goodenough, Progress in Solid State Chemistry (ed. H. Reiss), 5, 145 (1972).
2. J. B. Goodenough, Phys. Rev. 164, 789 (1968).
3. C. N. R. Rao and G. V. Subba Rao, phys. stat solidi 1a, 597 (1970).
4. G. V. Subba Rao, B. M. Wanklyn and C. N. R. Rao, J. Phys. Chem. Solids 32, 345 (1971).
5. P. M. Raccah and J. B. Goodenough, Phys. Rev. 155, 932 (1967).
6. V. G. Bhide, D. S. Rajoria, G. Rama Rao and C. N. R. Rao, Phys. Rev. B6, 1021 (1972).
7. R. A. Bari and J. Sivardiere, Phys. Rev. 5B, 4466 (1972).
8. G. H. Jonker and J. H. Van Santen, Physica 19, 120 (1953).
9. H. Watanabe, J. Phys. Soc. Japan 12, 515 (1957).
10. J. B. Goodenough, J. Phys. Chem. Solids 6, 287 (1958).
11. P. M. Raccah and J. B. Goodenough, J. App. Phys. 39, 1209 (1968).
12. R. J. H. Voorhoeve, J. P. Rameika, P. E. Freeland and B. T. Matthias, Science, 177, 353 (1972).
13. U. Shimony and J. M. Knudsen, Phys. Rev. 144, 361 (1966).
14. N. Menyuk, P. M. Raccah and K. Dwight, Phys. Rev. 166, 510 (1968).
15. P. W. Anderson, Phys. Rev., 79, 350, 705 (1950).
16. C. L. Zener, Phys. Rev. 82, 403 (1951).

^{61}Ni MÖSSBAUER STUDIES OF SUBSTITUTED Ni SPINELS*

John C. Love
Florida Institute of Technology, Melbourne, Fla. 32901

Felix E. Obenshain
Oak Ridge National Laboratory, Oak Ridge, Tenn. 37830

ABSTRACT

^{61}Ni nuclear gamma resonance (NGR) spectra were measured at 4.2K for the mixed spinels $NiFe_xCr_{2-x}O_4$ (0<x<2), and magnetic hyperfine fields $|H_{A,B}(x)|$ were derived for tetrahedral (A) and octahedral (B) sites. The B-site fields lie in the range (∿20-94)kOe, quite comparable with values obtained for many octahedrally-coordinated Ni compounds. For x<1, Ni ions partially occupy A-sites and the hf fields for these ions are anomalously large, varying from 440 kOe for x = 0 to ∿630 kOe for x = 0.5. This later field value, which is also found for the related compound $NiFe_{0.5}V_{1.5}O_4$, is the largest yet reported for ^{61}Ni. It is suggested that the composition dependence of the A-site field is due to large orbital contributions and supports an interpretation of Goodenough that spin-orbit interaction dominates Jahn-Teller interaction at A-site Ni^{2+} for the range ∿ $0.2 \le x < 1.0$.

INTRODUCTION

The ^{61}Ni nuclear gamma resonance (NGR) spectrum of the spinel $NiCr_2O_4$ shows the largest magnetic dipole hyperfine (hf) interaction yet reported.[1,2] The effective magnetic field of (±)440 kOe, for the tetrahedrally-coordinated (A-site) Ni^{2+} ion in the chromite may be compared with values obtained for octahedrally-coordinated Ni oxides, such as NiO (100 kOe)[3] and $NiFe_2O_4$ (94 kOe).[1] The source of the large field in $NiCr_2O_4$ has been ascribed to a large orbital contribution since the electronic ground state of A-site Ni^{2+} is orbitally degenerate in cubic symmetry. However, this ion displays a strong Jahn-Teller (J-T) effect, and tetragonal (c/a > 1) distortions below 310K lead to a singlet ground state and large reductions of orbital contributions to the g factor and the hf field.[4] Thus, any orbital moment that may be present at low temperatures results from a balance between J-T and spin-orbit interactions, and other effects, and is not easily estimated a priori.

In order to investigate experimentally the origin of this large field, we measured ^{61}Ni NGR spectra at 4.2K of powdered absorbers for the series of mixed spinels $NiFe_xCr_{2-x}O_4$ ($0 \le x \le 2$) which spans

---
*Research sponsored jointly by the U.S. Atomic Energy Commission under contract with Union Carbide Corporation, by Oak Ridge Assoc. Universities, and by the Research Corporation.

the range from the "anomalous" $NiCr_2O_4$ to the ferrite $NiFe_2O_4$. These materials have interesting magnetic properties and have been the subject of several investigations.[5-11] It is known that Cr^{3+} is found only on the B-sites whereas Fe^{3+} is found on both sites. It is believed that Ni^{2+} also tends to seek B-sites, but direct evidence is difficult to obtain because of the almost equal scattering factors of Fe and Ni for both neutrons and x-rays.

Low temperature x-ray studies[5,8] of this series show tetragonal distortions with the J-T signature (c/a>1) for the range 0<x<0.18, orthorhombic structure for compositions 0.2 < x < 0.28, and tetragonal (c/a<1) for $0.28 \le x < 1.0$. Goodenough[9] has interpreted this complicated structure as due to competitive J-T and spin-orbit interactions at A-site Ni^{2+}. The spin-orbit interaction produces a tetragonal (c/a<1) distortion of the site and stabilizes the ion in a ground state with unquenched orbital moment ($J = |L-S| = |1-1| = 0$), in contrast to the effect of J-T interaction. One may expect that the ^{61}Ni hyperfine field at A-site ions will exhibit large changes with composition in this region ($0 \le x < 1.0$); this provides the motivation for the present experiments.

EXPERIMENTAL DETAILS

The starting materials for the preparation of the samples were the nitrates of Ni and Fe and CrO_3, all of which had been analyzed to confirm metal ion content. Appropriate portions were weighed out (the CrO_3 was handled in a dry box) and mixed, melted under a heat lamp in the water of crystallization of the nitrates, well-stirred and dried until decomposition was complete. The dried mixture was heated further in Pt enclosures in a muffle furnace, powdered to 140 mesh, pressed into billets (10 tons, 1/2" diameter die) and fired overnight in air at 1300°C. The billets were cooled in the furnace to 900°C then quenched to room temperature ("quenched" samples); x-ray diffraction patterns were obtained and showed the presence of only the spinel phase. After the NGR data were obtained two samples with composition x = 0.5 were subsequently annealed at 600°C in air for 110 hours and additional spectra were obtained. Other experimental details have been described previously.[3]

RESULTS AND DISCUSSION

Representative spectra are shown in Figure 1. For x = 0, the single-valued field (±)440 kOe is measured and may be associated with the A-sites which are exclusively occupied by Ni. For both x = 1.5 and 2.0, only a small-field pattern is seen, and it is known that Ni occupies B-sites in these samples. For intermediate values of x, two distinct hf patterns are seen, with "small" field component growing in intensity at the expense of the "large" field component as x is increased. The large field is in fact a distribution of field values, as would be expected in mixed systems such as this, whose average value and width both increase as x increases. The average value for the large component reaches the value of H_A = (±)630 kOe for x = 0.5, and a similar result is found for the related compound $NiFe_{0.5}V_{1.5}O_4$

(see Figure 1(b)).

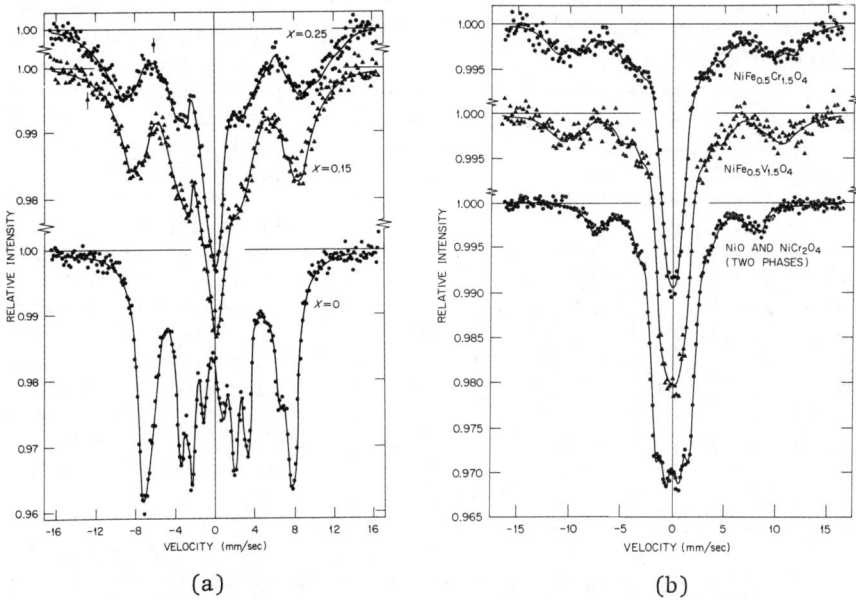

Fig. 1. ^{61}Ni NGR spectra for $NiCr_xFe_{2-x}O_4$ spinels. The absorption dip at zero velocity is the B-site component.

There are a large number of effects displayed by these spectra, including distributions of fields and quadrupole interactions for two sites, and we have not attempted least-squares fittings of sufficient flexibility to take them into account. Rather, rough estimates of average fields were obtained by fittings with two unique hf patterns, one for each site, with variable line widths. Such estimates are listed in Table I.

The large hf fields at A-site Ni^{2+} ions are difficult to understand without invoking orbital contributions. The contact field due to core polarization (cp) by the 3d spin has been estimated by Watson and Freeman[12] to be -332 kOe for the free ion and -275 kOe for a simple cubic environment (an octahedral array of point charge ligands). If we adopt the value H_{cp} = -300 kOe, as a rough estimate, then rather large additional contributions are needed to explain the values of (±)440 kOe (x = 0) and (±)630 kOe (x = 0.5); specifically, these are (+740, -140) kOe for x = 0 and (+930, -330) kOe for x=0.5, depending on the signs of the net fields. Clearly, additional experiments with applied fields are needed to determine these signs of $H_A(x)$ and provide a definitive choice. We may, however, eliminate several possible sources for these contributions, including variations in transferred hf interaction[13] due to the larger B-sublattice spin as Fe replaces Cr (one expects ΔH < 40 kOe due to this factor) and spin dipole fields (only the lattice contributes, therefore negligible). Covalency-overlap effects are harder to estimate, but

Table I. Lattice constants and ^{61}Ni hf fields at tetrahedral (A) and octahedral (B) sites in NiFe$_x$Cr$_{2-x}$O$_4$

x	H_A (kOe)	H_B (kOe)	Lattice constant (at R.T. in Å)
0	440	-	8.3162
0.15	506	20	8.3115
0.25	560	20	8.3167
0.50	617	48	8.3112
0.50	631	49	(annealed)
0.50	609	44	8.3057
0.50	623	48	(annealed)
1.50	-	84	8.3160
2.0	-	94	8.3375
- Preliminary results NiFe$_x$V$_{2-x}$O$_4$			
(0.5)	639	71	-
1.0	-	-	8.3568

probably tend to reduce the magnitude of H_{cp} by (say) 10% or less, since oxides of the spinel type are predominantly ionic.[14]

As Watson-Freeman point out, the core field is the result of partial cancellation of large, opposing, 2s and 3s terms, and, as such, it should be sensitive to environment. However, it has been shown empirically that the core field per unit spin is (almost) constant over the 3d series for a range of ionicities and ligand bond lengths.[15,16] Moreover, the W-F calculation has been shown to be in excellent agreement with data for several octahedrally-coordinated Ni^{2+} fluorides and oxides.[3,16]

For these cases of octahedral coordination, the orbital fields are accurately estimated from the shift of the electronic g value (from EPR data) from the spin-only value of 2.0023, because both quantities are due to second-order, spin-orbit, perturbations of the $^3A_{2g}$ orbital singlet ground state, and are given by the same matrix elements to within a constant factor. Such estimates yield large, positive, orbital fields (typically +200 kOe for $\Delta g \sim 0.28$) that nearly cancel the negative cp component and give net fields of the order of -100 kOe, such as the B-site fields reported here. Unfortunately, the $^3T_1$ ground state appropriate to the (cubic) A-site is orbitally degenerate and one expects no simple relation between g-shifts and orbital hf fields. We may estimate the g-values for A-site Ni^{2+} in NiCr$_2$O$_4$ for the cubic (high-T) and J-T distorted phases (low T) by comparing with the susceptibility data of NiRh$_2$O$_4$, a compound isostructural with the chromite in which Ni^{2+} is the only magnetic ion. From Blasse's[17] data, we obtain g(c/a=1) = 3.2 μ_B and g(c/a>1) = 2.5 μ_B. The latter value is then to be associated with the pure chromite (x = 0) and the former value is applied to A-site Ni in the composition x = 0.5, since the cubic and spin-orbit distorted A-site ions should have about the same values for $\langle L_z \rangle$. The large g-values would allow for the huge positive orbital fields that are needed if H_A is positive. On the other hand, one notes a linear relation between g-shift and H_A if H_A and the orbital fields are both negative.

It is an open question whether a negative orbital field may occur, but possible mechanisms for its occurrence are d-p hybridization and/or the substantial mixing by the crystal field of the $^3T_1$

state (derived from the ^3F free ion state) with terms of the same symmetry and multiplicity arising from the ^3P free ion state.[18] In both cases, one obtains contributions of opposite signs to the orbital part of the g-factor, the spin-orbit interaction, and the orbital part of the hyperfine interaction.

In summary, the large values of $|H_A|$ observed in the series $NiFe_xCr_{2-x}O_4$ are due to significant orbital contributions, but the size and even the sign of these terms may not be deduced until the signs of the experimental fields are determined.

We wish to thank D. LaValle for the preparation of the samples.

REFERENCES

1. H. Sekizawa, T. Okada, S. Okamota and F. Ambe, J. dePhysique 32, C326, Suppl. 2-3 (1971).
2. J. Goring, Z. Naturforsch. A26, 1929 (1971).
3. J. C. Love, F. E. Obenshain and G. Czjzek, Phys. Rev. B3, 2827 (1971).
4. F. S. Ham, Phys. Rev. 138A, 1727 (1965).
5. T. R. McGuire and S. Greenwald, Proc. International Conf. on Solid State Physics, Brussels 1958. (Academic Press, N. Y., 1960), Vol. 3.
6. F. C. Romeijn, Philips Res. Rep. 8, 304 (1953).
7. I. S. Jacobs, J. Phys. Chem. Solids 15, 54 (1960).
8. R. J. Arnott, A. Wold and D. B. Rogers, J. Phys. Chem. Solids 25, 161 (1964).
9. J. B. Goodenough, J. Phys. Soc. Japan 17, Supp. B-1, 185 (1962).
10. J. Chappert and R. B. Frankel, Phys. Rev. Lett. 19, 570 (1967).
11. V. F. Belov, et al., Sov. Phys. - Solid State 13, 747 (1971).
12. R. E. Watson and A. J. Freeman, Phys. Rev. 120, 1125 (1960); 120, 1134 (1960).
13. A. J. Heeger and T. W. Houston, Proc. Int. Conf. on Magnetism, Nottingham, England, 1964, p. 395.
14. D. S. McClure, J. Phys. Chem. Solids 3, 311 (1957).
15. R. E. Watson and A. J. Freeman, in Hyperfine Interactions, edited by A. J. Freeman and R. B. Frankel (Academic Press, 1967), p. 53.
16. S. Geschwind, in Hyperfine Interactions, edited by A. J. Freeman and R. B. Frankel (Academic Press, 1967), p. 225.
17. G. Blasse, Philips Res. Rep., Supp. (1964), No. 3, (see p. 66); independent data by S. Miyahara and S. Horiuti, Proc. Int. Conf. on Magnetism, Nottingham, England, 1964, p. 550.
18. F. S. Ham and G. W. Ludwig, in Paramagnetic Resonance, edited by W. Low (Academic Press, 1963), Vol. 1, p. 130.

SUPERTRANSFERRED HYPERFINE FIELDS AT Sb^{5+} IN INSULATING FERRITES: EFFECTS OF LOCAL ORDER AND ION-SPECIFIC PROPERTIES

B. J. Evans*
University of Michigan
Ann Arbor, Mich. 48104

L. J. Swartzendruber
National Bureau of Standards
Gaithersburg, Md. 20760

ABSTRACT

The supertransferred hyperfine fields at Sb^{5+} in $LiFe_5O_8$, $CoFe_2O_4$, and YIG have been determined using ^{121}Sb Mössbauer spectroscopy. In contrast to $CoFe_2O_4$, $NiFe_2O_4$, and YIG, the small, average hyperfine field of ~ 100 kOe at Sb^{5+} in $LiFe_5O_8$ requires the existence of significant local order and indicates that the clustering of Li^{1+} about Sb^{5+} is approximately ten times as large as that expected for a random intrasite cation distribution, in agreement with the known strong influence of Sb substitution in destroying the Li:Fe ordering. The decrement in the hyperfine field at Sb^{5+} due to an A-site Co^{2+} is also found to be larger than that due to Ni^{2+}.

INTRODUCTION

As a part of a continuing effort to establish the systematics of supertransferred hyperfine fields (STHF) at diamagnetic cations in insulating ferrites,[1] ^{121}Sb Mössbauer measurements have been made for Sb^{5+} substituted $CoFe_2O_4$, $LiFe_5O_8$, and $Y_3Fe_5O_{12}$(YIG) and are compared with previous measurements on Sb^{5+} and Sn^{4+} in other ferrites.

First of all, the general trends in the previously reported studies[1] have been further confirmed. Secondly, the assumption of random cation distributions-- an assumption commonly made in applying microstatistical considerations to STHFs-- is found to be a poor description of the actual Li^{1+} intrasite distribution in Sb substituted $LiFe_5O_8$. Evidence for the influence of the specific details of the spin and electron density distributions of neighboring ions on the STHF is found in the different effects of Ni^{2+} and Co^{2+} on the STHF at Sb^{5+}.

*Alfred P. Sloan Research Fellow. Partial support of this research by the Alfred P. Sloan Foundation is gratefully acknowledged.

EXPERIMENTAL

All of the samples were prepared according to the prescriptions of Blasse.[3] They were characterized using x-ray powder diffractometry and wet-chemical analytical techniques. All of the samples were single-phase and the analyzed compositions are given in Table I. The Mössbauer spectrometer, data collection and analysis techniques were identical to those described in detail previously.[1,2]

RESULTS

The spectra of the different samples at 100 K are shown in Fig. 1. From our experience with other Sb substituted ferrites,[1,2] it was obvious that all of the samples, except YIG, had a distribution of hyperfine fields at the Sb^{5+} ion. Therefore, the spinel ferrite spectra were fitted to a magnetic-plus-electric quadrupole pattern, with the addition of an assumed Lorentzian distribution of hyperfine fields about some mean value H_{sthf}. The width (FWHM) of the hyperfine field distribution, Γ_{sthf}, was one of the fitted parameters. The solid lines in Fig. 1 are the results of a least mean-squares fit of these assumptions to the spectra. The parameters from these fits are given in Table I. The electric quadrupole interactions were found to be negligibly small and are not listed in Table I.

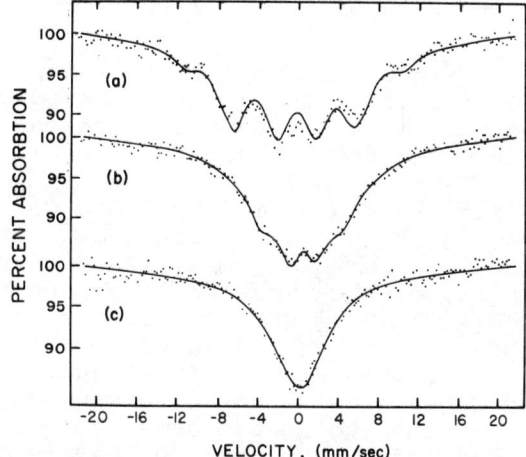

Fig. 1. ^{121}Sb Mössbauer spectra at 100 K for: (a) $Y_{2.9}Ca_{0.2}Fe_{4.9}Sb_{0.1}O_{12}$, (b) $Co_{1.1}Fe_{1.85}Sb_{0.1}O_4$, (c) $Li_{1.2}Fe_{4.6}Sb_{0.2}O_8$.

DISCUSSION

With the possible exception of Sb substituted $LiFe_5O_8$, the ^{121}Sb Mössbauer spectra of Fig. 1 evince appreciable magnetic hyperfine fields, as in other ferrimagnetic spinels and garnets.[1,2,4] Even though there is no resolved splitting in the spectrum of Sb substituted LiFe$_5$O$_8$ (Fig. 1c), the width of an assumed single-line would be ~7 mm/sec and three times that expected for an intrinsic, single-line ^{121}Sb pattern.[2] The possibility of large, unresolved quadrupole splitting can be dismissed since the electric quadrupole interaction has been found to be small in a variety of different spinel ferrites;[2] the spectrum also lacks the expected asymmetry which accompanies a large electric quadrupole splitting.

Table I. ^{121}Mössbauer parameters at 100 K of Sb substituted ferrites. $\langle H_{sthf} \rangle$ is the average value of an assumed Lorentzian distribution of hyperfine fields and Γ_{sthf} is the width (FWHM) of this distribution. Individual lines were assumed to have Lorentzian profiles and relative widths and intensities for a thin absorber.

Sample (Actual Compositions)	$\langle H_{sthf} \rangle$ (kOe)	Γ_{sthf} (kOe)	Isomer Shift[a] (mm/sec)
Li$_{0.60}$Fe$_{2.30}$Sb$_{0.11}$O$_4$	116(3)	149(10)	-0.2(1)
Co$_{1.10}$Fe$_{1.82}$Sb$_{0.06}$O$_4$	172(2)	63(6)	-0.6(1)
Co$_{1.21}$Fe$_{1.68}$Sb$_{0.12}$O$_4$	165(2)	60(6)	-0.6(1)
Y$_{2.9}$Ca$_{0.2}$Fe$_{4.9}$Sb$_{0.1}$O$_{12}$	267(5)	--	-0.1(1)

[a] w.r.t. BaSn(Sb)O$_3$ at 100 K

Since our samples were prepared in a manner identical to that of Blasse,[3] we may reasonably assume them to have very similar cation distributions. Sb^{5+} is located exclusively on the octahedral sites in every case. 6% of the A sites are occupied by Li in Li$_{0.6}$Fe$_{2.3}$Sb$_{0.1}$O$_4$; 26% of the A sites are occupied by Co in Co$_{1.2}$Fe$_{1.7}$Sb$_{0.1}$O$_4$; and the cation is not accurately known for Co$_{1.1}$Fe$_{1.85}$Sb$_{0.05}$O$_4$ but is assumed to be similar to that for pure CoFe$_2$O$_4$. In all of these materials except YIG, the Sb^{5+} ion has chemical and magnetic disorder among next-nearest-neighbor (NNN) cation environments.

If the value of the magnetic hyperfine field at Sb^{5+} in CoFe$_2$O$_4$ and LiFe$_5$O$_8$, having only A site Fe neighbors, is assumed to be the same as that in NiFe$_2$O$_4$,[2] we obtain values of 80 kOe and 400 kOe, respectively, for the decrement, ΔH, in the magnetic hyperfine field per non-Fe A site neighbor. These ΔH values differ from that of 50 kOe as found for NiFe$_2$O$_4$. While the <u>differences</u> between the ΔH values of 50 kOe for Sb^{5+} and 30 kOe for Sn^{4+} in YIG are believed to understood qualitatively, the results reported here for Sb^{5+} in CoFe$_2$O$_4$ and LiFe$_5$O$_8$ are not explicable in terms of the general systematics of ^{121}Sb

STHFs; and a consideration of the dominance of effects unique to the Co^{2+} and Li^{1+} ions and/or local structural details might be instructive.

Recent studies of the STHF at ^{57}Fe in $Co_{1-x}Zn_xFe_2O_4$[5] have shown that the ΔH associated with substituting Co for Fe on an A site has a value somewhat larger than that due to a diamagnetic cation such as Zn and also larger than that due to Ni.[2] If we assume a similar behavior for Co^{2+} on the STHF at Sb^{5+}, then the larger value of ΔH for Co^{2+} relative to that for Ni^{2+} is to be expected.

The small difference in the magnitudes of $\langle H_{sthf} \rangle$ and Γ_{sthf} for $Co_{1.1}Fe_{1.85}Sb_{0.05}O_4$ and $Co_{1.2}Fe_{1.7}Sb_{0.1}O_4$ (cf. Table I) vis-à-vis the diferences between these parameters for $Ni_{1.1}Fe_{1.85}Sb_{0.05}O_4$ and $Ni_{1.2}Fe_{1.7}Sb_{0.1}O_4$[2] follows directly from the small differences in the A site cation distributions for these two compositions of cobalt ferrites. The fractional occupation of the A sites by Co^{2+} in $Co_{1.1}Fe_{1.85}Sb_{0.05}O_4$ is at least 0.22,[5] and it increases by only 4% to 0.26 in $Co_{1.2}Fe_{1.7}Sb_{0.1}O_4$.[3]

In view of the small variations in the value of the STHF at Sb^{5+} in a variety of ferrites for a given local environment, the unusually small value of 100 kOe for $\langle H_{sthf} \rangle$ and large value of $\sim$400 kOe for ΔH at Sb^{5+} in $LiFe_5O_8$ are anomalous. However, in arriving at this value for ΔH a random, intrasite distribution of Li^{1+} ions on the A site s was assumed. There is reason to believe that this assumption is not correct. Blasse[3] has shown that Sb substitution in $LiFe_5O_8$ is quite effective in destroying the 1:3 Li-Fe ordering on the B site. Presumably, the disordering involves local clustering of Li^{1+} ions about the Sb^{5+} ions. Such clustering would involve Li^{1+} ions on both the A and B sites. Therefore, the number of Li^{1+} A-site ions that are NNN to a B-site Sb^{5+} ion is much greater than that expected on the basis of a random intra-site distribution of Li^{1+}. Once local clustering of Li^{1+} about Sb^{5+} is taken into account, the smaller value of $\langle H_{sthf} \rangle$ for Sb^{5+} in $Li_{1.2}Fe_{4.6}Sb_{0.2}O_8$ relative to that in $Ni_{1+2x}Fe_{2-3x}Sb_xO_4$ and $Co_{1+2x}Fe_{2-3x}Sb_xO_4$ is found to be qualitatively consistent with the systematics of STHFs at diamagnetic cations in spinel ferrites. Using previously developed data analysis techniques,[2] the average number of A site Li^{1+} ions that are NNN to a Sb^{5+} ion is estimated to be $\sim$4. This estimate was arrived at by employing a value of 50 kOe for ΔH and 300 kOe for H_{sthf} at a Sb^{5+} ion having only Fe A-site NNN. This result is to be compared with an average number of 0.4 Li^{1+} A-site ions about a Sb^{5+} ion expected for a random intra-site distribution of tetrahedral Li^{1+} ions in $Li_{1.2}Fe_{4.6}Sb_{0.2}O_8$.

The ^{121}Sb Mössbauer spectrum of Sb^{5+} in YIG was reported by other investigators[6] shortly after we completed our measurements. However, our sample composition is different from that of $Y_{2.5}Ca_{0.5}Sb_{0.25}Fe_{4.75}O_{12}$ reported on in the earlier study.[6] We obtain 267 kOe for the STHF at Sb^{5+} in $Y_{2.8}Ca_{0.2}Sb_{0.1}Fe_{4.9}O_{12}$ which is less than that of 290 kOe obained for Sb in $Y_{2.5}Ca_{0.5}Sb_{0.25}Fe_{4.75}O_{12}$. We can offer no explanation for the apparent discrepancy since the saturation value of the STHF is expected to be similar for these two materials.[2] In agreement with the known cation distribution in Sb-substituted YIG and the fact that the tetrahedral neighbors of a given Sb ion are all Fe ions, the ^{121}Sb Mössbauer spectrum shown in Fig. 1a can be adequately fitted with a single hyperfine field.

CONCLUSION

In conclusion, we note that our previous prediction[1] of the relative magnitudes of the STHFs at ^{121}Sb and ^{119}Sn in YIG has been verified[6] and that our interpretaion of the predominance of 5s covalent charge transfer to the supertransferred hyperfine interaction mechanisms seems to be firmly established for Sb^{5+} and isoelectronic Sn^{4+}. Further refinements in our understanding of the systematics of STHFs at Sb^{5+} and Sn^{4+} will require careful consideration of the chemical bonding properties and magnetic exchange interactions specific to each ion as well as a knowledge of the local and possibly non-random cation distributions.

REFERENCES

1. B. J. Evans and L. J. Swartzendruber, Phys. Rev. 6, 223 (1972).
2. B. J. Evans, in Mössbauer Effect Methodology, edited by I. J. Gruverman (Plenum, New York, 1968) Vol. 4, p.139.
3. G. Blasse, Philips Res. Repts. Suppl. 3, 1 (1964).
4. I. S. Lyubutin and Yu. S. Vishnyakov, Phys. Stat. Sol. (a) 12, 47 (1972).
5. G. A. Petitt and D. W. Forester, Phys. Rev. 4, 3912 (1971).
6. V. A. Golovnin, S. M. Irkaev, R. N. Kuz'min, and V. V. Mill, Sov. Phys. JETP Letters 11, 21 (1970).

A MÖSSBAUER STUDY OF Dy_2O_3 AND THE ISING ANTI-FERROMAGNET $DyPO_4$: RELAXATION, ANISOTROPY AND SITE DEPENDENCE.

D.W. FORESTER and W.A. FERRANDO[†], Naval Research Laboratory, Washington, D.C. 20375

$DyPO_4$ is a highly anisotropic antiferromagnet [$g_{\parallel}$=19.3 and $g_{\perp}$=0.5±0.5 as shown by Prinz and Wagner and by Wright and Moos]. Spectra of ^{161}Dy taken from below T_N=3.4K to 300K were computer-analyzed by using a complete diagonalization of $\mathcal{H}=f(t)\{AI_zS_z\}+B[S_+I_- + S_-I_+]+P[3I_z^2 - I(I+1)]$ in both the ground (g) and excited (e) nuclear states. A (and B) and P are magnetic and quadrupole parameters respectively and $f(t) = \pm 1$. The time evolution description uses a markoffian stochastic model. Both excited and ground state parameters were obtained with $A^g/2h$ = 799.4 Mc/sec and $B^g/A^g \cong .007$. If we use $\mathbf{\underline{A}} = k\mathbf{\underline{g}}$ for an isolated multiplet, then $g_{\perp}/g_{\parallel} \cong .007$ or $g_{\perp} \cong 0.14$. The relaxation is dominated below 30K by an Orbach process and above this by a T^9 Raman type process. In Dy_2O_3, two 16-line hyperfine spectra (hfs) are resolved with the intensity ratio 3:1 corresponding to Dy^{3+} at sites with C_2 and C_{3i} symmetries. The C_{3i} magnetic hfs is almost identical to that of $DyPO_4$ corresponding to a pure $|J_\zeta=\pm 15/2\rangle$ Kramers doublet and exhibits faster relaxation than the C_2 spectrum. The strong C_2 spectrum has reduced hfs parameters indicating a small crystal field mixing of states with $M_J<15/2$ into the ground state.

[†]NRL-NRC Research Associate

NUCLEAR MAGNETIC RESONANCE IN HIGH NEEL TEMPERATURE GARNETS

Hirotaka Yokoyama
Toshiba Research and Development Center,
Kawasaki Japan

ABSTRACT

V^{51} and Fe^{57} NMR of $Y_{3-2x}Ca_{2x}Fe_{5-x}V_xO_{12}$ garnets for x between 0.2 and 1.5 have been studied at 4.2°K by pulsed NMR method. This garnet system is known to retain a high Neel temperature for the increase of x right up to x = 1.5. The center of gravity of the V^{51} NMR line increases from 9.5 MHz to 15.6 MHz with increasing x from 0.2 to 1.5. The hyperfine field of V ion nuclei between 8.5 and 14 kOe is what one would expect in its order of magnitude for the supertransferred hyperfine fields at the diamagnetic V^{5+} ion nuclei on tetrahedral sites in iron garnets. The Fe^{57} NMR spectra of octahedral Fe^{3+} consist of several peaks with nearly equal spacing. Frequencies of these peaks are approximately given by $\nu_n = \nu_c + n\nu_T$, with n = 0, 1, 2, 3, $\cdots$, where $\nu_c \simeq 75.7$ MHz and $\nu_T \simeq 2.2$ MHz. Relative intensities of these peaks are calculated by a model in which Fe and V are distributed at random on the tetrahedral sites. Similar calculation is made for V^{51} NMR intensity distribution. From the comparison between the calculated and the observed NMR spectra, it is suggested that there exists a certain degree of short range ordering in the distribution of V ions on the tetrahedral sites. The NMR frequency of about 64.5 MHz for the tetrahedral site Fe^{57} is almost constant up to x = 1.5, which is in contrast with the NMR frequency of 62.5 MHz for the tetrahedral Fe^{57} of $Y_{3-x}Ca_xFe_{5-x}Si_xO_{12}$ with x = 1.5. If we assume that the decrease of core polarization hyperfine fields for the both garnets are roughly the same, this result leads to a conclusion that the supertransferred hyperfine field at tetrahedral Fe^{3+} in $Ca_3Fe_{3.5}V_{1.5}O_{12}$ is larger than that in $Y_{1.5}Ca_{1.5}Fe_{3.5}Si_{1.5}O_{12}$.

MAGNETIC ENVIRONMENT OF HYDROGEN IN FE FROM MUON PRECESSION MEASUREMENTS*

Neil Heiman,[†] M. L. G. Foy, and W. J. Kossler
College of William and Mary, Williamsburg, Va. 23185

C. E. Stronach
Virginia State College, Petersburg, Va. 23803

ABSTRACT

The behavior of hydrogen in metals is a matter of considerable technological importance and it is a matter of intrinsic interest as one of the simplest alloy problems. Because of its hydrogen-like character, the positive muon provides a useful tool for studying this problem, particularly in ferromagnetic metals. Using a technique similar to that used in our investigation of μ^+ in Ni, we stopped polarized μ^+ from the Space Radiation Effects Laboratory (SREL) in an ellipsoidal iron target and observed the precession of the μ^+ in a transverse magnetic field. From this data we determined both the internal field at the μ^+ site (B_μ) and the relaxation time of the μ^+ polarization (τ) from 77K to above the Curie point of Fe. As a function of T, B_μ approximated a Brillouin function with a saturation value of 4100G. This result indicates the conduction electron polarization in this region is less than expected. τ increased from 0.1 μsec at 300K to 0.5 μsec at 900K, then dropped to 0.1 μsec near T_c=1050K and rose to 0.2 μsec above T_c. This indicates that the relaxation is dominated by the static inhomogeneity to 900K at which point magnetization fluctuations become important.

INTRODUCTION

The study of hydrogen impurities in metal is attractive because it is one of the least complicated alloy problems. The simplicity of the hydrogen system is especially attractive for the study of magnetically ordered metals because the interaction of the hydrogen nucleus with the polarized conduction electrons is not complicated by such features as core polarization and volume overlap. Unfortunately obtaining experimental data on the microscopic environment of dilute hydrogen impurities is extremely difficult; however, from the solid state physics point of view, the positive muon is nothing more than a light hydrogen nucleus, and information about the

---

*Work performed at Space Radiation Effects Laboratory which is supported by NASA, NSF and the State of Virginia.
[†]Present Address: IBM Research Laboratory, San Jose, California 95193.

microscopic environment of the muon can be easily obtained by
implanting the muon in a sample and allowing it to precess in a
transverse magnetic field. The muon decays, emitting a positron,
and the precession can be observed by monitoring the time dependence
of the anisotropic distribution of the emitted positrons. The information thus obtained is equivalent to the information which would be
obtained from the NMR spectrum of hydrogen impurities if it were
possible to observe a hydrogen resonance. A number of recent
papers[1,2,3,4] have demonstrated that the implanted positive muon does
indeed behave as a hydrogen nucleus and can provide information in
such areas as lattice structure, chemical reactions, and the
electronic configuration of hydrogen in semiconductors. Our own
recent work[5] has shown that muons implanted in ferromagnetic metals
can be used to measure the internal fields and relaxation rates at
the hydrogen site. For the work reported in that paper we chose
nickel as a target because nickel had the advantages of a low Curie
point and an interstitial position of high symmetry. Iron may well
be a more interesting target material because of the existence of
more information on conduction electron polarization. Therefore,
using improved apparatus, we have studied muon precession in iron
from room temperature to above the Curie point. In this paper we
report the results of this study.

EXPERIMENTAL PROCEDURE

We implanted polarized positive muons from the Space Radiation
Effects Laboratory 600 MeV synchrocyclotron into an iron target.
The target was in the form of an oblate ellipsoid (7.5 cm diameter,
1 cm thick) and was mounted inside a furnace capable of achieving
1000°C. The target-furnace assembly was placed in the gap of a very
large electromagnet. The magnet with pole faces 75 cm by 35 cm
provided magnetic fields up to 8 kilogauss. The incoming muons and
the decay positrons were detected with an array of plastic scintillators. The experimental geometry is shown in Fig. 1.

A muon which stops in the target will pass through scintillators
1 and 2 but not reach scintillator 3. Consequently the signal that
a muon stopped in the target was defined as $12\overline{3}$ (meaning a coincidence
between 1 and 2 but no signal in 3). A signal from scintillators 4
or 5 indicated a positron had been detected. The stopped muon signal
($12\overline{3}$) was used to start a time to amplitude converter (TAC) and a
positron signal (4 or 5) was used to stop the TAC. The output of the
TAC was fed to a pulse height analyzer which directly produced a plot
of the number of decay positrons reaching either the 4 scintillator
or the 5 scintillator as a function of time after a muon stop. An
example of the data is shown in Fig. 2. The data were fitted to the
function:

$$N(t) = N_o \exp(-\lambda t) + N_o \exp(-\lambda t)[aP \exp(-t/\tau)\cos(\omega t+\phi)] + \text{Bgnd}$$

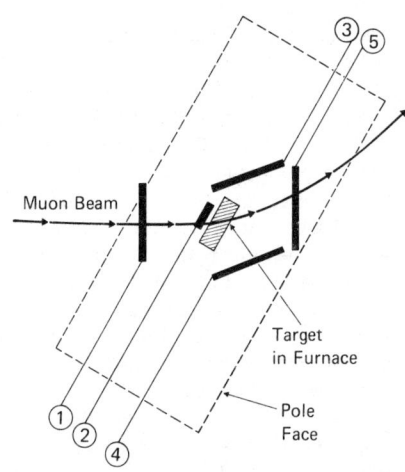

Fig. 1: EXPERIMENTAL GEOMETRY

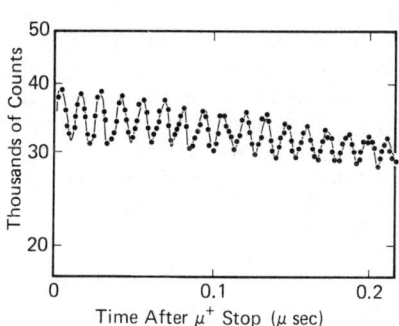

Fig. 2: NUMBER OF COUNTS IN NUMBER 4 SCINTILLATOR AS A FUNCTION OF TIME AFTER MUON STOPS IN FE TARGET AT 875K

The expression $N_o \exp(-\lambda t)$ accounts for the fact that the muons are radioactive particles and decay with a mean life $1/\lambda = 2.2$ μsec. The term in brackets describes the oscillating count rate in the detectors due to the precession of the muons. P is the initial degree of polarization of the stopped muons, a is the anistropy of the angular distribution of the emitted positrons, τ is the time constant characterizing the relaxation of the transverse polarization of the precessing muons. The angular precessing frequency ($\omega = 2\mu_\mu B_\mu/\hbar$, were μ_μ is the magnetic moment of the muon) provides a direct measure of the magnetic field (B_μ) at the muon site. The initial phase angle (ϕ) and a background term (Bgnd) are included as fitting parameters. Note that the expression in brackets is simply the Fourier transform of the muons magnetic resonance absorption spectrum.

RESULTS

Fig. 3 shows the behavior of B_μ as a function of temperature. B_μ as plotted in Fig. 3, has already been corrected for the external and demagnetizing fields. The correctness of subtracting these fields was verified by a run in which the iron target was part of a closed magnetic loop such that $H_{ext} \simeq 0$ and $H_{demag} \simeq 0$. The temperature dependence of B_μ approximates a Brillouin function with a 0°K intercept of 4100 gauss.

Contributions to the internal field at the muon site consists of only 3 terms: (1) the dipole moment sum H_d, (2) the Lorentz field $4\pi M/3$ and (3) the conduction electron polarization contact hyperfine field H_{cep}. Because of the low symmetry of the interstitial site in iron, the dipolar sum is not zero and in a polycrystalline sample would lead to such a broad distribution of B_μ values as to make the experimental determination of B_μ impossible.

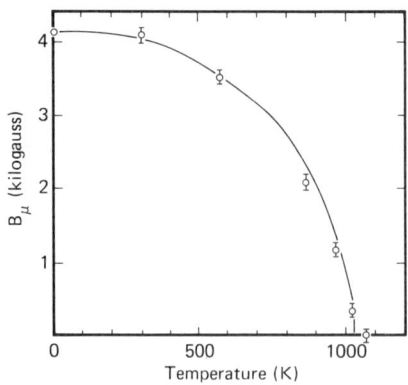

Fig. 3: B_μ AS A FUNCTION OF TEMPERATURE

Fortunately the muon hopping is so fast compared to the precession period that H_d is averaged to zero. The Lorentz field $4\pi M/3$ at 0°K is about 7000 gauss, thus indicating a conduction electron contribution $H_{cep}=-3000$ gauss. Using these numbers Mary Beth Stearns[6] was able to extend the conduction electron polarization curve for iron to interstitial distances. The extension resulted in a more negative polarization than was anticipated. It should be noted that the values observed for B_μ (+4100 gauss in Fe and +1500 gauss in Ni) are small and in fact only about 1/2 the magnitude of the Lorentz field values. The correct interpretation of B_μ therefore requires careful treatment of such contributions as the Lorentz field. Rasetti[7] (experimentally) and Wannier[8] (theoretically) have shown that unenergetic muons do not sense the magnetization of their host. Consequently one might suggest that the Lorentz field for the stopped muon is zero. If that is truly the case, the conduction electron polarization at interstitial distances may be more in line with expected values.

The relaxation time of the muons transverse polarization (τ) was also measured. A nearly fivefold increase in τ from 0.1 μsec at room temperature to over 0.5 μsec at 900°K seems to be understandable in terms of static local field inhomogeneity. A drop in τ to 0.1 μsec near T_c is most likely due to fluctuations. The smallness of τ in the paramagnetic region ($\tau \simeq 0.2$ μsec) seems hard to understand particularly when compared to paramagnetic Ni ($\tau \geq 4.0$ μsec).

REFERENCES

1. A. Schenck and K. M. Crowe, Phys. Rev. Lett. **26**, 57 (1971).
2. J. H. Brewer, K. M. Crowe, R. F. Johnson, A. Schenck, and R. W. Williams, Phys. Rev. Lett. **27**, 297 (1971).
3. J. Brewer, K. M. Crowe, F. N. Gygox, R. F. Johnson, B. D. Patterson, D. G. Fleming, and A. Schenck, Phys. Rev. Lett. **31**, 143 (1973).
4. Meeting on Muons in Solid State Physics, Burgenstock. 1-3 Sept. 1971 (unpublished).
5. M. L. G. Foy, N. Heiman, W. J. Kossler, and C. E. Stronach, Phys. Rev. Lett. **30**, 1064 (1973).
6. Mary Beth Stearns, private communication.
7. F. Rosetti, Phys. Rev. **66**, 1 (1944).
8. G. H. Wannier, Phys. Rev. **72**, 304 (1947).

HIGH TEMPERATURE EPR IN DENSE SOLID AND MOLTEN PARAMAGNETS*

Elmar Dormann[†] and Vincent Jaccarino
Department of Physics, University of California
Santa Barbara, CA 93106

ABSTRACT

To probe the nature of exchange correlations in the magnetically dense solid and molten paramagnet we have studied the linewidths ΔH of the EPR of certain Mn^{2+} salts in the temperature region 20° to 1200°C at 9.3 GHz. For the salts $XMnF_3$ (X=K,Rb,Cs) ΔH increases linearly with T until the vicinity of the melting point, T_m. Just below T_m a more rapid increase is observed followed by an abrupt doubling of ΔH as the crystals melt. Above T_m, ΔH rapidly decreases with increasing T. On the other hand, MnF_2 shows a continuous decrease of ΔH as T is raised with a most pronounced drop appearing as a precursor to melting and a continuing decline through the molten region. The relation of these observations to the effects that lattice vibrations, symmetry and liquid motion have on dipolar and exchange interactions are discussed.

INTRODUCTION

The structure and widths of EPR and NMR lines provide information on the amplitude and spectral density of the local field fluctuations from which certain characteristic properties of the electronic spin correlation functions in dense magnetic systems may be inferred. Such information for the high temperature solid and molten phase of paramagnets is scarce; EPR information for the molten state is mainly limited to diluted chlorides.[1,2,3] We began our investigations with the intent of understanding high temperature "motion" in the broadest sense - including the influence of lattice vibrations in the solid as well as liquid motion. We initially chose the fluoride salts of manganese - despite their chemical aggressivity at high temperatures - because the S-state character of Mn^{2+} simplifies the interpretation and because dipolar, Mn^{55} hfs as well as F^{19} super hfs interactions can be studied. However, we restrict this report to the EPR of the dense Mn-salts, where the Mn^{55} and F^{19} hfs, relative to the Mn^{2+} dipolar interactions, contribute negligibly to the linewidth. We investigated $KMnF_3$ and $RbMnF_3$ as examples of cubic structure, hexagonal $CsMnF_3$ and tetragonal MnF_2. One of our most interesting first findings was that even in the solid the high temperature EPR linewidths (dipolar broadened and extremely exchange narrowed) were far from being temperature independent; this being so not only in the fluorides of Mn^{2+}, but also in oxides and sulfides not reported here.

---

*Supported in part by the National Science Foundation.
†Fazit-Stiftung Fellow; on leave from II. Physikalisches Institut, Technische Hochschule, Darmstadt, Germany.

EXPERIMENTAL DETAILS

All measurements were made at about 9.3 GHz with a magic tee microwave bridge. An Impatt diode was used as constant frequency rf source.[4] Fiberfrax insulation of platinum strip coated sandblasted quartz tubes allowed our watercooled TE_{102}-cavity of the Singer, et al.[5] -type to be operated up to temperatures of about 1400°C. The thickness of the resistance heated Pt-strips was adjusted in order to minimize temperature gradients inside the cavity. Most temperature measurements were performed with Chromel/Alumel thermocouples. Temperature stability during measurement was better by far than ± 1°C; the error bars shown in the following figures give the inaccuracy in absolute value. The salts were contained in crucibles made of pyrolytic boron nitride (Boralloy, Union Carbide) closed with a plug of normal sintered BN. Closed crucibles were essential to prevent oxidation and decomposition despite the fact that precaution was taken to surround the heater tube with dry N_2. The noncubic crystals were oriented by x-rays and optical means to about ± 1°, but after transfer into crucible and heater, alignment with respect to the field was not better than ± 3.5°, as inferred from measurements of the angular dependence of the linewidth. The lines in both solid and molten states were Lorentzian in shape. The peak-to-peak linewidths (ΔH_{pp}) in the solid state were derived by comparison of the theoretical with the observed derivatives of absorption lines, recorded using 280 Hz field modulation. Especially in the lower temperature range, care had to be taken to prevent radiation induced broadening due to inadmissibly big sample size.[6] The molten salts are also quite conductive, with, e.g., $\sigma \approx 5(\Omega cm)^{-1}$ and a skin depth δ of about ¼ mm for MnF_2 at 9.3 GHz and 1000°C. Therefore, in order to observe well defined lineshapes, a sample size $d \gg \delta$ or $d \ll \delta$ must be used. We tried to realize the second extreme. Alternatively, two procedures were used which both gave the same values of ΔH_{pp} within experimental accuracy: a) the bridge was tuned for absorption, giving superimposed absorption-dispersion signals that are well-known in metal EPR; the recorded lines then were analysed using the diagrams given by Peter, et al.[7] or b) the bridge was tuned until just that mixture of absorption and dispersion was obtained which resulted in a simple absorption derivative being recorded and was then analysed as would be in the solid state. The use of "big" samples gave us a convenient access to supplementary information; we could observe and follow semiquantitatively how samples became conductive or lossy quite below T_m (e.g., MnF_2 begins to be quite lossy at 100° below T_m); a decrease of the asymmetry in the line profile in the molten state informed us of increasing sample oxidation, etc. Since higher melting temperature oxides can give, in many cases, linewidths that are quite similar to the ones reported here, temperature reversibility of the measurements played a crucial role (therefore polycrystalline data for the noncubic samples had to be taken too!).

RESULTS

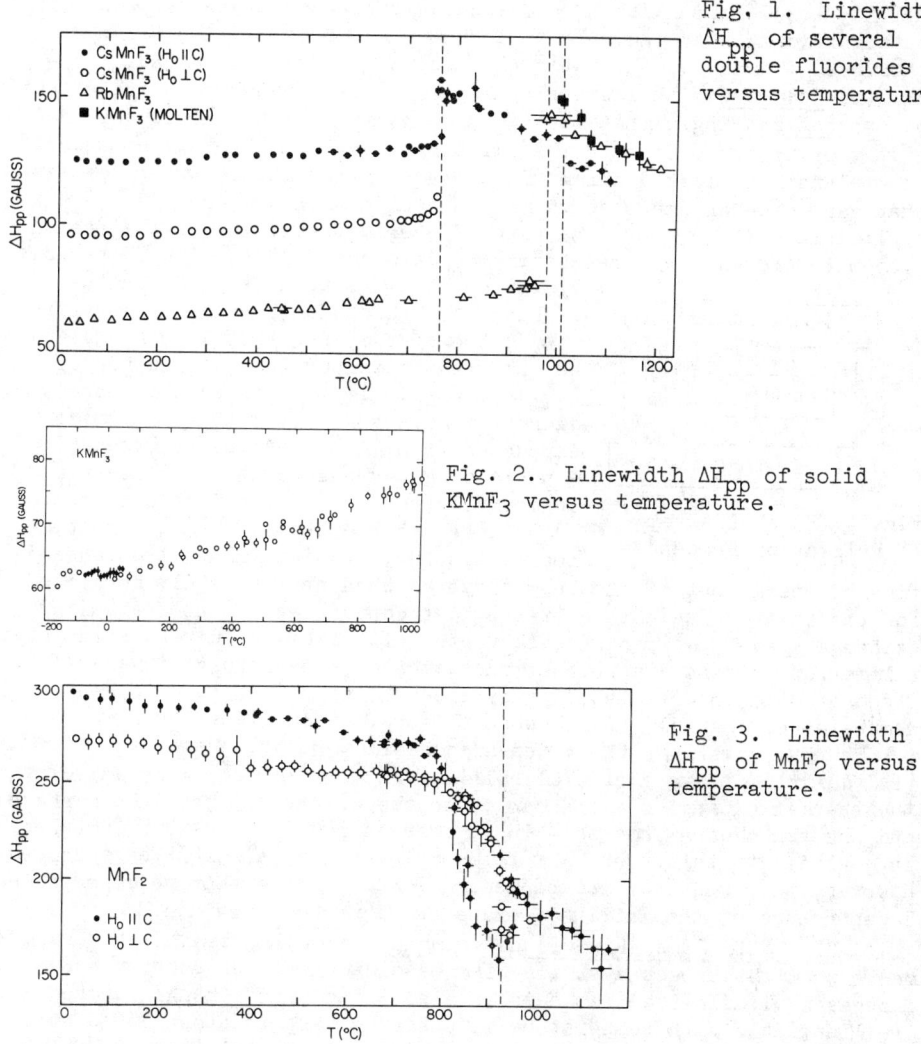

Fig. 1. Linewidth ΔH_{pp} of several double fluorides versus temperature.

Fig. 2. Linewidth ΔH_{pp} of solid $KMnF_3$ versus temperature.

Fig. 3. Linewidth ΔH_{pp} of MnF_2 versus temperature.

DISCUSSION

A. <u>Solid State Linewidths of the Double Fluorides $XMnF_3$</u>. Linewidths between 300 K and T_N for the $XMnF_3$ compounds were published before: in $RbMnF_3$[8] the linewidth is virtually constant with some narrowing very near T_N, for $KMnF_3$ [9] a constant linewidth was reported, and in the hexagonal $CsMnF_3$ ΔH appears to diverge as T approaches T_N.[10] However, our data, shown in Fig. 1 and 2, reveal a quite similar behavior for all three $XMnF_3$ compounds <u>above</u> room temperature; namely there is a linear increase of ΔH with T of about (14.3 ± 1.6)G /1000° for X=Rb, (13.8 ± 0.9) for X = K and (9.5 ± 1.8) and

(11.3 ± 1.3)G/1000° with $H_0 \parallel C$ and $\perp C$, respectively for X = Cs. It is significant that $d/dT(\Delta H)$ is larger the smaller is ΔH itself. Furthermore, in CsMnF$_3$, the ratio $\Delta H_{pp}(\parallel)/\Delta H_{pp}(\perp)$ as well as $\Delta H_{pp}(\parallel) - \Delta H_{pp}(\perp)$ decrease with increasing T. These two features indicate that we are not dealing with a linewidth mechanism which contributes additively or in proportion to the existing linewidth. When we consider the relaxation block diagram of Fig. 4, we find mainly two processes that can cause an increase of the linewidth with temperature. (δ_{ZL} is negligible.[11]) a) when the rate of energy transfer from the Zeeman system to the exchange reservoir (δ_{ZE}) is small compared with the rate by which energy is transferred from the exchange to the phonon(lattice) reservoir (δ_{EL}), the system is "unbottlenecked", and we observe the temperature dependence of $\delta_{ZE} \sim H_d^2/J$ [12]. Since the exchange narrowing decreases more strongly with the lattice expansion than does the dipolar broadening — neglecting harmonic contributions in J — an overall broadening with T is observed.

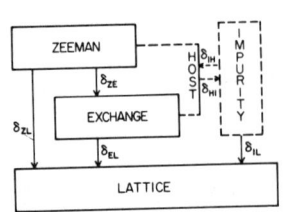

Fig. 4. Block Diagram of Relaxation Processes.

b) in the other extreme, for the "bottlenecked system", δ_{EL} determines the rate of flow of energy out of the Zeeman system when appropriately corrected for the relative weighting of the heat capacities of the Zeeman and exchange systems. Then we observe the effect of the direct spin lattice relaxation process via phonon modulation of the exchange interaction.[13] The one phonon process would vary linearly with T when $T > \Theta_D$ and would decrease very rapidly with T when $T \ll \Theta_D$. A quantitative comparison for RbMnF$_3$ shows that the r dependence of J, or the quantity $\lambda = |1/J \, dJ/dr|$ plays a crucial role. For $\lambda \geq 4 \cdot 10^8$ cm^{-1} or $J \sim 1/r^{\geq 18}$ the thermal expansion alone can cause the observed linewidth increase, and the two phonon process[13] in δ_{EL} has opened the possible "bottleneck". In contrast to this, for $\lambda \leq 1 \cdot 10^8$ cm^{-1} process b) might be observed. However, the frequency independence of ΔH_{pp} proved that we observe the T-dependence of the intrinsic exchange narrowed linewidth.

B. Solid State Linewidths in MnF$_2$. According to the above arguments we might expect a similar increase of the linewidths in MnF$_2$ with increasing T. Instead, as seen in Fig. 3, an even stronger decrease of the linewidths with temperature is observed. Up to about 750°C, the ratio $\Delta H_{pp}(\parallel C)/\Delta H_{pp}(\perp C)$ remains constant at about 1.081±0.008, in excellent agreement with the result of a moment analysis by Gulley[14] $(M_2^{3/2}/M_4^{1/2})_{\parallel}/(M_2^{3/2}/M_4^{1/2})_{\perp} = 1.083$. We believe that the decrease in linewidth is a consequence of the increase in exchange narrowing. Perhaps this is not surprising insofar as the nearest neighbor exchange in MnF$_2$, J_{nn} is surprisingly small and is almost certainly tied to the directional dependence of the superexchange interaction and the low symmetry of the F- sites. Hence, increasing lattice vibrations might well result in a larger r.m.s. J_{nn} contribution, with the respective increase in exchange narrowing more than compensating for the expected linear increase of ΔH_{pp} from the processes considered above. At more than 100° below T_m, ΔH_{pp} decreases yet more strongly and even the ratio $\Delta H(\parallel)/\Delta H(\perp)$ is interchanged! One possible interpretation, besides changes in the lattice constants, is that there is an appreciable concentration of Mn^{2+} vacancies whose rapid diffusion as a precursor to melting contributes further to the narrowing of the line. It is interesting to note that it is in this very same temperature region that the microwave conductivity appears to increase noticeably. Further supplementary measurements are clearly necessary before a detailed picture may be had.

C. <u>Linewidths of the Molten Fluorides</u>. Figs. 1 and 3 show that all of the molten fluorides behave, as might be expected, very similarly at higher temperatures. Just above T_m the double fluorides all have quite similar $\Delta H_{pp} \simeq 150\pm5G$, whereas MnF_2, with the higher density of magnetic moments, starts with $\Delta H_{pp} \simeq 194\pm14G$. Perhaps the most interesting feature associated with the solid-liquid transition is the abrupt initial increase in ΔH_{pp} of the double fluorides upon melting. One might argue that a substantial decrease in the local density upon melting would cause a more rapid decrease in the exchange than in the dipolar interactions - and hence the narrowing would be relatively suppressed. However, we believe that the abrupt increase in ΔH_{pp} is mainly a manifestation of the decrease of the <u>dynamic</u> superexchange interaction as a consequence of its strong directional dependence, most favorable for the 180° Mn-F-Mn configuration.

When the temperature increases above T_m, all linewidths decrease more or less linearly. This variation cannot be due to simple volume expansion, but must be caused by motional narrowing.[15] For motional narrowing along ΔH_{pp} should be proportional to η/T (η = viscosity.). Values of η are not available, but the slow decrease of η with T requested to fit our temperature dependence seems reasonable. The combined effect of exchange and motional narrowing has still to be solved, but the strong variation of ΔH_{pp} from T_m on points towards the prevailing influence of motion to the decay of the spin correlation functions.

ACKNOWLEDGEMENTS

We gratefully acknowledge the assistence of Mr. N. Nighman, who prepared the samples used in this work and who made accessible his crystal growth experience, and the help of Dr. A.R. King for many discussions and suggestions.

REFERENCES

1. L. Yarmus. M. Kukk, and B.R. Sundheim, J. Chem. Phys. <u>40</u>, 33(1964).
2. T.B. Swanson, J. Chem. Phys. <u>45</u>, 179(1966).
3. B.R. Sundheim, J. Flato, and L. Yarmus, J. Chem. Phys. <u>51</u>, 4132(1969).
4. R.D. Hogg, Rev. Sci. Instrum. <u>44</u>, 582(1973).
5. L.S. Singer, W.H. Smith, and G. Wagoner, Rev.Sci.Instrum.<u>32</u>,1249(1969).
6. M.S. Seehra, Rev. Sci. Instrum. <u>39</u>, 1044(1968).
7. M. Peter, D. Shaltiel, J.H. Wernick, H.J. Williams, J.B. Mock, and R.C. Sherwood, Phys. Rev. <u>126</u>, 1395(1962).
8. R.P. Gupta and M.S. Seehra, Phys. Lett. <u>33A</u>, 347(1970).
9. K. Horai and K. Saiki, J. Phys. Soc. Japan, <u>21</u>, 397(1966).
10. K. Lee, A.M. Portis, and G.L. Witt, Phys. Rev. <u>132</u>, 144(1963).
11. J.E. Gulley and V. Jaccarino, Phys. Rev. <u>B6</u>, 58(1972).
12. P.W. Anderson, P.R. Weiss, Rev. Mod. Phys. <u>25</u>, 269(1953).
13. R.B. Griffiths, Phys. Rev. <u>124</u>, 1023(1961).
14. J.E. Gulley, Ph.D. thesis, 1970, UCSB (unpublished).
15. N. Bloembergen, E.M. Purcell, and R.V. Pound, Phys. Rev. <u>73</u>, 679(1948).

EPR STUDY OF FeS_2:Ni, Co

R. N. Chandler and R. W. Bené
Department of Electrical Engineering and
Electronics Research Center, University
of Texas at Austin, Austin, Texas 78712

ABSTRACT

We report on an EPR study of the pyrite solid solutions $Co_xFe_{1-x}S_2$, $Ni_xFe_{1-x}S_2$, and $Co_xNi_yFe_{1-x-y}S_2$. The single-ion spectrum for dilute Ni^{2+} in a pyrite structure was fitted to a spin Hamiltonian of the form $H_s = g_z \beta S_z H_z + g_x \beta (S_x H_x + S_y H_y) + D S_z^2$, where $g_z = 2.13 \pm .005$, $g_x = g_y = 2.11 \pm .005$, and $D = -1.21$ cm^{-1}. The single-ion spectrum for Co^{2+} in FeS_2 showed evidence of dynamic Jahn-Teller activity, with a broad, isotropic line at $g = 2.18 \pm .02$. At metallic concentrations, FeS_2:Co displayed the classically asymmetric, isotropic line of a conduction electron but with an unusually high g-value of $g = 2.12 \pm .005$, indicating the partial ionic character of narrow band electrons. The system FeS_2:Ni:Co displayed both the broad lines characteristic of the itinerate Cobalt e_g electrons and the localized Ni^{2+} spectrum. The peculiar Ni^{2+} lineshapes resulting from exchange effects with the itinerate electrons are analyzed, and dependences upon power level are displayed.

EVIDENCE FOR EXCHANGE ENHANCEMENT BY IMPURITY-BAND ELECTRONS IN $Sm_{1-x}Gd_xS$

W. M. Walsh, Jr., L. W. Rupp, Jr., E. Bucher and L. D. Longinotti
Bell Laboratories, Murray Hill, N.J. 07974

ABSTRACT

The anomalously larger exchange-induced g-shifts previously reported for Gd^{3+} ions, compared with those arising from isoelectronic Eu^{2+} ions, as trace impurities in the samarium monochalcogenides suggested that the "extra" electron remains localized in the immediate vicinity of the donor. Such a model appears to be consistent with low temperature ESR studies of these ions at high concentrations in $Sm_{1-x}Gd_xS$ and $Sm_{1-x}Eu_xS$. An initial decrease of the Gd^{3+} g-shift for $x \sim 0.01$ is followed by a rapid increase up to $x = 0.15$ at which concentration a semiconductor-to-metal transition occurs. This behavior appears to arise from delocalization of the "extra" electrons into an impurity band with subsequent buildup of electronic density at higher concentrations. Some evidence for an overall antiferromagnetic Gd-Gd coupling is found. Resonance has not been observed in the metallic regime except near $x = 1$ where a $g = 2$ signal is observed above the Néel temperature of GdS. In contrast the g-shift of Eu^{2+} increases slightly up to $x \sim 0.05$ followed by a monotonic decrease to zero in pure ferromagnetic EuS.

INTRODUCTION

The samarium monochalcogenides have attracted considerable attention over the past few years because the divalent state ($4f^6, ^7F_0$) of the samarium ions produces strong VanVleck paramagnetism and is relatively easily disturbed by lattice compression leading to semiconductor-to-metal transitions.[2] Electron spin resonance (ESR) studies of dilute S-state probe ions have revealed striking exchange effects in the semiconducting states of these crystals.[3-6] The $Sm^{2+}-Sm^{2+}$ exchange interactions of the pure host crystals have been deduced from the low-temperature magnetic susceptibilities[4] and the resultant large hyperfine field at the chalcogen nuclei have been observed via NMR.[7] We report further ESR studies of Gd and Eu resonance g-shifts in SmS over the entire concentration range from very dilute to fully substituted. The results appear qualitatively consistent with a model suggested earlier[4] in which the "extra" electron introduced by Gd^{3+} doping remains closely associated with that ion in the dilute limit but is now inferred to delocalize into an impurity band at concentrations of $\sim$ 1 mole % GdS.

EXPERIMENTAL RESULTS

The two series of crystals $Sm_{1-x}Gd_xS$ and $Sm_{1-x}EuS$ were prepared by melting prereacted salts in sealed tantalum crucibles followed by quenching to preserve the NaCl phase. Thin [100] plates ~ 2mm × 2mm × 0.5mm were cleaved for ESR studies conducted principally at 17.4 GHz and 4.2 K. Considerable care was required at the higher concentrations to correct for demagnetization effects and spurious shifts due to image effects. In several cases spectra were also taken at 12.0 and 34.5 GHz and at higher temperatures. At concentrations appreciably greater than 1% line widths ranged from several hundred to as much as two thousand gauss so no great accuracy in the measured g-shifts can be claimed but the gross trends shown in Fig. 1 proved reproducible.

Fig. 1 The g-values measured at 4.2 K in the two series $Sm_{1-x}Gd_xS$ (black circles and ellipses) and $Sm_{1-x}Eu_xS$ (open triangles). The solid curves merely connect the data; the dashed curve is based on the model for Gd described in the text.

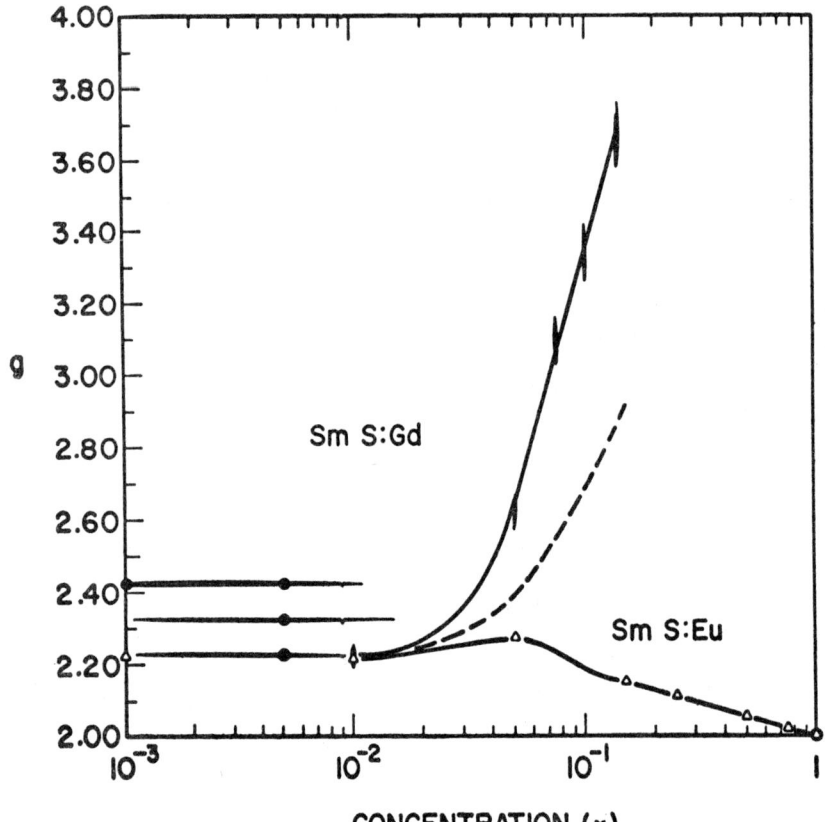

The simplest results were obtained for the Eu-doped series where the dilute-limit g value 2.23 is seen to increase slightly near 5% concentration then decrease monotonically to 2.00 in pure EuS. All samples appeared quite insulating at low temperature and strong magnetization effects were seen as the ferromagnetism of the end member developed

In the case of Gd one must distinguish several regimes: At very low concentrations ($\lesssim$ 0.1%) a cubic fine-structure spectrum centered at g = 2.43 is found. At somewhat higher concentrations complex spectra grow in with apparent g-values of 2.33 and 2.23. Finally near 1% a broad unstructured resonance appears at g = 2.23 and microwave loss is first observed. At higher concentrations this line broadens further and moves very rapidly to large g-values. These very large shifts decrease with increasing temperature but we will not attempt to deal with that phenomenon here. No resonance has been found at concentrations higher than 14% save for a simple g = 2.00 signal in pure GdS at temperatures well above the Néel point of 47 K. It should be stressed that at Gd concentrations greater than 15% the crystals are no longer black with lattice constants consistent with Sm^{2+} ions but have collapsed into an isostructural golden metallic phase due, very crudely, to the internal pressure of the smaller Gd^{3+} cores.[8] In the range 1%-14% appreciable microwave loss was observed and the resonance lineshapes developed the asymmetry characteristic of electrically conducting media.

DISCUSSION

The variation of the g-shift across the Eu-doped series appears consistent with decreasing availability of Sm^{2+} ions which produce δg = 0.23 in the dilute limit, though we are not aware of any calculations for this relatively simple interplay of exchange coupling between randomly situated Eu^{2+} local moments and Sm^{2+} singlet ions.

The larger g-shift δg = 0.43 for very dilute Gd^{3+} relative to Eu^{2+}, despite the smaller size of the trivalent ion, led to the suggestion[4] that the "extra" electron does not delocalize but occupies a cubically symmetric orbital around the donor and mediates the observed large exchange with the nearest-neighbor Sm^{2+} ions on which it principally resides. Such a wavefunction would be principally made up of s- and d-like excited states of the Gd^{3+} ion and of its twelve nearest Sm^{2+} neighbors. It would thus have a diameter of $\gtrsim$ 10 Å.

As the gadolinium concentration is increased from trace values the probability of interaction between such large entities increases rapidly. We therefore expect the complicated spectra exhibiting intermediate g-shifts (observed in the concentration range $10^{-3} < x \lesssim 10^{-2}$) to result from complexes (e.g., pairs, triplets, etc.) of Gd ions located near enough to each other for some overlap of their "extra"-electron wavefunctions to occur. Since overlap of even the outermost parts of the wavefunctions may be expected to allow further delocalization of the "extra" electrons the average density of such electrons on the samarium ions nearest to the gadolinium ions will be reduced and, in turn, a corresponding reduction in the exchange field felt by those ions will occur.

Finally at concentrations $x \gtrsim 0.01$ the degree of connectivity among the overlapping "extra"-electron wavefunctions becomes so great that an impurity band is a more meaningful description. Alternatively the phenomenon is known as a Mott transition. The impurity band involves essentially all the Sm^{2+} ions therefore the "extra"-electron density decreases from 1/12 per Sm^{2+} nearest-neighbor ion in the very dilute limit to $\sim 1/99$ for $x = 0.01$. This decrease produces the reduced g-shift. At concentrations greater than 1% the electron density per Sm^{2+} increases as the impurity band broadens and the g-shift correspondingly increases. The dashed curve of Fig. 1 is given by such a simple model in which undressed Gd^{3+} ions have a $\delta g = 0.23$, fortuitously equal to that for Eu^{2+} ions, and become dressed by impurity electrons as the concentration increases with the scale of that effect determined by the dilute-limit g-shift. A correction for the decreasing lattice constant[4,8] has also been included.

This model may be expressed as

$$\delta g(x) = \left[\delta g_{bare} + [\delta g(0.001) - \delta g_{bare}]\frac{12x}{1-x}\right]\left[\frac{a(o)}{a(x)}\right]^{21} \quad (1)$$

where the lattice constant dependence is taken from our pressure experiments[5] using the revised compressibility of SmS.[9] It should be recognized that treating the g-shift as the primary variable is superficial in the sense that one does not distinguish between changes in Sm^{2+}-Gd^{3+} exchange and the host Sm^{2+}-Sm^{2+} exchange which enters via renormalization of the $J = 0$, $J = 1$ energy difference (see Ref. 4). Such oversimplification is probably trivial with respect to the true complexity of the random mixed crystal, however. One particularly important question which we have not dealt with is the overall interaction between the exchange-enhanced Gd moments. Susceptibility measurements[10] indicate a net antiferromagnetic behavior for a sample with 7.5% Gd. It is quite likely that such antiferromagnetic coupling between randomly arranged exchange-enhanced Gd moments results in anisotropy fields which contribute significantly to the apparent g-shift. This may explain the increase in $\delta g(c)$ for $c \geq 0.01$ which is twice as rapid as our simple noninteracting dressing model could account for. The fact that these very large g-shifts decrease as the temperature is increased is at least qualitatively consistent with weakening of the Gd-Gd antiferromagnetic interactions.

ACKNOWLEDGMENTS

We have benefited from discussions with R. J. Birgeneau, A. Jayaraman and P. D. Dernier.

REFERENCES

1. E. Bucher, V. Narayanamurti and A. Jayaraman, J. Appl. Phys. $\underline{42}$, 1741 (1971).

2. A. Jayaraman, V. Narayanamurti, E. Bucher and R. G. Maines, Phys. Rev. Letters $\underline{25}$, 368 (1970) and $\underline{25}$, 1430 (1970).

3. F. Mehran, K. W. H. Stevens, R. S. Title and F. Holtzberg, Phys. Rev. Letters $\underline{27}$, 1368 (1971).

4. R. J. Birgeneau, E. Bucher, L. W. Rupp, Jr. and W. M. Walsh, Jr., Phys. Rev. $\underline{B5}$, 3412 (1972).

5. W. M. Walsh, Jr., L. W. Rupp, Jr. and L. D. Longinotti, AIP Conf. Proc. $\underline{10}$, 1560 (1973).

6. F. Mehran, J. B. Torrance and F. Holtzberg, Phys. Rev. $\underline{B8}$, 1268 (1973).

7. K. C. Brog and R. P. Kenan, Phys. Rev. $\underline{B8}$, 1492 (1973).

8. A. Jayaraman, E. Bucher, P. D. Dernier and L. D. Longinotti, Phys. Rev. Letters $\underline{31}$, 700 (1973).

9. E. Kaldis and P. Wachter, Sol. Sta. Comm. $\underline{11}$, 907 (1972).

10. E. Bucher (unpublished data).

MAGNETIC PROPERTIES OF GADOLINIUM CHALCOGENIDES WITH VARYING STOICHIOMETRY.

W. Beckenbaugh, G. Güntherodt, R. Hauger, E. Kaldis,
J.P. Kopp* and P. Wachter
Laboratorium für Festkörperphysik, ETH Zürich,
Hönggerberg, 8049 Zurich, Switzerland.

ABSTRACT

The magnetic and optical properties of GdS and GdSe compounds with varying stoichiometry have been studied. The results indicate an antiferromagnetic nearest neighbor exchange, which is dependent on charge carrier concentration, an effective moment enchanced over the free ion value and a possible second transition temperature below T_N. These effects are discussed in terms of polarization of the 5d conduction electrons which can result in a magnetic order within the electron system.

INTRODUCTION

There have been several investigations of how the magnetic properties of Eu chalcogenides are affected by the presence of charge carriers (1,2,3,4). This paper reports the study of the magnetic properties of the GdS and GdSe systems as a function of stoichiometry. The principal advantage of these systems is that a small change in the relative concentration of the Gd ion will cause a large change in the electron concentration without significantly changing the lattice parameters or the structure of the lattice. Additionally, one does not encounter difficulties when introducing foreign ions as dopants, either with or without a magnetic moment. In the past it has not been always realized that the wide range of Curie temperatures from -60 K to -120 K for GdS (5,6,7) and from -60 K to -90 K for GdSe (5,6) are connected with deviations from stoichiometry and thus varying carrier concentrations. In our investigation the relative charge carrier concentration has been measured optically (8,9) on the very same samples on which the magnetic measurements have been performed. In addition chemical analysis gave us not only the ratio between Gd and the chalcogen ion concentrations, but also the absolute concentration of ions per formula unit. The lattice constant and the cubic phase has been determined by X-ray analysis. The respective data are compiled in Table I.

ELECTRONIC STRUCTURE

The Gd chalcogenides are bright lustrous metals which exhibit spectacular changes in color with varying stoichiometry. In these compounds Gd is strictly trivalent and since the chalcogen ion can

---

*Permanent address: Department of Physics, Loras College, Dubuque, Iowa 52001, USA

Table I. Physical and Magnetic Properties of GdS and GdSe Compounds

Compound	No.	No. of ions Formula unit		a_o (Å)	free carriers $\times 10^{22}$ (cm)$^{-3}$	Θ_p (K)	T_N (K)		T_1 (K)	I_1/k (K)	I_2/k (K)
		Gd	S or Se				χ	χ$_o$			
GdS	612	0.99	1.05	5.5606	2.12	−97	45	38	23	−0.41	−0.71
GdS	834	1.01	0.95	5.5505	2.63	−102	48	42	20	−0.43	−0.76
GdS	542	1.02	0.90	5.555	(a)	−115	55	44	12	−0.48	−0.87
GdSe	540	0.92	1.08	5.762	1.4	−28	50	(b)	18	+0.17	−0.79
GdSe	144	0.98	1.02	5.766	1.8	−60	54	(b)	24	−0.05	−0.86
GdSe	907	1.02	0.97	5.7754	2.43	−92	58	(b)	18	−0.27	−0.92

(a) Surface quality did not allow accurate reflectivity measurements.
(b) χ_o apparatus was not adequately sensitive above 50 K.

accept only 2 electrons we are left with one conducting electron per Gd ion in stoichiometric samples. The problems which we want to discuss in this section are the character of the lowest conduction band and the relative energies of the respective s, p, d and f levels. If we look at the absorption spectra of Gd^{3+} ions in various host lattices, we realize that the 4f-5d transition energies are about 10 eV (10). Also there is about a 4 eV energy difference between the center of gravity of the lower lying 5d and the 6s states. The same relative sequence of energy levels is retained also in the solid, where the lowest conduction band is filled with one electron per Gd ion. The valence p band of the chalcogen is in between the localized 4f^7 levels and the 5d conduction band. Basically this energy level scheme has been confirmed by photoemission data (11). From optical spectra we derive the following statements (8,9). The energy difference 4f^7-Fermi level amounts to 9 eV. From a Kramers Kronig analysis of the reflectivity we derive a plasma resonance of the free carriers, which, depending on stoichiometry, is between 4 and 5.5 eV. The optical effective mass of the free carriers is between 1.2 and 1.6 m. The lustrous and different colors of GdS and GdSe are due to a plasma reflection edge in the visible part of the spectrum, which corresponds to coupled modes of a plasmon with a interband optical transition. From this visible plasma edge one cannot determine the concentration of free carriers or the effective mass. Such coupled modes have been observed for example in Ag where the interband transition involves d states (14). We are thus taking the coupled modes as strong evidence that the interband transition goes into the 5d-band and that the free carriers are mostly 5d electrons. Apparently there is an energy gap between the bottom of the 5d conduction band and the top of the p valence band (11). Deviations from stoichiometry create localized levels within this energy gap, which, at 300 K, can trap electrons. It is shown in Table I how the free carrier concentration varies with stoichiometry.

RESULTS OF MAGNETIC MEASUREMENTS

The systems being investigated have been studied magnetically by measuring the susceptibility (χ) using a Faraday balance in fields of 4 and 12 kOe and the initial susceptibility (χ_o) using a mutual induction method (12).

In the paramagnetic temperature range follows an exactly linear Curie-Weiss law with θ_p values which vary systematically with carrier concentration (see Table I). When the stoichiometry ratios are taken into account the measured effective magnetic moments are consistently higher than the free ion value ($P_{eff} = 8.2 \pm 0.1 \mu_B$ and $P_{free\ ion} = 7.94 \mu_B$).

As indicated by Fig. 1 it is difficult to determine a precise value for T_N. The values in Table I are determined according to the following criteria: (1) the existence of a high temperature maximum in χ_p, (2) a point of inflection in the $1/\chi$ curves, and (3) the invariance of the point of inflection in $1/\chi$ in repeated measurements on selected samples over a 2 month period of time. The values of T_N quoted here are in satisfactory agreement with (6).

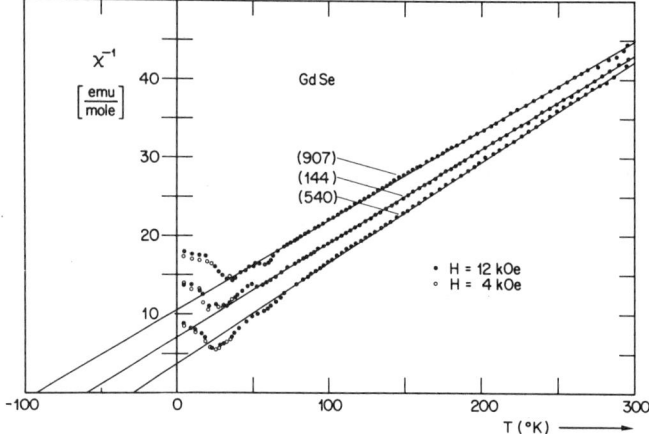

Figure 1. Reciprocal susceptibilities versus temperature for three GdSe compounds. See Table I for identification. The varying slopes indicate the raw data before correcting for varying Gd stoichiometry values.

The pattern which can be established in the behavior of the low temperature susceptibilities include the following observations: (1) the χ below T_N is critically field dependent (see Fig. 2), (2) in samples with higher carrier concentrations the χ_o differs more drastically from χ than in samples with excess chalcogens, (3) the low temperature χ changes with aging of the sample. Since the observed effects seem to correlate with the change in carrier concentration, it seems reasonable to assign the second maximum in χ_o below T_N to the 5d electrons. Possible interpretations of this result will be given in the following section.

ANALYSIS OF RESULTS

If the 5d electrons only have a weak magnetic interaction with the ion spin system they can be considered as sufficiently delocalized at high enough temperatures. If one assumes a very simple picture of the susceptibilities of these two systems adding, the effective moment would be

$$P^2_{eff} = g_1^2 J_1(J_1 + 1) + K^2 g_2^2 J_2(J_2 + 1) \tag{1}$$

The factor K is the ratio of conducting electrons to Gd ions and would be 1 for stoichiometric samples. With the following entities, $g_1 = 2$, $J_1 = 7/2$, $g_2 = 2$, and $J_2 = 1/2$ we obtain $P_{eff} = 8.13 \mu_B$ in agreement with the measured value $8.2 \pm 0.1 \mu_B$. That is, by assuming the orbital component of the conduction electrons is quenched, the 5d electron matrix could provide the additional moment of a "spin one-half" system weakly coupled to a spin system obeying a Curie-Weiss law.

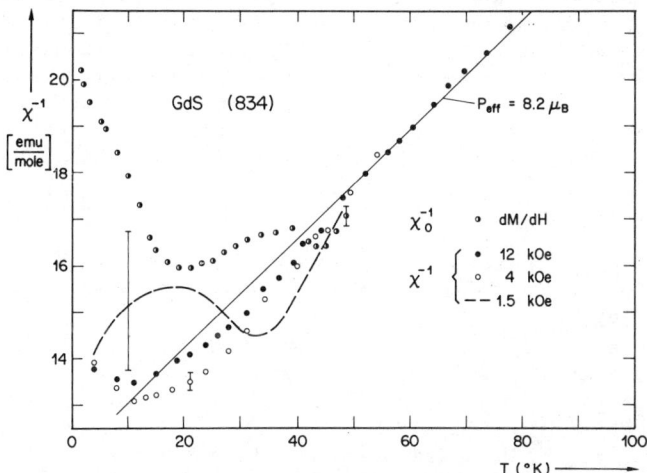

Figure 2. Low temperature reciprocal susceptibility of GdS (834) in applied fields of 0, 1.5, 4, and 12 kOe. The dashed curve for H = 1.5 kOe is the average of several data points at selected temperatures.

Generally we can expect three magnetic exchange interactions for the Gd chalcogenides: a superexchange between the $4f^7$ spins, an interaction between the 5d electrons and the $4f^7$ spin system, and a magnetic exchange amongst the conduction electrons (the RKKY interaction is generally considered invalid in the case of d-electrons). The third mechanism could provide us with a second magnetic ordering temperature. We are well aware of the problematics of applying a MFA to these systems, but making use of the observed AF_2 structure in GdS and GdSe at 4.2 K (6) and the values of T_N and θ_p we can extract the nearest (I_1) and the next nearest (I_2) exchange energies, which are quoted in Table 1. In general it is seen that I_2 increases in magnitude from the S to the Se, when comparing similar stoichiometries. The tendency towards anti-ferromagnetic I_1 correlates with the charge carrier concentration, especially for the case of GdSe.

Considering I_1 in Table I it is evident that the greater the density of carriers, either in sulfide compared with the selenide or within the stoichiometry range of one compound, the larger the tendency toward antiferromagnetic coupling. Even without using MFA it is obvious that the more electrons the larger the difference between $-\theta_p$ and T_N. We thus conclude that the 5d electrons couple antiferromagnetically. Theories favoring such an antiferromagnetic coupling between d-electrons have been presented by Arai (13).

Finally we offer some brief remarks with regard to the low temperature maximum in χ, labelled T_1 in Table I. This peak is strongly field dependent and it changes with aging of the sample. Here one is tempted to speculate about a second ordering process within the 5d matrix. Such ordering processes of electrons have been observed in (3) and possibly in (2). Here we would have to assume that the order of the electrons is of antiferromagnetic sign and that for T going to 0 K in stoichiometric compounds the ion spins are antiparallel, the 5d electron spins are antiparallel, but the electron and ion spins are parallel as demanded by Hund's rule. Another possibility which cannot be ruled out and must await confirmation by neutron diffraction is that between T_1 and T_N the magnetic structure is different from AF_2 and that we are in fact observing a second magnetic phase of the 4f moments.

REFERENCES

1. S. Methfessel and D. Mattis, Handbuch Phys. 28/1 (1968)
2. N. F. Oliveira, S. Foner, Y. Shapira and T. B. Reed, Phys. Rev. B 5, 2634 (1972)
3. J. Vitins and P. Wachter, Solid State Commun. 13,   (1973)
4. J. Schoenes and P. Wachter, Phys. Rev.   (1973)
5. F. Holtzberg, T. R. McGuire, S. Methfessel and J. C. Suits, J. Appl. Phys. 35, 1033 (1964)
6. T. R. McGuire, R. J. Gambino, S. J. Pickart and H. A. Alperin, J. Appl. Phys. 40, 1009 (1969)
7. T. R. McGuire and F. Holtzberg, AIP Conf. Proc. 5, 855 (1971)
8. W. Beckenbaugh, G. Guntherodt, E. Kaldis and P. Wachter, (1973) in print.
9. G. Guntherodt and P. Wachter, 19th Magnetism Conf., Boston (1973)
10. G. H. Dieke and H. N. Crosswhite, Appl. Optics 2, 673 (1963)
11. D. E. Eastman, and M. Kuznietz, J. Appl. Phys. 42, 1396 (1971)
12. The authors are grateful to Dr. O. Vogt, M. Landolt and Dr. F. Hulliger for making their apparatus available and for helpful discussions.
13. T. Arai, Phys. Rev. B 4, 216 (1971)
14. H. Ehrenreich and H. R. Phillipp, Phys. Rev. 128, 1622 (1962).

A SPHERICAL TENSOR OPERATOR DESCRIPTION OF THE EXCHANGE SPLITTINGS OF Gd^{3+} IN $GdCl_3$

R. S. Meltzer and R. L. Cone
Department of Physics and Astronomy, University of Georgia
Athens, Georgia 30602

ABSTRACT

Exchange contributions to the splittings of the Kramers doublet ^6P_J exciton levels at $\vec{k}=0$ have been determined. Both static and dynamic (dispersive) contributions to these splittings were considered in the analysis which used spherical tensor operators to include the effects of anisotropic exchange. The two-electron exchange parameters in the Hamiltonian were determined from the excited state data alone. While anisotropic exchange was dominant in the dynamic contribution, isotropic exchange gave the primary contribution to the static term for these states. The results show for the first time that the isotropic exchange parameter for a rare earth ion in an excited state is the same as that for the groundstate.

INTRODUCTION

Although the effective spin Hamiltonian method has proved very useful in describing the limited group of states spanned by the effective spin, the resulting exchange parameters are not generally applicable to other states. It is therefore desirable to formulate the exchange problem in terms of the <u>real</u> angular momenta of the electrons involved so that exchange matrix elements for different states can be related theoretically.

In this paper we analyze the exchange splittings of the ^6P_J states of Gd^{3+} in $GdCl_3$ using the two-electron tensor operator formulation of the exchange interaction developed by Levy[1]. This approach, which takes into account the orbital dependence of the interaction, allows a determination of the isotropic exchange parameter for the ^6P_J optical excited states which is in excellent agreement with the same parameter determined by Clover and Wolf[2] from studies of the groundstate.

THEORY OF THE $\vec{k}=0$ EXCITON SPLITTINGS

In a pure material such as $GdCl_3$, Frenkel excitons rather than single-ion product states are the true crystal eigenstates. Meltzer and Moos[3] have observed measurable dispersion in several states of the ^6P_J manifolds. This dispersion results in non-uniform shifts of the $\vec{k}=0$ energy levels which must be taken into account in analyzing the $\vec{k}=0$ exchange splittings in $GdCl_3$ and presumably in other pure compounds.

The exciton wavefunctions are determined in the usual way. Sublattice excitons are defined by

$$|\mu_{\vec{k}},p\rangle = N^{-\frac{1}{2}} \sum_{\ell=1}^{N} e^{-i\vec{k}\cdot\vec{T}_{\ell p}} |\mu_{\ell p}\rangle \qquad (1)$$

where N is the number of unit cells, $\vec{k}$ is a reciprocal lattice vector, and $\vec{T}_{\ell p}$ is a translation vector connecting the origin to the pth ion of the ℓth unit cell. The state $|\mu_{\ell p}\rangle$ is a single-ion product state in which ion p of unit cell ℓ is excited, all other ions being in the groundstate. (For the GdCl$_3$ structure there are two ions per unit cell.)

When the Hamiltonian matrix is formed in the $\vec{k}=0$ sublattice exciton representation defined by Eq. 1, two classes of matrix elements of the form $\langle \mu_{\ell p}|\mathcal{H}|\mu'_{\ell' p'}\rangle$ occur: those for which the labels ℓ p and ℓ'p' are identical and those for which one or more of the labels is different. (In this case only matrix elements for which $\mu=\mu'$ are important.) The latter lead to interionic energy transfer and are responsible for the dispersion of the states. These so-called "dynamic" matrix elements have been determined for GdCl$_3$ by Meltzer and Moos[3]. The first or "static" class of matrix elements, which we determine here, uniformly splits the exciton bands arising from the two components of a single-ion Kramers doublet.

Only the exchange interaction and magnetic dipole-dipole interactions must be considered, since electric multipole interactions are negligible[4]. The magnetic dipole-dipole contributions to the static terms are readily calculated using the measured Zeeman splitting factors and the calculated dipolar field. The contribution of that interaction to the dynamic terms is negligible[4].

Levy[1] has shown that the exchange interaction between two ions in states which are antisymmetrized only with respect to the n electrons on ion 1 and n' electrons on ion 2 may be written as

$$\mathcal{H}_{exch} = \sum_{i=1}^{n} \sum_{j=1}^{n'} \sum_{\substack{kk' \\ qq'}} -\Gamma^{kk'}_{qq'} u^{(k)}_{q}(i) u^{(k')}_{q'}(j) \left\{ \frac{1}{2} + 2\vec{s}_i\cdot\vec{s}_j \right\} \qquad (2)$$

where the $\Gamma^{kk'}_{qq'}$ are two-electron exchange parameters which are different for nearest-neighbor and next-nearest-neighbor ion pairs. The $u^{(k)}_q$ are spherical-tensor operators acting on the orbital angular momentum of individual electrons, and $\vec{s}_i$ and $\vec{s}_j$ are the spin operators of individual electrons on each ion.

Analysis of the dynamic matrix elements has been reported earlier[3-4]. Only Γ^{11}_{00} and $\Gamma^{11}_{1-1} = \Gamma^{11}_{-11}$ contribute, and dynamic effects occur only for $M_J = -7/2, -5/2$, and $-3/2$ due to spin selection rules.

For the static matrix elements, we are interested in the interactions of an ion in a ^6P_J excited state with its neighbors which are all in the ground state $^8S_{7/2}$ $M_J=-7/2$. Since the excited state and groundstate are essentially pure ^6P and ^8S, respectively, only Γ^{00}_{00} and Γ^{20}_{00} are involved. The Γ^{00}_{00} term is the familiar isotropic exchange which is usually written $-2J\vec{S}\cdot\vec{S}$. (The definition of $u^{(0)}_0$ determines $\Gamma^{00}_{00} = 7J$.) Since the static splittings are a first order effect, the contributions of the two nearest neighbors and six next-nearest neighbors are additive.

EXPERIMENTAL RESULTS AND ANALYSIS

The spectrum of Gd^{3+} was measured as a function of magnetic field along the easy axis. The data above 10kG was described using a linear least-square fit. The use of only "high-field" data guaranteed that the excited ion was interacting only with "aligned" Gd^{3+} ions. The transition energies were then extrapolated to zero-field to yield the total splitting of the $\vec{k}=0$ excitons arising from a single-ion Kramers doublet.

The splitting at some arbitrary external field consists of the contributions shown in Fig. 1. The "static" exchange and magnetic dipole-dipole interactions give rise to a symmetric splitting to which must be added the "dynamic" exchange in order to properly describe the total $\vec{k}=0$ splitting. The extrapolated zero-field $\vec{k}=0$ splittings ΔE_{obs} along with the individual contributions are listed in Table I for four components of the $^6P_{7/2}$ and $^6P_{5/2}$ manifolds of Gd^{3+}. The "dynamic" contributions $\Delta E_{dynamic}$ are those of Ref. 3. Since the splitting is a first order effect, the contributions are additive. The static contribution ΔE_{static} was thus determined by subtracting the dynamic

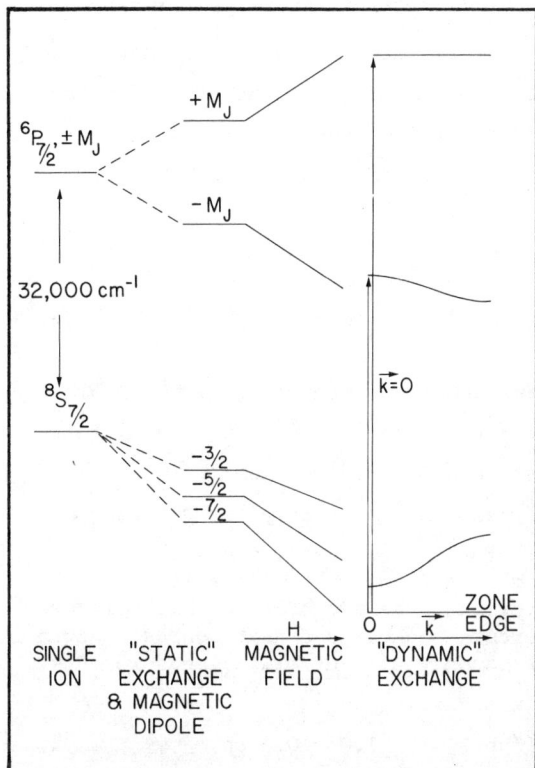

Fig. 1 Contributions to the $\vec{k}=0$ splitting of a Kramers doublet.

Table I Contributions to the extrapolated zero-field splittings of the k=0 Kramers doublet exciton levels in GdCl$_3$ at 1.2°K. All Energies are in cm^{-1}.

$|SLJM_J\rangle$	ΔE_{obs}	$\Delta E_{dynamic}$	ΔE_{dip}	ΔE_{static} obs	ΔE_{static} calc
$^6P_{7/2} \pm 7/2$	5.29±0.4	-0.30±0.26	1.61±0.16	4.0±0.8	4.1±1.1
$^6P_{7/2} \pm 5/2$	4.13±0.4	0.0±0.26	1.10±0.10	3.0±0.8	3.0±0.8
$^6P_{7/2} \pm 1/2$	1.24±0.4	0.0±0.1	0.26±0.03	1.0±0.5	0.6±0.2
$^6P_{5/2} \pm 1/2$	0.59±0.4	0.0±0.1	0.27±0.03	0.3±0.5	0.8±0.2

exchange and magnetic dipole-dipole contributions ΔE_{dip} from the observed splittings.

Relative contributions of the Γ_{00}^{00} and Γ_{00}^{20} terms were determined by evaluating the matrix elements of the spin and orbital operators[5] in Eq. 2. For the $^6P_{7/2} M_J = \pm 5/2$ splitting, the calculated coefficient of Γ_{00}^{20} was 0.3% that of Γ_{00}^{00}, so that the contribution of Γ_{00}^{20} could be quite reasonably neglected. This splitting thus yielded

$$\left\{ 2 \Gamma_{00}^{00}(nn) + 6 \Gamma_{00}^{00}(nnn) \right\} = 0.9 \pm 0.2 \text{ cm}^{-1}. \quad (3)$$

Coefficients of the Γ_{00}^{20} term for the other states listed in Table 1 were at most 15% of those of the Γ_{00}^{00} term. We cannot clearly rule out a contribution to the splittings of these states by Γ_{00}^{20}; however, by neglecting the Γ_{00}^{20} term and using only the Γ_{00}^{00} term of Eq. 3, a reasonable fit to all of the four measured states can be obtained as shown in the right-hand column of Table I.

Clover and Wolf[2] have determined the groundstate isotropic exchange parameters for both nearest-neighbor and next-nearest-neighbor Gd^{3+} ions in GdCl$_3$ from high-frequency susceptibility measurements. In the Γ_{00}^{00} notation, their values give

$$\left\{ 2 \Gamma_{00}^{00}(nn) + 6 \Gamma_{00}^{00}(nnn) \right\} = 1.0 \pm 0.1 \text{ cm}^{-1}. \quad (4)$$

There is thus excellent agreement between the isotropic exchange parameters determined <u>independently</u> from ground and excited state measurements.

CONCLUSIONS

While anisotropic exchange gives the dominant contribution to dynamic exchange effects such as exciton dispersion, isotropic exchange gives the dominant contribution to the static effects for the ^6P_J states. Moreover, the excellent agreement between the two-electron exchange parameters for ground and excited states has indicated the feasibility of a <u>unified</u> description of exchange interactions for a wide range of electronic states.

ACKNOWLEDGEMENT

The authors wish to thank Mr. Tom Lynch for his assistance with the measurements.

REFERENCES

1. P.M. Levy, Phys. Rev. <u>117</u>, 509 (1969).
2. R.B. Clover and W.P. Wolf, Solid State Commun. <u>6</u>, 331 (1968).
3. R.S. Meltzer and H.W. Moos, Phys. Rev. B <u>6</u>, 264 (1972).
4. R.L. Cone and R.S. Meltzer, Phys. Rev. Letters <u>30</u>, 859 (1973).
5. The orbital operators are defined by $(\ell \| u^{(k)} \| \ell) = (2k+1)^{-\frac{1}{2}}$

INTERACTION OF U^{3+} PAIRS IN $LaCl_3$*

T.C.L.G. Sollner and R.N. Rogers
Department of Physics and Astrophysics
University of Colorado, Boulder, Colorado 80302

ABSTRACT

We report the first measurement of actinide pairs in the well studied $LaCl_3$ system. Second shell U^{3+} pairs were observed by EPR, resulting in the determination of the anisotropic interaction tensor for 3rd and 6th nearest neighbors. As this yields a precise measure of the superexchange contribution, the results are discussed in terms of superexchange paths available to the pair in question. Comparison is made of these results to those of second shell Nd^{3+} pairs in $LaCl_3$.

INTRODUCTION

There has been considerable interest in recent years in interaction between ions with unquenched orbital angular momentum.[1] Non-dipolar interactions of lanthanide ion pairs beyond the second nearest neighbor has been studied only for Nd^{3+} in the halides[2], and no results for isolated actinide pairs have been published. Of the actinides, uranium would seem to be the most likely candidate for study, since its half-life is sufficiently long that radiation damage to the host lattice is negligible, and electronically U^{3+} is identical to Nd^{3+}, except that the principal quantum number of the outer shells is greater by one for U^{3+}, leading to the expectation of a larger exchange interaction. This work reports the measurement of long-range interactions for U^{3+} pairs in $LaCl_3$. Detailed measurements for Nd^{3+} in the same lattice are available for comparison[2-5].

THEORETICAL ASPECTS

The effective spin Hamiltonian for a pair of interacting spins $\vec{S}_i, \vec{S}_j$ in a magnetic field $\vec{H}$ is

$$\mathcal{H} = \beta\vec{H}\cdot\vec{g}_i\cdot\vec{S}_i + \beta\vec{H}\cdot\vec{g}_j\cdot\vec{S}_j + \vec{S}_i\cdot\vec{K}\cdot\vec{S}_j$$
$$= \beta\vec{H}\cdot\vec{g}_o\cdot(\vec{S}_i+\vec{S}_j) + 1/2\ \beta\vec{H}\cdot(\Delta\vec{g}_i+\Delta\vec{g}_j)\cdot(\vec{S}_i+\vec{S}_j)$$
$$+ 1/2\ \beta\vec{H}\cdot(\Delta\vec{g}_i-\Delta\vec{g}_j)\cdot(\vec{S}_i-\vec{S}_j) + J_{ij}\vec{S}_i\cdot\vec{S}_j + \vec{S}_i\cdot\vec{A}_{ij}\cdot\vec{S}_j\ . \quad (1)$$

The isotropic contribution J_{ij} has been separated since only the traceless part $\vec{A}_{ij}$ can be determined from the EPR spectrum. The

*Supported by the National Science Foundation.

quantity $\vec{g}_o$ is found from the isolated ion resonance as measured by Hutchison, et al.[6] If $(\Delta\vec{g}_i - \Delta\vec{g}_j)$ is small, the pair spectrum will yield $(\Delta\vec{g}_i + \Delta\vec{g}_j)$, which we denote as $2\Delta\vec{g}$.

The Hamiltonian is diagonalized in the Zeeman interaction, the other terms being treated as perturbations. For the magnetic field applied parallel and perpendicular to the c-axis, resonances due to pairs occur at ΔH from the isolated ion line at H_o given by[2]

$$\theta = 0°: \quad g_\parallel \beta \Delta H^\pm = -\Delta g_{zz} \beta H_o \mp \frac{3}{4} A_{zz} \tag{2}$$

$$\theta = 90°: \quad g_\perp \beta \Delta H^\pm = -(\Delta g_{xx}\cos^2\phi + 2\Delta g_{xy}\sin\phi\cos\phi + \Delta g_{yy}\sin^2\phi)\beta H_o$$

$$\mp \frac{3}{4}(A_{xx}\cos^2\phi + 2A_{xy}\cos\phi\sin\phi + A_{yy}\sin^2\phi), \tag{3}$$

where (θ,ϕ) are the polar angles of $\vec{H}$, and ΔH^+ corresponds to the transition $S_z = 0 \to +1$.

The interaction mechanisms which can contribute to $\vec{A}$ and $\Delta\vec{g}$ are magnetic multipole interactions (MMI), electric multipole interactions (EMI), virtual phonon exchange (VPE), and exchange, and in addition, crystal field distortion due to the presence of a neighboring impurity ion contributes to $\Delta\vec{g}$.[7,1]

For U^{3+} second shell neighbors in $LaCl_3$, the magnetic dipole-dipole interaction may be accurately calculated (see Table I), contributes only to $\vec{A}$, and dominates the MMI. Estimation of the effect of the remaining interactions indicates that, except for exchange, only VPE may contribute to $\vec{A}$ and only crystal field distortion affects Δg. Based on the considerations of Baker,[8] we conclude that VPE contributes not more than 10% of the observed, non-dipolar interactions, and probably much less. Thus any non-dipolar contribution to $\vec{A}$ is due almost entirely to superexchange.

Table I. Dipolar contribution to $\vec{A}$ for U^{3+} neighbors in $LaCl_3$. For each neighbor the coordinate system (xyz) is chosen so that the pair bond lies in the xz-plane. Units are 10^{-4} cm^{-1}.

n	1	2	3	4	5	6
A_{xx}	635.0	-196.1	-99.4	50.6	-26.8	79.4
A_{yy}	635.0	14.5	-27.7	69.3	7.6	79.4
A_{zz}	-1270	181.6	127.1	-120.0	19.2	-158.8

RESULTS

The experimental spectrum for $\theta = 90°$ is shown in Fig. 1. Magnetic susceptibility measurements[9] indicated that first and second neighbor interactions would be too large to be observed, as turned out to be the case. Resonances interpretable as either third neighbor (3n) or fifth neighbor (5n) were observed (Fig. 1).

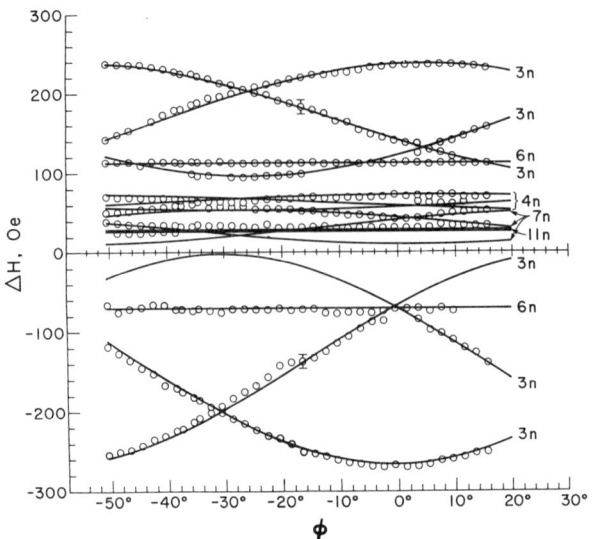

Figure 1. Angular variation about the z axis ($\theta = 90°$) of U^{3+}: $LaCl_3$ second shell neighbor pair resonances. The solid curves for 3n and 6n were fit to the data. Those labeled 4n, 7n and 11n are calculated from magnetic dipole-dipole alone.

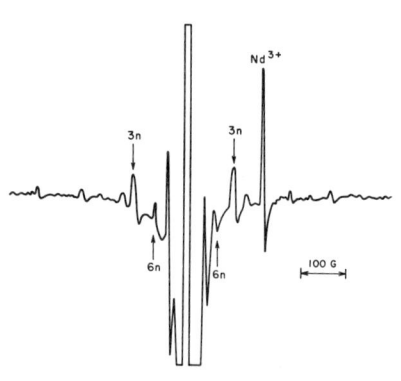

Figure 2. Derivative EPR spectrum near the large central resonance of isolated U^{3+}: $LaCl_3$ for $\theta = 0°$. The smaller, almost evenly spaced hyperfine lines of a trace of U^{235} are also visible.

These were assigned to 3n pairs on the basis of the size of the observed interaction. Identification of these resonances with 5n pairs would imply a 3n interaction $A_{zz} > 2$ cm^{-1}, which is extremely unlikely. That these are corresponding high- and low-field lines can be seen by their correct prediction of 3n resonances in Fig. 2, thus measurement at other frequencies was not required to separately determine both $\vec{\Delta g}$ and $\vec{A}$.

The 6n pair bond lies along the c-axis and thus possesses axial symmetry. Isotropic resonances which appear near the 6n dipolar values in Fig. 1 correctly predict the position of lines of the proper intensity in the $\theta = 0°$ spectrum, Figure 2, confirming the 6n assignment.

Additional resonances appear in the high field spectrum for which the low field counterparts were probably obscured by chips of twinned crystal. We have tentatively identified these as due to 4n, 7n and 11n because they fall close to the resonance expected from the corresponding purely dipolar interactions shown in Fig. 1. Dipolar values indicate that 5n, 8n, 9n and 10n will be obscured by the central transition.

553

Table II. Measured non-dipolar contributions for $\vec{A}$ in units of 10^{-4} cm^{-1} and measured $\vec{\Delta g}$ for U^{3+} pairs. Values of A_{zz} in parentheses were determined from $\theta = 0°$ data. Errors are in last digit quoted. Values in brackets [ ] are approximate results for Nd^{3+} from reference 2.

n	A_{xx}	A_{xy}	A_{yy}	A_{zz}	Δg_{xx}	Δg_{xy}	Δg_{yy}	Δg_{zz}
3	-138.5	-1	-19.0	157.5(157)	.002	-.0006	-.0070	.0078
	[1]	[0]	[-3]	[2]	[.008]	[-.005]	[.000]	[-.001]
5	[-20]	[0]	[-17]	[37]	[-.007]	[-.002]	[.006]	[.010]
6	7.6	--	7.6	-15(-13)	-.003	--	-.003	.007
	[-2]		[-2]	[5]	[-.005]		[-.005]	[.014]

Table II summarizes the results of our measurements on the now-assigned 3n and 6n interactions. Since the sign of the interaction is not determined, we have chosen the smaller of the two possible values of the non-dipolar interactions, a result which is at least consistent with the imputed small magnitude of the 4n and 5n interactions. Also listed are similar results for Nd^{3+} as found by Baker and Marsh[2].

DISCUSSION AND CONCLUSIONS

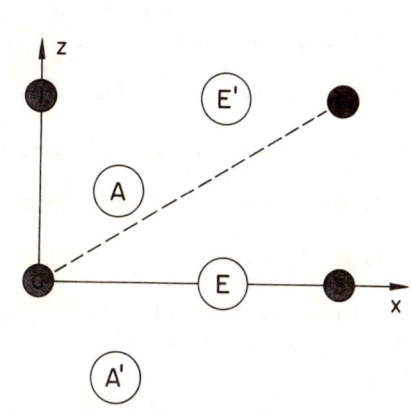

Figure 3. Position of the LaCl$_3$ ligands relevant to superexchange for 3n and 5n. All of the anions lie somewhat out of the plane defined by ions 0, 1, 3 and 5. 6n is above 1 on z-axis. (From ref. 2)

Comparison of Nd^{3+} and U^{3+} values in Table II shows that the Δg for U^{3+} are generally smaller than for Nd^{3+} in LaCl$_3$. This supports the assumption that the Δg originate in a crystal field distortion, since the U^{3+} ion is a better fit in LaCl$_3$ than is Nd^{3+}.

The larger non-dipolar values for 3n and 6n U^{3+} over Nd^{3+} confirms the suspicion that superexchange becomes more important for the actinides. The fact that $A_{xy} \simeq 0$ for 3n indicates that the exchange path is nearly symmetric about the xz-plane. Thus only paths O-(A,A')-E-3, and perhaps O-E-3 as discussed below, appreciably contribute to the interaction (see Fig. 3). Furthermore, the small non-dipolar 6n interaction suggests that paths which involve La^{3+} are relatively weaker paths for superexchange. These last two observations tend to confirm the suggestions of Baker and Marsh.

It was shown by Baker and Marsh that the exchange path for 5n was similar to that for 3n, except that it does not contain the sharp bend O-A-E shown in Fig. 3. They suggested that the relatively larger non-dipolar interaction found for 5n Nd^{3+} may be thus explained. For U^{3+} however, the very large non-dipolar interaction for 3n and the apparent smaller contribution to 5n indicates that when ligand overlap is sufficiently large, separation dependence predominates. In this case it may be that the path O-E-3 contributes substantially to the 3n interaction.

We feel that further theoretical work into superexchange could be fruitfully applied to the rare earth halides in light of these findings.

ACKNOWLEDGMENTS

We wish to thank Dr. J. G. Conway and his staff at Lawrence Berkeley Laboratory for their hospitality and able assistance while one of us (T.C.L.G.S.) utilized their crystal growing facility to prepare the samples used in this investigation, and Professor W. P. Wolf for a very helpful discussion.

REFERENCES

1. This subject has been thoroughly reviewed by J. M. Baker, Rep. Prog. Phys. 34, 109 (1971).
2. J. M. Baker and D. Marsh, Proc. Roy. Soc. A, 323, 341 (1970).
3. K. L. Brower, H. J. Stapleton and E. O. Brower, Phys. Rev. 146, 233 (1966).
4. J. M. Baker, J. D. Riley, and R. G. Shore, Phys. Rev. 150, 198 (1966).
5. J. D. Riley, J. M. Baker, and R. J. Birgeneau, Proc. Roy. Soc. A, 320, 369 (1970).
6. C. A. Hutchison, P. M. Llewellyn, E. Wong, and P. Dorian, Phys. Rev. 102, 292 (1956).
7. R. J. Birgeneau, M. T. Hutchings, J. M. Baker, and J. D. Riley, J. Appl. Phys. 40, 1070 (1969).
8. J. M. Baker, J. Phys. C: Solid St. Phys. 4, 1631 (1971).
9. P. Handler and C. A. Hutchison, Jr., J. Chem. Phys. 25, 1210 (1956).

Section 12.   Amorphous and Disordered Materials   555

EXCHANGE INTERACTION IN DILUTE YTTRIUM-RARE EARTH ALLOYS[*]

R. M. Nicklow and N. Wakabayashi
Solid State Division, Oak Ridge National Laboratory
Oak Ridge, Tennessee, 37830

ABSTRACT

Dilute alloys of the heavy rare earths Gd, Tb, Dy, and Ho in yttrium all exhibit the identical type of helical long-range magnetic order at low temperature. This fact suggests that the magnetic order in these alloys is determined predominantly by the RKKY exchange interaction $J(q)$ which is proportional to the susceptibility $\chi(q)$ for the conduction electrons of yttrium. In order to obtain information about $J(q)$, the magnetic excitations in the c-direction of Y(10%Tb) and Y(10%Ho) have been studied at 4.7°K by means of neutron inelastic scattering techniques. In spite of the low concentration of magnetic atoms well-defined propagating excitations were observed for both alloys. The data have been analyzed on the basis of a model in which the equations of motion for spins are averaged over randomly distributed sites of magnetic atoms with the assumption that $J(q)$, properly normalized, is common to both alloys. The results show this to be a reasonably good assumption, and the resulting $J(q)$, which may be interpreted to be characteristic of Y, shows a peak at the wave vector of the helical structure which is much larger than those measured in the pure rare earth metals.

INTRODUCTION

The principal mechanism responsible for the complicated magnetic order exhibited by the heavy rare earth metals is believed to be an indirect isotropic exchange interaction which takes place through the conduction electrons. Of central importance is the response of the conduction electrons to the exchange interaction I between the localized 4f spin moments and the conduction electron spins. This response is described by the susceptibility $\chi(q)$ which is related to the energy bands and Fermi surface of the conduction electrons. The Fourier transform of the exchange interaction $j(q)$ between rare-earth ions is given approximately by[1]

$$j(q) = I^2(q)\chi(q) . \qquad (1)$$

Yttrium while having no 4f electrons, does have an outer electron configuration and electronic energy band structure quite similar to that of the heavy rare earth metals. Thus although

---
[*]Research sponsored by the U. S. Atomic Energy Commission under contract with Union Carbide Corporation.

yttrium is itself non-magnetic, the conduction electron susceptibility provides a mechanism for the existence of a long range exchange interaction between rare earth ion impurities. Indeed rather dilute (e.g., ∼ 2% Tb) quantities of rare earth ions in Y show a long range ordering of their magnetic moments at low temperatures in spite of their random locations in the yttrium structure. Furthermore, the magnetic order of dilute alloys is essentially independent of the particular rare earth impurity in that all such alloys possess oscillatory antiferromagnetic structures which are described by the same periodicity or wave-vector $\vec{q}_0$.[2,3] Thus the susceptibilities for these alloys are probably very similar and perhaps essentially the same as that of pure yttrium.

As in the case of the pure rare-earth metals, information about $\chi(q)$ for yttrium-rare earth alloys may be inferred from the exchange interaction $j(q)$ which can normally be obtained from neutron inelastic scattering measurements of the spin-wave dispersion relations.[4] Of course the interpretation of the magnetic excitations in an alloy is itself a non-trivial problem. The substitution of impurities into magnetic insulators[5] and metals[4] can strongly perturb the spin waves, resulting in distortions of the host spin-wave dispersion relations as well as producing localized magnetic excitations. In the present case as a starting approximation, we shall ignore such problems. Although the host is non-magnetic and the concentration of magnetic impurities is rather low (∼ 10%) in the samples studied, a good description of the measurements is obtained with a mean lattice model. Since the exchange interaction is long range, the effects of the variations of the local environment around different magnetic ions may be small. Thus, in the magnetic Hamiltonian which we assum to be

$$H = B \sum_i^{mag} S_{i\zeta}^2 - \frac{1}{2} \sum_{\substack{ij \\ i \neq j}}^{mag} J_{ij} \vec{S}_i \cdot \vec{S}_j , \qquad (2)$$

the summations over the sites of the magnetic ions are replaced by summations over all atomic sites by merely taking into account the probability, c, that a site is occupied by a magnetic ion, viz,

$$H = cB \sum_i^{all} S_{i\zeta}^2 - c^2 \frac{1}{2} \sum_{ij}^{all} J_{ij} \vec{S}_i \cdot \vec{S}_j . \qquad (3)$$

Here J_{ij} is the exchange interaction, B defines the single-ion crystal field anisotropy, c is the concentration of magnetic ions, and ζ is in the c-direction of the (HCP) yttrium structure. The spin-wave energies along the c-direction are then given by

$$\hbar\omega(q) = S[\{cJ(q_0) - cJ(q) + 2B\}\{cJ(q_0) - \frac{1}{2}cJ(q_0+q)$$
$$- \frac{1}{2}cJ(q_0-q)\}]^{1/2} , \qquad (4)$$

where S is the total angular momentum so that $J(q) = (g-1)^2 j(q)$ and g is the Landé factor. This is the same expression usually used for the pure metals in their helical phases[4] except the exchange interaction is reduced by the factor c.

EXPERIMENT

A triple axis neutron spectrometer at the Oak Ridge HFIR was used to measure the spin-wave energies at 4.7°K along the c-direction for two yttrium-rare earth alloys; one containing 9.8 at.% Tb and one containing 9.5% Ho. The Néel temperatures were observed to be approximately 54°K for Y(Tb) and 20°K for Y(Ho). The measurements were performed with the energy of the scattered neutrons fixed at 12.8 meV and with Be crystals as monochromator and analyzer. Graphite filters were placed between the sample and the analyzer.

In spite of the low concentration of magnetic ions, rather well defined excitations were observed, having energy widths of typically 0.5 meV. The spin-wave dispersion curves obtained from the measurements are shown in Figs. 1 and 2. Compared to the results for the helical magnetic phases of the pure rare earth metals,[4,6] these curves have much larger initial slopes and much less dispersion at intermediate wave-vectors.

ANALYSIS AND DISCUSSION

In addition to our assumption that $\chi(q)$ for both alloys is nearly equal to that for pure Y, we also assumed that the interaction $I(q)$ in Eq. (1) is the same for both Tb and Ho. If these assumptions are valid, the spin-wave measurements for both alloys can be described by the same $j(q)$ function. Therefore, with $j(q)$ expressed in terms of interplanar exchange constants, j_ℓ, i.e.,

$$J(q) = (g-1)^2 j(q) = 2(g-1)^2 \sum_\ell j_\ell \cos(\pi\ell\zeta) \;, \tag{5}$$

the measured dispersion curves have been fitted by a least squares procedure to eq. (4) with j_ℓ's, taken to be common to both alloys, and the anisotropy constants B's as the fitting parameters. The overall agreement between the experimental results and the calculations, shown as the full lines in Figs. 1 and 2, is quite good thereby substantiating our assumptions.

The anisotropy constants obtained in the fitting analysis are in fair agreement with those measured for the pure rare earth metals.[4,6,7] Of particular interest, however, is the exchange function $j(q) = J(q)/(g-1)^2$ deduced from the measurements. This is shown in Fig. 3 together with earlier results for pure Ho,[6] which possesses the largest peak $J(q_0) - J(0)$ of the pure metals. Such a peak is necessary for the existence of a helical magnetic structure. It is apparently much larger in the yttrium-rare-earth alloys, indicating that $\chi(q_0) - \chi(0)$ is much larger for Y than for the rare earths.

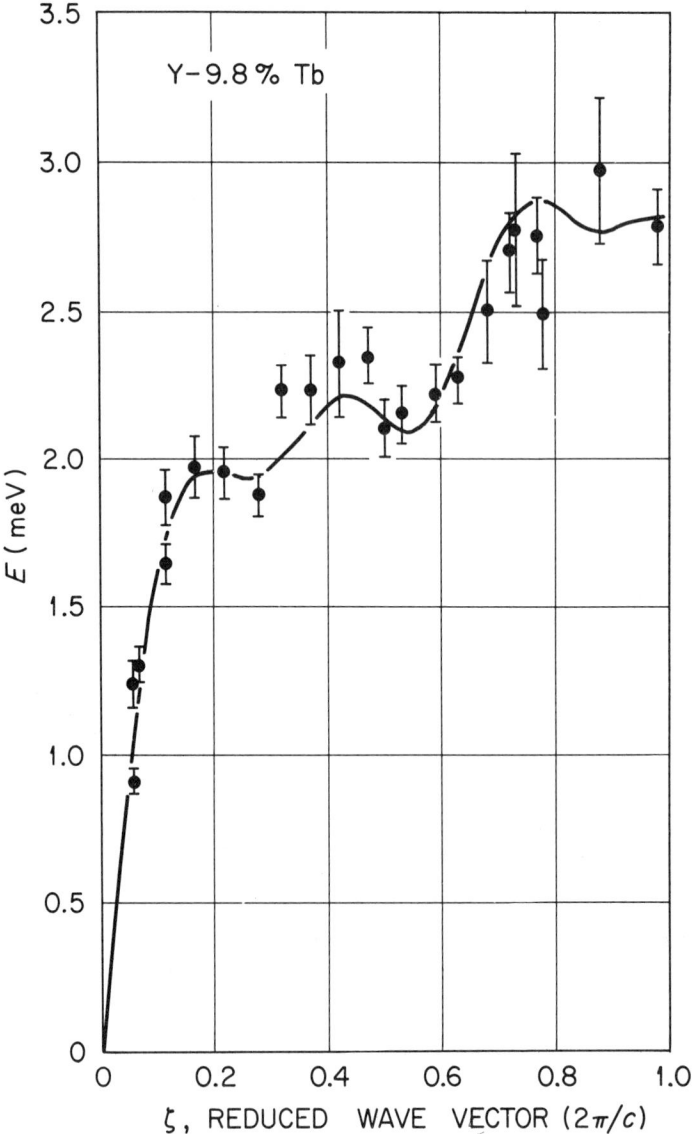

FIGURE 1. SPIN-WAVE ENERGIES MEASURED FOR Y(9.8 AT.% Tb) ALONG c-DIRECTION. THE FULL CURVE IS A FIT TO THE DATA AS DESCRIBED IN THE TEXT.

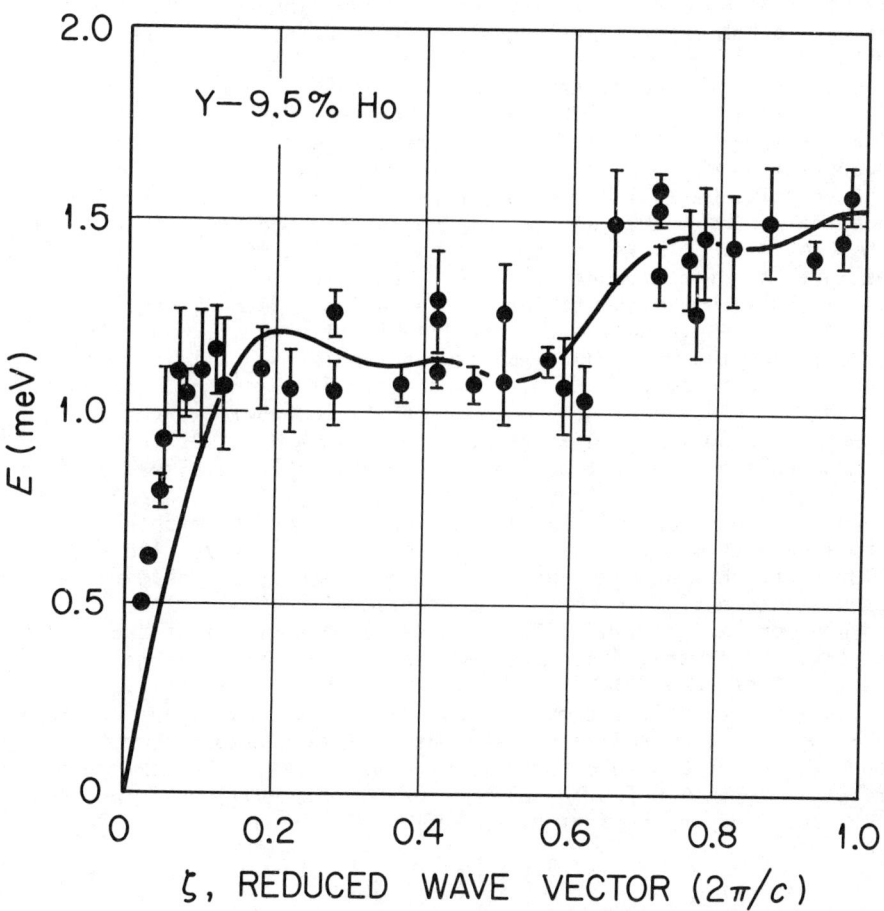

FIGURE 2. SPIN-WAVE ENERGIES MEASURED FOR Y(9.5 AT.% Ho) ALONG c-DIRECTION. THE FULL CURVE IS A FIT TO THE DATA AS DESCRIBED IN THE TEXT.

This conclusion is in good qualitative agreement with recent $\chi(q)$ calculations by Liu et al.,[8] and it also appears to be consistent with other experimental results. For example, when the rare-earths are diluted with yttrium the tendency toward helical magnetic ordering increases strongly with increasing yttrium concentration.[2] The ferromagnetic structures of Tb and Dy, both of which also have helical structures, are completely suppressed at Y concentrations of 24% and 5%, respectively. In Gd which is ferromagnetic only, helical ordering occurs in Gd-Y alloys at an Y concentration of $\sim$ 20%, and ferromagnetism is completely suppressed at $\sim$ 60 at.% Y.[3] It is tempting to explain such structure properties in terms of a concentration dependent $J(q)$, or $\chi(q)$, for which $\chi(q_o) - \chi(0)$ grows with yttrium concentration. Such a model would qualitatively explain also the results of Belov et al.[9] who find that although the ordering temperatures for Tb_cY_{1-c} alloys decrease as expected with decreasing c, the magnetic field which is required to transform their helical structures to ferromagnetic structures increases strongly with decreasing c for $0.36 \leq c \leq 1.0$.

One consequence of the mean lattice assumption employed here is that single-ion interactions such as the crystal field, are proportioned to c, whereas two ion interactions such as the exchange are proportional to c^2. In the limit of small c the crystal field contribution to the magnetic Hamiltonian can become comparable to the exchange interaction so that the conventional spin-wave theory we used may not be justified. To investigate this aspect of the data analysis, the crystal field energy levels for a point charge model in a mean field approximation were calculated for both Y(Tb) and Y(Ho) alloys at various temperatures and for various values of $J(q_o)$ and B which gave approximately the observed Néel temperatures. The ground states at 0°K were found to have essentially the maximum S_z for both alloys, $\sim$ 6 for Tb and $\sim$ 8 for Ho, thereby justifying the procedures used in the present analysis.

SUMMARY

Neutron inelastic scattering measurements on dilute ($\sim$ 10%) alloys of the heavy rare-earths Tb and Ho in yttrium metal show the existence of rather well defined propagating spin-waves. The resulting dispersion relations are satisfactorily described by a mean lattice model with the assumption that the exchange interaction, $j(q)$, is essentially proportional to the conduction electron susceptibility $\chi(q)$ of yttrium and, hence, is common to both alloys. The $j(q)$ function deduced from the data possesses a peak at the wave vector of the helical magnetic structures of the alloys which is much larger than those measured for the pure rare earth metals. This result is consistent with theoretical calculations of $\chi(q)$ from the electronic energy bands for yttrium and with measurements of the magnetic structures of yttrium-rare earth alloys as a function of composition, temperature, and applied magnetic field.

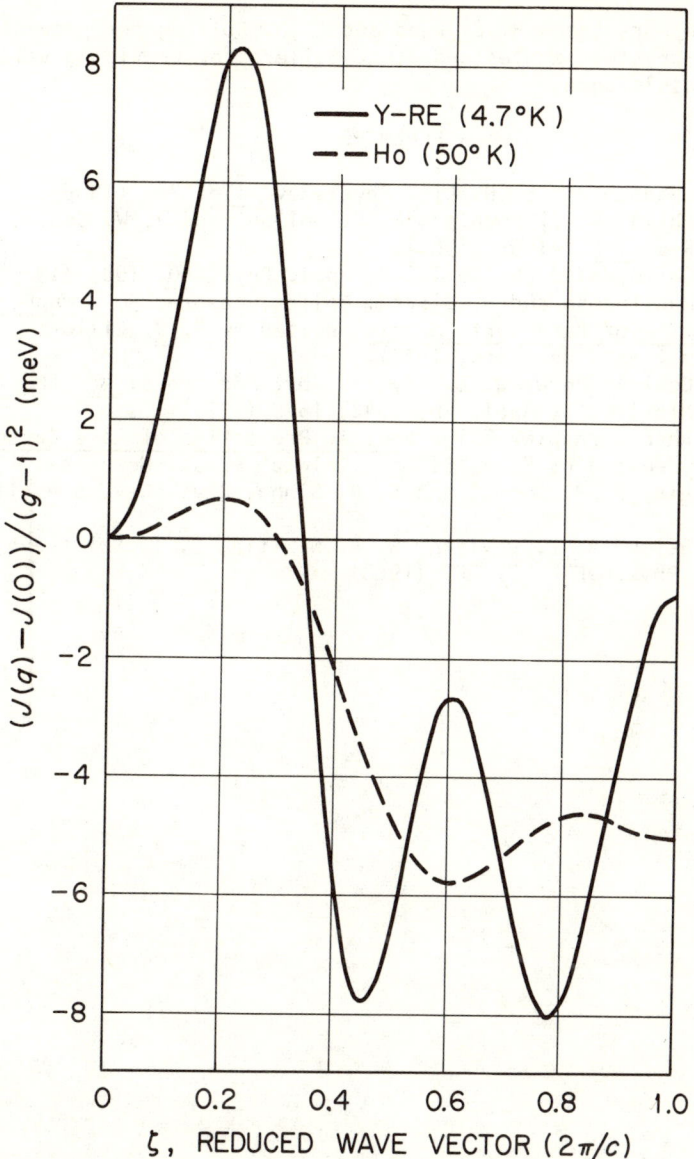

FIGURE 3. THE EXCHANGE INTERACTION $J(q)/(g-1)^2$ FOR DILUTE Y(RE) ALLOYS AND FOR Ho METAL.

ADKNOWLEDGEMENTS

The authors thank R. E. Reed and E. D. Bolling for providing the single crystal samples and J. L. Sellers for providing valuable technical assistance.

REFERENCES

1. W. E. Evensen and S. H. Liu, Phys. Rev. 178, 783 (1969).
2. H. R. Child, W. C. Koehler, E. O. Wollan, and J. W. Cable, Phys. Rev. 138, A1655 (1965).
3. H. R. Child and J. W. Cable, J. Appl. Phys. 40, 1003 (1969).
4. A. R. Mackintosh and H. Bjerrum Møller, chapter 5 in Magnetic Properties of Rare Earth Metals, edited by R. J. Elliott (Plenum Press, New York, 1972).
5. R. A. Cowley and W. J. L. Buyers, Rev. Mod. Phys. 44, 406 (1972).
6. R. M. Nicklow, J. Appl. Phys. 42, 1672 (1971).
7. J. J. Rhyne, chapter 4 in Magnetic Properties of Rare Earth Metals, edited by R. J. Elliott (Plenum Press, New York, 1972).
8. S. H. Liu, R. P. Gupta, and S. K. Sinha, Phys. Rev. B 4, 1100 (1971).
9. K. P. Belov, R. Z. Levitin, S. A. Nikitin, L. I. Solvtseva, Soviet Phys. JETP 27, 207 (1968).

MAGNETISM IN AMORPHOUS TERBIUM-IRON*

J. J. Rhyne, S. J. Pickart and H. A. Alperin
Naval Ordnance Laboratory, Silver Spring, MD 20910 and
National Bureau of Standards, Washington, DC 20234

ABSTRACT

Neutron diffraction and magnetization measurements on a sputtered specimen of 33% terbium, 66% iron have confirmed the existence of long range magnetic order and an amorphous spatial distribution of atoms and spin sites. Elastic diffraction data taken above and below the 409K Curie temperature have enabled the separation of nuclear and magnetic scattering contributions. The magnetic spin density function contains a very broad major peak, dominated by the Tb-Tb pair distribution function, with a maximum near 3.6 A. Longer range correlations are very much weaker. The principal features of the atomic structure can be accounted for by considering only near neighbor correlations between terbium and iron atoms in a random close-packed arrangement.

Inelastic scattering studies have revealed a significant shift in the magnetic density of states to lower energy in the amorphous material relative to the crystalline $TbFe_2$ counterpart. No discrete spin-wave excitations or critical scattering near T_c could be observed.

I. INTRODUCTION

The concept of long range ordered magnetism existing in a material which lacks a regular crystal structure, or is amorphous, is of relatively recent origin. Gubanov[1] in 1960 first gave the theoretical evidence, using a molecular field model, that translational invariance was not a requirement for ferromagnetic order and that the effect of the topological disorder was to introduce fluctuations into the exchange interaction which modify the Curie temperature and magnetization.

The first apparent experimental observation of amorphous ferromagnetism was the work of Brenner et al.[2] in 1950 who observed ferromagnetic behavior in an amorphous film of Co-P prepared by electrodeposition. However, it was not until the mid 1960's that metallic amorphous materials received much attention.[3] Since then several amorphous materials have been prepared, primarily by one of three techniques: (1) Rapid quenching from the liquid state "splat cooling" as for example in studies of Fe-Pd-P by Maitre-Pierre[4] and Sharon and Tsuei[5]; (2) Electrodeposition from solution as in the case of Ni-P[6] and Co-P[7] alloys studied by Cargill and others; (3) by evaporation onto a cooled substrate as used by Fujima[8] and Mader and Norwick[9] to prepare amorphous alloys at low temperatures which often recrystallize when heated to room temperature.

The materials prepared by these techniques (except for the low temperature evaporation) have several features in common. Among these is the universal presence of a "glass former" atom (e.g., C, P, or Si) which is required to produce the amorphous phase, presumably by filling in holes in the metal atom stacking arrangement. In general these materials are found to be amorphous for only a limited compositional range, typically approximately 75 to 85 atomic percent of the transition metal. In many cases the measured effective moment of the magnetic element is significantly reduced by the exchange fluctuations. As expected intuitively no magnetic anisotropy is observed and the alloys are magnetically "soft."

The recent availability of sputter-deposited rare earth-transition metal alloys (e.g., $TbFe_2$) has presented an amorphous magnetic system with properties radically different from the amorphous materials described above. These sputtered alloys have been prepared in both thin films by the IBM group[10] using a.c. sputtering and in millimeter thick bulk plates by the Battelle Northwest Laboratories[11] using a d.c. sputtering technique. These alloys contain no "glass forming" atoms and are strongly magnetic with Curie temperatures in general above room temperature. As discussed later, some possess a giant local magnetic anisotropy and a large low temperature coercivity. The alloys can be prepared in amorphous form over a wide range of composition. We have confirmed by neutron diffraction the absence of crystallinity in samples from the Tb_xFe_{1-x} system with $x = 11.1\%$ and $x = 75\%$ as well as $x = 33\%$ ($TbFe_2$). The precise mechanism behind the occurrence of the amorphous structure in these alloys is at best speculative; however, it is felt that the "cold" deposition of the atoms on the substrate in the sputtering process (compared to conventional evaporation techniques) plus the large atomic size difference between rare earth and transition metal atoms are key factors required for producing the amorphous phase.

II. SAMPLE PREPARATION

The R-Fe (R = rare earth) materials used for the studies reported here were all supplied by R. Allen of Battelle Northwest and were prepared by a high deposition rate d.c. triode sputtering system[11] using a water cooled copper target and sample substrate with a krypton sputtering gas. The target was made from compacted powders mixed according to the desired composition. No detectable variation in composition was found in the final sputtered sample. Care was taken in cleaning and baking the sputtering chamber to minimize oxygen contamination of the samples. The copper backing of the samples was removed by grinding before neutron measurements were made to eliminate contamination of the scattering pattern by the copper Bragg lines.

Most of the work reported here was done on a material of composition $Tb_{.33}Fe_{.67}$ prepared as a one-inch disk. Two samples were studied, one a one millimeter thick disk and a second three millimeters thick. No differences were detected between these two samples in neutron studies. The density of the amorphous material was 8.3 gm/cm^3.

compared to the crystalline x-ray density of 9.1 gm/cm^3 and 8.9 gm/cm^3 measured on a cast and annealed bulk sample of crystalline TbFe$_2$. Magnetization data on the amorphous TbFe$_2$ gave a Curie temperature of 409 K to be compared with 697 K[12] for the Laves phase crystalline TbFe$_2$. The room temperature resistivity is 220 μΩcm and remains essentially unchanged at 77 K due to the large temperature independent atomic disorder scattering. Room temperature resistivity for polycrystalline TbFe$_2$ is 58 μΩcm. The amorphous material has been shown to recrystallize into the TbFe$_2$ Laves phase on prolonged annealing above 600 K which also produces dramatic changes in the magnetic properties.[13]

III. AMORPHOUS NEUTRON DIFFRACTION STUDIES

Neutron scattering measurements were made on both two and three axis spectrometers at the National Bureau of Standards Reactor. The technique of neutron scattering provides a sensitive method of examining the atomic arrangement present in the materials and, due to the interaction of the neutron magnetic moment with that of the magnetic electrons, also allows a probe of the magnetic spin distribution. Unlike x-rays, the neutrons readily penetrate the entire sample, and thus the total scattering is proportional to the volume of the sample. The absence of Bragg scattering peaks provides a sensitive indication of the lack of clustered crystalline phases in an otherwise amorphous sample.

A neutron scattering experiment measures the intensity of neutrons of incident wavevector $\vec{k}_o$ scattered by the sample through an angle 2θ into a final state $\vec{k}_f$. It is convenient to describe the scattered intensity in terms of the change in neutron wavevector $\vec{Q} = \vec{k}_f - \vec{k}_o$ as I(Q). The scattering may be either elastic $|k_f| = |k_o|$ in which case $|Q| = \frac{4\pi \sin \theta}{\lambda}$ (λ = incident wavelength) or inelastic in which the neutron either gains or loses energy from creating or annilating an excitation in the solid (e.g. phonon or magnon). For this case $|k_o| \neq |k_f|$ and one defines the energy exchange of the neutron with the excitation ω = - ΔE$_{neutron}$ = E$_f$-E$_o$.

The total neutron scattering measured by a conventional two axis spectrometer contains both elastic and inelastic events -- the latter being integrated over all energy transfers which satisfy conservation of initial state momentum and energy. True elastic scattering processes (ΔE = 0) may be measured only with the help of an additional scattering from an analyzing crystal which selects out only those scattered neutrons whose energy has remained unchanged within an energy resolution width (typically a few tenths of a millivolt for incident energies of order 10 meV). In conventional crystalline diffraction measurements the distinction between total and elastic scattering is usually inconsequential, due to the dominant elastic cross-section. However, in these amorphous magnetic materials the difference is critical, particularly in determining the atomic structure or nuclear scattering from measurements above the Curie temperature as described in Section V.

In general the scattered neutron intensity is a measure of the scattering cross-section $\sigma(Q,\omega)$ which can be written in terms of a scattering function $S(Q,\omega)$ defined by

$$I(Q,\omega) \propto \sigma(Q,\omega) = \frac{\sigma_b}{4\pi} \frac{k_f}{k_o} S(Q,\omega) \tag{1}$$

where $\sigma_b = 4\pi b^2$ is the bare atomic scattering cross-section. In the case of amorphous materials, $S(Q,\omega)$ is the Fourier transform of the real space atomic pair correlation function or time dependent density function $G(r,t)$.

For the case of elastic scattering the differential scattering cross-section per unit solid angle for N atoms may be written[14]

$$\left(\frac{d\sigma}{d\Omega}\right)^{elas} = N\bar{b}^2 [1 + \int \rho_e(r) e^{-iQr} dr] e^{-2w} \tag{2}$$

where $\rho_e(r)$ is the equilibrium pair correlation function (corresponding to $t = \infty$) and e^{-2w} is the Debye-Waller temperature factor.

The expression for the total scattering cross-section is somewhat more complex to evaluate, and it is convenient to introduce the "static approximation" in which the energy change of the scattered waves is small compared to the incident energy. Such an approximation is clearly valid for x-rays and generally satisfactory for neutrons with high incident energies. Under this restriction the total scattering cross-section may be written

$$\left(\frac{d\sigma}{d\Omega}\right)^{total} = N\bar{b}^2 [1 + \int \rho_o(r) e^{-iQr} dr] \tag{3}$$

where $\rho_o(r)$ is the instantaneous ($t = 0$) pair correlation function. Thus within the static approximation and the assumption that the two pair correlation functions represent essentially the same distribution, then the total and elastic cross-section differ only by the inclusion of the Debye-Waller factor in the elastic scattering.

In the case of $TbFe_2$, which contains two atoms, there are really three pair correlation functions $\rho(r)$, one for each of the Tb-Tb, Tb-Fe, and Fe-Fe pair combinations. These will be lumped together in the observed scattering from the compound and can only be separated uniquely by studying isotopically substituted specimens with different atomic scattering amplitudes. Such experiments are in progress on amorphous $DyFe_2$.

IV. TOTAL NEUTRON SCATTERING

As we reported previously[15] the diffraction pattern from sputtered $TbFe_2$ above the Curie point as shown in Figure 1(A) shows an atomic distribution much like that of a liquid. The lack of significant structure in the scattering at higher Q compared to that observed in many amorphous semiconductors and conventional glasses indicates a relative lack of longer range atomic correlations. This

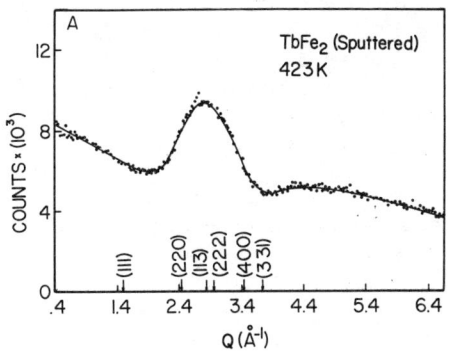

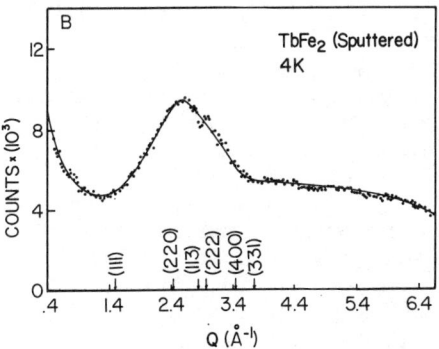

Figure 1. Total neutron diffraction intensity. (A) Above the magnetic ordering temperature. (B) At 4.2 K, as a function of scattering vector Q. The "blips" in the data near the peak are the result of remnant scattering from the Aluminum sample container.

will be discussed more fully later. The data of Figure 1 have been corrected for instrumental background but do include contributions from inelastic paramagnetic scattering which produce the envelope or form-factor-like decrease in intensity with increasing Q. The arrows in the figure indicate the positions of some of the Bragg reflections in crystalline $TbFe_2$. The absence of significant scattering from the sputtered $TbFe_2$ in the region of both high and low angle Bragg crystalline peaks (in particular the (331) and (111)) provides further indication of the absence of a micro-crystalline structure in these materials.

For the data reported, the scans were generally taken with 40 minute collimation; restricting this to 20 minutes did not materially change the observed scattering and thus instrumental resolution corrections were neglected.

Figure 1(B) shows the total neutron scattering at 4 K over the same Q range as in Figure 1(A). The diffuse peak is further broadened and shifted to lower Q. An additional low angle scattering contribution is observed which is not seen in the high temperature data. Both of these result from the added scattering from the microscopic spin polarization. The shift in peak position is principally a consequence of increased weight being placed on scattering from Tb as opposed to Fe atom and spin positions in the ordered magnetic state as discussed later. The low angle scattering is of purely magnetic origin and is interpreted as a broadening of the (000) forward scattering corresponding to the introduction of inhomogeneities in the ferromagnetic order several orders of magnitude smaller than conventional ferromagnetic domains.

V. NUCLEAR STRUCTURE DETERMINATION

The scattering intensity from amorphous $TbFe_2$ at temperatures higher than the Curie temperature is dominated by the nuclear scattering, which then provides a mechanism to probe the atomic distribution in the material in a manner analogous to x-ray diffraction, which of course has higher statistical precision. The neutron method, however, enjoys one significant advantage over x-rays -- namely the lack of an atomic form factor which suppresses the high angle x-ray intensity. Removing this form factor dependence from the data is one of the major uncertainties in the x-ray analysis, particularly in multi-element systems for which the form factor must be calculated. As shown in equations (2) and (3) of the previous section, the neutron scattering at large scattering angles (large Q) becomes a constant ($N\bar{b}^2$) which greatly simplifies and improves the accuracy of the data analysis.

As observed in Figure 1(A) the total scattering above T_c contains also a contribution from the paramagnetic scattering. This scattering is inelastic and of the form of a continuum energy spectrum. It has a magnetic form factor angular dependence (see Figure 7) with a fairly rapid fall off due to the relatively dispersed nature of the magnetic electrons and reaches zero essentially at $Q=7$ A^{-1}. To obtain the nuclear scattering this paramagnetic contribution must be removed. Due to its inelastic nature this is conveniently accomplished by using elastic (energy analyzed) data with relatively tight energy resolution. At large Q the form factor removes the paramagnetic contribution and then elastic and conventional total scattering are equivalent aside from the Debye-Waller Factor (eq. 2). This assumes negligible nonmagnetic inelastic scattering as shown in Section IX.

Figure 2 illustrates the nuclear scattering as a function of Q at 433 K. The data are a composite of two experiments (1) Elastic scattering from $Q = .4$ to 6.4 A^{-1} using an incident wavelength of 1.5A (35 meV) with an energy resolution of 0.5 meV and (2) Total scattering from $Q = 6.4$ A^{-1} to $Q = 14.4$ A^{-1} using a wavelength of 0.7 A. The data sets were individually corrected for sample absorption at each wavelength and instrumental background has been subtracted. No corrections were made for incoherent scattering or for multiple scattering. The former is independent of Q (except for the Debye-Waller factor) and acts only as an essentially constant additive factor. The multiple scattering is expected to be small because of the thin disk sample geometry.

The multiple pair distribution function $\rho(r)$ may be obtained by Fourier inversion of the data of Figure 2. The transform function of the intensity I(Q) is

$$\rho(r) = \frac{2r}{\pi} \int_0^{Q_{max}} Q[I(Q)-C]\sin Qr \, dQ \qquad (4)$$

where C is the constant high Q asymptote of the scattering (equal to b^2/atom if there is no incoherent scattering). The exact transform requires $Q_{max} = \infty$ and the effect of the smaller experimentally attainable Q_{max} is to broaden the peaks in the resulting $\rho(r)$ distribution, and to cause oscillations at small r.

Figure 3 shows the resulting multiple pair density function obtained from the data of Figure 2. Several features should be noted. The first major peak which reflects the shortest range atomic correlations

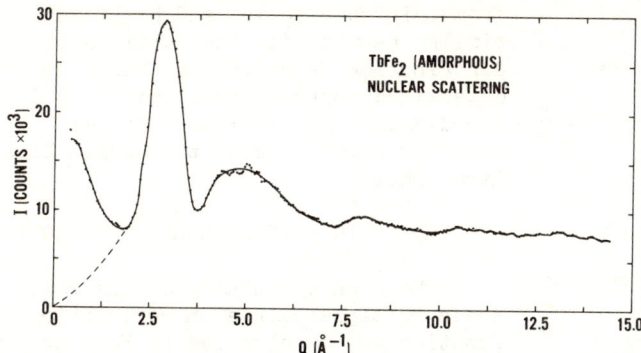

Figure 2. Nuclear scattering from amorphous TbFe$_2$ at 433 K. The dotted line indicates an extrapolation of the forward beam scattering to zero as required by the Fourier transform procedure.

in the material is nearly resolved into two subpeaks and a shoulder corresponding to nearest neighbor atom pairs of Fe-Fe, Fe-Tb and Tb-Tb. The relative amplitudes of the three peaks are in accord with their total scattering amplitude and relative numbers of atoms. The peak positions are strikingly close to the sums of the metallic radii (12-fold coordination radius for Tb=1.78 A and for Fe=1.27 A) for these atoms in contact as indicated by the arrows. It is to be noted that this is in direct contrast to the Laves phase crystalline TbFe$_2$ in which the Tb-Tb interatomic distance is significantly compressed from 3.6 A down to 3.2 A possibly as a result of charge transfer to the Fe d shell. Further confirmation that the Tb-Tb distance is indeed 3.6A is provided by the magnetic scattering (next section). Possible stacking arrangements of three and four atoms in contact are shown at positions of highest probability in the second peak in $\rho(r)$. The weaker higher order peaks in the $\rho(r)$ function do correspond to other possible planar and 3 dimensional stacking arrangements of Tb and Fe atoms; however, a more positive identification must await the isotope substitution experiments mentioned above. Atomic coordination numbers may be obtained from the first peak areas by scaling the results to a vanadium standard scatterer. The ripples observed at low r are the result of termination effects in the data transform.

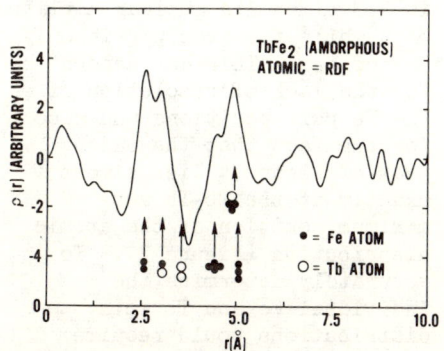

Figure 3. Multiple pair correlation function obtained from the data of Figure 2.

It is clear from these results that the predominant features of the Laves phase unit cell, namely the tetrahedra of Fe atoms and the near planar arrangement of compressed rare earths are absent in the sputtered material. Rather these correlations are replaced by a predominance of pair interactions based solely on statistical factors representative of the dense random packing sequences of Polk[16] and others. Cargill[17] has done x-ray scattering on a sputtered sample of GdFe$_2$

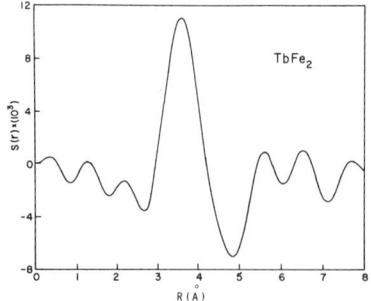

Figure 4. Magnetic spin correlation function obtained from the difference pattern of Figure 1(A) and 1(B).

from this same series and has found similar results for the multiple pair distribution function. He has determined approximate atomic coordination numbers which differ markedly from those of the crystalline Laves phase.

VI. MAGNETIC SPIN CORRELATION FUNCTION

The magnetic analog of the real space multiple pair atom correlation function may be obtained by Fourier transformation of the ordered magnetic moment scattering intensity. This is properly accomplished by transforming the difference pattern between the 4 K and 433 K data shown in Figures 1(A) and 1(B). The high temperature data contains the total magnetic intensity distributed into the paramagnetic "form factor." The subtraction procedure then removes the effects of absorption, incoherent scattering and phonon scattering to the extent that they are temperature independent. The resulting function is one which oscillates about zero and has built in the natural convergence factor provided by the magnetic form factor.

The transformed magnetic spin correlation function is shown in Figure 4. Note it possesses the same general features as the atomic pair distribution function except that the principal peak is not split. In contrast to the atomic case in which all three pair correlation functions have approximately equal weight, the magnetic scattering is dominated by the Tb-Tb spin pair correlation function due to its much larger magnetic moment and hence magnetic scattering amplitude. The magnetic scattering amplitude p_{Tb} (Q=0) for terbium at T=C K is 3.3 times its nuclear amplitude b_{Tb}, while for iron p_{Fe} is only .62 of b_{Fe}. This can account for the lack of resolution of the Fe pair positions and also for the fact that the major peak of Figure 4 lies almost exactly at the Tb-Tb submaximum position of the atomic distribution (Figure 3). To accurately determine the individual Fe and Tb spin distributions would require experiments with substituted nonmagnetic RE and transition metal atoms in a material

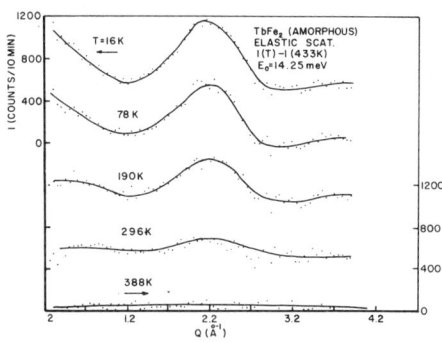

Figure 5. Magnetic elastic scattering at various temperatures obtained from subtracting elastic scattering at 433 K from elastic scattering at temperature T. Arrows indicate left or right intensity scales; other curves are displaced vertically one ordinate unit.

retaining the exact atomic distribution.

VII. MAGNETIC NEUTRON DIFFRACTION

In this section we present data representing the wavevector dependence of the long range ordered magnetism in amorphous TbFe$_2$ at various temperatures. The ordered magnetic component always contributes to the elastic neutron scattering, while the total scattering contains in additional inelastic contributions from any magnetic and phonon excitations plus the scattering from paramagnetic spins. Elastic diffraction is then required to study the long range magnetic order. Figure 5 shows the results of elastic diffraction studies at several temperatures using an incident neutron energy of 14.3 meV. In each case the elastic data at 433 K, which is above the Curie temperature and represents the nuclear or atomic scattering function, has been subtracted. The result is the pure elastic magnetic intensity. The Q=0 intercept, which is experimentally inaccessible, is proportional to the magnitude of the ordered ferromagnetic moment. For Q>0 the major intensity is seen to be concentrated in a broad maximum confirming the amorphous nature of the spin distribution as is the case of the atomic distribution.

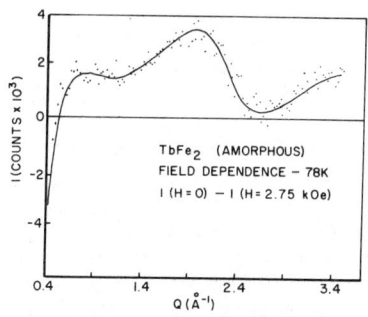

Figure 6. Magnetic field dependence of total scattering at 78 K.

The elastic as well as the total scattering shows the development of considerable low angle scattering in the magnetic intensity compared to the nuclear scattering (Figure 1(B)) and is thus part of the long range ordered ferromagnetic structure. This low angle scattering is quite sensitive to applied fields which may indicate magnetic inhomogeneities analogous to domains on the scale of the atomic coherence lengths. Figure 6 illustrates the change in magnetic scattering on application of a 3 kOe field parallel to the scattering vector. The low angle scattering is observed to increase strongly (the curve represents field-off minus field-on scattering) which is consistent with a decrease of the broadening of the forward beam, reflecting an increase in the size of the magnetic inhomogeneities. The high Q scattering is reduced by the field consistent with the rotation of spins into the scattering vector where they do not contribute to the scattering. Some selective redistribution of spins is observed since the maximum near 2 Å^{-1} in Figure 6 is shifted from the position of maximum magnetic intensity observed at 2.3 Å^{-1} (Figure 5.)

It is instructive finally to examine the total inelastic scattering integrated over all energies satisfying the energy and momentum conservation conditions. This can be obtained by subtracting the elastic scattering properly normalized to a vanadium standard at each temperature. The result is shown in Figure 7. At 433 K the inelastic scattering shows a Q dependence characteristic of a magnetic form factor and can be identified with paramagnetic inelastic scattering. The dotted line represents the square of a form factor constructed additively from Tb and Fe elemental form factors. No correction has been made for Boltzmann population factors in these data since the exact energy of the contributing excitation cannot be identified. It is noted that the inelastic scattering contains no evidence of a low angle tail as mentioned in the previous section. The inelastic scattering persists down to low temperatures and retains its diffuse character below the bulk T_c although its magnitude is reduced.

VIII. MAGNETIZATION OF AMORPHOUS $TbFe_2$

Magnetization measurements have been made[18] on a specimen of sputtered $TbFe_2$ from 4 K to 480 K in applied fields up to 18 kOe, and additional high field magnetization data have been taken at 4 K and 77 K[19] in fields of 100 kOe.

Figure 8(A) illustrates selected magnetization isotherms from 435 K (above the bulk Curie temperature of 409 K) down to 160 K. Below about 120 K an anomalous coercive force type behavior develops as shown in Figure 8(B). The coercive field reaches almost 30 kOe at 4.2 K[19]. Significant remanent magnetization appeared at 201 K and below and amounted to 74% of the saturation value at 82 K. A small time dependence of the remanence was observed reducing the remanent moment by an additional one percent after two minutes. The sample was annealed to above 250 K between each magnetization to remove the remanence.

The spontaneous magnetization was obtained by extrapolation of the linear high field portion of the magnetization back to the demagnetizing field and is given in Figure 9. The dotted portion indicates the region in which a reliable extrapolation cannot be made due to the anomalous magnetization curves. The square symbols represent the magnetization obtained from the neutron elastic magnetic intensity of Figure 5 normalized to the bulk magnetization at N_2 temperature. For convenience

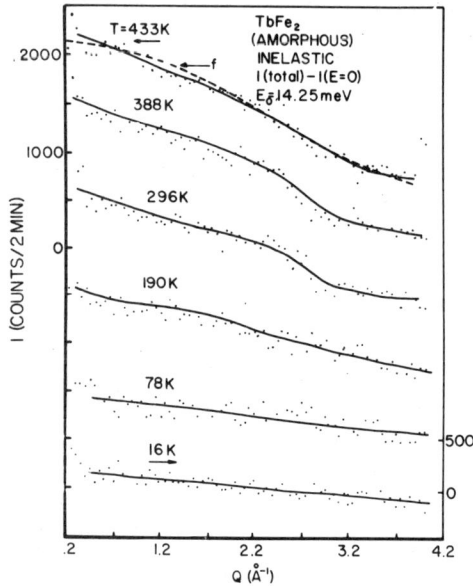

Figure 7. Total inelastic scattering at various temperatures obtained by subtraction of elastic scattering from the total scattering.

this magnetic intensity was obtained by integration under the broad distribution of Figure 5 which will be proportional to the square of the bulk magnetization assuming no redistribution of magnetic intensity over wavelength occurs with changing temperature. The Curie temperature obtained from the data of Figure 9 is 409 K which is somewhat higher than the 388 K reported earlier[15] using a different sample and different magnetization apparatus. The low temperature portion of the magnetization curve also differs from the previous result[15] due to incomplete demagnetization of the sample between each run in the earlier data. The 0 K intercept of Figure 9 corresponds to a spontaneous moment of 4.2 μ_B per formula unit. If the antiparallel arrangement of two iron moments and one terbium moment found in the Laves phase compounds is maintained here, then the minimum net obtainable magnetization would be either 5.6 μ_B or 4.6 μ_B assuming the full moment of 9 μ_B for Tb and respectively the 1.7 μ_B Fe moment of the Laves phase or the 2.2 μ_B of elemental iron. The low moment of the amorphous material implies either an alternate magnetic structure (e.g., one containing antiparallel Tb moments) or that a significant proportion of the magnetization is disordered. The response of such a loosely coupled system could account for the nearly constant susceptibility observed above technical saturation in the magnetization curves below T_c, and also for the inelastic scattering distribution shown in Figure 7. The observed susceptibility is nearly independent of temperature from 435 K down below 150 K which would not be the case if its origin lay in incomplete saturation due to an anisotropy effect.

The concept of an anisotropy energy in an amorphous material may at first appear to be a contradiction in terms; however, one must consider the Coulomb interaction of the highly non-spherical Tb 4f charge distribution located in an electric field of unknown and spatially varying symmetry produced by the neighboring ions. This

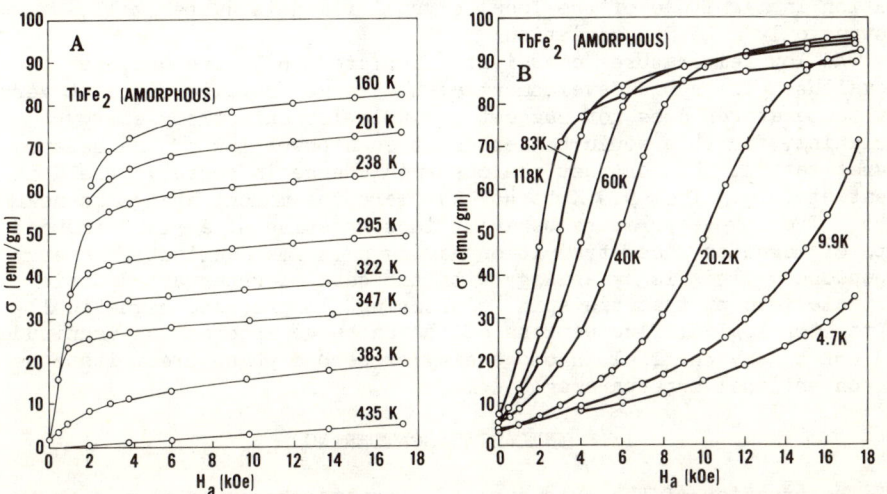

Figure 8. (A), (B). Magnetic moment as a function of applied field at temperatures indicated.

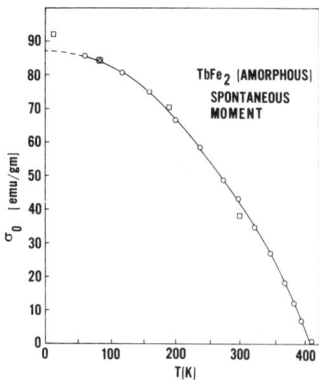

Figure 9. Spontaneous moment as a function of temperature, O obtained from isothermal magnetization data, and □ obtained from elastic neutron diffraction.

plus the strong L-S coupling, provide a local anisotropy which, together with a spatially fluctuating exchange interaction, determines the equilibrium direction of a spin or group of spins satisfying the condition of minimum free energy. The bulk manifestation is then a material for which conceptually all directions are hard magnetic directions and energy is required to align spins parallel along any single magnetization direction. Such a local anisotropy model has been examined theoretically by Harris, Plischke, and Zuckerman[20] who showed that the fluctuating anisotropy can produce a reduced zero Kelvin magnetization as well as a reduced Curie temperature. Their calculations were done for total momentum J = 1,2, and 3 and for a one atom ferromagnetic system; therefore their results are not quantitatively applicable to the $TbFe_2$ system. Harris et al. observed that the Curie temperature reduction of almost 40% in amorphous $TbFe_2$ would require unrealistically large fluctuations in exchange field if this was the only operable mechanism. The inclusion of the randomized crystal field interaction then reduces the required exchange field disorder and also introduces a local anisotropy of second order in the spin variables which is forbidden in crystalline $TbFe_2$ by the cubic symmetry. A surprising result of the calculation[20,21] is that, in spite of the randomized direction of the local field axes, the fluctuation in magnitude of the local crystal field is quite small, showing only a 2% RMS deviation.

The low temperature "coercivity" depicted in Figure 8(B) is anomalous. The rapid development of the large coercive field at very low temperatures does not suggest a single-ion anisotropy energy mechanism, for this would be dependent on a power law of the related magnetization. In contrast the observed change in coercive field is greatest (e.g., from 9.9 K to 4.7 K) where the moment change is nearly zero. The low temperature behavior is suggestive of a field-induced type of magnetic phase transition having a thermal activation energy dependence. This is by analogy with the case of superparamagnetic particles except that the spin "clustering" is produced magnetically by the topological fluctuations of the exchange interaction magnitude, and not by any chemical inhomogeneity or second phase precipitate as in conventional superparamagnetism.

IX. INELASTIC SCATTERING

The question of the existence of discrete phonon or magnon interactions in amorphous $TbFe_2$ is currently being studied using inelastic triple axis and time-of-flight methods. Only a brief account will be

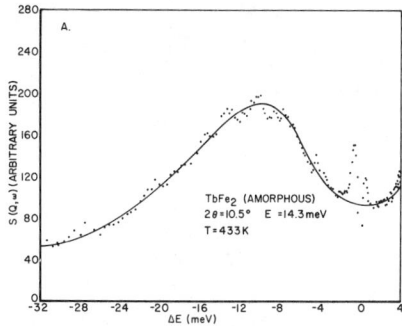
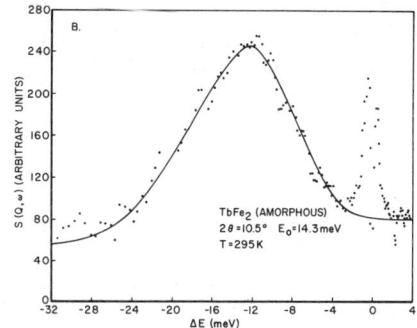

Figure 10. (A) Inelastic time-of-flight spectrum at $T > T_c$. (B) Inelastic spectrum at room temperature. The peak at $\Delta E = 0$ represents the elastic scattering. The Boltzman population factor was removed from the data for comparison of different temperatures.

given here, full details will be published later.[22]

Figures 10(A), 10(B) show the results for the scattering function $S(Q,\omega)$ at 433 K (above T_c) and at room temperature. The results have been corrected for the Boltzmann population factor, time of flight scale and instrumental constants. The wavevector Q was equal to .48 A^{-1} at the elastic position and increased to 1.1 A^{-1} at $\Delta E = -12$ meV neutron energy gain. The spectra show a broad inelastic distribution centered about 12 meV at room temperature and at a somewhat lower energy at 433 K. These excitations have been shown to be magnetic as opposed to phonon in origin as they exhibit a decrease in intensity with increasing Q (magnetic form factor). When analyzed using an expression for paramagnetic scattering[23] the 433 K data show a symmetric distribution about $\Delta E = 0$ which exhibits approximately the predicted Gaussian shape.

It is to be noted for the low T inelastic data that the topological disorder of the amorphous material destroys the uniqueness of the wavevector Q as a descriptor of an excitation, and this effect plus the expected broadening from spin wave life-time effects contributes to the smearing out of the magnetic excitations into a continuous spectrum reflecting the density of states. It is expected that compared to the crystalline material the short wavelength modes (higher energy if an acoustic branch) will be suppressed due to their inability to propagate in the disordered structure, whereas long wavelength modes will be present and produce the magnetic order. This expected shift in inelastic energy spectrum between amorphous and crystalline materials has been confirmed[24] by a time-of-flight experiment on polycrystalline $TbFe_2$ (Figure 11) which shows a much narrower excitation spectrum and one centered at 18 meV versus the 12 meV found for the amorphous material.

A preliminary search for critical magnetic scattering was carried out on the triple axis spectrometer at NBSR. Constant Q scans were made through $\Delta E = 0$ at 5 Kelvin temperature increments near the Curie temperature. The minimum Q attainable was 0.2 A^{-1} due to instrumental limitations. At this and larger Q no evidence for a change in either intensity of width of the $\Delta E = 0$ peak was observed as would normally accompany critical scattering at a magnetic phase transition. It is suggested that this scattering in analogy with the lack of discrete magnon excitations, may only occur at Q values too small to be experimentally observable in this preliminary experiment. Further experiments with better Q and temperature resolution are in progress.

ACKNOWLEDGMENTS

We have benefitted from discussions with S. Cargill, A. E. Clark, R. Harris, M. Zuckerman, and J. R. Cullen and are pleased to be able to quote some of their results in advance of publication as well as the results of some of our collaborative experiments with J. H. Schelleng of the Naval Research Laboratory and D. L. Price of Argonne National Laboratory. We also thank H. T. Savage of providing the resistivity data.

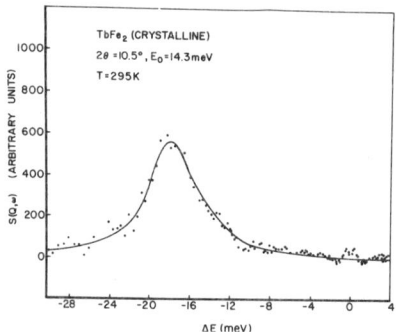

Figure 11. Inelastic time-of-flight spectrum of polycrystalline $TbFe_2$ at room temperature.

REFERENCES

*Work supported by NOL Independent Research Funds.
1. A. I. Gubanov, Fiz. Tver. Tela 2, 502 (1960).
2. A. Brenner, D. E. Couch and E. K. Williams, J. Res. National Bureau of Standards 44, 109 (1950).
3. A review of much of this work is contained in Amorphous Magnetism edited by H. O. Hooper and A. M. deGraaf, Plenum Press, New York, (1973).
4. P. L. Maitre-Pierre, J. Appl. Phys. 40, 4826 (1969).
5. T. E. Sharon and C. C. Tsuei, Phys. Rev. B 5, 1047 (1972); other similar materials were reviewed by C. C. Tsuei, loc. cit. Ref. 3, pp. 299; and by Pol. Duwez and S. C. H. Lin, J. Appl. Phys. 38, 4066 (1967).
6. G. S. Cargill, J. Appl. Phys. 41, 12 (1970).
7. G. S. Cargill and R. W. Cochrane, loc. cit. Ref. 3, pp. 313.

8. S. Fujime, Japan J. Appl. Phys. $\underline{6}$, 305 (1967).
9. S. Mader and A. S. Norwick, Appl. Phys. Lett. $\underline{7}$, 57 (1967).
10. Reviewed by R. J. Gambino, this conference; and P. Chaudhari, J. J. Cuomo, R. J. Gambino, IBM J. Res. and Dev. $\underline{17}$, 66 (1973).
11. R. W. Stewart, Battelle Northwest Laboratories Research Report (1972) unpublished.
12. E. Burzo, Zeit. fur Angewandt Physik $\underline{32}$, 127 (1971).
13. S. Pickart, J. Rhyne, H. Alperin, and H. Savage (to be published).
14. A. J. Leadbetter, A. C. Wright and A. J. Apling, <u>Amorphous Materials</u> edited by Douglas and Ellis, Wiley Press, New York, (1971).
15. J. J. Rhyne, S. J. Pickart, and H. A. Alperin, Phys. Rev. Lett. $\underline{29}$, 1562 (1972); J. J. Rhyne, S. J. Pickart and H. A. Alperin, loc. cit, Ref. 3, pp. 373.
16. D. E. Polk, J. Non-Cryst. Solids, $\underline{2}$, 381 (1973), based on the monatomic model of J. D. Bernal. Proc. Royal Soc. (London) $\underline{A280}$, 299 (1964).
17. G. S. Cargill (this conference).
18. J. J. Rhyne and J. H. Schelleng (to be published).
19. A. E. Clark, Appl. Phys. Lett. $\underline{23}$, 642 (1973).
20. R. Harris, M. Plischke, and M. J. Zuckerman, Phys. Rev. Lett. $\underline{31}$, 160 (1973), and private communication.
21. R. W. Cochrane, R. Harris and M. Plischke, Journ. Non-Cryst. Solids (to be published).
22. Detailed time-of-flight experiments are being performed in collaboration with D. L. Price of Argonne National Laboratory.
23. C. G. Windsor, Proc. Phys. Soc. $\underline{89}$, 825 (1966).
24. H. Alperin, J. Rhyne and S. J. Pickart, Proc. Intl. Conf. on Magnetism, Moscow 1973 (to be published).

AMORPHOUS MAGNETIC MATERIALS

R. J. Gambino, P. Chaudhari, and J. J. Cuomo
IBM Research Center, Yorktown Heights, N. Y. 10598

ABSTRACT

Magnetic ordering in atomically disordered systems is a subject of current experimental and theoretical interest. Amorphous magnetic materials have been known to exist for a number of years, but it has become clear only very recently that some of these materials have properties which make them useful for bubble domain devices and in thermomagnetic recording. In this paper, a brief review will be given of amorphous magnetic materials in general including structure, methods of fabrication and magnetic properties. The major emphasis, however, will be on the properties of amorphous rare earth transition metal alloys. The occurance of growth induced magnetic anisotropy in deposited thin films of amorphous alloys as well as the sources of coercivity in amorphous materials will be discussed.

INTRODUCTION

Alloys which are atomically disordered but magnetically ordered are of both scientific and technological interests. Amorphous materials have been shown to have properties which may make them useful in bubble devices,[1] thermomagnetic information storage[2] and as precursor materials for the preparation of permanent magnets.[3] Amorphous magnets and other disordered magnetic systems can be used as model systems to provide new insight into the nature of the exchange interactions in magnetic materials. Disorder in magnetic systems can take many forms (Table 1). For example, in a random solid solution of magnetic elements, the atomic positions are fixed by the crystal structure but the magnitude of the localized magnetic moment varies in an aperiodic, random fashion. An even simpler case is a spin glass which consists of a dilute solution of a magnetic element in a nonmagnetic host crystal. Again the atomic positions are fixed by the crystal lattice but the spin distribution is random. Other systems have been studied in which the spins are in an ordered lattice but have a disordered coordination sphere. Some examples are solid solutions of the type $GdN_{1-x}O_x$[4] or $Gd_{3-x}V_xS_4$.[5] In these systems, the spin system is periodic in magnitude and spacing but may be aperiodic in the strength of the exchange interaction. A theoretical treatment of this type of system has been given by Tahar-Kheli.[6]

Conceptually, amorphous magnetic elements constitute a class of systems with a still higher level of disorder. The magnitude of the atomic moment is fixed but its position is random and therefore the magnitude of the exchange is aperidoic. Pure amorphous magnetic elements are difficult to prepare and generally not expected to be stable at room temperature. Amorphous Ni has been reported.[7] Amorphous alloys of the type CoP_x approximate an amorphous element in some respects.

TABLE I
Disorder in Magnetic Systems

Class of System	Atomic Positions	Spin Positions	Spin Magnitude	Exchange Interaction	Examples
Crystalline:					
Element	—	—	—	—	
Spin glass	—	Disorder	—	Disorder	Dilute solution of M_n in Au
Solid solution alloy	—	—	Disorder	Disorder	Gd-Tb, Ni-Fe
Solid solution - compounds	Disorder	—	—	Disorder	$GdN_{1-x}O_x$, $Gd_{3-x}V_xS_4$
Amorphous:					
Element	Disorder	Disorder	—	Disorder	Ni
Alloys	Disorder	Disorder	Disorder	Disorder	Gd-Co

In an amorphous alloy, an additional disorder is introduced in that the magnitude of the spin varies from atom to atom. The positional disorder and exchange disorder are also present. Amorphous alloys, the most disordered case, represents the main concern of this paper. During the past decade it has been shown that amorphous magnetic alloys can be prepared by a variety of techniques. Mader and Nowick[8] studied evaporation and showed that a large class of alloys[9,10] can be made amorphous by evaporating the constituents onto a cold substrate. Duwez and co-workers[11,12] showed that selected compositions in certain transition metal - metalloid alloy systems can be made amorphous by rapidly quenching the molten mixture. Bremmer et al showed that amorphous magnetic alloys can be made by electro deposition and electroless deposition.[13] With these three methods of fabrication it is possible to design and prepare amorphous magnetic materials for specific applications and for testing theories of magnetism.

STRUCTURE

The structures of amorphous materials have been described by two principal models; dense random packing and microcrystallites.[14] Dense random packing is the structure obtained if hard spheres are forced together, for example, by hydrostatic pressure.[15] The radial distribution function (r.d.f.), free volume and density of such a structure have been calculated by a number of workers and are generally in good agreement with observed data.[16] The microcrystallite model[17] considers a system consisting of ordered regions a few atomic diameters in radius.

These regions are randomly oriented with respect to one another. The r.d.f. calculated on the microcrystallite model generally shows poorer agreement with the observed r.d.f.'s of amorphous materials.[16] Unfortunately, r.d.f.'s are relatively insensitive to the structural model assumed so it is usually impossible to find a unique structure for an amorphous material. Part of the motivation for studying amorphous magnetic materials is to elucidate the structure of amorphous materials in general. For example Mössbauer experiments[18] and neutron diffraction[19] can be used to probe the local environment of magnet atoms.

In any discussion of the structure of an amorphous material it is essential to consider phase separation. Phase separation into two amorphous solids has been observed in a number of amorphous systems. It manifests itself in the microstructure and the crystallization behavior of these materials. The material in the phase separated condition consists of amorphous regions which differ in composition. Phase separation can produce super paramagnetic clusters in some amorphous systems.[20] It can also influence, anisotropy, coercivity and magnetization. In most amorphous systems phase separation occurs on heat treatment, frequently, just below the crystallization temperature. The products of crystallization very often reflect the fact that the amorphous material had phase separated before crystallizing. Phase separation is most common in systems where the constituents differ considerably in bond character e.g. Fe-Ge, Gd-Gd_2O_3[21] and various silicate, phosphate and borate glasses. It is much less prevalent in amorphous metallic systems such as Gd-Co or Fe-P-C. The scope of this paper will be limited to amorphous systems which are not obviously phase separated. We will adopt the dense random packing model with composition fluctuations as a working structural model.

In any discussion of the magnetic properties of transition metals data on the electronic band structure is extremely valuable. Unfortunately, the electronic band structures of amorphous magnetic materials have not been investigated experimentally. However, it has been shown by photoemission studies that the d-band density of states of crystalline and liquid gold are very similar.[22] Apparently short range interactions common to the crystalline, amorphous and liquid states determine the d-band structure in metallic systems. In addition a large number of semiconductors have been studied in both the amorphous and crystalline states. We know from these studies that in disordered semiconductors, the major feature of the band structure of the crystalline solid are retained in the amorphous state. The exceptions to this generalization are cases where the local short range order (S.R.O.) in the amorphous phase is markedly different for the S.R.O. in the crystalline phase. We expect that in amorphous magnetic alloys with dense random packing, i.e. close packed S.R.O., that the major feature of the electronic density of states curve will be retained. On the other hand sharp structure in the density of states curve would be expected to be absent. It will be shown that the magnetic properties of amorphous alloys are consistent with this structural model.

PREPARATION

Amorphous magnetic alloys (a.m.a.) have been prepared by three major methods: liquid quench, vapor quench and electrochemical reaction. Liquid quenching usually involves splat cooling. A droplet of molten alloy is quenched between cold metal mandrels.[11,12] This method is only suitable for compositions which are relatively stable in the amorphous state such as Fe_3P type alloys.[23]

Vapor quench involves evaporating or sputtering the alloy onto a cold substrate.[10] This technique is applicable to a wide range of alloy compositions. There is in general a relationship between deposition rate and substrate temperature which determines whether an amorphous product will be obtained.[24] High deposition rate and low substrate temperature favor the formation of an amorphous phase. Indeed, if the substrate is cooled to liquid helium temperatures, virtually any vapor condensed at a rate of a few angstroms per second will probably yield an amorphous product. Whether or not the product will crystallize on heating to room temperature can be predicted from considerations of atomic size and composition range.[10]

Electrodeposition and electroless deposition have been used to produce a.m.a. for many years. These techniques have only been applied to a few compositions, however. Ion radiation damage is a well established method for producing amorphous layers in covalently bonded semiconductors and insulators.[25] This method has not been applied extensively to a.m.a. as yet.

MAGNETIC ORDERING

Magnetic ordering in amorphous alloys can be classified according to the dominant exchange mechanism in the corresponding crystalline phase. (Table II) For example, in the magnetic insulators, the garnets, spinels and hexagonal ferrites, the dominant-exchange mechanism is believed to be super exchange via the anion. When these materials are prepared in the amorphous form they invariably have shown super paramagnetic behavior.[26,27,28] The absence of long range magnetic order suggests that the sign and magnitude of the exchange is sensitive to structural features such as bond angle and distance. Since these features are random, long range order is absent.

Transition metal alloys, on the other hand, clearly show ferromagnetic as well as antiferromagnetic behavior.[29] In the corresponding crystalline transition metal alloys the exchange mechanism can best be described in terms of band magnetism. If we consider, as discussed above, that the major features of the d-band structure in the crystalline phase are usually present in the amorphous phase, it is not surprising the long range magnetic order is present.

Alloys containing the rare earths are a somewhat special case in that the exchange mechanism is via the conduction electrons. The concentration of conduction electrons (as well as the spin-spin distance) can be altered by the structure. Since the spin polarization is an oscillator function of the distance between spins we might expect the exchange interaction to vary in sign and magnitude in an amorphous material and superparamagnetic behavior to result as in the super exchange case, If the nearest neighbor exchange dominates however and the average nearest neighbor distance is near a maximum in the RKKY oscillating function, we might expect long range order. It would be interesting to study a disordered system near the cross-over

TABLE II

MAGNETIC ORDER IN AMORPHOUS MATERIALS

Exchange Mechanism in the Crystalline State	Observed Magnetic Order in the Amorphous State	Example Material	Method of Fabrication
Super-exchange	Super-paramagnetism	Rare earth iron garnets	Sputtering
		Ferrite spinels	Evaporation
Direct exchange	Ferromagnetism or ferrimagnetism	Co-P alloys	Electro-deposited
		Ni-Fe-P alloys	Liquid quench
		Co-Au alloys	Evaporated
Indirect exchange	Ferromagnetism or ferrimagnetism	Gd-Fe alloys	Evaporated
		Gd-Co alloys	Sputtered
		Tb-Fe alloys	Sputtered
		Pr-Co alloys	Sputtered

from a positive to a negative exchange in an RKKY system. One would anticipate super paramagnetic behavior in this case. The $GdN_{1-x}O_x$ system may be an example of this behavior.

In the metallic systems studied, all types of magnetic ordering have been observed; ferromagnetism, antiferromagnetism and ferrimagnetism. The implications of a comparison between the magnetic behavior of amorphous metals and amorphous insulators is that quasi-free electrons present in a metal may be essential to mediate the exchange which provides long range magnetic order in amorphous systems.

Gd-Co ALLOYS

Amorphous rare earth transition metal alloys constitute an interesting class of materials. The corresponding crystalline intermetallic compounds have been studied because of their potential as permanent magnet materials. Amorphous alloys in the Gd-Fe system are readily prepared in a composition range from at least 10 to 90 atomic percent Fe by a vapor quench.[30] In the Gd-Co system amorphous alloys have been prepared in the range 33 to 95 atomic % Co. The antiferromagnetic coupling between Gd and Co is shown by the composition dependence of the magnetizatin. The magnetization at room temperature for example shows a minimum at approximately 80 atomic percent cobalt.[31] The ferrimagnetic behavior is also clearly evident in the temperature dependence of the magnetization; a compensation point is observed in most compositions. (Fig. 1) It has been shown that in Gd-Co intermetallic compounds the Gd and Co sublattices are opposed. Clearly, antiparallel ordering of Co and Gd moments is preserved in the amorphous state.

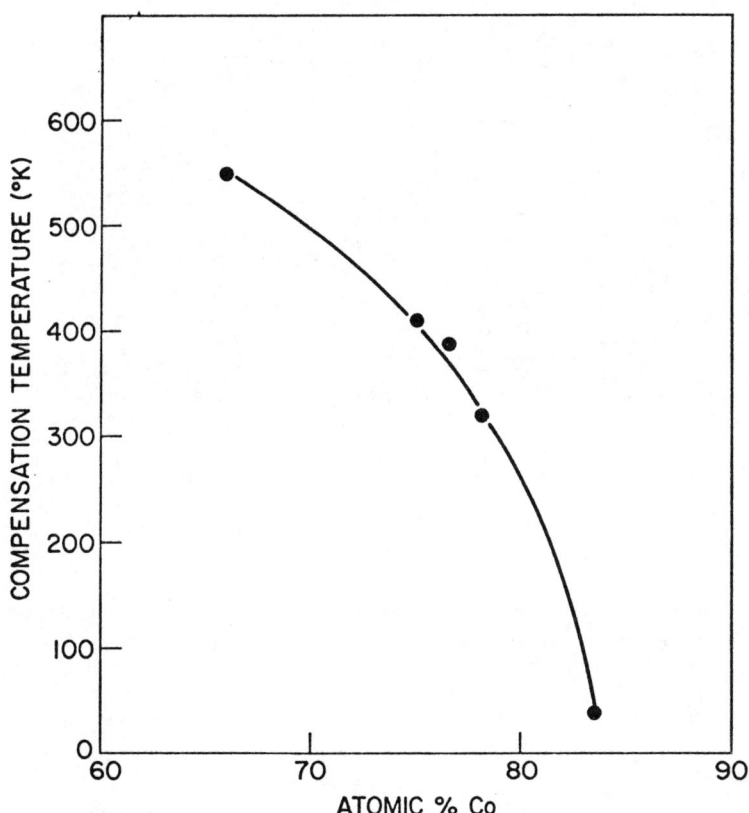

Fig. 1. Compensation temperature vs composition in amorphous Gd-Co alloys.

The approach to saturation of sputtered amorphous Gd-Co films is typical of a ferromagnetic thin film with a uniaxial anisotropy perpendicular to the plane of the film. The anisotropy has also been confirmed by FMR data.[32] The domain structure is also clearly that of a perpendicular uniaxial material showing stripes and bubble domains. (Fig. 2) Studies of the domain structure have been used extensively to monitor the magnetic properties of these films.

By measuring the stripe width in the demagnetized state (w_s) and the film thickness (h) it is possible to determine[33] the characteristic length parameter (ℓ)[34] which is defined as:

$$\ell \equiv \frac{\sigma_\omega}{4\pi M_s^2}$$

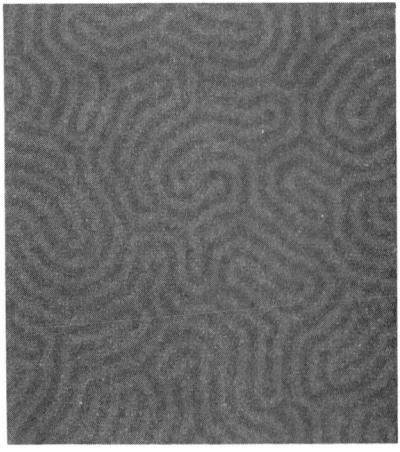 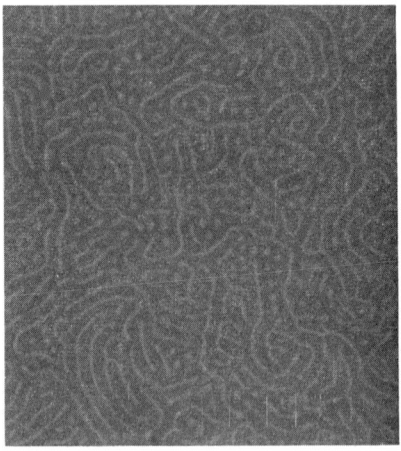

Fig. 2. a) Stripe domains in an a.c. demagnetized GdCo film. (ω_s = 1.8μm)
b) Bubble domains and short stripes in a GdCo film in an applied perpendicular field.

The saturation field (H_s) can be obtained on a polarizing microscope equiped with an electromagnet by observing the perpendicular field required to saturate the film to a single domain. Knowing H_s, ℓ and h it is possible to calculate the saturation magnetization (M_s)[33,35] and then the wall energy (σ_ω) from:

$$\sigma_\omega = \ell(4\pi M_s^2) .$$

The wall energy in a uniaxial film of this type is given by:

$$\sigma_\omega = 4\sqrt{AKu}$$

where A is the exchange constant and Ku is the anisotropy constant. Thus the domain structure can be used to monitor the anisotropy constant as a function of composition, deposition conditions and post deposition treatment.

What is the source of uniaxial anisotropy in an amorphous material? Anisotropy in magnetic materials in general can arise from four principle mechanisms: magneto crystalline, stress, shape or atomic short range ordering. Magneto crystalline anisotropy is the magnetic manifestation of the long range atomic order in a crystal. Since the Gd-Co alloys are amorphous this mechanism is ruled out. Stress anisotropy arise from a macroscopic planar stress which aligns the magnetization via the magnetostriction coefficient. We have shown that the strip width of Gd-Co films is independent of substrate constraint. Films were grown on NaCl substrates and their strip width measured and the films floated off the substrate. The strip width was unchanged in the free standing film. (Fig. 3) Since the strip width (ω_s) at constant films thickness (h) is proportional to (ℓ) the characteristic length parameter[34] and ℓ is given by:

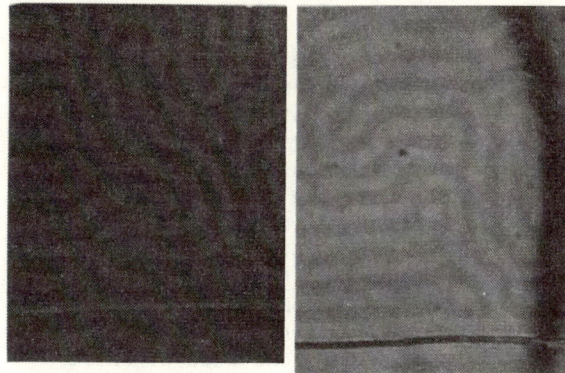

$$\ell = \frac{\sqrt{AK_u}}{\pi M_s^2}$$

it is clear that Ku is not sensitive to substrate constraint. Stress anisitropy is a possible mechanism in amorphous magnetic materials with a large magnetostriction and may dominate in particular compositions such as $TbFe_2$.[19]

Shape and atomic short range order are also possible mechanisms in amorphous magnetic materials. Shaped voids have been reported in amorphous semiconductors for example.

Fig. 3. Stripe domains before (left) and after (right) removing the film from its rock salt substrate. The line near the bottom of the photograph is a cleavage step used to identify the region of the sample.

Anisotropic optical properties have been observed in amorphous oxide films and have been attributed to film surface topography.[36] Also, shape effects are well documented as a source of anisotropy in polycrystalline cubic films such as permalloy evaporated at oblique incidence.[37] We have concluded that shape anisotropy is not an important mechanisms in amorphous Gd-Co films. Structural studies in thin films examined by transmission electron diffraction and microscopy show perpendicular domains by Lorentz microscopy but no evidence of shaped voids or surface topography.[38]

By the process of elimination, we conclude that atomic short range ordering is the main source of anisotropy in sputtered amorphous Gd-Co films. This conclusion is supported by two additional pieces of evidence. In a series of ion radiation damage experiments, it was shown that a relatively modest damage level (10^{14} argon ions/cm^2 at 2 MeV) was sufficient to destroy uniaxial anisotropy when the damage was sustained in zero field.[39] The level required is calculated to be sufficient to cause displacement of approximately half the atoms in the sample but certainly not sufficient to fill voids or level surface structure. It was also shown that a comparable dosage was capable of inducing perpendicular anisotropy in an in-plane film when the damage was sustained in a perpendicular magnetic field of 1.6 KG. In crystalline NiFe it has been shown that radiation damage in a magnetic field induces uniaxial anisotropy in the field direction.[40] The anisotropy was shown to arise from preferential ordering of Ni and Fe on alternating planes.

The anisotropy of Gd-Co films can also be adjusted by annealing the film below its crystallization temperature in a saturating perpendicular field. (Fig. 4) Annealing in zero field at the same temperature causes a decrease in anisotropy. Similar results have been obtained by B. Berry of our laboratory in splat cooled amorphous

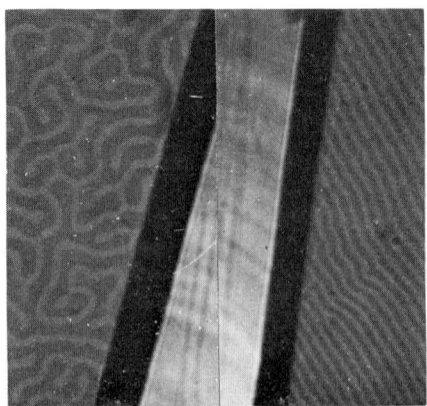

Fig. 4. Stripe domains in a Gd-Co film before (left) and after (right) annealing 18 hours at 200°C in a 3 KOe perpendicular field.

alloys.[41] The annealing temperature is apparently sufficient to cause single atomic jumps but not sufficient for long range diffusion which would cause crystallization. The jump probability in various directions is influenced by the orientation of the magnetization so as to favor the formation of short range atomic order of a particular orientation.

The simplest kind of short range order to consider in a binary alloy is single atomic pairs, for example, Co-Co pairs in amorphous Gd-Co alloys. Studies of hcp cobalt suggests that the easy direction associated with a Co-Co pair will be perpendicular to the pair axis and the anisotropy will have a magnitude of the order of 10^{-15} to 10^{-16} ergs/pair. The number of pairs required to account for the observed anisotropy in amorphous Gd-Co e.g. (10^5 erg/cm^3) can be readily calculated to be of the order of 10^{20} to 10^{21} pairs/cm^3. In other words, it is only necessary for a small fraction of pairs to be preferentially oriented in order to give the observed anisotropy.

How does this preferential orientation come about? In a thin film vapor deposition process, the adatom arrives at the film surface with considerable kinetic energy. It quickly loses its kinetic energy to the substrate and equilibrates to the substrate temperature. If the substrate temperature is sufficiently high the adatom will move by surface diffusion to an equilibrium sticking site, usually the growth step of a crystal nucleus.

If the substrate temperature is low so that surface diffusion is slow and if the impingment rate is high the adatom essentially sticks where it lands and an amorphous film is produced. This is the atomistic interpretation of the relation between deposition rate and substrate temperature for producing amorphous films.

Let us consider a somewhat intermediate case, for example, assume that the adatom can undergo a jump to any of its nearest neighbor (nn) surface sites. The jump probability for each site would be expected to be proportional to the energy difference between the initial site and the nn surface site. In these circumstances, the adatom would be expected to preferentially occupy lower energy sticking sites but not equilibrium crystal sites. The energy differences between the initial site and the nn surface site can arise from chemical, geometric or magnetic considerations. For example the bond energy of a Co-Co bond is undoubtedly considerably different from that of a Gd-Co bond. Geometric considerations may also be important because Gd and Co differ in atomic size by about 30%. The magnetic energy difference is apparently important in the bulk amorphous solid as evidence by the presence of magnetic annealing effects and may be even more important in the ordering of adatoms at the surface.

A special case of thin film vapor deposition is bias sputtering. In bias sputtering, the growing film is made a secondary target and is bombarded with low energy (e.g. 50eV) ions which cause some fraction of the surface atoms to sputter away. The probability that a surface atom will be resputtered should be proportional to its site energy. Since atoms in high energy sites are preferentially removed, bias sputtering can greatly enhance the probability of incorporated atoms residing in lower energy sticking sites. In the Gd-Co case, resputtering should greatly enhance the probability of incorporating Gd bonded to Co.

It should be mentioned that motion of a surface atom does not necessarily arise from thermal vibrations. Consider an adatom arriving at a film surface at a glancing angle of incidence. It can not transfer its kinetic energy to the substrate very efficiently so it may undergo several jumps before it sticks (if indeed it sticks at all since sticking coefficients are low for glancing angle adatoms). The net effect of glancing angle deposition is to create a situation where the surface of the growing films behaves as though it were much hotter than the bulk which is at the substrate temperature. The enhanced surface mobility in glancing angle deposition may account for many of the anisotropy effects observed in films deposited in this manner.

The model for the growth induced anisotropy in sputtered amorhous Gd-Co alloys emerges as follows. The uniaxial perpendicular anisotropy arises from Co-Co pairs in the plane of the film in excess of a random distribution. More complex short range order clusters may be more important but Co-Co pairs are the simplest cluster which will explain the data. The excess of pairs in the plane is caused by the higher resputtering probability of an adatom which is bonded to a like surface atom, i.e. Co-Co or Gd-Gd bonds are less stable than Gd-Co bonds. The statistical probability of a Gd-Co pair oriented perpendicular to the plane of the film is high, therefore, the concentration of Co-Co pairs oriented perpendicular is low. The excess Co-Co pairs in the plane results from the decreased concentration perpendicular to the plane.

The observations which support this model are the following. The anisotropy of a film is relatively independent of temperature in the vicinity of room temperature indicating that it is associated with the Co sublattice. The anisotropy induced during growth increases with increased resputtering as the model would predict. At high deposition rates or low substrate temperatures, conditions which would be expected to reduce site selection, the growth induced anisotropy is reduced. The observation that the anisotropy increases with thickness then reaches a limiting value suggests that a process akin to magnetic annealing is taking place during growth. (Fig. 5) The initial layers develop a small perpendicular component by the processes described above. The component of magnetization perpendicular to the plane of the film provides an exchange field which decreases the stability of vertical Co-Co pairs. The excess Co-Co

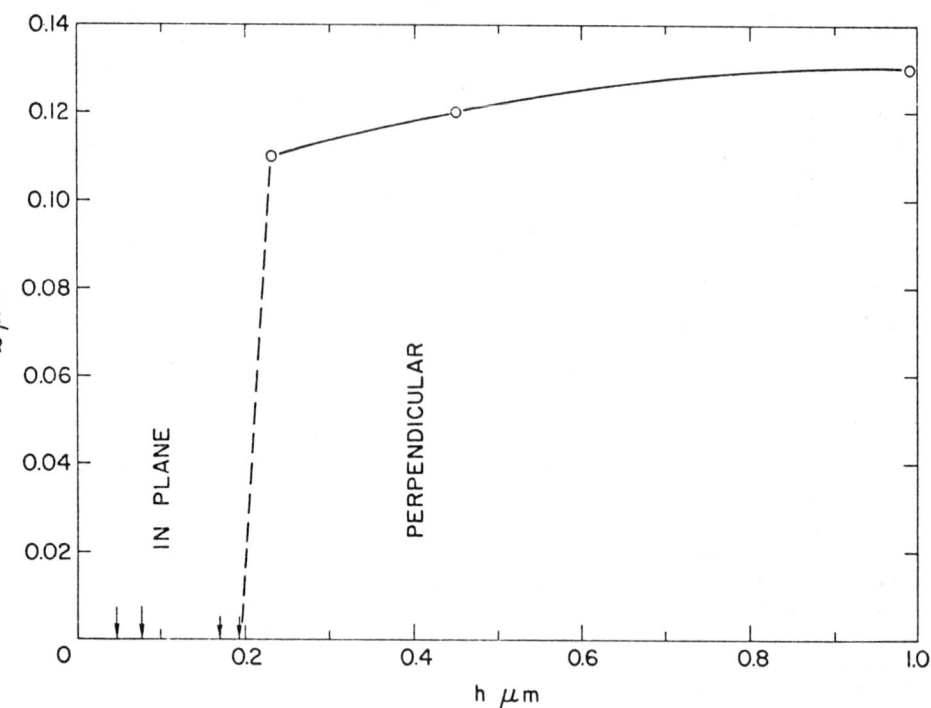

Fig. 5. Characteristic length parameter (ℓ) as a function of thickness for amorphous Gd-Fe films deposited under constant conditions. Note that for a fixed composition ℓ is proportional to $Ku^{1/2}$.

pairs in subsequent layers contribute an increased exchange field which further destabilizes vertical Co-Co pairs. A limiting value of anisotropy is reached because once $Ku > 2\pi M_s^2$, the full $4\pi M_s$ magnetization is oriented on the vertical direction so that further increases in Ku no longer increase the probability of in-plane Co-Co pair formation. One would expect Ku to be a function of M_s and to pass through a minimum at the compensated composition; this is found to be the case.

Let us turn out attention to coercivity in amorphous materials. It has been observed in a number of systems that the coercivity of an amorphous material is drastically lower than the coercivity of a corresponding polycrystalline material. In fact it can be said that with respect to coercivity, an amorphous material is more like a single crystal than like a polycrystalline material. This situation is analogous to the scattering of light in a transparent solid; amorphous fused quartz and a quartz crystal are more alike in this respect than a sintered polycrystalline aggregate of quartz crystals, for example.

In magnetic coercivity as in the scattering of light, discontinuities at grain boundaries make a major contribution. Having eliminated grain boundaries in an amorphous material what other discontinuities or fluctuations are there which can contribute to coercivity? Consider a magnetic bubble domain moving through an amorphous solid. What determines its coercivity? For convenience we will assume a square bubble of edge length D and height h (the thickness of the film). If the bubble moves distance dx the change in volume is 2Dhdx. The work done is

$$\Delta E_i = 2MH2Dhdx \tag{1}$$

where M and H are the magnetization and external field respectively. If the domain wall is in a local potential well it can be shown that the energy to move it is $\Delta E = \frac{\partial E}{\partial x} dx$. We now equate ΔE_i and ΔE to obtain the H field at which the bubble moves; this field is H_c.

$$4MH_c \, Dhdx = \frac{\partial E}{\partial x} dx \tag{2}$$

The energy of the wall is:

$$E_x = 4\sigma_w hD = 16(AKu)^{1/2} hD \tag{3}$$

Taking the appropriate partial derivitives we can express $\frac{\partial E}{\partial x}$ in terms of $\frac{\partial Ku}{\partial x}$, $\frac{\partial A}{\partial x}$ etc and arrive at the following expression for H_c:

$$H_c = \frac{\sigma_w}{M} \left[1/2 \left(\frac{\partial Ku}{Ku \partial x} + \frac{\partial A}{A \partial x} \right) + \frac{\partial D}{D \partial x} + \frac{\partial h}{h \partial x} \right] \tag{4}$$

To obtain some physical insight into this expression we assume a sawtooth shaped potential well of half width δ which is equal to the domain wall width. The latter assumption is valid since if δ is very much greater than the width of the potential wall, the wall will not be greatly perturbed by it. We rewrite equation 4 to obtain:

$$H_c = \frac{\sigma_w}{M}\left[\frac{1}{2}\left(\frac{\Delta Ku}{\delta Ku} + \frac{\Delta A}{\delta A}\right) + \frac{\Delta D}{\delta D} + \frac{\Delta h}{\delta h}\right] \quad (5)$$

$$= \frac{\sigma_w}{M\delta}\left[\frac{1}{2}\left(\frac{\Delta Ku}{Ku} + \ldots\right)\right]$$

$$\frac{\sigma_w}{\delta} = \frac{4\sqrt{AKu}}{\sqrt{A/Ku}} = 4Ku$$

Substituting 4Ku for σ_w/δ in equation 5 leads to:

$$H_c = \frac{2Ku}{M}\sum_i \frac{\Delta x_i}{x_i} \quad \text{where } x_i = Ku, A, D, h \quad (6)$$

We are now in a position to consider sources of coercivity in an amorphous material. The fluctuation terms $\frac{\Delta M}{M}$ and $\frac{\Delta A}{A}$ would be expected to be small in an amorphous material which is not significantly phase separated. We can estimate the mean square fluctuations in anisotropy from simple statistical mechanical considerations.

$$\frac{\Delta Ku}{Ku} = \frac{2\sqrt{n}}{n}$$

Where n is the number of pairs in a wall width region. Assuming 10^{20} pairs/cm^3 and typical values of Ku and M for a 2μm bubble in a Gd-Co film, we calculate a coercivity of the order of 0.1 Oe. The observed coercivity of as deposited Gd-Co films is typically 0.5 to 1.0 Oe. The coercivity can be reduced to less than 0.5 Oe by annealing at 200°C in a saturating field or by ion radiation damage with 2.5 meV argon ions at a dosage of 10^{12} to 10^{13} ions/cm^2. In accordance with equation 6, H_c is reduced by decreasing Ku as evidenced by a decrease in strip width. However, part of the decrease in H_c is not reflected in a decrease in Ku and is believed to be the result of a decrease in the ΔKu/Ku term. This suggests a greater fluctuation in anisotropy in the as deposited film than one would predict from a purely random process. Possibly these fluctuations are associated with the early stages of growth when local regions with a higher anisotropy develop the limiting anisotropy at an earlier stage by the exchange coupling mechanism.

SUMMARY

Amorphous magnetic alloys have magnet properties which make them potentially useful bubble domain materials. The magnetic anisotropy of sputter deposited amorphous Gd-Co films is understood in terms of atomic short range ordering which occurs during the deposition process. The coercivity in amorphous materials can be described in terms of fluctuations in magnetic properties over distances comparable to a domain wall width. Amorphous magnetic alloys are a class of materials which provide a great deal of flexibility in materials design for device applications and for the study of fundamental magnetic phenomena.

ACKNOWLEDGEMENTS

The authors are grateful to R. Hasegawa, H. Lilienthal, L. J. Tao, and J. Zeigler for data on coercivity, magnetization and ion

implantation prior to publication. They also thank S. M. Kane and
W. W. Molzen for film fabrication and F. Cardone, B. Chider and J. D.
Kuptis for microprobe analysis.

REFERENCES

1. P. Chaudhari, J. J. Cuomo and R. J. Gambino, IBM J. of Res. & Dev. <u>11</u>, 66 (1973).
2. P. Chaudhari, J. J. Cuomo and R. J. Gambino, Appl. Phys. Letters <u>22</u>, 337 (1973).
3. H. J. Garrett and W. G. D. Frederick, AIP Conf. Proceedings, No. 10, Part I Magnetism and Magnetic Materials-1972, Editors C. D. Graham, Jr. and J. J. Rhyne American Institute of Physics, New York (1973), p. 582.
4. H. Alperin, S. Pickard, R. J. Gambino and T. R. McGuire, J. Appl Phys.
5. S. von Molnar and F. Holtzberg, AIP Conf. Proceedings, No. 10 Part II (1973) p. 1259.
6. Tahar-Kheli, Proceedings of the Intn. Symposium on Amorphous Magnetism, Plenum, New York (1973), p. 373.
7. K. Tamara and H. Endo, Physics Letters <u>29A</u>, 52 (1969).
8. S. Mader and A. S. Nowick, Appl. Phys. Letters <u>7</u>, 57 (1965).
9. S. Mader and A. S. Nowick, IBM J. Res. & Dev. <u>9</u>, 358 (1965).
10. S. Mader in The Use of Thin Films in Physical Investigations edited by J. C. Anderson, Academic Press, New York (1966), p. 433.
11. P. Duwez and S. C. H. Lin, J. Appl. Phys. <u>38</u>, 4096 (1967).
12. C. C. Tsuei, G. Longworth and S. C. H. Lin, Phys. Rev. <u>170</u>, 603 (1968).
13. A. Brenner, D. E. Cough and E. K. Williams, J. of Res. of the N.B.S. <u>44</u>, 109 (1950).
14. S. Mader, J. Vac. Sci. and Tech. <u>2</u>, 35 (1965).
15. D. E. Polk, Acta Met. <u>20</u>, 485 (1972).
16. A. K. Sinha and P. Duwez, J. Phys. Chem. Solids <u>32</u>, 267 (1971).
17. M. L. Rudee, Phys. Stat. Sol. Short Notes(b) <u>46</u>, K1 (1971); M. L. Rudee and A. Howie, Phil. Mag <u>25</u>, 1001 (1972).
18. C. C. Tsuei, G. Longworth and S. C. H. Lin, Phys. Rev. <u>170</u>, 603 (1968).
19. J. J. Rhyne, S. J. Pickart and H. A. Alperin, Proceedings of the International Symposium on Amorphous Magnetism, Plenum, New York (1973)
20. H. O. Hooper, AIP Conf. Proceedings No. 10, Part I, Editors C. D. Graham, Jr. and J. J. Rhyne, AIP, New York (1973), p. 702.
21. O. Bostanjoglo and K. Rühkel, Phys. Stat. Sol. <u>11</u>, 161 (1972).
22. D. E. Eastman, Phys. Rev. Letters <u>26</u>, 1108 (1971).
23. C. C. Tsuei, Proceedings of the International Symposium on Amorphous Magnetic Materials, Plenum, New York (1973), p. 299.
24. I. H. Khan, Chapter 10, Handbook of Thin Film Technology Edited by L. I. Maissel and R. Glang, McGraw-Hill, New York (1970).
25. J. R. Parsons, Phil. Magaz. <u>12</u>, 1159 (1965).
26. J. J. Cuomo, V. Sadagopan, J. DeLuca, P. Chaudhari, and R. Rosenberg, Appl. Phys. Lett. <u>21</u>, 581 (1972).
27. M. H. Francombe, J. E. Rudisill and R. L. Cohen, J. Appl. Phys. <u>34</u>, 1215 (1963).

28. R. K. MacCrone, Proceedings of the International Symposium on Amorphous Magnetic Materials, Plenum, New York (1973).
29. R. Hasegawa, Phys. Rev. B5, 1631 (1971).
30. J. Orehotsky and K. Schröder, J. Appl. Phys. 43, 2413 (1972).
31. L. J. Tao, R. J. Gambino, S. Kirkpatrick and H. Lilienthal, (These proceedings).
32. D. C. Cronemeyer, (These Proceedings).
33. D. C. Fowlis and J. A. Copeland, AIP Conf. Proceedings Series 5, 240 (1972).
34. A. A. Thiele, J. Appl. Phys. 41, 1139 (1970).
35. H. Callen and R. M. Josephs, J. Appl. Phys. 42, 1977 (1971).
36. E. Neugebauer and C. V. Fragstein, Optik 29, 150 (1969).
37. I. H. Khan, Chapter Handbook of Thin Film Technology, Edited by L. I. Maissel and R. Glang, McGraw-Hill, New York (1970).
38. S. R. Herd and P. Chaudhari, Phys. Stat. Sol(a) 18, 603 (1973).
39. R. J. Gambino, J. Ziegler and J. J. Cuomo, (To be published).
40. J. Paulere, D. Dautreppe, J. Langier and L. Neel, Compt. rend. 254, 965 (1962).
41. B. Berry (To be published).

THEORETICAL APPROACHES TO SPIN WAVES IN DIS-
ORDERED MAGNETS

A. Brooks Harris
Department of Physics
University of Pennsylvania, Philadelphia, Pa. 19174

ABSTRACT

Insulating magnetic alloys, such as Zn-doped MnF_2, are favorable systems in which to study elementary excitations in disordered alloys because they are susceptible to inelastic neutron scattering experiments and their material parameters can be determined experimentally. We discuss four theoretical methods which have been used to treat the simplest such system, the dilute isotropic Heisenberg ferromagnet at zero temperature. The methods are 1) the coherent potential approximation (CPA) used by Tahir-Kheli[1] for the model where interactions ("bonds") are removed to form the diluted system, 2) the "site" CPA used by Harris, Leath, Elliott, and Nickel[2] to discuss the alloy formed by removing spins, 3) the construction of the spin-wave line shapes from their frequency moments as done by Nickel, and 4) a direct numerical evaluation of the line shapes for finite random ensembles. Methods 3 and 4 are essentially exact, but the former is superior, due to its simplicity. For vacancy concentrations up to 40% the site CPA is superior to the bond CPA. For higher concentrations both CPA's fail to describe properly the low-frequency resonance due to weakly bound spins and also they underestimate the effects of configurational fluctuations. Unlike method 3, method 4 is useful for mixed magnetic alloys and for the dilute antiferromagnet. Such calculations are being undertaken by Holcomb and the author. Most of this work is described in Reference 2.

1. R. A. Tahir-Kheli, Phys. Rev. B <u>6</u>, 2808 (1972).
2. A. B. Harris, P. L. Leath, B. G. Nickel, and R. J. Elliott, J. Phys. C (to be published).

MAGNETIZATION STUDIES OF AN AMORPHOUS ANTIFERROMAGNET

E.J. Friebele* and N.C. Koon
Naval Research Laboratory
Washington, D.C. 20375

ABSTRACT

The magnetization of the amorphous antiferromagnetic system x MnO-(1-x) P_2O_5, where 0.46 x 0.61, has been studied as a function of magnetic field and temperature. At $1.75^\circ K$ the magnetization vs. field data displayed considerable curvature toward the field axis, while at a given temperature and field the magnetization increased as the manganese concentration decreased. Although the curvature as a function of field could be due to formation of superparamagnetic precipitates or to a substantial number of structurally isolated paramagnetic spins, these mechanisms are ruled out because the data becomes almost independent of temperature below $1.75^\circ K$. The results are therefore interpreted in terms of a two spin model of amorphous antiferromagnetism where the probability of a given exchange interaction is calculated using crystalline $Mn_2P_2O_7$ as a structural model. With relatively small changes in the assumed exchange interactions for different compositions, the model can fit both the field and composition dependence of the low temperature data. Using the same parameters, the calculated low field susceptibility vs. temperature data, is in good agreement with the experimental data.

*National Research Council, Resident Research Associate

MAGNETIC BEHAVIOR OF BARIUM TITANIUM SILICATE GLASS

J. J. Santiago and H. W. Swenson Jr.
Aerospace Research Laboratories
Wright-Patterson AFB, Ohio 45433

ABSTRACT

Magnetic susceptibility by the Faraday method and EPR parameters were measured on samples prepared from the glass forming region of the system $BaO-TiO_2-SiO_2$. The material showed high resistivity (> 10^9 Ω.cm) and was of sufficient purity to satisfy the NBS requirements for the production of optical glass. The purpose of this work was to determine if antiferromagnetic exchange is observed in analogy to the non-stoichoimetric titanium oxides (Magneli phases). Antiferromagnetic interactions have indeed been observed in our samples. The room temperature susceptibility of the as formed glass (diamagnetic) is -0.29×10^{-6} emu/g and after heat treatment in a reducing atmosphere of 6% H_2-94% N_2 at 1600°C, the samples become paramagnetic with a specific susceptibility of $+0.59 \times 10^{-6}$ emu/g after the background diamagnetism has been subtracted. Temperature dependence of the susceptibility was measured in the range 77°K $\leq$ T $\leq$ 298°K. Using the Curie-Weiss law, $\chi = c/(T-\theta)$, and extrapolating from the linear part of the inverse susceptibility vs temperature plot, a paramagnetic Curie temperature of $-25 \pm 2°K$ was observed. In the as formed glass the Ti cation is in the tetravalent state (S = 0), reduction in a H_2 atmosphere induces the formation of Ti^{+3} ions which are involved in the coupling. The as formed glass (diamagnetic phase) shows no temperature dependence. Electron spin resonance was used to calculate the number of spins per gram according to the method of Wyard. This number for the reduced glass is $\sim 3.7 \times 10^{18}$ spins/g. The measured room temperature g-value is 1.9339.

INTRODUCTION

Since the work of Nicolau[1] is 1939 on the magnetic susceptibility of nickel sulphate aqueous solutions, researchers in this country and abroad have shown evidence for the existence of strong antiferromagnetic coupling in amorphous or highly disordered structures containing magnetic ions. Recently, Schinkel and Rathenau[2] reported magnetic susceptibility studies of borate glasses doped with manganese ions. These materials showed large paramagnetic Néel temperatures ($\theta = -85°K$). The same behavior was observed in rare-earth oxide-ferric oxide amorphous materials by Simpson and co-workers.[3,4] This same group also studied the transition element-phosphorus pentoxide glass system, reporting antiferromagnetic interactions in all the samples studied (doped with Mn^{+2}, Fe^{+3}, Ni^{+2}, Co^{+2} and Cu^{+2}). They also showed evidence of antiferromagnetic transition (Neel temperature) in the system of glasses containing Fe^{+3}. Another glass system studied which is more germane to our research is the vandium-phosphorous pentoxide glasses, extensively investigated by Wilson, Friebele and Kinser.[5,6] As in the work by Simpson, the high temperature inverse susceptibility obeys a Curie-Weiss law with a negative paramagnetic Neel temperature revealing the universal existence of antiferromagnetic interactions between transition metal ions in the glass systems they studied. The alkaliborate and aluminosilicate glasses doped with iron group ions were studied by Hooper and his coworkers. The results of their comprehensive study[7] show the existence of antiferromagnetic Neel temperature (T_N) in some of the aluminosilicate glasses and also some super-paramagnetic behavior due to a small amount of devitrification.

EXPERIMENTAL

The glass samples used in this study were grown in a globar furnace at ~ 1,600°C in a nitrogen atmosphere. The reduced samples were grown under similar conditions except for a flow of 6% hydrogen - 94% nitrogen through the system. The as grown material is a clear glass, slightly yellow. After reduction it turns black and totally opaque. The softening point is around 610°C and devitrification commences at a temperature of about 600°C. All glass samples are grown in a molybdenum-platinum crucible and all precautions are taken to keep the surroundings free of alkali metals which may unintentionally doped the sample with ionic charge carriers. The resistivity of the reduced glass is estimated to be ~ 10^{11} Ω-cm. For the as grown material is much larger than for the reduced and therefore difficult to estimate.

Magnetic measurements were performed using an electronic microbalance and Faraday type magnet pole pieces. The magnetic field intensity at the sample position was measured to be 7.9 KOe. A gas flow type temperature controller was used and the temperature range covered was from 77K to 298K.

The electron spin resonance (ESR) of the reduced glass samples was performed using a Varian model v-4560 x-band spectrometer. The g-value of the samples was determined by comparison to Holden DPPH value.[8] To determine the number of spins per gram the method of Wyard,[9] that is direct comparison between the resonance area of DPPH and the sample, was used.

The EPR results indicate spin concentration of about 3.7×10^{18} per gram in the reduced glass. The measured g-value at room temperature (298K) is 1.9339 $\pm$ 0.0003. The resonance first derivative was markedly broad (80 Gauss) suggesting some interaction between the paramagnetic centers.

Measurements on the clear, as grown glass reveals no paramagnetism in this material. Both static and dynamic techniques used in the temperature range from 77K to room temperature show a diamagnetic material with a specific susceptibility of -0.29×10^{-6} cmu/g. The field dependence of the susceptibility was also studied to discover the presence of ferromagnetic impurities incorporated into the material. Honda-Owens type plots reveals a constant χ_g vs H^{-1} profile implying a negligible amount of ferromagnetic impurities.

The dark, reduced glass shows a strong temperature dependent paramagnetism. When the inverse susceptibility corrected by subtracting the background diamagnetism is plotted versus temperature, the graph shows a linear relationship following closely the Curie-Weiss law $\{\chi = c/(T-\theta)\}$. Extrapolating a paramagnetic Curie temperature (θ) of $-25 \pm 2K$ is obtained.

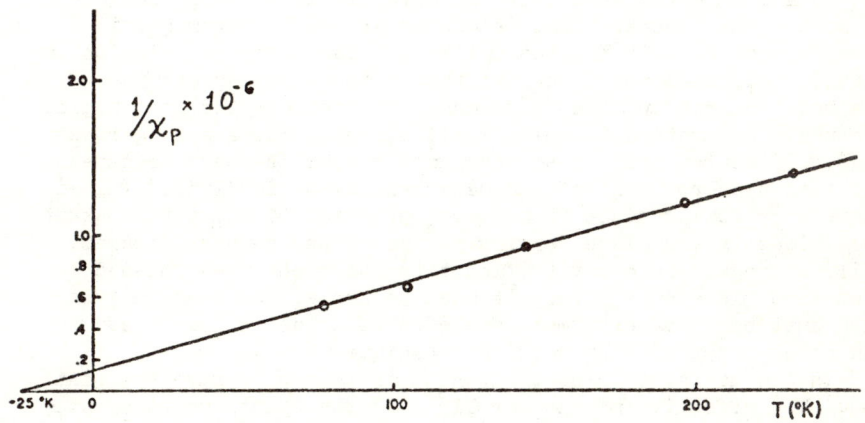

Fig. 1. Temperature Dependence of Susceptibility

The effective magnetic moment is difficult to obtain with any precision due to our ignorance of the true stoichiometry of the glass. Using an estimated empirical stoichiometry, from the Curie-Weiss relationship a $\mu_{eff} = 2.5 \pm 0.2$ Bohr magnetons was obtained.

DISCUSSION

The empirical stoichiometry of this glass (as grown) is roughly $TiBa_2Si_4O_{12}$. This formula, with some imagination, can be viewed as $Ba_2TiO_4 \cdot 4SiO_2$ or in other words as a mixture of non-stoichiometric barium titanate and excess silica. For the purpose of bonding the cations can be viewed as preserving an intact first sphere of coordination of the perovskite barium titanate and of quartz in the amosphous state. Barium in the perovskite lattice is coordinated by 12 oxygens and titanium by 6 (octahedral coordination, the same coordination of Ti in the compound rutile). It might be that by some form of stearic hindrance in the amorphous material the coordination of two cations is not exactly 6 and 12 but it will probably be reasonably close to those numbers. When the glass is reduced, most likely the homopolar bond between silicon and oxygen remains intact and the titanium-oxygen bond or the barium-oxygen bond, which are both more ionic in character, will be attacked by the hydrogen. If this is the case the metals first sphere of coordination will be disrupted, allowing the possibility of close proximity and therefore direct coupling between the transition element cations.

It has been called to our attention, that if one uses the spin density as measured by electron spin resonance, that is $\sim 3.5 \times 10^{18}$ spins/gm. one can deduce that less than 1% of the titanium ions are reduced to Ti^{+3} assuming a single state of oxidation. From the empirical stoichiometry of $TiBa_2Si_4O_{12}$ it appears that the ratio of the number of oxidized to reduced (Ti^{+3}) titanium atoms is less than 10^{-3}. So it seems that on the average the Ti^{+3} ions are separated by at least 10 non-magnetic atoms, if the material is assumed homogeneous. With such a large separation, it can be argued, a paramagnetic Curie temperature of $\theta = -25°K$ seems unlikely. The magnetic dilution implicit in this reasonong, suggest that some phase segregation of the glass has occurred and the Ti^{+3} ions have agglomerated in a region where their separation is quite small allowing for a stronger exchange. Of course this phase segregation would be hard to detect, since less than about 1% of the material may be involved. This is a plausible reasoning since it has been observed by the authors that compositions in the glass forming region of the ternary diagram readily segregate into devitrified titanium rich areas imbeded in a silica rich glass substrate. It must be emphasized that the segregation must be in a very small scale as no evidence for it can be found using x-ray line broadening techniques.

In the as-formed glass, titanium is in its tetravalent state (s=o). The effective moment for Ti^{3+} (in the spectroscopic state $^2D_{3/2}$) is 1.73 Bohr magnetons. The higher effective moment found in this study may be due to higher states of reduction of the titanium ion as some evidence for this has been obtained in this laboratory using ESCA (10). At this point it will be worthwhile to mention the relevance of this work to the studies of the titanium "magneli phases"[11] or reduced rutile materials. These solids show Mott type semiconductor to metal transitions and interesting magnetic phenomena such as thermal hysterisis of the susceptibility. Adler[12] mentioned in his review that magnetic ordering in these transition metal oxides is still unresolved. Some of these Magneli phases, over all the highly reduced ones possessed a highly disordered, open structure not too unlike the structure we envision for our glass samples.

CONCLUSION

The results of the present study suggest the possibility of nearest-neighbor antiferromagnetism coupling between Ti^{+3} ions in an inhomogeneous glass matrix. No evidence for phase segregation was discovered so the EPR line broadening is attributed to interactions between paramagnetic centers. Evidence for antiferromagnetic interaction was obtained although a Néel transition was not observed in the temperature range covered.

REFERENCES

1. A. Nicolau, Compt. Rend. 205, 557 (1937).
2. C. J. Schinkel and G. W. Rathenau, Physics of Non-Crystalline Solids (North Holland, 1965).
3. T. Egami, O. A. Sacli, A. W. Simpson, A. L. Terry and F. A. Wedgwood, Amorphous Magnetism, H. O. Hooper and A. M. de Graaf eds., Plenum Press, N. Y. (1973).
4. T. Egami, O. A. Sacli, A. W. Simpson, A. L. Terry and F. A. Wedgwood, J. Phys. (C) 5, L261 (1972).
5. L. K. Wilson, E. J. Friebele and D. L. Kinser, Amosphous Magnetism, H. O. Hooper and A. M. de Graaf eds. Plenum Press (1973).
6. E. J. Friebele, L. K. Wilson and D. L. Kinser, J. Amer. Cer. Soc. 55, 164 (1972).
7. H. O. Hooper, G. B. Beard, R. M. Catchings, R. R. Bukrey, M. Forrest, P. F. Kenealy, R. W. Kline, T. J. Moran, J. G. O'Keef, R. L. Thomas and R. A. Verhelst, Amorphous Magnetism, H. O. Hooper and A. M. de Graaf eds., Plenum Press (1973).
8. A. N. Holden, C. Kittel, F. R. Merritt and W. A. Yager, Phys. Rev. 77, 142 (1950).
9. S. J. Wyard, J. Sci. Inst. 42, 769 (1965).
10. M. Luciano and N. Fernelius, Personal Communication.
11. S. Asbrink and A. Magneli, Acta Cryst 12, 575 (1959).
12. D. Adler, Solid State Physics 21, 100 (1968).

FIELD DEPENDENCE OF MAGNETIZATION IN RANDOM FCC FERROMAGNETS*

Peter M. Richards
Sandia Laboratories, Albuquerque, New Mexico 87115

ABSTRACT

The magnetization $\bar{R}$ of a random ferromagnet is calculated for the fcc lattice using mean field theory. The treatment differs from previous works in that correlation between the magnetizations of neighboring sites is explicitly accounted for. The initial susceptibility and $\bar{R}^2$ vs. $H/\bar{R}$ plots are discussed, the latter showing a concave upward behavior which is probably masked by critical spin fluctuations in experimental situations. The magnetization vs. temperature is calculated for several concentrations in $Eu_xCa_{(1-x)}O$ and compared with experimental data. It is suggested that such a system should provide a better test than, say, NiCu alloys since it has local moments whose magnitude does not depend on the environment.

INTRODUCTION

A random ferromagnet is characterized by variations throughout the crystal of the number of magnetic neighbors. This causes fluctuations in the local exchange field apart from those due to spin disorder (critical fluctuations). Here we examine the consequences of the former type of fluctuations on the magnetization within the context of mean field theory (MFT). It is, of course, known that MFT cannot account for the latter spin fluctuations, but it does give a good overall picture of the behavior of the magnetization except in the immediate vicinity of the critical point T_c. Further, MFT is likely to be more successful in the presence of an applied field, as considered here, which tends to reduce the importance of spin fluctuations.

The main difference between our treatment of MFT and that of other workers[1] is that we account for the strong correlation in the fcc lattice between the numbers of magnetic neighbors of two atoms which are themselves nearest neighbors. This same feature has also been used[2] to explain clustering of magnetic moments in an fcc alloy.

FORMALISM

The MFT equation for a random system may be written as

$$R_0 = B_S\left(\frac{\lambda}{T} \sum_{\langle j \rangle} R_j e_j + \frac{H}{T}\right) \qquad (1)$$

*Work supported by the U. S. Atomic Energy Commission.

in which R_0 is the relative magnetization at an occupied site 0 (an occupied site is one which has a magnetic moment); $\varepsilon_j = 1$ or 0 for occupied or unoccupied sites, respectively; H is proportional to the applied field; T is the absolute temperature; λ is proportional to a nearest-neighbor exchange constant; and B_S is the Brillouin function for spin S. All occupied sites are assumed to have the same value of local moment, independent of the environment, and all unoccupied sites are taken to have zero moment. The summation in (1) is over the Z = 12 nearest neighbor (nn) sites of 0.

The average relative magnetization $\overline{R}$ is determined by averaging (1) over all possible configurations of occupied and unoccupied sites. We adopt the following assumptions to make the problem manageable: R_0 and R_j are assumed to depend only on N, the number of occupied nn sites of 0, for a given configuration (as well as depending on the average number of occupied nn sites $\overline{N} = cZ$ where c is the concentration). By symmetry we then take $R_j = R_1$ to be the same for all j; so that (1) reduces to

$$R_0(N) = B_S(\lambda N R_1(N)/T + H/T) , \qquad (2)$$

and the average magnetization is

$$\overline{R} = \sum_{N=0}^{Z} P_N R_0(N) \qquad (3)$$

in which $P_N = c^N (1-c)^N Z!/N!(Z-N)!$ is the probability of finding N occupied nn sites. The problem is not solved until an expression is obtained for $R_1(N)$. As mentioned in ref. 1, the only previous attempts of which we are aware set $R_1(N) = \overline{R}$, which, for a given N, neglects any correlation between the magnetization at 0 and that at a nn site. (Another method[1] is to let $R_1(N) = R_0(N) = R_0(Z\tilde{c})$ where $\tilde{c}$ is a concentration which varies throughout the sample in a random manner. Within each region of concentration $\tilde{c}$, however, the magnetization obeys conventional MFT for an atomically ordered system.)

We include the effect of correlation by considering the equation for $R_1(N)$, which is of the same form as (1) and involves a sum of over the nn sites k of 1,

$$\sum_{\langle k \rangle} R_k(N) \varepsilon_k(N) = R_0(N) + R_1(N) \sum_{\langle k \rangle}^{o1} \varepsilon_k(N) + \sum_{\langle k \rangle}' R_k(N) \varepsilon_k(N) \qquad (4)$$

where $\sum_{\langle k \rangle}^{o1}$ indicates a summation over those sites which are nn's both of 0 and 1, and $\sum_{\langle k \rangle}'$ is a summation over the remaining sites, exclusive of 0, which are nn's of 1. In (4) we have noted that all nn's of 0 are assumed to have magnetization $R_1(N)$, for a given N, and that $\varepsilon_0(N) = 1$ since the site 0 is assumed to be occupied.

The system of equations are now closed by assuming that $R_k(N)$ in Eq. (4), for which k is two nn steps removed from O, can be expressed in terms of $R_0(N)$, $R_1(N)$, and $\bar{R}$. In particular, we let

$$R_k(N) = a_k R_0(N) + b_k R_1(N) + c_k \bar{R} \qquad (5)$$

where $a_k + b_k + c_k = 1$. The choice $c_k = 1$, $a_k = b_k = 0$ would neglect any further correlation. Some improvement may be obtained by assuming that a_k, b_k, and c_k are proportional to the number of nn sites in common with O, in common with 1, and common to neither O nor 1, respectively. The basic rationale for this is that the magnetization at a site is primarily determined by the nn's of that site. If, in the hypothetical extreme case, two sites have the same nn's, they should have the same magnetization.

The final approximation is to neglect fluctuation of the quantities $e_k(N)$ in Eq. (4) and replace them by their average values for a given N. With these considerations, Eq. (4) becomes

$$\sum_{\langle k \rangle} R_k(N) = \left(1 + \frac{17}{12}c\right) R_0 + \left(\frac{4}{11}(N-1) + \frac{20}{12}c\right) R_1 + \frac{47}{12} c \bar{R} \quad . \qquad (6)$$

Eq. (6) is then used in the equation for $R_1(N)$ and the result combined with (2) and (3) to form a closed system of equations which can be solved numerically. The replacement of $e_k(N)$ by its average for a given N is not regarded as a serious approximation since the major effect of atomic fluctuations (variation of N) has been accounted for. Also, the choice of a_k, b_k and c_k is not critical since the results for $c_k = 1$, $a_k = b_k = 0$ ($R_k(N) = \bar{R}$ if k is not a nn of O or O itself) are quite similar to those obtained with the values discussed above.

RESULTS

The spontaneous magnetization $\bar{R}(0)$ was reported in ref. 1 to persist well above $\bar{T}_c = \lambda Zc(S+1)/3S$, the critical temperature in MFT for an atomically ordered system with the same concentration. Here we find that the initial susceptibility $d\bar{R}/dH$ is smaller in the random system than in the ordered one for the same value of $\bar{R}(0)$. One way to visualize this result physically is that the fluctuation $(N - Zc)R_1(N)$ already acts as an extra field to magnetize the random system, and thus the added external field is relatively less important. A further consequence, shown in Fig. 1, is that the standard plot of $\bar{R}^2$ vs. $H/\bar{R}$, which is a straight line for $\bar{R}$, $H/T \ll 1$ in standard MFT, becomes concave upward in this model. A quantity of interest, shown in Fig. 2, is the extrapolated spontaneous magnetization $\bar{R}^*$ defined as

$$\bar{R}^* = \bar{R}(H) - H d\bar{R}/dH \quad , \qquad (7)$$

which is often used by experimenters to estimate the spontaneous magnetization from high field measurements. $\bar{R}^*$ of course depends on the value of H and can be quite larger than the true $\bar{R}(0)$. But it is useful to have calculations of $\bar{R}^*$ since this quantity can be obtained experimentally without having to worry about the measurements in very small fields required for $\bar{R}(0)$.

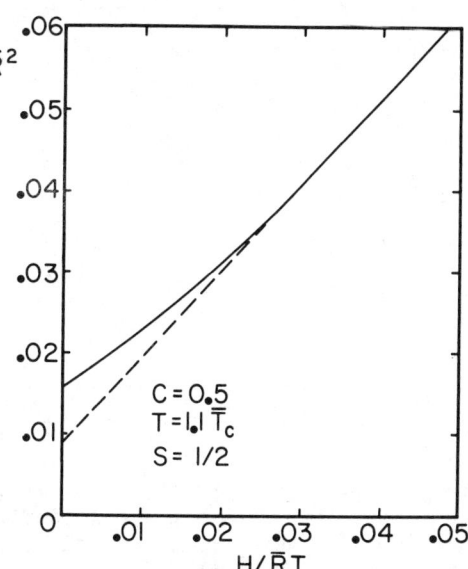

Fig. 1. Theoretical $\bar{R}^2$ vs. $H/\bar{R}T$ curve, showing concave upwardness. Dashed portion is continuation of straight line.

COMPARISON WITH EXPERIMENT AND COMMENTS

The predicted concave upwardness of $\bar{R}^2$ vs. $H/\bar{R}$ has not been observed to our knowledge. In fact, data both on pure Ni[3] and NiCu alloys[4] are concave downward, presumably because of the effect of spin fluctuations which invalidate MFT near T_c. Since the theoretical concave upwardness occurs only for $H/T \ll 1$ near T_c, it may be that critical fluctuations will always remove the effect.

A problem with NiCu alloys is the presence of giant moments[5] which prevent meaningful comparison of $d\bar{R}/dH$ between the alloys and pure Ni, since the effective local moment varies with concentration. It is preferable to study systems whose magnetization is due to purely local moments which do not change with environment. One such fcc material may be $Eu_xCa_{(1-x)}O$, which has been studied somewhat.[6] Data were reported for three high fields in the vicinity of 16 kOe from which estimates of the "spontaneous magnetization" $\bar{R}^*$ were made by Eq. (7). We have therefore computed $\bar{R}^*$ at 16 kOe and the experimental temperatures ($T_c \approx 70°K$ for $x = 1$) for $S = 7/2$ (Eu^{++}). Comparison with the data of ref. 6 is given in Fig. 2, where the experimental data have been normalized to the theoretical curves at $T/T_c = 0.8$ (where all curves have nearly the same value $\bar{R} = .62$) so that the shapes can be compared. Theoretical curves are shown only for $x = c = 1.0$ and 0.18 to avoid cluttering. Curves for intermediate values of x fall more-or-less uniformly between the ones shown. The data, on the other hand, seem to have nearly the same shape for all $x \leq 0.45$.

The qualitative differences between the data at low and high concentrations are certainly reproduced by the theoretical curves,

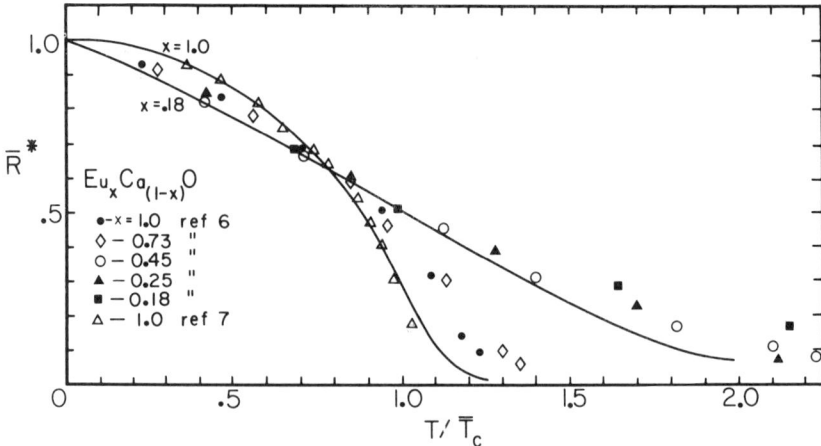

Fig. 2. Theory (solid curves) and experiment for $Eu_xCa_{(1-x)}O$ ($S=7/2$). Experimental temperature scale is defined such that $T = 0.8\,\bar{T}_c$ at temperature for which observed relative "spontaneous" magnetization $\bar{R}^*$, Eq. (7), is equal to 0.62.

but quantitative agreement is not particularly good. We do note, however, that the lack of agreement is no worse at $x = 0.18$ than at $x = 1.0$, where the theory is the same as normal MFT for an ordered system; so the problem may not be solely with applicability of our calculation to random magnets. It also should be observed that the data of Junod and Levy[7] for $x = 1.0$ (EuO) do fit MFT fairly well; so there is some experimental uncertainty, and a thorough study of $Eu_xCa_{(1-x)}O$ or similar local moment ferromagnet would be helpful.

REFERENCES

1. P. M. Richards, Physics Letters **44A**, 389 (1973), and references therein.
2. J. Perrier, B. Tissier, and R. Tournier, Phys. Rev. Lett. **24**, 313 (1970).
3. J. S. Kouvel and M. E. Fisher, Phys. Rev. **136**, A1626 (1964).
4. J. S. Kouvel and J. B. Comly, unpublished data, kindly supplied by Professor Kouvel.
5. T. J. Hicks, B. Rainford, J. S. Kouvel, G. G. Low, and J. B. Comly, Phys. Rev. Lett. **22**, 531 (1969).
6. A. A. Sumokhvalov et al., Soviet Phys.-Solid State **9**, 555 (1967).
7. P. Junod and F. Levy, Physics Letters **23**, 624 (1966).

DILUTE ANTIFERROMAGNETIC SYSTEMS IN FCC AND BCC LATTICES

Hiroshi Sato
Scientific Research Staff, Ford Motor Company,
Dearborn, Michigan 48121

Ryoichi Kikuchi
Hughes Research Labs., Malibu, California 90265

ABSTRACT

We contribute to the study of amorphous systems with antiferromagnetic (AF) interaction with the emphasis that the AF ordered state is sensitive to the crystalline (or non-crystalline) structure, while the AF disordered state is not. We calculate dilute AF states in both fcc and bcc structures, in which magnetic atoms are distributed randomly and interact at the nearest neighbor distance. Such a system resembles an amorphous structure as far as magnetic atoms are concerned. The tetrahedron approximation of the cluster-variation method is used. In a fcc crystal which can be divided into four sublattices, the AF transition is of the first order, has a lower Néel point and is harder to reach than in the bcc which is composed of two sublattices; the minimum concentration of magnetic atoms required to create the AF ordered state is 0.445 for fcc while it is 0.212 for bcc, and 0.156 and 0.212, respectively, for ferromagnetic case. This tendency suggests that in amorphous systems the AF disordered phase can persist up to a large density of magnetic atoms. The magnetic susceptibility in the AF disordered phase behaves qualitatively the same in fcc and bcc lattices. This behavior also explains well that observed in amorphous systems.

INTRODUCTION

The magnetic behavior of disordered systems with antiferromagnetic (AF) interactions (AF system) is attracting attention. Examples include amorphous systems containing magnetic atoms,[1] more or less disordered, dilute spin-glass alloys,[2] etc. There is, however, an intrinsic difficulty in visualizing an antiferromagnetically ordered state (AF ordered state) in such disordered systems. In addition to this difficulty, there can be a reasonable doubt if the AF ordered state can anyhow be established in amorphous systems. In contrast to the ferromagnetically ordered state (F ordered state) where all the magnetic moments align in a parallel fashion, the AF ordered state is very sensitive to the symmetry of the crystalline structure since the macroscopic cancellation of magnetic interactions, or misfit, can be involved in ordering. The situation has been investigated in detail for the order-disorder (or AF) transformation in the fcc lattice by the present authors.[3,4] In this case, the lattice is composed of four interpenetrating sublattices, and it is not possible to surround a plus spin by minus spins only or vice versa in the ordered state (misfit). Due to this situation, the transition point from the disordered state to the ordered state in an AF system is suppressed to a far lower temperature than that of a F system with equivalent strength of interactions, and the transition becomes first order. On the other hand, in the bcc lattice which is composed of two sublattices, there is no misfit problem and the behavior of both the F and the AF systems are equivalent.

In amorphous structures, we expect a certain, if not infinite, degree of misfit and, therefore, the suppression of the AF ordered state like that found in the fcc lattice occurs. The suppression of the AF ordered state also means that the minimum concentration to create the AF ordered state can be far larger than that of a corresponding F ordered state in dilute magnetic systems of certain crystalline structures. In order to investigate this situation quantitatively, we calculate here dilute AF states in both fcc and bcc structures.

RESULTS

We are interested in a system in which a fraction, ρ_B, of the lattice points of a given crystal is occupied randomly by magnetic atoms (species B). We assume further an interaction only between magnetic atoms which are nearest neighbors. We adopt the Ising model and the tetrahedron approximation of the cluster-variation (c-v) method is used. It has been shown that at least the tetrahedron approximation is required in order to obtain qualitatively correct results to calculate the AF ordered state in the fcc lattice. The phase transition in the fcc lattice is second order for ferromagnets and is first order for antiferromagnets in this approximation. Transition temperatures of Ising model ferromagnets and antiferromagnets calculated by several different approximations of the c-v method are given in previous papers.[3,4]

The calculation of transition points as a function of ρ_B was made by the natural iteration (NI) method developed by one of the authors.[5,12] The results are shown in Fig. 1. As ρ_B decreases, the transition points of both the ferromagnet and the antiferromagnet decrease linearly with ρ_B. The nature of the transition does not

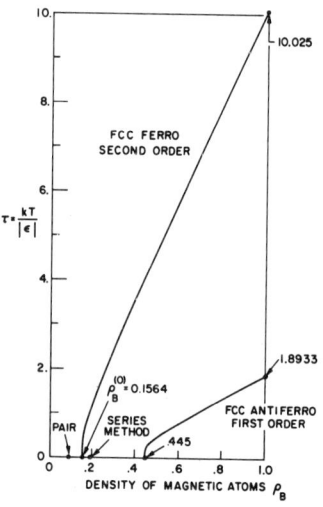

Fig. 1. Concentration dependence of the critical point for an Ising model ferromagnet and an Ising model antiferromagnet in a magnetically dilute fcc system, calculated by the tetrahedron approximation of the c-v method. The energy parameter ϵ is defined as $2\epsilon = \epsilon_{+-} - \epsilon_{++}$.

change by the decrease of ρ_B in both cases. The transition temperatures reach zero rather abruptly at certain concentrations and the limiting value $\rho_B^{(0)}$ for the ferromagnet is 0.1564 while for the antiferromagnet, it is 0.445. Let us call $\rho_B^{(0)}$ the "percolation limit" including the case of antiferromagnets. It is to be noted that the "percolation limit" for the antiferromagnet is much higher than that of the ferromagnet in the fcc lattice as expected. In the bcc lattice, the "percolation limit" is 0.2118 in the tetrahedron approximation for both the ferromagnet and the antiferromagnet. The values of the "percolation limit" calculated by several approximations of the c-v method are given in Table I.

Table I. Percolation Limits, $\rho_B^{(0)}$, of Ising Model Ferromagnets and Antiferromagnets in fcc and bcc Structures

Approximation	fcc Structure		bcc Structure
	Ferromagnet	Antiferromagnet	Ferromagnet Antiferromagnet
Point	0	0	0
Pair	1/11 = 0.909	no transition	1/7 = 0.143
Tetrahedron	0.1564	0.445	0.2118
(Series expansion[11])	0.195 ± 0.005		0.243 ± 0.010

The temperature dependence of inverse magnetic susceptibility in the disordered phase of a fcc AF system is calculated for several values of ρ_B and is plotted in Fig. 2. The change in slope of the

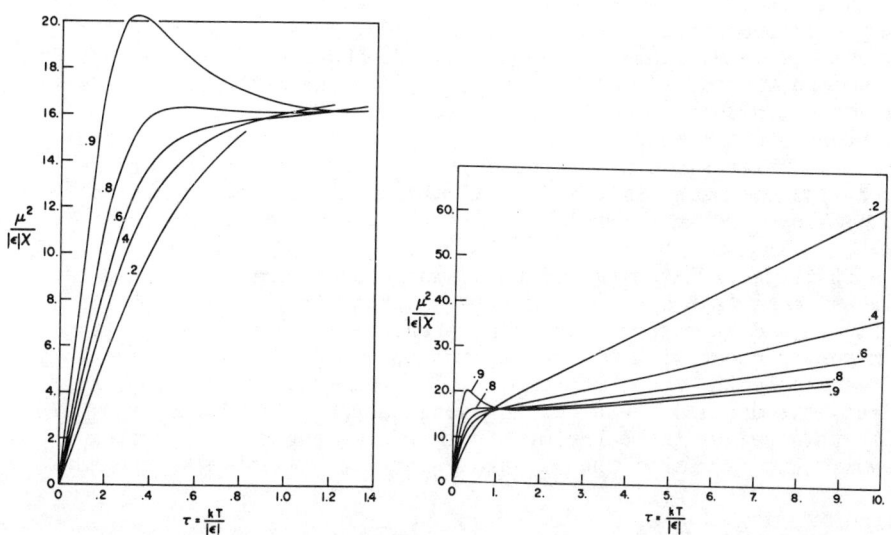

Fig. 2. The reduced inverse susceptibility per atom as a function of reduced temperature for several concentration of magnetic atoms in a disordered, fcc AF system calculated by the tetrahedron approximation of the c-v method.

$1/\chi$ vs. T curves, or the upward deviation from the Curie-Weis law, at low temperatures can be quite pronounced for large ρ_B. This is, however, a Schottky-type curve due to the short range AF ordering and should not be confused with a cooperative AF transition. On the other hand, the decrease in $1/\chi$ towards $T \to 0$ is due to the existence of magnetically isolated magnetic atoms. This behavior can then be suppressed by the inclusion of long range magnetic interactions. Very similar results are obtained for the bcc lattice. In contrast to the AF ordered state, the magnetic behavior of AF system in the disordered state is insensitive to crystalline structures.

DISCUSSION

Some time ago, the present authors calculated the "percolation limit", both for ferromagnets and antiferromagnets, in the bcc lattice in the Ising model with the pair approximation of the c-v method and obtained the value $1/(z-1)$, where z is the coordination number of the lattice.[6] Similar calculations[7] were also carried out for ferromagnets in the Heisenberg model and similar results were obtained. That the "percolation limit" should be of the order of $1/(z-1)$, whether it be a ferromagnetic or an antiferromagnetic, has since been quoted in many places. It is clear from the present calculation, however, that the "percolation limit" can be entirely different for a ferromagnet and for an antiferromagnet depending on the symmetry of the lattice.

That the "AF percolation limit" is high for the fcc lattice is important in understanding some behavior of spin-glass alloys with fcc structure. Copper rich Cu-Mn alloys can be taken as one of such examples.[8] In spite of the existing strong antiferromagnetic interaction between Mn atoms at the nearest neighbor distance, alloys do not become AF ordered until very high Mn concentration ($\sim 65\%$ Mn) is reached.[9] This is in qualitative agreement with our prediction. Therefore, in the composition range we are dealing with in ordinary spin-glass alloys, it is not necessary to consider the AF ordered state. At the same time, this indicates that the face centered cubic alloys are in general a favorable system for spin-glass studies.

It is generally expected that amorphous structures should have at least a certain degree of misfit and that the "AF percolation limit" should be rather high. Therefore, for ordinary concentration of magnetic atoms with AF interaction in amorphous substances[1,10] it is reasonable to expect that no AF long range order should exist. Indeed, the magnetic behavior in these amorphous substances[1,10] are well explained by the behavior of disordered phase of AF system shown in Fig. 2 rather than by that with a cooperative AF transition.

REFERENCES

1. A. W. Simpson and J. M. Lucas, J. Appl. Phys. $\underline{42}$, 2181 (1971). T. Egami, O. A. Sacli, A. W. Simpson, A. L. Terry and F. A. Wedgwood, J. Phys. (c) $\underline{5}$ L 261 (1972).
2. V. Cannella, Amorphous Magnetism, Proceedings of the International Symposium on Amorphous Magnetism, 1972. Eds. H. O. Hooper and A. M. de Graaf, Plenum Press, New York (1973), p. 195.
3. H. Sato and R. Kikuchi, AIP Conference Proceedings, No. 10, (1973) p. 505.
4. R. Kikuchi and H. Sato, to be published in Acta Met.
5. R. Kikuchi, J. Chem. Phys., to be published.
6. H. Sato, A. Arrott and R. Kikuchi, J. Phys. Chem. Solids $\underline{10}$, 19 (1959).
7. R. J. Elliot and B. R. Heap, Proc. Roy. Soc. $\underline{A265}$, 264 (1961).
8. H. Sato, S. A. Werner and R. Kikuchi, J. de Physique, to be published.
9. G. E. Bacon, I. W. Dunmur, J. H. Smith and R. Street, Proc. Roy. Soc. $\underline{A241}$, 223 (1957).
10. L. K. Wilson, E. J. Friebele and D. Kinser, Amorphous Magnetism, Eds. H. O. Hooper and A. M. de Graaf, Plenum Press, New York (1973) p. 65.
11. J. W. Essam, Phase Transition and Critical Phenomena, Vol. II, Eds. C. Domb and M. S. Green, Academic Press (1972), p. 224.
12. The NI method is a technique which is used in solving simultaneous equations appearing in the c-v method. It works far better than the usual Newton-Raphson iteration method as applied to the present problem. The details are reported in Ref. 5.

RANDOM MAGNETIC ALLOYS

R. A. Tahir-Kheli[*]
Department of Physics, Temple University
Philadelphia, Pennsylvania 19122

L. C. M. Miranda[†] and S. M. Rezende[†],
Instituto de Fisica, Universidade Federal
de Pernambuco, Recife, PE. Brazil

ABSTRACT

A mean-field model (MF) of a Heisenberg ferromagnet is generalized to apply to a two component alloy. When the spin-spin interactions are approximated according to a certain MFA, randomness can be treated exactly. We give numerical results for the magnetization for a few cases of positive exchange integrals and applied fields.

INTRODUCTION

While the spin-wave regime is of central importance in the study of ferromagnetism, the behavior of random alloy at general temperatures is also of interest. In the present study, we make a simple attempt to analyze such a behavior in terms of an admittedly very crude but hopefully a qualitatively useful model of random ferromagnet.

Random ferromagnetic alloy has recently been alalyzed in the spin-wave approximation by several authors[1-4]. The objective of these studies has been to study a model system, which in the non-random (i.e., 'pure') limit is linearized to be exactly soluble. In this manner, the mathematical and physical complications which arise due to an interplay between the exchange interactions and the system randomness are avoided and a somewhat simpler (though still intractable) problem manifesting only the randomness remains.

Recognizing the long established usefulness of the molecular field approximation (MFA) in providing a zeroth order qualitative understanding of exchange coupled spin systems, in the following we generalize the FMA model of a Heisenberg ferromagnet to the case of random ferromagnetic alloy. Due to the single particle nature of the MFA, the generalized model remains soluble even with arbitrary randomness.

FORMALISM

For a random ferromagnetic alloy,

$$\mathcal{H}(\text{alloy}) = - H \sum_{i,\lambda} \mu^{\lambda} \sigma_i^{\lambda} S_{i,\lambda}^{z} - \sum_{i,j,\lambda,\lambda'} J^{\lambda,\lambda'}(ij) \sigma_i^{\lambda} \sigma_j^{\lambda'} \vec{S}_{i,\lambda} \cdot \vec{S}_{j,\lambda'}, \quad (1)$$

[*]Supported by NSF grant # GH39023
[†]Supported by the C.N. Pq., Brazil.

where λ indicates the magnetic species, A, B, C, etc., and where σ_i^λ denotes the relevant occupation operator. The appropriate MFA reduction of (1) is

$$\mathcal{H}^{MFA}(\text{alloy}) = -\sum_{i,\lambda}[\mu^\lambda H + 2\sum_{j,\lambda'} J^{\lambda,\lambda'}(ij)\sigma_j^\lambda M_{\lambda'}]S_{i,\lambda}^z \sigma_i^\lambda \quad (2)$$

(Here $M_{\lambda'}$ is the magnetization for the λ' spin species).

Formal expressions for M_λ, from which a self-consistent solution for the magnetizations is to be obtained, can now be found straightforwardly, i.e.

$$2M_\lambda = (2S_\lambda + 1)\left\langle \coth[\beta(2S_\lambda + 1)\Gamma_i^\lambda]\right\rangle_{(i,\lambda)} - \left\langle \coth(\beta\Gamma_i^\lambda)\right\rangle_{(i,\lambda)} \quad (3)$$

$$2\Gamma_i^\lambda = \mu^\lambda H + 2\sum_{j,\lambda'} J^{\lambda,\lambda'}(ij)\sigma_j^\lambda M_{\lambda'} \quad (4)$$

Here for simplicity we consider only positive exchange integrals, i.e. $J^{\lambda,\lambda'} > 0$, S_λ denotes the magnitude of the λ spins in the Dirac units, $\beta^{-1} = kT$ and $\langle \ldots \rangle_{(i,\lambda)}$ is a conditional configurational average subject to the requirement that site (i) is occupied by a λ spin. Note that such a configurational average has to be carried out over the occupancy of all the sites j that are connected to i via the exchange integrals $J^{\lambda,\lambda'}(ij)$. For the case of nearest neighbor only exchange integrals, $J^{\lambda,\lambda'}$, this amounts to considering the entire neighboring shell of z atoms.

The Curie temperature, T_c, is determined by expanding the above expressions in powers of (Γ_i^λ) when $\mu^\lambda = 0$ and retaining only the linear terms in the small parameters, $M_{\lambda'}$. For a two component system, with relative concentrations c_A and c_B, this leads to the following expressions:

$$kT_c = z\Lambda + z[4(J^{AB})^2 c_A c_B X_A X_B + (\Delta)^2]^{\frac{1}{2}},$$

$$\Lambda = c_A X_A J^{AA} + c_B X_B J^{BB}; \quad \Delta = c_A X_A J^{AA} - c_B X_B J^{BB},$$

$$X_\lambda = S_\lambda(S_\lambda + 1)/3. \quad (5)$$

For determining the behavior of M_λ close to T_c, we need to retain the next leading order, i.e. $0\left\langle (\Gamma_i^\lambda)^3 \right\rangle_{(i,\lambda)}$, in the series expansion for Eq. (3). Because $\sigma_i^A + \sigma_i^B = 1$ and $\sigma_i^\lambda \sigma_i^{\lambda'} = \delta_{\lambda,\lambda'} \sigma_i^\lambda$ we find after some algebra

$$\langle (\Gamma_i^A)^3 \rangle_{(i,\lambda)} / z^3 = (M_A)^3 (J^{AA})^3 [(c_A)^3 + \frac{3}{z}(c_A)^2 c_B + (\frac{1}{z})^2 c_A c_B (c_B - c_A)]$$

$$+ 3M_A^2 M_B (J^{AA})^2 J^{AB} c_A c_B [c_A + \frac{1}{z}(1 - 3c_A) + (c_A - c_B)(\frac{1}{z})^2]$$

$$+ 3(M_B)^2 M_A J^{AA} (J^{AB})^2 c_A c_A [c_B + \frac{1}{z}(1 - 3c_B) + (c_B - c_A)(\frac{1}{z})^2]$$

$$+ (M_B)^3 (J^{AB})^3 [c_B^3 + \frac{3}{z}(c_B)^2 c_A + (\frac{1}{z})^2 c_A c_B (c_A - c_B)] \qquad (6)$$

The relevant expression for the B terms is obtained by interchanging A and B in Eq. (6). The result for the magnetization is thus found to be:

$$M_\lambda = D_\lambda (1 - T/T_c)^{\frac{1}{2}} \qquad (7)$$

$$D_B/D_A = \frac{2X_B c_B J^{AB}}{(kT_c - 2X_B c_B J^{BB})} \qquad (8)$$

RESULTS

In Fig. 1 we have plotted the relative magnetizations, $(M/S)_\lambda \equiv m_\lambda$ as a function of the ratio T/T_c for the case $J^{BB} = 3$, $J^{AB} = 2$ and $J^{AA} = 1$. Section a (top-left) refers to $S_A = S_B = 0.5$, Section b (top right) to $S_A = 0.5$, $S_B = 1$,
Section c (bottom left) to $S_A = 0.5$, $S_B = 7/2$ and Section d (bottom right) to $S_A = S_B = 7/2$. Full curves represent m_A and broken curves m_B.

For cases with equal spin magnitudes, i.e. Sections a and d, the lower of the full and of the broken curves correspond to $c_A = 0.2$ while the upper curves correspond to $c_A = 0.8$. The suppression of m_A curves below the 'normal' Brillouin curve (not shown) is larger for the case where the randomness is greater, i.e. for which c_A is smaller. Because of the somewhat greater stiffness of the B spins, the curves for m_B lie above the 'normal' curve, the case of larger randomness, i.e. $c_A = 0.8$, lying farther above it. (The normal curve lies just a little below the lower of the broken curves). Note that these departures from the normal curve are a shade smaller

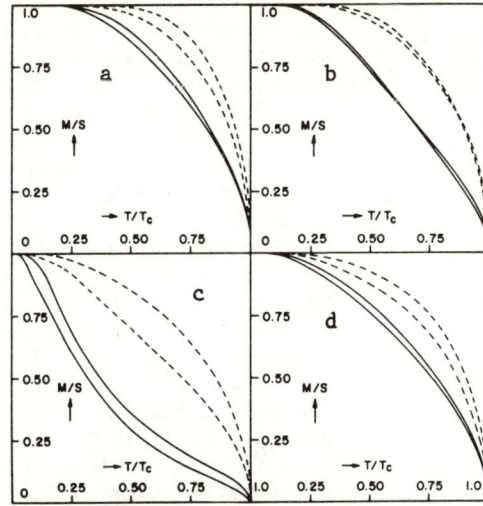

Figure 1, Sections a-d

in Section d where the spins are larger in magnitude.

The situation for the mixed spin magnitudes is more complicated. Looking at Section c, where the relevant effects are accentuated, we note first that the upper (of the broken and the full) curves correspond to $c_A = 0.2$. Moreover for m_B, the $c_A = 0.8$ curve is now below the normal curve. For the B spins, the presence of a larger number of atoms with much smaller stiffness causes a very large perturbation. Consequently, the B spin stiffness gets self-consistently pulled down.

Clearly, the coupled behavior of the A and the B magnetizations is very sensitive to the magnitude of exchange integrals, spins and the concentrations. Indeed, for any given set of these parameters, an a priori guessing of the behavior of m_λ versus T/T_c is not always possible. For example, in Section b of Fig. 1 the $c_A = 0.2$ and 0.8 curves for m_A cross over (as do the relevant curves for m_B) roughly half way to T_c.

To get a feel for the effects of external field we show in Fig. 2, the behavior of the mixed spin alloys ($S_A = ½$, $S_B = 7/2$) for two different field strengths. Sections a and c refer to $J^{AA} = 1$, $J^{AB} = 2$, $J^{BB} = 3$. Sections b and d refer to a similar alloy with larger AB interaction, i.e. $J^{AB} = 4$. The field strengths for Sections a and b are such that $z\mu^A H = z\mu^B H = 0.1$. For Sections c and d the field is ten times stronger. Initially all the m_B (broken) and the m_A (full) curves for the lower A concentration, i.e. $c_A = 0.2$, lie higher. However, along the temperature range, the smaller and larger c_A

Figure 2, Sections a-d

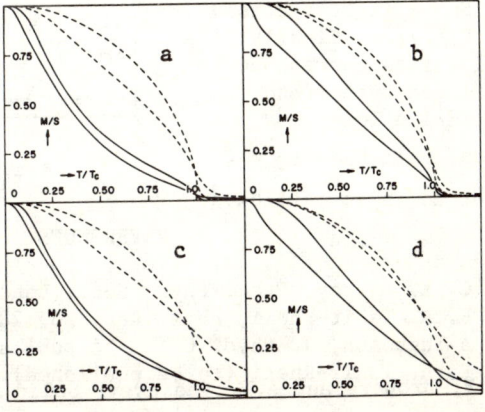

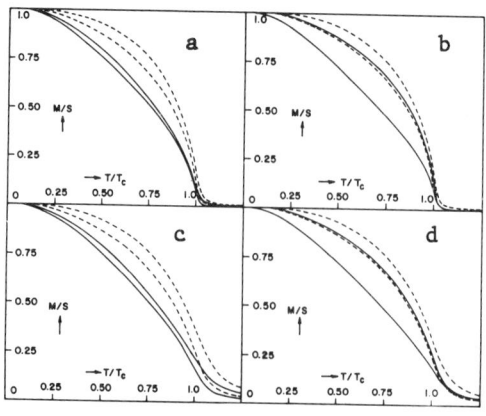

Figure 3, Sections a-d

curves cross over (T_c here refers to the corresponding alloy without external field). Differences in the behavior of the alloys with different J^{AB} exchange and different magnetic fields are again rich in their subtle variety.

To demonstrate the spin dependence of the magnetizations, in Fig. 3 we show the corresponding results for the same alloy described in Fig. 2 with the only difference that now $S_A = S_B = 7/2$. Here in Sections a and c, the relative disposition of the $c_A = 0.2$ and the $c_A = 0.8$ curves is inverted, i.e. the $c_A = 0.8$ results now lie higher. On the other hand, in Sections b and d, the behavior of m_B is inverted (with respect to c_A) but not that of m_A.

These results demonstrate the importance of interdependence of the two sets of magnetizations. Finally, it may be mentioned that this theory is clearly superior to the virtual crystal approximation which does not lead to any dependence of m_A upon the ratio (J^{AB}/J^{AA}) for the symmetrical case $c_A = c_B$ and $J^{AA} = J^{BB}$ when plotted against (T/T_c). The present theory, on the other hand, predicts an important dependence on this ratio. For instance, in Fig. 4 we show (M/S) for $J^{AB}/J^{AA} = 2$ (broken curves) and $= 0.01$ (full curves) for $S = \frac{1}{2}$ (left section) and $S = 7/2$. Clearly, the full curves correspond to a diluted system and should lie lower.

[Note: All the figures refer to an sc lattice.]

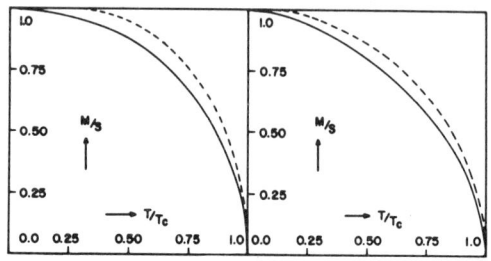

Figure 4, Section a and b

REFERENCES

1. G. A. Murray, Proc. Phys. Soc. (London) **89**, 87 (1966).
2. R. A. Tahir-Kheli, Phys. Rev. B**6**, 2808 (1972).
3. A Theumann, J. Phys. C (to be published).
4. R. A. Tahir-Kheli (to be published).
5. F. Brouers and A.V. Vedyayev, Solid St. Commun. **9**, 1521(1971)

STUDIES IN EXCHANGE AND ANISOTROPY DRIVEN AMORPHOUSNESS IN FINITE SIZE SPIN SYSTEMS

J. D. Patterson[†]
South Dakota School of Mines and Technology
Rapid City, South Dakota 57701

R. A. Tahir-Kheli[†*]
Temple University, Philadelphia, Pa. 19122

ABSTRACT

Finite size spin systems, consisting of n spins ($n \leq 5$), coupled through anisotropic Heisenberg exchange interactions and experiencing single ion anisotropy, are solved exactly by numerical methods. The results for the internal energy, the specific heat, the magnetization, and the magnetic susceptibility are computed for various configurations involving different exchange and anisotropy field strengths. To achieve an amorphous state, averages over many (randomly chosen) configurations are taken. Both the spin one-half and spin one cases are considered. Specific results for averaging over 100-1000 different choices of exchange and anisotropy parameters for the $n = 3$, $S = 1$ and $n = 5$, $S = \frac{1}{2}$ cases are represented. The results for the specific heat indicate an increase at low temperatures whenever the exchange, or the anisotropy, fields become random. Physically, this behavior can be thought to correspond to the low frequency increase of the density of states for random magnetic systems.[1]

---

[1] R. A. Tahir-Kheli, Phys. Rev. $\underline{B6}$, 2808 (1972); also Amorphous Magnetism, Hooper et al Editors, Plenum Press, 1973.

[†] Work begun at Universidade Federal de Pernambuco, Recife, Brazil.

[*] Supported by the N.S.F. Grant GH39023.

MOESSBAUER STUDIES IN NON-CRYSTALLINE MAGNETS

F.J. Litterst, G.M. Kalvius and A.J.F. Boyle[+]
Physikdepartment der Technischen Universität
München, D-8046 Garching, Germany

ABSTRACT

Non-crystalline (nc) layers of the known antiferromagnets (T_N) FeF_2 (78 K), $FeCl_2$ (24 K) and $FeBr_2$ (11 K) have been studied from 3-350 K. In contrast to the crystalline (crys) ferrous halides we find a single well-defined magnetic ordering temperature $T_m \approx 21$ K for all nc layers. Spectra below 21 K show a randomly oriented effective hyperfine field (H_{eff}) and electric field gradient (efg) to be present. H_{eff} is not sharply defined and shows a Gaussian distribution with a relative width of ~25%. This width is smaller than assumed in most theories of amorphous magnets. The rise of H_{eff} between $0.4 \cdot T_m$ and T_m is less steep than that calculated for any Brillouin function. The saturation values of H_{eff} deviate significantly from those for the crys ferrous halides.

INTRODUCTION

Present theoretical models [1-5] of nc magnets treat the changes in local magnetic properties either by fluctuations in the exchange integral or by a distribution of random local anisotropy fields. All theories give a slower temperature dependence of the magnetization in the nc state and predict either an increase or decrease in magnetic transition temperature. Thus macroscopic measurements at present cannot easily discriminate between the different models. Moessbauer spectroscopy gives selective microscopic information on the magnetic atom and its surrounding and delivers directly the actual distribution of hyperfine (hf) - fields in the nc solid.

EXPERIMENTAL

We present Moessbauer data on ^{57}Fe in FeF_2, $FeCl_2$ and $FeBr_2$ condensed from a molecular beam on a cold (<10K) substrate (30 μ Al-foil. About 50 mg of the anhydrous ferrous halides enriched in ^{57}Fe were heated in

---

[+]Permanent address: University of Western Australia, Nedlands 6009

a molecular beam oven (alumina) to 800-1200 K in a
vacuum better than 10^{-8} Torr. About one hour was needed
to deposit an absorber with a thickness on the order of
5000 Å (estimated from attenuation of X-rays during evaporation, assuming that the mean density is the same as
for the crys substances). The absorber plane and the
direction of γ-rays formed an angle of 50°. The absorber
temperature could be varied from 3-350 K. Absolute temperature calibration is good to ±0.5 K. Temperature control was better than ±0.2 K per day. Some data for $FeCl_2$
have been reported earlier [6].

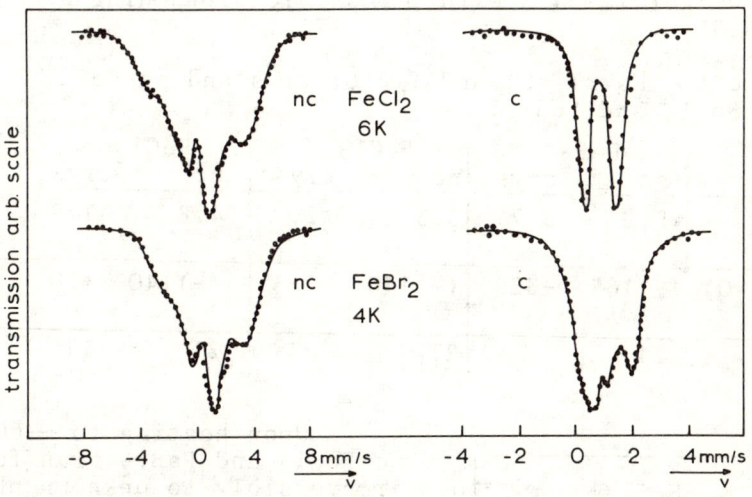

Fig. 1 Moessbauer spectra of nc and crys $FeCl_2$
and $FeBr_2$

RESULTS

Above 21 K the spectra of nc FeF_2, $FeCl_2$ and $FeBr_2$
are pure quadrupole doublets (see Table I). Figs. 1 and
2 show spectra found at lower temperatures together
with the spectra for the crys ferrous halides. The spectra of the nc absorbers are complex and indicate a magnetic hf splitting in addition to the electric quadrupole
splitting. Our data on $FeCl_2$ are identical with the
spectra of ref. 6, where the γ-rays and the absorber
plane formed an angle of 90°. From this we conclude that
H_{eff} is randomly oriented. We further assume that the
efg forms random angles with H_{eff}. Data analysis was performed by comparing the experimental data with computer
simulated spectra using a program, which is similar to

that given in ref. 7. From the broadened resonance lines in the regime of pure quadrupole interactions we deduce a Gaussian distribution in e^2qQ with a width of ~20%. For H_{eff} a similar distribution with a width of ~25% results. The solid lines in Fig. 1 show simulated spectra calculated under these assumptions. While for nc $FeCl_2$ and $FeBr_2$ it is not necessary to introduce an assymmetry parameter η for the efg, we have to use $\eta \approx 0.4$ for FeF_2 (Fig. 2). In Fig. 3 we have plotted the normalized mean effective hf magnetic fields $\langle H_{eff}(T)\rangle / \langle H_{eff}(T=0)\rangle$ against the normalized temperature T/T_m. The magnetic ordering temperature T_m at which H_{eff} vanishes, was determined by thermal scanning [8]. For all three nc ferrous halides we get $T_m \approx 21$ K with a possible broadening of about ±0.3K.

Table I e^2qQ, $H_{eff}(T=0)$ and T_m for crys and nc FeF_2, $FeCl_2$ and $FeBr_2$.

	FeF_2		$FeCl_2$		$FeCl_2$	
	nc	crys	nc	crys	nc	crys
e^2qQ (mm/s)	5.3	5.7	5.3	2.	4.8	1.9
$H_{eff}(T=0)$ (kOe)	(-)165	-329	(-)140	~+3	(-)140	~+30
T_m (K)	21	78	21	24	21	11

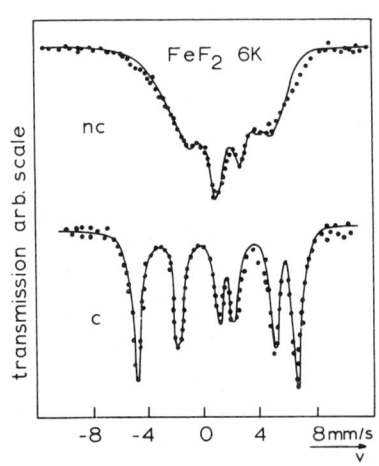

Fig. 2 Moessbauer spectra of nc and crys FeF_2

Upon heating to ~60K nc $FeCl_2$ and $FeBr_2$ transform irreversibly to unknown phases, which can be characterized by their quadrupole splittings. Complete crystallization is achieved at ~350 K. The absorbers of FeF_2 do not crystallize at room temperature. X-ray powder patterns show broad and diffuse rings, which are typical for nc solids. For crystallization ~800 K are needed. These crystallized layers show the well-known hf parameters for the crys ferrous halides [9,10].

DISCUSSION

Crys FeF_2, $FeCl_2$ and $FeBr_2$ are antiferromagnets. Their magnetic properties are sensitively related to their crystal structure [11]. For FeF_2 we find only little change in e^2qQ between a nc and a crys absorber, which leads to the conclusion that the local symmetry in the nc layer is close to that of a rutile-type lattice. In contrast, e^2qQ increases by a factor of ~2.5 by going from the crys to the nc structure in $FeCl_2$ and $FeBr_2$. This shows that the hexagonal layer structure of the crys phases with nearly octahedral halide coordination around the Fe is destroyed, favoring a symmetry which is more similar to that of FeF_2. The observed distribution in e^2qQ is caused by a spread in nearest neighbor bonding angles and distances.

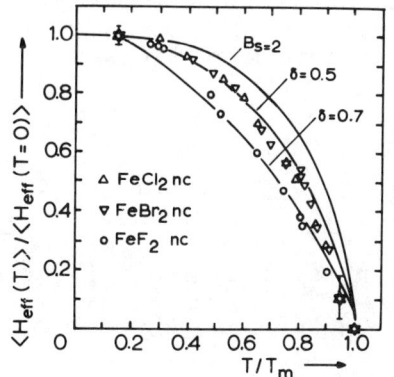

Fig.3 Temperature dependence of hf magnetic field in nc FeF_2, $FeCl_2$ and $FeBr_2$

The magnitude of H_{eff} in the ferrous halides is very sensitive to crystalline electric field interactions. In crys $FeCl_2$ and $FeBr_2$ the orbital hf field is incompletely quenched. In crys FeF_2 we have the usual strong quenching. From the changes in $H_{eff}(T=0)$ we conclude that in all the three nc ferrous halides the quenching of H_{orb} is incomplete, somewhat inbetween the cases of crys FeF_2 and $FeCl_2$. The observed distribution in hf-fields is rather small, particularly if compared to the model of ref.2. The relative width of the distribution in H_{eff} does not change significantly with temperature. The behaviour at T=0 indicates that the effective spin of the Fe ions is only slightly affected by the fluctuations in local order.

The rise of $H_{eff}(T)$ below T_m is clearly less steep than that calculated for a Brillouin function for S=2 as expected from theory. In Fig. 3 we have also plotted the curves given by a molecular field model for nc ferromagnets [1], where δ is a measure of the fluctuations in the exchange integral.

Whereas most theories only allow T_m either to increase or to decrease, a general model for a nc antiferromagnet [12] predicts both possibilities depending on the size and sign of the fluctuations in the exchange integrals. Although we do not know the magnetic structure of our nc layers, we discuss our results in terms of this model, which starts from a crys antiferromagnet that is made nc. The reduction in T_m for FeF_2 (for which the local symmetry has not drastically changed) can then be related to fluctuations of the order of 70% in the exchange integrals of nearest and next nearest Fe ions, which reduce the antiferromagnetic coupling. The interpretation for $FeCl_2$ and $FeBr_2$ is complicated by the change in local symmetry, which will also affect T_m. Our results show the magnetic coupling to be very similar in all nc ferrous halides. If we assume the rutile-like coordination and use the mean exchange integrals for the nearest neighbors from ref.13, it can be shown for nc $FeCl_2$, that exchange coupling cannot have changed drastically. For $FeBr_2$, however, the absolute mean exchange integral must have increased.

REFERENCES

1. K. Handrich, phys. stat. sol. 32, K55 (1969).
2. A.W. Simpson, phys. stat. sol. 40, 207 (1970).
3. S. Kobe, phys. stat. sol. 41, K13 (1970).
4. A.W. Simpson, D.R. Brambley, phys. stat. sol. (b) 49, 685 (1972).
5. R. Harris, M. Plischke, M.J. Zuckermann, Phys. Rev. Lett. 31, 160 (1973).
6. A.J.F. Boyle, G.M. Kalvius, D.M. Gruen, J.R. Clifton, R.L. McBeth, J. Phys. (Paris), Colloq. 32 Suppl., C1-224 (1971).
7. J.R. Gabriel, S.L. Ruby, Nucl. Instr. and Meth. 36, 23 (1965).
8. B.D. Dunlap, J.G. Dash, Phys. Rev. 155, 460 (1967).
9. G.K. Wertheim, D.N.E. Buchanan, Phys. Rev. 161, 478 (1967).
10. D.J. Simkin, Phys. Rev. 177, 1008 (1969).
11. J.S. Smart, in Magnetism III (Rado, Suhl ed., Academic Press, N.Y., 1963), pp. 63-144.
12. R. Hasegawa, phys. stat. sol. (b) 44, 613 (1970).
13. J. Kanamori, Progr. Theor. Phys. 20, 890 (1958).

ON THE DILUTE ANISOTROPIC HEISENBERG FERROMAGNET IN THE COHERENT POTENTIAL APPROXIMATION

T. Horiguchi
Ohio University, Athens, Ohio 45701

ABSTRACT

The dilute anisotropic Heisenberg ferromagnet of spin 1/2 with nearest neighbor exchange integrals is studied by using the coherent potential approximation within the random phase approximation of the two-time Green's function. The density of spin wave states is calculated for the sc, bcc, and fcc lattices.

INTRODUCTION

The impure Heisenberg ferromagnet has been investigated by several authors on the density of spin wave states(DSWS), on the concentration dependence of Curie temperature and so on. Montgomery et al.[1] studied the effects of distributed disorder on the properties of an isotropic Heisenberg ferromagnet by using the perturbation expansion for the two-time Green's function. Their theory was extended by Foo[2] to the anisotropic Heisenberg ferromagnet to show that such a system exhibits split bands which are not obtained in the isotropic case. The coherent potential approximation(CPA) was shown to be a very useful method for studing the disordered systems.[3,4] The application of the CPA to the isotropic Heisenberg ferromagnet has been done by Foo and Wu[5] to investigate the DSWS and the concentration dependence of Curie temperature for the sc lattice and by Bose and Foo[6] to investigate the DSWS for the linear lattice. In this paper, the theory given by Foo et al. for the isotropic Heisenberg ferromagnet is extended to the anisotropic Heisenberg ferromagnet by means of Morita and Chen's[7] scheme of generalizing the coherent potential approximation. The DSWS is studied in detail for the dilute magnetic system of the sc, bcc, and fcc lattices.

COHERENT POTENTIAL APPROXIMATION

We consider a dilute ferromagnet of the type $A_{1-x}B_x$. A and B are, respectively, magnetic and non-magnetic atoms and x is the concentration of non-magnetic component. The system is assumed to be described by the anisotropic Heisenberg Hamiltonian of spin 1/2 with nearest neighbor exchange integrals:

$$H = - \sum_{i,j}{}' [J_\perp(i,j)(s_i^x s_j^x + s_i^y s_j^y) + J_\parallel(i,j) s_i^z s_j^z] \quad (1)$$

where the prime on the summation means that the summation is taken over the sites on which magnetic atoms are located. In order to remove this restriction on the summation, we introduce an additional parameter, α_i, which takes the value 1 for the site i occupied by a magnetic atom and the value 0 for the site i occupied by a non-magnetic atom. Then we have

$$H = - \sum_{i,j} [I_\perp(i,j)(s_i^x s_j^x + s_i^y s_j^y) + I_\parallel(i,j) s_i^z s_j^z] \qquad (2)$$

where $I_\perp(i,j) = J_\perp(i,j)\alpha_i\alpha_j$ and $I_\parallel(i,j) = J_\parallel(i,j)\alpha_i\alpha_j$. The summation is now taken over all nearest neighbor pairs of sites i and j. $J_\perp(i,j)$ and $J_\parallel(i,j)$ represent the nearest neighbor exchange integrals and take a non-zero value $J_\perp^{AA}$ or $J_\parallel^{AA}$, respectively, only when the sites i and j are occupied by two A atoms. By using the states $|i\rangle$ which represents a single spin deviation state at site i following Kaneyoshi[8] and Bose and Foo[6], the Hamiltonian (2) can be expressed in the random phase approximation as follows:

$$H = \frac{1}{2} \sum_{i,j} [I_\parallel(i,j)(|i\rangle\langle i| + |j\rangle\langle j|) - I_\perp(i,j)(|i\rangle\langle j| + |j\rangle\langle i|)]. \qquad (3)$$

The Fourier time transform of the two time Green's function in the random phase approximation is obtained by using Hamiltonian(3) as follows:

$$G_{\ell m} = \langle \ell | \frac{1}{E - H} | m \rangle. \qquad (4)$$

Here $E = \omega/2\langle s^z \rangle$ where ω is the energy parameter of spin waves and is assumed to have a positive imaginary part.

In the CPA, the effective Hamiltonian is defined as follows:

$$\overline{G}_{\ell m} = \overline{\langle \ell | \frac{1}{E - H} | m \rangle} = \langle \ell | \frac{1}{E - H_{eff}} | m \rangle. \qquad (5)$$

In order to indicate the configurational average of a quantity, the bar is placed above the quantity. H_{eff} is assumed to be expressed within our CPA as follows:

$$H_{eff} = \frac{1}{2} \sum_{i,j} [I_\parallel^o(|i\rangle\langle i| + |j\rangle\langle j|) - I_\perp^o(|i\rangle\langle j| + |j\rangle\langle i|)] \qquad (6)$$

where generally $I_\parallel^o$ and $I_\perp^o$ are complex numbers. The summation is restricted over the nearest neighbor pairs of sites i and j. The Green's function defined by Eq.(4) is expressed in terms of a T-matrix and the Green's function defined by Eq.(5) when we formally use H_{eff} as the unperturbed Hamiltonian. Following Morita and Chen[7], the T-matrix is assumed to be expanded by means of the clusters of lattice sites. From the pairwise nature of our Hamiltonian(3), it is reasonable to assume that the expansion of the T-matrix begins with the cluster composed of two lattice sites. Taking into account only the effect of nearest neighbor pairs in the T-matrix as the lowest order approximation, we have the following equation to determine $I_\parallel^o$ and $I_\perp^o$:

$$\sum\sum_{\mu\nu} P_{\mu\nu} [\frac{1}{2}(I_\parallel(\mu,\nu) - I_\parallel^o) - \Delta_{\mu\nu} \overline{G}_{00}]/D_{\mu\nu} = 0, \qquad (7)$$

$$\sum\sum_{\mu\nu} P_{\mu\nu} [-\frac{1}{2}(I_\perp(\mu,\nu) - I_\perp^o) + \Delta_{\mu\nu} \overline{G}_{01}]/D_{\mu\nu} = 0, \qquad (8)$$

$$D_{\mu\nu} = \Delta_{\mu\nu}(\overline{G}_{00}^2 - \overline{G}_{01}^2) - (I_\parallel(\mu,\nu) - I_\parallel^o)\overline{G}_{00} + (I_\perp(\mu,\nu) - I_\perp^o)\overline{G}_{01} + 1, \qquad (9)$$

$$\Delta_{\mu\nu} = [(I_\parallel(\mu,\nu) - I_\parallel^o)^2 - (I_\perp(\mu,\nu) - I_\perp^o)^2]/4. \qquad (10)$$

μ and ν take one of species of atoms which is A or B. $P_{\mu\nu}$ is assumed to be expressed as $P_\mu P_\nu$ and $P_A=1-x$ and $P_B=x$. These equations are reduced to the one obtained by Foo et al.[5,6] for the isotropic Heisenberg ferromagnet by putting $I_\parallel(\mu,\nu)=I_\perp(\mu,\nu)$ and $I_\parallel^0=I_\perp^0$.

DENSITY OF SPIN WAVE STATES

$I_\parallel^0$ and $I_\perp^0$ in the effective Hamiltonian H_{eff} are now numerically obtained as the self-consistent solution of Eqs.(7) and (8). Then the density of spin wave states is obtained by taking the imaginary part of the Green's function for $\ell=m$ in Eq.(5). Figures of the DSWS given in this paper have their meaning only at low temperature where $<s^z>$ is nearly equal to 1/2. At first we investigate the behavior of the DSWS in the case of the isotropic Heisenberg ferromagnet. In Figs.1,2 and 3, we show that the DSWS for the sc, bcc and fcc lattices. For the low concentration of non-magnetic atoms, the shape of the DSWS depends upon the lattice structure. There do not appear localized impurity modes split from the main band. This is consistent with the fact that a non-magnetic impurity atom does not cause spin wave impurity levels[9]. As the concentration increases, the shape of the DSWS becomes independent of the lattice structure. For the concentration greater than a critical value x_c, a peak of δ-function appears at $E=0$ in the DSWS, that is, infinitely many states are found at $E=0$. The onset of the δ-function at $E=0$ makes Curie temperature equal to zero for reasons explained below and this critical value gives the critical concentration $c(=1-x_c)$ of magnetic atoms. The real part of the Greens function($Re\bar{G}_{00}$) is related to the imaginary part of the Green's function($Im\bar{G}_{00}$) in terms of the Kramers and Kronig's formula. If $Im\bar{G}_{00}$ has a δ-function at $E=0$, then $Re\bar{G}_{00}$ diverges as $1/E$ at $E=0$. The Curie temperature is inversely proportional to the value of $Re\bar{G}_{00}$ at $E=0$, so that if the DSWS has a δ-function at $E=0$, the system does not exhibit a transition to magnetic order. Next we investigate the effect of the anisotropy in the exchange integrals on the DSWS. In the pure system, the band width becomes narrow and the hight of the DSWS increases as the value $J_\perp^{AA}$ decreases and the DSWS has a δ-function at $E=zJ_\parallel^{AA}/2$ (z: the number of nearest neighbor lattice sites) in the limit of $J_\perp^{AA}=0$. For $x=0.1$, the DSWS is shown in Fig.4 for the sc lattice with the several values of anisotropy parameter $J_\perp^{AA}/J_\parallel^{AA}$. The thing different from the pure system is that the band width remains finite in the limit of $J_\perp^{AA}=0$. The DSWS for the sc lattice with $J_\parallel^{AA}=1$ and $J_\perp^{AA}=0.5$ is shown in Fig.5.

CONCLUDING REMARKS

In our figures for the DSWS, the curves drawn for the concentrations of non-magnetic atoms greater than x_c lose their meaning because the energy parameter $E=\omega/2<s^z>$ can not be defined any more. However, when we start with the Hamiltonian defined by Eq.(3), those curves have still their meaning and can be used to check the validity of our approximation in the high concentration. Let us consider two magnetic atoms which are on the nearest neighbor sites of each other and isolated from the other magnetic atoms by non-magnetic atoms. Then the energy eigenvalue of those two atoms are $(J_\parallel^{AA} \pm J_\perp^{AA})/2$. The peaks in the DSWS in high concentration exist sufficiently close to at $E=(J_\parallel^{AA}+J_\perp^{AA})/2$ and

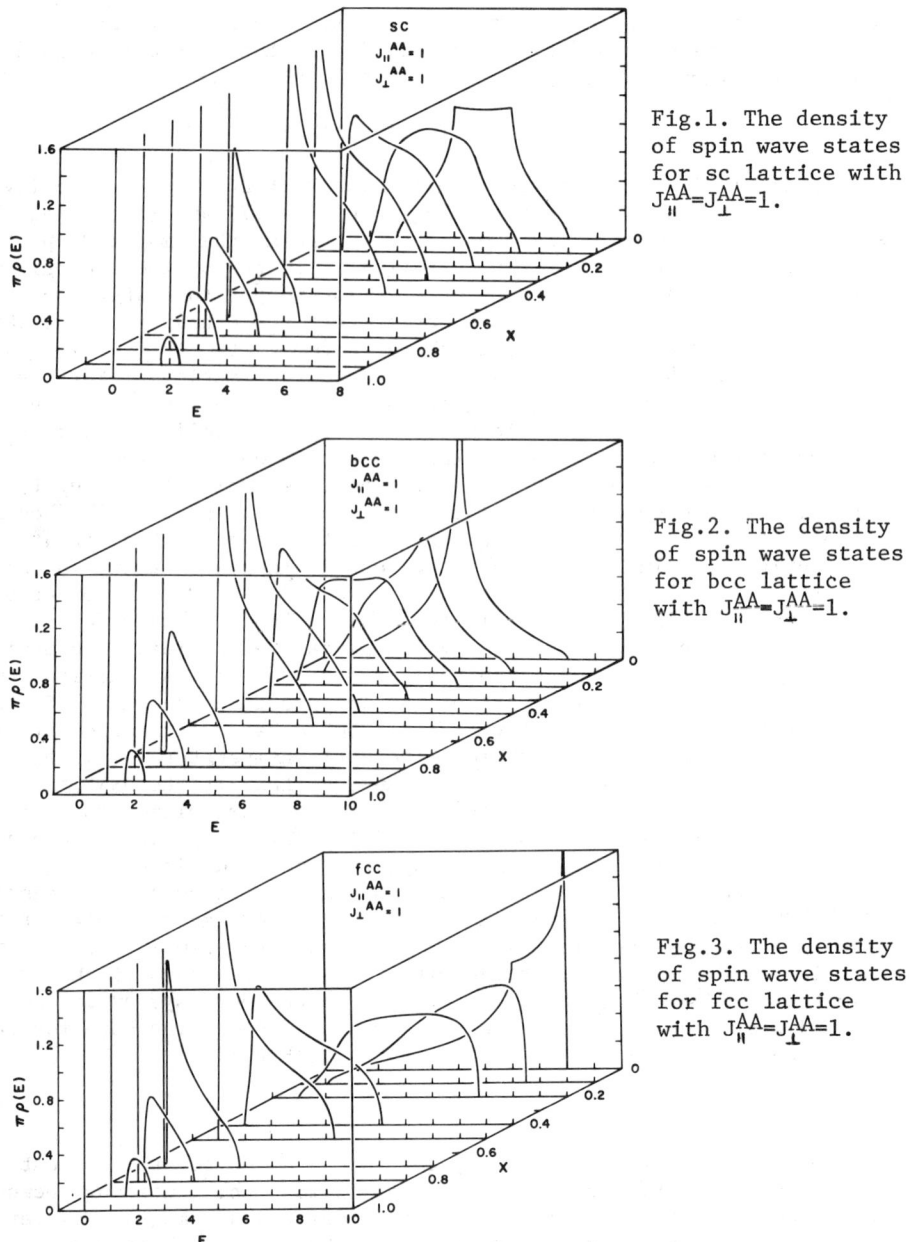

Fig.1. The density of spin wave states for sc lattice with $J_{\parallel}^{AA}=J_{\perp}^{AA}=1$.

Fig.2. The density of spin wave states for bcc lattice with $J_{\parallel}^{AA}=J_{\perp}^{AA}=1$.

Fig.3. The density of spin wave states for fcc lattice with $J_{\parallel}^{AA}=J_{\perp}^{AA}=1$.

at $E=(J_{\parallel}^{AA}-J_{\perp}^{AA})/2$. The critical concentration $c(=1-x_c)$ is given by Foo and Wu[5] for the isotropic Heisenberg ferromagnet. In the limit of $J_{\perp}^{AA}=0$, which corresponds to the Ising ferromagnet, the critical concentration is equal to $\sqrt{1/z}$. Detailed discussion on the concentration dependence of Curie temperature for the anisotropic Heisengerg

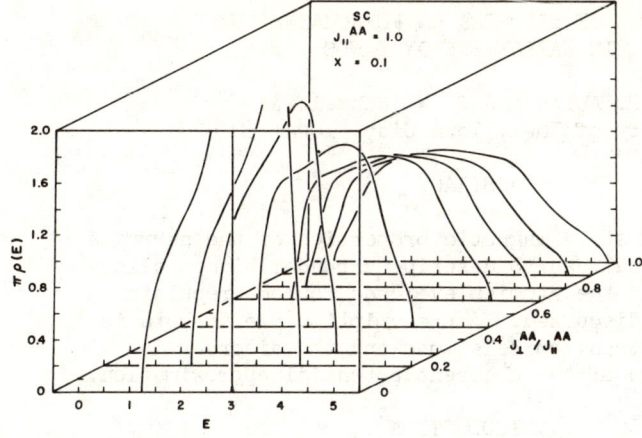

Fig.4. The density of spin wave states for sc lattice for x = 0.1.

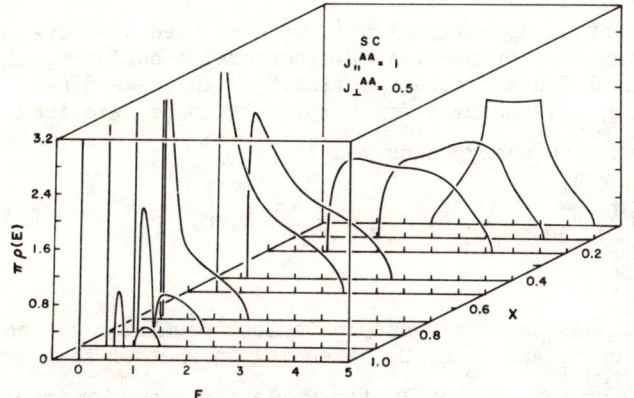

Fig.5. The density of spin wave states for sc lattice with $J_\parallel^{AA}=1$ and $J_\perp^{AA}=0.5$.

ferromagnet will be given on another occasion. Last of all, a mention is given about the form of H_{eff}(6) that the effect of the single site coherent potential in H_{eff} was taken into account for the case of the dilute isotropic Heisenberg ferromagnet in the precalculation to check the results obtained in Ref.5 but the author could not find any single site effect. Then the form of H_{eff} given in (6) was assumed here.

The author would like to express his sincere thanks to Professor T. Tanaka and Professor T. Morita for valuable discussions and comments.

REFERENCES

1. C. G. Montgomery, J. I. Krugler and R. M. Stubbs, Phys. Rev. Letters 25, 669 (1970).
2. E-Ni Foo, Solid State Commun. 10, 995 (1972).
3. P. Soven, Phys. Rev. 156, 809 (1967).
4. D. W. Taylor, Phys. Rev. 156, 1017 (1967).
5. E-Ni Foo and D-H Wu, Phys. Rev. B5, 98 (1972).
6. M. S. Bose and E-Ni Foo, J. Phys. C. 5, 1082 (1972).
7. T. Morita and C. C. Chen, J. Phys. Soc. Japan 34, 1136 (1973).
8. T. Kaneyoshi, Progr. Theoret. Phys. 42, 477 (1969).
9. S. Takeno, Progr. Theoret. Phys. 30, 731 (1963).

EFFECTS OF DISORDER ON FERROMAGNETISM IN NARROW ENERGY BANDS

G. F. Abito and J. W. Schweitzer
University of Iowa, Iowa City, Iowa 52242

ABSTRACT

The ground state magnetic properties of the narrow s band model generalized to describe a random binary alloy with disorder in the atomic level and Coulomb repulsion parameters are discussed. A prescription due to Roth is used to treat strong correlations and the disorder is treated by means of the coherent potential approximation.

INTRODUCTION

The coherent potential approximation (CPA)[1] has been used recently to study the magnetic properties of disordered substitutional binary alloys from the itinerant point of view.[2-4] In these discussions an alloy $A_x B_{1-x}$ is assumed for simplicity to be described by the nondegenerate narrow energy band model[5]

$$H = \sum_{\substack{i,j,\sigma \\ i \neq j}} t_{ij} c_{i\sigma}^+ c_{j\sigma} + \sum_{i,\sigma} \epsilon_i n_{i\sigma} + \tfrac{1}{2} \sum_{i,\sigma} U_i n_{i\sigma} n_{i-\sigma} \qquad (1)$$

where the atomic level and the intra-atomic Coulomb repulsion ϵ_i and U_i take on values ϵ_A or ϵ_B and U_A or U_B depending on whether the i-th site is occupied by a type A or B atom while the transfer integal t_{ij} is assumed independent of the species of atoms occupying the i and j sites. In most treatments[2] the Coulomb repulsion is approximated by the Hartree-Fock self-consistent field. This neglect of correlations is justified only if the Coulomb repulsion is weak, but that is not an interesting case to study since the stable phase is certainly the paramagnetic phase. For strong Coulomb repulsion the Hartree-Fock approximation does not provide a stringent enough condition on ferromagnetism.

Although the problem of strong correlations in the narrow energy band model is not satisfactorily understood, it is known from extensive studies of this model without disorder that correlations play a crucial role in determining the magnetic properties when the Coulomb repulsion is large. In order to get an indication of correlation effects in the CPA alloy generalization, Esterling and Tahir-Kheli[3] treated correlations in the spirit of the Hubbard I approximation.[5] More recently Fukuyama and Ehrenreich[4] applied the Hubbard III approximation.[6] In the case of a pure system ferromagnetism is not generally possible within both these approximations and this result remains when one introduces disorder. These approximations appear to provide a much too stringent condition on ferromagnetism.

In the present investigation we treat correlations within an approximation developed and used by Roth[7] to discuss this narrow band model in the case of a pure system. This approximation takes into account an energy shift which makes the stability of the ferromagnetic state more likely. Using this approximation Meyer and Schweitzer[8] obtained ground state magnetic properties in the limit of infinite Coulomb repulsion for cubic lattices with nearest neighbor hopping in complete agreement with the Nagaoka results[9] for nearly half-filled bands. Therefore this approximation seems to provide a suitable description of correlations in the strongly correlated limit for the investigation of magnetic properties, and furthermore it is one which can be straightforwardly extended to the alloy problem within the CPA framework.

PROCEDURE

The model as described by Eq. (1) is a special case of a model previously considered by Faulkner and Schweitzer[10] within the spirit of the Roth treatment of correlation effects. Here we cite several of the results of reference 10 where one can find a detailed discussion of the approximations involved. Within this scheme the one-electron Green function in the Wannier representation is found to be the solution of the equation

$$G_{ij}^{\sigma}(\omega) = G_i^{\sigma}(\omega)\, \delta_{ij} + G_i^{\sigma}(\omega) \sum_{\ell \neq i} t_{i\ell}\, G_{\ell j}^{\sigma}(\omega) \quad, \tag{2}$$

where $G_i^{\sigma}(\omega)$ is in general a quite complicated expression which depends on various expectation values that must be determined self-consistently. However for the two special cases we will discuss, U_i either zero or infinite, $G_i^{\sigma}(\omega)$ is given by

$$(U_i = 0) \quad G_i^{\sigma}(\omega) = [\omega - \epsilon_i]^{-1} \tag{3}$$

$$(U_i = \infty) \quad G_i^{\sigma}(\omega) = (1 - \langle n_{i-\sigma}\rangle)[\omega - \epsilon_i - E_{i\sigma}]^{-1} \tag{4}$$

where

$$E_{i\sigma} = -(1 - \langle n_{i-\sigma}\rangle)^{-1} \sum_{j \neq i} t_{ij} \langle c_{i-\sigma}^{\dagger} c_{j-\sigma}\rangle \quad. \tag{5}$$

For comparison the Hartree-Fock treatment for arbitrary U_i is obtained from

$$(H - F) \quad G_i^{\sigma}(\omega) = [\omega - \epsilon_i - U_i \langle n_{i-\sigma}\rangle]^{-1} \quad. \tag{6}$$

The single-site character of our treatment of correlations as clearly seen in Eq. (2) permits a straightforward application of the CPA treatment[1] of disorder. $G_i^\sigma(\omega)$ is assumed to depend only on the type of atom occupying the i-th site. Hence the average number of electrons with spin σ at the i-th site $\langle n_{i\sigma} \rangle$ and the spin-dependent energy shift $E_{i\sigma}$ appearing in Eq. (4) must be self-consistently determined in such a manner that this assumption is satisfied. This restricts our discussion to paramagnetism and ferromagnetism.

The coherent potential $\Sigma_\sigma(\omega)$ for a concentration x of A atoms is determined by the equation[1]

$$x\tilde{\epsilon}_{A\sigma}(\omega) + (1-x)\tilde{\epsilon}_{B\sigma}(\omega) - \Sigma_\sigma(\omega) = \left[\tilde{\epsilon}_{A\sigma}(\omega) - \Sigma_\sigma(\omega)\right] \times \left[\tilde{\epsilon}_{B\sigma}(\omega) - \Sigma_\sigma(\omega)\right] F_\sigma(\omega) \ . \tag{7}$$

where

$$\tilde{\epsilon}_{i\sigma}(\omega) = \omega - \left[G_i^\sigma(\omega)\right]^{-1} \ , \quad i = A, B \tag{8}$$

$$F_\sigma(\omega) = N^{-1} \sum_k \left[\omega - \epsilon_k - \Sigma_\sigma(\omega)\right]^{-1} \tag{9}$$

$$\epsilon_k = N^{-1} \sum_{\substack{i,j \\ i \neq j}} t_{ij}\, e^{i\vec{k}\cdot(\vec{R}_i - \vec{R}_j)} \ . \tag{10}$$

The quantities $\langle n_{i\sigma}\rangle$ and $E_{i\sigma}$ are evaluated by standard Green function techniques. For a given density of states associated with ϵ_k and a choice of the parameters x, ϵ_i, and the Fermi energy, one can calculate the one-electron properties. Of particular interest are $n_{A\sigma}$ and $n_{B\sigma}$, the average number of electrons with spin σ at an A and a B site, respectively, and the component densities of states $\rho_{A\sigma}(\epsilon)$ and $\rho_{B\sigma}(\epsilon)$.

RESULTS

We illustrate some of the effects of disorder and strong correlations on the ground state magnetic properties within our approximation by looking at two special cases in detail. Since we are interested in qualitative aspects rather than applications to specific alloys, we assume for simplicity the following form for the density of states associated with ϵ_k.

$$\rho^{(0)}(\epsilon) = \frac{2}{\pi}(1-\epsilon^2)^{\frac{1}{2}} \ , \quad |\epsilon| \leq 1$$

where the energy is in units of one half the bandwidth.

Case A: $\epsilon_B - \epsilon_A \gg 1$. Here the disorder splits the band into a band near ϵ_A and another near ϵ_B. The Fermi level is taken to be in the lower ϵ_A band so that the B sites are unoccupied. The CPA yields for the component density of states

$$\rho_{A\sigma}(\epsilon) = \frac{2}{\pi x} \{x - [\epsilon - \tilde{\epsilon}_{A\sigma}(\epsilon)]^2\}^{\frac{1}{2}}$$

which differs with the density of states for the pure A type crystal by having its width narrowed by a factor $x^{\frac{1}{2}}$.

The ferromagnetic solutions obtained as one varies the Fermi energy across the band are shown in Fig. 1 where the moment of an A atom, $m_A = n_{A\uparrow} - n_{A\downarrow}$, is plotted as a function of $n_A = n_{A\uparrow} + n_{A\downarrow}$.

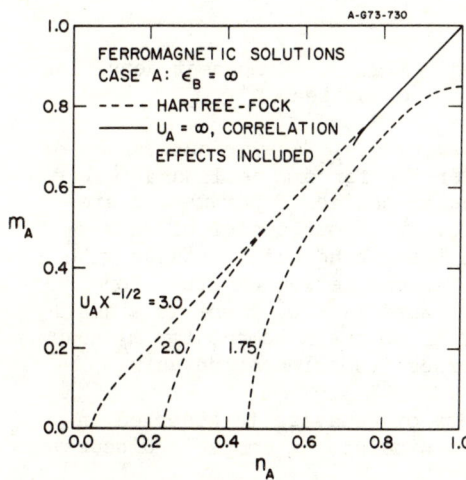

Fig. 1. Magnetization as a function of electron density.

The dashed curves are Hartree-Fock results for different values of $U_A x^{-\frac{1}{2}}$, which is the single parameter that characterizes the H-F solutions. For $U x^{-\frac{1}{2}} < \pi/2$ there are no ferromagnetic solutions. In contrast our treatment of correlations with U_A taken to be infinite yields the solid curve which is independent of the concentration x. For $n_A \leq 0.72$ there are no ferromagnetic solutions. The moment as a function of increasing n_A appears discontinuously with a finite value near the saturation value. The ferromagnetic solution is energetically stable relative to the paramagnetic solution with the same n_A.

The results that the curve in Fig. 1 is independent of x is particular to our choice for $\rho^{(o)}(\epsilon)$; however, to the extent that the major effect of the disorder is to narrow the band with dilution rather than to distort its shape, the solutions will be insensitive to x in our treatment with U_A infinite. Also the interesting discontinuous onset of ferromagnetic solutions is not a general result. In the previous study[8] of pure systems with nearest-neighbor tight-binding band structures the onset of magnetic solutions was continuous.

Case B: $\epsilon_B - \epsilon_A = 0.8$, $U_A = \infty$, $U_B = 0$. In this case the disorder parameter $\epsilon_B - \epsilon_A$ is sufficiently small that the electrons have access to the B sites yet large enough to significantly modify the density of states. Also we assume no Coulomb correlation on the B

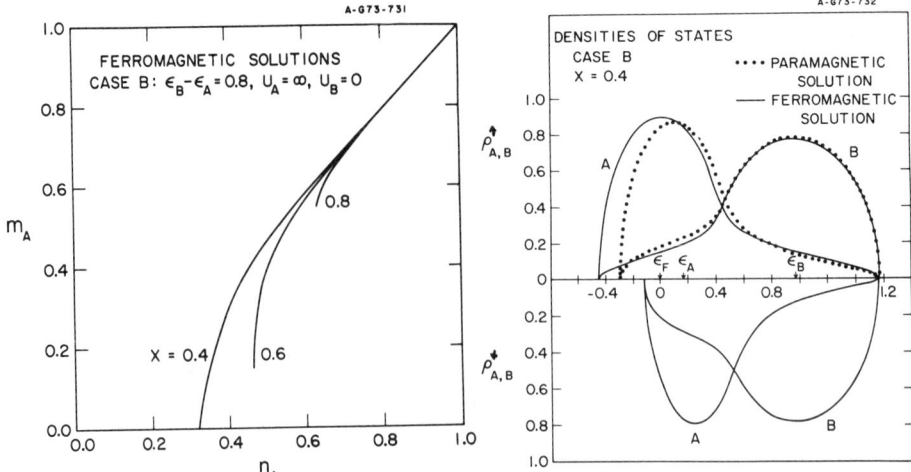

Fig. 2. Magnetization as a function of electron density for A sites.

Fig. 3. Component densities of states.

sites. Figure 2 shows curves of m_A versus n_A for various values of x. Component densities of states for the ferromagnetic and paramagnetic solutions for a particular choice of parameters are shown in Fig. 3 as a typical example. With our choice of $\epsilon_B > \epsilon_A$ the disorder causes a distortion in the A band which extends with decreasing x the range of n_A where ferromagnetism exists. For $\epsilon_A > \epsilon_B$ the range would be reduced. When $m_A \neq 0$, there is a small moment at the B site which is parallel to the A moment for m_A near 1 but decreases and eventually reverses its orientation as m_A decreases.

In summary a complicated variety of behavior is observed in this case and in other cases we have studied. A complete discussion will be presented elsewhere.

REFERENCES

[1] P. Soven, Phys. Rev. 156, 809 (1967); B. Velicky, S. Kirkpatrick, and H. Ehrenreich, Phys. Rev. 175, 747 (1968).
[2] For review see H. Fukuyama in AIP Conference Proceedings No. 10, Magnetism and Magnetic Materials - 1972, p. 1127.
[3] D. M. Esterling and R. A. Tahir-Kheli in Amorphous Magnetism edited by H. O. Hooper and A. M. de Graaf (Plenum Press, New York, 1973), p. 161.
[4] H. Fukuyama and H. Ehrenreich, Phys. Rev. B7, 3266 (1973).
[5] J. Hubbard, Proc. Roy. Soc. A276, 238 (1963).
[6] J. Hubbard, Proc. Roy. Soc. A281, 401 (1964).
[7] L. M. Roth, Phys. Rev. 184, 451 (1969).
[8] J. S. Meyer and J. W. Schweitzer, Phys. Rev. B7, 4253 (1973).
[9] Y. Nagaoka, Phys. Rev. 147, 392 (1966).
[10] D. Faulkner and J. W. Schweitzer, J. Phys. and Chem. Solids 33, 1685 (1972).

SHORT-RANGE ORDER IN AMORPHOUS GdFe$_2$*

G. S. Cargill III
Department of Engineering and Applied Science
Yale University, New Haven, Ct. 06520

ABSTRACT

X-ray scattering measurements on an amorphous sputtered GdFe$_2$ alloy indicate a short-range order significantly different from that of its crystalline Laves phase structure. Three prominent nearest neighbor distances for the amorphous alloy are 2.54Å, 3.04Å, and 3.47Å±.05Å. Associating these with Fe-Fe, Fe-Gd, and Gd-Gd pairs indicates that Fe atoms have an average of 6.3±.5 Fe and 3.3±.3 Gd nearest neighbors; Gd atoms have an average of 6.7±.6 Fe and 6.±1. Gd nearest neighbors. Corresponding coordination numbers in the Laves phase are 6, 6, 12, and 4. Although positions of Fe-Fe and Fe-Gd contributions are nearly identical both to those of the Laves phase and to distances expected from Goldschmidt atomic diameters, the observed Gd-Gd spacing is significantly larger than the Laves phase spacing (3.18Å) and somewhat smaller than the Gd Goldschmidt diameter (3.60Å). Binary dense random packing of hard spheres is proposed as a structural model for amorphous GdFe$_2$ and related alloys.

INTRODUCTION

This paper summarizes results of an x-ray scattering study of atomic arrangements in a sputtered amorphous GdFe$_2$ alloy. Although the structure of amorphous alloys containing approximately 80 at.% noble or transition metals (Fe, Ni, Co, Pd, Au) and 20 at.% metalloid elements (B, C, Si, P, Ge) has been well characterized,[1] this is the first report of detailed results for atomic arrangements in an amorphous rare earth--transition metal alloy.

Amorphous Gd-Co and Gd-Fe alloys are of current technological interest because of possible bubble domain applications.[2] Amorphous TbFe$_2$ has been extensively studied by elastic and inelastic neutron scattering, but with emphasis on magnetic effects rather than on details of atomic arrangements in the alloy.[3] Crystalline forms of GdFe$_2$, GdCo$_2$, and TbFe$_2$ all have the MgCu$_2$ Laves structure.[4] Results reported here indicate that short-range order in amorphous GdFe$_2$ differs significantly from that of its crystalline form.

EXPERIMENTAL

The amorphous alloy studied was prepared in bulk form (2.5 x 0.1 cm disk) at

---

* Supported by National Science Foundation.

Battelle Pacific Northwest Laboratories by d.c. krypton sputtering from a pressed powder target and was kindly made available by Dr. A. E. Clark. X-ray scattering measurements, at room temperature, employed a G.E. diffractometer with diffracted beam graphite monochromator and MoK$_\alpha$ radiation. Stability and statistical accuracy were better than 2%. Data analysis was similar to that in ref. 1.

The reduced interference function $F(k)=k[I(k)-1]$ is shown in Fig. 1; $k=4\pi(\sin\theta)/\lambda$. $F(k)$ was extrapolated linearly from $k=1.5\text{Å}^{-1}$ to zero at $k=0$. Fourier transformation of $F(k)$ with $k_{max}=17\text{Å}^{-1}$ yielded the reduced radial distribution function $G(r)=4\pi r[\rho(r)-\rho_o]$ shown in Fig. 2; ρ_o is the average atomic density. Oscillations in $G(r)$ for $r<2\text{Å}$ are attributed to termination effects and to possible slowly varying errors in the experimental scattering measurements. Small r oscillations of similar magnitude have also been found for amorphous Ni-P alloys.[1]

Calculating the radial distribution function $RDF(r)=4\pi r^2\rho(r)$ from $G(r)$ requires a value for the density of the amorphous alloy. Although the density of amorphous $GdFe_2$ has not been measured, Rhyne[5] found that the density of amorphous $TbFe_2$ was 93% of its crystalline density. The RDF of Fig. 3 was obtained using 93% of the crystalline $GdFe_2$ density.

Distribution functions obtained from scattering measurements on binary alloys are weighted sums of three partial distribution functions; $\rho_{ij}(r)=$ the number of j-type atoms per unit volume at distance r from an i-type atom, averaged over all i-type atoms. Weighting factors depend upon atomic scattering factors $f_i(k)$ and $f_j(k)$ and atomic fractions c_i and c_j.

$$\rho(r)=c_{Gd}(f_{Gd}^2/<f>^2)\rho_{GdGd}(r)+2c_{Gd}(f_{Gd}f_{Fe}/<f>^2)\rho_{GdFe}(r) \\ +c_{Fe}(f_{Fe}^2/<f>^2)\rho_{FeFe}(r) \qquad <f>^2=(c_{Gd}f_{Gd}+c_{Fe}f_{Fe})^2 \qquad (1)$$

the quotients of scattering factors are assumed to be independent of k. The actual k dependence of the coefficients of the $\rho_{ij}(r)$ may lead to broadening or narrowing of peaks in $\rho(r)$.[6] Termination of the fourier transform at finite k_{max} also contributes to peak broadening. The coefficients given in Table I were obtained by replacing f_{Gd} and f_{Fe} by the corresponding atomic numbers Z_{Gd} and Z_{Fe}.

DISCUSSION

The first true maximum in $G(r)$ appears to consist of three distinct contributions. This is more clearly illustrated in the corresponding maximum of $RDF(r)$ in Fig. 3. Also shown is the division of the maximum into three gaussian components, centered at 2.54Å, 3.04Å, and 3.47Å. The first of these is very close to the Fe-Fe distance in crystalline $GdFe_2$ and to the Fe Goldschmidt diameter, as shown in Table I. The second occurs very close to the crystalline Fe-Gd distance and to the Fe-Gd distance expected from the two Goldschmidt diameters. The third occurs at a significantly greater distance than the Gd-Gd spacing in crystalline GdFe$_2$, but at somewhat less than the Gd Goldschmidt diameter, as indicated in Table I.

If the three components in Fig. 3 are attributed to Fe-Fe, Fe-Gd, and Gd-Gd nearest neighbor pairs, their areas can be used to obtain the coordination numbers given in Table I. The Gd-Gd coordination number is much less certain than the others because Fe-Fe almost-nearest neighbor pairs are expected to contribute to RDF(r) in this region. A rough estimate of this contribution, shown in Fig. 3, was used in obtaining the Gd-Gd value. These experimental coordination numbers are consistent with local geometrical constraints imposed by atomic sizes of Gd and Fe. However, they are very different from those for crystalline GdFe$_2$, which are also given in Table I for comparison. In the amorphous alloy there appears to be a greater tendency for Fe-Fe and Gd-Gd nearest neighbor pairs than in crystalline GdFe$_2$.

Widths of the gaussian components of Fig. 3, after correction for termination effects, reflect distributions of nearest neighbor distances produced by structural and thermal disorder. The corrected full widths at half maximum were approximately 0.37Å.

Several weak maxima in G(r) are evident between 4Å and 7Å. Second neighbor distances in crystalline GdFe$_2$ occur at 4.78Å (Gd-Fe) and 5.20Å (Gd-Gd and Fe-Fe),[4] but no special features are found in G(r) at these distances. It may be possible to explain the observed structure in terms of dense random packing of hard spheres (DRPHS). This type of model has been used to interpret experimental results for amorphous Ni$_3$P and similar alloys.[1,7] Maxima associated with special linear and triangular configurations occur in RDF's for DRP of single sized hard spheres.[8] Introduction of two sizes of spheres for Gd (1.75Å radius) and Fe (1.27Å radius) complicates construction or generation of DRPHS models, particularly with regard to details of short-range chemical order. However, it is reasonable to expect that special configurations like those for DRP of single size spheres will produce maxima in binary DRPHS distribution functions.[9-11] The arrows in Fig. 2 indicate positions of expected maxima, although construction or generation of actual binary models is needed to verify the validity of this generalization and to determine widths and relative heights of the maxima.

Table I
Comparison of short-range order in amorphous and crystalline GdFe$_2$.

	Coef. of ρ_{ij}	Nearest Neighbor Distances			Coord. Numbers	
		Amorph.	Cryst.	From Goldschmidt Diameters	Amorph.	Cryst.
Fe-Fe	0.30	2.54±.05Å	2.60Å	2.54Å	6.3±.5	6
Fe-Gd	---				3.3±.3	6
Gd-Fe	0.74	3.04	3.05	3.07	6.7±.6	12
Gd-Gd	0.91	3.47	3.18	3.60	6 ± 1	4

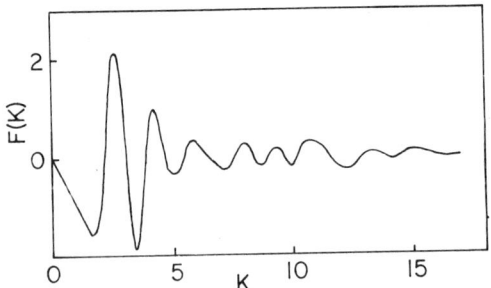

Fig. 1. Reduced interference function for amorphous $GdFe_2$.

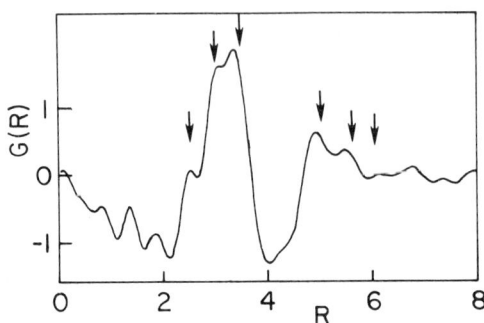

Fig. 2. Reduced radial distribution function for amorphous $GdFe_2$. Arrows indicate positions of maxima expected for binary DRPHS of radii 1.75Å and 1.27Å.

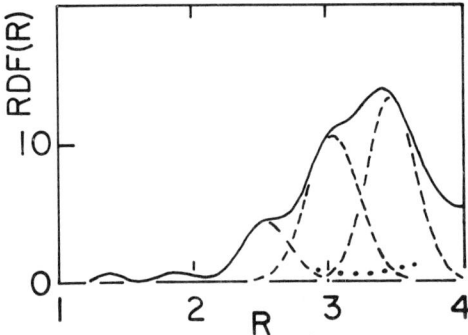

Fig. 3. Nearest neighbor portion of RDF for amorphous $GdFe_2$ (solid line), three gaussian components obtained by least-squares fit for 2.0Å <r<3.6Å (dashed line), and estimated Fe-Fe almost-nearest-neighbor contribution (dots).

CONCLUSIONS

Experimental data clearly indicate that well defined nearest neighbor short-range order exists in this amorphous $GdFe_2$ alloy, that the Gd-Gd nearest neighbor spacing is closer to the Gd Goldschmidt diameter than to the spacing in crystalline $GdFe_2$, and that the average coordination numbers are significantly different from those of crystalline structure. Binary DRPHS is proposed as a structural model for amorphous $GdFe_2$ and for other amorphous rare earth--transition metal alloys.

REFERENCES

1. G. S. Cargill III, J. Appl. Phys. 41, 12 and 2249 (1970).
2. P. Chaudhari, J. J. Cuomo, and R. J. Gambino, IBM J. Res. Dev. 17, 66 (1973).
3. J. J. Rhyne, S. J. Pickart, and H. A. Alperin, in Amorphous Magnetism, H. O. Hooper and A. M. deGraff, Eds. (Plenum Press, New York, 1973), p. 373.
4. J. H. Wernick, in Intermetallic Compounds, J. H. Westbrook, Ed. (Wiley, New York , 1967), p. 197.
5. J. J. Rhyne, private communication.
6. R. Kaplow, S. L. Strong, and B. L. Averbach, in Local Atomic Arrangements Studied by X-Ray Diffraction, J. B. Cohen and J. E. Hilliard, Eds. (Gordon and Breach, New York, 1966), p. 159.
7. G. S. Cargill III and R. W. Cochrane, "Amorphous Cobalt-Phosphorus Alloys: Atomic Arrangements and Magnetic Properties", to be published in Journal de Physique, January, 1974.
8. C. H. Bennett, J. Appl. Phys. 43, 2727 (1972).
9. D. E. Polk, J. Non-Cryst. Solids 11, 381 (1973).
10. J. F. Sadoc, J. Dixmier, and A. Guinier, J. Non-Cryst. Solids 12, 46 (1973).
11. R. W. Cochrane, R. Harris, M. Plischke, and M. J. Zuckermann, "A Model for Magnetism in Amorphous Metals," to be published in Journal de Physique, January, 1974.

MÖSSBAUER EFFECT STUDIES IN AMORPHOUS
$TbFe_2$, $DyFe_2$, $HoFe_2$ and $ErFe_2$

D. Sarkar and R. Segnan
American University, Washington, D.C. 20016

A. E. Clark*
U.S. Naval Ordnance Laboratory, White Oak, Md. 20910

ABSTRACT

Mössbauer measurements have been performed on sputtered $TbFe_2$, $DyFe_2$, $HoFe_2$ and $ErFe_2$. Neutron diffraction data indicates that sputtered $TbFe_2$ is structurally and magnetically amorphous. At room temperature, sputtered $TbFe_2$ exhibits a Mössbauer pattern which is interpreted as due to a broad distribution of hyperfine fields. A similar distribution of hyperfine fields is observed in $DyFe_2$ and $HoFe_2$ at 77K. $DyFe_2$, $HoFe_2$ and $ErFe_2$ show a resolved doublet at room temperature due to well defined quadrupolar interaction at the iron nuclei. Low temperature measurements on sputtered $HoFe_2$ show the broad pattern at 145K and the quadrupolar doublet at 230K. Our data indicate that the iron moments experience an exchange field of random orientation and magnitude and that the crystal field on the iron ions is relatively uniform.

INTRODUCTION

Neutron diffraction and magnetization measurements[1,2] indicate that sputtered RFe_2 compounds are structurally and magnetically amorphous. The Curie temperature of amorphous $TbFe_2$ has been found to be 388K[1] about 50% lower than that for crystalline $TbFe_2$ value of 697K[3]. The Curie temperature of amorphous $DyFe_2$, $HoFe_2$ and $ErFe_2$ have not been determined yet. The coercive force in these materials is very high[2].

---

*Supported by the Naval Ordnance Laboratory IR Fund and the Office of Naval Research.

EXPERIMENTAL RESULTS

Mössbauer spectroscopy on Fe^{57} nuclei has been used to investigate the properties of sputtered $TbFe_2$, $ErFe_2$, $HoFe_2$ and $DyFe_2$ samples[†], which were used by Clark[2] for his magnetization measurements and by Rhyne et al.[1] for the neutron diffraction measurements. We reduced these samples to powder form to use as Mössbauer absorbers.

Room temperature measurements of sputtered $TbFe_2$ shows a broad absorption curve (see Fig. 1), while those for sputtered $HoFe_2$, $ErFe_2$ and $DyFe_2$ exhibit a partially resolved doublet as shown for $DyFe_2$ in Fig. 2. At 77K $HoFe_2$ and $DyFe_2$ (both sputtered) show a broad absorption curve.

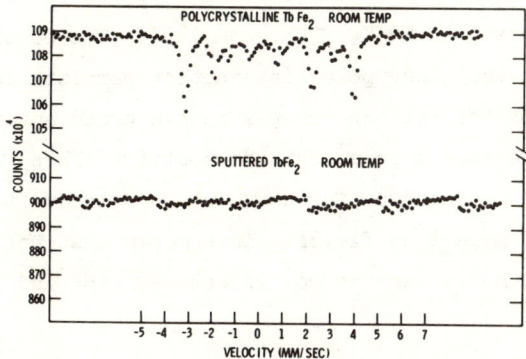

Figure 1
Mössbauer Spectra of $TbFe_2$ at Room Temperature

Since the Curie temperature of sputtered $TbFe_2$ is 388K, the alloy is magnetically ordered at room temperature. However, we interpreted the presence of the broad absorption spectrum as due to the presence of many nonequivalent Fe sites, giving rise to a broad distribution of hyperfine fields. A similar interpretation is inferred for $DyFe_2$ and $HoFe_2$ at 77K.

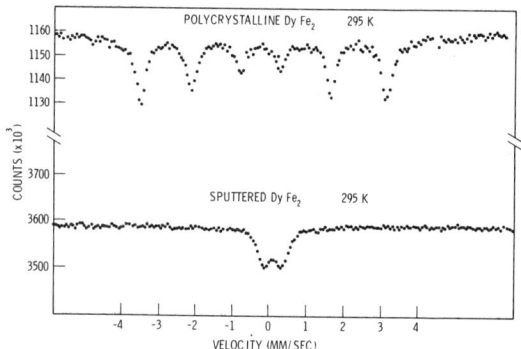

Figure 2
Mössbauer Spectra of DyFe$_2$ at Room Temperature

In Figure 3, we show the spectra for sputtered HoFe$_2$ at three different temperatures 295K, 230K, and 145K. The partially resolved doublet, due to the quadrupolar interaction persists from 295K to 230K but at 145K the pattern changes into a broad absorption spectrum (of the same type as for the sputtered TbFe$_2$ at room temperature). The spectrum for HoFe$_2$ at 145K is due to the broad distribution of hyperfine fields. This means that the Curie temperature for HoFe$_2$ (sputtered) is between 145K and 230K.

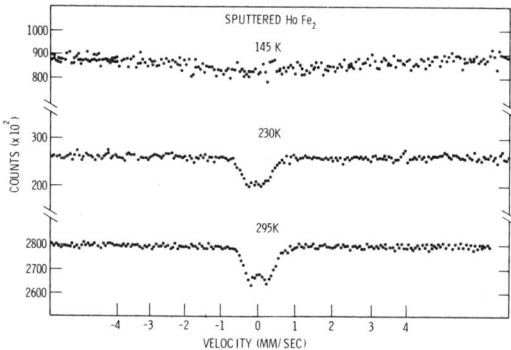

Figure 3
Mössbauer Spectra of Sputtered HoFe$_2$

From the data for amorphous $DyFe_2$ and $ErFe_2$ we conclude that their Curie temperatures are below room temperature. While the hyperfine fields in amorphous $HoFe_2$ in the magnetic state are distributed between 0 and about 190kOe, the quadrupole splitting has a well defined value of about 0.45 mm/sec. The corresponding values for the crystalline $HoFe_2$ are 201kOe and 0.40 mm/sec for the hyperfine field and the quadrupole splitting respectively as measured by Kimball et al.[4]

DISCUSSION AND CONCLUSION

Our experimental results shed some light on the nature of the exchange and crystal field in these amorphous materials.

The Hamiltonian can be written as

$$H = \sum_i V_i + \sum_{ij} J_{ij} \vec{J}_i \cdot \vec{J}_j$$

where V_i is the local crystal field acting on the ith ion and J_{ij} is the exchange constant between two ions i and j. A well defined quadrupolar splitting together with a broad distribution of hyperfine fields lead us to believe that the iron ions experience a crystal field which is relatively uniform but an exchange field of random orientation and magnitude. Our results seem to be consistent with the theory given by Harris et al.[5] who assume a Heisenberg model where each ionic spin experiences a local anisotropy field of random orientation.

ACKNOWLEDGMENT

The authors wish to thank Earl Callen and James Cullen for valuable discussions.

REFERENCES

1. James J. Rhyne, S. J. Pickart and H. A. Alperin, Phys. Rev. Lett. $\underline{29}$, 1562 (1972).
2. A. E. Clark, Appl. Phys. Lett. $\underline{23}$, 642 (1973).
3. E. Burzo, Z. Angew, Physik $\underline{32}$, 127 (1971).
4. C. W. Kimball, A. E. Dwight, R. S. Preston and S. P. Taneja, paper presented in this conference.
5. R. Harris, M. Plischke and M. J. Zukermann, Phys. Rev. Lett. $\underline{31}$ 160 (1973).
† The samples were prepared by rapid sputtering techniques by Pacific Northwest Laboratories, Batelle Memorial Institute, Richland, Washington.

MAGNETIC PROPERTIES OF AMORPHOUS GdCo FILMS

L. J. Tao, R. J. Gambino, S. Kirkpatrick, J. J. Cuomo and H. Lilienthal

IBM Thomas J. Watson Research Center
Yorktown Heights, New York 10598

ABSTRACT

We have measured the spontaneous magnetization of amorphous $Gd_{1-x}Co_x$ films with $0.36 \leq x \leq .96$, in the temperature range 4.2 - 600 K, and compare the properties of the films with those of six ordered Gd-Co compounds which form within this composition range. Like the crystals, the films appear to be collinear ferrimagnets, with Gd and Co spins ordered antiparallel, and display Gd-dominated, compensated, and Co-dominated magnetization curves as the Co concentration is increased. A compensation point is present over a wider range of compositions in the films than in crystals. There is a range (roughly $0.6 < x < 0.8$) in which T_C and the Co moment are greater in the films than in the crystals, in contrast to previous experience with disordered magnets. These changes in the magnetic properties of the Co atoms can be qualitatively accounted for as an effect of charge transfer.

INTRODUCTION

Intermetallic compounds consisting of rare earths and the transition metals Fe and Co can be made to be simultaneously amorphous and magnetic.[1,2] These systems are of particular value for fundamental study of amorphous magnetism because there exist a large number of well-studied ordered phases[3] which can be compared with amorphous materials of the same atomic composition.

For example, we recently reported[4] that the Curie temperature, T_C, observed in amorphous $Gd_{.33}Co_{.67}$ films was 30% higher than in the ordered compound, $GdCo_2$, in contradiction to the common belief that disorder should lower T_C. It was suggested[4] that charge transfer from the rare earth to the cobalt atoms has the dominant influence on the Co moment and the ordering temperature. Thus the lower density of the amorphous phase could result in less charge transfer into the Co d-bands, with the observed result that the Co moment and T_C increase. The results reported here provide further confirmation of this interpretation.

This paper presents a study of the magnetic properties of a series of amorphous Gd-Co films, comparing them with the known ordered compounds. The properties of the films--T_C, magnetization, compensation temperature and anisotropy--were all found to be smoothly-varying functions of film composition. By contrast, the properties of the ordered compounds vary in a less systematic way. Our results suggest that the unusual differences between the properties of crystals and amorphous films arise from special features of the crystalline, not the amorphous, state.

RESULTS

The spontaneous magnetization (extrapolated to zero field) was measured by the Faraday method over the temperature range 4.2 to 600 K for amorphous $Gd_{1-x}Co_x$ films with $0.36 \leq x \leq 0.95$. Films about 2μ thick were prepared by rf sputtering onto glass substrates, as described in Ref. 2. There were two limitations on the measurements. Our ability to determine the absolute moment per atom was limited by the roughly 5% accuracy of the physical measurements of film density. Curie temperatures greater than about 600 K, as occur in the Co-rich phases, could not be observed directly because of the onset of slow irreversible changes in film properties at about 550 K. These changes appeared to be associated with Gd oxidation, so the actual crystallization temperature of the films probably exceeds 600 K.

Six ordered compounds--Gd_4Co_3, $GdCo_2$, $GdCo_3$, Gd_2Co_7, $GdCo_5$, and Gd_2Co_{17}--occur in the composition range studied. The magnetic order in all six has been interpreted[3,5] as a collinear ferrimagnetic alignment with Gd and Co spins oppositely directed. Only Gd_2Co_7 shows a compensation point. In Gd_4Co_3, $Gd\ Co_2$, and $Gd\ Co_3$ at all temperatures the observed moment appears to be the magnetization of the Gd atoms minus that of the Co atoms while the reverse holds for $Gd\ Co_5$ and Gd_2Co_{17}. The latter two compounds have a large magnetocrystalline anisotropy,[6] which complicates interpretation of the magnetization measurements.

The films also exhibited Gd-dominated, then compensated, and finally Co-dominated behavior as the Co content was increased. Typical examples of the three regimes are shown in Fig. 1. The minimum value of M(T) at a compensation point was found to vary from sample to sample for fixed composition. The deep sharp minimum seen in the plot (Fig. 2) of the spontaneous magnetization

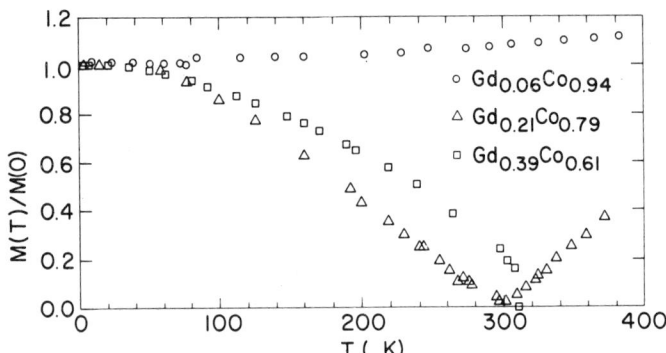

Fig. 1 Spontaneous magnetization as a function of temperature for three amorphous films.

Fig. 2 Low temperature spontaneous magnetization of the films, plotted against composition.

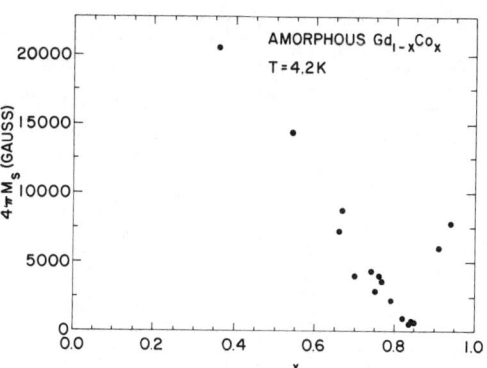

M_s(4.2 K) as a function of x also indicates that the observed moment is the difference of well-ordered Gd and Co moments. The fact that the Gd-Co films can be 'tuned' smoothly through this compensation region is of interest for bubble device applications, where a magnetically well-ordered system with small net magnetization is desired.

A compensation temperature was observed over a much wider range of compositions in the films than in crystals, as evidenced by the values plotted in Fig. 3. Either an increased Co magnetization, a weakened coupling between d and f spins, or decreased Gd magnetization could have caused the 150 K difference between T_{comp} in Gd_2Co_7 and in the corresponding films. However, the magnetization of a $Gd_{.25}Co_{.75}$ film was still increasing at the Curie temperature of the corresponding $Gd Co_3$ crystal, so it seems unlikely that the contribution to the internal field seen at a Co site (from the product of the square of the d-f coupling and the Gd magnetization) has decreased. The most reasonable interpretation of Fig. 3 is that M_{Co} and/or T_C are greater in the films than in the crystals for $x \lesssim .8$, and less for $x \gtrsim .8$.

Values of T_C for the ordered systems, compared when possible with film data, are plotted in Fig. 4. For the two films of lowest Co content studied, $x = .34$ and $x = .61$, T_C agreed with the trend of crystalline values. In the range $.6 \lesssim x < .8$, T_C for the films is indeed higher.

A comment about densities is required before discussing M_{Co} for the crystalline and amorphous phases. Densities of the amorphous films appeared to vary smoothly with x. To within measurement error, all lie on the linear interpolation between 12-fold coordinated pure Co(8.9 gm/cm^3) and pure Gd(7.95 gm/cm^3). This interpolated density,

$$\rho_{film} = 7.95 + .95 \, x \text{ gm/cm}^3, \qquad (1)$$

was used in extracting absolute moments from measurements on

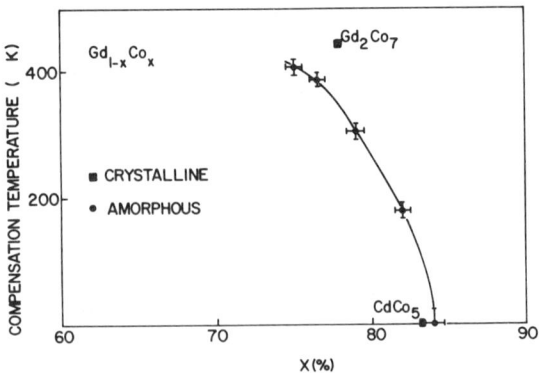

Fig. 3   Compensation temperatures of films with $.75 \leq x \leq .84$, and the ordered compound Gd_2Co_7. Magnetization in ordered $Gd\,Co_5$ is Co-dominated at all temperatures.

amorphous samples of known dimensions. By contrast, ordered compound densities differ erratically from each other. Gd_4Co_3, $Gd\,Co_5$, and Gd_2Co_{17} are only 1 - 3% denser than Eq. 1, but $Gd\,Co_2$ (9.6 gm/cm^3), $Gd\,Co_3$ (9.26 gm/cm^3) and Gd_2Co_7 (9.53 gm/cm^3) exceed Eq. 1 by 12, 7 and 10% respectively.

In the crystals, M_{Co} is nearly constant at about $1.6\mu_B$ for Gd_2Co_7 and all higher Co concentrations, but drops abruptly to $1.0\mu_B$ in $Gd\,Co_2$ and $0.65\mu_B$ in Gd_4Co_3. In the amorphous films, analyzing the 4.2 K data with the assumption that the Gd moment is perfectly aligned, we find M_{Co} tends to $1.4\mu_B$ at high Co concentrations, is about $1.45\mu_B$ in $Gd_{.33}Co_{.67}$, then drops at lower x, but still remains greater than in the crystals. M_{Co} for nearly compensated films (.65 < x < .8) appears to be greater than for .9 ≲ x, although the data is scattered. This may be due to incomplete alignment of the Gd spins, which would cause us to underestimate M_{Co} at large Co concentrations, and overestimate it at intermediate compositions.

DISCUSSION

The observation that the concentration range in which T_C and M_{Co} in the films exceed the crystalline values is also the region of unusually high crystalline densities confirms the role of charge transfer in the structure dependence of the magnetic properties of this material.

Recent x-ray structure studies on $Gd_{.33}Fe_{.67}$[7] show that some charge transfer persists in the amorphous phase. The data can be interpreted[8] in terms of a random packing of two sizes of hard spheres with a Gd radius of 1.73 Å, more than the 1.59 Å required to fit into the $GdFe_2$ Laves structure, but still less than the metallic radius of 1.80 Å.

If this small size and the charge transfer it entails are general features of the films, the Gd atoms will be forced to form some sort of open structure, maximizing Gd-Co contacts. Such a structure, formed at metallic densities, provides directional

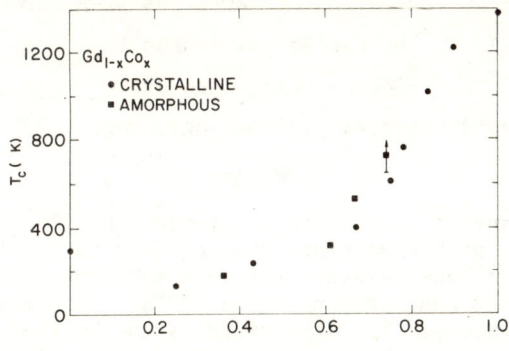

Fig. 4 Curie temperatures of amorphous films and ordered compounds. As described in the text, T_C for the film with x = .75 was not observed directly, but is estimated to lie above 700 K.

constraints needed to keep the material from crystallizing, yet is general enough to be consistent with the observed smooth dependence of the density on composition.

ACKNOWLEDGEMENTS

We would like to thank W. Molzen and S. Kane for technical assistance.

REFERENCES

1. J. Orehotsky and K. Schroeder, J. Appl. Phys. 43, 2413 (1972); J. J. Rhyne, S. J. Pickart and H. A. Alperin, Phys. Rev. Letters 29, 1562 (1972).
2. P. Chaudhari, J. J. Cuomo and R. J. Gambino, IBM J. Res. Dev. 17, 66 (1973); S. R. Herd and P. Chaudhari, phys. stat. sol. (a) 18, 603 (1973).
3. K. N. R. Taylor, Adv. in Phys. 20, 551 (1971); E. Burzo, Phys. Rev. B6, 2882 (1972).
4. L. J. Tao, S. Kirkpatrick, R. J. Gambino and J. J. Cuomo, Solid State Comm., to appear.
5. R. Lemaire, Cobalt 32, 132 (1966); 33, 201 (1966).
6. B. Barbara, D. Gignoux, D. Givord, F. Givord and R. Lemaire, Int. J. Magnetism 4, 77 (1973).
7. G. S. Cargill, III, this conference, talk 8E-1.
8. S. Kirkpatrick, unpublished.

MAGNETIC PROPERTIES OF AMORPHOUS Ni-P ALLOYS*

D. Pan and D. Turnbull

Gordon McKay Laboratory

Harvard University, Cambridge, Mass. 02138

ABSTRACT

The magnetic properties of electrodeposited amorphous Ni-P alloys, with P contents ranging from 12-23 a/o, were measured on a Foner vibrating-sample magnetometer in fields up to 8 KOe between 70 and 370°K. Ferromagnetic Curie temperatures, T_c, determined by the "method of thermodynamic parameters", were found to decrease from 212 to 75°K as x_p increases from 12.41 to 16.77 a/o. Paramagnetic Curie temperatures, T_p, were deduced from the temperature dependence of initial susceptibilities, χ_i, above T_c. The differences $(T_p - T_c)$ were of the order of 100°K for all specimens and several times larger than for crystalline ferromagnetic materials. At high field the low temperature magnetization decreases with x_p, and ferromagnetism disappears at about 17 a/o P, in close agreement with the T_c measurement. Our results indicate that at $x_p \sim 16-18$ amorphous Ni-P alloys behave as weak itinerant ferromagnets.

I. INTRODUCTION

The occurrence of ferromagnetism in amorphous materials was predicted by Gubanov in 1960[1], and it has been demonstrated experimentally for a number of amorphous alloys[2-9]. There have been several studies of the ferromagnetic behavior of amorphous Co-P alloys. Bagley and Turnbull[2] reported that amorphous Ni-P alloys, in contrast with Co-P alloys, are non-ferromagnetic at room temperature at P concentrations of 20-25 a/o. However Simpson and Brambley[7] have reported ferromagnetism in alloys at 15 a/o P. We report here an investigation of the dependence of the magnetic behavior of amorphous Ni-P alloys on P concentration over the composition range 12-23 a/o P.

II. EXPERIMENTAL PROCEDURE

Ni-P samples with 12-23 a/o P were electrodeposited from baths described by Brenner[10]. Copper strips 0.25-mm thick were used as cathodes and Ni rods with diam. 0.2" and purity 99.9999% were used as anodes. Samples were deposited in rectangular shapes of 2x3-cm. The deposition was carried out at a temperature of 70 ± 2°C with a current density of 10 amp/dm^2 in a period of 7 to 8 hours. Details

*Research sponsored by the National Science Foundation and the Office of Naval Research.

of the cathode preparation were described in Ref. 9. The thicknesses of the deposited samples were about 400 μ. All of the samples used exhibited the diffuse diffraction haloes characteristic of amorphous alloys[11] when examined on a G.E. diffractometer with Mo-Kα radiation. The magnetic specimens were cut out from the central 1x2-cm area of the deposited samples.

The amorphous specimens transformed to a stable mixture of crystalline Ni and Ni_3P after heating up to 600°C for a period of 15-20 minutes[12]. This permitted determination of the composition by the magnetic method described in detail in Ref. 9. This analysis utilizes the finding that Ni_3P is non-ferromagnetic at room temperature.

The magnetic measurement was carried out on a Foner-type vibrating-sample nagnetometer with a sensitivity of 1×10^{-4} emu[13], in fields up to 8 KOe in the temperature range of 70-370°K. The accuracy of temperature measurement was within 1-2°K. A pure nickel sphere sample was used for calibration.

III. RESULTS AND DISCUSSION

A. T_c Measurement

Magnetization curves at various temperatures typical for alloys with $x < 16.50$ a/o are shown in Fig. 1 for a specimen with 14.95 a/o P. In contrast with the amorphous Co-P alloys[8,9] complete saturation was not reached at fields up to 8 KOe at 77°K. The ferromagnetic Curie temperatures of five samples with composition between 12-17 a/o of P were determined by the "method of thermodynamic parameters"[14], in which H/σ is plotted against σ^2 as in Fig. 2. T_c was found to decrease with increasing P content and it vanishes at the composition of 17 a/o P as is clear from Fig. 3. For the 15 a/o P composition, our T_c (=190°K) is far below Simpson's reported value of 390°K[7]. This disparity may reflect the partial crystallization of his sample. Our T_c measurement is consistent with Bagley's observation[2].

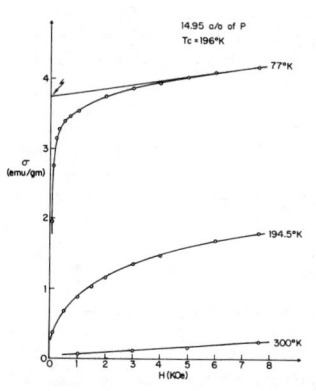

Fig. 1. Field dependence of magnetization at different temperatures for amorphous Ni-P with 14.95 a/o P.

B. χ_i above T_c and Determination of T_p

The temperature dependence of the initial susceptibility, χ_i, above T_c of specimens containing 14.95, 16.42 and 16.77 a/o P also was measured (see Fig.4). $1/\chi_i(T)$ deviated significantly from Curie-Weiss (C.W.) behavior at temperatures up to 150°K above T_c. At still higher temperatures, $1/\chi_i(T)$ became linear T, as shown in Fig.4, in accordance with C.W. law. The paramagnetic Curie temperatures, T_p, were determined by extrapolation of the linear parts of the curves. For all the samples T_p was about 100°K higher than T_c (see Fig. 3). The differences, T_p-T_c are several times or more larger than for crystalline ferromagnetic

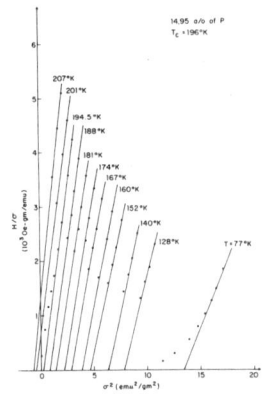

Fig. 2. $H/\sigma - \sigma^2$ plot for amorphous Ni-P with 14.95 a/o P.

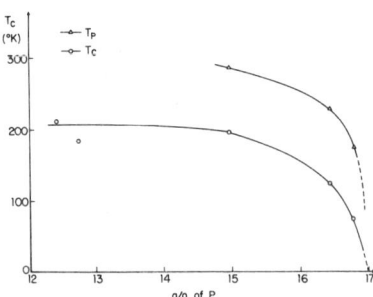

Fig. 3. Compositional dependence of T_c and T_p.

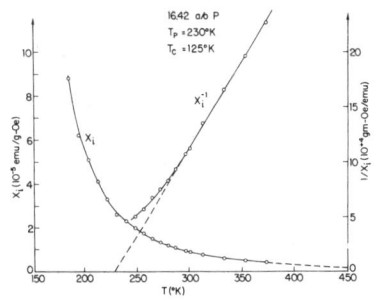

Fig. 4. Temperature dependence of χ_i and $1/\chi_i$ above T_c for amorphous Ni-P with 16.42 a/o P.

materials. This behavior is a unique feature of amorphous ferromagnetic material and is due to its strong magnetic short range order relative to its long range order.

C. Compositional Dependence of Magnetization at Low Temperature

The specific magnetization was measured as a function of composition at T = 77°K and H = 7.6 KOe which was still insufficient to achieve saturation. σ(7.6 KOe, 77°K) decreases linearly with composition between 12-17 a/o P as shown in Fig. 5. Above 17 a/o P, the samples were paramagnetic at 77°K, consistent with the results reported in the preceding section. The theoretical line was calculated assuming that amorphous nickel has the same electronic structure as its crystalline state, i.e., magnetic moment per atom n = 0.6 Bohr magneton, and spectroscopic splitting factor g = 2, independently of composition. Also the rigid band model, with all 5 valence electrons of P transferred to fill the Ni d-band was assumed. Linear extrapolation of the experimental data to zero P content indicates that $\sigma \simeq 40$ emu/gm which is only about a reduction to 2/3 that of pure crystalline Ni. This result is consistent with those of recent F.M.R. studies on amorphous nickel thin films at low temperatures.[15]

D. Amorphous Ni-P Alloys as Itinerant Ferromagnets

It is interesting to consider whether the ferromagnetism of amorphous Ni-P is consistent with the localized moment or itinerant model. According to Rhodes and Wahlfarth[16], $q_c/q_{s1} \simeq 1$ for the first type of ferromagnet while $q_c/q_s > 1$ for the second type, where q_c is the number of magnetic carriers per atom deduced from the Curie-Weiss constant, and q_s is the number of

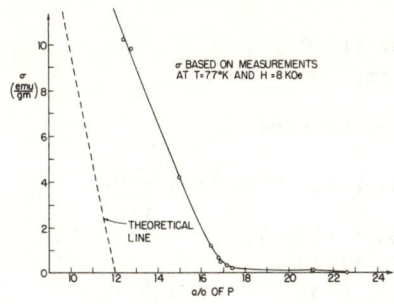

Fig. 5. Compositional dependence of low temperature magnetization at high fields.

carriers deduced from the saturation moment at 0°K. Calculation of q_c, q_s and their ratios for two samples are listed below. q_s is based on measurements at 7.6 KOe and 77°K.

a/o of P	q_c	q_s	q_c/q_s
14.95	0.1±0.02	0.041	2.4±0.5
16.42	0.13	0.011	12

The results are most consistent with the itinerant model. Moreover, the large values of q_c/q_s, particularly of the sample with 16.42 a/o P, suggests very weak itinerant ferromagnetism like single phase $ZrZn_2$[17,18] and compositionally disordered Cu-Ni alloys with compositions near 60 a/o of Cu[18,19]. According to Edwards and Wohlfarth[18], the following relation holds for weak itinerant ferromagnets:

$$\sigma(H,T)^2 = \sigma(0,0)^2 \{1 - (T/T_c)^2 + 2\chi_o \frac{H}{\sigma(H,T)}\}$$

From our H/σ-σ^2 plots, one of which is displayed in Fig. 2, the isotherms are approximately parallel except for those measurements at low fields where domain rotation still occurs. The intercepts on σ^2-axis, $I(T)$, are plotted against T^2 for the sample with 16.42 a/o P, in Fig. 6. A linear relation was found to hold below T_c. It meets the straight line joining two data points above T_c on T^2-axis at about the value of T_c^2. The two linear sections have different slopes, presumably due to short range order above T_c[18]. The following values were obtained from this sample.

$\sigma(0,0) = 0.90$ emu/gm $\qquad \chi_o = (0.7\pm0.3)\times10^{-4}$ emu/gm-Oe

The $I(T)$-T^2 plot for the sample with 14.95 a/o P, however, showed a small but clear curvature toward the T^2-axis at $T < T_c$.

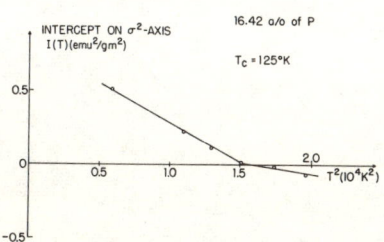

Fig. 6. $I(T)$-T^2 plot for amorphous Ni-P with 16.42 a/o P.

REFERENCES

1. A.I. Gubanov, Fiz.Tverd. Tela 2, 502 (1960).
2. B.G. Bagley and D. Turnbull, Bull. Am. Phys. Soc. 10, 1101 (1965).
3. S. Mader and A.S. Nowick, Appl. Phys. Letters 7, 57 (1965).
4. C.C. Tsuei, G. Longworth, and S.C.H. Lin, Phys. Rev. 170, 603 (1968).
5. J.G.M. de Lau, J. Appl. Phys. 41, 5355 (1970).
6. R. Hasegawa, J. Appl. Phys. 41, 4096 (1970).
7. A.W. Simpson and D.R. Brambley, Phys. Stat. Sol. (b) 49, 685 (1972).
8. G.S. Cargill and R.W. Cochrane in Amorphous Magnetism, edited by H.O. Hooper and A.M. de Graaf (Plenum, N.Y., 1973), P. 313; Proceedings of Conference on "Disordered Metallic Systems" to be published by Journale de Physique.
9. D. Pan and D. Turnbull, Bull. Am. Phys. Soc. [11] 18, 771 (1973); to be published in J. Appl. Phys.
10. A. Brenner, Electrodeposition of Alloys (Academic Press, New York and London, 1963) Vol. II, P. 457.
11. G.S. Cargill, J. Appl. Phys. 41, 12 (1970).
12. B.G. Bagley and D. Turnbull, J. Appl. Phys. 39, 5681 (1968); Acta Met. 18, 857 (1970).
13. S. Foner, Rev. Sci. Instr. 30, 548 (1959); G. Nowlin, Ph.D. Thesis, Harvard Univ., 1960.
14. K.P. Belov and A.N. Goriaga, Fiz. Metal, i, Metalloved 2, 3 (1956); A. Arrott, Phys. Rev. 108, 1394 (1957).
15. Y. Ajiro, K. Tamura and H. Endo, Phys. Letters 35A, 275 (1971); G. Mather, Jr., Phys. Lett. 38A, 37 (1972).
16. P. Rhodes and E.P. Wohlfarth, Proc. Roy. Soc. 273, 247 (1963).
17. S. Ogawa and N. Sakamoto, J. Phys. Soc. Japan 22, 1214 (1967); S. Foner, E.J. McNiff and V. Sadagapan, Bull. Am. Phys. Soc. 12, 311 (1967).
18. D.M. Edwards and E.P. Wohlfarth, Proc. Roy. Soc. A303, 127 (1968); E.P. Wohlfarth, J. Appl. Phys. 39, 1061 (1968).
19. E.P. Wohlfarth, Proc. Roy. Soc. A195, 434 (1949).

LOW FIELD MAGNETIC SUSCEPTIBILITY OF NOBLE METAL-TRANSITION METAL SPIN GLASS ALLOYS*

V. Cannella
Department of Physics, Wayne State University
Detroit, Michigan 48202

J. A. Mydosh
Institüt fur Festkörperforschung, KFA
517 Jülich 1, West Germany

ABSTRACT

To study the interactions between magnetic moments in spin glass alloys we have completed systematic studies of the temperature dependence of the low field ($\sim$5 oe) magnetic susceptibility $\chi(T)$ of noble metal-transition metal spin glass alloys including AgMn, AuMn, CuMn, AuFe, AuCr, AuCo, and CuFe in the concentration range .1 at.% $\leq C \leq$ 10 at.%. $\chi(T)$ for these alloys is characterized by a sharp peak at a magnetic "freezing" or ordering temperature T_0, and a Curie-Weiss law at higher temperatures, while approaches a non-zero value as $T \to 0$. We find that $T_0 \sim C^m$ where $m \simeq 2/3$, where C is the concentration of magnetic atoms, despite the differences in the clustering characteristics of the various alloys. At present theoretical models are unable to account completely for this behavior.

INTRODUCTION

In recent studies of AuFe,[1] CuMn,[2] and AuMn[2] alloys with concentrations C between 1 and 10 at.% of the magnetic impurity we have found sharp peaks in the temperature dependence of the very low field ($\sim$5 gauss) magnetic susceptibility $\chi(T)$. These sharp peaks in $\chi(T)$ occur at temperatures in good agreement with the magnetic "ordering" temperatures indicated by the onset of hyperfine field splitting in available Mössbauer data.[3,4] The existence of a sharp peak in $\chi(T)$ at well defined temperatures, T_0, coupled with the Curie-Weiss behavior of $\chi(T)$ above T_0 and a non-zero intercept of χ in the limit $T \to 0$, are ordinarily characteristic of antiferromagnetic ordering. Neutron diffraction studies,[5,6] however, have given no evidence for long range antiferromagnetic order in the usual sense, and Mössbauer spectra taken in polarizing fields correspond to a complete distribution of spin orientations.[3] Furthermore, there have been no observed critical anomalies in the specific heat[7] at T_0 to corroborate a cooperative phase transition. Studies of the electrical resistivity ρ have shown an ambiguous

---

*Work supported in part by Air Force Office of Scientific Research under Grant AFOSR-71-2002.

maximum in $d\rho/dT$ near T_0,[8] but no normal critical divergence. The origin, therefore, of the sharp peaks in $\chi(T)$ is not clear. Because these sharp peaks in $\chi(T)$ become rounded in moderate magnetic fields (~500 gauss), earlier susceptibility measurements found broad rounded maxima in $\chi(T)$.[9] These broad maxima in $\chi(T)$, and the absence of critical anomalies in the specific heat and transport properties, gave rise to the "spin glass" model of these alloys.[10] In this model the orientations of the magnetic spins gradually freeze into fixed positions due to the increased importance of RKKY interactions or various anisotropy mechanisms[11] as the temperature is lowered. Klein[12] and others[13] have developed mean field treatments of this model which predict that $\chi(T)$ will exhibit broad maxima whose magnitudes are nearly independent of concentration C and which occur at T_0 proportional to C.

In this paper we report briefly the completion of an extensive investigation of the low field $\chi(T)$ in noble metal-transition metal spin glasses which has expanded the concentration range studied down to 0.1 at.%, and added AuCr and AgMn, and CuFe, along with the less typical AuCo system to the list of alloys studied. Isolated solute atoms are not magnetic in AuCo alloys[14] where the Kondo or spin fluctuation temperatures of both isolated Co atoms, and of Co-Co near-neighbor pairs are high. Thus longer range interactions in AuCo are dependent upon short range interactions for the appearance of a moment on the magnetic impurities.

$\chi(T)$ was measured in an rms field of ~5 gauss using a low frequency (17 Hz-155 Hz) a-c mutual inductance technique. Alloys of AgMn, CuMn, and AuMn were annealed in a vacuum for 24 hrs. at 900°C and slowly cooled. Alloys of AuCo, AuCr, AuFe and CuFe were homogenized in a H_2 atmosphere for 6 hrs. at 850°C, then quenched in ice water and stored in liquid N_2 to randomize the distribution of the solute in the matrix.

EXPERIMENTAL RESULTS

A cross section of $\chi(T)$ curves for lower concentration alloys is shown in Fig. 1. Data points were taken every .1°K near the peaks, and every .2° or .5° elsewhere. The scatter was less than the thickness of the drawn curves.

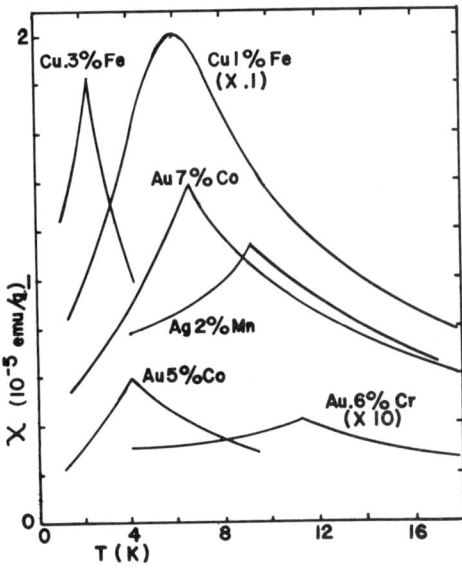

Fig. 1. $\chi(T)$ for various spin glass alloys. Note that the magnitudes of the data for Cu 1% Fe and Au .6% Cr have been multiplied .1 and 10 respectively.

Except for the Cu 1% Fe alloy, the curves show the sharp peak in $\chi(T)$ and the non-zero limit for $\chi(T=0)$ which is characteristic of AgMn, CuMn, AuMn, AuCr, AuFe, and AuCo with .1 at.% $\leq$ C $\leq$ 10 at.%. In AgMn, CuMn, AuMn, and AuFe alloys a Curie-Weiss analysis of $\chi(T)$ above T_0 gives values of the effective moment, p_{eff}, which increase with concentration from a low concentration (single magnetic atom) limit. This indicates a type of superparamagnetic clustering where local atomic arrangements of the magnetic atoms interact ferromagnetically. The details of this behavior depend upon the particular alloy and will be discussed at length in a future publication. In AuCr the absence of superparamagnetic clustering indicates that neither local nor longer range interactions are ferromagnetic.

Several obvious effects of superparamagnetism were observed. Values of the paramagnetic Curie-Weiss θ were found to be close to zero for more dilute alloys (C $\leq$ 1 at.%), but took on small positive values ($\theta \leq$ 10°K) for alloys where superparamagnetism was evident. The ratio of χ_0, the limit of χ as $T \to 0$, to χ_{max} was $\simeq$ .6 for more dilute alloys while χ_0/χ_{max} decreased as p_{eff} increased. The obvious rounding of the peak in $\chi(T)$ for Cu 1% Fe was the extreme example of the slight rounding found in higher concentrations of other alloys[1,2] as superparamagnetic clustering became large. This reflects the excessive clustering due to the poor solubility of Fe in Cu. As a result Cu 1% Fe was the highest concentration of this alloy studied.

The concentration dependence of T_0 found for these alloys is shown on a log-log plot in Fig. 2. (The data points for AuFe[1] and AuMn[2] have been published previously and are omitted here because they overlap the curves for CuMn and AgMn respectively.) For all of these alloys except AuCo we find $T_0 \propto C^m$ where m $\simeq$ 2/3. This $C^{2/3}$ dependence disagrees with the mean field theories[12,13] which predict $T_0 \propto C$. At first AuCo would seem an exception to the $C^{2/3}$ dependence of T_0, but the complications of moment formation on a Co atom in Au must be considered. It has been shown by Boucaï and co-workers[14] that neither isolated Co atoms nor first near neighbor pairs of Co atoms are magnetic in the Au matrix. However, first

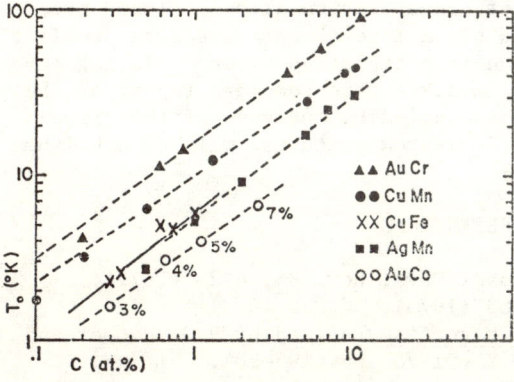

Fig. 2. T_0 vs. C for AuCr, CuMn, AgMn, CuFe, and AuCo. The points for AuCo are plotted for the concentration of magnetic Co atoms and the actual atomic concentration of Co is written beside each point.

near neighbor groups of three or more Co atoms are magnetic. The solubility of Co in Au is adequate to assure a random distribution of Co atoms in the Au matrix for carefully quenched samples. Assuming a random distribution of Co atoms in the Au matrix the concentration of actual magnetic Co atoms can be calculated by finding the statistical probability of Co atoms occurring in first near neighbor clusters of three or more. Using the same technique as in previous results for face centered cubic lattices[15] we find the following conversions from the atomic concentration of Co to the concentration of magnetic atoms in AuCo alloys: 2 at.% → .09%, 3 at.% → .3%, 4 at.% → .64%, 5 at.% → 1.1%, 7 at.% → 2.4%. Fig. 2 shows that by using the actual concentration of magnetic Co atoms in AuCo alloys, T_o follows the same $C^{2/3}$ dependence found for other alloys. The broad validity of this $C^{2/3}$ power law for T_o is particularly surprising since the magnetic species in the alloys vary from single magnetic atoms in more dilute alloys to superparamagnetic clusters which are estimated to average twenty or more magnetic atoms for higher concentrations of AuMn and AuFe.

The relatively large values of p_{eff} found for Cu 1 at% Fe alloys support the fact that considerable clustering of Fe atoms is occurring even in this carefully quenched 1% sample. Since the concentrations of the CuFe alloys are nominal and have not been analyzed, the scatter in the T_o vs. C curve is probably due to small errors in the concentration.

CONCLUSIONS

The similarities in the behavior of the low field $\chi(T)$ for these noble metal-transition metal spin glass alloys indicate that similar freezing phenomena or longer range interactions occur despite the differences in the short range clustering characteristics found for the various alloys. There exist sharp peaks in $\chi(T)$ at freezing or ordering temperatures T_o proportional to $C^{2/3}$, where C is the concentration of magnetic atoms. These temperatures T_o, coincide with the onset of hyperfine field splitting found from Mössbauer studies.[3,4] $\chi(T)$ follows a Curie-Weiss Law for $T > T_o$, with $\chi(T=0)/\chi_{max} \simeq .6$ for lower concentration alloys. These properties indicate that there exists a type of spin freezing peculiar to disordered alloys which cannot be explained by any existing mean field theory for spin glass alloys. A more complete report of the results of these investigations, including analyses of the effective moments and paramagnetic Curie temperatures, will be published in the future.

REFERENCES

1. V. Cannella and J.A. Mydosh, Phys. Rev. B6, 4220 (1972); A.I.P. Conf. Proc. 10, 785 (1973).
2. V. Cannella in Amorphous Magnetism, edited by H.O. Hooper and A.M. de Graaf (Plenum, N.Y., 1973) pp. 195-206.

3. For $\underline{Au}Fe$ Mössbauer studies and references see R.J. Borg, Phys. Rev. $\underline{B1}$, 349 (1970); B. Window, Phys. Rev. $\underline{B6}$, 2013 (1972).
4. For $\underline{Cu}Mn$ Mössbauer studies and references see B. Window, J. Phys. $\underline{C3}$, 922 (1970).
5. P. Wells and J.H. Smith, J. Phys. F: Metal Phys. $\underline{1}$, 763 (1971).
6. H. Sato, S. Werner, R. Kikuchi, Strasbourg Conference, 1973, and private communication.
7. J.E. Zimmerman and F.E. Hoare, J. Phys. Chem. Solids $\underline{17}$, 52 (1960); J. Souletie and R. Tournier, J. Low Temp. Phys. $\underline{1}$, 95 (1969).
8. J.A. Mydosh, M.P. Kawatra, J.I. Budnick, T.A. Kitchens, and R.J. Borg, in Proceedings XIth Conference on Low Temperature Physics (St. Andrews University Press, St. Andrews, Scotland, 1969), Vol. 2, p. 1324.
9. O.S. Lutes and J.S. Schmit, Phys. Rev. $\underline{134}$, A676 (1964); J.L. Tholence and R. Tournier, J. Phys. (Paris) 32, C1-211 (1971).
10. See, for example, B.R. Coles, in Amorphous Magnetism, edited by H.O. Hooper and A.M. de Graaf (Plenum, N.Y., 1973) pp. 169-184.
11. J.S. Kouvel, J. Phys. Chem. Solids $\underline{24}$, 795 (1963). See also R. Harris, M. Plischke, and M.J. Zuckermann, Phys. Rev. Lett. $\underline{31}$, 160 (1973); J.L. Tholence and R. Tournier, Strasbourg Conference 1973, to be published in J. Phys. (Paris), Jan. 1974.
12. M.W. Klein, Phys. Rev. $\underline{136}$, A1156 (1964); $\underline{173}$, 552 (1968); $\underline{188}$, 933 (1969).
13. K.H. Bennemann, J.W. Garland, and F.M. Mueller, Phys. Rev. Lett. $\underline{23}$, 1503 (1969); J. Souletie and R. Tournier, J. Phys. (Paris) $\underline{32}$, C1-172 (1971).
14. E. Boucaï, B. Lecoanet, J. Pilon, J.L. Tholence, and R. Tournier, Phys. Rev. $\underline{B3}$, 3834 (1971).
15. K.J. Duff and V. Cannella, in Amorphous Magnetism, edited by H.O. Hooper and A.M. de Graaf (Plenum, N.Y., 1973) pp. 207-214.

VARIATIONAL DEFINITION OF ANDERSON'S LIGAND FIELD STATES*

Nilton P. Silva[†] and T.A. Kaplan
Department of Physics, Michigan State University
East Lansing, Michigan 48824

ABSTRACT

We point out that Anderson's Hartree-Fock definition of his ligand field states, even as modified by Fuchikami, is unsatisfactory. We propose a new definition which is variational and which overcomes the objections raised, and we discuss our initial exploration of its consequences.

This paper is an examination of the concept of the ligand field problem which is crucial to Anderson's theory of exchange in insulators.[1] It has been noted[2,3] that on theoretical grounds Anderson's definition, which involved the Hartree-Fock approximation leads to incorrect results; we point out that Fuchikami's modification[2] is also unsatisfactory. We propose a new definition which overcomes the objections raised. This definition is motivated to a considerable extent by the work of Mattheiss,[4] the objective being to reduce the arbitrariness of his choice of potential. Our definition represents a solution to the long-standing problem of how to go beyond the Hartree-Fock approximation in a variational and controlled way, at least such that the results are correct in the narrow-band region. The relation of our proposal to Hubbard's derivation of the Hubbard Hamiltonian[5] and a possible relationship to the problem of understanding the behavior of NMP-TCNQ[6] has been discussed elsewhere.[7]

The ligand field problem was defined[1] as that of determining "the exact one-electron states" for the crystal as a whole, <u>excluding</u> the effects of exchange interactions between the magnetic ions. These states are Bloch functions $|k\rangle$ plus their energies ε_k; by their definition they are to be "non magnetic" (spin independent). From these states one is to construct "the exact Wannier functions" and then do perturbation theory for the full crystal many-electron problem considering interionic overlap to be small. This procedure leads, according to Anderson, to the dominant contribution to the exchange integral being the "kinetic exchange", $-2b^2/U$, where b is the transfer integral calculated from the above ε_k, and U is an appropriate energy.[1] Furthermore Anderson defined $|k\rangle$ and ε_k ("most exactly") as eigenfunctions and energies of the Hartree-Fock (HF) operator in which the spins of all the magnetic ions in the crystal are parallel.

Anderson motivated this HF definition by saying that certain desirable properties are expected to follow (for example the nonmagnetic electrons in the system, e.g. from F^- ions, can then be treated as core electrons whose wavefunctions will not change very much with magnetic excitations; again, the perturbation theory is rapidly convergent when use is made of "the exact localized functions"). While there is an intuitive appeal to this, we note that these desirable properties were not shown[1] to occur.

Recently it was explicitly (and independently) shown[2,3] that something was wrong with Anderson's definition in certain 2-site models. In the simpler model[3] having two electrons and two spatial orbitals (like H_2), the low-lying singlet-triplet splitting according to Anderson's prescription was found to differ (within the kinetic exchange) from the exact results by terms of the same order in overlap as those kept. It was found[2,3] that if the electron spins associated with localized states in the HF operator were set anti-parallel, then the correct splitting resulted. Contrary to Fuchikami,[2] one cannot conclude from this special model that the HF operator for the antiferromagnetic state is correct in general (one can conclude only that using parallel spins is not correct in general). In fact, in certain cases (e.g. MnO) it is impossible to have all the nearest-neighbor pairs be antiparallel. Thus it would be impossible there to satisfy Anderson's requirement that the exact HF eigenstates be solutions for the crystal as a whole, simultaneously with this antiparallelism of all near-neighbor pairs. Furthermore in any such HF scheme, the states do not satisfy Anderson's requirement of spin-independence. For example, even in the parallel-spin case the eigenvalue for a down spin $\varepsilon_{k\downarrow} \equiv (k\downarrow|F(\uparrow\uparrow\uparrow...)|k\downarrow)$ is not equal to $(k\uparrow|F(\uparrow\uparrow\uparrow...)|k\uparrow) \equiv \varepsilon_{k\uparrow}$, the eigenvalue for an up-spin ($\varepsilon_{k\downarrow}-\varepsilon_{k\uparrow} \sim$ intra-atomic Coulomb interaction); here $F(\uparrow\uparrow\uparrow...)$ is the HF operator for the ferromagnetic configuration. We conclude that the HF definition, despite its being variational, is unsatisfactory.

We then must start anew if we wish to find a mathematical definition of "the ligand field problem" which satisfies many of Anderson's requirements. The related work of Mattheiss[4] provides us with additional motivation. He made band calculations[4] using Slater's $\rho^{1/3}$ exchange for a variety of materials including the metal ReO_3 and the insulator $KNiF_3$. Whereas his one-electron theory made sense in terms of experimental properties of the metal, it did not[4] for the insulator; he then added a Hubbard type of Coulomb repulsion[5] U within the 3d-band, thereby causing the drastic types of changes needed. Although this was sensible in the context of his $\rho^{1/3}$ model, it is ad hoc (Slater's exchange was derived for nearly free electrons, whereas U was greater than the 3d-bandwidth[4]). To overcome the arbitrariness of this one-electron operator, it would be natural to define the one-electron states in this context in the framework of the variationally best trial Hamiltonian of the form (tentatively)

$$\tilde{H}_{ten} = \sum_{\nu ij\sigma} b_{\nu i,\nu j} \tilde{c}^{\dagger}_{\nu i\sigma} \tilde{c}_{\nu j\sigma} + \sum_{\nu i} U_\nu \tilde{N}_{\nu i\uparrow} \tilde{N}_{\nu i\downarrow} \quad ; \tag{1}$$

ν is a band index, i,j label atomic sites, $\tilde{c}^{\dagger}_{\nu i\sigma}$ creates an electron in a Wannier function $\tilde{w}_{\nu i}$, with spin σ, $\tilde{N}_{\nu i\sigma} = \tilde{c}^{\dagger}_{\nu i\sigma}\tilde{c}_{\nu i\sigma}$, and $b_{\nu i,\nu j}$, U_ν are variational parameters; further the $\tilde{w}_{\nu i}$ are taken as orthonormal linear combinations of a basis set of one-electron functions (which can be chosen arbitrarily), the coefficients also being variational parameters. In other words, we take $\tilde{H}_{ten}$ of the form finally assumed by Mattheiss. But instead of choosing the band-structure Hamiltonian arbitrarily, and presumably in a way associated with nearly-free electrons, we choose it (namely $\tilde{H}_{ten}^{(1)}$, the one-electron operator in (1)) variationally, treating it on the same footing as the U-terms. The variational principle used is[8]

$$F[\tilde{H}] \equiv \mathrm{tr}\ \tilde{\rho}(H-\tilde{H}) - kT\ \ln\ \mathrm{tr}\ \exp[-\beta(\tilde{H}-\mu N)] \geq F[H] \qquad (2)$$

with $\qquad \tilde{\rho} \equiv \exp[-\beta(\tilde{H}-\mu N)]/\mathrm{tr}\ \exp[-\beta(\tilde{H}-\mu N)] \qquad ,\qquad (3)$

for all Hermitean $\tilde{H}$. Here H is the "exact" Hamiltonian which is being approximated by $\tilde{H}$, μ and $T=1/k\beta$ being the chemical potential and temperature. By (2), the best estimate of the free energy attainable from $F[\tilde{H}]$ is its minimum, and this is the criterion for determining the "best" parameters in (1).

The resulting one-electron Hamiltonian $\tilde{H}_{ten}^{(1)}$ is by definition variational, it is a property of the crystal as a whole and it is non-magnetic. Further we will see that (1) gives the correct behavior for single band models (a special example of which is that of ref. 3 referred to above) provided the kinetic exchange dominates the potential exchange.[9] Therefore, for such models the (one-electron) eigenstates of $\tilde{H}_{ten}^{(1)}$ could provide a definition of the ligand field states which represents a major improvement over the Hartree-Fock definition.[1,2] However, the limitation to cases with dominant kinetic exchange is unsatisfactory (there are cases where this is expected to be violated[10]). Further, for more than one narrow band (e.g. a "degenerate" band such as a 3d band in NiO), (1) will be unsatisfactory on the additional ground that Hund's rule would be violated![11]

These difficulties should be overcome by adding potential exchange terms to (1); also generalizing the one-electron terms (with no additional serious increase in difficulty) we obtain as our trial Hamiltonian

$$\tilde{H} = \Sigma\ b_{\nu i,\mu j}\tilde{c}_{\nu i\sigma}{}^{\dagger}\tilde{c}_{\mu j\sigma} + \Sigma\ U_{\nu}\tilde{N}_{\nu i\uparrow}\tilde{N}_{\nu i\downarrow}$$
$$- \tfrac{1}{2}\ \Sigma'_{\nu i,\mu j}\ \Sigma_{\sigma\sigma'}\ U_{\nu i,\mu j}\ \tilde{c}_{\nu i\sigma}{}^{\dagger}\tilde{c}_{\nu i\sigma'}\tilde{c}_{\mu j\sigma'}{}^{\dagger}\tilde{c}_{\mu j\sigma} \qquad (4)$$

where $U_{\nu i,\mu j}$ are additional variational parameters which physically appear as effective potential exchange matrix elements. It seems to us that it would be sufficient for most purposes to put equal to zero all parameters connecting magnetic to non-magnetic bands. However we prefer to regard this suggestion as tentative since we have so far studied only the single band problem, taken for simplicity of discussion to be an s-band, to which we limit ourselves in the remainder of this paper.

For the single band model, the so-called exact Hamiltonian is

$$H = \Sigma\ h_{ij}c_{i\sigma}{}^{\dagger}c_{j\sigma} + \tfrac{1}{2}\ \Sigma\ v_{ijkl}c_{i\sigma}{}^{\dagger}c_{j\sigma'}{}^{\dagger}c_{l\sigma'}c_{k\sigma} \qquad (5)$$

where h is the kinetic energy plus the electron Coulomb interaction with the ions, and h_{ij} are the matrix elements of h between Wannier functions; v_{ijkl} are the matrix elements of the electron-electron Coulomb interactions. And the trial Hamiltonian becomes

$$\tilde{H} = \Sigma\ b_{ij}\tilde{c}_{i\sigma}{}^{\dagger}\tilde{c}_{j\sigma} + U\Sigma\tilde{N}_{i\uparrow}\tilde{N}_{i\downarrow} - \tfrac{1}{2}\ \Sigma_{\sigma\sigma'}\ \Sigma'_{ij}U_{ij}\tilde{c}_{i\sigma}{}^{\dagger}\tilde{c}_{i\sigma'}\tilde{c}_{j\sigma'}{}^{\dagger}\tilde{c}_{j\sigma} \qquad (6)$$

The Bloch functions $\psi_k(r)$ are physically invariant for a single band (aside from a phase factor $e^{i\gamma k}$), but the Wannier functions,

$$w_i(r) \equiv N_s^{-\frac{1}{2}} \Sigma_k \exp(-ik \cdot R_i + i\gamma_k) \psi_k(r) \qquad (7)$$

are not invariant, because of γ_k. Here R_i is the position of site i, N_s the number of sites, and Σ_k goes over a Brillouin zone. Then we should ask for the best w_i in the sense of (2), leading to many variational parameters. However it is natural to demand that w_i, in common with the atomic 1s function a_i, be real and invariant under inversion through R_i; and also to demand that w_i go continuously to a_i as the interatomic separation goes to infinity. Then one can prove that γ_k is constant, leaving no physical arbitrariness in the w_i, and reducing the parameters only to b_{ij}, U and U_{ij}.

To find the best $\tilde{H}$ we must solve the stationarity equation $(\partial \tilde{F}/\partial \lambda_i)=0$, where λ_i stands for the variational parameters. For small interatomic overlap, b_{ij} and U_{ij} for $i \neq j$ are expected to be small; thus one can use expansion techniques. We have investigated two cases, in both of which the average number of electrons $\bar{N}$ is one per site, namely $U_{ij}=0$ and $U_{ij} \neq 0$; we will limit our detailed discussion here to calculations made in the grand canonical ensemble, at low $T(<<|b_{ij}|/k)$, with nearest neighbor b_{ij} only.

In the first case ($U_{ij}=0$), for arbitrary N_s we restrict the traces in $\partial \tilde{F}/\partial \lambda_i$ to the eigenstates G of $\tilde{H}$ resulting from the splitting of the (low-lying) singly-occupied-site eigenstates of the atomic limit of $\tilde{H}$. Here we use second order degenerate perturbation theory in order to handle the operator $\exp(-\beta \tilde{H})$ since the Hubbard Hamiltonian $\tilde{H}$ is not diagonalizable in practice for general N_s. Although this degenerate perturbation theory leads in turn to another insoluble problem, that of the Heisenberg spin-Hamiltonian, we do not need the eigenvalues and eigenstates of the latter - for carrying out the variational calculation it turns out that it is sufficient to know that the eigenvalues of the Hubbard Hamiltonian corresponding to G are those of the Heisenberg Hamiltonian with $J_{ij} = -2b_{ij}^2/U$. The results are:

$$U = (V-v)(1+X_p/X_k) \quad , \quad b_{ij} = t_{ij}(1+X_p/X_k) \qquad (8)$$

where $V=v_{ii,ii}$, $v=v_{ij,ij}$, $X_p=v_{ij,ji}$, $X_k=-2t_{ij}^2/(V-v)$,

$$t_{ij} \equiv h_{ij} + v_{ii,ji} + \Sigma_{l \neq i,j} v_{il,jl} \quad ,$$

i and j being nearest neighbors. These results hold provided $|X_k| > X_p$. Also b_{ij} does not follow from this procedure due to the restriction of the traces.

The low T results are seen to give the Heisenberg exchange interaction, appropriate to (6) with $U_{ij}=0$, of $-2b_{ij}^2/U = X_k + X_p$. This interaction is identical to the nearest neighbor exchange interaction appropriate[3] to the <u>exact</u> Hamiltonian (5), despite the lack of U_{ij} in $\tilde{H}$. Hence we have accomplished Anderson's objective of defining a <u>non-magnetic</u> one-electron Hamiltonian whose transfer integral leads via the kinetic exchange to the correct low-lying exchange splittings.

However, the above approach fails to give a physically acceptable approximation to the properties of H when $X_p > |X_k|$. This failure suggests that we try to improve our approximation by including U_{ii} terms in $\tilde{H}$ (as in (2)). We have begun an investigation of this problem by considering the case of $N_s=2$. We found that for low T (and small overlap)

$$b_{12} = \frac{V}{V-v}t_{12} \quad , \quad U_{12} = X_p - \frac{v}{V-v}X_k \quad , \quad b_{11} = h_{11}+v \quad , \quad U = V . \qquad (9)$$

These results lead to $U_{12}-2b_{12}^2/U = X_p+X_k$ so that the approximate exchange splitting is again identical to the exact splitting but now for all $X_p/|X_k|$. We point out that (9) could not be obtained uniquely by following the same procedure used for $U_{ii}=0$, since the restriction of the trace discussed there now leaves not only b_{11} undetermined, but due to the additional parameter U_{ii} it also leaves the others not uniquely determined. The results (9) were obtained by including in the traces the first excited states other than G, namely those for which $N=N_s\pm1$. It is interesting to try to find an effective one-electron potential $W(r)$ such that $b_{ij}=(w_i|p^2/2m +W|w_j)$. We have not been able to do that; but we note that the diagonal element is the same as that ($b_{ii}=h_{ii}+v$) obtained with the Wigner-Seitz potential,[1] W_{ws}. However the off-diagonal element does not appear to be that of W_{ws}.

Other implications of the present work will be discussed elsewhere. We are also considering the generalization of the above single-band results to arbitrary N_s, and to models which have more than one band (e.g. cases in which both magnetic and non magnetic ions are present).

* Work supported in part by NSF Grant GH-34565
† Permanent address, Depto de Fisica-ICEX-UFMG-Belo Horizonte, MG, Brazil.

REFERENCES

1. P.W. Anderson, Solid State Physics 14, 99 (1963)
2. N. Fuchikami, J. Phys. Soc. Japan 28, 871 (1970)
3. T.A. Kaplan, unpublished lecture notes for a course in Magnetism given at Michigan State University in 1971 and 1972.
4. L. Mattheiss, Phys. Rev. B2, 3918 (1970)
5. J. Hubbard, Proc. Roy. Soc. (London) A276, 238 (1963)
6. T.A. Kaplan, Bull. Am. Phys. Soc. 18, 399 (1973)
7. Nilton P. Silva and T.A. Kaplan, Bull. Am. Phys. Soc. 18, 450 (1973)
8. A. Huber, in Mathematical Methods in Solid State and Superfluid Theory, Clark & Derrick, Eds. (Plenum Press, Inc. N.Y., 1968)
9. See Ref. 1 for this terminology
10. Nai Li Huang Liu, R. Orbach, AIP Conf. Proc. No. 10, Magnetism and Magnetic Mat'ls (1972) p. 1238
11. Hund's rule was implicitly assumed in ref. 4.

POLAR SINGLET-GROUND-STATE:
APPLICATION TO WURSTER'S BLUE PERCHLORATE*

Robert A. Bari
Physics Department, SUNY, Stony Brook, N.Y. 11790

ABSTRACT

Wurster's Blue Perchlorate (TMPD-ClO$_4$) undergoes a first-order phase transition at $T_c \simeq 190°$K. The susceptibility, χ, increases with T but then exhibits a jump at T_c. Above T_c, χ has a Curie-Weiss behavior. Kommandeur and coworkers[1] suggested that the system can be characterized as having a singlet-ground-state with alternating TPMD^o and TMPD^{++} ions. The system also shows a doubling of the unit cell below T_c. A detailed model of such a transition has been given by us earlier[2]. Here we calculate χ for the model and compare it with experiment. The model yields either a first- or second-order transition depending on the Coulomb repulsion, I_1, between the two holes on TMPD^{++}. For I_1 such that the transition is second-order, the theory fits the data well except near T_c. On the other hand, for I_1 chosen to fit the jump in χ at T_c, the theory does not fit as well away from the transition region.

---

*Work supported in part by the National Science Foundation, Grant No. H040884

1) J. Kommandeur et al, J. Chem. Phys. 47, 391, 395, 401, 408 (1967)

2) R. A. Bari, Phys. Rev. B 3, 2662 (1971).

PERSISTENCE OF EXCHANGE SPLITTING
IN METALLIC FERROMAGNETS ABOVE T_c*

J. B. Sokoloff
Physics Department, Northeastern University
Boston, Massachusetts 02115

Recently Mook, Lynn, and Nicklow observed using inelastic neutron diffraction that the point at which the spinwave mode enters the Stoner mode continuum does not come down in energy as the temperature is raised through T_c, as Stoner-Hartree-Fock theory would predict! The photoemission work of Pierce and Spicer supports this view[2]. In this talk, it is shown that the standard RPA one electron self-energy, used in spin fluctuation or paramagnon theory, leads to a splitting of the bands due to short range order of a ferromagnetic metal just above T_c or a nearly ferromagnetic metal near $T = 0$, no matter how weak the electron-electron exchange interaction. This implies that the occurrence of a band splitting which does not go to zero in a metallic ferromagnet as T goes through T_c is not surprising result, and in fact, it should be the usual occurrence. Although the RPA should only be valid for weak interactions, since work on the Hubbard model implies a splitting[3], of the bands exists for all temperatures for sufficiently strong interaction, it is not surprising that the Stoner mode continuum edge does not come down as T goes through T_c. It is also shown that below T_c the conventional Stoner spin split bands go continuously into the bands that occur above T_c. The susceptibility is also calculated in this approximation and found to have spin wave poles just above T_c. More details are contained in reference 4.

References

1. H. A. Mook, J. W. Lynn and R. M. Nicklow, Phys. Rev. Lett. 30, 556 (1973).
2. D. T. Pierce and W. E. Spicer, Phys. Rev. Lett. 25, 581 (1970) and Phys. Rev. B6, 1787 (1972).
3. J. Hubbard, Proc. Roy. Soc. A276, 738 (1963) and Proc. Roy. Soc. A281, 401 (1964).
4. J. B. Sokoloff, Phys. Rev. Lett. 31, 1421 (1973).

*Work supported by ONR contract number N00014-68-A-0207-0004

FINITE TEMPERATURE PROPERTIES OF THE HUBBARD MODEL: PHASE SEPARATION*

P. B. Visscher[†]

Physics Department and Materials Research Laboratory,
Univ. of Illinois, Urbana, Illinois 61801

ABSTRACT

A new non-perturbational method involving no <u>ad hoc</u> approximations has been developed for calculating the finite temperature properties of lattice systems. An exact system of coupled first-order differential equations is written for the many-body correlations as functions of temperature. These equations are then integrated numerically down from infinite temperature. Numerical results for the Hubbard model are presented. These support the conventional assumption of antiferromagnetism for a half-filled band, but suggest an anomalous separation into coexisting antiferromagnetic and ferromagnetic phases for the non-half-filled band. Independent evidence for the existence of this separation is presented.

INTRODUCTION

In this paper, I describe a new method for calculating properties of many-body systems, and apply it to the Hubbard model.

The method can be derived very simply. We wish to calculate the equilibrium properties of an arbitrary many-body system in the Grand Canonical Ensemble (GCE)[1]. If the second-quantized Hamiltonian of the system is H and the number of particles is variable, we must add a chemical potential term to H, obtaining

$$H_{GCE} = H + \mu N \tag{1}$$

where μ is the chemical potential and N (an operator) is the particle number.

It is well known that the expectation value of any operator A is then

$$\langle A \rangle = \frac{\text{tr } A \exp(-\beta H_{GCE})}{\text{tr } \exp(-\beta H_{GCE})} \tag{2}$$

---

*Supported in part by the NSF under Grant GH-33634.
†Present address: Department of Physics, University of California at San Diego, LaJolla, California 92037.

where $\beta = 1/kT$ is the inverse temperature and tr is the trace (ie. the sum of the diagonal matrix elements of an operator over all possible many-particle states). From the expectation values of such operators any equilibrium property of the system may be obtained.

The new method consists in calculating the expectation values from a system of coupled differential equations with respect to the inverse temperature β. If we observe that

$$(d/d\beta) \, A \, \exp(-\beta \, H_{GCE}) = - A \, H_{GCE} \, \exp(-\beta \, H_{GCE}) \tag{3}$$

we may differentiate Eq. (2) with respect to β. Using Eq. (2) again to express the results in terms of expectation values, we get

$$(d/d\beta)<A> = - <A \, H_{GCE}> + <A> \, <H_{GCE}> \tag{4}$$

The initial conditions (the expectation values at infinite temperature) are in general easily calculated (for $\beta=0$, $<A>=\text{tr } A$). Thus Eq. (4) determines the $<A>$'s for all temperatures.

The exact method whereby this system of equations is integrated numerically for a lattice model (Hubbard model, Ising model, etc.) is described elsewhere.[2] It involves changing variables from the $<A>$'s to cumulants (the apparent infinities on the right in Eq. (4) cancel). Here I will give only some results of applying the new technique to the Hubbard model.

The present method has the great advantage over an ad hoc approximation scheme that by observing the convergence of the solution as more terms are taken in Eq. (4), a reliable estimate of the error may be obtained. Another technique which (in principle at least) has this property involves extrapolating from finite systems.[3] Unfortunately, this seems to be useful only in one dimension.

THE HUBBARD MODEL

The Hubbard model[4] is the simplest model of a system of interacting electrons on a lattice. The electrons are allowed to occupy one s-orbital per lattice site. Their motion is described by the tight-binding Hamiltonian except that their interactions are taken into account through a short-ranged repulsive potential (which acts only between two electrons on the same site). The Hubbard Hamiltonian is thus

$$H = I \sum_{\vec{R}} n_{\vec{R}\uparrow} n_{\vec{R}\downarrow} - t \sum_{\vec{R},\vec{R}';\sigma} c^+_{\vec{R}\sigma} c_{\vec{R}'\sigma} \tag{5}$$

where t is the tight-binding overlap integral (t>0) and I is the intra-atomic Coulomb repulsion.

This Hamiltonian was originally introduced ten years ago as a model for the d-electrons in a transition metal, with the objective of explaining the magnetic behavior (the paradoxical formation of local magnetic moments by otherwise itinerant electrons, presumably due to strong local correlations). Typical values of the parameters in such a metal might be I ≈ 10 volts, t ≈ a few tenths of a volt. The Hubbard model has also been proposed as a model for the metal-insulator transitions which occur in transition-metal oxides and sulfides.[5] It has been suggested[6] that the one-dimensional Hubbard model may explain some of the behavior (both magnetic and metal-insulator transition) of the charge-transfer salt NMP-TCNQ.

In all of these applications of the Hubbard model, the band width is much smaller than the Coulomb repulsion (so t <<I). Except possibly in NMP-TCNQ, we are interested in temperatures lower than the parameters t and I. Thus the region of most physical interest is described by kT < t << I.

NUMERICAL CALCULATION

The present technique is basically local. That is, we calculate expectation values of local operators (products of creation and destruction operators for local orbitals). It therefore must be expected to become intractable at low temperatures, if the system undergoes an ordering transition (inducing long-ranged correlations).

It is worth noting that Hubbard's original calculation[4] is also essentially a high temperature approximation, though he did not so label it.[7] He asserted that it is valid in the two limiting cases: atomic limit (t ≈ 0, I finite) and noninteracting limit (t finite, I ≈ 0), and should therefore be reasonable in between. Closer analysis shows that the real variables involved are βt and βI; thus the two limits correspond to t<<kT, t<<I and I<<kT, I<<t. These are both high temperature limits; it is unclear whether the Hubbard solution should be trusted in the most interesting (kT<t<<I) region.

The present method has the advantage that its accuracy may be easily estimated; it is still good for kT ≈ t. It is possible that the calculation can be carried to much lower temperature by replacing the local operators with Bloch waves. This should reach the degenerate Fermi gas (kT<<t) but ordered phases may still be inaccessible due to critical correlations.

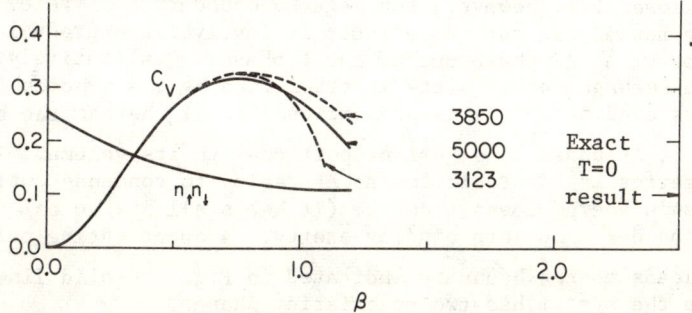

1. Specific heat (C_v) and $<n_\uparrow n_\downarrow>$ for 1-dimensional Hubbard model (half-filled band, t=1, I=4)

I can give here only a few numerical results to indicate what the technique can do. Fig. 1 gives results for the one-dimensional Hubbard model. The three branches of C_v correspond to taking into account different numbers of the coupled equations (4); the number is roughly indicated next to the curve. It is apparent that for $\beta \leq 0.5$, convergence is extremely good. In Fig. 2 the corresponding 3-dimensional result is given.

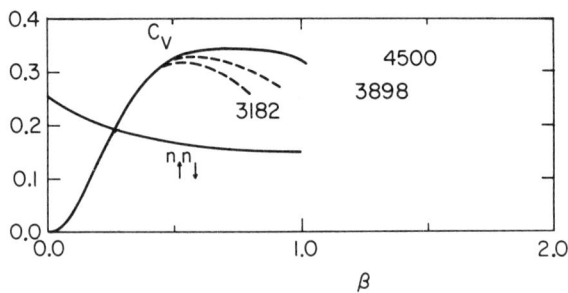

2. Same as Fig. 1, for simple cubic lattice.

PHASE SEPARATION

High temperature calculations for the non-half-filled band[2] (not given here) suggest the possibility of a condensation phenomenon at low temperatures. It is proved in detail elsewhere[8] by zero temperature arguments that a peculiar condensation in fact occurs for t<<I and densities near but not at the half-filled band.[9] Here we can only describe the phenomenon, starting from Nagaoka's exact result that for $t/I \to 0$ slightly off the half-filled band, the ground state is ferromagnetic (F). On the phase diagram (Fig. 3) n is the number of holes in an otherwise half-filled band (N electrons on N atoms). As we increase t/I from zero at fixed n/N the system must cease to be F at some critical value of t/I. Nagaoka suggested the dotted line in Fig. 3 as the boundary, because a spin-wave instability would cause a transition to an antiferromagnetic (AF) state there. If some other effect leads to the instability of the F state at lower t/I, however, the Nagaoka boundary is irrelevant.

I have found just such an effect; it involves the gradual condensation of an AF phase out of the F phase. Qualitatively, it occurs because Nagaoka's F state is stabilized by the n holes (and for n<<N its binding energy is proportional to n) whereas the binding energy of the AF phase[10] is just proportional to its volume. Thus it is energetically favorable for an AF region to condense out: the F region's energy doesn't change (it keeps all its holes) and the AF region has a nonzero binding energy. A quantitative calculation[8] leads to the boundary indicated in Fig. 3 (solid line) above which the system has two co-existing phases.

Clearly phase separation will not occur in a real system; it is a consequence of the Hubbard model's neglect of interatomic

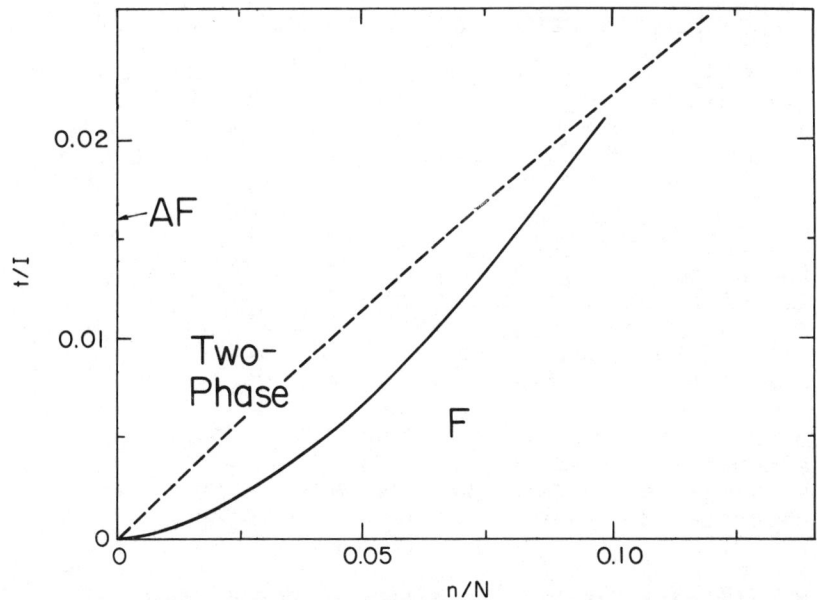

3. Phase diagram for simple cubic Hubbard model at T=0, for small t/I and hole concentration n/N. Dashed line is Nagaoka boundary. System is purely AF on n/N=0 line, F below solid curve, and phase-separated in between.

Coulomb repulsion. Thus for the choices of parameters yielding condensation this model is unrealistic. Although this includes some physically interesting cases, it should be emphasized that it does not include the most interesting case, namely the half-filled band.

REFERENCES

1. K. Huang, Statistical Mechanics (Wiley, New York, 1963).
2. P. B. Visscher, to be published.
3. H. Shiba and P. Pincus, Phys. Rev. B$\underline{5}$, 1966 (1972); D. Cabib and T. A. Kaplan, Phys. Rev. B$\underline{7}$, 2199 (1973).
4. J. Hubbard, Proc. Roy. Soc. (London) A$\underline{276}$, 238 (1963).
5. D. Adler, Rev. Mod. Phys. $\underline{40}$, 714 (1968).
6. A. J. Epstein, S. Etemad, A. F. Garito, and A. J. Heeger, Phys. Rev. B$\underline{5}$, 952 (1972).
7. Related observations have been made by D. M. Esterling and R. V. Lange, Rev. Mod. Phys. $\underline{40}$, 796 (1968).
8. P. B. Visscher, to be published.
9. Y. Nagaoka, Phys. Rev. $\underline{147}$, 392 (1966).
10. P. W. Anderson in Solid State Phys., Vol. 14, ed. F. Seitz and D. Turnbull (Academic Press, N. Y., 1963).

SINGLE SITE APPROXIMATION IN THE HUBBARD MODEL

L. M. Roth, SUNY Albany, Albany, New York 12222

ABSTRACT

We consider a single site approximation for the Hubbard Model for electron correlation in narrow energy bands. An effective field is assumed to operate on every site except one, on which the Coulomb interaction is allowed to operate between electrons. The effective field is evaluated self-consistently in analogy with the coherent potential approximation. Several Green's function decoupling schemes are used, and it appears that the single site approximation does not give saturated feromagnetism in the "Nagaoka limit."

In his classic work on electron correlation in narrow energy bands, Hubbard[1,2] developed several approximations for the one band Hamiltonian given in second quantized form by

$$\mathcal{H} = \sum_i U n_{i\uparrow} n_{i\downarrow} + \sum_{ij} t_{ij} c_{i\sigma}^\dagger c_{j\sigma} \qquad (1)$$

One of these, the alloy analogy, consists of regarding the down spin electrons as fixed on a fraction $n_\downarrow$ of the sites and examining the scattering of up spin electrons by them. Our understanding of this view is enormously increased by more recent work on binary alloys, and in particular the coherent potential approximation[3,4] (CPA) of Soven. From the standpoint of binary alloys the "Hubbard I" approximation[1] corresponds to what has been termed the average T-Matrix approximation[4], but which is more properly viewed as an average locator approximation[5]. The "Hubbard III" result[2] which includes only his "scattering correction" corresponds to the CPA, and is often referred to as the alloy analogy result. While the alloy analogy has had some success in dealing with the metal-insulator transition[2], the above methods do not give good results for magnetic problems. Thus, the Hubbard I result is extremely reluctant to give ferromagnetism[1], and the alloy analogy as extended to binary alloys has been shown not to give a ferromagnetic instability[6].

Improvements upon Hubbard I[7,8,9] have introduced an energy shift which deals approximately with the dynamics of the down spin electrons. However, this author's first result[8] does not include the scattering correction, so that the bands are always split by the correlation. A second result[8] does so only for the case of one reversed spin in an otherwise aligned band. In Hubbard III, there is a result including both the scattering and "resonance broadening" corrections, but which does not appear to include the energy shift.

In the present work we explore the possibility of including Hubbard's scattering correction but treating the dynamics of the reversed spin electrons by a method similar to our previous work. We shall explicitly introduce a single site approximation, so that the work gives further insight into the relation between the CPA and Hubbard's work. We shall discuss the result in several limits.

We use the method of retarded and advanced Green's functions[10], and we refer the reader to references 1, 2 and 8 for definitions and notation. Then, developing our single site approximation, we assume that site 0 has a Coulomb interaction as in Eq.(1), but that the rest of the sites are described by a spin and frequency dependent self energy Σ_σ. We shall determine Σ_σ self consistently by requiring that the one electron Green's function exhibit no scattering from site 0, which is analagous to the CPA condition. Thus, the Hamiltonian is assumed to be given effectively by

$$\mathcal{H} = \sum_i n_{i\sigma} \Sigma_\sigma + \sum_{ij} t_{ij} c_{i\sigma}^\dagger c_{j\sigma} + U n_{o\uparrow} n_{o\downarrow}$$

$$-\Sigma_\uparrow n_{o\uparrow} - \Sigma_\downarrow n_{o\downarrow}$$
(2)

Let us first derive the alloy analogy result by decoupling the Green's function equations of motion. The equation of motion for the one electron Green's function $G_{ij\sigma} = \langle\langle c_{i\sigma}; c_{j\sigma}^\dagger \rangle\rangle$ is

$$\omega G_{ij\sigma} = \delta_{ij} + \Sigma_\sigma G_{ij\sigma} + \sum_\ell t_{i\ell} G_{i\ell\sigma}$$
$$+ \delta_{i0} [U \Gamma_{oj\sigma} - \Sigma_\sigma G_{oj\sigma}]$$
(3)

where $\Gamma_{ij\sigma} = \langle\langle n_{0-\sigma} c_{i\sigma}; c_{j\sigma}^\dagger \rangle\rangle$ is a higher order Green's function. For Γ we have a further equation of motion

$$\omega\Gamma_{ij\sigma} = \langle n_{0-\sigma}\rangle \delta_{ij} + \Sigma_\sigma \Gamma_{ij\sigma} + \Sigma_\ell t_{i\ell} \Gamma_{\ell j\sigma}$$
$$+ \delta_{io}(U-\Sigma_\sigma)\Gamma_{oj\sigma} \qquad (4)$$
$$+ \Sigma_j t_{oj} \langle\langle (c^\dagger_{o-\sigma}c_{j-\sigma} - c^\dagger_{j-\sigma}c_{o-\sigma})c_{i\sigma}; c^\dagger_{j\sigma}\rangle\rangle$$

Suppose we make a decoupling approximation on the last term, replacing the quantity in parenthesis by its average value, which is zero. The two equations then become a closed set, and we can solve for the one electron Green's function $G_{ij\sigma}$. The result clearly separates into a term due to the uniform coherent potential Σ_σ, which Fourier transforms into

$$G^o_{\vec{k}\sigma} = (\omega - \varepsilon_{\vec{k}} - \Sigma_\sigma)^{-1} \qquad (5)$$

where $\varepsilon_{\vec{k}} = \Sigma_j t_{oj} e^{i\vec{k}\cdot\vec{R}}$, and a scattering term which, when set equal to zero, yields the condition

$$\Sigma_\sigma = n_{-\sigma}U + \Sigma_\sigma(U-\Sigma_\sigma)F_\sigma \qquad (6)$$

$$F_\sigma = N^{-1}\sum_{\vec{k}} (\omega - \varepsilon_{\vec{k}} - \Sigma_\sigma - W_{\vec{k}-\sigma})^{-1}. \qquad (7)$$

Where $W_{\vec{k}\sigma}=0$ at this point. This is exactly the alloy analogy result.

We now treat this by our method[8], which is a generalized random phase approximation based on a stationary principle. We choose trial operators A_n (such as $c_{i\sigma}$) and construct energy and normalization matrices

$$E_{nm} = \langle [[A_n, \mathcal{H}], A^\dagger_m]_+\rangle \qquad (8)$$

$$N_{nm} = \langle [A_n, A^\dagger_m]_+\rangle \qquad (9)$$

Then the result is that the best Green's function for this set is

$$\langle\langle A_n; A^\dagger_m \rangle\rangle = [N(N\omega-E)^{-1}N]_{nm} \qquad (10)$$

This is well defined but involves statistical averages which can hopefully be obtained from the Green's function by standard methods[8].

Applying this to our case, we use the operators $c_{i\sigma}$ and $n_{i-\sigma}c_{i\sigma}$ and the Hamiltonian, Eq.(2). Again, the result for the one electron Green's function consists of a coherent potential term given by Eq.(5), and a scattering term whose vanishing now yields a condition for Σ_σ given by Eqs.(6) and (7) with

$$n_\sigma(1-n_\sigma)W_{\vec{k}\sigma} = (2n_{\vec{k}-\sigma}-1)N^{-1}\sum_{\vec{k}} \varepsilon_{\vec{k}} n_{\vec{k}\sigma} \qquad (11)$$

$W_{\vec{k}\sigma}$ is a level shift which is essentially the average excitation energy of the reversed spin in the scattering process. We should remark that we have made an approximation in evaluating the quantity $<(c_{o-\sigma}^\dagger c_{j-\sigma}^\dagger c_{j-\sigma} c_{o-\sigma}) c_{\ell\sigma}^\dagger c_{i\sigma}>$ in that we have neglected the scattering, i.e. we have factored it into up and down spin parts. Here also we write $n_{\vec{k}\sigma} = <c_{\vec{k}\sigma}^\dagger c_{\vec{k}\sigma}> = \int f_o(\omega)\{\frac{1}{\pi}\,\text{Im}\,G_{\vec{k}}(\omega)\}d\omega$ where f_o is the Fermi function.

The result appears to give us a relatively simple method of including the energy shift term[9] which was found to be important in allowing the system to become ferromagnetic. The calculation involves the self consistent solution of two simultaneous equations. We now consider the result in several limits. First, in the dilute limit we find an effective interaction parameter $U_{eff} = U\Delta/(U+\Delta)$ where $\Delta^{-1} = N^{-1}\sum_{\vec{k}}(\varepsilon_{\vec{k}}-2\varepsilon_b)^{-1}$ and where ε_b is the band edge energy. This is closer to the exact Kanamori result $\Delta^{-1} = N^{-1}\Sigma(2\varepsilon_{\vec{k}}-2\varepsilon_b)^{-1}$ than our previous result of $\Delta = 2\varepsilon_b$ but still is not exact in the low density limit. It is interesting to note that in the 2-electron scattering process the present method replaces the energy of one of the electrons by its average value (zero), while our previous result did this for <u>both</u>.

A somewhat disturbing result occurs when we examine the theory for $U\to\infty$ and $n<1$, in which limit Nagaoka[13] showed that under certain conditions the ferromagnetic state should be stable. In Fig.1(a) we sketch the bands for this region assuming all the spins are up. The center of the reversed spin band is pushed up toward the top of the band by the energy shift, as in our previous work. (See Fig. 1 in Ref.8). However, in the present case the width of the band is proportional to $\sqrt{1-n\uparrow}$, a typical CPA result. For a parabolic band edge, this is $\propto(\varepsilon_t-\varepsilon_F)^{3/4}$ which puts the reversed spin band edge below the Fermi level, as indicated in Fig.1(b).

Thus our theory is not consistent with Nagaoka's result in the nearly half filled band region. Our previous work for one reversed spin however[8], together

Fig. 1 a) Density of states
b) reversed spin band edge
versus Fermi energy.

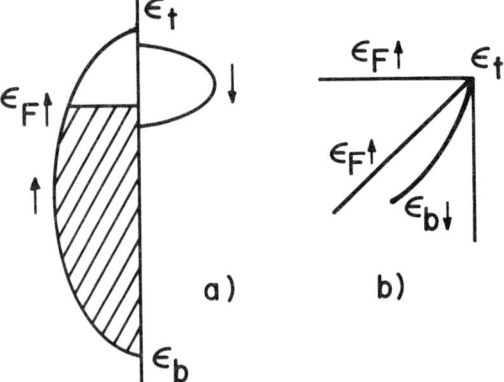

with results for spin wave stability[13] strongly suggests that the saturated ferromagnetic state should be stable to order $(1-n_\uparrow)$ under Nagaoka's conditions[14]. An examination of the theory shows that the discrepancy is a direct consequence of the single site approximation which allows the reversed spin states to be too broad.

The appeal of the alloy analogy and the present single site approximation is that they can be readily incorporated into the CPA for the case of ferromagnetic alloys. Fukayama and Ehrenreich[6] showed that the alloy analogy does not give ferromagnetism for the Hubbard Model. Our improvement does not give saturated ferromagnetism in the Nagaoka limit, although it may yet give partial spin alignment. It appears that it will be necessary to go beyond the single site approximation to treat correlation in ferromagnetic alloys.

1. J. Hubbard, Proc. Roy. Soc. A276, 738 (1963).
2. J. Hubbard, Proc. Roy. Soc. A281, 401 (1964).
3. P. Soven, Phys. Rev. 156, 809 (1967).
4. B. Velicky et.al., Phys. Rev. 175, 747 (1968).
5. L.M. Roth, Phys. Rev. B7, 4321 (1973).
6. H. Fukuyama & H. Ehrenreich, preprint.
7. A.B. Harris & R.V. Lange, Phys. Rev. 157, 295 (1967).
8. L.M. Roth, Phys. Rev. 184, 451; 186, 428 (1969).
9. R. Tahir-Kheli & H. Jarrett, Phys. Rev. 180, 544 (1968).
10. D.N. Zubarev, Soviet Phys. Usp. 3, 320 (1960).
11. J. Kanamori, Prog. Theoret. Phys. (Kyoto) 30, 276 (1963).
12. T. Nagaoka, Phys. Rev. 147, 392 (1966).
13. L.M. Roth, J. Phys. Chem. Solids 28, 1549 (1967).
14. This view is not universally accepted. Nagaoka's proof actually holds only for $1-n = 1/N$. See L.C. Bartel and H.S. Jarrett, the following paper.

MAGNETIC PROPERTIES OF THE HUBBARD HAMILTONIAN INCLUDING THE RESONANCE-BROADENING TERMS: COMPARISON TO THE CPA

L. C. Bartel*
Sandia Laboratories, Albuquerque, NM 87115

H. S. Jarrett †
E. I. du Pont de Nemours and Co., Wilmington, DE 19898

ABSTRACT

In this paper the resonance-broadening terms are included along with the spin-disorder scattering term to study the zero temperature magnetic properties of the Hubbard model Hamiltonian and to compare results with those using the coherent potential approximation (CPA). Although our solution is valid for all strengths of the Coulomb repulsion term, we restrict our discussion here to the strong correlation limit. We find that the inclusion of the resonance-broadening terms (in particular the spin-flip scattering term) leads to qualitative differences with calculations using only the spin-disorder scattering term (CPA), and thus casting serious doubt onto the validity of those simpler calculations.

INTRODUCTION

In recent years it has become quite popular to discuss the Hubbard model Hamiltonian using the coherent potential approximation (CPA) which has been shown equivalent to Hubbard's alloy analogy.[1,2] For example, Hasegawa and Kanamori[3] discussed the Fe-Ni alloy system, Fukuyama and Ehrenreich[4] discussed strong correlations in disordered systems, and Levin and Bennemann[5] discussed the local moments in paramagnetic metals all using the CPA. In the ordinary CPA or alloy analogy only one scattering term, spin-disorder scattering in the magnetic case, is considered while the resonance-broadening terms of Hubbard[1] are neglected. The resonance-broadening terms include the scattering of a σ spin electron into a $-\sigma$ spin hole and the spin-flip scattering.

Recently Jarrett[6] has examined the effect of electron scattering on the ferromagnetic band width for the half-filled band case (one electron per site, n=1). He found that the inclusion of resonance-broadening terms, in particular the spin-flip scattering term, affects the pseudoparticle density of states significantly. In a forthcoming publication[7] we shall examine in some detail the magnetic and thermal properties of the Hubbard Hamiltonian for fractional occupancy where the resonance-broadening terms are included.

The purpose of this paper is to compare the possible ferromagnetic states when all three of Hubbard's scattering terms are in-

---

* Work at Sandia Labs supported by the U.S. Atomic Energy Commission.

cluded to those using the simpler CPA. Although our solution is valid for all strengths of the Coulomb repulsion term, we restrict ourselves in this paper to discussing the strong correlation case and defer to a future publication the weak correlation limit. We find that the inclusion of the resonance-broadening terms has a very significant effect, thus casting serious doubt onto the validity of the simpler CPA solutions at least in the strong correlation limit. From Hubbard's comment,[1] we would expect that in the weak correlation limit the importance of the resonance-broadening terms would become even more pronounced.

HUBBARD HAMILTONIAN PLUS A SYMMETRY BREAKING FIELD

The Zener model as applied to transition metals, referred to here as the modified Zener model (MZM), has been discussed at some length by Arai and Parrinello[8] using the simplest of Hubbard's approximations in which all scattering is omitted and by Bartel[9] using a Hartree-Fock type approximation. In brief, the MZM considers the Hund'a rule coupling of an itinerant electron to a localized spin which is itself a result of Hund's rule coupling. The model Hamiltonian is[8,9]

$$H = \Sigma_{ij\sigma} t_{ij} c_{i\sigma}^+ c_{j\sigma} + \Sigma_{i\sigma} T_0 n_{i\sigma}$$
$$+ I \Sigma_i n_{i\uparrow} n_{i\downarrow} - 2J \Sigma_i \vec{\sigma}_i \cdot \vec{S}_i \quad , \tag{1}$$

where i and j denote lattice sites with t_{ij} the energy of an electron hopping from site j to site i; the C's are the electron operators with $n_{i\sigma} = c_{i\sigma}^+ c_{i\sigma}$; T_0 is the atomic binding energy; I is the Coulomb repulsion energy of electrons of opposite spin localized on the same site; J is the Hund's rule energy coupling parameter; $\vec{\sigma}$ and $\vec{S}$ are the electron and localized spin operators, respectively. Using a molecular field approximation, the last term can be written as $-\Sigma_{i\sigma} \sigma J \bar{S} n_{i\sigma}$, where the magnetization per site is $m = n_+ - n_-$ (+ and - denote up and down spins, respectively) and $\bar{S}$ is the average localized spin per site.

In the atomic limit the energy of a (+) spin electron on an otherwise empty site is $T_0 - J\bar{S}$ while for a site occupied by both (+) and (-) spins the energy is $T_0 + I - J\bar{S}$. For the (-) spin electron the corresponding energies are $T_0 + J\bar{S}$ and $T_0 + I + J\bar{S}$. The electrons in the presence of the molecular field behave as though $J\bar{S}$ represented an externally applied magnetic field.

CALCULATIONS

We have discussed elsewhere[7] our solution to the Hubbard model Hamiltonian using a locator technique[6,10], where we have taken into account all three of Hubbard's scattering terms, spin-disorder, σ spin electron into a $-\sigma$ spin hole, and spin-flip scattering. For the sake of brevity we will only outline our results. The pseudoparticle density of states is given by

$$\rho_\sigma(E) = -\pi^{-1} \operatorname{Im} g_\sigma(E) \quad , \tag{2}$$

where for a single-particle density of states given by

$$\rho^{(0)}(E) = (8/\pi\Delta^2)\left[(\Delta/2)^2 - E^2\right]^{1/2} \quad , \tag{3}$$

with Δ the band width; g_σ satisfies the equation

$$-\frac{1}{8} g_\sigma^3 + \left[\frac{3}{4}(E + \sigma J\bar{S}) + \frac{1}{4}\Gamma_{-\sigma}\right] g_\sigma^2 + \left[(I/2)^2 - (E + \sigma J\bar{S})^2 - \frac{1}{2}\right] g_\sigma$$

$$+ E + \sigma J\bar{S} - \Gamma_{-\sigma} - \frac{1}{16} g_\sigma^2 g_{-\sigma} + \frac{1}{4}\left[E + \sigma J\bar{S} + \Gamma_{-\sigma}\right] g_\sigma g_{-\sigma}$$

$$- \frac{1}{4} g_{-\sigma} = 0 \quad , \tag{4}$$

where

$$\Gamma_\sigma = (I/2)(1 - 2n_\sigma) \quad . \tag{5}$$

Eq. (4) represents coupled equations for g_σ and $g_{-\sigma}$, and the resulting equation for say g_+ is ninth-order. The above solution, Eq. (4), is valid for all values of I and $n = n_+ + n_-$, where n is the average number of electrons per site. In this paper, we restrict ourselves to the strong correlation limit, $I/\Delta \geq \sqrt{3}/2$, the value at which the Mott-Hubbard gap disappears.

We find that in zero field ($J\bar{S} = 0$) the only self-consistent solutions are when $m = 0$ and $m = n$ (except for $n = 1$[6]) with the $m = 0$ (nonferromagnetic) solution having the lower energy. Because of the spin-flip scattering term the energies of the σ and $-\sigma$ spin electrons are symmetric, and without a symmetry breaking field one does not expect nor obtain a self-consistent ferromagnetic solution for $0 < m < n$. Self-consistent ferromagnetic solutions for $0 < m < n$ are possible when a symmetry breaking field ($J\bar{S} \neq 0$) is included. Of course, the $m = n$ ferromagnetic solution is possible when $J\bar{S} \neq 0$, and it does have the lower energy. Although the $0 < m < n$ ferromagnetic solutions may not have a lower energy than the $m = n$ solution in a magnetic or molecular field, it is still interesting to examine these solutions.

To illustrate the zero temperature behavior of m vs. n, we show our results in Fig. 1a at various fields with $I/\Delta = 10$. In Fig. 1b we show an illustrative example of $\rho_\sigma(E)$ for $I/\Delta = 10$ and $J\bar{S}/\Delta = 0.1$.

In Fig. 2a the results of zero temperature calculations of m vs. n are shown at various I/Δ with $J\bar{S} = 0$ using the CPA. In our CPA calculations we used $\rho^{(0)}(E)$ given by Eq. (3) and Eqs. (3.17), (3.18), and (4.29 - 4.32) of Ref. 2 to self-consistently solve for m. An illustrative $\rho_\sigma(E)$ is shown in Fig. 2b. In the CPA, the behavior of m vs. n in the presence of a field is similar to that in zero field.

Comparison of the solution for m vs. n which includes the three scattering terms (Fig. 1a) to the CPA solution (Fig. 2a) shows quite graphically the importance of the resonance-broadening terms and in

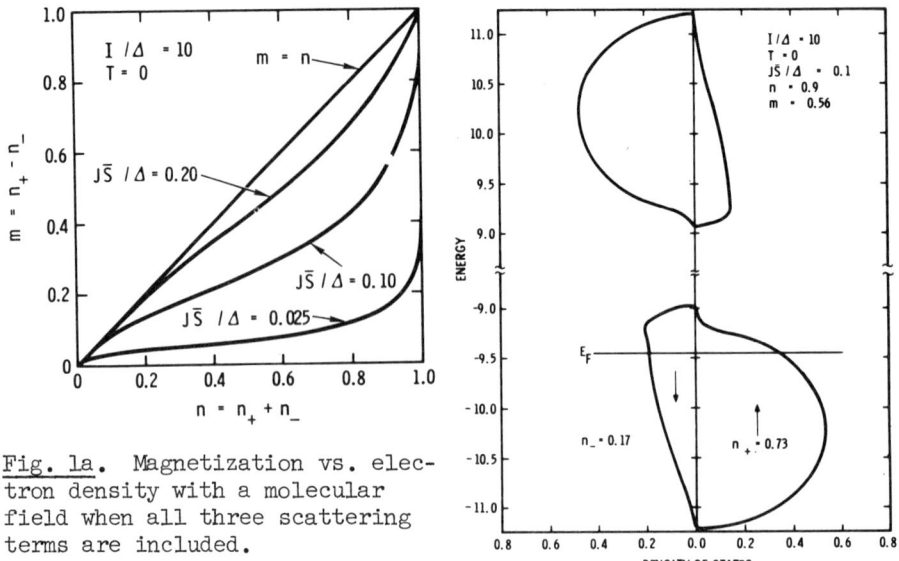

Fig. 1a. Magnetization vs. electron density with a molecular field when all three scattering terms are included.

Fig. 1b. Typical density of states, in units of $(\Delta/2)^{-1}$, as a function of energy, in units of $\Delta/2$, with a molecular field when all three scattering terms are included.

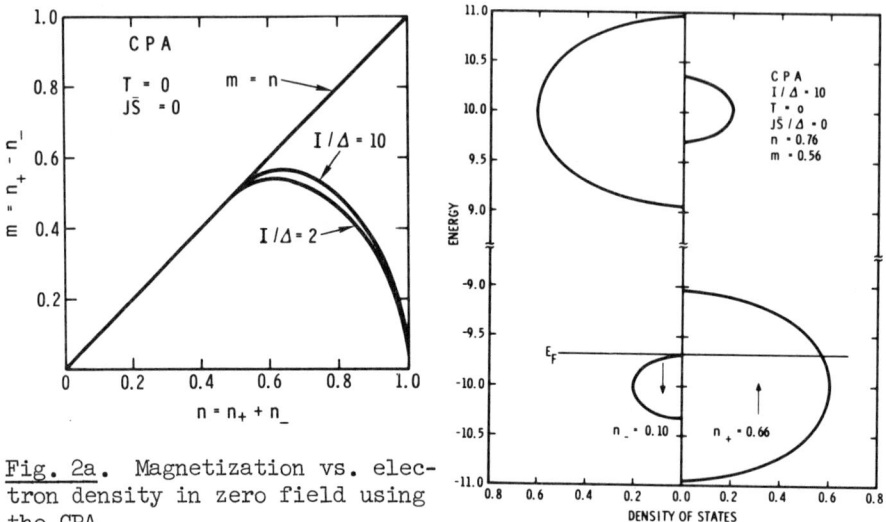

Fig. 2a. Magnetization vs. electron density in zero field using the CPA.

Fig. 2b. Typical density of states, in units of $(\Delta/2)^{-1}$, as a function of energy, in units of $\Delta/2$, in zero field using the CPA.

particular the spin-flip scattering term. Calculations of density of states and m vs. n involving spin disorder and electron-hole scattering only yield quite similar results to those using the CPA.[6,11] In zero field ($J\bar{S} = 0$) the self-consistent solutions when all three scattering terms are included are m = 0 and m = n with m = 0 (non-ferromagnetic) solution having the lower energy at least in the strong correlation regime ($I/\Delta > \sqrt{3}/2$). A symmetry breaking field, such as an externally applied magnetic field or an internal molecular field, is required to obtain a ferromagnetic solution in which the m < n or m = n solution has a lower energy than the m = 0 solution. On the other hand, the CPA solution gives self-consistent ferromagnetic solutions in zero field with the ferromagnetic solution having the same or lower energy than the m = 0 solution.

CONCLUSIONS

The spin-disorder and the σ spin electron into a $-\sigma$ spin hole scattering do not couple the motions of the σ spin and $-\sigma$ spin electrons. However, the inclusion of the spin-flip scattering term couples the motions of the σ spin and $-\sigma$ spin electrons. The results of calculations of such items as density of states and m vs. n considering all three scattering mechanisms differ significantly from those using the CPA or those omitting the spin-flip scattering. It is quite clear that the CPA is not a good approximation in discussing the magnetic properties of the Hubbard Hamiltonian. A very important scattering mechanism is the spin-flip scattering term. We feel that in order to obtain meaningful results one must include all three of Hubbard's scattering mechanisms and that results obtained using the CPA are of doubtful validity.

REFERENCES

† Contribution No. 2093
1. J. Hubbard, Proc. Roy. Soc. A281, 401 (1964).
2. B. Velicky, S Kirkpatrick, and H. Ehrenreich, Phys. Rev. 175, 747 (1968).
3. H. Hasegawa and J. Kanamori, J. Phys. Soc. Japan 31, 382 (1971).
4. H. Fukuyama and H. Ehrenreich, Phys. Rev. B 7, 3266 (1973).
5. K. Levin and K. H. Bennemann, Phys. Rev. B 5, 3770 (1972).
6. H. S. Jarrett, in Proceedings of the Eighteenth Annual Conference on Magnetism and Magnetic Materials, No. 10, ed. by C. D. Graham and J. J. Rhyne (AIP, New York, 1973), pp. 521-525.
7. L. C. Bartel and H. S. Jarrett, to be published.
8. T. Arai and M. Parrinello, Phys. Rev. Letts. 27, 1226 (1971).
9. L. C. Bartel, Phys. Rev. B 7, 3153 (1973).
10. H. Shiba, Prog. Theor. Phys. 46, 77 (1971).
11. H. S. Jarrett, unpublished calculations.

PROJECTION OPERATOR FORMALISM: APPLICATION TO THE HUBBARD MODEL*

A.J. Fedro and R.S. Wilson
Northern Illinois University, DeKalb, Illinois 60115
and Argonne National Laboratory, Argonne, Illinois 60439

ABSTRACT

We present a method of doing many body theory based on the commutator projection operator discussed by Kim and Wilson. Such a method immediately yields the Dyson equation leaving one the task of evaluating the self energy term. Such a term was previously handled by ordinary perturbation theory. We find, on the other hand, an exact set of differential equations involving the self energy and its time derivatives in Zubarev-like fashion which can be solved self-consistently in every order of perturbation in contrast to chain breaking procedures which are not in general perturbationally correct. We apply our method to the Hubbard model and find the Esterling and Lange (EL) type result. That is, the form of EL is correct but their $T - L(\vec{k})$ expression is not complete. It is worthwhile to point out here that our calculation is done with finite hopping so we do not encounter the degeneracy problem of EL.

INTRODUCTION

In this short paper we attempt to outline a procedure for handling many body Hamiltonians based on the use of projection operators originally developed by Zwanzig[1] and Mori[2]. In particular, we focus our attention on the commutator projection operator used by Kim and Wilson[3]. This choice of projection operator immediately yields the Dyson equation and will be a convenient starting point. We point out that previous calculations based on this idea have been done in standard perturbation fashion with no attempt to be self-consistent in the Zubarev sense. We, on the other hand, developed a chain of equations for the self energy term and its time derivatives in a Zubarev fashion and, as in chain breaking, our approximation lies in neglecting commutators with appropriate parts of the Hamiltonian which are internal to the particular averages involved, but only after explicitly insuring a perturbationally correct result. We apply our method to the Hubbard model and obtain the Esterling and Lange type result[4], that is, their form is correct but their T-L(k) expression is not complete.

*Supported in part through the auspices of the U.S. Atomic Energy Commission.

THEORY AND DISCUSSION

In this section we present a brief sketch of a projection operator technique that can be used to yield solutions of a many body Hamiltonian which are self-consistent as well as being perturbationally correct at every stage of the calculation. We apply the method to the Hubbard Hamiltonian which can be written as follows:

$$H = H_I + H_\mu + H_t , \qquad (1)$$

where

$$H_I \equiv \frac{I}{2} \sum_{j,\sigma} N_{j,-\sigma} N_{j\sigma} , \qquad (2)$$

where

$$H_\mu \equiv -\mu \sum_{j,\sigma} N_{j\sigma} , \qquad (3)$$

and where

$$H_t \equiv \sum_{i,j,\sigma} t_{ij} C_{i\sigma}^\dagger C_{j\sigma} ; \quad t_{jj} = 0 . \qquad (4)$$

Here μ is the chemical potential, $C_{j\sigma}^\dagger$ and $C_{j\sigma}$ are the Fermi creation and destruction operators at site R_j with spin σ and $N_{j\sigma} \equiv C_{j\sigma}^\dagger C_{j\sigma}$. I is the intraatomic Coulomb repulsion parameter and t_{ij} the hopping matrix elements. We also define the Liouville operators L_p in the following manner.

$$L_p \overline{X} \equiv [H_p, \overline{X}]_- \qquad (5)$$

for $p = I, \mu$ or t and $\overline{X}$ an arbitrary operator. Here $[\ ,\ ]_-$ is the ordinary commutator and we define $L \equiv L_I + L_\mu + L_t$. The theory to be presented will be independent of the particular form of the projection operator $P_{\ell\sigma}$ used, but, for convenience, we will use the specific one introduced by Kim and Wilson,[3] that being

$$P_{\ell\sigma} \overline{X} \equiv C_{\ell\sigma}^\dagger <[C_{\ell\sigma}, \overline{X}]_+> \qquad (6)$$

with $\overline{X}$ arbitrary. Here $[\ ,\ ]_+$ is the anticommutator and $<\cdots>$ is the grand canonical average over H of Eq. (1). This particular form of $P_{\ell\sigma}$ immediately yields the Dyson equation and will be a convenient starting point.

In this paper we will focus our attention on the one particle Green's function G defined as follows:

$$G_{ij\sigma}(t) \equiv i <[C_{i\sigma}, C_{j\sigma}^\dagger(t)]_+> , \qquad (7)$$

with

$$C_{j\sigma}^\dagger(t) \equiv e^{iHt} C_{j\sigma}^\dagger e^{-iHt} = e^{iLt} C_{j\sigma}^\dagger ; \quad C_{j\sigma}^\dagger \equiv C_{j\sigma}^\dagger(t=0) . \qquad (8)$$

From Eqs. (7) and (8) we find the equation of motion for G to be

$$\frac{\partial}{\partial t} G_{ij\sigma}(t) = - <[C_{i\sigma}, L C_{j\sigma}^\dagger(t)]> . \qquad (9)$$

The essence of the projection operator formalism originally developed by Zwanzig and Mori is to write $C_{j\sigma}^\dagger(t)$ in the following way:

$$C_{j\sigma}^\dagger(t) = (\sum_\ell P_{\ell\sigma}) C_{j\sigma}^\dagger(t) + (1 - \sum_\ell P_{\ell\sigma}) C_{j\sigma}^\dagger(t) \qquad (10)$$

and solve for $(1-\sum_\ell P_{\ell\sigma}) C_{j\sigma}^\dagger(t)$ in terms of $(\sum_\ell P_{\ell\sigma}) C_{\ell\sigma}^\dagger(t)$. From the equations of motion for $C_{j\sigma}^\dagger(t)$, Eq. (8), and the definition of $P_{\ell\sigma}$ in Eqn. (6) we can write the exact expression:

$$(1-\sum_\ell P_{\ell\sigma}) C_{j\sigma}^\dagger(t) = i\int_0^t dt'\, K_\sigma(t')(1-\sum_{\ell_1} P_{\ell_1 \sigma})L(\sum_\ell P_{\ell\sigma})C_{j\sigma}^\dagger(t-t'), \quad (11)$$

where we have defined

$$K_\sigma(t) \equiv \exp[it(1-\sum_\ell P_{\ell\sigma})L] \quad . \quad (12)$$

Use of Eqs. (6), (8), (10) and (11) in Eq. (9) yields

$$\frac{\partial}{\partial t} G_{ij\sigma}(t) = i\sum_\ell \Omega_{i\ell\sigma} G_{\ell j\sigma}(t) - \sum_\ell \int_0^t dt'\, \gamma_{i\ell\sigma}(t') G_{\ell j\sigma}(t-t') \quad (13)$$

with

$$\Omega_{ij\sigma} \equiv \langle [C_{i\sigma}, L\, C_{j\sigma}^\dagger]_+\rangle \quad , \quad (14)$$

and

$$\gamma_{ij\sigma}(t) \equiv \langle [C_{i\sigma},\, L\, K_\sigma(t)(1-\sum_\ell P_{\ell\sigma})\, L\, C_{j\sigma}^\dagger]_+\rangle \quad (15)$$

Equation (13) is easily solved by Laplace and Fourier transforming it as follows:

Let

$$A_{ij\sigma}(\omega) \equiv \int_0^\infty A_{ij\sigma}(t)\, e^{-[\varepsilon+i\omega]t}\, dt \;;\; \varepsilon = 0^+ \quad , \quad (16)$$

and

$$A_{\vec{k}\sigma}(\omega) \equiv \sum_{(i-j)} A_{ij\sigma}(\omega)\, e^{-i\vec{k}\cdot(\vec{R}_i-\vec{R}_j)} \quad (17)$$

for arbitrary A. Use of Eqns. (16) and (17) in Eq. (13) yields the Dyson equation

$$G_{\vec{k}\sigma}(\omega) = \frac{1}{\bar{\omega}-\Omega_{\vec{k}\sigma}-i\,\gamma_{\vec{k}\sigma}(\omega)} \quad (18)$$

with $\gamma_{\vec{k}\sigma}(\omega)$ the self energy (damping like) term and $\bar{\omega}\equiv\omega-i\varepsilon$. The problem, of course, lies in trying to evaluate $\gamma_{\vec{k}\sigma}(\omega)$ since use of Eqns. (14) and (17) give the exact result

$$\Omega_{\vec{k}\sigma} = \varepsilon_{\vec{k}} + I\langle N_{j-\sigma}\rangle - \mu \quad , \quad (19)$$

with $\varepsilon_{\vec{k}}$ the Fourier transform of t_{ij}. The method of evaluating $\gamma_{\vec{k}\sigma}(\omega)$ is the key to our formalism. Instead of the standard perturbation treatment of expanding $K_\sigma(t)$ of Eq. (12) in powers of L_t and replacing $\langle\ldots\rangle$ by $\langle\ldots\rangle_0$ everywhere it appears, where $\langle\ldots\rangle_0$ is the average in the absence of H_t, we develop a chain of equations involving $\gamma_{ij\sigma}(t)$ and its time derivatives in a Zubarev-like fashion and find that the exact result for $G_{\vec{k}\sigma}^{-1}(\omega)$ can be written as follows:

$$G_{\vec{k}\sigma}^{-1}(\omega) = \tilde{\bar{\omega}} - \varepsilon_{\vec{k}} - \frac{I\langle N_{j-\sigma}\rangle\tilde{\bar{\omega}}}{\tilde{\bar{\omega}}-I(1-\langle N_{j-\sigma}\rangle)} + \frac{I^2[B_{\vec{k}\sigma}-i\,D_{\vec{k}\sigma}(\omega)]}{[\tilde{\bar{\omega}}-I(1-\langle N_{j-\sigma}\rangle)]^2} \quad (20)$$

where $\tilde{\omega} \equiv \bar{\omega} + \mu$, where

$$B_{\vec{k}\sigma} \equiv \sum_i t_{ij} \langle C_{i\sigma}^{\dagger} C_{j\sigma} \rangle - \sum_{(i-j)} e^{-i\vec{k}\cdot(\vec{R}_i-\vec{R}_j)} t_{ij} [\langle \Delta N_{i-\sigma} \Delta N_{j-\sigma} \rangle + \langle C_{i\sigma} C_{i-\sigma}^{\dagger} C_{j-\sigma} C_{j\sigma}^{\dagger} \rangle]$$

$$+ \sum_{(i-j)} e^{-i\vec{k}\cdot(\vec{R}_i-\vec{R}_j)} t_{ij} \langle C_{j\sigma}^{\dagger} C_{j-\sigma}^{\dagger} C_{i-\sigma} C_{i\sigma} \rangle - \sum_i t_{ij} \langle (C_{i-\sigma}^{\dagger} C_{j-\sigma} + C_{j-\sigma}^{\dagger} C_{i-\sigma}) N_{j\sigma} \rangle, \quad (21)$$

where

$$D_{\vec{k}\sigma}(\omega) \equiv \frac{1}{I^2} \sum_{(i-j)} e^{-i\vec{k}\cdot(\vec{R}_i-\vec{R}_j)} \int_0^{\infty} e^{-[\epsilon+i\omega]t} dt \langle [L_t L_I C_{i\sigma}, K_\sigma(t)(1-\sum_{\ell} P_{\ell\sigma}) L_t L_I C_{j\sigma}^{\dagger}]_+ \rangle, \quad (22)$$

and where $\Delta N_{j-\sigma} \equiv N_{j-\sigma} - \langle N_{j-\sigma} \rangle$. To bring $G_{\vec{k}\sigma}^{-1}(\omega)$ in this form is not trivial and requires one to be rather adept with projection operator manipulations. The points to notice are as follows. First, all the the terms of $B_{\vec{k}\sigma}$ involve two site averages and are explicitly linear in t_{ij}. In fact, the first two terms on the right hand side of Eq. (21) represent the Esterling and Lange result while the first of the remaining two terms of $B_{\vec{k}\sigma}$ reflects "double hopping", that is, where two electrons on a given site hop onto an empty site and the other reflects "single hopping" in the presence of an electron of opposite spin. These particular averages in $B_{\vec{k}\sigma}$ are of order $(t_{ij})^2/I$ and can be neglected to first order. However, we are left with our unknown function $D_{\vec{k}\sigma}(\omega)$ of Eq. (22). This term is explicitly proportional to the square of the hopping matrix element since it involves taking two commutators with respect to H_t. At first glance, one might tend to neglect such a term in a calculation correct only to the first power of t_{ij}, but this term is frequency dependent and could yield a contribution that is first order in t_{ij} if it had the form

$$D_{\vec{k}\sigma}(\omega) \sim \frac{A_1}{\tilde{\omega}} + \frac{A_2}{\tilde{\omega}-I} \quad (23)$$

with A_i for $i = 1,2$ explicitly proportional to $(t_{ij})^2$ and the poles of $G_{\vec{k}\sigma}^{-1}(\omega)$ occurring at $\tilde{\omega}$ or $\tilde{\omega}-I \approx O(t_{ij})$. A consistent approximation to first order in $D_{\vec{k}\sigma}(\omega)$ would be to replace L in the exponential terms, $K_\sigma(t)$, by $L_I + L_\mu$, that is neglect commutators with H_t from the exponential operator, but always leaving the resulting average over full H alone, to see if the form, Eq. (23), arises. Such a calculation is difficult because $K_\sigma(t)$ contains the projection operator; however, one can find an exact expression for $D_{\vec{k}\sigma}(\omega)$ that does not involve projection operators in which one can do the calculation outlined above and indeed obtain the form Eq. (23). This is not trivial and, due to space limitations, will be repeated elsewhere. We can point out though that $D_{\vec{k}\sigma}(\omega)$ involves mainly three site averages. That the structure of $D_{\vec{k}\sigma}(\omega)$ is of the form, Eq. (23), indicates that our method, when applied to the Hubbard model, has the same inherent difficulties pointed out by Bari[5] and later by Esterling[6] that, in a moment-like expansion in powers of t_{ij}, the neglected term, $D_{\vec{k}\sigma}(\omega)$ in our case, is really of the same order as $B_{\vec{k}\sigma}$ due to the frequency dependence in the neighborhood $\tilde{\omega}$, $\tilde{\omega}-I \approx O(t_{ij})$. Thus one still has not collected all the terms of $O(t_{ij})$ in this neighborhood. This feature is peculiar to the Hubbard Hamiltonian; such difficulties do not arise in other problems where expansion about the Hartree-Fock is a valid one.

In summary we point out that the averages are always in the presence of t_{ij} and the only approximations made involve neglect of commutators with respect to H_t in $K_\sigma(t)$ so that the degeneracy problem of Esterling and Lange is avoided. Of course, in the end one must be able to evaluate the averages appearing in $G_{k\sigma}^{-1}(\omega)$ and this is done in the same fashion as for $G_{\vec{k}\sigma}(\omega)$ by selection of appropriate projection operators depending on the particular average needed. The specific details will appear in a later paper.

REFERENCES

1. R. Zwanzig, J. Chem. Phys. 33, 1338 (1960).
2. H. Mori, Prog. Theor. Phys. 33, 423 (1965).
3. S.K. Kim and R.S. Wilson, Phys. Rev. 7, 1396 (1973).
   S.K. Kim and R.S. Wilson, Phys. Rev. B15, (1973).
4. D.M. Esterling and R.V. Lange, Rev. Mod. Phys. 40, 796 (1968).
5. Robert A Bari, Phys. Rev. B2, 2260 (1970).
6. D.M. Esterling, Phys. Rev. B2, 4686 (1970).

ELECTRON CORRELATION IN A NARROW BAND METAL
A NEW METHOD OF CALCULATION

Samuel P. Bowen
University of Wisconsin, Madison, WI. 53706

ABSTRACT

A generalization of Feenberg's Perturbation Theory is shown to give exact expressions for the temperature dependent retarded greens functions for a many body problem. The theorem establishing this relationship is sketched and the result briefly applied to the Hubbard model.

In 1948 Feenberg[1] developed formulas for calculating the matrix elements of the resolvent operator $(\omega-H)^{-1}$ for a Hamiltonian $H=H_o+V$ which acts on a Hilbert space with an inner product $<\phi|\psi>$. With respect to an orthonormal basis ϕ_n of the Hilbert space, the diagonal matrix element of the resolvent is

$$<\phi_n|(\omega-H)^{-1}|\phi_n> = (\omega-E_n-\Sigma(n,\omega))^{-1}, \quad (1)$$

where $E_n = <\phi_n|H|\phi_n>$ and the self energy $\Sigma(n,\omega)$ is defined in terms of further off diagonal matrix elements of the potential $V_{\ell\ell'}=<\phi_\ell|V|\phi_{\ell'}>$. The expression for the self energy is given by a series of terms, the first two of which are

$$\Sigma(n,\omega) = \sum_{\ell \neq n} \frac{|V_{n\ell}|^2}{\omega-E_\ell-\Sigma'(\ell,\omega)} + \sum_{\substack{\ell \neq n \\ \ell' \neq n}} \frac{V_{n\ell}V_{\ell\ell'}V_{\ell'n}}{(\omega-E_\ell-\Sigma'(\ell))(\omega-E_{\ell'}-\Sigma''(\ell',\omega))} + \ldots \quad (2)$$

The reduced self energy expressions $\Sigma'(\ell)$ and $\Sigma''(\ell')$ appearing in the denominator have effectively the same definition except that the summations are further restricted to not include any states n or ℓ which have occurred previously in the sum.

These expressions are exact, but somewhat cumbersome. In the thermodynamic limit many of the summation restrictions can be ignored and for some models these formulas can yield tractable approximations in this limit.

The major result of this paper is to note that these Feenberg formulas can be shown rigorously to allow the exact calculation of thermally averaged retarded greens functions for any many body Hamiltonian. The theorem relating retarded greens functions to the Feenberg formulas will be sketched below and then it will briefly be used to discuss some exact properties of the one electron self energy of the Hubbard model.

For convenience the discussion will be limited to the one electron retarded greens function.

$$G^{(r)}(\vec{k}\uparrow,t) = -i\theta(t) \langle\{c_{\vec{k}\uparrow}(t), c^{\dagger}_{\vec{k}\uparrow}(0)\}\rangle, \quad (3)$$

where $\langle...\rangle$ is the exact thermal average $\{,\}$ is an anticommutator, and $\theta(t)$ the unit step function.

For complex ω in the upper half frequency plane the fourier transform of the retarded greens function $G(\vec{k},\omega)$ can be expressed in terms of the spectral function $A(\vec{k}\uparrow,\omega)$ by the usual expression

$$G^{(r)}(\vec{k}\uparrow,\omega) = \int \frac{d\bar{\omega}}{2\pi} \frac{A(\vec{k},\bar{\omega})}{\omega - \bar{\omega}}. \quad (4)$$

Making use of the operator equations of motion in the Heisenberg representation, one can easily derive the following formal large $|\omega|$ expansion of $G(\vec{k},\omega)$

$$G^{(r)}(\vec{k},\omega) = \sum_{n=0}^{\infty} \frac{1}{\omega^{n+1}} \langle\{\mathbf{L}^n c_{\vec{k}}, c^{\dagger}_{\vec{k}}\}\rangle, \quad (5)$$

where the operator $\mathbf{L}$ is defined using the Hamiltonian H and the commutator,

$$\mathbf{L} c_{\vec{k}} = [c_{\vec{k}}, H]. \quad (6)$$

Note that repeated application of $\mathbf{L}$ on the one electron operator generates terms involving products of odd numbers of electron fermion creation and annihilation operators. A. Lonke[2] has observed that one can form an abstract Hilbert space whose elements e_j are products of an odd number of creation and/or annihilation operators

$$e_j = c^{(\dagger)}_{\vec{k}_{i_1}} c^{(\dagger)}_{\vec{k}_{i_2}} \ldots c^{(\dagger)}_{\vec{k}_{i_{2n+1}}}, \quad n = 0,1,2,\ldots \quad (7)$$

The scalar product on this abstract Hilbert space is defined in terms of the thermal average $\langle...\rangle$ and the anticommutator,

$$((e_i, e_j)) = \langle\{e_j, e^{\dagger}_i\}\rangle. \quad (8)$$

The elements of this abstract Hilbert space are the products of the creation and annihilation operators themselves. On this space the operator playing the role of a Hamiltonian is the $\mathbf{L}$ introduced above. If the Hamiltonian H commutes with the density operator ρ which defines the average $\langle...\rangle$ it is straightforward to see that $\mathbf{L}$ is Hermitian on this Hilbert space. By comparing the Neumann expansion of the resolvent of $\mathbf{L}$ on this abstract Hilbert space, one immediately sees that the resolvent

$$((c_{\vec{k}\uparrow}, (\omega - \mathbf{L})^{-1} c_{\vec{k}\uparrow})) \quad (9)$$

is exactly the fourier transform of the retarded greens function for ω in the upper half plane.

$$G^{(r)}(\vec{k},\omega) = ((c_{\vec{k}\uparrow},(\omega-\mathbf{L})^{-1} c_{\vec{k}\uparrow})) \tag{10}$$

Since the Feenberg formulas apply for any Hilbert space with an inner product and a self adjoint Hamiltonian they can be applied directly to the calculation of the retarded greens function. The matrix elements in this generalized Hilbert space are temperature dependent and often must be determined using self consistency conditions.[3]

The major point of this theorem is that the derived expression for the self energy is exact for any Hamiltonian. The construction of approximations to the infinite series which defines the self energy will involve various "summations" of the terms to correspond to correct matrix approximations which will be discussed elsewhere.

As a brief illustration of these results, let us examine the leading terms of the one electron self energy for the Hubbard Hamiltonian, with particular emphasis on the non-locality emphasized by Esterling and coworkers.[4]

The main task in the application of the generalized Feenberg formulas is the construction of an orthonormal basis among the elements "e_j" of the generalized Hilbert-space. The one electron "states" $c_{\vec{k}\uparrow}^j$ and $c_{\vec{k}+\vec{q}\uparrow}$ form an orthonormal basis of the one electron subspace

$$((c_{\vec{k}\uparrow},c_{\vec{k}+\vec{q}\uparrow})) = \langle\{c_{\vec{k}+\vec{q}\uparrow},c_{\vec{k}\uparrow}^{\dagger}\}\rangle = \delta_{\vec{q},0} \tag{11}$$

If $\varepsilon_{\vec{k}\uparrow}$ is the one electron band energy, then the diagonal energy $E_{\vec{k}}$ is

$$E_{\vec{k}} = \varepsilon_{\vec{k}} + U\langle n\downarrow\rangle \tag{12}$$

The contribution to the self energy involving the one particle subspace can have a contribution of the form

$$U^2 \sum_{G'\neq 0} \frac{|\langle\tilde{\rho}_{G'\downarrow}^{\dagger}\rangle|^2}{\omega-\varepsilon_{k+G'\uparrow}-U\langle n\downarrow\rangle-\Sigma'(k+G')} \tag{13}$$

where

$$\langle\rho_{\vec{G}'\downarrow}^{\dagger}\rangle = \frac{1}{N}\sum_{\ell} e^{-i\vec{G}'\cdot\vec{R}_\ell} \langle n_{\vec{\ell}\downarrow}\rangle. \tag{14}$$

The vectors G' correspond to reciprocal lattice vectors of a possibly antiferromagnetic spin distribution which occur within the Brillouin zone of the crystal structure. For those cases where the Fermi energy occurs at a favorable point these amplitudes can be determined in a self consistent fashion from band structure effects.

A "state" from the 3 particle subspace which is orthogonal to the single particle subspace is

$$(\rho_{\vec{q}\downarrow} - \langle\rho_{\vec{q}\downarrow}\rangle) c_{\vec{k}+\vec{q}\uparrow}.$$

The contribution to the self energy from this "state" will be of the form

$$\left(\frac{U}{N}\right)^2 \sum_{\vec{q}} \frac{(\langle\rho^\dagger_{\vec{q}\uparrow}\rho_{\vec{q}\downarrow}\rangle - \langle\rho^\dagger_{\vec{q}\downarrow}\rangle\langle\rho_{\vec{q}\downarrow}\rangle)}{\omega - \varepsilon_{\vec{k}+\vec{q}\uparrow} - U(1-\langle n\downarrow\rangle) - \Sigma''(\vec{k}+\vec{q}\uparrow,\omega)}. \quad (15)$$

Only the $\vec{q} = 0$ part of this numerator is what was retained in this term in the Hubbard alloy analogy and related treatments.[5] The full $\vec{q}$ dependent numerator involves, as may be seen by writing out a Wannier representation exactly the nonlocal correlation which has been emphasized by Esterling and others.

Methods utilizing the average of the anticommutator as a means of orthogonalizing fermion operators have been presented before. Notably, the work of L. Roth on the Hubbard model[6] and R. H. Parmenter on the Anderson model[7] are two examples. The generalization of the Feenberg formula to thermal green's functions presented above may be regarded as a rigorous formulation of some aspects of the decoupling procedures of the above authors. Neither of the above methods systematically dealt with the construction of an orthonormal basis of the operator Hilbert space. As will be discussed elsewhere, the careful attention paid to a proper orthonormalization will uniquely determine contributions to the effective matrix elements which are of physical importance and would be easily missed in a less rigorous method.

A more detailed discussion of this generalized Feenberg method and its application to the Hubbard Model will be discussed elsewhere.

REFERENCES

1. E. Feenberg, Phys. Rev. 74, 206 (1948).
2. A. Lonke, J. Math. Phys. 12, 2422 (1971).
3. S. P. Bowen, to be published.
4. D. M. Esterling, Phys. Rev. B2, 4686 (1970).
   D. M. Esterling and H. C. Dubin, Phys. Rev. B6, 4276 (1972).
5. H. Fukayama and H. Ehrenreich, Phys. Rev. B7, 3266 (1973).
6. L. M. Roth, Phys. Rev. 184, 451 (1969).
   L. M. Roth, Phys. Rev. Letters 20, 1431 (1968).
7. R. H. Paramenter, Phys. Rev. B5, 2704 (1972).

DETERMINATION OF ERROR MINIMIZED GREEN'S FUNCTIONS FOR THE S-D MODEL

Philip E. Bloomfield
City College of CUNY, N.Y., N.Y. 10031

Edward B. Brown
Manhattan College, Bronx, N.Y. 10471

ABSTRACT

We construct a functional, J, which is a positive definite measure of the net error introduced in response functions (RF) which have been truncated in an equations of motion hierarchy. J is also a measure of the net error in the positions of the poles (approximate excitation energies) of these approximate RF. Minimizing J yields a closed set of equations for error minimized RF which for the s-d model of dilute magnetic impurities may be considered a generalization of those arising from the cumulant decoupling method.

FORMALISM

We consider the frequency dependent correlation function

$$(A;B^+)_\Omega = \int_{-\infty}^{\infty} dt\, e^{-i\Omega t} (e^{iHt} A e^{-iHt} ; B^+) \tag{1}$$

where H is the Hamiltonian of interest and the scalar product $(C;D^+)$ is required to satisfy the conditions

$$(C;C^+) \geq 0, \quad ([B,H];C^+) = (B;[H,C^+]), \quad (B;C^+) = (C;B^+)^*. \tag{2}$$

Two scalar products of particular interest satisfy (2),

$$(C;D^+) = \langle \{C,D^+\} \rangle = Tr\,\rho\{C,D^+\} \tag{3}$$

where $\rho = e^{-\beta H}/Tr\, e^{-\beta H}$

which produces the spectral function when used in (1) and

$$(C;D^+) = \int_0^\beta d\lambda \langle e^{-\lambda H} C e^{\lambda H} D^+ \rangle \tag{4}$$

which produces the response function when used in (1).
For an operator set $\{A_\mu\}$, which is to be speci-

fied later, we form the hermitean matrices $\eta_{\mu\nu}(\Omega)$ and $\eta_{\mu\nu}^{(0)}$ where

$$\eta_{\mu\nu}(\Omega) = (A_\mu; A_\nu^\dagger)_\Omega, \tag{5}$$

$$\eta_{\mu\nu}^{(m)} = \int_{-\infty}^{\infty} d\Omega \, \Omega^m (A_\mu; A_\nu^\dagger)_\Omega. \quad m=0,1,2,\ldots \tag{6}$$

Since $\eta_{\mu\nu}^{(0)}$ is hermitean, it can be diagonalized by a unitary matrix T, i.e.

$$T^\dagger \eta^{(0)} T = N. \tag{7}$$

Note that the eigenvalues of the matrix $\eta^{(0)}$ are positive, i.e. $N_{\eta\eta} \geq 0$. By ignoring the equality we assume the existence of an inverse for $\eta^{(0)}$. Since N is thus positive definite we may write

$$N = N^{1/2} N^{1/2} \tag{8}$$

$$N^{-1} = N^{-1/2} N^{-1/2} \tag{9}$$

and define the hermitean matrix $\bar{M}$ by

$$2\bar{M} = N^{-1/2} T^\dagger M T N^{1/2} + N^{1/2} T^\dagger M^\dagger T N^{-1/2} \tag{10}$$

where the matrix M is to be specified later.
The unitary matrix diagonalizing $\bar{M}$ is S, i.e.

$$S^\dagger \bar{M} S = \bar{\lambda} \tag{11}$$

The hermitean matrix $\mathcal{F}(\Omega)$ is defined by

$$\mathcal{F}(\Omega) = (\Omega - \bar{\lambda}) S^\dagger N^{-1/2} T^\dagger \eta(\Omega) T N^{-1/2} S (\Omega - \bar{\lambda}) \tag{12}$$

and the unitary matrix diagonalizing $\mathcal{F}(\Omega)$ is $\mathcal{P}(\Omega)$, i.e.

$$\mathcal{P}(\Omega)^\dagger \mathcal{F}(\Omega) \mathcal{P}(\Omega) = f(\Omega) \tag{13}$$

where $f(\Omega)$ is a diagonal matrix whose elements are the eigenvalues of $\mathcal{F}(\Omega)$.
Now consider the properties of the functionals

$$E_\nu(\Omega) = \sum_{\lambda\eta} \sum_{\chi'\eta'} (\mathcal{P}(\Omega)^\dagger S^\dagger Q)_{\nu\lambda} (\Omega \delta_{\lambda\lambda'} - m_{\lambda\lambda'}) \eta_{\lambda'\eta'}(\Omega)$$
$$\times (\Omega \delta_{\eta\eta'} - m_{\eta\eta'}^*)(Q^\dagger S \mathcal{P}(\Omega))_{\eta\nu} \geq 0 \tag{14}$$

where the inequality follows from $(B;B^+)_\Omega \geq 0$ and the matrices m and Q are given by

$$2m = 2M + n^{(0)}M^+(n^{(0)})^{-1}, \qquad Q = N^{-1/2}T^+. \qquad (15)$$

Using the definitions of the matrices in $E_\nu(\Omega)$ we find

$$E_\nu(\Omega) = f_\nu(\Omega). \qquad (16)$$

Thus, the <u>approximation</u> generated by setting

$$E_\nu(\Omega) = 0. \qquad (17)$$

for all ν, Ω is equivalent to setting all of the eigenvalues of the hermitean matrix $\mathcal{F}(\Omega)$ to zero, i.e. to setting

$$\mathcal{F}(\Omega) = 0. \qquad (18)$$

In terms of $n(\Omega)$ and m, (18) becomes

$$(\Omega - m) n(\Omega)(\Omega - m^+) = 0. \qquad (19)$$

We denote the retarded correlation function corresponding to $n_{qv}(\Omega)$ by $\mathcal{A}_{qv}(\Omega)$, i.e.

$$\mathcal{A}_{qv}(\Omega) = -i\int_{-\infty}^{\infty} dt\, e^{-i\Omega t}\, \Theta(t)\, (e^{iHt} A_q e^{-iHt}; A_\nu^+) \qquad (20)$$

where $\Theta(t)$ is the step function, and find, upon using (19) in the hierarchy for $\mathcal{A}_{qv}(\Omega)$, the approximation

$$(\Omega - m)\mathcal{A}(\Omega) = n^{(0)} + \left\{ \int_{-\infty}^{\infty} d\Omega' \left[((\Omega' - m)\mathcal{A}(\Omega'))_{\Omega' + i\epsilon} - ((\Omega' - m)\mathcal{A}(\Omega'))_{\Omega' - i\epsilon} \right] \right\} (\Omega - m^+)^{-1} \qquad (21)$$

which has the solution

$$(\Omega - m)\mathcal{A}(\Omega) = n^{(0)}. \qquad (22)$$

From (17), the approximation generating (22), the net error in the approximation (22) is obtained by summing the positive quantities $E_\nu(\Omega)$ over all ν and Ω. That is, denoting the error in (28) by E,

$$E = \sum_\nu \int d\Omega\, E_\nu(\Omega). \qquad (23)$$

In terms of $\{A_q\}$ and M, (23) is

$$E = T_r (n^{(2)} - m n^{(1)} - n^{(1)}m^+ + m n^{(0)} m^+)(n^{(0)})^{-1} \qquad (24)$$

We will choose $\{A_q\}$ and M to minimize (24).

One can also demonstrate the inequality

$$E \geq \sum_\lambda (\Omega_\lambda - \bar{m}_\lambda)^2 \qquad (25)$$

where $\bar{m}_\lambda$ is the λ th eigenvalue of $\mathcal{M}$ and Ω_λ is the nearest exact excitation energy. Since the poles of $\mathcal{A}(\Omega)$, i.e. our approximate excitation energies, are at the eigenvalues of $\mathcal{M}$, (25) shows that minimizing E also minimizes the net error in the positions of our approximate excitation energies.

Since only RF of the form $(A_\mu; A_\nu^+)_\Omega$ have been error-minimized in our scheme, we require that the correlations which arise in the specification of $\{A_\mu\}$ and $\mathcal{M}$ be determined in terms of these RF only. No external information can be introduced without destroying the error-minimizing characteristics of our scheme. This condition can be satisfied by a suitable choice of ansatz for $\{A_\mu\}$ and $\mathcal{M}$ which, in contrast to other schemes, are <u>both</u> subject to direct variation in our scheme. This feature is demonstrated in our treatment of the s-d model.

S-D MODEL

The s-d model Hamiltonian is given by[1]

$$H = \sum_{k\sigma} \epsilon_k c_{k\sigma}^+ c_{k\sigma} - \frac{J}{2N} \sum_{kp\sigma} (\sigma c_{k\sigma}^+ c_{p\sigma} S^z + c_{k\sigma}^+ c_{p\bar{\sigma}} S^{\bar{\sigma}}) + \frac{V}{N} \sum_{kp\sigma} c_{k\sigma}^+ c_{p\sigma} . \qquad (26)$$

As our trial operator set $\{A_\mu\}$ we use the particularly simple ansatz

$$A_{k\sigma}(1) = c_{k\sigma}, \quad A_{k\sigma}(2) = c_{k\sigma} S^z, \quad A_{k\sigma}(3) = c_{k\bar{\sigma}} S^{\bar{\sigma}}. \qquad (27)$$

while as our ansatz for $M_{k\sigma,k'\sigma'}$ we take

$$M_{k\sigma,k'\sigma'} = \begin{pmatrix} \epsilon_k + K_{k\sigma}(1,1) & K_{k\sigma}(1,2) & K_{k\sigma}(1,3) \\ K_{k\sigma}(2,1) & \epsilon_k + K_{k\sigma}(2,2) & K_{k\sigma}(2,3) \\ K_{k\sigma}(3,1) & K_{k\sigma}(3,2) & \epsilon_k + K_{k\sigma}(3,3) \end{pmatrix} \qquad (28)$$

Ansatzes (27) and (28) are then used to construct our functional to measure, for this case, the net error in our solutions for the Green's functions $\langle\!\langle A_{k\sigma}(i); A_{k'\sigma'}^+(j)\rangle\!\rangle_\Omega$. Extremizing determines the elements $K_{k\sigma}(i,j)$ to be

$$K_{k\sigma}(1,1) = 0, \quad K_{k\sigma}(1,2) = K_{k\sigma}(1,3) = -J/2N, \quad K_{k\sigma}(31) = 2K_{k\sigma}(21),$$

$$K_{k\sigma}(2,1) = -\frac{J}{2N}\left(\frac{1}{4} - L_{k\sigma}\right), \quad K_{k\sigma}(2,2) = -\frac{J}{2N}\left(\frac{4}{\sigma}\right)\frac{1}{\frac{1}{2} + \frac{1}{N}\sum_k L_{k\sigma}}\left\{\left(\frac{n_{k\sigma}}{2} - \frac{1}{4}\right)\frac{1}{N}\sum_k L_{k\sigma} + T_{k\sigma}\left(\frac{1}{2} - \frac{2}{N}\sum_k L_{k\sigma}\right)\right\}$$

$$K_{\kappa\sigma}(2,3) = -\frac{J}{2N}\left(\frac{4}{\alpha}\right)\frac{1}{\frac{1}{2}+\frac{1}{N}\sum_{\kappa}L_{\kappa\sigma}}\left\{\left(\frac{n_{\kappa\sigma}}{2}-\frac{1}{4}\right)\frac{1}{4} - \frac{T_{\kappa\sigma}}{2}\left(\frac{1}{2}-\frac{2}{N}\sum_{\kappa}L_{\kappa\sigma}\right)\right\},$$

$$K_{\kappa\sigma}(3,2) = 2K_{\kappa\sigma}(2,3), \quad K_{\kappa\sigma}(3,3) = K_{\kappa\sigma}(2,2) + K_{\kappa\sigma}(2,3). \tag{29}$$

where $\alpha = 1 - \frac{4}{N}\sum_{\kappa}L_{\kappa\sigma}$, $Q_{\kappa\sigma} = \frac{1}{N}\sum_{p p \kappa'}<C_{\kappa'\sigma}^{+}C_{p\sigma}C_{q\bar{\sigma}}^{+}C_{\kappa\sigma}S^{\sigma}>$,

$$T_{\kappa\sigma} = (L_{\kappa\sigma} - 2Q_{\kappa\sigma})/2 - (\alpha-1)/2. \tag{30}$$

and the remaining quantities are defined in ref. 1.

Our results for the Green's functions thus determined can all be expressed in terms of the t matrix

$$t(\Omega) = \left(\frac{J}{2N}\right)^2 \frac{3E(\Omega)}{\left(1-\frac{V}{N}F(\Omega)\right)^2 + \frac{2}{\alpha}\left(1-\frac{V}{N}F(\Omega)\right)G(\Omega) - 3\left(\frac{J}{2N}\right)^2 F(\Omega)E(\Omega)} \tag{31}$$

where

$$F(\Omega) = \sum_{\kappa}\frac{1}{\Omega-\epsilon_{\kappa}}, \quad G(\Omega) = \sum_{\kappa}\frac{(n_{\kappa\sigma}-1/2)}{\Omega-\epsilon_{\kappa}}, \quad E(\Omega) = \sum_{\kappa}\frac{(1/4-L_{\kappa\sigma})}{\Omega-\epsilon_{\kappa}} \tag{32}$$

It should be noted that all the averages which appear in (29) can be determined from t. This a direct consequence of the form of our ansatz for $M_{\kappa\sigma,\kappa'\sigma'}$. Also, (31) reduces to the t determined by the cumulant decoupling technique if in (31) one sets $\alpha=1$, its high temperature value.

REFERENCES

1. P.E. Bloomfield, et al, Phys. Rev. **B2**, 3714 (1970)

MICROSCOPIC THEORY FOR THE FLUCTUATION-DRIVEN PHASE TRANSITION IN WEAK ITINERANT MAGNETS*

K. K. Murata
Brookhaven National Laboratory, Upton, New York 11973

ABSTRACT

The microscopic basis for a recent phenomenological theory for the thermodynamics near T_c of weak itinerant magnets is presented. Calculation of dynamic renormalization effects within paramagnon theory justifies the previous assumption of weakly temperature dependent Ginsburg-Landau α and β coefficients. The cutoff on the fluctuation phase space is found to go as $q_m \sim (T_c)^{1/3}$ rather than the usual inverse coherence length. In spite of the phase space reduction, as $T_c \to 0$, the phenomenological description becomes exact in a temperature region of order T_c. However, it is found that the α and β coefficients must be determined phenomenologically since no small parameter enters the calculation of their overall magnitude.

INTRODUCTION

Recent work[1,2] on spin fluctuations has explained why the random phase approximation for the Hubbard model $U \sum_i n_{i\uparrow} n_{i\downarrow}$ for a conduction band gives incorrect predictions for the experimental susceptibilities of extremely weak itinerant magnets. One faces the problem in deriving the Ginsburg-Landau functional

$$\mathcal{F} = \frac{1}{2} \sum_q^{q_m} |h_q|^2 (\alpha_F + \mu_F^2 q^2) + \frac{\beta_F \tilde{T}}{4!N} \sum_{\{q\}}^{q_m} h_q h_{q'} h_{q''} h_{-q-q'-q''} \quad (1)$$

where $\tilde{T} = TN(\epsilon_F)$ is the reduced temperature, h is a unit normalized field representing magnetization fluctuations, and the partition function $Z/Z_0 = \langle e^{-\mathcal{F}} \rangle$ is given as an ensemble average over the h fields.

The comparison with experiment fails if one assumes, as in the theory of superconductivity, that the order parameter susceptibility $\chi \propto \langle |h_0|^2 \rangle$ is given by α^{-1} calculated in the ladder approximation (RPA). The point is that $\alpha_{RPA} = 1 - 2UN(\epsilon_F)$ contains temperature dependence due only to the thermal average $\overline{N(\epsilon_F)}$ of the density of states. This results in the extremely weak dependence $\alpha_{RPA} = \alpha_{RPA}(T=0) + \mathcal{O}(\tilde{T}^2)$. Now in weak magnets χ^{-1} is observed to have linear[3] rather than quadratic dependence on T over several T_c. Further the Curie coefficient obtained by expanding α_{RPA} around its zero is proportional to ϵ_F/T_c, whereas experimentally this number should be more like unity.

* Work performed under the auspices of U.S. Atomic Energy Commission.

The suggestion of Murata and Doniach[1] (hereafter called I) is that the leading temperature dependence in χ^{-1} may come from the β_F term. With the Hartree factorization

$$\sum_{\{q\}} h^4 \approx 6 \sum_{qq'} |h_q|^2 <|h_{q'}|^2> - 3 \sum_{qq'} <|h_q|^2><|h_{q'}|^2>$$

of the β_F term, the model is solvable, and we obtain

$$\chi_q^{-1} \sim D_q \equiv <|h_q|^2> = \{\alpha_F + \beta_F T \xi/2 + \mu_F^2 q^2\}^{-1}$$

with $\xi = N^{-1} \sum_{q<q_m} D_q$. Not too close to T_c, ξ is roughly $\xi_0 = \Omega_a q_m^2/2\pi^2 \mu_F^2$, with Ω_a the atomic volume. The β_F term can thus provide the required Curie temperature dependence in χ provided q_m is sufficiently large.

The magnitude of q_m is not given, as one might expect, by the inverse of the finite or zero temperature coherence length. Rather, it is argued below that the momentum sum defining ξ encompasses the thermal modes. If these modes are assumed to be overdamped paramagnons with width $\Gamma(q) \sim q^3$, one immediately obtains $q_m \sim \tilde{T}^{1/3}$, a result obtained from less obvious means below. The anomalously large q_m and partial success of the fluctuation model is thus due to the extremely "shallow" q^3 paramagnon dispersion.

LOWEST ORDER DYNAMIC COUPLING

The free energy (1) does not omit the finite (Matsubara) frequency modes. These have been integrated out and because of coupling effects renormalize α_F and β_F. To lowest order in this coupling the dynamic susceptibility is given by

$$D_q(\omega \pm i\delta) = [\eta + \mu_F^2 q^2 \mp i\omega/\gamma q]$$

where $\eta = \alpha_0 + \alpha^{(1)} + \beta_0 \tilde{T}\xi/2$, $\gamma = 2V_F/(2\pi \ UN(\epsilon_F))$, and α_0 and β_0 are nominal values given by $\alpha N(\epsilon_F)$ and $\beta N(\epsilon_F)$ in the notation of I. Here $\alpha^{(1)}$ is a functional of $D_{qn} = D_q(\omega_n)$ and the electron Green's function $G_{kn} = [\omega_n - \epsilon_k]^{-1}$, where ω_m is the Matsubara frequency:

$$\alpha^{(1)} = \frac{2U^2 T^2}{N^2} \sum_{kqnm}' D_{qm} \left[2G_{k-q,n-m}^3 G_{kn} + G_{k-q,n-m}^2 G_{k,n}^2 \right]$$

The primed sum implies for m=0, only values of $q > q_m$ are taken; the rest are included in ξ. The q_m is determined variationally.

Converting the Matsubara sums in the standard fashion to integrals in the complex plane, we deal with terms in the <u>temperature derivative</u> of $\alpha^{(1)}$ as follows: 1) The derivative with respect to explicit Fermi factors generates terms of order $\tilde{T}$ and $\tilde{T}\ln\tilde{T}$. 2) The term containing the boson factor $n(\omega)$ can be written, after some manipulation, as:

$$\alpha_{Boson}^{(1)} = - \frac{U^2 N''(\epsilon_F)}{N} \left\{ \frac{1}{\pi} \sum_q \int d\omega' \left[n(\omega) - \frac{T}{\omega'} - \frac{1}{2} \right] ImD_q(\omega+i\delta) + T \sum_{q > q_m} ReD_q(0) \right\} \qquad (2)$$

The cutoff q_m is determined by requiring that $\left[\frac{\partial}{\partial T} + \frac{\partial q_m}{\partial T} \frac{\partial}{\partial q_m} \right] \alpha_{Boson}^{(1)}$ vanish. As shown below, this condition reduces to $q_m \sim T^{1/3}$. 3) The implicit dependence on η generates a term $\frac{\partial \alpha^{(1)}}{\partial \eta} \frac{\partial \eta}{\partial T}$. The explicit form for $\frac{\partial \alpha^{(1)}}{\partial \eta} \approx \frac{\partial \alpha^{(1)}}{\partial \eta}\bigg|_{\eta=0} < 0$ is found to be

$$\frac{d\alpha^{(1)}}{d\eta} = \frac{U^2 N''(\epsilon_F)}{N} \left[\sum_q (\gamma q)^2 / \pi \Gamma(q) + \sum_q \gamma^2 q \, \ell n \left(\frac{qV_F}{\Gamma(q)} \right) \right] \qquad (3)$$

where $\Gamma(q) = \gamma q (\eta + \mu_F^2 q^2)$. We see that (3) is approximately constant and of order unity. Thus

$$\frac{d}{dT} \eta \approx \frac{d}{dT} (\alpha_0 + \beta_0 \tilde{T} \xi / 2) + \frac{\partial \alpha^{(1)}}{\partial \eta}\bigg|_{\eta=0} \frac{d}{dT} \eta + \mathcal{O}(\tilde{T} \ell n \tilde{T}) \, .$$

Integrating this and defining $\eta = \alpha_F + \beta_F \tilde{T} \xi / 2$, we find to order $\tilde{T}^2 \ell n \tilde{T}$:

$$\alpha_F = (\alpha_0 + \alpha_0^{(1)})/(1 - \frac{\partial \alpha^{(1)}}{\partial n}\bigg|_{\eta=0})$$

$$\beta_F = \beta_0/(1 - \frac{\partial \alpha^{(1)}}{\partial n}\bigg|_{\eta=0}) \qquad (4)$$

Here $\alpha_0^{(1)}$, which is also found to be of order one, is $\alpha^{(1)}$, evaluated for $\eta=0$ and $T=T_c$. Equation (4) shows the qualitative way the dynamic renormalization effects enter. Comparable corrections are expected from higher order couplings since $UN(\epsilon_F) \sim 1$ appears to be the relevant parameter.

DETERMINATION OF q_m

This is rather delicate. Writing

$\text{Im}D_q(\omega+i\delta) = (\gamma q/2) d\ell n(\Gamma^2(q)+\omega^2)/d\omega$, we integrate (2) by parts. We next split up the weight of

$$\frac{1}{\pi} \frac{\partial}{\partial \omega} [n(\omega)-T/\omega-1/2] = \frac{T/\pi}{\omega^2+T^2} + \frac{1-\pi}{\pi\omega} g(\frac{\omega}{T}) \qquad (5)$$

into a Lorentzian part and a function g of unit weight and asymptotic behavior $g(x)/x \to x^{-4}$. Substituting (5) and forming the derivative $[\frac{\partial}{\partial T} + \frac{dq_m}{dT} \frac{\partial}{\partial q_m}]$, we obtain a sum of terms proportional to

$$\int_0^{2k_F} \frac{dq\gamma q^2}{\Gamma(q)+T} + \frac{1-\pi}{\pi} \int_0^{2k_F} dq\gamma q \int \frac{dx}{x^2} \frac{xg(x)T}{\Gamma^2(q)+x^2T^2}$$

$$+ T \frac{q_m^2}{\eta+\mu_F^2 q_m^2} \frac{dq_m}{dT} - \int_{q_m}^{2k_F} \frac{q^2 dq}{\eta+\mu_F^2 q^2} \qquad (6)$$

In (6) one may set $\eta=0$ to good approximation. One then obtains a sum of the form $aq_m+bT^{1/3}+cT\, dq_m/dT$, which has the solution $q_m \sim T^{1/3}$. The coefficient can be obtained numerically from (6) if necessary.[4]

DISCUSSION

It was assumed in I that the temperature dependence of α_F and β_F is relatively weak. The above suggests fairly weak $\tilde{T}^2 \ell n \tilde{T}$ dependence, with a coefficient which presumably cannot be readily calculated. However, at some point for sufficiently low T_c, we expect the $T^{4/3}$ dependence from $\beta_F T \xi$ to dominate and the model (1) to become appropriate.

Criteria for the validity of the Hartree approximation to (1) are given in I in terms of the two small parameters T/T_0 and $\epsilon \sim T^2/(T-T_c) \cdot T_0$. Since $T_0 \sim q_m$, the Hartree theory becomes valid at sufficiently low T_c over a range determined essentially by the constancy of α_F and β_F. This is presumably $|T-T_c|/T \sim \mathcal{O}(1)$.

As mentioned in I, Sc_3In with $T_c \sim 6K$ fits the model, whereas $ZrZn_2$ with $T_c \sim 25K$ does not. Systems with considerably lower T_c's are presumably required for extremely good fits to the Hartree theory. However, in the event such systems are found, one may be able to test for the $\tilde{T}^{1/3}$ dependence by including a temperature-dependent q_m in the Hartree theory.

Unfortunately the above results seem to indicate that contrary to the suggestion of Moriya and Kawabata[2] a proper calculation of α_F and β_F from given bandstructure is not feasible because the expansion parameter $UN(\epsilon_F)$ is not small.

Full details of the calculation will be published elsewhere.

REFERENCES

1. K.K. Murata and S. Doniach, Phys. Rev. Letters **29**, 285 (1972)
2. T. Moriya and A. Kawabata, J. Phys. Soc. Japan **34**, 639 (1973) and Technical Report of the Institute for Solid State Physics, University of Tokyo, Series A, No. 574 (1973).
3. See, for example, the experimental references in Ref. 1.
4. The crossover behavior in the numerical work of Moriya and Kawabata indicates that q_m can be of order k_F even for $T/\epsilon_F \sim 0.01$.

HYDRODYNAMICS OF A FERROMAGNETIC ELECTRON GAS WITH SPIN-ORBIT COUPLING

J. Tilley and A. Luther*
Harvard University, Cambridge, Mass. 02138

ABSTRACT

We derive the hydrodynamic equations for the transverse magnetization of a ferromagnetic electron gas, including spin-orbit coupling and a finite electronic mean free path. The electrons are assumed to be described by tight-binding t_{2g} functions and the short range intraband exchange is treated in the RPA. Spin-orbit coupling removes the spin rotational invariance, introduces a gap into the spin-wave spectrum, and mixes the transverse and longitudinal susceptibilities. The latter is diffusive for wavelengths longer than the electronic mean free path. We examine the structure of the spectral function and find the mixing of the spin-waves with this diffusive mode.

INTRODUCTION

Ferromagnetic resonance studies[1,2] of the transition metals have raised interesting questions about the hydrodynamics of an itinerant ferromagnet. It has been suggested recently[3,4] that the spin-orbit interaction in such metals can account for the observed temperature dependence of the damping of spin-waves in nickel and cobalt. We study here the proper hydrodynamic description of spin-waves including both spin-orbit coupling and a finite electronic mean free path.

We introduce a specific model for clarity, and attempt to choose features of this model which are characteristic of ferromagnetic nickel. The bulk of the density of states at the Fermi surface in nickel is formed from electrons with wavefunctions of t_{2g} symmetry. The e_g bands are separated in energy from the t_{2g} bands because of ligand field effects and do not play a significant part in the magnetic properties of nickel. Accordingly, we have chosen to study a model with two triply degenerate bands of electrons, with the usual spin-splitting of the bands due to the short range Coulomb exchange interaction. To introduce an electron lifetime, we include a contact interaction of the electrons with a

---

*Work supported in part by National Science Foundation Grant NSF-GH-32774.

low concentration of impurities located randomly in the electron gas. The model Hamiltonian which we use is

$$H = \sum_{\vec{k},\mu,\sigma} \varepsilon(\vec{k}) c^+_{\mu\sigma}(\vec{k}) c_{\mu\sigma}(\vec{k}) + v_c \sum_{\vec{k},\mu,\sigma} n_{\mu,\sigma}(\vec{k}) n_{\mu,-\sigma}(-\vec{k})$$

$$+ u_{imp} \sum_{\vec{x}_i} \sum_{\substack{\vec{k},\vec{k}' \\ \mu,\sigma}} e^{i(\vec{k}-\vec{k}')\cdot \vec{x}_i} c^+_{\mu\sigma}(\vec{k}) c_{\mu\sigma}(\vec{k}')$$

$$+ v_s \sum_{\substack{\vec{k}, \\ \mu,\sigma,\mu',\sigma'}} M_{\mu\sigma,\mu'\sigma'} c^+_{\mu\sigma}(\vec{k}) c_{\mu'\sigma'}(\vec{k}) \qquad (1)$$

In Equation (1), $\varepsilon(\vec{k})$ gives the band structure, v_c is the screened Coulomb parameter, u_{imp} represents the coupling to the impurities located at positions $\vec{x}_i$, and v_s is the spin-orbit coupling constant. For the spin-orbit interaction, we have used the site-diagonal matrix elements $M_{\mu\sigma,\mu'\sigma'}$ between pairs of t_{2g} wavefunctions with band indices μ,μ' and spin indices σ,σ', as have been calculated by Abate and Asdente.[5] The band calculations of Hodges and Ehrenreich[6] indicate that such treatment of the spin-orbit forces is sufficient to do justice to the band structure of nickel.

CALCULATION OF THE SUSCEPTIBILITY

We define $\chi_{\uparrow\downarrow}(v_c = 0)$ to be the transverse intraband susceptibility including the effects of impurity scatterings and spin-orbit interactions but with no Coulomb scatterings other than the Hartree-Fock contributions to the band structure (resulting in spin-split bands). Then, in the RPA for the exchange, we have

$$\chi_{\uparrow\downarrow} = \frac{\chi_{\uparrow\downarrow}(v_c = 0)}{1 - v_c \chi_{\uparrow\downarrow}(v_c = 0)} \qquad (2)$$

The averaging over random impurities is carried out in the usual manner. It is important to note that this procedure leads to effective interband impurity scatterings. Since the model has assumed s-wave scattering from the impurities, it is clear that the electron-impurity coupling will be the same for all of the t_{2g} electrons. This means that the impurity ladder corrections to the parallel spin but band off-diagonal intermediate states will result in diffusive behavior just as in the analysis for a paramagnetic electron gas.[7] We treat the spin-orbit interaction as weak and determine the corrections to the "mass operator" for the magnons to second order in v_s.

By using the "density-of-states" approximation appropriate to a ferromagnetic electron gas with a spherical Fermi surface and small spin-splitting of the bands, we arrive at the final form of our result.

$$\chi_{\uparrow\downarrow}(q,\omega) = \frac{-N(0)\Delta}{\omega - Jq^2 + i\frac{v_c v_s^2}{2\pi\Delta}\frac{J_L}{1-uJ_L}} \quad (3)$$

where

$$J_L = -\frac{i\pi N(0)}{v_F q} \ln\left[\frac{-i/\tau - \omega + v_F q}{-i/\tau - \omega - v_F q}\right] \quad (4)$$

Here u is the effective two-body impurity scattering potential, τ is the electronic lifetime, Δ is the spin-splitting of the bands, v_F is the Fermi velocity, and Jq^2 is the usual magnon dispersion. The expression for the susceptibility is valid if ω, $v_F q$, $1/\tau \ll \Delta$.

DISCUSSION

Outside the hydrodynamic region, we can let $\tau \to \infty$ giving the result

$$\chi_{\uparrow\downarrow}^{\tau=\infty} = \frac{-N(0)\Delta}{\omega - Jq^2 - \omega_{gap}\left(1 + \frac{1}{3}\frac{v_F^2 q^2}{\omega^2}\right)} \quad (\omega/v_F q \gg 1)$$

$$\chi_{\uparrow\downarrow}^{\tau=\infty} = \frac{-N(0)\Delta}{\omega - Jq^2 - i\frac{\pi}{2}\frac{\omega_{gap}}{v_F q}\omega} \quad (\omega/v_F q \ll 1)$$

(5)

where $\omega_{gap} = v_s^2/\Delta$.

The spectral function shows a peak at the gap energy and another peak at

$$\omega_{max} = \left(\frac{1}{1 + \frac{\pi^2}{4}\frac{\omega_{gap}^2}{v_F^2 q^2}}\right)^{\frac{1}{2}} Jq^2 \quad (6)$$

Our calculation of the specific heat of this pure system

shows that for $T \lesssim J\omega_{gap}^2/k_B v_F^2 = T_o$, the leading term is proportional to $T^{4/3}$ while for $T \gg T_o$, the usual spin-wave $T^{3/2}$ law is valid. We estimate T_o to be typically 3 mK and therefore expect this effect to be difficult to observe experimentally.

In the hydrodynamic region, our results do not agree with other calculations. We find impurity vertex corrections to be very important in the susceptibility calculation. If we were to omit these "impurity ladders", our results would be in agreement with others. For $\omega/v_F q \ll 1$ we have

$$\chi_{\uparrow\downarrow}(q,\omega) = \frac{-N(0)\Delta}{\omega - Jq^2 - \omega_{gap}\left[\dfrac{i\omega}{i\omega - Dq^2}\right]} \quad (7)$$

where $D = \frac{1}{3}v_F^2 \tau$ is the diffusion constant. A peak in the spectral function occurs at

$$\omega_{max} = \left(\frac{1}{1 + \dfrac{\omega_{gap}^2}{D^2 q^4}}\right)^{\frac{1}{2}} Jq^2 \quad (8)$$

If $\omega_{gap}/Dq^2 \gg 1$, the peak lies at DJq^4/ω_{gap} while if $\omega_{gap}/Dq^2 \ll 1$, the maximum occurs at the magnon energy Jq^2.

CONCLUSION

Insofar as the qualitative details of our model Hamiltonian correctly describe nickel, our conclusions about the hydrodynamic region also apply. At first appearance it seems that our results suggest an increasing spin-wave damping constant for decreasing mean free path. It is more accurate to state that the picture of a damped spin-wave collective mode is not appropriate for wavelengths larger than $\lambda_\infty = v_F/\omega_{gap}$ in a pure system or for wavelengths larger than $\lambda_\ell = (D/\omega_{gap})^{\frac{1}{2}}$ in a system with electronic mean free path ℓ. (This latter condition is reasonable only when the system is hydrodynamic, i.e. if $\ell/\lambda_\ell \lesssim 1$ or equivalently, if $\ell \lesssim \frac{1}{3}\lambda_\infty$.) At such large wavelengths there are two collective modes - a mode at approximately the gap frequency with almost the full spectral weight and a low-lying coupled spin-orbital mode with practically no spectral weight. These features imply that the comparison of FMR data with the Landau-Lifshitz-Gilbert equation for the relaxation of the magnetization will be correct if the skin depth of the sample at the rf frequency used in the experiment is smaller than the appropriate λ_ℓ or λ_∞. However, this does not appear to be the case in typical FMR experiments.

ACKNOWLEDGMENT

The authors acknowledge many helpful discussions with Professor P. C. Martin.

REFERENCES

1. S. M. Bhagat and L. L. Hirst, Phys. Rev. <u>151</u>, 401 (1966).
2. S. M. Bhagat, J. R. Anderson and Ning Wu, Phys. Rev. <u>155</u>, 510 (1967).
3. R. E. Prange and V. Korenman, J. Mag. Res. <u>6</u>, 274 (1972).
4. V. Korenman and R. E. Prange, Phys. Rev. <u>B6</u>, 2769 (1972).
5. E. Abate and M. Asdente, Phys. Rev. <u>140A</u>, 1303 (1965).
6. H. Ehrenreich and L. Hodges, Methods in Comp. Phys. <u>8</u>, 149 (1968).
7. P. Fulde and A. Luther, Phys. Rev. <u>170</u>, 570 (1968).

UNIFIED ITINERANT-LOCALIZED THEORY OF MAGNETISM

B. H. Brandow

Battelle Memorial Institute, Columbus, Ohio 43201

ABSTRACT

A general procedure is described for deriving Heisenberg hamiltonians in a fully itinerant and quantitative manner. Strong orbital rearrangement plays an essential role. The resulting picture resolves basic conceptual problems in magnetic materials as diverse as metallic iron, Mott insulators, and solid ^{3}He. This scheme also provides a localized-orbital description for Mott insulators which is logically satisfactory and intuitively appealing.

INTRODUCTION

One of the oldest and most persistent problems in the quantum theory of magnetism has been that of reconciling the itinerant and localized models of the ferromagnetic transition metals. This has often been approached as a problem of choosing between one or the other of these two models. However, it is now clear on experimental grounds, at least for iron, that this either/or approach is no longer tenable. Detailed band calculations[1,2] have accounted quite well for most features of the de-Haas-van-Alphen data of iron,[2] as well as for the specific heat and for the X-ray and neutron scattering form factors. On the other hand, there is abundant evidence[3] for the existence of local moments in the high temperature regime, $T > T_c$. One can no longer avoid the conclusion that the itinerant and localized models are <u>both</u> valid within their proper domains, and that the proper task of theory is to work out a reconcilation. We shall now describe how the localized electron features can be <u>derived</u>, in a quantitative manner (at least in principle), starting from a fully itinerant Hartree-Fock model.

Before doing so, however, we should note that there is a closely related problem of itinerancy <u>vs</u>. localization for a very different class of magnetic materials, namely the Mott insulators. Basic theory suggests that the ground states of these materials should be describable by means of band theory, in this case the antiferromagnetic band theory of Slater.[4] On the other hand, Slater's theory seems to require that the insulating band gap and the local moments should both disappear at the Neel temperature, contrary to experiment. The electrons in Mott insulators behave phenomenologically as if they are localized in some quite literal manner. A similar conceptual problem arises in understanding the magnetism of solid ^{3}He,[5] since the ^{3}He <u>atoms</u> remain localized on a periodic lattice at temperatures far greater than T_N. This situation should also occur in Wigner's electron solid at low densities.

LOCALIZED ELECTRONS?

These observations led us to examine whether the Hartree-Fock equations for a periodic system could ever lead to localized (Wannier-like) canonical eigenfunctions, that is, eigenfunctions which diagonalize the matrix λ of Langrange multipliers. The most general form of the Hartree-Fock equations is

$$\mathcal{F} \psi_i = \sum_j \lambda_{ji} \psi_j , \qquad (1)$$

where $\mathcal{F}$ is the usual Fock operator (kinetic plus ionic plus 2-body potential terms). It is well known that off-diagonal λ terms are introduced when the HF equations for the usual Block eigenfunctions are expressed in terms of the corresponding Wannier orbitals. A <u>canonical</u> Wannier solution (one with $\lambda_{ji} \equiv 0$ for $j \neq i$) would therefore necessarily be <u>physically different</u>, even in the case of filled bands. If the resulting total energy should turn out to be lower than that of the usual Bloch solution (or Slater antiferromagnetic band solution), say for a low-density lattice of hydrogen atoms, this could have important practical consequences for the understanding of Mott insulators. This existence question was raised as early as 1940, in Seitz's textbook (§53), but it had never received a satisfactory answer.*

We have carefully studied this existence question, and have concluded that localized canonical HF orbitals are <u>not</u> possible.[5] This is an inherent consequence of the structure of the exchange terms in the HF potential. The Hartree equations, where all exchange terms except the self-exchange are absent, can easily produce localized orbitals, but inclusion of the "other particle" exchanges destroys this localization. (However, the <u>combined</u> effect of the off-diagonal λ's and the corresponding exchange terms does suffice to produce localization, since the resulting Wannier HF equations are formally equivalent to the canonical Bloch solution.) This negative result brings the subject of Mott insulators closer to that of the ferromagnetic transition metals, since it is now quite clear that formally correct treatments of these (and other) magnetic materials <u>must</u> be based on de-localized canonical eigenfunctions. A "literal" localization of the electrons is definitely inconsistent with HF theory.

---

*The only investigation that we are aware of, prior to Ref. 5, is that of Kaplan and Argyres.[6] Unfortunately their non-existence proof is invalid, for the following reason: They begin by assuming the validity of first-order perturbation theory to estimate the elements of a certain matrix "P". Under this assumption it must be legitimate to insert the <u>exact</u> eigenfunctions into their perturbation formula (3.55) for P, since any resulting differences must (formally) be of second order in P. Thanks to the assumed validity of their canonical eigenvalue equation (3.46), however, Eq. (3.55) now takes on the indeterminate form (zero/zero). This result does not necessarily involve any logical inconsistency, so their analysis is inconclusive.

The ferromagnetic transition metals and the Mott insulators both exhibit "localized electron behavior" in the sense that their magnetic phenomenology can be described (for the most part) by Heisenberg hamiltonians,

$$\mathcal{H} = \sum_{i<j} J_{ij}\, \vec{S}_i \cdot \vec{S}_j \;. \tag{2}$$

We shall now describe how such effective hamiltonians can be obtained from HF theory, regardless of whether the system is metallic or insulating, ferromagnetic or antiferromagnetic.

THEORY

We begin with Slater's band picture of antiferromagnetism.[4] Slater pointed out that the net spin polarization at any given site will tend to attract electrons of the same spin to this site, via the exchange terms of the HF potential. Electrons with the opposite spin will not experience as strong an exchange attraction, thus the net spin polarizations will tend to maintain themselves self-consistently. Our proposal is based on the observation that <u>this mechanism for the self-consistency of the local moments is really quite unrelated to the periodicity of the antiferromagnetic ground state</u>. Self-consistent HF solutions must therefore exist for <u>arbitrary</u> assignments (up or down) for the net spin polarizations at the various sites,[5,7] provided the intra-atomic exchange is sufficiently strong compared to the hopping tendency. (Most spin assignments will require large unit cells, often as large as the entire crystal sample, but this feature does <u>not</u> introduce any <u>fundamental</u> problem for HF theory.) Let us imagine, now, that complete self-consistent calculations had been carried out for all 2^N of these spin assignments. It should then be possible to parametrize the <u>differences</u> in the total HF energies of these 2^N different states by means of a Heisenberg hamiltonian (2), at least to a good approximation. (The main neglects are terms involving more than two sites, bi-quadratic exchange terms, and the weakening of the moments due to one-electron thermal excitations. The latter neglect will invalidate this approach for systems with weakly formed moments, such as chromium, $ZrZn_2$, and NiS.)

In practice, one would do complete self-consistent calculations for only a few spin arrangements, and use the resulting total energy differences to determine the J_{ij}'s for near neighbors. Three calculations, for example, would suffice to determine the J's for first and second neighbors. One would naturally choose these spin arrangements to preserve as much symmetry as possible, for calculational convenience.

An essential characteristic of the present scheme, which sharply distinguishes it from Slater's, is the presence of strong orbital rearrangement. Whenever one alters the directions (up or down) of some of the moments, all of the de-localized (Bloch-like) eigenfunctions are strongly modified in the vicinities of these altered sites. It can readily be seen[5,7] that this rearrangement incorporates the so-called strong correlations needed to explain the persistence of the band gap for $T > T_N$ in Mott insulators.

The present conceptual picture has been hinted at in the past, in one way or another, by a number of investigators,[†] but a clear and forthright presentation has been lacking. There appears to have been a general timidity arising from doubts that HF theory can support local moments when these moments are arranged in disordered configurations. Such doubts are without any objective foundation.[††]

TRANSITION METALS

For the magnetic transition metals, Friedel and Blandin[8] have suggested roughly the same picture, but their discussions consider far less realistic detail. The present approach is completely non-perturbative, and it can incorporate <u>all</u> of the realistic details (crystal structure, degeneracy, hybridizations, etc.) that are treated in conventional band calculations, as well as all the intra-atomic interactions, including the Hund's rule coupling. Of course it remains to be seen whether it is actually feasible to calculate the very small total energy differences associated with the J_{ij}'s. In any event, however, we believe that the present picture is of considerable conceptual value. It provides a clear and straight-forward reconciliation of the itinerant and localized electron features seen experimentally. It also shows that the J_{ij}'s arise from a 3d hopping mechanism (necessarily of short range) of the type described by Alexander and Anderson,[9] in addition to the well known RKKY contribution[10] from the 4s electrons.

MOTT INSULATORS

Adler[7] had previously considered this approach for Mott insulators, but he rejected this because it did not appear to account for the insulating band gap in CoO. We have observed[5] that the insulating aspect of the Mott insulators can be associated with the concept of filled bands <u>even for disordered spin arrangements</u>. This allows the introduction of a self-consistent (but spin-order dependent) Wannier representation based on (1), with off-diagonal λ's included of course.* This representation accounts for the main characteristic features of Mott insulators in a simple and intuitive manner. For

---

†See especially Refs. 7, 8, p. 541 of Ref. 4, and p. 611 of Ref. 14.

††To illustrate this, we have obtained some well-behaved numerical solutions for the periodic one-dimensional model of Ref. 5, with a 7-site unit cell in which the spin arrangement is +++-+--. Note that this configuration has no internal symmetries other than the 7-site periodicity. Details of these calculations will be presented elsewhere.

*It should be noted that the localized orbitals of Ref. 5 differ in several respects from those of Ref. 11. The latter are all space-orthogonal and spin-independent; furthermore the doubly-occupied orbitals of that scheme remain de-localized or Bloch-like.

example, the insulating states all have the __same__ number of Wannier orbitals centered on each cation, regardless of orbital degeneracy or choice of spin arrangement. In some cases the L-S coupling will require this HF approach to be generalized to a __multi-determinant__ description, but this is conceptually straightforward when one starts with the present Wannier picture. (This generalization allows for orbital contributions to the magneton numbers; it also resolves Adler's difficulty with CoO.) The "crystal field" absorption peaks (Frenkel excitons)[12] are also readily understood from this picture. This representation also makes it clear that there will be a strong band-narrowing effect when "U/t" is large (especially for the pure antiferromagnetic spin arrangement), in qualitative agreement with Hubbard's result[13]. The usual cluster approach[11] for calculating the J_{ij}'s in these insulators is quite consistent with the present general picture. It must be acknowledged that the cluster approach offers some practical advantages for quantitative studies[11] of the J_{ij}'s, such as the feasibility of calculating correlation corrections to the pure HF result, by means of perturbation theory, and also the prospect of obtaining bi-quadratic exchange terms. Nevertheless, essentially all of the basic qualitative characteristics of Mott insulators are properly described by the present Wannier HF picture. In its direct relation to the physics of Mott insulators, this localized representation is analogous to the familiar relation between Bloch orbitals and the properties of metals. (Incidentally, it seems appropriate to classify __any__ magnetic insulator at a Mott insulator, if it continues to insulate and have local moments above T_N.)

REFERENCES

1. A. V. Gold, L. Hodges, P. T. Panousis, and D. R. Stone, Intern. J. Magnetism __2__, 357 (1971).
2. S. Wakoh and J. Yamashita, J. Phys. Soc. Japan __21__, 1712 (1966); K. J. Duff and T. P. Das, Phys. Rev. __B3__, 192 and 2294 (1971); R. A. Tawil and J. Callaway, Phys. Rev. __B7__, 4242 (1973).
3. M. B. Stearns, Phys. Rev. __B8__ (November, 1973).
4. J. C. Slater, Phys. Rev. __82__, 538 (1951).
5. B. H. Brandow, Ann. Phys. (N.Y.) __74__, 112 (1972); see Appendix C.
6. T. A. Kaplan and P. N. Argyres, Phys. Rev. __B1__, 2457 (1970).
7. D. Adler, Solid State Physics __21__, 1 (1968); see p. 11.
8. J. Friedel, G. Leman, and S. Olszewski, J. Appl. Phys. __32__, 325S (1961); A. Blandin and P. Lederer, Proceedings of the International Conference on Magnetism (Nottingham, 1964), p. 71.
9. S. Alexander and P. W. Anderson, Phys. Rev. __133__, A1594 (1964); T. Moriya, Progr. Theoret. Phys. __33__, 157 (1965).
10. A. W. Overhauser and M. B. Stearns, Phys. Rev. Letters __13__, 316 (1964); M. B. Stearns, Phys. Rev. __B4__, 4081 (1971).
11. K.-I. Gondaira and Y. Tanabe, J. Phys. Soc. Japan __21__, 1527 (1966); S. Freeman, Phys. Rev. __B7__, 3960 (1973).
12. R. Newman and R. M. Chrenko, Phys. Rev. __114__, 1507 (1959); G. W. Pratt and R. Coelho, Phys. Rev. __116__, 281 (1959).
13. J. Hubbard, Proc. Roy. Soc. __A281__, 401 (1964).
14. J. des Cloizeaux, J. Phys. Radium __20__, 601 and 751 (1959).

Theory of Motional Narrowing of EPR Spectral Density of Magnetic Ions

George Reiter
The City College of New York, New York, New York 10031

Abstract

A theory of the EPR lineshape of a magnetic impurity coupled to a lattice is presented that is capable of treating both the "slow" and "fast" regimes i.e. the case that the characteristic frequency of the fluctuations is comparable to or smaller than the linewidth that they produce, as well as the motional narrowed limit, where the characteristic frequency is large compared to the linewidth. A self consistent equation for the propagator of the ion motion is obtained, valid in both regimes, under the assumption that the fluctuations of the lattice are not affected by the presence of the impurity, for arbitrary spin Hamiltonians. The equation is shown to correspond to the choice of the lowest order term in a renormalized perturbation theory for the "self energy" operator of the system, and is equivalent to the Kubo-Tomita theory in the fast limit. This lowest order approximation is analogous to the mode coupling, or independent mode approximation in extended systems. A diagrammatic representation of the perturbation theory is given which one may readily use to include higher order corrections, and which makes manifest the role of the static distribution of the fluctuating fields in determining the lineshape. The theory may be generalized to arbitrary finite dimensional systems coupled to a bath.

EFFECTS OF NEAREST NEIGHBOR FOUR-SPIN CORRELATION UPON THE CRITICAL PROPERTIES OF SPIN-1/2 HEISENBERG FERROMAGNET

Tomoyasu Tanaka
Ohio University, Athens, Ohio 45701

Louis F. Libelo
U.S. Naval Ordnance Laboratory, White Oak, Md. 20910

ABSTRACT

A spin-1/2 f.c.c. isotropic Heisenberg ferromagnet is treated by the cluster variation method in which the nearest neighbor pair, triangle and tetrahedral correlations are taken into account. The Helmholtz free energy, which is expressed in terms of eight quantities, $<S^z>$, $<S_1^z S_2^z>$, $<S_1^- S_2^+>$, $<S_1^z S_2^z S_3^z>$, $<S_1^- S_2^+ S_3^z>$, $<S_1^z S_2^z S_3^z S_4^z>$, $<S_1^- S_2^+ S_3^z S_4^z>$ and $<S_1^- S_2^- S_3^+ S_4^+>$, is minimized with respect to those expectation values. Above the Curie temperature three of these quantities vanish and there are only two linearly independent variables. Below T_c the eight variables satisfy non-trivial, non-linear transcendental equations. The critical data obtained are: $<S^z> = <S_1^z S_2^z S_3^z> = <S_1^+ S_2^- S_3^z> = 0$, $2<S_1^+ S_2^-> = 4<S_1^z S_2^z> = 1/5$, $6<S_1^+ S_2^+ S_3^- S_4^-> = 24<S_1^+ S_2^- S_3^z S_4^z> = 16<S_1^z S_2^z S_3^z S_4^z> = 0.081802$, and $T_c = 0.67347 * T$ (Weiss). It is found that $<S^z>^3$ is more linear than $<S^z>^2$ over a wide range of temperature below T_c.

INTRODUCTION

It is well known that the Weiss molecular field type approach describes essential features of second order phase transition although in detail there are many undesirable features of thermodynamical quantities displayed near the transition temperature. In particular, the critical indices predicted are known to be incorrect. When the effect of the nearest neighbor pair correlation is included (Bethe-Peierls-Weiss approximation) the value of the critical temperature and the behavior of the specific heat are substantially improved, although there is almost no improvement in the critical exponents at this stage. It is not clear, however, whether or not such behaviors are common to all molecular field type approaches. It would, therefore, be worthwhile to push statistical approximation a few more steps in succession in order to find out if there is any sign of convergence towards the correct behaviors and if there is any minimal number of spins for which correlations must be included in order to find a nearly correct set of critical indices. There are, of course, unmatchable advantages of the molecular field type approaches, i.e., the thermodynamic quantities are obtained both below and above the transition temperature. As one of the first attempts towards these goals we shall formulate a theory of second order phase transition in which the nearest neighbor four-spin correlation f function is included. As the basic lattice structure we take the face centered cubic, and the four-spin cluster is chosen to be the smallest tetrahedron. Because of the geometrical symmetry of the four-spin cluster, calculations are carried out semi-analytically near the transition temperature and relatively small amount of numerical computations is needed.

FORMULATION

The physical system treated in this formulation is the face centered cubic, nearest neighbor coupled, isotropic, spin $S = 1/2$, Heisenberg ferromagnet. The Weiss critical temperature, T_w is given by $2JS(S+1)z/3k$. If the values of $S = 1/2$ and $z = 12$ are introduced one obtains $T_w = 6J$. The energy per lattice point, measured in the units of the Weiss critical temperature, is given by

$$E = -2(<S_i^+ S_j^-> + <S_i^z S_j^z>) \tag{1}$$

where $<S_i^+ S_j^->$ and $<S_i^z S_j^z>$ are the nearest-neighbor transverse and longitudinal pair correlation functions, respectively. The entropy per lattice point, in the four-spin tetrahedral approximation is given by[1]

$$-S/kN = \gamma^{(1)} + z_2 \gamma^{(2)} + z_3 \gamma^{(3)} + z_4 \gamma^{(4)}, \tag{2}$$

where z_2, z_3, z_4 are the numbers of nearest neighbor pairs, equilateral triangles, and regular tetrahedra, per lattice point respectively. Some simple geometrical analysis yields

$$z_2 = 6, \quad z_3 = 8, \quad z_4 = 2. \tag{3}$$

$\gamma^{(1)}$, $\gamma^{(2)}$, $\gamma^{(3)}$ and $\gamma^{(4)}$ are given in terms of the reduced density matrices:

$$\Gamma^{(1)} = \text{tr } \rho^{(1)} \ln \rho^{(1)} = \gamma^{(1)} \tag{4}$$

$$\Gamma^{(2)} = \text{tr } \rho^{(2)} \ln \rho^{(2)} = 2\gamma^{(1)} + \gamma^{(2)} \tag{5}$$

$$\Gamma^{(3)} = \text{tr } \rho^{(3)} \ln \rho^{(3)} = 3\gamma^{(1)} + 3\gamma^{(2)} + \gamma^{(3)} \tag{6}$$

$$\Gamma^{(4)} = \text{tr } \rho^{(4)} \ln \rho^{(4)} = 4\gamma^{(1)} + 6\gamma^{(2)} + 4\gamma^{(3)} + \gamma^{(4)}. \tag{7}$$

In the tetrahedral approximation all the four spins enter symmetrically into the above density matrices, and hence there is no need to indicate which spins are referred to in respective density matrices.

When eqs. (4)-(7) are substituted into eq. (2) one finds the entropy per lattice point to be given by

$$-S/kN = 5\Gamma^{(1)} - 6\Gamma^{(2)} + 2\Gamma^{(4)} \tag{8}$$

At this point one should observe the fact that $\Gamma^{(3)}$ does not appear in the entropy at all. In the general formulation of the cluster variation method[2] this situation is stated in geometrical or topological terms, i.e., any diagram which is not a common part of two neighboring parent diagrams does not contribute to the entropy. In the present case of the tetrahedral approximation, none of equi-lateral triangles can be a common part of two neighboring tetrahedra and hence the triangles ($\gamma^{(3)}$ or $\Gamma^{(3)}$) do not contribute to the entropy.

Our next task is to diagonalize the two-spin and four-spin density matrices. Since S = 1/2, diagonalization of the two-spin density matrix is trivial. Due to a complete geometrical symmetry of our regular tetrahedron the diagonalization of the four-spin density matrix can be achieved analytically and one finds the following eigenvalues:

$$\rho^{(4)}(1) = (1/16)(1 + 4x_1 + 6x_2 + 4x_3 + x_4) \qquad g_1 = 1$$

$$\rho^{(4)}(2) = (1/16)(1 - 4x_1 + 6x_2 - 4x_3 + x_4) \qquad g_2 = 1$$

$$\rho^{(4)}(3) = (1/16)(1 - x_4 + 12x_5 + 12x_7 + 2x_1 - 2x_3 + 24x_6) \qquad g_3 = 1$$

$$\rho^{(4)}(4) = (1/16)(1 - x_4 + 12x_5 + 12x_7 - 2x_1 + 2x_3 - 24x_6) \qquad g_4 = 1$$

$$\rho^{(4)}(5) = (1/16)(1 - x_4 - 4x_5 - 4x_7 + 2x_1 - 2x_3 - 8x_6) \qquad g_5 = 3$$

$$\rho^{(4)}(6) = (1/16)(1 - x_4 - 4x_5 - 4x_7 - 2x_1 + 2x_3 + 8x_6) \qquad g_6 = 3$$

$$\rho^{(4)}(7) = (1/16)(1 - 2x_2 + x_4 - 16x_5 - 16x_7 + 16x_8) \qquad g_7 = 1$$

$$\rho^{(4)}(8) = (1/16)(1 - 2x_2 + x_4 - 8x_5 + 8x_7 + 16x_8) \qquad g_8 = 2$$

$$\rho^{(4)}(9) = (1/16)(1 - 2x_2 + x_4 - 16x_8) \qquad g_9 = 3$$

$$\cdots \cdots \quad (9)$$

where g_i's (i = 1, 2, ... 9) are the multiplicity factors and $x_1, \ldots, x_8$ are the correlation functions as defined by

$$x_1 = 2\langle S^z \rangle \qquad x_5 = \langle S_1^+ S_2^- \rangle$$

$$x_2 = 4\langle S_1^z S_2^z \rangle \qquad x_6 = 2\langle S_1^x S_2^+ S_3^- \rangle$$

$$x_3 = 8\langle S_1^z S_2^z S_3^z \rangle \qquad x_7 = 4\langle S_1^z S_2^z S_3^+ S_4^- \rangle$$

$$x_4 = 16\langle S_1^z S_2^z S_3^z S_4^z \rangle \qquad x_8 = \langle S_1^+ S_2^+ S_3^- S_4^- \rangle$$

$$\cdots \cdots \quad (10)$$

At temperature T, reduced in the units of the Weiss critical temperature, the Helmholtz free energy per lattice point is given as

$$F = -(1/2)x_2 - x_5 + T(5\Gamma^{(1)} - 6\Gamma^{(2)} + 2\Gamma^{(4)}) \quad (11)$$

where

$$\Gamma^{(1)} = (1/2)[(1 + x_1) \ln (1 + x_1) + (1 - x_1) \ln (1 - x_1)] \quad (12)$$

$$\begin{aligned}\Gamma^{(2)} = (1/4)[&(1 + 2x_1 + x_2) \ln (1 + 2x_1 + x_2) \\ &+ (1 - x_2 + 4x_5) \ln (1 - x_2 + 4x_5) \\ &+ (1 - x_2 - 4x_5) \ln (1 - x_2 - 4x_5) \\ &+ (1 - 2x_1 + x_2) \ln (1 - 2x_1 + x_2)]\end{aligned} \quad (13)$$

$$\Gamma^{(4)} = \sum_{j=1}^{9} g_j \, \rho^{(4)}(j) \ln \rho^{(4)}(j). \quad (14)$$

The equilibrium values of the correlation functions are found in such a way that the Helmholtz free energy is a minimum with respect to variations of these correlation functions. The minimum conditions of the free energy yield 8 coupled non-linear equations and these must be solved numerically. However, in an immediate neighborhood of the critical temperature these equations become algebraic and in particular one finds a few simple relations. These are

$$2\langle S_1^z S_2^z S_3^z \rangle = 3\langle S_1^z S_2^+ S_3^- \rangle, \quad (15)$$

$$\langle S_1^z S_2^z \rangle = 1/20 \quad (16)$$

x_4 is the real solution of a cubic equation

$$27(11 + 5x_4)(1 - 25x_4)^2 = (9 - 25x_4)^3 \quad (17)$$

$$x_4 = 16 \langle S_1^z S_2^z S_3^z S_4^z \rangle = 0.08180185$$

$$2\langle S_1^z S_2^z \rangle = \langle S_1^+ S_2^- \rangle = 1/10 \quad (18)$$

$$3\langle S_1^+ S_2^+ S_3^- S_4^- \rangle = 12\langle S_1^z S_2^z S_3^+ S_4^- \rangle = 8\langle S_1^z S_2^z S_3^z S_4^z \rangle \quad (19)$$

$$1/kT_c = \ln \frac{1}{27} \frac{(5x_4 + 11)}{(25x_4 - 1)} \quad (20)$$

CONCLUSION

The spin $S = 1/2$ f.c.c. Heisenberg ferromagnet is treated in the tetrahedral approximation. This is the crudest four-spin approximation because all the correlation functions included contain only the nearest neighbor distance. The value of the transition temperature, the spontaneous magnetization, and the specific heat near the transition temperature are substantially improved. In particular, the third power of the spontaneous magnetization is found to be more linear than the second power of the same quantity over a wide range of temperature below the transition temperature. We have so far only three cases of the molecular field type approximation, i.e., one-spin, two-spin, and four-spin approximations. In order to be able to test the capability of the molecular field type approximation at least two more cases, e.g., 6-spin and 8-spin treatments, would be necessary. These approximations are under consideration and the results will be reported in future.

REFERENCES

1) T. Morita and T. Tanaka, Phys. Rev. $\underline{145}$, 288 (1966).
2) T. Morita, Journ. Math. Phys. $\underline{13}$, 115 (1972).

Some Consequences of the Non Mixing Behaviour of the Four Spin Correlation Function in Heisenberg Paramagnets

Boyd Foster
University of California at Irvine, Irvine, Cal.

George Reiter
The City College of New York, New York, New York 10031

Abstract

A simple argument based on energy conservation shows that the correlation function of the Heisenberg paramagnet,

$$\langle \bar{S}(q,t) \cdot \bar{S}(-q,t) \, \bar{S}(q') \cdot \bar{S}(-q') \rangle - \langle \bar{S}(q) \cdot \bar{S}(-q) \rangle \langle \bar{S}(q') \cdot \bar{S}(-q') \rangle$$

does not vanish as $t \to \infty$, for any dimension of the lattice. The time dependence of this quantity has been evaluated by solving the kinetic equations for the one dimensional Heisenberg chain numerically, and it is found that this function approaches a limit $K(q, q')$. As a consequence, the fourier transform of this correlation function has a singularity at zero frequency. In a real system, where the spin system can relax through coupling to the environment, this singularity will be replaced by more complicated structure, which in the simplest approximation, can be assumed to be a simple pole on the imaginary axis at the relaxation frequency for the energy of the system. We conjecture that this behaviour occurs in two and three dimensions also, and is the explanation for the acoustic attenuation measurements of Moran and Luthi in $RbMnF_3$, where the energy decay is due to the spin lattice coupling. The low frequency singularity which is a result of the relaxation of the internal structure of the system, should be observable in, EPR, APR, and light scattering experiments, and the conditions for observing it will be discussed.

DYNAMIC FORM FACTOR AND NONEQUILIBRIUM BEHAVIOR OF THE
$S=\frac{1}{2}$ XY MODEL ON CUBIC LATTICES

M. Howard Lee
Department of Physics and Astronomy
University of Georgia, Athens, Ga. 30602

ABSTRACT

The $S=\frac{1}{2}$ XY model order parameter $S^x(k)$ is not a constant of motion at any wavelengths. Hence, if an external perturbation is turned off, $S^x(k,t)$ is expected to relax to an equilibrium value. In a molecular-field approximation, the time behavior of the order parameter can be analytically given. The resulting dynamic form factor satisfies the standard sum rules but not the spin diffusion equation. With the fluctuation-dissipation theorem, the dynamic susceptibility is obtained, which shows that our result is exact in the high-frequency limit and that there exists an undamped collective mode which is formally similar to the plasmon. The approximate time behavior of the order parameter can be expressed in terms of the generalized relaxation function, whose Laplace transform is given by a continued fraction. The MFA result then follows when the relaxation function is given by a two-pole approximation. The more exact time behavior follows when the relaxation function is given by a higher multipole approximation.

INTRODUCTION

Present theoretical attempts to understand dynamic behavior of lattice models of spin exchange systems such as the Heisenberg paramagnet are largely semiphenomenological.[1] They are more or less limited to relating the fluctuation spectrum to the first two moments of the spectral shape function which can be exactly calculated at $T=\infty$. Dynamic behavior is usually looked at in low and high frequency regimes. In the low frequency regime the dominant behavior is believed to be a diffusion process. Central to this notion is that the total spin is a constant of motion, as is for the Heisenberg paramagnet.

The $S=\frac{1}{2}$ XY paramagnet serves as an interesting model for studying dynamic behavior since the total spin $M(x)$ or $M(y)$ is not a constant of motion. Thus, the dynamics of the XY paramagnet may prove to be very different from that of the Heisenberg paramagnet. In particular in the low frequency regime the dominant process may not be diffusionlike. Also since the total spin is not conserved at long wavelengths, one can distinguish in the XY model what is purely kinematic from what is dynamic.

In general the dynamic form factor cannot be given an analytic expression since the time-dependence of $S_k^x(t)$ cannot be reduced to a tractable form. We introduce here a procedure by which the time-dependence can be obtained approximately. First expand $S_k^x(t)$ in terms of successive zero-time commutators and consider $[\mathcal{H}, \dot{S}_k^x]$ in a perturbative way such that in zeroth order it represents a certain truncation of the commutator. Within zeroth order we find an analytic expression for $S_k^x(t)$. The result is then used to obtain the dynamic form factor and with aid of the fluctuation-dissipation theorem the dynamic susceptibility. Our result for the dynamic susceptibility is found to be exact in the high-frequency limit.

DYNAMIC FORM FACTOR AND TRUNCATION OF COMMUTATORS

The XY paramagnet is described in the XY Hamiltonian

$$\mathcal{H} = -2J \sum S_r^x S_{r'}^x + S_r^y S_{r'}^y , \qquad (1)$$

where the sum is over n.n. pairs and J is the exchange integral. Since the Hamiltonian does not have spin exchange in the longitudinal direction, the transverse and longitudinal total spins M(x) and M(z), respectively, have very different dynamical and static behavior.

The dynamic form factor is defined as

$$\mathcal{S}^{xx}(k\omega) = \int_{-\infty}^{\infty} dt\, e^{-i\omega t} <S_{-k}^x(o)\, S_k^x(t)> , \qquad (2)$$

where

$$S_k^x(t) = (\exp i\mathcal{H}t)\, S_k^x\, (\exp -i\mathcal{H}t). \qquad (3)$$

The above eq. can be written as

$$S_k^x(t) = S_k^x + it\, [\mathcal{H}, S_k^x] + \tfrac{1}{2}(it)^2\, [\mathcal{H},[\mathcal{H},S_k^x]] + \cdots$$

$$\equiv \sum (it)^n\, c_n/n! . \qquad (4)$$

If c_1 is defined by

$$[\mathcal{H}, S_k^x] = y_k , \qquad (5)$$

by working out the commutator it can be shown that

$$y_k = i \sum J(rr')\, S_r^z\, S_{r'}^y\, (\exp ik.r). \qquad (6)$$

Using the above result, we find for c_2 that

$$[\mathcal{H}, y_k] = qJ^2\, S_k^x + \psi_k + \theta_k , \qquad (7)$$

where q is the coordination number,

$$\psi_k = \sum J(rr')\, J(r'r'')\, S_r^z\, S_{r'}^z\, S_{r''}^x\, \exp ik.r , \qquad (8)$$

and

$$\theta_k = \sum J(rr')\, J(r'r'')\, S_r^y \cdot (S_{r'}^x, S_{r''}^y - S_{r'}^y, S_{r''}^x) \cdot \exp ik.r' . \qquad (9)$$

Observe in (7) that apart from the constant factor qJ^2 the first term of c_2 is exactly c_o. The second term ψ contains essentially the z-spin pair correlation which in the XY model is weak compared with the x-spin correlation.[2] Hence, here we can regard ψ as a kind of perturbation to the first term. If the restriction on the lattice sum(rr') is slightly relaxed, the third term θ is identically zero. When one calculates the dynamic form factor, it enters into the calculation in the form tr $\theta \mathcal{H}^n$, which becomes very small in the large-n limit. Next consider c_3. The first term of c_2 reproduces c_1 exactly. Hence, if (7) is truncated keeping only the first term $qJ^2 S^x$, it leads to a simple summable series for the expansion (4). This is our zeroth order approximation. As to the truncated portion ψ and θ, $[\mathcal{H},\theta]$ produces terms which are all proportional to θ. Thus, as a start it is reasonable to neglect the contribution of θ in all orders of expansion. In contrast, $[\mathcal{H},\psi]$ generates nonlinear terms. By regarding ψ formally as a perturbation to S^x, one can also make the expansion summable. The summability is found to be related to some interesting physical properties of the system.[3] The inclusion of ψ in eq. 4 and neglecting terms proportional to θ will be our first-order approximation. Finally, the contribution of θ is to be considered as a still-higher-order correction.

In zeroth order then we truncate the commutator (7) and write

$$[\mathcal{H}, y_k] = \omega_0^2 S_k^x , \quad (10)$$

where $\omega_0 = \sqrt{q}\, J$. Then, using (5) and (10) we can evaluate the successive commutators to all orders of expansion and obtain

$$S_k^x(t) = S_k^x \cos \omega_0 t + i\, y_k \sin \omega_0 t/\omega_0 . \quad (11)$$

Hence, the dynamic form factor is

$$2 \mathcal{S}^{xx}(k\omega) = \left(Y^{xx}(k) + U^{xx}(k)/\omega\right)\left(\delta(\omega-\omega_0) + \delta(\omega+\omega_0)\right) , \quad (12)$$

where

$$Y^{xx}(k) = \langle S_{-k}^x S_k^x \rangle - \langle S_{-k}^x \rangle \langle S_k^x \rangle , \quad (13)$$

and

$$U^{xx}(k) = \langle S_{-k}^x y_k \rangle . \quad (14)$$

The dynamic form factor satisfies the static form-factor sum rule, the f-sum rule, and the compressibility sum rule. The static susceptibility is

$$\chi^{xx}(k) = Y^{xx}(k)\, \sinh\beta\omega_0/\omega_0 + U^{xx}(k) \cdot (1-\cosh\beta\omega_0)/\omega_0^2 . \quad (15)$$

In the high-temperature limit, the susceptibility and fluctuation behave identically:

$$\beta^{-1} \chi^{xx}(k) = Y^{xx}(k) + O(\beta) . \quad (16)$$

We have already shown[2] that in the long-wavelength limit the fluctuation and susceptibility have the same critical behavior. Also, the diffuse scattering cross section shows that it is sharply peaked at $\omega = \pm\omega_0$ but it has no central peak. This is a consequence of the fact that the XY-model dynamic form factor does not satisfy the diffusion equation. Transverse spins do not diffuse but propagate. This is in contrast to the cross section for the Heisenberg paramagnet which is centrally peaked.[1]

DYNAMIC SUSCEPTIBILITY AND RELAXATION FUNCTION

The imaginary part of the dynamic susceptibility follows from the fluctuation-dissipation theorem

$$-2\text{Im } \chi^{xx}(k\omega) = \pi (1 - \exp - \beta\omega) \cdot (Y^{xx}(k) + U^{xx}(k)/\omega)$$
$$\times (\delta(\omega-\omega_0) + \delta(\omega+\omega_0)). \qquad (17)$$

Now since the real and imaginary parts of the dynamic susceptibility is related by a Kramers-Kronig relation, we obtain

$$\text{Re } \chi^{xx}(k\omega) = \omega_0 (\omega_0^2 - \omega^2)^{-1} (Y^{xx}(k) \sinh \beta\omega_0$$
$$+ U^{xx}(k) (1-\cosh \beta\omega_0)/\omega_0). \qquad (18)$$

From (15) and (18) we see that

$$\text{Re } \chi^{xx}(k,\omega\to 0) = \chi^{xx}(k) + O(\omega^2) \qquad (19)$$

and

$$\text{Re } \chi^{xx}(k,\omega\to\infty) = -\omega_0^2 \chi^{xx}(k)/\omega^2 + O(\omega^{-4}). \qquad (20)$$

The above result is valid for all values of k. Thus, (19) implies that in the long-time limit the dynamic (Kubo or isolated) susceptibility becomes identical to the isothermal susceptibility. The longitudinal component does not satisfy this relation. That is, Re $\chi^{zz}(k=0,\omega\to 0) = 0$, whereas the isothermal susceptibility $\chi^{zz}(k=0) \ne 0$. Also, from (18) we observe that Re $\chi^{xx}(k\omega)$ has two real poles. When the first order corrections are introduced, the poles can become complex whose imaginary part is then related to the width in the diffuse cross section.

The relaxation function R^{xx} can be obtained from the relation

$$\omega R^{xx}(k,\omega) = (1 - \exp(-\beta\omega)) \mathscr{J}^{xx}(k\omega) \qquad (21)$$

Hence,

$$R^{xx}(k,t) = (\cos \omega_0 t) \cdot \chi^{xx}(k). \qquad (22)$$

Observe that the relaxation function satisfies the required relations[1]

$$R^{xx}(k,t=0) = \chi^{xx}(k) \qquad (23)$$

and

$$\frac{\partial}{\partial t} R^{xx}(kt) = -\chi^{xx}(kt). \qquad (24)$$

DISCUSSION

While the validity of our approximation cannot be discussed here, we can give a limited physical picture of the nature of our approximation. Since (10) depends only on commutation relations, one can find an equivalent Hamiltonian $\widetilde{\mathscr{H}}$ for which our approximation is exact. Such a Hamiltonian can be found if we, as in molecular field approximation, let $S_r^z = \lambda$ for all r, where λ is a constant. Then, in terms of some new operators we find that the equivalent Hamiltonian is

$$\widetilde{\mathscr{H}} = \widetilde{E}_0 - \Sigma_k \Omega_k n_k, \qquad (25)$$

where n_k and $\widetilde{E}_0$ are, respectively, the number operator and the ground state energy (provided that $\Omega_k < 0$--that is, antiferromagnetic). The equivalent Hamiltonian satisfies the familiar commutation relations for harmonic oscillators. This result recalls the RPA solution of a dense electron gas problem where Ω_k would represent the single-particle and collective excitations.[4] For small k, most of the correlation energy is taken up by the collective mode, which is essentially independent of k and is undamped. Thus, this plasma oscillation corresponds to our ω_0-mode in the XY model.

The first order approximation generalizes the zeroth-order expression (11) in the form

$$S_k^x(t) = \Xi_k(t) S_k^x + \phi_k(t) Y_k ,\qquad(26)$$

where Ξ and ϕ are, respectively, the generalized relaxation function and its time derivative. This is similar to the time-correlation function of Mori derived from the Langevin equation using a continued-fraction representation.[5] In fact, a two-pole approximation reduces (26) to (11). By combining our approach with that due to Mori, we can extend the calculation of the relaxation function beyond a 3-pole approximation, which is a present limit.[6] We can obtain the singularities of the relaxation function, where the imaginary part of the dominant singularity is the decay rate, and also describe the summability and the validity of a multi-pole approximation via theorems associated with the convergence of the Padé approximants.

REFERENCES

1. W. Marshall and R.D. Lowde, Rept. Prog. Phys. 31, 705 (1968).
2. M.H. Lee, Phys. Rev. B 8, 1203 (1973).
3. M.H. Lee, Phys. Rev. B Oct. 1973 (in press).
4. K. Sawada, Phys. Rev. 106, 372 (1957); and M.H. Lee, Prog. Theor. Phys. 40, 990 (1968).
5. H. Mori, Prog. Theor. Phys. 34, 399 (1965).
6. S.W. Lovesey and R.A. Meserve, J. Phys. C 6, 79 (1973).

RECIPROCITY RELATIONS FOR SUSCEPTIBILITIES AND FIELDS IN MAGNETOELECTRIC ANTIFERROMAGNETS

George T. Rado
Naval Research Laboratory, Washington, D.C. 20375

ABSTRACT

A clarified formulation of the generalized Onsager reciprocity relation for susceptibilities is proposed for antiferromagnets. No net magnetic moments and no biasing fields are assumed to be present. A detailed quantum mechanical treatment of time reversal is used to show explicitly that the Onsager relation for the matrix [S] of the generalized susceptibility does remain valid in a magnetoelectric, antiferromagnetic medium provided the medium (and hence its [S]) is time-reversed whenever [S] is transposed. By combining the resulting reciprocity relation for the magnetoelectric susceptibility (i.e., the magnetoelectric part of [S]) with Maxwell's equations, a modified reciprocity theorem is then derived for electromagnetic fields and circuits. The essence of this theorem is that the usual reciprocity theorem of electromagnetic theory does remain valid in a magnetoelectric, antiferromagnetic medium provided the medium is time-reversed whenever an electromagnetic source and a detection apparatus are interchanged. The full text of this paper will appear in <u>The Physical Review B1</u>.

LINEAR MOMENTUM IN THE REST FRAME OF A FERROMAGNET AND THE ELECTRODYNAMIC CONSEQUENCES*

F. R. Morgenthaler
Department of Electrical Engineering
and Center for Materials Science and Engineering
Massachusetts Institute of Technology
Cambridge, Massachusetts 02139

ABSTRACT

A new energy-momentum tensor for a moving ferromagnet is presented that contains large-signal linear momentum due to spin precession even when the lattice is at rest. The associated energy density and power flow are interpreted as arising from precession of the intrinsic angular momentum. Therefore, it is at least plausible that the quantum-mechanical exchange of electrons between sites that are unequal energetically can cause kinetic momentum flow--quite apart from lattice motion.

In the generally accepted formulation of magnetostatic stress and force density in a ferromagnet,[1] there is no large-signal material momentum. Therefore, although the second-order components of large-signal power and energy connect directly with the corresponding small-signal quantities, that are based upon the linearized equations of motion, the same cannot be true of stress and momentum. Because all four quantities are interrelated through the transformation properties of the energy-momentum tensor, this fact seems inconsistent. A new formulation of energy-momentum conservation in a moving ferromagnet is presented that does connect with the small-signal theorem in the expected way; moreover, the new model lends itself to a simple physical interpretation. Consequently, the previous interpretation of the small-signal stress and momentum[2] as physically significant force producing quantities seems justified.

We choose to work in a four-dimensional representation because, although we are not interested in relativistic terms in v^2/c^2, connection with the electromagnetic fields requires that all equations be Lorentz invariant. The magnetostatic approximation--valid when the wave group velocity is nonrelativistic--is contained as a special case.

If each of the four-vector magnetic moments $\mathcal{Q}$ that comprise a ferromagnet is conserved, the governing equation of motion must be expressible (in MKS units) as

$$\frac{\partial \mathcal{Q}}{\partial \tau} = -\frac{i}{c} \gamma_g' \mu_o (\mathcal{Q} \times \mathbb{H}^E)^{\dagger} \cdot \mathbb{V} + \mathcal{Q} \cdot \frac{\partial \mathbb{V}}{\partial \tau} \frac{\mathbb{V}}{c^2} \qquad (1)[3]$$

where $\gamma = (1 - v^2/c^2)^{-1/2}$, c the free space velocity of light, $\gamma_g' = \gamma_g \gamma$, the rest frame gyromagnetic ratio (negative)

$$\mathcal{Q} = (\bar{a} + \gamma^2 (\frac{\bar{a} \cdot \bar{v}}{c^2}) \bar{v}, \; i\gamma^2 \frac{\bar{a} \cdot \bar{v}}{c}) \qquad (2)$$

*This research was sponsored by the NSF under Contract GH-33635.

τ the proper time ($\frac{\partial}{\partial \tau} = \gamma \frac{d}{dt}$) and

$$\mathbb{V} = (\gamma \overline{v}, i\gamma c) \qquad (3)$$

the four-vector velocity associated with the motion of $\overline{a}$.

In terms of three-space vectors, (1) is equivalent to

$$\frac{d\overline{a}}{dt} + \frac{\overline{a_3} \cdot \overline{v}}{c^2} \frac{d\overline{v}}{dt} = \gamma_g \mu_o \overline{a_3} \times \overline{H}^E \qquad (4)$$

where

$$\overline{H}^E = \frac{1}{\gamma} \overline{\mathbb{H}}_3^E + \gamma \frac{\overline{v}}{c} (\frac{\overline{\mathbb{H}}_3^E \cdot \overline{v}}{c} + i \mathbb{H}_4^E) \qquad (5)$$

and $\overline{a_3}$ and $\overline{\mathbb{H}}_3^E$ are respectively the three-space components of $\mathcal{Q}$ and $\mathbb{H}^E$. Notice that $\mathcal{Q} \cdot \mathcal{Q} = \overline{a_3} \cdot \overline{a} = a^2$.

For $\mathbb{H}^E$ arising only from isotropic (in the rest frame) exchange, it can be shown[4] that

$$\mathbb{H}^E = \gamma \frac{1}{n} \frac{\partial}{\partial x_i} (n \Lambda_{ij} \frac{\partial \mathcal{Q}}{\partial x_j}) \qquad (6)$$

where

$$\Lambda_{ij} = \Lambda(\delta_{ij} + \frac{\mathbb{V}_i \mathbb{V}_j}{c^2}) \qquad (7)$$

Here Λ is the exchange constant, n the density of magnetic moments and δ_{ij} the four-dimensional Kronecker delta. Indices take on the values 1,2,3 and 4; summation over repeated indices is assumed.

The tensor

$$\mathbb{T}_{ij} = \begin{cases} -\mu_o n \frac{\partial \mathcal{Q}}{\partial x_i} \cdot \Lambda_{jk} \frac{\partial \mathcal{Q}}{\partial x_k} \\ + \frac{i}{c} \frac{n}{2\gamma \gamma_g'} \frac{}{\mathcal{Q} \cdot \mathcal{Q}} \mathbb{U} \cdot (\mathcal{Q} \times \frac{\partial \mathcal{Q}}{\partial x_i})^\dagger \cdot \mathbb{V} \mathbb{V}_j \\ - \frac{1}{2} \mu_o n \frac{\mathcal{Q} \cdot \mathcal{Q}}{\gamma \mathcal{Q} \cdot \mathbb{U}} \mathbb{U} \cdot \mathbb{H}^E \delta_{ij} \end{cases} \qquad (8)$$

where $\mathbb{U}(\overline{r},t)$ is an arbitrary unit four-vector, has the interesting property that if $\mathbb{V}$, $\mathbb{U}$, and all material parameters are uniform in space and time, then correct to fourth-order in the cone angle (θ) that may exist between $\overline{a}$ and its axis of precession(if the latter is along $\mathbb{U}$),

$$\Box \cdot \mathbb{T} = \frac{\partial}{\partial x_j} (\mathbb{T}_{ij}) = 0 \qquad (9)$$

Thus, $\mathbb{T}$ has the form of an energy-momentum tensor for a closed system; moreover it can be shown[4] to connect with the corresponding electromagnetic tensor in the event that the four-vector magnetic field

$$\mathbb{H} = [\gamma(\overline{H} - \epsilon_o \overline{v} \times \overline{E}), i\gamma \frac{\overline{H} \cdot \overline{v}}{c}] \qquad (10)$$

is added to $\mathbb{H}^E$. Consequently, we postulate (8) to have such physical significance when $\mathbb{U}$ is taken to be the precession-axis.

In the exchange only magnetostatic limit, all normal electromagnetic effects are negligible, Galilean Relativity applies and the total energy density, momentum density, Poynting Flux and stress including the contributions due to the continuum kinetic energy are

$$W = \tfrac{1}{2}nmv^2 + \tfrac{1}{2}\mu_o n\Lambda \frac{\partial \bar{a}}{\partial x_j} \cdot \frac{\partial \bar{a}}{\partial x_j} + (\bar{G} - nm\bar{v}) \cdot \bar{v} \tag{11}$$

$$G_i = nmv_i + \frac{n}{2\gamma_g \bar{a} \cdot \bar{u}} \bar{a} \times \frac{\partial \bar{a}}{\partial x_i} \cdot \bar{u} \tag{12}$$

$$S_i = \tfrac{1}{2}nmv^2 v_i - \mu_o n\Lambda \frac{\partial \bar{a}}{\partial x_i} \cdot \frac{\partial \bar{a}}{\partial t} - \frac{n}{2\gamma_g \bar{a} \cdot \bar{u}} \bar{a} \times \frac{\partial \bar{a}}{\partial t} \cdot \bar{u} \, v_i \tag{13}$$

and

$$T_{ij} = -G_i v_j - \mu_o n\Lambda \frac{\partial \bar{a}}{\partial x_i} \cdot \frac{\partial \bar{a}}{\partial x_j} - \tfrac{1}{2}\mu_o n \frac{a^2}{\bar{a}\cdot\bar{u}} \bar{u} \cdot \bar{H}^{ex} \delta_{ij} \tag{14}$$

where nm is the mass density of the crystal. The second-order components of (11) through (14) agree with the corresponding small-signal quantities derived from the linearized equations.

Eq. (11) can be rewritten as

$$W = \tfrac{1}{2}nmv^2 - \frac{n}{2\gamma_g \bar{a}\cdot\bar{u}} \bar{a}\times\frac{\partial \bar{a}}{\partial t}\cdot\bar{u} - \tfrac{1}{2}\mu_o n \frac{a^2}{\bar{a}\cdot\bar{u}} \bar{u}\cdot\bar{H}^{ex} \tag{15}$$

and the second term identified[5] as excess kinetic energy acquired by the precessing angular momentum vector $\bar{a}/(-\gamma_g)$ when the precession cone angle θ is nonzero; the last term as potential energy associated with the precessional motion.

For a single plane wave of frequency ω, wave vector $\bar{k}$ and constant cone angle θ traveling in a uniform crystal, the large signal equations of motion gives

$$\omega = -\gamma_g \mu_o \Lambda a k^2 \cos\theta + \bar{v}\cdot\bar{k} \tag{16}$$

$$W = \tfrac{1}{2}nmv^2 + \tfrac{1}{2}n^*m^*(v_{group}^2 - v^2) \tag{17}$$

$$\bar{G} = nm\bar{v} + n^*m^*(\bar{v}_{group} - \bar{v}) \tag{18}$$

$$\bar{S} = \tfrac{1}{2}nmv^2\bar{v} + \tfrac{1}{2}n^*m^*(v_{group}^2 - v^2)\bar{v}_{group} \tag{19}$$

and

$$\bar{\bar{T}} = -nm\bar{v}\,\bar{v} - n^*m^*(\bar{v}_{group} - \bar{v})\bar{v}_{group} \tag{20}$$

where the magnon density and effective mass are, respectively,

$$n^* = \frac{na}{-2\gamma_g \hbar} \frac{\sin^2\theta}{\cos\theta} \qquad (21)$$

and

$$m^* = \frac{\hbar}{-2\gamma_g \mu_o \Lambda a \cos\theta} \qquad (22)$$

Evidently the excitation of magnons raises the kinetic energy density of the continuum by an amount that is positive definite even when the lattice is at rest ($\bar{v} = 0$). It is this energy borrowed from the reservoir of orbital and/or spin cooperative motion, that is available to carry exchange power.

For the single plane wave, the force density created when the RHS of (9) is nonzero, is basically $n^* \vec{\nabla} \hbar \omega(\vec{k},\vec{r})$ where ω is the dispersion relation in the nonuniform crystal. This is the reaction to the force $d(\hbar \vec{k})/dt$ that accelerates each magnon of the wave packet.[6] Evidently the particle representation transcends the small signal regime for which $\sin\theta \approx \theta$ and $\cos\theta \approx 1$. The nonzero fourth-order terms mentioned previously are present when θ is non-uniform because according to (16) ω is proportional to $\cos\theta$.

Electromagnetic effects are included through use of the four-vector fields $\mathbb{E}$ and $\mathbb{H}$, given by

$$\mathbb{E} = [\gamma(\bar{E} + \mu_o \bar{v} \times \bar{H}), i\gamma \frac{\bar{E} \cdot \bar{v}}{c}] \qquad (23)$$

and Eq. (10). One can show[4] that the Chu formulation (discussed extensively in Ref. 1) of Maxwell's Eqs. is equivalent to

$$\square \cdot [-\frac{i}{c}(\mathbb{H} \times \mathbb{V})^\dagger - \mathbb{D} \times \mathbb{V}] = \mathbb{J}_f = 0 \qquad (24)$$

and

$$\square \cdot [-\frac{i}{c}(\mathbb{E} \times \mathbb{V})^\dagger + \mathbb{B} \times \mathbb{V}] = 0 \qquad (25)$$

where $\mathbb{D} = \varepsilon_o \mathbb{E} + \mathbb{P}$, $\mathbb{B} = \mu_o(\mathbb{H} + \mathbb{M})$ and $\mathbb{M} = \frac{n}{\gamma}\mathbb{Q}$; we concentrate on the magnetic issues by setting $\mathbb{P} = 0$. The appropriate additions to $\overline{\overline{T}}$ appear to be $\varepsilon_o \mathbb{E} \mathbb{E} + \mu_o \mathbb{H}(\mathbb{H} + \mathbb{M}) - \frac{1}{2}\mu_o \mathbb{H} \cdot (\mathbb{M} - \mathbb{M}^o)(\mathbb{I} + 2\frac{\mathbb{V}\mathbb{V}}{c^2})$
$+ \frac{i}{c^3}\{[\mathbb{E} \times (\mathbb{H} + \mathbb{M} - \mathbb{M}^o)]^\dagger \cdot \mathbb{V}\mathbb{V} + \mathbb{V}(\mathbb{E} \times \mathbb{H})^\dagger \cdot \mathbb{V}\} + \frac{1}{2}\mu_o \mathbb{H} \cdot \mathbb{M}^o \mathbb{I}$

where $\mathbb{M}^o = \frac{1}{2}(\frac{\mathbb{M} \cdot \mathbb{M}}{\mathbb{M} \cdot \mathbb{U}} + \mathbb{M} \cdot \mathbb{U})\mathbb{U}$. When the elastic and magnetoelastic contributions are also added, the entire system of coherent energy-momentum (if lossless) satisfies

$$\square \cdot (\rho_o \mathbb{V}\mathbb{V} - \overline{\overline{T}}) = 0 \qquad (26)$$

where ρ_o is the rest frame mass density of the crystal.

Dedicated to the memory of Lan Jen Chu

REFERENCES

1. The 1964-1965 theories of H. F. Tiersten, D. E. Eastman and W. F. Brown, Jr. are referenced, discussed and rederived by means of the "principle of virtual power" developed by P. Penfield, Jr. and H. A. Haus in their important book, Electrodynamics of Moving Media, Research Monograph No. 40 MIT Press, Cambridge, Mass., (1967). Refer especially to Sec. 4.9, p. 89, but note that their sign convention for material and electromagnetic

stress is opposite to that taken here.
2. F. R. Morgenthaler, "Exchange Energy, Stress and Momentum in a Rigid Ferromagnet", J. Appl. Phys. $\underline{38}$, 1069, (1967). "Dynamic Magnetoelastic Coupling in Ferromagnets and Antiferromagnets", IEEE Trans. Magnetics MAG-8 130, March 1972.
3. The cross product of the four-vectors $\mathbb{A} = (\bar{A}, A_4)$ and $\mathbb{B} = (\bar{B}, B_4)$ is the antisymmetric tensor $(\mathbb{AB} - \mathbb{BA})$ which is made up of six independent terms that are the components of $\bar{A} \times \bar{B}$ and $B_4 \bar{A} - A_4 \bar{B}$. The operation $(\mathbb{A} \times \mathbb{B})\dagger$ simply interchanges the roles of the latter two three-vectors to form the dual.
4. F. R. Morgenthaler, MIT Microwave and Quantum Magnetics Group Technical Report 28, July, 1973.
5. This interpretation is based upon consideration of the following elementary model used to represent the angular momentum vector $\bar{L} = \bar{a}/(-\gamma_g)$ when it precesses about a fixed axis (taken here along $-z$) with frequency ω_p.

Assume that $\bar{L}$ arises from orbital precession of a single point charge of mass m. The orbit $x = r\cos\omega t$, $y = -r\sin\omega t$, $z = -r\tan\theta\cos(\omega+\omega_p)t$ creates

$$\bar{L} = -L_o[\sin\theta(\bar{i}_x \cos\omega_p t + \bar{i}_y \sin\omega_p t) + \cos\theta \bar{i}_z]$$

provided $mr^2\omega = L_o\cos\theta$ and $\omega \gg \omega_p/\cos\theta$. The nonrelativistic kinetic energy of the particle can be time-averaged over the very high frequency ω to give

$$\frac{1}{2}mv^2 = \frac{1}{2}L_o\omega \frac{1+\cos^2\theta}{2\cos\theta} + \frac{1}{2}L_o \frac{\sin^2\theta}{\cos\theta}\omega_p$$

Correct to fourth-order in θ, the <u>excess</u> kinetic energy (over the $\theta = 0$ value) is simply

$$\frac{1}{2}L_o \frac{\sin^2\theta}{\cos\theta}\omega_p = \frac{-1}{2\gamma_g \bar{a}\cdot\bar{i}_z}(\bar{a}\times\frac{d\bar{a}}{dt})\cdot\bar{i}_z$$

On a per unit volume continuum basis the connection with the second term of Eq. (15) is evident.

Notice that

$$-\frac{1}{2}\mu_o n\bar{a}\cdot\bar{H}^E = \frac{n}{2\gamma_g \bar{a3}\cdot\bar{u}}\bar{a}\times(\frac{d\bar{a}}{dt} + \frac{\bar{a3}\cdot\bar{v}}{c^2}\frac{d\bar{v}}{dt})\cdot\bar{u}$$

$$-\frac{1}{2}\mu_o n \frac{a^2}{\bar{a3}\cdot\bar{u}}\bar{u}\cdot\bar{H}^E$$

(where $\bar{u}$ is an arbitrary unit vector) is an identity based upon Eq. (4). However, physical significance in connection with the model enters only when $\bar{u}$ is taken to be the axis of precession.
6. The force leads to an exchange radiation pressure $p = \mu_o na^2 \Lambda k^2 \theta_o^2/2$ exerted on a free surface by a normally-incident reflected magnon, where θ_o is the precession cone angle at the surface. For yttrium iron garnet and k in cm^{-1}, $p = 4 \times 10^{-7} k^2 \theta_o^2$ dynes/cm^2.

AUTHOR INDEX

Abeledo, C.R.	380	Budnick, J.I.	307,437
Abito, G.F.	626	Buhrer, C.F.	139
Adam, J.D.	1279	Bui-Thieu, T.	1284
Adour, D.L.	362	Bullock, D.C.	232
Aharony, A.	863	Burch, T.J.	307,437
Ahn, K.Y.	1103	Buttry, R.W.	432
Ahrenkiel, R.K.	1118	Buyers, W.J.L.	1305
Akselrad, A.	949		
Alben, R.	770	Cable, J.W.	426
Albert, P.A.	1103	Callen, E.	1360
Aldred, A.T.	336	Camp, W.J.	878
Alperin, H.A.	563	Campagna, M.	1388
Anderson, E.E.	362	Campbell, C.C.M.	319
Anderson, P.W.	895	Campbell, T.G.	759
Anderson, W.E.	1385	Cannella, V.	307,651
Arajs, S.	362	Cannon, J.A.	437
Archer, J.L.	115,116,139	Cargill III, G.S.	631
Arrott, A.S.	396	Carlo, J.T.	1268
Asai, K.	1044	Carnall, Jr., E.	1118
Asama, K.	110	Carrara, P.	341
Asano, A.	262	Chandler, R.N.	534
Asik, J.R.	1349	Chang, T.S.	776
Aton, T.J.	252	Chaudhari, P.	247,578,896
		Chen, C.W.	432
Baberschke, K.	984	Chen, E.Y.	329,334
Bachmann, K.	1168	Chen, T.	1222
Bahurmuz, A.A.	1059	Chen, T.T.	115
Baird, D.H.	139	Chen, Y.S.	100
Baker, G.L.	318	Chin, G.Y.	1159
Bambakidis, G.	324	Chipman, D.R.	1187
Bari, R.A.	661	Chock, E.P.	989
Barker, R.C.	1284,1285	Clark, A.E.	636,1015
Barmatz, M.	754	Clark, P.E.	1412
Barrett, P.H.	438	Clinton, J.R.	316
Bartel, L.C.	673	Coburn, T.J.	1118
Beckenbaugh, W.	540	Coleman, J.A.	152
Bekebrede, W.R.	167	Coleman, R.V.	292
Belt, R.F.	954	Collins, J.H.	1279
Bene, R.W.	534	Colpa, J.H.P.	806
Benedict, V.	335	Cone, R.L.	545
Benz, M.G.	1173	Coren, R.L.	122,1287
Berger, L.	918	Cowen, J.A.	391
Bergman, D.J.	771	Cox, D.E.	386
Bernasconi, J.	1182	Craig, R.S.	1207,1297
Bertram, H.N.	1113	Cronemeyer, D.C.	75,85
Bhatia, S.N.	335	Crowell, J.M.	427
Bhide, V.G.	504	Cuomo, J.J.	578,641
Birgeneau, R.J.	1066,1299	Cusick, J.P.	324
Bloembergen, P.	806	Cuthill, J.R.	1039
Bloomfield, P.E.	687		
Blume, M.	893	Das, D.K.	1149
Bohning, O.D.	139	Das, T.P.	1049
Bongianni, W.L.	1005	Date, M.	1078
Bonner, J.C.	1311	Davidov, D.	989
Bonner, W.A.	68	Davidson, G.R.	381,386
Bonyhard, P.I.	100	Davidson, J.B.	396
Borg, R.J.	317	Davis, H.L.	1068
Bouricius, W.G.	132	De Jongh, L.J.	806
Bowen, S.P.	683	Deffeyes, R.J.	1108
Boyce, J.B.	252	de Leeuw, F.H.	199
Boyle, A.J.F.	616	Della Torre, E.	127
Brandow, B.H.	702	Dentz, D.J.	954
Bremer, J.C.	970	Deyoung, T.F.	362
Brezin, E.	849	Dillon, J.F.	329,334
Bridger, N.J.	357	Dixon, G.S.	1073
Brittain, J.O.	272	Doane, D.A.	1192
Brodsky, M.B.	357	Donoho, P.L.	1263
Brooks, M.S.S.	1258,1300	Doran, J.C.	980
Brown, E.B.	687	Dormann, E.	529
Brun, T.O.	426	Doyle, W.D.	152
Bucher, E.	535,1298	D'Silva, T.	1253
Buchner, E.	748	Dunk, P.	162

Dunlap, B.D.	366,381	Guenzer, C.S.	1292
Dwight, A.E.	1242	Guggenheim, H.J.	329
		Guyot, M.	1382
Eagen, C.F.	301	Gyorgy, E.M.	70
Eastman, D.E.	903,1030		
Edwards, L.R.	401	Haddad, J.B.	964
Egami, T.	1258,1300	Hagedorn, F.B.	222
Eibschutz, M.	381,386	Hanabusa, M.	1286
Elliott, S.S.	1273	Hankey, A.	776
Endoh, Y.	1065	Harbus, F.	776
Engel, U.	984	Harmer, R.S.	1217
England, W.B.	1054	Harris, A.B.	593
Entin-Wohlman, O.	771	Harrold, W.J.	1340
Erbudak, M.	1030	Hartings, M.F.	1163
Erickson, N.E.	1039	Hata, H.	1078
Erickson, R.A.	411,416	Hauger, R.	540
Evans, B.J.	518	Heilner, E.J.	1232
		Heiman, N.	525
Fairbank, W.M.	1335	Heinz, D.M.	194
Fedro, A.J.	678,1055	Heller, P.	764,893
Felcher, G.P.	426	Henmi, Z.	730
Feldman, M.	764	Henry, R.D.	194
Ferer, M.	883	Hibiya, T.	80
Ferrando, W.A.	523	Hirano, M.	110
Fert, A.R.	341	Hoch, M.	1197
Feuersanger, A.E.	105	Holden, T.M.	1305
Fickett, F.R.	740	Holt, S.L.	376
Fisher, M.E.	888	Holtzberg, F.	340,478,902,908,1030
Fitchen, D.B.	929	Honda, S.	741
Flanders, P.J.	1354	Horiguchi, T.	621
Flannery, W.E.	152	Hornfeldt, S.P.	426
Flood, D.J.	1345	Hsin, C.H.	122
Foner, S.	1362	Hu, H.L.	213
Forester, D.W.	523	Hubbell, W.C.	1263
Foster, B.	713	Huber, D.L.	892,894
Foy, M.L.G.	525	Huberman, B.A.	351
Frankel, R.B.	380	Hubert, A.	178
Frederick, W.G.D.	1197	Huck, F.B.	969
Freeouf, J.	1030	Hudis, J.	267
Friebele, E.J.	594	Huffman, G.P.	277
Friedberg, S.A.	1311,1316	Hüfner, S.	984
Fritz, I.J.	401	Hutchens, R.D.	1212
Fujiwara, S.	157		
Fulde, P.	1	Igarashi, H.	1253
		Iida, S.	913,1044
Gambino, R.J.	578,641	Imry, Y.	771
Gangulee, A.	1103	Irons, H.R.	90,1383
Gardner, J.A.	970	Isaacs, L.L.	312
Gaunt, P.	1202	Ishikawa, Y.	1065
Gelard, J.	341		
George, P.K.	115,116	Jaccarino, V.	529,1083
Geusic, J.E.	69	Jacobs, J.T.	1122
Giessen, B.C.	282	Jao, J.K.	940
Glinka, C.J.	1060	Jarrett, H.S.	673
Globus, A.	1382	Jennings, L.D.	1187
Goer, D.A.	416	Johnson, G.H.	929
Goldberg, N.	1138	Johnson, J.W.	1306
Golding, B.	754	Johnson, Jr., C.E.	1108
Gonsalves, A.	1361	Jones, G.A.	162
Gorter, F.W.	806	Josephs, R.M.	227
Gottlieb, A.M.	764	Judy, J.H.	1108
Graham, Jr., C.D.	1192		
Greedan, J.E.	749,1212	Kadar, G.	421
Green, D.C.	896	Kahan, H.M.	746
Green, M.L.	1159	Kaldis, E.	540
Greiner, J.D.	432	Kalvius, G.M.	616
Grinstein, G.	876	Kandel, L.D.	380
Grobman, W.D.	903	Kaplan, B.Z.	1330
Grodkiewicz, W.H.	1232	Kaplan, T.A.	656
Gruzalski, G.R.	282	Karnezos, M.	1316
Guentherodt, G.	540,1034	Kasevich, R.S.	1340

Kasper, H.M.	748	Malozemoff, A.P.	242
Kassai, M.	730	Maple, M.B.	447
Kawashima, M.	1363	Marti, J.	1377
Kehler, T.P.	1287	Martin, D.L.	1173
Keller, D.A.	1207	Martin, J.J.	1073
Kennedy, T.A.	287	Matcovich, T.J.	122
Kenner, G.W.	1361	Matsumi, K.	80
Kestigian, M.	167	Mayer, A.M.	1361
Kikuchi, R.	605	McAlister, A.J.	1039
Kim, H.M.	989	McClure, Jr., J.C.	1355
Kimball, C.W.	1242	McCollum, D.C.	376
King, A.R.	1093	McGuire, T.R.	903
Kinne, R.W.	930	Mee, J.E.	63
Kinsner, W.	127	Meier, D.	1316
Kirkpatrick, S.	641	Melcher, C.L.	1263
Kitchens, T.A.	959	Meltzer, R.S.	545
Kiwi, M.	347	Menth, A.	1177,1182
Kjems, J.K.	1066,1299	Mezei, F.	406
Klein, H.P.	1177,1182	Michaelis, P.C.	140
Klokholm, E.	75	Miedema, A.R.	806
Knott, H.W.	352	Mildrum, H.F.	1163
Kobayashi, T.	172,1284	Miller, A.E.	1253
Koch, E.F.	1144	Minkiewicz, V.J.	1060
Koon, N.C.	302,594,1247	Minnaja, N.	1123
Koopmann, G.	984	Miranda, L.C.M.	610
Kopp, J.P.	540	Mizushima, K.	1044
Kossler, W.J.	525	Montano, P.A.	371,438
Kramer, J.J.	1367	Moody, J.W.	237
Kren, E.	421	Mook, H.A.	781,1068
Krishnan, R.	1227	Moore, E.B.	193
Kryder, M.H.	213	Morgenthaler, F.R.	720,940,1268
Kushida, T.	1286	Morris, R.C.	292
Kusuda, T.	741	Morris, T.M.	242
		Morrish, A.H.	1412
Laforce, R.P.	1173	Mueller, M.H.	352,442
Lam, D.J.	366,442	Mullen, M.E.	1298
Landau, D.P.	819	Murata, K.K.	692
Lander, G.H.	352,442	Murphy, J.C.	1286
Lawson, A.W.	759	Murphy, J.J.	437
Lazzari, J.P.	990	Mydosh, J.A.	651
Lecraw, R.C.	188	Myron, H.W.	1054
Lee, K.	747		
Lee, K.H.	376,1408	Narasimhan, K.S.V.	1212,1248
Lee, M.H.	714	Nelson, D.R.	888
Lee, P.A.	895	Nelson, T.J.	95
Leiper, W.	319	Nereson, N.G.	1297
Leroux-Hugon, P.	897	Nesbitt, E.A.	1159
Lessoff, H.	1292	Neurath, P.W.	1322
Levy, P.M.	923	Neurath, V.H.	1322
Lewis, J.L.	928	Newbury, D.E.	1372
Libelo, L.F.	708	Newman, P.	391
Lilienthal, H.	641	Ng, G.	127
Lilioussis, K.T.	1237	Nicklow, R.M.	555,781
Litterst, F.J.	616	Nicoli, D.F.	896
Little, G.R.	411	Nishida, H.	172
Littman, M.	764	Nobile, M.	1123
Liu, S.H.	1054	Noordermeer, L.	806
Liu, Y.J.	1284,1285	Novak, R.E.	949
Livingston, J.D.	1144		
Longinotti, L.D.	535,1298	O'Sullivan, W.J.	930
Longo, R.T.	287	Obenshain, F.E.	513
Love, J.C.	513	Ogawa, S.	1361
Love, N.D.	1073	Okada, M.	730
Lund, R.E.	1133	Okuda, K.	1078
Luo, H.L.	316	Oosterhuis, W.T.	1326
Lurie, N.A.	893	Opfer, J.E.	1335
Luther, A.	697,876	Owens, J.M.	1279
Luthi, B.	1298		
Lynn, J.W.	781	Pal, L.	421
		Pan, D.	646
Mahoney, J.V.	1297	Paquette, D.	1093
Makino, H.	80	Passell, L.	1060

Patterson, D.L.	949
Patterson, J.D.	615
Patton, C.E.	318
Paul, D.	1377
Pearlman, D.	1118
Peirce, R.J.	140
Penney, T.	908
Perlman, M.L.	267
Petzinger, K.G.	8
Pickart, S.J.	563
Pierce, D.T.	1393
Pifer, J.H.	287
Plaskett, T.S.	75
Pointon, A.J.	1237
Popma, T.J.A.	944
Preston, R.S.	1242
Prinz, G.A.	928
Puttbach, R.C.	954
Quinn, R.K.	297
Rado, G.T.	719
Ramachandran, K.	1098
Rao, C.N.R.	504
Rao, K.V.	362
Rao, P.	1144
Rao, V.U.S.	1297
Ratnam, D.V.	1154
Ravishankar, K.	923
Ray, A.E.	1143
Ray, R.	282
Reddy, J.F.	352,442
Reiter, G.	707,713
Rettori, C.	989
Rezende, S.M.	610,1083
Rhyne, J.J.	563
Rice, T.M.	895
Richards, P.M.	600
Riedel, E.K.	834
Rijnierse, P.J.	199
Rives, J.E.	335
Rivoire, M.	1227
Robertson, J.M.	944
Rogers, R.N.	550
Roth, L.M.	668
Roy, G.	1202
Rupp, Jr., L.W.	535
Ryan, J.F.	930
Sablik, M.J.	923
Salama, K.	1263
Sandercock, J.R.	935
Sandfort, R.M.	237
Sandin, T.R.	1222
Sankar, S.G.	1207
Santiago, J.J.	595
Sarachik, M.P.	964
Sarkar, D.	636
Sasaki, T.	730
Sato, H.	605
Sattler, K.	1388
Saul, D.M.	769
Savage, W.R.	969
Sawatzky, E.	1122
Scalapino, D.J.	8
Schindler, A.I.	301
Schlömann, E.	183
Schneider, T.	891
Schroder, K.	1355
Schwartz, B.B.	1362

Schwartz, L.H.	262
Schwee, L.J.	1383
Schweitzer, J.W.	626,969
Scott, B.A.	896
Scott, J.F.	930
Searle, C.W.	1197
Segnan, R.	636
Sellmyer, D.J.	282
Semura, J.S.	894
Sery, R.S.	90
Shafer, M.W.	1060
Shaffer, J.C.	1055
Shanfield, Z.	438
Shanley, C.W.	1217
Shanley, J.	1253
Shapero, D.	1360
Shaw, R.W.	237
Sherwood, R.C.	1159
Shiga, M.	463
Shiles, E.	1317
Shirane, G.	1065,1299
Showalter, A.B.	974
Shulman, R.G.	1361
Shumate, P.W.	140
Siegel, E.	735
Siegmann, H.C.	1388,1393
Silberglitt, R.S.	794
Silva, N.P.	656
Simon, W.J.	1133
Singal, C.M.	1049
Singh, S.K.	121
Skjeltorp, A.T.	770
Slichter, C.P.	252
Smarra, N.	1326
Smeggil, J.G.	1144
Smith, A.B.	167
Smith, D.H.	69,173
Smith, F.W.	975
Smith, J.L.	100
Smith, K.	1361
Soares, E.	1083
Sokoloff, J.B.	342,662
Sollner, T.C.L.	550
Soohoo, R.F.	1098
Spartalian, K.	1326
Spence, R.D.	391
Spronken, G.	1284
Stalinski, B.	490
Stanley, H.E.	776
Stearns, F.S.	63
Stearns, M.B.	257
Stein, B.F.	48,227
Steingroever, E.	725
Stoll, E.	891
Strassler, S.	1182
Street, G.B.	747
Streever, R.L.	1088
Strnat, K.J.	1163
Stronach, C.E.	525
Stubbs, D.P.	746
Sturge, M.D.	70
Stutius, W.E.	1222
Sugita, Y.	172
Suhl, H.	33
Suits, J.C.	747
Sullivan, D.B.	740
Sun, J.N.	1367
Swartzendruber, L.J.	518,735
Swenson, Jr., H.W.	595
Symko, O.G.	980
Szofran, F.R.	282

Taggart, G.B.	1317
Tahir-Kheli, R.A.	610,615,1317
Tajima, K.	1065
Takahashi, K.	110
Takahashi, M.	172
Takashima, K.	1363
Tanaka, M.	1044
Tanaka, T.	708
Taneja, S.P.	1242
Tang, C.H.	1340
Tao, L.J.	340,641,908
Taylor, R.D.	959
Thiele, A.A.	173
Tilley, J.	697
Toccio, L.R.	115,139
Tommet, T.	892
Torok, E.J.	1133
Torrance, J.B.	340,896,902
Trainor, J.	376
Treves, D.	1122
Tu, K.N.	1103
Turk, R.A.	1403
Turnbull, D.	646
Tyler, E.H.	316
Uffer, L.F.	923
Umemura, S.	913
Valadez, L.	1377
Van Dyke, J.P.	878
Van Uitert, L.G.	69,70,381,1299
Vella-Coleiro, G.P.	69,217
Viens, N.P.	1340
Visscher, P.B.	663
Vittoria, C.	1278,1292
Voegeli, O.	193
Von Molnar, S.	908
Voorhoeve, R.J.H.	19
Vytal, J.J.	139
Wachter, P.	540,1034
Wada, T.	1363
Wakabayashi, N.	555
Walker, J.C.	427
Wallace, W.E.	1207,1212,1248,1297
Walsh, Jr., W.M.	535
Wang, I.	437
Wang. Y.Y.	1408
Wang, Y.L.	1306
Warren, R.G.	63
Watson, R.E.	267,1039
Weaver, H.T.	297
Weber, M.A.	380
Weihrauch, P.	1149
Weiner, R.A.	1398
Wells, M.G. H.	1154
Welsh, L.B.	272
Werner, S.A.	396
Wettling, W.	935
Weymouth, J.W.	282
Whitcomb	63
White, H.W.	376,1408
Wieder, H.	1128
Wigen, P.E.	1403
Wikswo, Jr., J.P.	1335
Willhite, J.R.	272
Williams, C.M.	1247
Williams, M.L.	1039
Wilsey, N.D.	1278
Wilson, R.S.	678
Wittekoek, S.	944
Wohlleben, D.	447
Wolf, W.P.	334,770
Wolfe, R.	188
Wollan, E.O.	426
Woolhouse, G.R.	247
Wortis, M.	769
Yakowitz, H.	1372
Yamane, T.	1361
Yamomoto, M.	913
Yelon, A.	1284,1285
Yokoyama, H.	524
Yoshimi, K.	157
Yoshitomi, T.	272
Young, B.W.	292
Young, C.Y.	342
Yu, J.T.	1403
Zrudsky, D.R.	974
Zuckermann, M.J.	347,1059

AIP Conference Proceedings

		L.C.Number	ISBN
No. 1	Feedback and Dynamic Control of Plasmas (Princeton 1970)	70-141596	0-88318-100-2
No. 2	Particles and Fields - 1971 (Rochester)	71-184662	0-88318-101-0
No. 3	Thermal Expansion - 1971 (Corning)	72-76970	0-88318-102-9
No. 4	Superconductivity in d- and f-Band Metals (Rochester 1971)	74-188879	0-88318-103-7
No. 5	Magnetism and Magnetic Materials - 1971 (2 parts) (Chicago)	59-2468	0-88318-104-5
No. 6	Particle Physics (Irvine 1971)	72-81239	0-88318-105-3
No. 7	Exploring the History of Nuclear Physics (Brookline 1967, 1969)	72-81883	0-88318-106-1
No. 8	Experimental Meson Spectroscopy - 1972 (Philadelphia)	72-88226	0-88318-107-X
No. 9	Cyclotrons - 1972 (Vancouver)	72-92798	0-88318-108-8
No.10	Magnetism and Magnetic Materials - 1972 (Denver)	72-623469	0-88318-109-6
No.11	Transport Phenomena - 1973 (Brown University Conference)	73-80682	0-88318-110-X
No.12	Experiments on High Energy Particle Collisions - 1973 (Vanderbilt Conference)	73-81705	0-88318-111-8
No.13	π-π Scattering - 1973 (Tallahassee Conference)	73-81704	0-88318-112-6
No.14	Particles and Fields - 1973 (APS/DPF Berkeley)	73-91923	0-88318-113-4
No.15	High Energy Collisions - 1973 (Stony Brook)	73-92324	0-88318-114-2
No.16	Causality and Physical Theories (Wayne State University, 1973)	73-93420	0-88318-115-0
No.17	Thermal Expansion - 1973 (Lake of the Ozarks)	73-94415	0-88318-116-9
No.18	Magnetism and Magnetic Materials - 1973 (Boston)	59-2468	0-88318-117-7
No.19	Physics and the Energy Problem - 1974 (APS Chicago)	73-94416	0-88318-118-5

QC
761
C6
1973
pt.1

FEB 10 1975